Fundamentals of Dispersed Multiphase Flows

Dispersed multiphase flows are at the heart of many geophysical, environmental, industrial, and energy applications. Volcanic eruptions, rain formation, powder snow avalanches, sediment transport, and dust storms are some classic examples from the environment, while industrial applications include fluidized beds, slurry transport, fuel injection, cyclone separators, and plasma coating, to name a few. Although each application is unique, they share significant commonalities in the underlying dispersed multiphase-flow physics that govern their dynamics.

This book takes a rigorous approach to explaining the complex interconnected physical processes that are at play, before developing different classes of mathematical models and numerical techniques that are appropriate for different regimes of dispersed multiphase flows. Containing many examples and over 100 exercises, it is suitable for use as a graduate-level textbook as well as a reference for researchers who want to model and simulate a multiphase flow phenomenon in their application.

S. Balachandar is Newton C. Ebaugh Professor of Mechanical & Aerospace Engineering at the University of Florida. He is a fellow of the American Physical Society and the American Society of Mechanical Engineers. He has received the Thermal Fluids Engineering Award from the American Society of Thermal Fluids Engineers, the Gad Hetsroni Senior Researcher Award from the International Conference on Multiphase Flow, the Freeman Scholar Award from the American Society of Mechanical Engineers, and the Francois Naftali Frenkiel Award from the American Physical Society. He is co-editor-in-chief of the *International Journal of Multiphase Flow*.

Fundamentals of Dispersed Multiphase Flows

S. BALACHANDAR

University of Florida

Shaftesbury Road, Cambridge CB2 8EA, United Kingdom

One Liberty Plaza, 20th Floor, New York, NY 10006, USA

477 Williamstown Road, Port Melbourne, VIC 3207, Australia

314–321, 3rd Floor, Plot 3, Splendor Forum, Jasola District Centre, New Delhi – 110025, India

103 Penang Road, #05–06/07, Visioncrest Commercial, Singapore 238467

Cambridge University Press is part of Cambridge University Press & Assessment,
a department of the University of Cambridge.

We share the University's mission to contribute to society through the pursuit of
education, learning and research at the highest international levels of excellence.

www.cambridge.org
Information on this title: www.cambridge.org/highereducation/isbn/9781009160469

DOI: 10.1017/9781009160452

First published 2024

Printed in the United Kingdom by CPI Group Ltd, Croydon CR0 4YY

A catalogue record for this publication is available from the British Library

A Cataloging-in-Publication data record for this book is available from the Library of Congress

ISBN 978-1-009-16046-9 Hardback

Additional resources for this publication at www.cambridge.org/bala_multiphase

Contents

	Preface	*page* ix
1	**Introduction**	1
	1.1 Scope	2
	1.2 A Multiscale Approach	5
	1.3 Dispersed Multiphase Flow Processes	8
2	**Scales, Mechanisms, and Parameters**	18
	2.1 Dispersed and Continuous-Phase Length Scales	18
	2.2 Multiphase Mechanisms	20
	2.3 Illustration with Examples	23
	2.4 Time Scales and Nondimensional Parameters	28
3	**Description of the Dispersed Phase**	46
	3.1 Characterization of a Realization	49
	3.2 Volume and Ensemble Averages	51
	3.3 Uniform Distribution and Poisson Statistics	55
	3.4 Statistical Description of the Dispersed Phase	59
	3.5 Statistical Description of Multiphase Dynamics	73
	3.6 Micro, Meso, and Joint Ensembles	81
	3.7 Polydispersity	84
4	**Isolated Rigid Particle in an Unbounded Ambient Flow**	88
	4.1 Five Steady Stokes Flow Problems	91
	4.2 Basset–Boussinesq–Oseen Equation of Motion	111
	4.3 Maxey–Riley–Gatignol Equation of Motion	124
	4.4 Force on a Particle in the Inviscid Limit	129
	4.5 Extension to Finite Reynolds Number	132
5	**Lift Force and Torque in Unbounded Ambient Flows**	145
	5.1 Saffman Shear Lift	146
	5.2 Lift Force in Other Linearly Varying Flows	150
	5.3 General Representation	154

5.4	Particle Rotation, Spin-Induced Torque, and Lift	159
5.5	Unsteady Lift	165
6	**Heat and Mass Transfer from an Isolated Sphere**	**168**
6.1	Steady Heat Transfer in the Small Péclet Number Limit	168
6.2	BBO-Like Equation of Heat Transfer	170
6.3	MRG-Like Equation of Heat Transfer	173
6.4	Heat Transfer at Finite Pe	174
6.5	Unsteadiness Due to Velocity Variation	178
7	**Particle–Turbulence Interaction in the Dilute Limit**	**181**
7.1	Flow Around an Isolated Stationary Particle with Increasing Re	182
7.2	Flow and Dynamics of a Freely Moving Particle	195
7.3	Pseudo Turbulence at the Dilute Limit	200
7.4	Effect of Turbulence on Particle Settling Velocity	204
7.5	Turbulent Diffusion of Particles	212
7.6	Direct Effect of Turbulence on Finite-Size Particle Force	233
8	**Particle–Wall Hydrodynamic Interactions**	**242**
8.1	Preliminary Discussion	243
8.2	Zero Reynolds Number Stokes Limit	244
8.3	Small Reynolds Number Results	257
8.4	Finite-Re Results – in Contact with the Wall	265
8.5	Finite-Re Results – Not in Contact with the Wall	274
8.6	Wall Effect on Added-Mass and History Forces	287
8.7	Particle Interaction with Near-Wall Turbulence	288
8.8	Wall Deposition in Turbulent Flows	293
9	**Particle–Particle Interactions**	**296**
9.1	Particle–Particle Collision	297
9.2	Particle–Particle Lubrication Interaction	306
9.3	Flow-Mediated Particle–Particle Interaction	310
9.4	Volume Fraction Effect on Drag in the Stokes Flow Limit	319
9.5	Finite-Re Volume Fraction Effects on Drag Law	331
9.6	Volume Fraction Dependence of Added-Mass and History Forces	338
9.7	Pseudo Turbulence at Finite Volume Fraction	341
9.8	Volume Fraction Effect on Heat Transfer	349
10	**Collisions, Coagulation, and Breakup**	**352**
10.1	Stochastic Collision Model	355
10.2	Collision Kernel	363
10.3	Combined Effects of Turbulence, Inertia, and Gravity	371
10.4	Particle Number Density Distribution	377
10.5	Other Considerations	381

11   **Filtered Multiphase Flow Equations**   385
       11.1 Spatial Filter Operation   388
       11.2 Filtered Continuous-Phase Equations   396
       11.3 Filtered Dispersed Phase Equations   414

12   **Equilibrium Particle Fields**   423
       12.1 Criterion for Uniqueness   425
       12.2 Other Examples of Non-unique Particle Velocity   428
       12.3 Equilibrium Eulerian Approximation   430
       12.4 Multiphase Physics Explored with the Equilibrium Model   441

13   **Multiphase Flow Approaches**   450
       13.1 Introduction to Different Computational Approaches   451
       13.2 How to Choose the Appropriate Computational Approach   456

14   **Particle-Resolved Simulations**   469
       14.1 Particle-Resolved Methods   470
       14.2 Direct Forcing Immersed Boundary Methodology   472
       14.3 Variations and Improvements of Direct Forcing   478
       14.4 Techniques Employing Body-Fitted Grids   483

15   **Euler–Lagrange Approach**   487
       15.1 Chapter Plan Based on Length Scales   489
       15.2 EL Governing Equations – Small Particle Limit   491
       15.3 Closure Relations of EL-DNS – Small Particle Limit   495
       15.4 Closure Relations of EL-LES – Small Particle Limit   510
       15.5 Multiphase Effects and Further Simplifications   514
       15.6 Implementation Details: Small Particle Limit   517
       15.7 EL Approach for Large Particles   521
       15.8 Force Coupling Method   543

16   **Euler–Euler Approach**   548
       16.1 EE Governing Equations from Volume Filtering   550
       16.2 Closure Relations of the EE Approach   553
       16.3 Kinetic Granular Theory   569
       16.4 Simplified Formulation – Mixture Approach   587
       16.5 Simplified Formulation – Sedimentation Approximation   590
       16.6 Ensemble-Averaged Equations   594

Appendix A   **Index Notation**   605

Appendix B   **Vector Calculus**   609

Appendix C   **Added Dissipation of an Isolated Particle**   612

Appendix D   **Solution of the Helmholtz Equation**   614

Appendix E   **Derivation of the Perturbation Force of the BBO Equation**   616

Appendix F   **Derivation of the MRG Equation with Reciprocal Theorem**   619
      F.1   Force Expression in the Time Domain   621

References   623

Index   659

Preface

The last 30 years have been an exciting period for our understanding of multiphase flows. I have been fortunate enough to witness first-hand and contribute to this, and to see how dispersed multiphase flows in particular are central in applications ranging from volcanic eruptions to flocculation of cohesive sediment or aluminum oxide smoke in solid rocket motors. I have benefitted immensely, both professionally and personally, from this decades-long journey of exploration. Though the different applications have subtle features that make them special in their own way, it is striking that the underlying physics is the same, and that it provides a strong, unifying theme. This book is my tribute to the field that has delighted me and continues to do so to this day.

I have written the book for those who have just entered the fascinating field of multiphase flow and who have a scientific or technological curiosity. For such a reader, this book will introduce the fundamental concepts that underpin multiphase-flow physics and build the way toward solution strategies that are both accurate and affordable.

Dispersed multiphase flows are everywhere around us and even inside us. There are many environmental applications of interest, and these are typically at very large scales with quite immense destructive power. Examples include dust storms, powder snow avalanches, and volcanic eruptions. At smaller scales, there are numerous industrial applications such as fluidized beds, fuel sprays, and phase-change heat exchangers. At still smaller scales, biological applications range from blood flow to shock-assisted needless drug delivery. The range of spatial and temporal scales at which these flows occur is huge: whether heat transfer enhancement with nanoparticles or the powerful explosion of a supernova on a galactic scale, both are dispersed multiphase applications.

By focusing on the underlying physics, I have tried my best to make this book accessible to the entire range of scientists and engineers, whether they be mechanical, aerospace, chemical, environmental, biological, or coastal engineers, or volcanologists, geomorphologists, sedimentologists, or astrophysicists. Although examples from the different disciplines are given here and there, the downside of my decision to focus on the common ground is that this book will not address any applications in great detail. Students who want to study, for example, fluidized beds or volcanic eruptions can start with this book in order to learn the common underlying multiphase physics before consulting more specialized books.

There are a few good books that focus on particulate, droplet, or bubbly flows. Here, we mark out a different course in the following five respects. First, we adopt a *scientific* rather than a *technological* viewpoint. As a result, we emphasize basic understanding and theoretical results, including their rigorous derivation, instead of presenting a long list of empirical correlations that address a wide range of complex situations. In this sense, this book is intended to be a *textbook*, and not a *reference book*. For example, I do not want readers to simply accept and use classic results such as the Basset–Boussinesq–Oseen equation of motion or the Saffman lift force. I believe that anyone with a deeper interest needs to know how these classic results arose. Of course, this book is not just one derivation after another. But my philosophy has been not just to present the results, but also to discuss how they came about and what the implications are.

The second distinguishing feature of the book stems from my experience with a variety of dispersed multiphase flow problems. The research literature has an amazing number of different versions of the governing equations that have been formulated for multiphase flows and a wide range of numerical methods (interface-resolved, Euler–Lagrange, and Euler–Euler) that have been employed to solve them. Any young researcher faced with the challenge of solving a multiphase flow will have to make difficult decisions as to what optimal set of governing equations to solve and what numerical methodology to use. Oversimplifying the governing equations will result in important effects being neglected. On the other hand, it is not necessary to always solve the most complicated set of governing equations, as it can be a pointless waste of precious resources. Knowing how to make these decisions can be a formidable challenge, which we address by providing researchers with analytical estimates and scaling relations that can be used to make judicious decisions given the multiphase problem at hand.

Third, this book enjoys the advantage of being more recent and therefore provides a state-of-the-art picture in terms of multiphase flow modeling. One particular emphasis is the highlighting of significant gaps in our understanding and modeling. These are important opportunities for researchers to pursue.

Fourth, by focusing only on a limited set of self-contained topics, I have tried to provide in-depth coverage. Of course, no book is complete, and this one is no exception. There are several additional topics that I would have liked to cover – exciting ones such as shock–particle interaction, droplet breakup/agglomeration, phase-change heat transfer, and bubbly flows, to mention just a few. But I refrained from making the book any longer. For readers who need to go beyond and seek input on more complex multiphase phenomena, I have tried to provide pointers to excellent books and review articles.

Finally, I have created about 60 or so video recordings, each about 30 minutes in length, to go with the book. These recordings are informal, and they cover almost all of the chapters of this book. In particular, each video recording calls out the specific sections of the book that are covered, and therefore the book and the recordings can be used hand-in-hand. You can find these recordings at www.cambridge.org/bala_multiphase.

The trigger for writing this book was a graduate-level course on multiphase flow I taught at the University of Florida (UF). Though other books covered different aspects of what I wanted to teach, what I needed was scattered over different sources. I started writing a book because I wanted to assemble all the fundamental information into one single source. It turned out to be a significant undertaking and took nearly eight long years. During this period, I have given the course at UF several times, and I want to thank my students for being guinea pigs. They helped me to focus on the chosen topics and also spotted many typos and errors. I also thank my family for their sustained support and encouragement, which allowed me to engage in this intellectual pursuit. In particular, I want to thank my daughter Anjana for the cover design using the OpenAI tool Dall-E.

1 Introduction

Multiphase flow is a branch of fluid mechanics that has grown rapidly over the past few decades. The term *phase* in "multiphase" refers to the solid, liquid, or gaseous state of matter. Thus, a multiphase flow is one that involves more than one phase. Multiphase flow can be a *gas–solid* flow, as in the case of a sand storm or pneumatic transport of powder. It can be a *liquid–solid* flow, as in the case of transport of sediments by a river or slurry flow in a pipe. *Gas–liquid* multiphase flow can be in the form of liquid droplets in a gas flow, with examples being rain, mist, and fuel spray. Gas–liquid flows can also be in the form of bubbles in a liquid flow, with examples being geysers and boilers.

The above examples are what can be considered as conventional multiphase flows. Under extreme conditions you can even have a multiphase flow of a solid phase around gas or liquid inclusions!! For example, in the context of an intense shock wave propagating through a sample of condensed-phase explosive, if the post-shock pressure is larger than the yield stress of the explosive, then the solid condensed phase begins to flow. If there are void regions within the condensed phase, then we have a *solid–gas* multiphase flow. The presence of voids within a condensed-phase explosive leads to local hot spots, which have been known to trigger ignition and subsequent detonation of the explosive. Shock-induced ignition is now an active area of research in multiphase flow.

In the case of immiscible fluids, their mixture can result in a *liquid–liquid* multiphase flow, as in the case of an emulsion of oil droplets in water. We have so far considered examples of gas–solid, liquid–solid, gas–liquid, and liquid-liquid flows, and all these examples involve only two different materials. Often such multiphase flows are also referred to as *two-phase flows*. There are more complex multiphase flows that involve three or more phases, as in the case of a bubbly flow with suspended sediment particles or oil droplets.

Multiphase flows are all around us. Mother Nature can be credited with creating some of the most fascinating multiphase flows. Volcanic eruptions, such as those at Mount St. Helens in the United States, or Pinatubo in the Philippines, or Eyjafjallajökull in Iceland, put on a spectacular show with explosive ejection of enormous amounts of ash, dust, and debris into the sky. They are catastrophic examples of multiphase flow. Pyroclastic flows, or similarly powder snow avalanches down a mountain slope, are multiphase flows on a somewhat smaller scale than a full-blown volcanic eruption, but nevertheless can be quite destructive. Sediment transport by rivers and streams, sediment-laden turbidity currents in a submarine canyon, and wave or tide-driven sediment transport in coastal regions are multiphase flows (also known as *sediment-laden flows*) that have immense potential to slowly, but steadily, change not only our coastlines but also the entire Earth surface. Dust storms, rain clouds, and geysers are other stunning examples of multiphase flows.

Blood flow is an example of multiphase flow within us. Medicinal inhalers, lithotrispy, and contrast-enhanced ultrasound with micro-bubbles are some of the modern medical technologies that rely on multiphase-flow physics. Industrial or technological uses of multiphase flow are numerous: fluidized beds, cyclone separators, spray drying, pneumatic or slurry transport, fuel injection, electrostatic precipitation, to mention but a few. Finally, multiphase flows occur even on galactic scales. Eagle nebula columns and starburst galaxies are some of the awesome examples of astrophysical multiphase flows. Check them out on the Internet. It is quite clear that multiphase flows occur over a wide range of length scales, from micrometers to light years. The associated time scales of these multiphase flows also vary, from microseconds to years.

1.1   Scope

Multiphase flows can be broadly classified as **dispersed multiphase flows** or **separated multiphase flows**. In both, there are two or more phases that remain distinct at the continuum level (i.e., there is no molecular-level mixing between phases). In the former, the phases are divided into the **continuous phase** and the **dispersed phase**. By definition, the continuous phase will remain connected over the entire domain of the flow, while the dispersed-phase material will remain scattered as individual particles, droplets, or bubbles. For example, in the case of an aerosol flow, the aerosol particles are disconnected and therefore form the dispersed phase, while the surrounding air remains connected and forms the continuous phase. In a sediment-laden flow, the sediment particles are the dispersed phase while the continuous phase (water in this case) is the continuous phase.

In contrast, in a separated multiphase flow, both phases remain connected and the boundary between the two phases forms the interface. A wavy air–water interface over a large body of water, such as a lake or sea, is an example of separated multiphase flow. In a separated multiphase flow, the two phases interact only at the interface, but the dynamics of the interface is typically very complicated and often the subject of great interest. In a dispersed multiphase flow, the dispersed phase deeply penetrates throughout the volume of the continuous phase, and thus at the macroscale the interaction between the phases can be thought of as volumetric, although the true interaction is still at the interface between the phases (which is dispersed over the volume of the continuous phase).

Unfortunately, not all applications can be nicely classified as either dispersed or separated multiphase flow. Consider the case of an air–water interface under energetic conditions of wave breaking. The air side of the air–water interface will now include sea spray. Similarly, the plunging waves can entrain air and the water side of the interface will include air bubbles. This is a very complex multiphase flow where there is a large-scale air–water interface, however on either side of the interface we now have a dispersed multiphase flow.

This book will primarily be concerned with dispersed multiphase flows. Special considerations are required in case of separated multiphase flows, dealing with the interface dynamics. These include the physics of interfacial instabilities, the role of surfactants and other chemical additives, and numerical techniques such as level-set and volume-of-fluid methodologies that have been developed to track the interface. These additional topics will not be addressed in this book. There are excellent textbooks and review articles that the reader can refer to: Hirt and Nichols (1981); Scardovelli and Zaleski (1999); Sethian (1999); Sethian and Smereka (2003); Osher and Fedkiw (2006); Tryggvason et al. (2011). In the area of dispersed multiphase flow, the classic books are Brennen (2005), Crowe et al. (2011), and Zhu et al. (2021). These books can be supplemented with recent edited volumes and review articles on various aspects of dispersed multiphase flows: Prosperetti and Tryggvason (2007); Balachandar and Eaton (2010); Fox (2012); Subramaniam (2013); Tenneti and Subramaniam (2014); Loth (2016); Subramaniam and Balachandar (2022).

There is an increasing body of literature in fluid mechanical journals on multi-material flows, many of them in the context of modern numerical methods that can handle multimaterial flows. The terms "multimaterial flows" and "multiphase flows" must be contrasted. First it must be noted that their difference is not clear cut, since a multimaterial flow can clearly be multiphase flow as well. Their difference is mainly based on their current common usage. The focus of computational methodologies for multimaterial flows has been to accurately resolve-in-space and evolve-in-time the interface between immiscible fluids, and to accurately propagate waves, such as shock waves and expansion fans, across material interfaces. Many of these recent developments in multimaterial flows are directly relevant to multiphase flows as well. *This book will not address special numerical issues that must be considered in the context of multimaterial interfaces.* The reader should consult Fedkiw et al. (1999), Saurel and Abgrall (1999), and Braeunig et al. (2007).

1.1.1 One, Two, Three, and Four-Way Coupling

The essence of a dispersed multiphase flow is in the coupling between phases. The continuous phase influences the dispersed phase through processes such as dispersion, preferential accumulation, agglomeration, and breakup. The dispersed phase modulates the continuous phase through back coupling of mass, momentum, and energy. As a result of this back coupling, the continuous-phase dynamics in a multiphase flow will differ substantially from that of single-phase flow.

Four fundamental interactions can be identified in a dispersed multiphase flow: (1) interaction among the continuous-phase fluid elements represented by the continuous-phase Navier–Stokes equation; (2) continuous-to-dispersed-phase forward coupling, which accounts for the effect of the continuous phase on the dispersed phase; (3) dispersed-to-continuous-phase coupling, which accounts for the effect of the dispersed

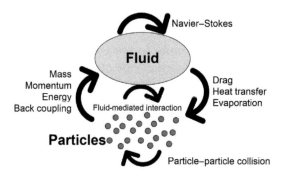

Figure 1.1 Four fundamental ways of coupling among and between the continuous and dispersed phases. Depending on the level of coupling, we can have one-way, two-way, or four-way coupled dispersed multiphase flow.

phase back on the continuous phase; (4) interaction among the dispersed-phase elements, which accounts for the interaction between the particles, droplets, or bubbles, either through direct collisions or as mediated by the continuous phase. See Figure 1.1.

A dispersed multiphase flow is said to be **one-way coupled** when only the first two interactions are of importance. The dispersed multiphase flow is **two-way coupled** when dispersed-to-continuous-phase back coupling is also of importance. When interactions among the dispersed-phase elements are important, and when this interaction is primarily mediated through the continuous phase, the flow is **three-way coupled**. In this regime of multiphase flow, for example, the wake behind a particle will affect the dynamics of the neighboring particles. When interactions among the dispersed-phase elements are important and are in the form of collisions between the dispersed-phase elements, the multiphase flow problem is considered to be **four-way coupled** (Elghobashi, 1991, 1994).

Here and henceforth, throughout the book, we will use the term "particle" to denote a solid particle or droplet or bubble, where this difference does not matter. In situations where the difference matters, we will explicitly refer to the dispersed-phase element as a droplet or bubble. Similarly, we will use the term "fluid" to refer to the continuous phase.

The dispersed-phase volume fraction is a parameter that plays a key role in determining the importance of three and four-way coupling. With increasing dispersed-phase volume fraction, the mean distance to the nearest neighbor decreases. At very low dispersed-phase volume fraction, particle–particle interaction, even as mediated by the fluid, becomes rare and multiphase flow is either one-way or two-way coupled. With increasing volume fraction the multiphase flow first becomes three-way coupled. When interparticle collisions become sufficiently frequent that they matter in the overall momentum exchange, the flow becomes four-way coupled. But there are other regimes of multiphase flow. With further increase in volume fraction, the dispersed multiphase enters the **collisional regime** where interparticle collisions dominate and dictate the entire flow.

In the near close-packing limit, the dispersed multiphase flow becomes so dense that it enters the **contact regime**, where each dispersed-phase element is in enduring contact with one or more of the neighboring elements, as opposed to frequent, but momentary, collisions with neighbors in the collisional regime. Under extreme conditions, such as when an intense shock propagates over a dense-packed bed of spherical particles, one can define a **compaction regime**, where the dispersed phase begins to flow (i.e., under extreme pressure of the surrounding flow the particles compress and deform) and as a result the local volume fraction can even exceed the close-packing limit for spheres. The collisional regime is well described by the granular kinetic theory and its recent extensions apply to a contact-dominated regime as well. *This book will primary focus on dispersed multiphase flows that are one to four-way coupled.* The books by Gidaspow (1994), Rao et al. (2008), and Brilliantov and Pöschel (2010) are excellent references on multiphase flows in the collisional regime. In the compaction regime of dispersed multiphase flow, the formulation by Baer and Nunziato (1986) and its variants (Saurel and Abgrall, 1999; Kapila et al., 2001; Abgrall and Saurel, 2008) are recommended.

It can be noted that all dispersed multiphase flow examples considered are turbulent in nature. But this need not be the case. There are many examples of low Reynolds number multiphase flows that have been actively studied, often under the general title of **suspensions**. These flows are viscous dominated and thus have allowed fundamental theories such as the **suspension balance model** (Nott and Brady, 1994; Morris and Brady, 1998; Nott et al., 2011) and sophisticated numerical approaches based on **Stokesian dynamics** (Brady and Bossis, 1988). *This book will be mainly concerned with inertia-dominated multiphase flows, where turbulence is often the key feature.*

Turbulence still remains an unsolved grand challenge problem in classical physics and mathematics, even in the context of single-phase flows. Needless to say, multiphase turbulence poses an even grander challenge. This challenge is what makes inertial multiphase flows a fascinating subject to study. Clearly, most flows that we will consider in this book are inertia dominated and turbulent at large scales. Even in such turbulent multiphase flows, the dispersed-phase elements may be so small that the local flow at the microscale around them may be viscous dominated. Thus, the vast body of knowledge gained from the study of low Reynolds number multiphase flows is of great value in our understanding and modeling of multiphase flows at the microscale. This aspect of dispersed multiphase flows will be addressed in this book.

1.2   A Multiscale Approach

Multiphase flows often involve a very wide range of length and time scales. Let us consider the example of sediment-laden river water flowing into an ocean. Sometimes the density of the sediment-laden warm river water is lighter than the clear, cold, saline water of the ocean. In this case, the muddy river flow initially rides on top and spreads on the surface of the ocean – such flows are called hypopicnal flows. Suppose the

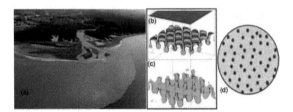

Figure 1.2 An example of sediment-laden hypopicnal river flow into the ocean at three different length scales. (a) Macroscale aerial photograph of sediment-laden Elwha River entering the straight of Juan de Fuca (photo by Tom Roorda, Roordaaerial.com). (b) Mesoscale schematic of the double-diffusive instability that occurs at the interface between the sediment-laden river flow (green) moving over the bottom clear water. In this vertical plane the sediment fingers are due to 5 μm clay particles and fingers of millimeter width. (c) As (b) but for 20 μm silt particles. Now the instability is due to gravitational Rayleigh–Taylor instability and the fingers are of centimeter width (Yu et al., 2013, 2014). (d) A microscale schematic of the sediment-laden flow, where the sediment diameter ranges from a few microns (clay) to tens of microns (silt).

river water contains a very dilute 0.4% volume concentration of 20 μm silt particles. The corresponding mean interparticle spacing can be estimated to be about 100 μm. Each silt particle settles at a single particle settling velocity of about 0.36 mm/s. An instability analysis of this flow shows that as the sediments settle from the top layer of sediment-laden river water into the bottom ocean water, a denser intermediate later develops, which undergoes gravitational Rayleigh–Taylor instability. This results in the formation of centimeter-sized sediment fingers (Yu et al., 2013, 2014). Most importantly, the sediment fingers settle at a much faster velocity of about 1 cm/s. Thus, because of the instability, the effective settling velocity of the silt particles is more than 25 times larger. The sediment-laden flow typically extends over hundreds of meters to kilometers into the ocean. Twenty microns to kilometers represents an eight orders of magnitude variation in length scale, with intermediate length scales generated by flow instabilities and turbulence. These length scales are associated with a wide range of time scales as well. Clearly this is a multiscale problem whose various scales are shown in Figure 1.2. This feature is shared by many other multiphase flows.

It is virtually impossible to solve all these length and time scales in their entirety from first principles. The standard strategy for such multiscale problems is *divide and conquer*. In other words, we need to divide the wide range of scales into micro, meso, and macroscales, each of which is limited in length and time, so that they can be studied in greater detail. The real challenges are then to (i) meaningfully define these micro, meso, and macroscales, (ii) develop mathematical models that best describe the multiphase flows at these respective scales, and (iii) at each scale appropriately represent the effect of the neglected smaller and larger scales.

Multiphase flows present a significant opportunity to develop a universal framework for solving multiscale problems. In most dispersed multiphase flow problems there is a natural way to separate the wide range of scales into micro, meso, and macroscales. In this book we will consistently consider the following definitions of the three different scales defined in terms of the dispersed-phase elements, multiphase instabilities, and the physical problem at hand. These definitions will then be used to address multiphase flow processes, their mathematical modeling, numerical simulations, and analysis at the three different scales.

- *Microscale.* We define the microscale as the smallest scale on which the dispersed and the continuous phases interact. In the limit of very low dispersed-phase volume fraction, when only one and two-way couplings are important, the microscale can be at the level of an isolated particle. When three and four-way interactions are important, the microscale will be defined as the scale on which the continuous-phase flow interacts with a representative cluster of particles. Investigations of multiphase flows at the microscale can be detailed and fully resolved. Such microscale investigation typically limits the number of dispersed-phase elements (particles, droplets, or bubbles) to about $O(10^4)$, so that their interaction with the continuous phase and among themselves can be studied in much greater detail.

- *Mesoscale.* The mesoscale is defined as the length scale on which the instabilities of the dispersed multiphase flow can be explored. At the length scale of the instability, the multiphase flow may involve millions of particles. Thus, at the mesoscale, the collective action of a very large number of dispersed-phase elements becomes far more important. In this sense, the dispersed phase behaves like a second fluid at the mesoscale. The detailed flow physics that occurs at the microscale is not directly important at the mesoscale, but the net effect of this microscale interaction must be accurately taken into account at the mesoscale. In the earlier example of the sediment-laden hypopicnal river flow into the ocean, a mesoscale investigation of the gravitational Rayleigh–Taylor instability of the interface requires a domain of about $10 \times 10 \times 10$ cm^3 (see Figure 1.2b and c taken from Yu et al., 2013, 2014). Such a cubic domain will contain about 238 million silt particles at 0.4% volume fraction. Clearly, mesoscale instability is of critical importance in explaining the rapid settling of the silt particles. This order of magnitude increase in the settling velocity of a large layer of particles over the settling velocity of an isolated particle cannot be explained at the microscale by considering only $O(10^4)$ particles.

  We shall discuss more about multiphase instabilities in Section 1.3.4, each of which introduce a characteristic length scale that defines the mesoscale of the problem. One important outcome of these instabilities is that they introduce volume fraction variation by locally redistributing the particles. Another meaningful way to define the mesoscale is based on particle volume fraction variation. Thus, the mesoscale of a multiphase flow is the length scale on which the volume fraction begins to vary. With this definition, the particle volume fraction can be taken to be uniform at the microscale.

- *Macroscale.* The macroscale is defined as the entire region of interest of dispersed multiphase flow. It can be an entire fluidized bed, or a river delta, or the exhaust cloud of a solid rocket booster. Depending on the problem size, the macroscale may contain billions, trillions, or an even larger number of particles, droplets, or bubbles. In many applications, the macroscale is clearly many orders of magnitude larger than the mesoscale.

At the microscale, the motion and thermal evolution of particles depend on the flow around them. In return, the particles modify the local flow by the formation of momentum and thermal wakes. Particle structures, such as fingers, layers, chains, and clusters, can spontaneously form due to wake–wake, particle–wake, and particle–particle interactions. Thus, the focus at the microscale is on accurate representation of the details of the interaction between the phases. In the case of bubbles and droplets, this will include time-dependent deformation of the interface and complex processes such as collision, agglomeration, and breakup.

At the mesoscale, suspended particles are responsible for flow instabilities, which eventually lead to multiphase turbulence at the meso/macroscale. Flow instabilities can lead to large-scale structures in particle distribution. Furthermore, due to inertial interaction with continuous-phase turbulence, particles preferentially accumulate (Squires and Eaton, 1991; Wang and Maxey, 1993). These non-uniformities have a profound collective back influence on the flow. Thus, the primary issues of interest at the mesoscale are the volumetric distribution of the dispersed phase, how the non-uniform distribution evolves over time and space, and the manner in which the dispersed-phase elements (particles, droplets, or bubbles) collectively interact with the surrounding continuous phase. In the study of dispersed multiphase flows at the mesoscale, attention is not on the evolution of the interface between the phases at the microscale level of an individual dispersed-phase element.

At the macroscale (or system scale), the geometric details of the problem influence the coupling between the particles and the continuous phase. This may be the scale of interest for many application-driven scientists and engineers. This is the scale on which problems of practical interest are often solved with commercial codes. However, just like in experiments, where one goes from table top, to pilot plants, to production units, a thorough understanding of the physics of the problem at the microscale – leading to a better understanding of the mesoscale physics – will lead to better models that can be used in macroscale investigation.

1.3   Dispersed Multiphase Flow Processes

Some flow physics is unique to multiphase flows, which makes them far more interesting and challenging than a single-phase flow. Clearly, there are a number of unresolved questions even in the context of single-phase flows. These questions become far more complex in the context of multiphase flows. A case in point is fluctuations arising from turbulence. In the context of dispersed multiphase flows, even under otherwise laminar conditions, due to the random location of the dispersed-phase elements, there are fluctuations in the motion of both the dispersed and the continuous phases (this is called **pseudo turbulence**). Thus, in a turbulent multiphase flow, classical single-phase turbulence of the continuous phase gets modulated, and this in turn modifies the pseudo turbulence. This is just an example of an interesting new physics that arises only in multiphase flows. In this section, we will present an overview of some of the more interesting multiphase flow processes that we will discuss in greater detail over the rest of the book.

1.3.1 Preferential Particle Accumulation

Traditional theories of turbulent particle dispersion assume that the continuous-phase turbulence applies seemingly random forcing to the particles, producing behavior similar to diffusion of a scalar contaminant. These theories predict that local concentration peaks will always reduce in time and that if the concentration distribution ever becomes statistically uniform, it will always remain so in the absence of deterministic external forces applied on the particles. It is now well known that this view is incorrect (Maxey, 1987; Squires and Eaton, 1991; Eaton and Fessler, 1994). Particles that are initially uniformly distributed in space may be drawn into dense clusters by turbulence, and particle dispersion is often caused by deterministic eddy motions which produce highly non-uniform concentration distributions. These effects are now called **preferential concentration**, because in many instances it is found that the particles are concentrated within particular flow structures. This phenomenon is also called inertial clustering, because it is caused by the difference in inertia between a particle and an equivalent element of the continuous phase. Preferential concentration can occur when particles are denser than the continuous phase, as in particle or droplet-laden gas flows. It can also occur when the particles are lighter than the continuous phase, as in bubbly flows.

Particle inertia is usually expressed in terms of a particle response time constant. The important parameter in preferential concentration is the Stokes number, which is the ratio of the particle time scale to the time scale of turbulence (we will discuss their precise definitions in the next chapter). Particles of very small Stokes number will faithfully follow all of the coherent motions of ambient turbulence. Just as fluid elements are well mixed by turbulence, these small particles will be dispersed to a uniform distribution. At large Stokes number, the particles are very sluggish. A particle encountering an eddy will have only a small response during the eddy's lifetime. In this case, the turbulence can be modeled as providing small random impulses to the particles. At intermediate $O(1)$ Stokes number, turbulent eddies induce significant coherent motion to the particles and we expect preferential concentration to be most active.

Most of the reported work on preferential concentration is for flows in which the density of the particles is much greater than the continuous phase. Vortex structures can preferentially concentrate dense particles because the particles' inertia prevents them from following highly curved fluid streamlines. Particles which are initially distributed uniformly are driven in the tangential direction by the vortex. Particles cannot follow the curved streamlines and spiral outward as they move out of the vortex core. Particles end up collected in a highly concentrated ring around the vortex. Even more intense concentrations of particles may be formed at convergence zones, where multiple vortices interact to form a saddle-point flow. There is strong experimental

(Aliseda et al., 2002) and computational (Squires and Eaton, 1991; Wang and Maxey, 1993) evidence that the local concentration of particles and droplets can even be as large as 10 times or more the mean value.

A completely different type of preferential concentration occurs in bubbly flows, where bubbles are drawn toward the core of vortices in the same way that heavy particles are spun out of vortices. This can easily be observed when rowing a boat, where bubbles are entrained into vortex rings shed during the impulsive motion of the oar. We now mention two areas where preferential concentration may play an important role in the natural world. Shaw et al. (1998) discussed the potential impact of preferential concentration on the formation and evolution of clouds. Preferential concentration may affect both the spatial distribution of water saturation and droplet coalescence rates.

Cuzzi et al. (2001) pointed out that many meteorites are composed of uniform size $O(1 \text{ mm})$ particles called chondrules. These meteorites may have formed due to preferential concentration in the protoplanetary nebula that eventually formed our solar system. Cuzzi et al. concluded that preferential concentration by a factor as large as 105 may have led to the coalescence of massive stable clusters of chondrules forming much larger objects, and that this mechanism could play an important role in the formation of terrestrial planets as well.

There are other mechanisms by which the dispersed phase tends to accumulate in certain regions of the flow. In a turbulent pipe or channel flow it is observed that heavier-than-fluid particles migrate toward the walls. This migration is due to the tendency of particles to move in the direction of decreasing mean turbulence level and, in analogy to thermophoresis, was termed **turbophoresis** by Reeks (1983).

There are also other clustering mechanisms. For example, Kline et al. (1967) observed that microbubbles released uniformly in the near-wall region of a turbulent boundary layer quickly accumulated in the low-speed streaks. Similarly, Rashidi et al. (1990), Pedinotti et al. (1992), and Rouson and Eaton (2001) observed nearly linear clusters of particles near the wall in a turbulent boundary layer. The particles that settle to the wall are swept into low-speed streaks by the longitudinal vortices that formed the low-speed streaks. This accumulation can be explained by the fact that under a stationary state, the upward turbulent flux of particles must balance the downward settling flux. This demands that there be a positive correlation between particle concentration and upward velocity fluctuation. This accumulation of particles in the near-wall upwelling regions of low-speed streaks has the potential to greatly alter turbulence production and has been shown to even totally kill turbulence (Cantero et al., 2012; Shringarpure et al., 2012). We will consider physical and mathematical modeling of preferential accumulation and turbophoretic migration in Section 12.4.

1.3.2   Settling Velocity

Settling of heavier-than-fluid particles and droplets, and rising of bubbles through turbulent flows, are common phenomena of significance in many industrial, environmental, and geophysical applications. In the atmosphere, settling of particles naturally suspended by dust storms or volcanic eruption, and settling of man-made pollutants, such as smoke and ash particles, is of importance to air quality. In cloud physics, the settling rate of water droplets plays a critical role in their growth process and in the eventual precipitation as rain (Shaw, 2003). Erosion and settling of mud and sand particles in rivers, beaches, and submarine turbidity currents play an important role in their dynamics (Peakall et al., 2000). In the oceanic context, settling of biological particles from the mixed top layer to the deep ocean is an important mechanism for vertical transport of carbon. In all the above examples (and many others), the turbulent nature of the ambient flow has been known to significantly alter the mean settling velocity of the particles from their still-fluid settling velocity.

If the continuous phase has a mean nonzero vertical velocity, clearly it will add to the still-fluid settling velocity and contribute to a decrease or increase in the net particle settling velocity. More interestingly, even in the absence of a mean continuous-phase velocity, turbulence has been established to substantially influence the mean particle settling velocity through several mechanisms. First, in case of sufficiently large particles, whose Reynolds number based on relative velocity is larger than unity, the effect of a nonlinear relation between drag coefficient and particle Reynolds number is to reduce the mean settling velocity. Tunstall and Houghton (1968) demonstrated this nonlinear drag effect for the case of particles settling in a spatially uniform flow that oscillated over time about a zero mean. Their results were confirmed in later experiments (Schöneborn, 1975).

Even in the linear drag regime, the mean settling velocity can be influenced by the *loitering effect*. Consider a particle settling through a cellular flow where part of the time, when the particle is in the down-flow region, settling is enhanced and part of the time, when the particle is in the up-flow region, settling is opposed. If we assume the particle trajectory to equi-partition and pass through up and down-flow regions with equal probability, it can be argued that the particle will spend more time in the up-flow region, since the settling velocity is lower there, than in the down-flow region, where settling is enhanced. This temporal bias toward the up-flow region is termed the "loitering effect" (Nielsen, 1993).

Trapping of particles in regions of strong updraft is another mechanism that can result in substantial reduction in the mean settling velocity. Such trapping can be considered as an extreme case of the loitering effect. The possibility of permanent and temporary trapping of particles in regions where the downward buoyancy force is balanced by the upward drag force has been well recognized for a long time (Stommel, 1949; Nielsen, 1984), and the resulting closed particle trajectories are often referred to as the *Stommel retention zone*. These trapped particles, when taken into account in the calculation of the mean settling velocity, clearly contribute to a reduction in the mean settling velocity. While particles can be permanently trapped in steady cellular flows of sufficient strength, in a time-dependent cellular flow, particles can be temporarily trapped within the cells in a chaotic manner (Fung, 1997).

In his pioneering work, Maxey (1987) demonstrated an important mechanism of interaction between the particles and the turbulent eddies, which can result in a substantial increase in particle settling velocity. Two physical processes were observed to drive this mechanism. First, due to preferential accumulation, heavier-than-fluid particles tend to spin out of vortices and concentrate in the high-strain-rate regions on the periphery of the vortices. Second, due to the combined effect of particle inertia and local flow field, particles tend to prefer down-flow regions than up-flow regions (Wang and Maxey, 1993). The net effect is that the fluid velocity averaged over the particle trajectory is not the same as the volume (or ensemble)-averaged fluid velocity. Since the particles preferentially sample down-moving fluid, the net settling velocity of particles is higher than the still-fluid settling velocity. This mechanism is referred to as *trajectory bias* or *fast-tracking* (Nielsen, 1993) and is most active when the particle Stokes number is of order one. Such particles respond best to turbulent eddies and are able to follow the trajectory bias. Much smaller particles better follow the fluid and behave more like tracer particles, and as a result do not fast-track. On the other hand, bigger particles of much larger Stokes number fail to respond to the turbulent eddies and simply follow a near-vertical path.

Recent experiments on particle settling in isotropic turbulence (Aliseda et al., 2002; Yang and Shy, 2003, 2005) have measured the increase in settling velocity due to trajectory bias and confirmed the theoretical prediction that the mechanism is most effective for particles of $O(1)$ Stokes number. The results of Aliseda et al. (2002) have further shown the effect of trajectory bias to increase with increasing particle concentration. The mechanistic reason for this particle concentration effect is simple. When a region of higher particle concentration is surrounded by lower particle concentration, the density variation of the mixture (continuous phase plus particles) begins to play a role. In other words, the collective effect of the excess particles in the high-concentration region is to induce an additional vertical motion. Bosse et al. (2006), in their simulations, explored this two-way coupling effect on particle settling and observed it to be effective for mean particle volume fractions in excess of 10^{-5}. It is to be noted that this two-way coupling effect is primarily through the role of gravity.

Of the five mechanisms discussed – namely, nonlinear effect, loitering effect, particle trapping, trajectory bias, and two-way coupling – the first three contribute to a reduction in settling velocity, while the later two result in an increase in settling velocity. The three primary parameters that influence these effects are the particle Stokes number, the particle Reynolds number, and the particle volume fraction. Depending on the location in this parameter space, the relative importance of the above mechanisms varies and as a result past experiments and computations have shown conflicting influences on settling velocity, ranging from a substantial increase (Maxey, 1987) to a modest increase or even a reduction in settling velocity (Fung, 1997). Davila and Hunt (2001) have systematically considered different (Eulerian average, Lagrangian average, and bulk average) definitions of settling velocity and their inter-relation to illustrate how such differences in the definition of mean settling velocity could partially account for the observed and computed differences in settling velocity. We will consider the effect of ambient turbulence on mean settling velocity in greater detail in Section 7.4.

1.3.3   Turbulence Modulation

The simplest dispersed multiphase flows contain very small volume fractions of very small particles, allowing the flow to be treated as a single-phase flow contaminated by widely scattered particles. In such instances, the classical mechanisms of single-phase turbulence dominate, where the continuous-phase turbulence is produced by Reynolds stresses acting on the mean shear and is dissipated by viscosity acting on the smaller-scale motions. This is the classic case of one-way coupling, in which the continuous-phase turbulence is exactly the same as the equivalent single-phase flow. However, as the particle volume fraction and mass loading increase, or particle diameters exceed the size of turbulent eddies, several additional mechanisms of turbulence production, distortion, and dissipation become important. The turbulent stresses can be either reduced or increased by the addition of particles. This effect can be very strong in situations where the single-phase production rate is fairly small. For example, reductions in the turbulent kinetic energy by more than a factor of five have been observed near the centerline of turbulent channel flows. Conversely, particles falling or bubbles rising through a laminar flow produce active turbulence, which may trigger significant mean-flow changes.

We first address the effect of the dispersed phase on the mean motion of the continuous phase. There are particle-laden flows where the mean flow is significantly altered due to the particles. Experiments by Tsuji et al. (1984) provide a good example that illustrates the dramatic effect the dispersed phase can have on the flow. Two of their figures are reproduced here in Figure 1.3, showing the mean velocity profile in the presence of 500 μm and 1 mm-sized particles for varying mass loading ranging from 0 to 3.5. The larger particles have only a small effect on the mean velocity profile, but for the same mass loading the 500 μm particles dramatically alter the mean velocity distribution – the peak velocity is not at the pipe center but has moved halfway between the center and the pipe wall. Interestingly, particles smaller then 500 μm do not influence the flow as much. These results clearly illustrate the importance of the back-effect of particles on the flow and the need to consider simultaneously the mechanism of preferential accumulation. Otherwise, the critical dependence on particle size cannot be explained.

Particles can have an even more dramatic effect on the fluctuating turbulent quantities. Turbulence modulation is important to understand because it can be so large as to qualitatively change the behavior of natural or engineering systems. For example, for small particles dispersed in gas in a vertical duct flow such as a fast fluidized bed, it is likely that the particle dispersion, heat transfer rates, and mixing/reaction rates will be much lower than might be expected if turbulence attenuation is ignored. In liquid systems, rising bubbles can induce turbulence and may produce large changes in chemical mixing rates.

If the mean velocity is substantially altered due to the presence of the particles, then turbulence production that depends on mean velocity gradients will be altered as well. In this case, large differences in the rms fluctuation and higher-order turbulence statistics can be expected. There are particle-laden flows where the mean flow may

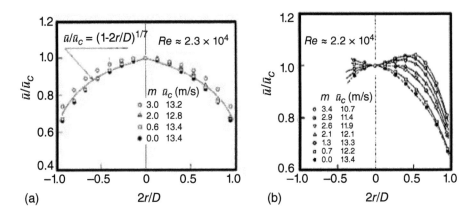

Figure 1.3 The results of Tsuji et al. (1984) on the normalized turbulent mean velocity, measured in a pipe flow of Re = 2.2 × 10⁴ for particle mass loading ranging from 0.0 to 3.4 for two different sized particles: (a) 1 mm and (b) 500 μm. Also plotted in frame (a) as a solid line is the expected single-phase turbulent mean flow velocity. The dramatic effect of the inclusion of the particles for the case of 500 μm particles can clearly be seen.

not be significantly altered by particles, however turbulent fluctuation is substantially influenced. The interaction between particles and suspending fluid remains complex and poorly understood. The conditions under which turbulence will be enhanced or suppressed remain to be fully determined. Based on a collection of experimental measurements, Crowe et al. (1996) suggested that turbulence modulation is dictated by the ratio of particle diameter to the characteristic size of the energy-containing eddies. If this ratio is greater than 0.1, then turbulence is augmented and otherwise suppressed. On the other hand, Elghobashi and Truesdell (1993) observed turbulence enhancement even for particles of diameter comparable to the Kolmogorov scale of turbulence. Hetsroni (1989) argued that the vortex-shedding process is responsible for turbulence augmentation and therefore suggested Re > 400 to be the criterion for turbulence enhancement, where Re is the Reynolds number based on particle diameter and relative velocity between the particle and the surrounding flow. Bagchi and Balachandar (2004), however, have observed the shedding process to initiate at a much lower Re, triggered by free-stream turbulence.

On the other hand, the dispersed phase can also contribute to extra dissipation. A turbulent eddy in a flow laden with unresponsive (high Stokes number) particles experiences a force field which opposes its motion. Energy is transferred from the continuous-phase turbulence into very small-scale (sub-Kolmogorov scale) flow around each particle, where it is presumably dissipated very rapidly and does not reappear as turbulent kinetic energy. Clearly, the back-effect of particles on the flow is perhaps the most challenging of all the mechanisms. It is of fundamental importance, without which we cannot answer the basic question of whether the continuous-phase turbulence is augmented or suppressed by the inclusion of the dispersed phase. For further discussion of turbulence modulation in dispersed multiphase flows, see Balachandar and Eaton (2010) and Brandt and Coletti (2022).

There is ample evidence that bubbles can cause turbulence modulation in liquid flows. Lance and Bataille (1991) found that Taylor microscale-sized bubbles strongly augmented grid turbulence, while Druzhinin and Elghobashi (1998) found the opposite effect in simulations of grid turbulence with very small bubbles. However, there remain numerous open questions about the accuracy of simulations of bubbly flows, and experiments are very difficult because of the high bubble volume fraction in most flows. Therefore, most studies have focused on gas and liquid flows laden with solid particles, where there is a wealth of experimental and numerical simulation results.

A less appreciated mechanism of turbulence modulation occurs at the mesoscale, due to the collective action of the dispersed phase. In many systems, the differential density between the dispersed and the continuous phases leads to self-stratification of the dispersed phase. The resulting vertical variation in the volume fraction of the dispersed phase in combination with gravity contributes to the substantial buoyancy effect at the mesoscale. For example, in the context of a sediment-laden turbidity current, sediment concentration increases with depth, resulting in a stable density stratification of the current. Similarly, in a bubbly flow through a horizontal pipe, one can expect a higher concentration of bubbles at the top of the pipe, again resulting in a mean stable stratification. The effect of this stable stratification on turbulence production and transport can be quite strong (see Turner, 1979). As observed by Cantero et al. (2012) and Shringarpure et al. (2012), even modest stratification can lead to total suppression of turbulence. Interestingly, even in the absence of differential settling, stratification occurs, with an associated strong influence on continuous-phase turbulence (Richter and Sullivan, 2014; Richter, 2015). These turbulence modulations are clearly mesoscale mechanisms and they do not depend on microscale mechanisms such as vortex shedding and enhanced dissipation in the wakes of individual particles. These mesoscale turbulence modulation mechanisms dominate most geophysical and environmental multiphase flows. Although generally ignored in engineering multiphase flow problems, it is likely that these mesoscale turbulence modulation mechanisms are important even in these problems. In this book we will consider turbulence modulation both at the microscale and at the mesoscale in Chapter 7.

1.3.4   Mesoscale Instabilities

There are many situations where a front of dispersed multiphase flow propagates into an ambient. Some of these are displayed in Figure 1.4. For example, in a turbidity current, sediment-laden water propagates into ambient clear water. In the case of a dust storm or a powder snow avalanche, sand or snow-laden air moves through unladen ambient air. In these examples, the ambient fluid happens to be the same as the continuous phase of the dispersed multiphase flow. But this is not always the case. In the case of volcanic eruption, an energetic mixture of hot gases, steam, ash, and particulate matter expands into the ambient air. Also shown in Figure 1.4 is a picture taken during explosive dispersal of particles. In this experiment a cylinder of explosive was surrounded by an annular bed of iron and glass particles of about 100 μm. Immediately after detonation of the explosive, the resulting high-pressure products of detonation rapidly push the bed of particles out. The leading front of the resulting multiphase flow can be seen to undergo instability and form long coherent spikes of suspended particles. Thus, in all the cases shown in Figure 1.4, the dispersed multiphase flow front undergoes instability. The length scale of these instabilities is much larger than both the diameter of the dispersed-phase elements and the interparticle spacing. These instabilities have their origin in the mesoscale processes.

Here we will list some of the more common mesoscale instabilities of the dispersed multiphase flow. As seen in Figure 1.4a, as the sediment-laden current moves under the ambient fluid, a stratified shear layer forms which can undergo classic **Kelvin–Helmholtz instability** and **Holmboe instability**. These instabilities that occur at the interface can be seen on the top surface of the current. The **lobe and cleft instability** can be seen at the front of the current. As the current moves forward, the sediment-laden fluid moves over a thin layer of clear ambient fluid that remains attached to the bottom boundary due to the no-slip condition. This appears as an overhanging nose of the current as it moves forward. This gives rise to the classic heavy-over-light **Rayleigh–Taylor instability**, which in the context of the propagating front is called the lobe and cleft instability, named after the shape that forms as a result of the instability.

Instead of the bottom-propagating heavy current, if we consider the sediment-laden hypopicnal flow that travels along the top surface of the ambient, we can identify two types of instability. The first is shown in Figure 1.2c. As mentioned before, this is due to the gravitational Rayleigh–Taylor instability. In this case, as the sediment settles down from the sediment-laden top layer into the cold saline bottom layer, an intermediate layer of cold saline water with sediment forms and it is heavier than both the layers above and below it. This heavy-over-light configuration is unstable and undergoes strong Rayleigh–Taylor instability, which is the dominant mode of instability for larger silt-sized sediments.

In the case of fine clay particles, their single-particle settling velocity is quite small and as a result the interface undergoes **double-diffusive instability**. This instability arises because of the difference in the diffusivity of the sediments and salinity. Since sediment diffusivity is much smaller than that of salinity, again an intermediate layer of unstable stratification develops, leading to double-diffusive instability. The resulting instability fingers are shown in Figure 1.2b, but their wavelength is finer at the scale of a millimeter.

Finally we address the instabilities seen in the explosive dispersal of particles. Here, immediately following detonation, the high-pressure gas rapidly accelerates the bed of particles. But as the bed expands, the velocity of both the dispersed and the continuous phases within the bed continues to decrease over time, due to radial expansion. Thus, if we consider the particle-laden layer of fluid as the heavy fluid and the ambient into which it is moving as the lighter fluid, we have the situation of a heavy fluid decelerating into a lighter fluid. This gives rise to Rayleigh–Taylor instability, which can be conjectured to be the dominant instability mechanism responsible for the particulate jets seen in Figure 1.4d. In addition, in the case of cylindrical detonation, the expansion fan becomes a second shock, reflects off the origin, and eventually moves over the expansing layer of particles. This gives rise to the classic **Richmeter–Mechkov instability**.

These dispersed multiphase flow instabilities are often of fundamental importance as they control the propagation of these multiphase flows. These instabilities are responsible for the level of mixing with the ambient. In the case of turbidity currents, instability-induced turbulence at the interface is of paramount importance, as it controls entrainment of ambient flow and the downstream evolution of the current. In the case of explosive dispersal of metal particles, the instability of the front and the level of mixing with the ambient air control the burn rate of the metal particles. Thus, any mathematical model that attempts to describe the dispersed multiphase flow at the macroscale must be capable of accurately capturing these mesoscale instabilities and the resulting turbulent mixing.

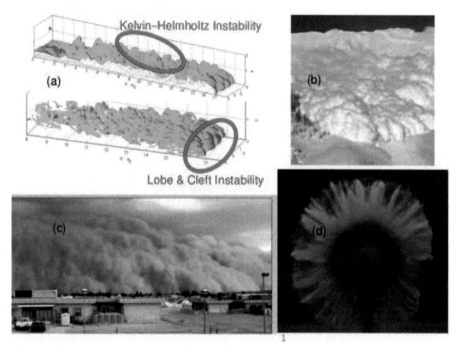

Figure 1.4 Examples of mesoscale instabilities in dispersed multiphase flows. (a) Results from a numerical simulation of sediment-laden current where the sediments are so fine that their settling can be ignored (Cantero et al., 2007). The top figure shows the turbulent interface between the sediment-laden current and the ambient clear fluid. The bottom figure shows the turbulent vortical structures as extracted by swirling strength (Zhou et al., 1999; Chakraborty et al., 2005). The Kelvin–Helmholtz or Holmboe instability of the interface and the lobe and cleft instability of the advancing front can be seen. (b) Picture of a power snow avalanche that shows the instabilities of its top surface. (c) Instabilities in an advancing front of a sand storm. (d) Still image of the explosive dispersal of a particle from an experiment of Frost et al. (2012), reprinted with the permission of AIP Publishing. A cylindrical explosive was surrounded by an annulus of heavier iron and lighter glass particles. Finger-like instability of the expanding glass particle front can be seen as the lighter color. The front of the heavier iron particle (black material near the center) is lagging behind and its instability is less pronounced.

2   Scales, Mechanisms, and Parameters

The multiscale nature of dispersed multiphase flows makes their characterization challenging. A single-phase flow may be reasonably characterized in terms of nondimensional parameters, such as the Reynolds number, Mach number, or Rayleigh number. But characterization of a multiphase flow requires additional parameters that describe the dispersed phase and its relation to the continuous phase. In this chapter we will introduce mathematical definitions of some basic quantities and explain how they characterize the dispersed multiphase flow.

2.1   Dispersed and Continuous-Phase Length Scales

We start with a discussion of the dispersed and the continuous-phase length scales and how they relate to each other. The two main length scales associated with the dispersed phase are the particle diameter d_p and the mean distance between two closest neighbors l_d. For simplicity, we will consider the dispersed phase to be spherical and monodispersed. That is, all the particles will be taken to be of the same diameter d_p. The distance between any particle and its closest neighbor is a random variable, whose distribution can be expressed in terms of a probability density function. But for now we consider l_d to be the characteristic length scale that defines interparticle distance.

Both d_p and l_d are dispersed-phase quantities at the microscale and therefore they play an important role in the microscale processes. For example, mass, momentum, and energy exchange between the phases depend on the particle Reynolds number and thus on the particle diameter. The importance of particle–particle interaction, either through direct collision or through interaction of one particle with the wake of another particle, depends on interparticle distance. There can be longer length scales associated with the dispersed phase, based on the manner in which the particles are distributed within the domain of interest. Particle distribution measured in terms of number density or volume fraction can show spatial variation at length scales larger than l_d. As in the hypopicnal flow example considered in Figure 1.2, these mesoscale length scales arise from instabilities.

To understand multiphase flow, it is useful to start the discussion with the corresponding *single-phase limit*, where the continuous-phase flow is in the absence of the dispersed phase. This limit can also be called the *globally undisturbed limit*, since the flow is fully undisturbed (or unperturbed) by the dispersed phase. The flow in this single-phase limit may be quiescent, laminar, or turbulent in nature. If the single-phase flow is turbulent, then it will be characterized by a range of length scales from the Kolmogorov scale (η_c) to the integral scale ($\mathcal{L}_c$). With the addition of the dispersed phase to the continuous phase, the multiphase flow can differ substantially from the corresponding single-phase limit.

Two different regimes can be envisioned, as depicted in Figure 2.1, and they differ in terms of the relative length scales of the dispersed phase and the globally undisturbed continuous-phase turbulence. In the regime depicted in Figure 2.1a, the spectrum of scales associated with the flow around the particles and in the wakes can be considered microscale and represented by the dashed line. The solid line represents the energy spectra of the flow in the single-phase limit (i.e., spectra of globally undisturbed turbulence). Note that in this schematic the two are well separated. If the dispersed multiphase flow is one-way coupled, then the main part of the continuous-phase energy spectrum will remain the same as that of the single-phase limit (solid line) and the effect of the particles is limited only to the microscale. If the multiphase flow is two-way coupled, then particles' influence on the continuous phase extends beyond the boundary layers and wakes around the particles. The main spectra of continuous-phase turbulence will itself be modified by the dispersed phase and this two-way coupled energy spectrum is depicted as the dashed–dotted line in Figure 2.1a.

In the regime where the particle size and the interparticle separation are of the same order as the scales of globally undisturbed turbulence, there will be no scale separation between undisturbed turbulence and microscale flow features around the particles. In this regime, the multiphase flow is always two-way coupled. The energy spectrum of the globally undisturbed turbulence and that of the dispersed multiphase flow are shown as solid and dotted lines in Figure 2.1b. From a modeling perspective, this regime is significantly more complicated since the microscale processes cannot be clearly separated from those at larger scales. Nevertheless, one can conceptualize the multiphase energy spectrum (dotted line) to be separated into a spectrum due to the collective action of the particle (dashed–dotted line) and one due to individual particles (dashed line).

Furthermore, from an individual particle's perspective, the scenario depicted in Figure 2.1a is far simpler than that presented in Figure 2.1b. In the former, since the particle is much smaller than all the scales of turbulence, the ambient flow encountered by each particle can be taken to be spatially smooth. As will be discussed in Chapter 4, the momentum and energy coupling between the particle and such a well-behaved ambient flow can be accurately modeled as drag and heat transfer coefficients. By contrast, in the scenario depicted in Figure 2.1b, each particle is subjected to a turbulent flow that varies on the scale of the particle. This presents formidable challenges for the accurate modeling of momentum and energy coupling between the phases.

2.2   Multiphase Mechanisms

In both regimes depicted in Figure 2.1, the effect of the dispersed phase is quite clear in terms of the substantial difference between the energy spectrum of the actual multiphase turbulence and that of the single-phase limit. In this section, we will discuss five different multiphase mechanisms by which the presence of the dispersed phase changes the flow from the single-phase limit. These mechanisms are *inertia effect*, *body-force effect*, *volume effect*, *thermodynamic/transport effect*, and *slip effect*. To explore these mechanisms, let us consider the following sequence of four problems whose end members are the multiphase flow problem under investigation and its single-phase limit, respectively.

Single-Phase (SP) Limit

As discussed above, in this limit the continuous-phase flow is in the absence of the dispersed phase. If the single phase is driven by external driving forces that are maintained identically as in the multiphase problem, them the resulting flow is the single-phase limit.

Point-Particle Dusty Gas (PPD) Limit

In this limit, each particle is considered to be a point mass that accounts for the excess mass of the actual particle over that of the displaced continuous phase. However, at all times each particle's velocity is maintained to be the same as that of the local fluid. Thus, there is no relative velocity between the phases. Such a flow is often called *dusty gas* flow. Since particle motion is perfectly synchronized with the local fluid, it is sufficient to think of the multiphase flow as a single fluid which is a mixture of continuous and dispersed phases. However, the density of the mixture depends on the local volume fraction of the dispersed phase. Thus, this will be similar to a single-phase flow, but with variable density. The dusty gas limit accounts for the *inertia effect* of the dispersed phase through this variable density. For example, if a pressure difference is suddenly applied, then the acceleration of a heavier particle-laden mixture will be slower than that of the corresponding single-phase limit. The dusty gas limit also accounts for the *body-force effect* (or the gravity effect) of the dispersed phase. For example, the dusty gas limit of a non-uniform suspension of heavier-than-fluid particles will result in a buoyancy-driven flow purely due to the spatial variation in the density of the fluid–particle mixture.

Finite-Particle Dusty Gas (FPD) Limit

In this limit, the finite size of the particle is recognized in addition to the density effect of the particle. However, as in the PPD limit, at all times each particle's velocity is maintained to be the same as that of the local fluid. Due to the finite size of particles, two additional multiphase effects influence the flow in this limit. The first is the *volume effect* that arises from the fact that with finite-sized particles the fluid is forced to occupy only the region outside of the particles. As the particle volume fraction varies over space and time, the mass of continuous phase per unit volume (or the effective density of the continuous phase) varies accordingly, even when the fluid is incompressible and its material density is a constant. Thus, the volume effect gives the continuous phase an effective compressibility.

The second multiphase mechanism due to the finite size, or the finite volume occupied by the particles, is the *thermodynamic/transport effect*. In essence, the thermodynamic and transport properties of the fluid–particle mixture are different from their single-phase counterpart. Consider a volume of fluid with spherical particles of volume fraction ϕ_d randomly distributed within it. Since the particles are moving with the fluid in the FPD limit, we will consider the entire volume of mixture to be static. Some of the thermodynamic properties, such as the heat capacity of the mixture, can be expressed as a simple sum of the heat capacity of the fluid and the particle, weighted by their respective mass, per mass of the mixture. On the other hand, transport properties, such as the effective thermal conductivity and viscosity of the mixture, are not simple weighted sums of the individual components. The effective transport properties depend on the manner in which the dispersed phase is distributed within the fluid. But the evaluation of the effective transport properties of a multiphase mixture is a well-studied problem. For example, the effective thermal conductivity of a medium with a random distribution of spherical particles goes back to Maxwell (1873), and the effective viscosity of a multiphase mixture to Einstein (1906) (see Section 4.1.4 for a detailed discussion).

Multiphase (MP) Limit

Finally, we consider the multiphase flow with none of the above simplifications. The key difference between this and the FPD limit is that now the particles are not restricted to move at the same velocity as the local fluid and this allows the multiphase flow to account for the important *slip effect*. In fact, the slip effect can be generalized. The temperature of the particle is allowed to be different from the local fluid. The angular velocity of the particle can be different from that of the local fluid. Also, the pressure within the dispersed phase can be different from that of the surrounding continuous phase. If the translational velocity, temperature, angular velocity, and pressure of each particle are the same as those of the surrounding fluid, then there is no significant slip effect. Otherwise, as a result of the relative translational and rotational motion of the particle and the difference in temperature, there will be a perturbation flow around each particle in the form of boundary layers and wake flow. Depending on the particle Reynolds number (which will be defined in Section 2.4.1), the wake flow around each particle may be laminar or turbulent.

Thus, a multiphase flow differs from its single-phase counterpart due to these five different mechanisms: inertia effect, body-force effect, volume effect, thermodynamic/transport effect, and slip effect. Later, when we discuss Euler–Lagrange and Euler–Euler methods of multiphase flow in Chapters 15 and 16, we will consider how each of these effects are modeled and under what conditions some of these effects can be ignored.

The perturbation from the single-phase limit induced by the above five effects can be classified into two different groups. The inertia and the body-force effects that are captured by the difference between the SP and the PPD limits represent the collective effect of the dispersed phase. The corresponding perturbation flow is primarily at the mesoscale, whose length scales are dictated by the scales of non-uniformities in the spatial distribution of the dispersed phase.

In contrast, the perturbation flow induced by the volume, thermodynamic/transport, and slip effects is primarily at the level of individual particles. The relative velocity between the particle and the surrounding fluid (i.e., the slip effect) results in the formation of a boundary layer around the particle and a recirculating wake region behind the particle. The individual volume, thermodynamic/transport, and slip effects, when aggregated over all the particles, lead to an overall perturbation to the continuous-phase flow. Owing to the random distribution of particles, this overall perturbation flow will be chaotic and this is what we earlier referred to as *pseudo turbulence*.

In Figures 2.1a and b, the spectra of pseudo turbulence is denoted by the dashed line, while the perturbation to the meso/macroscale flow due to inertia and body-force effects contributes to the deviation of the dashed–dotted line from the single-phase spectrum given by the solid line. The importance of both sources of turbulence must be appreciated. In the case of the sediment-laden hypopicnal flow shown in Figure 1.2, turbulence is entirely driven by inertia and body-force effects of the suspended sediments. In this problem the role of particle-induced pseudo turbulence is not important, since the volume, thermodynamic/transport, and slip effects are very small. Whereas, in the case of the particle-laden pipe flow shown in Figure 1.3, pseudo turbulence is very important and in this case perhaps even more important than the inertia and body-force effects. Modeling and simulation of multiphase flows must carefully consider these aspects of the problem at hand and direct efforts to focus only on the relevant physics.

Since the primary source of pseudo turbulence is at the level of individual particles, the length scales associated with the pseudo turbulence of each particle are dictated by the wake flow, which is of the order of d_p and l_d. When particles are randomly distributed in space with uniform probability without any mesoscale structure, the corresponding pseudo turbulence will mainly be at the microscale, schematically represented by the red dashed lines in Figures 2.1a and b. Unfortunately, often the situation is not so simple. As discussed in Section 1.3, particles preferentially accumulate, generating number density variations at the mesoscale. Furthermore, mesoscale instabilities result in mesoscale variations in particle distribution. As a result, pseudo turbulence due to the collective action of all the particles will also show mesoscale variation. In other words, the spectrum of pseudo turbulence (depicted by the dashed line in Figure 2.1) can be much broader than what is shown in the schematic.

We will now briefly discuss the nature of pseudo turbulence. Like in classical single-phase turbulence, the perturbation flow around a random distribution of particles in chaotic motion will exhibit a range of length and time scales. This will be the case even when the flow around each individual particle is laminar. Superposition of randomly located and oriented laminar wakes can appear to be turbulent (Parthasarathy and Faeth, 1990; Mizukami et al., 1992). However, the spectra of such a pseudo turbulence will not follow classical turbulence scaling relations. In many applications, the particle

Reynolds number remains modest and the perturbation flow around each particle remains laminar. However, in applications where the particle Reynolds number is order 300 or more, the wake downstream of the particle transitions to turbulence (Wu and Faeth, 1993, 1994a). With further increase in particle Reynolds number, the transition to turbulence moves closer to the particle, with the shear layer and attached boundary layer around the particle becoming progressively more turbulent. In such cases, the nature of pseudo turbulence is quite complex indeed.

To firmly fix these ideas, we will consider some concrete examples and discuss them in detail in Section 2.3.

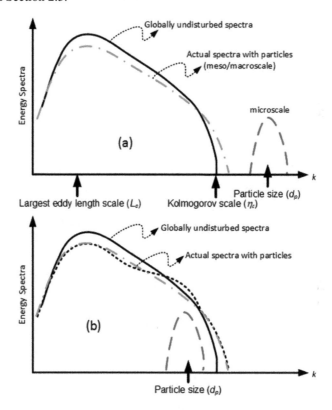

Figure 2.1 Schematic of the energy spectra of the continuous phase: (a) for the case of scale separation, where particles are much smaller than the undisturbed flow scales; (b) when there is no scale separation between the particle and the undisturbed flow.

2.3   Illustration with Examples

In the preceding sections we introduced several new concepts related to continuous and dispersed-phase length scales and multiphase flow mechanisms. In this section, we will consider these concepts in the context of four simple examples, in order to help us better grasp these new concepts. These examples are chosen to illustrate an important point. Among the multiphase mechanisms – inertia, body-force, volume, thermodynamic/transport, and slip effects – not all are important in every problem. Depending on the physics of the problem at hand, only one or more of these mechanisms play a role and must be included in the mathematical description of the multiphase flow.

This gives rise to the single biggest challenge that can vex a young researcher of multiphase flow: *There are so many different formulations of multiphase flow with varying governing equations, which one should I use in my problem?* While there are Navier–Stokes equations for the single-phase flow,[1] one can find a wide variety of multiphase flow equations in the literature. Depending on which combination of multiphase flow mechanisms are included, the resulting multiphase flow governing equations have been simplified and tailored to the problem at hand. In the following examples, we will only focus on identifying which mechanisms are important and which ones can be ignored. Only later in this book will we show how these decisions translate into simplified multiphase flow formulations and governing equations.

As you will see below, the first two environmental multiphase flow examples are simple in the sense that they are dominated only by the body-force effect. But they are nevertheless very complicated due to their wide range of scales, and intense turbulence, among other reasons. The next two examples, drawn from engineering applications, provide a different picture of multiphase flow. Virtually everything around you, from the microscale to the astronomical scale, involves multiphase flow of some form. So the following examples provide an extremely limited picture.

Example 1: Hypopicnal Flow

A simple example is to consider the flow generated by a very dilute distribution of very small sedimenting particles in an otherwise very large body of stagnant fluid. Each sedimenting particle has its own wake flow. Randomness in the distribution of particles contributes to mesoscale density variation, which leads to mesoscale convective instabilities such as those shown in Figures 1.2b and c. These instabilities grow and often develop into large-scale turbulence. Note that in this sedimentation example, the single-phase limit is stagnant fluid. Thus, in this example, mesoscale turbulence is entirely created (not just modified) by the dispersed phase. Due to the small size of the sedimenting particles, in this example, the large-scale turbulence at the mesoscale is well separated from the pseudo turbulence due to the individual particles. If interest is limited to the mesoscale behavior of sedimentation, then the problem can be considered at the level of the mesoscale.

We can now examine the importance of the different multiphase flow mechanisms in this example. As stated above, and as we will show in Section 2.4.3, only a very dilute distribution of particles is needed to create strong instability-driven flows. Since the volume fraction is very low (typically much less than 1%), the volume and thermodynamic/transport effects are unimportant. The sedimentation process is of importance from the particle's perspective, but from the perspective of the continuous phase the

---

[1] There again we have compressible vs. incompressible equations, inviscid Euler equations, potential flow equations, etc.

associated slip effect and pseudo turbulence are unimportant and can be ignored. Not only is the volume fraction of suspended sediments low, even the density difference between the sediment-laden and the clear fluid is quite small. In other words, the density variation in this flow can be so small that the inertia effect is also not important. This leaves the body-force effect as the only remaining multiphase mechanism. It is the sole source of instability and the resulting complex flow. See Table 2.1 for a summary of the relevant mechanisms.

Note that the body-force effect is also due to the density variation caused by the random distribution of suspended sediments. Thus, while we ignore the density variation as being small in the inertia effect, we have argued the density variation to be important in the body-force term of the Navier–Stokes equation. This is called the *Boussinesq approximation*, where the density variation is important only when multiplied by the acceleration due to gravity. In essence, with the Boussinesq approximation of the multiphase flow: (i) the water–sediment mixture can be treated as an effective variable density single fluid; (ii) the density variation is, however, important only in the body-force term; and (iii) the velocity of the sediment phase can simply be taken to be the sum of the local fluid velocity plus the sediment settling velocity. The Boussinesq approximation thus substantially simplifies the multiphase model and this approach has been taken by several researchers (see, e.g., Bosse et al., 2006 and Yu et al., 2013, 2014).

Example 2: Turbidity Current

Let us consider another example where the flow is entirely driven by the dispersed phase and thus there is no flow or turbulence in the single-phase limit. The turbidity current shown in Figure 1.4a is an example. In a typical turbidity current, the sediments are of $d_p \approx O(100 \, \mu\text{m})$ and at a very low volume fraction. But the largest turbulent scales are much larger, on the order of the depth of the current, which can be several centimeters to meters. Depending on the intensity of the current, the Kolmogorov scale can be either larger than l_d, in which case we are in the regime depicted in Figure 2.1a, or, if the conditions are such that $\eta_c < l_d$, we are in the regime depicted in Figure 2.1b. In either case, it is fascinating that the dispersed phase, through collective action, creates mesoscales much larger than d_p and l_d, which was the case even in the sedimentation example.

Here again, the influence of pseudo turbulence due to volume, thermodynamic/transport, and slip effects is negligible, and the dynamics of the turbidity current can be well studied without the complications of these multiphase flow mechanisms. Once again, if the turbidity current is characterized by a very dilute suspension of sediments, then the inertia effect can also be ignored. This leads to the Boussinesq model where the body-force effect is the only multiphase flow mechanism, which in this case is the sole source of flow and turbulence. Note that even though pseudo turbulence due to the slip effect is ignored from the perspective of the fluid, the settling velocity is important from the sediment perspective and is taken into account in keeping track of the sediment concentration (see Table 2.1). We shall see more about Boussinesq approximation and the resulting simplified governing equations in Chapter 15. You

Table 2.1 The importance of different multiphase mechanisms in the four examples considered. The reader should be careful in reading this table. As discussed in the text, there are many parameters that determine the importance of the different multiphase mechanisms. The table is for a certain choice of values for these parameters and therefore must be considered only as an illustration of possibilities. Here, the asterisk indicates that the effects are unimportant at lower particle-to-fluid mass ratios.

	Inertia Effect	Body-Force Effect	Volume Effect	Strain Effect	Slip Effect	Comment on Particle Velocity
Example 1	–	✓	–	–	–	Fluid + settling
Example 2	–	✓	–	–	–	Fluid + settling
Example 3	*	*	–	–	✓	More complex
Example 4	*	*	–	–	✓	More complex

can refer to Necker et al. (1984), Cantero et al. (2009), Meiburg and Kneller (2010), or Shringarpure et al. (2012) for use of the Boussinesq model in the study of turbidity currents.

Example 3: Multiphase Stirred Tank

Next we consider a multiphase stirred tank where the fluid is stirred by tangential jets, as opposed to by an impeller (without the impeller the problem becomes geometrically simpler, but the behavior of the multiphase flow to be discussed below will hold for a tank stirred with an impeller). With proper design of jets, away from the walls of the tank and the injectors, the flow can be considered to well approximate isotropic turbulence (Variano and Cowen, 2008; Carter et al., 2016). In this problem, the single-phase limit is turbulent, and its range of scales will be dictated by the specifications of the tangential jets and the tank. We can now consider the effect of introducing particles and study the difference between single-phase and multiphase isotropic turbulence, for the same external forcing.

This example is chosen to contrast the extreme simplicity of the first two examples. The two most important parameters of the multiphase problem are the ratio of particle-to-fluid length scale and the ratio of particle-to-fluid time scale. The length scale ratio is given by d_p/η_c, where η_c is the Kolmogorov scale of ambient turbulence. The time scale ratio is given by the particle Stokes number St, which we will study in greater detail in Section 2.4.5. For now, the Stokes number measures the relative inertia of the particles with respect to the surrounding flow. If St is larger than unity, then the particles will not follow the fluid and there will be slip between the particles and the surrounding flow. The length and time scale ratios, along with the amount of particles in the system, control the importance of the different multiphase mechanisms.

If the volume fraction occupied by the particles is small, then the volume and thermodynamic/transport effects can be ignored. In addition, if the mass of particles per unit volume is much smaller than the continuous-phase mass within that volume, then inertia and body-force effects can be neglected as well. Here a subtle point must be noted. The mass ratio is the source of density variation and the density variation was small even in Examples 1 and 2. However, in those examples the small density variation was the sole source of multiphase flow and therefore cannot be neglected. In contrast, in this example, turbulence is driven by other external mechanisms and compared to them the body-force effect can be neglected.

The remaining slip effect depends on the length and time-scale ratios. In the limit where particles cannot be considered small (i.e., when $d_p \gtrsim \eta_c$), there is no scale separation between the particle and the single-phase spectra. As discussed before, this makes the problem quite complex and the slip effect along with local pseudo turbulence cannot be ignored. In the regime of very small particles (i.e., when $d_p \ll \eta_c$), the importance of the slip effect depends on the Stokes number. For St $\gtrsim$ 1, each particle's motion will substantially differ from that of the local fluid and the slip velocity will be the source of substantial local pseudo turbulence. Even when there is pseudo turbulence generated by individual particles, its effect on meso and macroscale turbulence will depend on how many particles there are per fluid volume (i.e., on volume fraction).

Example 4: Particle-Laden Pipe Flow

Finally, we consider the problem of particle-laden flow in a horizontal pipe driven by oscillatory pressure gradient. At sufficiently large Reynolds number, the flow within the pipe will be turbulent even in the single-phase limit. Here again, the different multiphase effects will be discussed first in the simpler limit of small-sized particles, with the understanding that the problem will be quite complicated when the particle size becomes comparable to the Kolmogorov scale. For a specified oscillatory pressure gradient, the response of the particle-laden flow will be different from that of the single-phase limit due to the modified mass of the multiphase mixture (inertia effect). In a horizontal pipe, the particles experience a vertical settling flux toward the bottom of the pipe. This downward flux is compensated by an upward turbulent flux which tends to resuspend the particles in the pipe. The net effect of these two fluxes is to generate a vertical variation in the mean number density of particles, with a higher concentration near the bottom wall and progressively lower concentration with increasing height. This density stratification can have a strong effect on the turbulence, thus differentiating single and multiphase turbulence at the mesoscale. In addition, there is pseudo turbulence due to the slip, thermodynamic/transport, and volume effects, which in turn depend on both the particle Stokes number and the particle size relative to the ambient flow scales.

The above examples illustrate the complexities associated with even canonical multiphase flows. Many parameters such as particle volume/mass fraction, particle-to-flow length scale ratio, particle Stokes number, and particle Reynolds number influence the different multiphase mechanisms, which we will discuss in greater length over the course of the rest of this chapter. It is also clear that not all the mechanisms are equally important in all the problems. Depending on the situation, one or more mechanisms play an important role. Such understanding will allow us to choose the right multiphase flow model, best suited for a given problem.

2.4   Time Scales and Nondimensional Parameters

In this section we will follow the presentation of Section 2.1 on length scales and discuss the key time scales and velocity scales of the dispersed and continuous phases. We will also define the particle Reynolds number and particle Stokes number, which – along with the volume fraction and dispersed-phase mass concentration – control the importance of the various multiphase flow mechanisms discussed in Section 2.2.

2.4.1   Translational Time Scale

The translational time scale of a particle defines how fast it responds to a change in the velocity of the surrounding continuous phase. In addition to the translational time scale, we will also define other time scales associated with particle temperature and pressure. However, the translational time scale is often the most critical and therefore it is simply referred to as the *particle time scale*. The particle time scale depends primarily on the properties of the dispersed and continuous phases. In order to obtain the particle time scale, consider a particle injected at velocity V_0 into a quiescent fluid. Due to hydrodynamic drag, the particle velocity will decay and approach the zero ambient fluid velocity. The time scale on which this decay occurs is the particle time scale. The equation of motion of the particle can be written as

$$\rho_p \mathcal{V}_p \frac{dV}{dt} = -3\pi \mu_f d_p V \, \Phi - \frac{1}{2} \rho_f \mathcal{V}_p \frac{dV}{dt}, \qquad (2.1)$$

where ρ_f and μ_f are the constant density and dynamic viscosity of the continuous phase. The left-hand side is the mass times acceleration of the particle, and on the right-hand side the first term is the quasi-steady drag, which includes the finite Reynolds number nonlinear correction, $\Phi(\mathrm{Re})$. The Reynolds number is defined as $\mathrm{Re} = d_p V / \nu_f$, where $\nu_f = \mu_f / \rho_f$ is the kinematic viscosity of the fluid. In the low Reynolds number limit (i.e., $\mathrm{Re} \to 0$), the correction $\Phi \to 1$ and the first term reduces to the Stokes drag on a sphere. The second term on the right is the added-mass force, where the factor $1/2$ is the added-mass coefficient of a sphere. We shall consider the equation of motion more in Chapter 4, but the above is sufficient to establish the particle time scale.

Using the definition of the volume of a sphere ($\mathcal{V}_p = \pi d_p^3 / 6$), Eq. (2.1) can be rewritten in terms of particle time scale (τ_p) as

$$\frac{dV}{dt} = -\frac{V}{\tau_p} \quad \text{where} \quad \tau_p = \frac{d_p^2 (1 + 2\rho)}{36 \, \nu_f \, \Phi}, \qquad (2.2)$$

where $\rho = \rho_p / \rho_f$ is the particle-to-fluid density ratio. In the dual limit of a heavy particle (i.e., $\rho \gg 1$) and $\mathrm{Re} \to 0$ (i.e., $\Phi \to 1$), we recover the conventional definition that $\tau_p = d_p^2 \rho / (18 \nu_f)$.

This conventional definition does not capture the behavior of a lighter-than-fluid particle. For example, in the case of a bubble, as $\rho \to 0$, Eq. (2.2) correctly predicts the particle time scale to be $d_p^2 / (36 \nu_f)$. Since the mass of a bubble is negligible, the time scale of the bubble is dictated by the added mass of the surrounding fluid. This is the reason why we include the added-mass force in Eq. (2.1).

At finite Reynolds number a commonly used correction to the Stokes drag is

$$\Phi(\text{Re}) = 1 + 0.015\,\text{Re}^{0.687}\,. \tag{2.3}$$

The drag formula represented by the first term on the right-hand side of Eq. (2.1) with this correction is often called the *standard drag law* (Clift et al., 1978). Thus, it can be seen that nonlinear drag can have a strong effect on particle time scale. For example, at Re = 1000 the particle time scale obtained with the standard drag is substantially lower than that predicted with the Stokes drag by a factor of about 18.26 due to nonlinear drag.

Note that in the particle time-scale formula given in Eq. (2.2), the particle diameter, particle-to-fluid density ratio, and kinematic viscosity of the fluid are parameters that are often time independent. If we further assume Φ to be nearly a constant (which is a good assumption in the limit of small Reynolds number), then τ_p can be taken to be time independent as well. We can then integrate Eq. (2.2) in time to obtain the solution

$$V(t) = V_0 \exp(-t/\tau_p), \tag{2.4}$$

where V_0 is the initial velocity of the particle. It can readily be seen that over a particle time scale (i.e., for $t = \tau_p$), the particle velocity decreases by a factor $1/e$, or to 36.8% of the initial value.

2.4.2   Thermal Time Scale

The thermal time scale of a particle similarly defines how fast it thermally responds to a change in the temperature of the surrounding continuous phase. To obtain the thermal time scale, consider a stationary particle in a uniform ambient flow of Reynolds number Re. Let the flow be fully developed and let the particle be at the same temperature as the ambient fluid temperature T_c. The temperature of the fluid is also taken to be spatially uniform. Consider suddenly changing the temperature of the particle to T_{p0} at $t = 0$, and subsequently allowing the particle temperature to relax to the ambient temperature through heat transfer with the surrounding fluid. The time scale on which this change occurs is the thermal time scale of the particle. The equation of thermal evolution of the particle can be written as

$$\rho_p \mathcal{V}_p C_p \frac{dT_p}{dt} = 2\pi k_f d_p (T_c - T_p)\Phi_T\,, \tag{2.5}$$

where k_f is the constant thermal conductivity of the continuous phase and C_p is the specific heat of the particle. On the right-hand side, $2\pi k_f d_p (T_c - T_p)$ is the heat transfer from a sphere in the Stokes limit of zero Reynolds number and Φ_T is the finite Reynolds number correction to heat transfer. Here the correction is written as analogous to the finite Reynolds number correction to the Stokes drag. Traditionally, heat transfer is expressed in terms of the *Nusselt number*, and is related to the correction as Nu $= 2\Phi_T$. The Nusselt number is the ratio of convective to conductive heat transfer and in the zero Reynolds number limit Nu $= 2$ for a spherical particle in a uniform flow. At finite Reynolds number the two commonly used empirical expressions for the Nusselt number or Φ_T are given by

$$
\mathrm{Nu} = 2\Phi_T = \begin{cases} 2 + (0.4\,\mathrm{Re}^{1/2} + 0.06\,\mathrm{Re}^{2/3})\,\mathrm{Pr}^{0.4} & \text{Whitaker (1972)}, \\[2mm] 2 + 0.6\,\mathrm{Re}^{1/2}\,\mathrm{Pr}^{1/3} & \text{Ranz and Marshall (1952)}. \end{cases}
$$

$$(2.6)$$

Thus, heat transfer depends not only on Re, but also on the Prandtl number, which is defined as $\mathrm{Pr} = \nu_f/\kappa_f$, where κ_f is the thermal diffusivity of the continuous phase. Note that there is no added-mass-like effect in the case of heat transfer and therefore there is only one term on the right-hand side of Eq. (2.5) – there is no term analogous to the second term on the right-hand side of Eq. (2.1). We shall consider the question of thermal evolution of a particle more in Chapter 6.

We now use the definition $k_f = \kappa_f\,\rho_f\,C_{pf}$, where C_{pf} is the specific heat at constant pressure of the continuous phase, to rewrite Eq. (2.5) in terms of the thermal time scale τ_T as

$$
\frac{dT_p}{dt} = \frac{T_c - T_p}{\tau_T} \quad \text{where} \quad \tau_T = \frac{d_p^2 \rho C_r}{12\,\kappa_f\,\Phi_T},
$$

$$(2.7)$$

with $C_r = C_p/C_{pf}$ the particle-to-fluid specific heat ratio. Again, if we approximate Φ_T to be a constant, which is a good assumption in the limit $\mathrm{Re} \to 0$, and let the ambient fluid temperature remain steady, Eq. (2.7) can be solved to obtain

$$
T_p(t) = T_c + (T_{p0} - T_c)\exp(-t/\tau_T).
$$

$$(2.8)$$

The particle temperature, which starts at T_{p0}, thus evolves exponentially toward the ambient fluid temperature. As before, we can observe that over a thermal time scale (i.e., for $t = \tau_T$), the particle temperature decreases by a factor $1/e$ toward the ambient temperature.

We now investigate typical values of the translational and thermal time scales of the dispersed phase under the following three scenarios:

- An aerosol particle or water droplet of density 1000 kg/m^3 falling through air of approximate density 1 kg/m^3.
- A sediment particle of density 2650 kg/m^3 falling through a column of water.
- A spherical bubble of density 1 kg/m^3 rising through a water column.

The three scenarios cover respectively very large, $O(1)$, and very small density ratios. In each case, the size of the dispersed phase is varied and the results are plotted in Figure 2.2. The quasi-steady drag given by the first term on the right-hand side of Eq. (2.1) and the quasi-steady heat transfer given by the right-hand side of Eq. (2.5) are for a rigid particle and do not strictly apply for a droplet or bubble. With internal circulation inside the droplet or bubble, the drag force differs from the Stokes drag on a rigid particle, and is given by the *Hadamar–Rybczynski drag law*. We will consider this problem in greater detail in Chapter 4. Here we merely note that the time scales of a droplet or bubble with internal circulation are different by a factor of $2/3$ at most from those obtained from the above relations for a rigid particle that ignores internal circulation. Furthermore, in the case of bubbles in normal water (which invariably is not ultra pure), the bubble's drag force is closer to that of a rigid particle of the same size. Thus, the results presented in Figure 2.2 are quite useful even in the case of droplets and bubbles.

The translational and thermal time scales defined in Eqs. (2.2) and (2.7) are plotted in Figures 2.2a and b for the three different scenarios. These plots are evaluated at the terminal free-fall velocity or free-rise velocity in case of a bubble. The Reynolds numbers in the expressions (2.3) and (2.6) for Φ and Φ_T are evaluated at the terminal velocity. As the particle diameter is varied from 0.1 μm to 0.1 mm, the translational and thermal time scales increase as d_p^2. As we will see below, in this range the particle Reynolds number remains smaller than unity and thus $\Phi \sim 1$ and $\Phi_T \sim 1$. This, along with the time-scale expressions given in Eqs. (2.2) and (2.7), explains the quadratic dependence on d_p. For larger diameter, the nonlinear effect due to finite Reynolds number comes into play and the increase in time scale slows down. The time scales of water droplets are larger than those of sedimenting particles, which in turn are larger than those of air bubbles. We caution that these figures must be used only as a guideline, since other physical effects will significantly influence the actual values. For example, large droplets and bubbles will not remain spherical.

Also of interest is the ratio τ_T/τ_p, since it measures how fast a particle will reach the surrounding temperature compared to reaching the surrounding velocity. This ratio can be expressed as

$$\frac{\tau_T}{\tau_p} = 3 \Pr \frac{\rho C_r}{1 + 2\rho} \frac{\Phi}{\Phi_T}, \qquad (2.9)$$

and is plotted in Figure 2.3 for the different cases. Depending on the scenario, the ratio can be as small as 10^{-4} or as large as 1.5. The ratio is small for an air bubble in water, indicating that a bubble will thermally equilibrate with the ambient faster than its translational motion. In the case of a water drop in air, the translational and thermal time scales are comparable and their ratio is nearly independent of the droplet size.

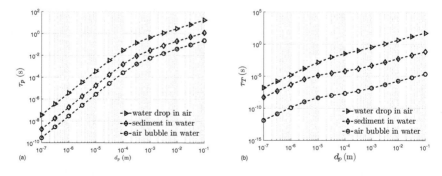

Figure 2.2 (a) Plot of translational time scale τ_p of the particle as a function of particle diameter for the three different scenarios: (i) water droplet in air; (ii) sand particle in water; and (iii) air bubble in water. (b) Plot of thermal time scale vs. d_p for the three different scenarios.

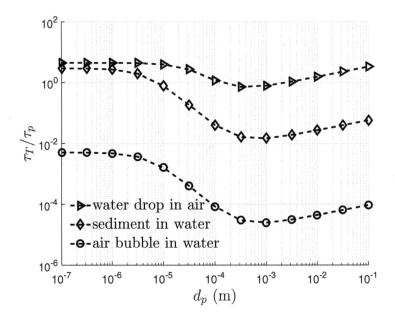

Figure 2.3 Plot of the time-scale ratio τ_T/τ_p as a function of d_p.

2.4.3   Settling Velocity and Reynolds Number

We now investigate the terminal velocity at which an isolated particle will fall in still fluid. The equation of motion includes the effect of gravity (third term on the right-hand side):

$$\rho_p \mathcal{V}_p \frac{dV}{dt} = -3\pi\mu_f d_p V \, \Phi - \frac{1}{2}\rho_f \mathcal{V}_p \frac{dV}{dt} + (\rho_p - \rho_f)\mathcal{V}_p \, g \,, \qquad (2.10)$$

where g is the acceleration due to gravity and V is the particle velocity in the downward direction. When released from rest, V will increase from 0 and accelerate toward the terminal velocity (V_s); the time scale of this change in particle velocity is τ_p. The final steady state of the falling particle is given by $dV/dt = 0$. Using this in Eq. (2.10), the terminal velocity is obtained from a balance of the first and third terms on the right-hand side:

$$V_s = \frac{d_p^2}{18\nu_f \, \Phi(\mathrm{Re}_s)}(\rho - 1)g = \frac{\rho - 1}{\rho + 1/2}\tau_p \, g \,, \qquad (2.11)$$

where $\Phi(\mathrm{Re}_s)$ is the correction factor evaluated at the Reynolds number of the final terminal velocity. The terminal velocity of a heavy particle (i.e., as $\rho \to \infty$) is $\tau_p \, g$ and at finite values of density ratio this value is modified by the factor $(\rho - 1)/(\rho + 1/2)$.

The particle Reynolds number is defined in terms of particle diameter as

$$\mathrm{Re} = \frac{|u_c - V| \, d_p}{\nu_f} \,, \qquad (2.12)$$

and it depends on the relative velocity. In the case of a particle settling at its terminal velocity, the relative velocity is known, and the Reynolds number of a particle falling at terminal velocity can be obtained as

$$\mathrm{Re}_s = \frac{V_s \, d_p}{\nu_f} . \tag{2.13}$$

The terminal settling velocity and the corresponding Reynolds number for the three different scenarios are presented in Figure 2.4. From the Reynolds number plot it is clear that the difference between the different scenarios is small. The most important observation is that in all three scenarios the Reynolds number based on settling velocity transitions from $\mathrm{Re}_s < 1$ to $\mathrm{Re}_s > 1$ at about a diameter of 0.1 mm. This transition was earlier observed in the time-scale plots and can also be observed in the settling velocity plot shown in Figure 2.4a.

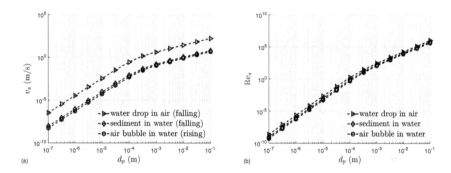

Figure 2.4 (a) Plot of terminal velocity of the particle as a function of particle diameter for the three different scenarios: (i) water droplet in air; (ii) sand particle in water; and (iii) air bubble in water. (b) Particle Reynolds number at terminal velocity.

Problem 2.1 In Figures 2.2, 2.3, and 2.4, the physical parameters that were used in the calculation are as follows:

$$\text{Air} \begin{cases} \rho = 1 \, \text{kg/m}^3, \; \nu = 1.568 \times 10^{-5} \, \text{m}^2/\text{s}, \\ \kappa = 2.2 \times 10^{-5} \, \text{m}^2/\text{s}, \; C_p = 1005 \, \text{J/(kg K)}, \end{cases}$$

$$\text{Water} \begin{cases} \rho = 1000 \, \text{kg/m}^3, \; \nu = 1.0 \times 10^{-6} \, \text{m}^2/\text{s}, \\ \kappa = 0.143 \times 10^{-6} \, \text{m}^2/\text{s}, \; C_p = 4181 \, \text{J/(kg K)}, \end{cases}$$

$$\text{Sand} \begin{cases} \rho = 2650 \, \text{kg/m}^3, \; C_p = 1381 \, \text{J/(kg K)}. \end{cases}$$

(a) Use these values and calculate the settling velocity for the three scenarios and the corresponding particle Reynolds number.

(b) Use the Reynolds number to calculate Φ and Φ_T. Also evaluate τ_p and τ_T for the three scenarios to reproduce the plots given in Figures 2.2, 2.3, and 2.4. At finite Reynolds number you need to solve a nonlinear algebraic equation.

2.4.4   Velocity Due to Collective Action

In Section 1.3.2 we considered several factors that modify the effective settling velocity of the dispersed phase. Here we present a powerful demonstration of the collective action of the dispersed phase at work. Consider a spherical cloud of particles released into an otherwise quiescent surrounding fluid, as shown in Figure 2.5a. Let the individual particles be of diameter d_p and the volume fraction of the cloud ϕ_d be uniform. Let the cloud be a large sphere of radius $\mathcal{D}$. If we take the continuous-phase density to be a constant, then the density of the two-phase mixture within the cloud becomes

$$\rho_m = \rho_p\,\phi_d + \rho_f\,(1 - \phi_d) = \rho_f + \rho_f(\rho - 1)\phi_d\,. \tag{2.14}$$

If we consider the cloud to be a sphere of this mixture density, then its settling velocity can be estimated from Eq. (2.11) as

$$V_{cl} = \frac{\mathcal{D}^2}{18\nu_f\,\Phi(\mathrm{Re}_s)}\left(\frac{\rho_m}{\rho_f} - 1\right)g = \frac{\left(\mathcal{D}\sqrt{\phi_d}\right)^2}{18\nu_f\,\Phi(\mathrm{Re}_s)}(\rho - 1)g\,. \tag{2.15}$$

Thus, the cloud will settle with an effective settling velocity corresponding to that of a rigid solid particle of diameter $\mathcal{D}\sqrt{\phi_d}$. Now consider a sediment-laden cloud of diameter 10 cm in water. Let the sediment particles be sand grains of density 2650 kg/m^3 and diameter 100 μm at 1% volume fraction. From the above argument we see that the cloud will settle as if it is a particle of diameter 1 cm. From Figure 2.4 it can be seen that the settling velocity of an individual 100 μm sediment grain is less than 1 cm/s, while that of the cloud is 1 m/s.

Admittedly the analysis for the cloud is crude, as it ignores (i) flow between the particles within the cloud at the microscale, (ii) the large-scale circulation that will develop within the cloud, and (iii) the change in shape of the cloud as it moves down, among other approximations. Nevertheless, an order of magnitude increase in settling velocity is observable in nature and is due to the collective action of the dispersed phase. This mechanism is the cause of the significant increase in settling velocity observed in the sediment fingers in Figure 1.2.

Also from Figure 2.4 we can estimate the Reynolds number of the flow around an individual 100 μm grain to be about 1. The Reynolds number of the flow around a 1 cm particle is about 10^4. The Reynolds number of the sediment cloud will be even larger, since the diameter of the cloud is 10 cm, while its effective settling velocity is that of a 1 cm particle. In this example, the 10 cm cloud contains 1% by volume particles and 99% by volume fluid, and it moves down relative to the surrounding fluid at around 1 m/s. Within the cloud, at the microscale, particles may move relative to the local fluid at a much slower settling velocity of 1 cm/s.

Another example of this collective action is the sediment-laden gravity current shown in Figure 2.5b. If the height of the current is H, then the velocity of the sediment-laden front propagating into the ambient clear water of density ρ_f can be estimated as

$$V_{fr} = \sqrt{\left(\frac{\rho_m}{\rho_f} - 1\right) g H} = \sqrt{(\rho - 1)\,\phi_d g\,H}\,. \tag{2.16}$$

Consider a turbidity current of height 1 m containing finer silt particles of 30 μm size at a much lower volume fraction of 0.1%. While the settling velocity of an individual grain is about 1 mm/s, the current itself will be moving at about 135 mm/s. What is impressive about this example is that the same excess weight of an individual grain that was the source of its settling velocity, when working collectively with all the other particles within the current, is able to drive the current at a velocity two orders of magnitude larger.

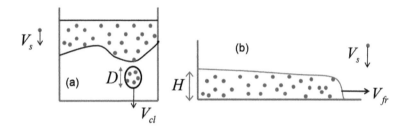

Figure 2.5 Schematics showing the power of collective action of the dispersed phase at the mesoscale. (a) The settling velocity of a single isolated particle, u_{pf}, is given in Eq. (2.11), while that of a spherical cloud is given in Eq. (2.15). (b) Another example of a sediment-laden gravity current. Here the current velocity is perpendicular to the direction of gravity and can be much larger than the terminal fall velocity of an isolated particle.

2.4.5 Stokes Number

The Stokes number is a parameter of great importance in dispersed multiphase flows. It captures the response of a particle to changes in the surrounding fluid. In Figure 2.6, flow through a converging nozzle is shown, where the fluid velocity increases as it goes through the nozzle. It is easy to envision an inertial (sluggish) particle taking time to respond to this change in fluid velocity. The centerline fluid velocity changes from an initial value of u_i to the final post-nozzle velocity of u_e at the end of the nozzle. This change in fluid velocity is due to the change in cross-sectional area across the nozzle. A particle initially traveling along the centerline at u_i will also eventually reach the post-nozzle velocity of u_e. But the downstream distance over which this change occurs increases with increasing particle time scale (see Figure 2.6b).

We define a fluid time scale $\tau_c \approx 2h/(u_i + u_e)$, where h is the length of the nozzle and $(u_i + u_e)/2$ is the average fluid velocity. The Stokes number is then defined as

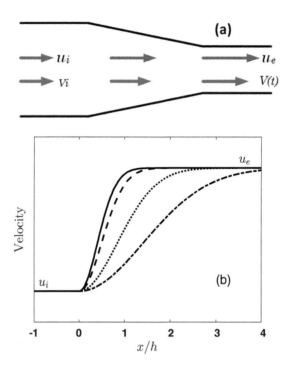

Figure 2.6 (a) Schematic nozzle where the blue arrows indicate fluid velocity which increases from u_i to u_e across the nozzle of length h. The red arrows correspond to the particle velocity V_i which was the same as that of the fluid at the entrance to the nozzle. Due to the particle inertia, its increase in velocity across the nozzle is slow. At the exit of the nozzle, the particle velocity V_e is smaller than u_e but will eventually approach u_e further downstream. (b) The streamwise increase in fluid velocity (solid line) is compared with that of the particle. Schematics of three different particle Stokes numbers are shown. The dashed line corresponds to small Stokes number, which nearly moves with the fluid. The dotted line is intermediate Stokes number and the dashed–dotted line is much larger Stokes number.

$$\text{St} = \frac{\tau_p}{\tau_c} . \tag{2.17}$$

If the Stokes number is small, then the particle is able to quickly adjust to the change in fluid velocity and the particle velocity across the nozzle will closely follow that of the fluid. In the other limit of very large Stokes number ($\text{St} \gg 1$), the particle is so sluggish that its velocity will remain close to u_i even long after the fluid velocity has changed to u_e. At modest values of St, the particle velocity will change from u_i to u_e over a distance longer than h.

In Figure 2.6 we only provided a schematic of the inertia effect of the particle; the actual lag in velocity between the particle and the local fluid depends on the details of how the fluid velocity changes within the nozzle. Here we will consider an even simpler setup that illustrates the same physics, but whose solution can easily be obtained.

Particle Falling Through a Linear Shear Flow

Consider a particle of diameter d_p released from rest at $t = 0$ from the top of a shear flow that extends downward from $y = 0$ (see Figure 2.7; note that positive y points downward). In this simple linear shear flow, the fluid velocity is given by $u(y) = Gy$, where G (the shear magnitude) is positive. Due to gravity, the particle will fall down into the shear flow and the streamwise fluid velocity will accelerate the particle along the horizontal direction. Our quest is to find the vertical and horizontal velocities of the particle and relate them to those of the local fluid. This will allow us to compute the slip velocity between the particle and the surrounding fluid.

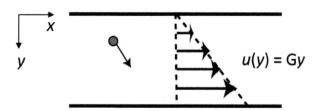

Figure 2.7 Schematic of a particle falling down a linear shear flow. As the particle falls down at its terminal velocity, its streamwise velocity will increase. But it will lag the local fluid velocity due to its inertia and the lag depends on the Stokes number.

Similar to Eq. (2.1), the governing equations of motion for the particle along the vertical and horizontal directions can be written as

$$(\rho_p + \tfrac{1}{2}\rho_f)\mathcal{V}_p \frac{dV_x}{dt} = 3\pi\mu_f d_p(u(y) - V_x)\,\Phi,$$

$$(\rho_p + \tfrac{1}{2}\rho_f)\mathcal{V}_p \frac{dV_y}{dt} = (\rho_p - \rho_f)\mathcal{V}_p\, g - 3\pi\mu_f d_p V_y\,\Phi,$$

(2.18)

where g is the acceleration due to gravity. On the left-hand side we have the mass plus the added mass of the particle times the acceleration of the particle along the x and y-directions. On the right-hand side of the x equation of motion, the only force on the particle is the drag force due to the relative motion between the particle and the surrounding fluid. In the vertical direction, in addition to the drag force, we also have the weight of the particle minus the buoyancy force.

We first solve the particle motion in the y-direction. If we introduce the particle time scale τ_p as defined in Eq. (2.2) and the particle settling velocity as in Eq. (2.11), the y equation of motion can be rewritten as

$$\frac{dV_y}{dt} = \frac{V_s - V_y}{\tau_p},$$

(2.19)

whose solution for an initially stationary particle is

$$V_y(t) = V_s(1 - e^{-t/\tau_p}).$$

(2.20)

Here and for the rest of this discussion we will assume τ_p to be time invariant, which is appropriate in the Stokes flow limit. Thus, the vertical fall velocity of the particle will increase exponentially and approach the terminal fall velocity. As is to be expected, the time scale of this exponential increase is the particle time scale.

For the investigation of the particle motion in the x-direction, let us simplify the analysis by ignoring the exponential term in the vertical velocity (the full analysis with the exponential term is also possible and is left as an exercise for you to do). The vertical position of the particle as a function of time is then $Y_p(t) = V_s t$ and the fluid velocity at the particle position can be expressed as $u(Y_p) = GV_s t$. With these, the x equation of motion can be rewritten as

$$\frac{dV_x}{dt} = \frac{GV_s t - V_x}{\tau_p},$$
(2.21)

whose solution can be written as

$$V_x(t) = V_s Gt - V_s G\tau_p (1 - e^{-t/\tau_p}).$$
(2.22)

We again consider $t \gg \tau_p$ and ignore the transient effect represented by the exponential term. Since we can express time in terms of the vertical location of the particle, we rewrite the above as

$$V_x(Y_p) = u(Y_p) - V_s \text{ St}.$$
(2.23)

Here the Stokes number $\text{St} = \tau_p G$ and G can be interpreted as the inverse time scale of the shear flow. From the above, it is clear that as the particle falls through the linearly increasing shear flow, at times much longer than the particle time scale, the particle velocity will lag the local fluid velocity by an amount V_s St, which is the streamwise slip velocity. This relative velocity is, in addition to the vertical slip, represented by the settling velocity.

This lag is due to the inertia of the particle. For small St, as it falls down the particle is able to adjust to the ever-increasing fluid velocity and be very close to it. This is just like a particle of small Stokes number being able to stay with the fluid and increase in velocity across the nozzle in the previous example. Meanwhile for large St the streamwise slip velocity will be large. This example nicely illustrates the role of the Stokes number in capturing the inertia effect of a particle.

Note that if we replace the heavier particle by a lighter-than-fluid oil droplet or bubble, there will be a corresponding rise in velocity and it is easy to see that the bubble's horizontal velocity will lead that of the local fluid. This is because the bubble will arrive at any y location from below, from a region of higher horizontal fluid velocity. The bubble will remember this past history and this inertia effect results in a positive slip velocity of the bubble.

Problem 2.2 (a) First, verify Eqs. (2.18).

(b) Follow the steps outlined above and solve both the x and y equations of motion exactly, without ignoring the term e^{-t/τ_p}. You still need to assume the time scale τ_p to be a constant. In the end, if you ignore the transient terms represented by the decaying exponentials, show that you still obtain Eq. (2.23).

Particle in a Vortex

Another example shown in Figure 2.8 is a particle trajectory in a vortex. While the fluid streamlines go round and round, a heavier-than-fluid particle will spiral out due to its inertia. A lighter-than-fluid bubble will migrate toward the center of the vortex. This behavior is also captured by the Stokes number, which is now defined as the ratio of the particle time scale to the time scale of the vortex. If we take the vortex to be in solid body rotation with angular velocity ω, then the vortex time scale can be taken to be $1/\omega$ and the Stokes number is given by St $= \omega\tau_p$. If St is small, then the particle will closely follow the circular fluid streamlines. Heavier-than-fluid particles of St $\gg 1$ will spiral out rapidly, while bubbles will spiral inward to the vortex center.

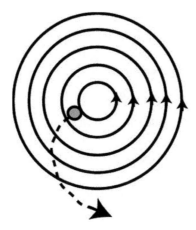

Figure 2.8 Inertial motion of a heavy particle in a vortex of radially dependent circumferential velocity $u_\theta(r)$. While the fluid streamlines are circular, the particle will have an outward radial velocity and thus will spiral out. The degree of outward spiraling will depend on the Stokes number.

In order to mathematically obtain the above results, we start with the radial component of the particle equation of motion

$$\left(\rho_p + \frac{1}{2}\rho_f\right)\mathscr{V}_p\frac{dV_r}{dt} = \rho_p\mathscr{V}_p\frac{V_\theta^2}{R_p} - \rho_f\mathscr{V}_p\frac{u_\theta^2}{R_p} - 3\pi\mu_f d_p V_r \,\Phi, \qquad (2.24)$$

where on the left-hand side we have the mass plus the added mass of the particle times the radial acceleration of the particle. On the right-hand side the first term is the centripetal force on the particle, where V_θ is the circumferential velocity of the particle and R_p is the radial location of the particle. The second term arises from the radial pressure gradient in the vortex, which from the r component of the Navier–Stokes equation becomes $dp/dr = \rho_f u_\theta^2/r$. This radial increase in pressure from the center of the vortex generates an inward-pointing pressure gradient force on the particle. Note that for a solid-body vortex $u_\theta = \omega R_p$ corresponds to the circumferential velocity of the vortex at the radial location of the particle R_p. Thus, the first two terms on the right are analogous to the gravitational and buoyancy forces in the lower of Eqs. (2.18), with the acceleration V_θ^2/R_p or u_θ^2/R_p playing the role of g. The last term on the right is the radial drag on the particle.

In order to solve the above equation, we first make the assumption that the circumferential velocity of the particle is nearly the same as the local fluid (i.e., $V_\theta \approx u_\theta$). As we will see at the end of the solution, the difference is $O(\text{St}^2)$ and therefore can be taken to be very small in case of $\text{St} \ll 1$. We introduce the density parameter

$$\beta = \frac{\rho - 1}{\rho + 1/2} \tag{2.25}$$

and rewrite Eq. (2.24) as

$$\frac{dV_r}{dt} = \beta \omega^2 R_p - \frac{V_r}{\tau_p}. \tag{2.26}$$

Assuming τ_p to be a constant, the above can be solved along with the definition $V_r = dR_p/dt$. For small Stokes number, the effect of the left-hand side becomes small and the following solution can be obtained:

$$R_p = R_{p0} \exp(\beta\, \text{St}\, \omega t) \quad \text{and} \quad V_r = \beta\, \text{St}\, u_\theta, \tag{2.27}$$

where R_{p0} is the particle location at $t = 0$. The radial velocity is proportional to both St and the local circumferential velocity of the vortex. Note that the density parameter β is zero for a neutrally buoyant ($\rho = 1$) particle, positive for a heavier-than-fluid particle, and negative for a lighter-than-fluid particle. As ρ ranges from 0 to ∞, β varies from -2 to 1. Thus, radial migration of a particle is similar to the particle falling down the linear shear flow due to gravity. Heavier particles spiral out, while lighter particles spiral in.

We can also anticipate the circumferential velocity of the particle to be different from the local fluid velocity due to the radial migration of the particle. To obtain an expression for the circumferential velocity, we start with the θ equation of motion of the particle

$$\left(\rho_p + \frac{1}{2}\rho_f\right)\mathcal{V}_p\frac{dV_\theta}{dt} = 3\pi\mu_f d_p(u_\theta(R_p) - V_\theta)\,\Phi, \tag{2.28}$$

where the only circumferential force on the particle is the drag force given on the right-hand side. Evaluating the fluid velocity $u_\theta = \omega R_p$ at the particle position given in Eq. (2.27) and substituting in the above, we obtain

$$\frac{dV_\theta}{dt} = \frac{\omega\, R_{p0} \exp(\beta\, \text{St}\, \omega t) - V_\theta}{\tau_p}. \tag{2.29}$$

The above equation can readily be solved. We then consider the small Stokes number limit and ignore the transient terms involving e^{-t/τ_p}, which yields the following simplified expression for the circumferential particle velocity:

$$V_\theta(R_p) = \frac{u_\theta(R_p)}{1 + \beta\, \text{St}^2} = u_\theta(R_p)(1 - \beta\, \text{St}^2 + \cdots). \tag{2.30}$$

Here, in the second expression we have expanded the denominator for $\text{St} \ll 1$. Thus, a heavier-than-fluid particle ($\beta > 0$) will lag the fluid because as it travels radially out it constantly moves from a region of lower to a region of higher circumferential fluid velocity. In contrast, a lighter-than-fluid particle ($\beta < 0$) will lead the fluid because as it travels radially inward it constantly moves from higher to lower circumferential fluid velocity. The similarity between this problem and the previous problem of particle settling through a shear flow becomes clearer if we use the expression for V_r from Eq. (2.27) in the above to get

$$V_\theta(R_p) = u_\theta(R_p) - V_r(R_p)\,\mathrm{St}, \qquad (2.31)$$

which is analogous to Eq. (2.23). The only difference is that, in the previous example, the vertical velocity of a settling particle is a constant, while in the case of a particle spun out by a vortex, the radial velocity increases with radial position and accordingly circumferential slip increases with radial position. Finally, we can also justify the approximation $V_\theta \approx u_\theta$ made at the beginning, since the difference is $O(\mathrm{St}^2)$.

Problem 2.3 (a) Verify Eqs. (2.24) and (2.28).

(b) Then, follow the steps and obtain the results in Eqs. (2.27) and (2.30).

The above examples clearly illustrate the role of particle inertia and its parameterization through the Stokes number. In all the examples, in the limit $\mathrm{St} \to 0$ the particle follows the fluid faithfully. At finite inertia, departure of the particle path from the fluid streamlines increases with St. In the case of gravitational settling of particles, the primary slip velocity is in the vertical direction, but it also includes an $O(\mathrm{St})$ slip in the streamwise velocity. Similarly, in the case of a particle in a vortex, the primary slip is in the radial direction, which in turn induces a slip in the circumferential velocity. The $O(\mathrm{St})$ slip velocity is a general feature of particle motion and we will formally obtain this as an asymptotic solution of the particle equation of motion in Chapter 12.

In a turbulent flow, there are a range of flow scales from the integral scale eddies to the Kolmogorov eddies. Correspondingly, there are a range of time scales from the integral time scale $\mathcal{T}$ to the Kolmogorov time scale τ_k. So, a range of Stokes numbers can be defined: $\mathrm{St}_I = \tau_p/\mathcal{T}$ is the smallest Stokes number and it measures the response of the particle to large integral-scale eddies. The Stokes number based on the Kolmogorov scale, $\mathrm{St}_k = \tau_p/\tau_k$, characterizes the response of the particle to small Kolmogorov-scale eddies. Note that $\mathrm{St}_I \ll \mathrm{St}_k$ since the time scale of the Kolmogorov-scale eddies is much smaller than that of the integral time scale. Thus, three different particle regimes can be identified. (i) Very small particles with $\mathrm{St}_I \ll \mathrm{St}_k \ll 1$, indicating that they respond well to all turbulent eddies. Such particles faithfully follow the turbulent fluid quite well. (ii) Medium-sized particles, with $\mathrm{St}_I \ll 1$ but $\mathrm{St}_k \gg 1$, which respond well and follow the large eddies of turbulence, but do not follow the smaller scales of motion. (iii) Very large particles ($1 \ll \mathrm{St}_I \ll \mathrm{St}_k$), which are ballistic and do not respond to any of the turbulent eddies. We will consider these regimes in greater detail in Chapter 12.

2.4.6   Rotational and Pressure-Equilibrium Time Scales

In Section 2.2 we discussed the slip effect as an important source of perturbation to the single-phase limit. The translational slip velocity between the particle and the surrounding fluid is the primary source of momentum exchange between the particle and the surrounding fluid, and it is parameterized as the drag force. If the slip velocity is zero, then the particle is perfectly synchronized with the local fluid and the slip-induced drag force is zero. As we will see throughout this book, the essence of multiphase flow is in the understanding and modeling of mass, momentum, and energy exchanges between the phases. The time scales associated with these exchanges are of great importance.

In Sections 2.4.1 and 2.4.2 we discussed translational and thermal time scales, which correspond to linear momentum and energy exchange between the particle and the fluid phases. There are two other exchanges, which are in general somewhat less important for the evolution of the multiphase flow. For completeness, here we will discuss the time scales associated with these exchanges.

Rotational Time Scale

The first of these other exchanges concerns the rotational motion of the particle and the corresponding local angular velocity of the fluid. The problem of rotational motion of a particle was first considered by Basset (1888) and later improved by Feuillebois and Lasek (1978). Consider a spherical particle that is held stationary in a quiescent fluid for time $t \leq 0$. For $t > 0$, a constant torque T is applied, which results in the rotation of the particle about, say, the z-axis. The particle begins to accelerate angularly, and its angular velocity Ω_p as a function of time is given as

$$I_p \frac{d\Omega_p}{dt} = T - \pi \mu_f d_p^3 \Omega_p \Phi_R, \tag{2.32}$$

where I_p is the moment of inertia of the particle and Φ_R is the nonlinear correction factor that depends on the rotational Reynolds number defined as $\mathrm{Re}_R = |\Omega_p| d_p^2 / \nu_f$. In the Stokes limit of $\mathrm{Re}_R \to 0$, there is no need for a nonlinear correction and $\Phi_R \to 1$. In the above, the time rate of change of angular momentum of the particle given on the left-hand side is set equal to the external torque applied on the particle minus the hydrodynamic torque applied by the surrounding stagnant fluid resisting the relative rotation of the particle.

Over time, the particle will reach its terminal angular velocity

$$\Omega_{p,s} = \frac{T}{\pi \mu_f d_p^3 \Phi_R}, \tag{2.33}$$

which is obtained by balancing the two terms on the right-hand side. The constant torque T plays the role of gravitational force and as a result the terminal angular velocity is the analog of the terminal settling velocity. Note that there is no term in Eq. (2.32) that is analogous to the added-mass force. In other words, there is no added moment of inertia. One may define a rotational time scale τ_r as the time it takes for the sphere to reach $(e - 1)/e$ of its terminal value (which is the same as the difference decreasing by a factor $1/e$). For a solid sphere, the moment of inertia is $I_p = m_p d_p^2 / 10$, where m_p is the mass of the particle. With this we obtain the time scale of rotation to be

$$\tau_r = \frac{1}{60} \frac{d_p^2 \rho}{\nu_f \Phi_R} .$$
(2.34)

The rotational time scale is the time it takes for a particle's angular velocity to adjust to the local angular velocity of the fluid. The time-scale ratio τ_r / τ_p can be evaluated and, in the Stokes limit, when $\Phi = \Phi_R = 1$, τ_r / τ_p depends only on the particle-to-fluid density ratio. The time-scale ratio is listed in Table 2.2 for different density ratios. It is clear that over the entire range of density ratio the particle will approach rotational equilibrium faster than the translational equilibrium.

Table 2.2 Ratio of rotational to translational time scale in the Stokes limit.

ρ	τ_r / τ_p
0.1	0.050
0.5	0.150
1	0.200
2	0.240
5	0.273
10	0.286
50	0.297
100	0.299
200	0.299
500	0.299

Pressure Equilibrium

Finally, we address the process and time scale on which the dispersed-phase pressure equilibrates with the continuous-phase pressure. The pressure equilibrium is most relevant for the case of a bubble and therefore in this subsection we will use the term *bubble* in place of *particle*. In general, the state of the fluid around a bubble is given by fluid velocity, pressure, and temperature. If the bubble velocity and temperature are different from those of the local fluid, then the force exchange and heat transfer between the bubble and the fluid will tend toward equilibrium. Similarly, if the bubble pressure is different from that of the surrounding, then the bubble size will change over time in order to reach an equilibrium pressure.

The key aspect of the pressure equilibrium process is that it is oscillatory. As discussed in the previous sections, if a particle's velocity is different from the constant velocity of the surrounding fluid or if a particle's temperature is different from the constant temperature of the surrounding fluid, then the particle's velocity or temperature will monotonically change toward those of the surrounding fluid. This is not the case with pressure equilibrium. If a bubble's initial pressure is different from that of the surrounding medium, then its radius will undergo oscillation and correspondingly the pressure within the bubble will overshoot and undershoot the ambient pressure on its

way to equilibrium. Without going through the derivation, here we will simply present
the natural frequency of oscillation of the bubble to be

$$\omega_p = \left[\frac{3k(p_\infty - p_{b0})}{\rho_f R_{eq}^2} \right]^{1/2}, \tag{2.35}$$

where p_∞ is the constant ambient pressure, p_{b0} is the initial bubble pressure, R_{eq} is
the final equilibrium radius of the bubble when the bubble pressure is equal to the
ambient pressure, and k is a constant equal to 1 when the bubble is isothermal and
γ under adiabatic conditions (γ is the specific heat ratio of the bubble material). The
pressure-equilibration time scale is the inverse of the frequency of oscillation (i.e.,
$\tau_{pr} = 1/\omega_p$).

The pressure equilibrium is important in the case of droplets and particles when they
are subjected to very high-speed flows such as shocks. For example, when a particle
is subjected to an intense shock, the fluid pressure around the particle rapidly changes
from the pre-shock pressure to the post-shock pressure. In response, the pressure within
the particle also rapidly changes and this equilibration happens on the acoustic time
scale, defined as the ratio of the particle diameter to the speed of sound within the
particle, d_p/a_p. For a 100 μm particle subjected to an air shock, this time scale is
0.3 μs. Thus, under ordinary conditions, pressure equilibration is very rapid and the
time scale of pressure equilibration is not critical in case of droplets and particles.

2.4.7   Interparticle Collisional Time Scale

Another important feature of dispersed multiphase flows is collisions between the
dispersed-phase elements. The frequency of these collisions increases with increasing
volume fraction occupied by the dispersed phase, and also with increasing relative
motion between the particles. Here the objective is to get a very simple estimate of the
time scale on which these collisions occur.

Consider a reference particle marked A in Figure 2.9 surrounded by a random
distribution of other particles. For simplicity, consider the relative velocity of the
reference particle with respect to all other particles to be u_r. Over time Δt, the reference
particle can be pictured to trace out a cylinder of diameter $2d_p$ and length $u_r \Delta t$ (see
Figure 2.9). If we ignore any close-range hydrodynamic effect between the particles,
then all the particles marked B whose center lies within the traced cylinder can be
considered to have collided with the reference particle during time Δt. Thus, the
number of collisions during the period Δt can be estimated as

$$\pi d_p^2 \, \bar{n} \, u_r \, \Delta t, \tag{2.36}$$

where $\bar{n}$ is the average number density of particles in the cylindrical region traced out by
the reference particle. From the above, the time between collisions can be estimated as

$$\tau_{co} = \frac{1}{\pi d_p^2 \, \bar{n} \, u_r}, \tag{2.37}$$

which is the collisional time scale.

The above estimate is crude and involves several approximations. Most importantly, it does not recognize the fact that a random distribution of particles obeys the Poisson statistics (this aspect will be discussed in Section 3.3). As a result, the collisional time is a random variable whose mean value can be carefully evaluated. But the above crude estimate is sufficient for us to evaluate the importance of collisions in a multiphase flow. Note that even if the two colliding particles have been following the surrounding fluid closely before collision, immediately after collision both their motions will deviate from the local fluid velocity. If $\tau_p > \tau_{co}$ then collisions are happening more frequently than particles are able to readjust their velocities to the surrounding fluid. Such a scenario can be termed a *collision-dominated regime* or *dense regime*. In this regime, the particle velocity will be determined by the collisions and will not be dictated by the surrounding fluid flow directly. On the other hand, if $\tau_p < \tau_{co}$ then collisions are happening less frequently and the multiphase flow can be considered not dense.

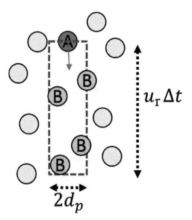

Figure 2.9 The reference red particle is falling at a relative velocity u_r with respect to all other particles. Over time Δt it will trace out a cylinder of diameter $2d_p$ and will collide with all particles whose center is within the cylinder. This simple picture of course ignores the motion of all the particles due to their hydrodynamic interaction. This picture of particle–particle interaction can easily be extended to collisions between particles of varying diameter.

3 Description of the Dispersed Phase

In this chapter our attention will primarily be restricted to the dispersed phase. Clearly the continuous phase is also important, but in this chapter we will discuss the state or evolution of the continuous phase only as needed in the context of characterizing the state of the dispersed phase. Consider the case of a turbulent multiphase flow with a random distribution of monosized spherical particles (or droplets or bubbles) within it. Imagine taking pictures of the particle distribution in an experiment (i.e., in one realization) without recording the details of the flow surrounding the particles. Each picture records the state of the dispersed phase at one time instant, and a sequence of pictures records the evolution of the dispersed phase in that realization. Repeating this process for several runs of the experiment will yield an extensive record of the ensemble. In this chapter we address the question of how best to mathematically describe or characterize the state of the dispersed phase as seen in an individual picture, as seen in an ensemble, and the time evolution of the dispersed phase.

The crucial point we want to stress from the beginning is the fact that a multiphase flow is chaotic by its very nature. In order to appreciate the inherent randomness of a multiphase flow, let us discuss three simple related scenarios. The first is the familiar single-phase laminar pipe flow with a parabolic velocity profile. From elementary fluid mechanics we know that in this laminar flow only the streamwise or x-component of velocity is nonzero and is a function of only the radial position, while the other two components of velocity are identically zero.

$$\text{Laminar pipe flow:} \quad u_x(r), \ u_r \equiv 0, \ u_\theta \equiv 0. \tag{3.1}$$

The second is a single-phase turbulent flow through the same pipe. Now we are faced with a complicated flow whose three velocity components are all functions of the three coordinates and also time dependent. Thus, the turbulent flow is characterized by

$$\text{Laminar pipe flow:} \quad u_x(r,\theta,x,t), \ u_r(r,\theta,x,t), \ u_\theta(r,\theta,x,t). \tag{3.2}$$

Now consider a particle-laden flow through the same pipe, such as the one shown in Figure 3.1, under conditions which we would have called laminar in the single-phase limit. In other words, in the single-phase limit, or in the absence of particles, let the flow be laminar. Even then, a velocity probe placed within the pipe will measure a random time-dependent velocity. Each particle within the pipe will not simply be moving with the local fluid. As we will see later in Chapter 4, due to local fluid shear, the particle will experience a lift force which will move the particle radially. As we saw in Figure 2.7, this transverse motion will also create a streamwise slip between the particle and the surrounding fluid. In addition, since particles whose center is closer to the pipe centerline will mover faster than particles further away from the centerline, there will be collisions between particles, which will result in complex post-collision motion of the colliding particles. In essence, due to the presence of the particles, the flow within the pipe will become complex. The velocity at the probe may be nearly steady when there are no particles around it, but every time one or more particles go nearby there will be velocity fluctuation. This is what we earlier referred to as pseudo turbulence. Thus, even in a slow-moving multiphase flow, the velocity field is characterized by its complete space and time dependence as given in Eq. (3.2). In this sense, a multiphase flow is chaotic due to the random location of particles and their chaotic motion.

In a turbulent multiphase flow we can identify two distinct sources of fluctuations. The first is the continuous-phase turbulence, which can exist even in the absence of the dispersed phase. The second is pseudo turbulence, which arises from the random position, finite size, and relative motion of the particles, which can exist even in the absence of continuous-phase turbulence. Typically, both these sources of fluctuations co-exist.

Following traditional turbulence literature, we define an individual dispersed multiphase flow as a *realization*. For example, a single run of a laboratory experiment of a lock-release turbidity current (Sequeiros et al., 2009) or an individual experiment of explosive dispersal of particles as in Figure 1.4d is a realization. Repeating these flows again and again, under nominally identical conditions, will generate an *ensemble* of realizations. In the example of a pipe flow with particles, Figure 3.1 shows three different realizations, which are quite different from each other in the distribution of particles, although they are statistically from the same distribution, where the concentration is higher near the pipe centerline and lower as we move toward the pipe walls. Many such realizations will form an ensemble. The ensemble is a conceptual tool, since it is impractical to repeat any multiphase flow more than a few times. Each realization is unique and differs from any other realization due to both continuous-phase turbulence and pseudo turbulence.

The experience that scientists have gained over the years in single-phase turbulence is of vital importance as we proceed to consider multiphase flows. The following are some valuable single-phase lessons that have been learned, which we plan to exploit in our description of multiphase flows. (i) The velocity and pressure fields of an individual realization are necessarily space and time dependent in a very complex manner. Therefore, it is advantageous to study the statistical properties of ensemble averages. (ii) Many problems present exact or approximate symmetries that can be exploited to simplify the space-time dependence of ensemble averages. (iii) Governing equations of the ensemble-averaged quantities can be derived and solved. This, however, gives rise to a closure problem. (iv) Direct numerical simulation of an individual realization yields detailed understanding of turbulence that cannot be obtained by considering only the ensemble-averaged problem.

The rest of the chapter is organized as follows. In Section 3.1 we start with an individual realization, where tracking the time-dependent motion of each and every particle within the system is sufficient for complete characterization of the dispersed phase. This section will highlight the need for averaging in order to address the complexities of practical multiphase flows. In Section 3.2 we then introduce two kinds of average that have been wisely used – spatial averaging and ensemble averaging. In this section we will also discuss how these two different averages address statistical symmetries of the problem with some simple examples. In Section 3.3 we then address an important property of the dispersed phase that arises due to its discrete nature. We note that even a random distribution of particles that has been distributed with uniform probability over space will have number density or volume fraction variation. While the time evolution of the dispersed phase in an individual realization can be very instructive, in itself it is not sufficient. Different realizations may evolve in substantially different ways, and thus knowing the evolution of one does not allow us to predict, or even estimate, the evolution of another realization. So, we next turn our attention to statistical characterization of the dispersed phase in the ensemble of realizations as a whole. Before addressing the time evolution of an ensemble, in Section 3.4 we introduce N-particle probability density functions for the complete statistical characterization of the ensemble at one instant in time. This statistical description of the state of the dispersed phase is then generalized for the time evolution of the ensemble in Section 3.5. With the preliminary understanding built up in the previous sections, in Section 3.6 we present different kinds of ensemble that have been specifically adapted to better address micro and mesoscale physics. Finally, in Section 3.7 we briefly address issues pertaining to polydisperse distribution of particles.

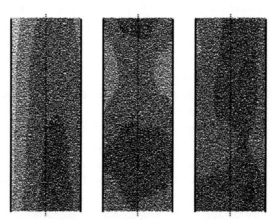

Figure 3.1 Three realizations of the particle distribution with a vertically oriented pipe flow. On average, the particle concentration is higher near the pipe centerline and lower near the walls, with no variation along the pipe axis. Nevertheless, in this schematic, particle concentration differs substantially from realization to realization.

3.1   Characterization of a Realization

Before we consider the ensemble average and its statistical properties, let us first look at the complete characterization of an individual realization and the system of coupled governing equations that govern its dynamics. This will be the basis of the direct numerical simulations of individual realizations that was alluded to at the end of the previous section.

Let us consider the mth realization of a dispersed multiphase flow consisting of N_m rigid particles of diameter d_p, where the subscript m indicates that the total number of particles is a random variable that varies from realization to realization. Henceforth, we will drop the subscript m and leave it to the reader to understand that appropriate quantities are specific to a realization. The position, velocity, and angular velocity of the particles are given by

$$ \mathbf{X}_j(t), \quad \mathbf{V}_j(t), \quad \mathbf{\Omega}_j(t) \quad \text{for} \quad j = 1, 2, \ldots, N, \tag{3.3} $$

and they fully characterize the dynamics of the dispersed phase. Similarly, the continuous phase is fully characterized by the time-dependent velocity $\mathbf{u}(\mathbf{x}, t)$ and pressure $p(\mathbf{x}, t)$ fields.

We now define a dispersed-phase indicator function as

$$ I_d(\mathbf{x}, t) = \begin{cases} 1 & \text{if inside a particle,} \\ 0 & \text{otherwise,} \end{cases} \tag{3.4} $$

while the continuous-phase indicator function is given by $I_c(\mathbf{x}, t) = 1 - I_d(\mathbf{x}, t)$. Clearly the continuous-phase velocity, pressure, and so on are defined only where $I_c(\mathbf{x}, t) = 1$. The time evolution of the continuous phase is governed by the Navier–Stokes equations, which in the incompressible form can be expressed as

$$ \nabla \cdot \mathbf{u} = 0, $$
$$ \frac{\partial \mathbf{u}}{\partial t} + \mathbf{u} \cdot \nabla \mathbf{u} = -\frac{1}{\rho_f} \nabla p + \nu_f \nabla^2 \mathbf{u}, \tag{3.5} $$

where ρ_f and ν_f are the density and kinematic viscosity of the fluid. The time evolution of the jth particles is given by the particle equation of motion

$$ \frac{d\mathbf{X}_j}{dt} = \mathbf{V}_j, $$
$$ m_p \frac{d\mathbf{V}_j}{dt} = \int (\nabla \cdot \boldsymbol{\sigma}) \, dA, \tag{3.6} $$

where m_p is the mass of a particle, the integral is over the surface of the jth particle, and $\boldsymbol{\sigma} = -p\mathbf{I} + \mu_f(\nabla \mathbf{u} + \nabla \mathbf{u}^\mathrm{T})$ is the total stress in the continuous phase. Thus, the right-hand side of the second equation is the net hydrodynamic force on the jth particle. If there are other external forces that act on the dispersed and continuous phases, they must be added to the respective momentum equations. The Navier–Stokes equation requires boundary conditions, which are the matching of the dispersed and continuous-phase velocities at the interface.

The above governing equations that govern the evolution of an individual realization are in principle the simplest, since they do not involve any kind of averaging. The difficulty with this *particle-resolved formulation* is that even with state-of-the-art computational power, one can only consider $O(10^4)$ to $O(10^5)$ particles. Even this number will have to decrease with increasing particle Reynolds number, since each particle will demand higher resolution. Thus, fully resolved simulations are not suitable for the investigation of a problem at the system scale (or at the macroscale). Even investigations of mesoscale physics, such as mesoscale instabilities, are currently out of reach as they typically require millions of particles. Fully resolved simulations are an excellent option for the detailed investigation of the interaction between the dispersed and continuous phases at the microscale. We will discuss particle-resolved simulations in greater detail in Chapter 14.

Fully resolved simulation of a realization is a powerful concept, even if such a simulation at the meso and macroscales is beyond the current computational capability. It can serve as a conceptual framework, since it provides mental access to the "true" fully resolved solution even at the meso and macroscales. With such a conceptual framework we can proceed in two complementary directions to obtain simplified or averaged formulations that will allow us to investigate complex multiphase flows at the meso and macroscales. Such simplifications are essential since fully resolved simulations are well-nigh impossible. The first direction is to spatially average the fully resolved flow over a suitable length scale that is typically much larger than the length scales associated with the particles (i.e., larger than d_p and l_d). The intent of this spatial averaging process is to filter all the details of the multiphase flow at the microscale (i.e., filter out the microscale pseudo turbulence) so that the problem can be studied at or above the mesoscale. The advantage of such a study is that it still retains all the mesoscale structures and their dynamics. This approach of spatial averaging of an individual realization will be introduced in the next section, but will form the basis of the Euler–Lagrange and Euler–Euler methods to be discussed in Chapters 15 and 16.

The second direction is to consider an ensemble of fully resolved realizations, for example by repeating an experiment over and over again. The conceptual ensemble of realizations can then be used to define a hierarchy of ensemble-averaged statistics. Governing equations for these ensemble-averaged statistics can be derived as well. Fundamental symmetries of the problem can be exploited to greatly simplify the ensemble-averaged statistics and their governing equations. In the following section we will also introduce the ensemble average, which will then be reconsidered in greater detail in Chapter 16.

It should also be pointed out that the formulation presented in Eqs. (3.5) and (3.6) is for rigid particles and it becomes somewhat more complicated in the case of droplets and bubbles. In the case of droplets and bubbles, due to their internal flow, the rigid-body Eqs. (3.6) must be replaced by Navier–Stokes equations inside the dispersed-phase elements. Furthermore, at the interface between the phases, both velocity and stress must match. Additional interfacial forces such as surface tension and other effects must be taken into account.

3.2   Volume and Ensemble Averages

From the complete characterization of an individual realization, we can obtain other derived quantities through local volume averaging (also known as spatial averaging). The local particle volume fraction around a point $\mathbf{x}$ can be defined as

$$\phi_d(\mathbf{x}, t) = \frac{1}{\mathcal{V}} \int_{\mathcal{V}} I_d(\mathbf{x}', t) d\mathbf{x}' , \qquad (3.7)$$

where $\mathcal{V}$ is the reference volume chosen around the point $\mathbf{x}$ and it can be chosen following considerations that are typically used in continuum mechanics.[1] That is, $\mathcal{V}$ must be large enough to contain sufficient number of particles to yield smooth enough volume fraction variation, but not so large as to average out meaningful large-scale variation. The local fluid volume fraction can then be obtained from

$$\phi_c(\mathbf{x}, t) + \phi_d(\mathbf{x}, t) = 1 . \qquad (3.8)$$

The number density of particles can be obtained from the volume fraction as

$$n_d(\mathbf{x}, t) = \frac{\phi_d(\mathbf{x}, t)}{\mathcal{V}_p} , \qquad (3.9)$$

where $\mathcal{V}_p = \pi d_p^3 / 6$ is the volume of a particle.

An important point must be stressed here. In any realization of the multiphase flow, the separation between the dispersed and the continuous phases is discrete. For any given $(\mathbf{x}, t)$ there can only be either a dispersed or a continuous phase – they cannot co-exist. Thus, definitions such as volume fraction or number density as a function of $(\mathbf{x}, t)$ are possible only upon some kind of *averaging*. In the above definitions of volume fraction and number density, we have employed averaging over space, also known as *volume averaging* over the reference volume. The simple volume average defined above is sufficient for this purpose. For example, the volume-averaged volume fractions of particles shown in the three realizations of Figure 3.1 are shown in Figure 3.2. It is clear that the volume average differs from realization to realization due to differences in the mesoscale features of the dispersed-phase distribution.

We now proceed to define a few other quantities that will prove useful in future discussion. If we consider the density of each particle to be ρ_p, then the effective density of the dispersed phase becomes (see Crowe et al., 2011)

$$\rho_d(\mathbf{x}, t) = \frac{1}{\mathcal{V}_d} \int_{\mathcal{V}} \rho_p I_d(\mathbf{x}', t) d\mathbf{x}' , \qquad (3.10)$$

where $\mathcal{V}_d = \int_{\mathcal{V}} I_d(\mathbf{x}', t) d\mathbf{x}'$ is the volume occupied by the dispersed phase. If all the particles are of the same density, then $\rho_d = \rho_p$. The effective density of the continuous phase defined over the reference volume becomes

$$\rho_c(\mathbf{x}, t) = \frac{1}{\mathcal{V}_c} \int_{\mathcal{V}} \rho_f I_c(\mathbf{x}', t) d\mathbf{x}' , \qquad (3.11)$$

---

[1] The volume-averaging process will be carried out rigorously in Chapter 11, where we will derive the resulting averaged equations.

where $\mathcal{V}_c = \int_{\mathcal{V}} I_c(\mathbf{x}',t)d\mathbf{x}'$ is the volume occupied by the continuous phase. In a constant-density flow $\rho_c = \rho_f$. The bulk densities of the dispersed and continuous phases are then $\rho_d(\mathbf{x},t)\,\phi_d(\mathbf{x},t)$ and $\rho_c(\mathbf{x},t)\,\phi_c(\mathbf{x},t)$, respectively, and the mixture density within the reference volume $\mathcal{V}$ is

$$\rho_m = \rho_p\,\phi_d + \rho_c\,\phi_c\,. \tag{3.12}$$

From the above definitions, the dispersed-phase mass concentration is defined as (Crowe et al., 2011)

$$C = \frac{\rho_d\,\phi_d}{\rho_c\,\phi_c}, \tag{3.13}$$

which is the ratio of dispersed to continuous-phase mass within the reference volume. From the definition it is clear that in the case of $O(1)$ particle-to-fluid density ratio, C depends mainly on the particle volume fraction. However, in the case of heavier-than-fluid particles, such as sand particles in air, C can be substantial, even at very small volume fraction. Thus, both volume fraction ϕ_d and mass loading C are independent parameters that critically control multiphase behavior.

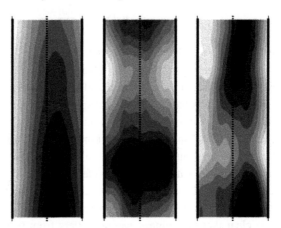

Figure 3.2 The volume average of the three realizations of the particle distribution shown in Figure 3.1. Upon volume averaging, all the microscale detail of how individual particles are distributed is erased. But the mesoscale variations are still preserved. The corresponding ensemble average will show no variation along the pipe axis and will be symmetric about the pipe centerline, with larger concentration along the centerline.

3.2.1   Ensemble Average

In this subsection we will introduce the ensemble average as an average over an ensemble of realizations. Suppose we want the ensemble-averaged particle volume fraction at a point $\mathbf{x}$ at time t. This can be obtained with the following definition:

$$\langle\phi_d\rangle(\mathbf{x},t) = \lim_{M\to\infty}\left[\frac{1}{M}\sum_{m=1}^{M}I_{d,m}(\mathbf{x},t)\right], \tag{3.14}$$

where $I_{d,m}$ is the particle indicator function of the mth realization. Thus, we sort through all the realizations and count only those realizations in which there is a particle at $(\mathbf{x}, t)$. The fraction of realizations with particles gives the ensemble-averaged particle volume fraction, which is denoted with an angle bracket. Analogous to Eq. (3.8) we have $\langle \phi_d \rangle + \langle \phi_c \rangle = 1$. Ensemble-averaged particle density can similarly be defined as

$$\langle \rho_d \rangle (\mathbf{x}, t) = \frac{1}{\langle \phi_d \rangle} \lim_{M \to \infty} \left[\frac{1}{M} \sum_{m=1}^{M} \rho_m I_{d,m}(\mathbf{x}, t) \right], \qquad (3.15)$$

where ρ_m is the density of the particle that happens to be located at $(\mathbf{x}, t)$, if there is one, in the mth realization. Clearly, this and other ensemble-based definitions are analogous to those defined with the volume average.

But the differences between the two averages are profound and important. This is illustrated in the multiphase pipe flow example. If an average over all the realizations such as those shown in Figure 3.1 were carried out, the resulting ensemble average volume fraction would be much simpler than the volume averages shown in Figure 3.2. The ensemble average will have no variation along the pipe axis, and will be left–right symmetric about the pipe axis. In this particular example, the radial dependence will be such that the particle volume fraction will be larger along the pipe axis and will reduce toward the pipe walls. The volume-averaged particle volume fraction has only removed the microscale features of individual particles, but retains the three-dimensional and time-dependent nature of the mesoscale variation. As a result, volume-averaged dispersed-phase quantities will vary in space and time, and from one realization to another. Only in the limit where the reference volume within which the quantities are averaged becomes very large does the volume average cease to be space and time dependent.

In contrast, since the ensemble average is over an infinite number of realizations, in a fully developed pipe flow $\langle \phi_d \rangle$ will not depend on the streamwise or circumferential direction. It will be time independent as well. Thus, $\langle \phi_d \rangle$ is only a function of the radial location within the pipe. Thus, the advantage of the ensemble average is that it erases not only the microscale details, but also the mesoscale variations that may exist in individual realizations. Clearly, for analytical treatments, the simplification brought about by ensemble averaging is very attractive. On the other hand, the volume average presents a more vivid picture of what happens in an individual realization. Thus, both these view-points serve different, but important, purposes.

As we saw in the case of pipe flow, statistical symmetries of a problem greatly simplify the ensemble average. The following three classes of symmetry find extensive application in single and multiphase turbulent flows.

- **Temporal stationarity.** First and foremost, stationarity does not imply that the flow is steady. As discussed earlier in this chapter, a multiphase flow is never steady, since particles will come and go at any observation point. Thus, stationarity pertains to ensemble-averaged statistics. It implies that the one-time statistics is independent of time. For example, in a stationary flow, the ensemble-averaged ϕ_d or number density will be time independent. Stationarity presents another important advantage:

an average over time can be used to approximate the ensemble average. This is a significant advantage, since taking measurements and averaging over a long enough time is far simpler than repeating the experiment many times. This replacement of the ensemble average with the time average, however, assumes ergodicity. We will not pursue this matter further here. The interested reader is referred to books on turbulence.

- **Spatial homogeneity.** Homogeneity means *translational invariance* along one, two, or all three spatial directions. One-point statistics are independent of the homogeneous directions.[2] In the example of the pipe flow shown in Figure 3.1, the streamwise and circumferential directions are homogeneous. In other words, the turbulent pipe flow is translationally invariant along the x and θ-directions. Therefore, the ensemble average is only a function of r. Homogeneity also has the advantage that an average along homogeneous directions can be used to approximate the ensemble average.

- **Spatial isotropy.** Isotropy implies *rotational and reflectional equivariance* of the statistics in space. For statistics to be isotropic, it must be homogeneous. But the reverse is not necessary – statistics can be homogeneous but not isotropic. Isotropy does not affect ensemble averages of one-point scalar quantities. But ensemble averages of one-point vector and tensor quantities must obey rotational equivariance. We will discuss this further in the context of higher-order statistics, such as the radial distribution function. Isotropy can also be used to improve statistical convergence by approximating the ensemble average with an average over space.

The symmetries, while they were helpful in simplifying the ensemble averages, do not simplify the spatial average. For example, in Figure 3.2a, the volume-averaged volume fraction varies along x, θ, and t, despite the statistics being stationary and homogeneous along the axial and circumferential directions. This preservation of spatio-temporal complexity is perhaps the most attractive feature of the volume average, since it is well suited to capture the complex dynamics of multiphase flow at the mesoscale and above. However, volume-averaged quantities are not simple enough for statistical analysis. Thus, the results of volume-averaged simulations are often further averaged over time (in case of stationarity) and over space (along a homogeneous direction). With these additional averages, the two approaches begin to converge.

3.2.2   Notational Convention to be Followed

Here are a few important points on the notation to be followed throughout this book.

- We will use subscript p to denote the physical properties of the particle (e.g., ρ_p, d_p) and subscript f to denote the thermodynamic and transport properties of the fluid (e.g., ρ_f, ν_f). These properties are generally taken to be constants, since we mainly deal with incompressible flows with monodispersed particles.

---

[2] As in classical turbulence, two-point and two-time statistics can be introduced. In stationary turbulence, two-time statistics will only depend on the time separation. In homogeneous turbulence, two-point statistics will only depend on the distance between the two points along the homogeneous direction.

- We will use subscripts d and c to denote dispersed and continuous-phase quantities that are obtained from averaging (or coarse-graining) over a volume $\mathscr{V}$. Because of averaging over length scales larger than the dispersed-phase scales d_p and l_d, dispersed-phase quantities also become Eulerian field variables (they lose their Lagrangian character). *We will use the subscripts d and c to imply that these are averaged mesoscale quantities* (e.g., ϕ_d, ϕ_c, ρ_c, $\mathbf{u}_c$, $\mathbf{u}_d$).
- Capital letters are generally reserved for Lagrangian quantities associated with individual particles (e.g., $\mathbf{V}$ is the velocity of a particle). Note that, on the other hand, when averaged over all the particles inside the reference volume, we obtain

$$\mathbf{u}_d(\mathbf{x}, t) = \frac{1}{\mathscr{V}_d} \int_{\mathscr{V}} \mathbf{V} \, I_d(\mathbf{x}', t) d\mathbf{x}', \tag{3.16}$$

where the particle velocity $\mathbf{u}_d$ is an Eulerian field that depends on $(\mathbf{x}, t)$, and thus is denoted by a lowercase letter.
- Variables without the subscript d or c define dispersed and continuous-phase quantities as appropriate (e.g., fluid and particle velocities $\mathbf{u}$ and $\mathbf{V}$). Unsubscripted quantities imply full resolution. They involve no averaging and thus include microscale variation.
- Ensemble averages are denoted by angle brackets.

Occasionally we may need to stray from the above notation. If so, either it should be obvious from the context or it will be explained explicitly.

3.3   Uniform Distribution and Poisson Statistics

In this section we will establish an important fact about uniformly distributed particles such as the two realizations that are shown in the top row of Figure 3.4 later. Clearly, particles are randomly distributed and as a result no two realizations are identically the same. However, the realizations are all statistically the same in the sense that particles are distributed with uniform probability and the eye cannot discern any preferential accumulation of particles. The ensemble average volume fraction $\langle \phi_d \rangle$ is a constant. That is, at every point within this box, $\langle \phi_d \rangle$ is the same. However, this is not true in the case of volume average. Only when the reference volume is large will the volume average approach $\langle \phi_d \rangle$. For small-sized reference volumes (but still bigger than d_p and l_d), even though the dispersed phase is uniformly distributed, we will see that ϕ_d and n_d are random fields. This was also seen in the pipe flow example shown in Figures 3.1 and 3.2.

The behavior of the volume average as a function of the reference volume can be established under dilute conditions. For this, we follow the discussion in Chandrasekhar (1943) and define a large cloud of volume Ω consisting of N_Ω particles. The particles are randomly distributed within the cloud and thus the average number of particles within the reference volume centered around a point $\mathbf{x}$ is a random number. In some realizations the number of particles $N_{\mathscr{V}}$ inside the volume $\mathscr{V}$ will be a bit smaller and in some realizations a bit higher, and so on. When this random variable $N_{\mathscr{V}}$ is averaged over an ensemble of realizations, we obtain the average number of particles within the reference volume as $\langle N_{\mathscr{V}}\rangle = N_\Omega \mathscr{V}/\Omega$. We emphasize that this is a deterministic quantity and not a random variable.

Since $N_{\mathscr{V}}$ is a random variable, we now proceed to evaluate its probability distribution. The probability of finding a particle within the volume $\mathscr{V}$ is

$$\frac{\mathscr{V}}{\Omega} = \frac{\langle N_{\mathscr{V}}\rangle}{N_\Omega}, \tag{3.17}$$

and the probability of not finding a particle in this volume is $1 - (\langle N_{\mathscr{V}}\rangle/N_\Omega)$. Hence, the probability of finding *exactly* k particles within the reference volume follows the Bernoulli distribution:

$$P_k = \frac{N_\Omega!}{k!(N_\Omega - k)!}\left(\frac{\langle N_{\mathscr{V}}\rangle}{N_\Omega}\right)^k\left(1 - \frac{\langle N_{\mathscr{V}}\rangle}{N_\Omega}\right)^{N_\Omega - k}. \tag{3.18}$$

We now wish to consider the limit where the reference volume is much smaller than the volume of the domain (i.e., $\Omega \gg \mathscr{V}$) and as a result $N_\Omega \gg \langle N_{\mathscr{V}}\rangle$. In order to take this limit, we rewrite Eq. (3.18) as

$$\begin{aligned} P_k &= \frac{\langle N_{\mathscr{V}}\rangle^k}{k!} N_\Omega(N_\Omega - 1)\ldots(N_\Omega - k + 1)\left(\frac{1}{N_\Omega}\right)^k\left(1 - \frac{\langle N_{\mathscr{V}}\rangle}{N_\Omega}\right)^{N_\Omega - k} \\ &= \left(\frac{\langle N_{\mathscr{V}}\rangle^k}{k!}\right)(1)\left(1 - \frac{1}{N_\Omega}\right)\ldots\left(1 - \frac{k-1}{N_\Omega}\right)\left(1 - \frac{\langle N_{\mathscr{V}}\rangle}{N_\Omega}\right)^{N_\Omega - k}. \end{aligned} \tag{3.19}$$

Now we take the limit $N_\Omega \to \infty$ and obtain the *Poisson distribution*

$$P_k = \frac{\langle N_{\mathscr{V}}\rangle^k \exp(\langle N_{\mathscr{V}}\rangle)}{k!}. \tag{3.20}$$

The quantity P_k represents the probability that there are k particles within the reference volume and it is only a function of the average number of particles $\langle N_{\mathscr{V}}\rangle$. Sample Poisson distributions for $\langle N_{\mathscr{V}}\rangle = 5$ and $\langle N_{\mathscr{V}}\rangle = 10$ are shown in Figure 3.3a. It can be observed that the probability for $\langle N_{\mathscr{V}}\rangle = 5$ peaks around $k = 5$, and becomes virtually zero for $k > 14$, indicating that the chance of having more than 14 particles within the volume is virtually nil. In comparison, the probability distribution for $\langle N_{\mathscr{V}}\rangle = 10$ is more Gaussian-like and centered about $k = 10$. The mean value of the above Poisson distribution is $\langle N_{\mathscr{V}}\rangle$, which is the expected number of particles within the reference volume $\mathscr{V}$.

The standard deviation of the Poisson distribution is $\sqrt{\langle N_{\mathscr{V}}\rangle}$. Thus, to one standard deviation, the number of particles within the reference volume can be between $\langle N_{\mathscr{V}}\rangle - \sqrt{\langle N_{\mathscr{V}}\rangle}$ and $\langle N_{\mathscr{V}}\rangle + \sqrt{\langle N_{\mathscr{V}}\rangle}$. These fluctuations can be viewed in two different ways. In an individual realization of a large cloud of uniformly distributed particles, if one were to choose the reference volume $\mathscr{V}$ at various points within the cloud, the number of particles within those reference volumes would show the above variation, since the spatial distribution of particles within the cloud follows a Poisson distribution. On the other hand, if one considers a single reference volume chosen around a fixed point, but different realizations of the cloud, then the realization-to-realization variation in the number of particles within the reference volume will also follow the above Poisson distribution. In both cases, as the average number of particles within the reference volume increases, the relative level of fluctuation ($\sqrt{\langle N_{\mathcal{V}} \rangle}/\langle N_{\mathcal{V}} \rangle$) decreases.

From the number of particles within the volume $\mathcal{V}$, we can calculate the number density and volume fraction as $n_d = N_{\mathcal{V}}/\mathcal{V}$ and the local particle volume fraction as $\phi_d = n_d \mathcal{V}_p$. These are random variables as well. That is, as we vary the location of the reference volume within an individual realization, or when we consider the same reference volume in different realizations, both the number density and the volume fraction will exhibit variation. Ensemble averages of number density and volume fraction can be evaluated as

$$\langle n_d \rangle = \frac{\langle N_{\mathcal{V}} \rangle}{\mathcal{V}} = \frac{N_\Omega}{\Omega} \quad \text{and} \quad \langle \phi_d \rangle = \langle n_d \rangle \mathcal{V}_p . \tag{3.21}$$

The standard deviation of number density and particle volume fraction variation within the reference volume can also be easily obtained.

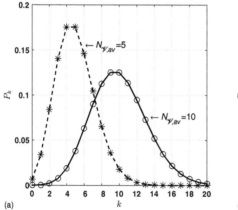
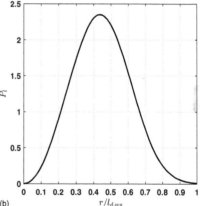

Figure 3.3 (a) Sample Poisson distributions given in Eq. (3.20) for $\langle N_{\mathcal{V}} \rangle = 5$ and $\langle N_{\mathcal{V}} \rangle = 10$. (b) Probability distribution (see Eq. (3.27) later) as a function of nearest-neighbor distance.

In Eq. (3.17) of the above derivation, the probability of finding a particle within the reference volume was taken to be independent of all other particles. This is a good approximation as long as $\mathcal{V}_p \ll \Omega$ and the average volume fraction of particles within the volume is very small. In this limit, a realization of the cloud can be built by adding one particle at a time, with each new particle randomly placed within the volume independent of the position of all earlier particles. This picture of building a random cloud of particles changes as their volume fraction within the cloud increases. Each new particle will now be restricted to occupy only the volume excluded by all other earlier placed particles. This complicates the random distribution process significantly and the resulting distribution will not be strictly Poisson. In fact, the mean volume fraction alone is not sufficient to characterize the cloud; additional information is needed in terms of the *radial distribution function*, which governs how pairs of particles are correlated. In other words, one must go beyond single-particle statistics and start to consider two-particle statistics. We will consider this further in Section 3.4.

3.3.1   Interparticle Spacing

We now consider the distance between particles within a cloud, which can be precisely defined as the distance between the center of a particle and the center of its nearest neighbor. As discussed in the previous subsection, this distance to the nearest particle is a random variable. We will obtain the probability distribution of this distance to the nearest neighbor. But before that, let us obtain some simple geometric measures assuming the particles to follow a certain simple regular geometric structure.

A simple measure of interparticle distance can be obtained by first defining the *Wigner–Seitz radius* as the ball of volume associated with each particle. Again we consider a cloud of volume Ω with N_Ω particles and thus on average each particle is associated with a volume $\Omega/N_\Omega = 1/\langle n_d \rangle$. If we define the average nearest-neighbor distance to be twice the Wigner–Seitz radius, we obtain

$$l_{d,ws} = 2 \left(\frac{3}{4\pi \langle n_d \rangle} \right)^{1/3}. \tag{3.22}$$

Another simple measure is to consider the particle to be arranged in a cubic lattice with the sides of the cube being $l_{d,cu}$. Since the number density is the inverse of the volume of this cube:

$$l_{d,cu} = \left(\frac{1}{\langle n_d \rangle} \right)^{1/3}. \tag{3.23}$$

These different measures of nearest-neighbor distance are a reflection of the basic fact that there is no unique value, since the particles can be arranged in different ways. However, the main conclusion to be drawn from the above two estimates is that the mean nearest-neighbor distance scales as $1/(\langle n_d \rangle)^{1/3}$.

The above estimations are for a structured distribution of particles and therefore there is a unique distance between the neighbors. In a random distribution, the distance to the nearest neighbor is a random variable. For the Poisson distribution of particles, the nearest-neighbor distance can be calculated as follows. With a particle at the origin, the probability of a particle falling within an annular region of radius r and thickness dr is given by the volume ratio $4\pi r^2 dr/\Omega$. Thus, for large N_Ω, the probability of finding any one particle within the annulus becomes

$$\frac{4\pi r^2 dr}{\Omega}. \tag{3.24}$$

No other particles must lie within this sphere of radius r for the nearest-neighbor distance to be r. For each particle, the probability of this exclusion is $1 - (4\pi r^3/3\Omega)$. Thus, the overall probability of the nearest-neighbor distance being r is

$$P_l(r)dr = N_\Omega \frac{4\pi r^2 dr}{\Omega} \left(1 - \frac{4\pi r^3}{3\Omega}\right)^{N_\Omega - 1}, \tag{3.25}$$

which can be rewritten in terms of $l_{d,ws}$ as

$$P_l(r)dr = 6\left(\frac{2r}{l_{d,ws}}\right)^2 \left(1 - \left(\frac{2r}{l_{d,ws}}\right)^3 \frac{1}{N_\Omega}\right)^{N_\Omega - 1} \frac{dr}{l_{d,ws}}. \tag{3.26}$$

Finally, we take the limit $N_\Omega \to \infty$ and obtain the nearest-neighbor distance probability as

$$P_l(r)dr = 6\left(\frac{2r}{l_{d,ws}}\right)^2 \exp\left[-\left(2r/l_{d,ws}\right)^3\right] \frac{dr}{l_{d,ws}}. \tag{3.27}$$

Also shown in Figure 3.3 is the above distribution of nearest-neighbor distances. The peak of the probability occurs at a distance of $(1/12)^{1/3}l_{d,ws}$, or equivalently $(1/(2\pi\langle n_d\rangle))^{1/3}$. From the probability distribution, the mean nearest-neighbor distance can be calculated as

$$\langle l_d\rangle = \frac{\Gamma(4/3)}{2}l_{d,ws} = \Gamma(4/3)\left(\frac{3}{4\pi}\right)^{1/3}\frac{1}{(\langle n_d\rangle)^{1/3}} = \frac{0.554}{(\langle n_d\rangle)^{1/3}}. \tag{3.28}$$

Thus, the mean nearest-neighbor distance of a uniform random distribution is little more than half of the cubic arrangement. This illustrates that one must have a clear probabilistic picture in mind when it comes to the dispersed phase.

3.4   Statistical Description of the Dispersed Phase

Here we will consider an ensemble of random distributions of N particles at a particular instant in time. For now we will not worry about the dynamics. We are not concerned about how the distribution of particles got there or how it will change in the future. In other words, we consider a distribution of particles such as those shown in Figure 3.4 and systematically consider their statistical description. Fortunately, this problem of characterizing the micro-structural details of a random distribution of dispersed elements in a continuous surrounding has been well studied. The reader is referred to the excellent books by Stoyan and Stoyan (1994) and Torquato (2013), which go well beyond the random distribution of spherical particles to be considered here.

We start with the N-particle *specific probability density function* $G_N(\{\mathbf{x}\}^N)$, where $\{\mathbf{x}\}^N$ represents the N position vectors $\{\mathbf{x}_1, \mathbf{x}_2, \ldots, \mathbf{x}_N\}$. Here, $G_N(\{\mathbf{x}\}^N)d\{\mathbf{x}\}^N$ represents the probability of finding the first particle in a volume element $d\mathbf{x}_1$ around the location $\mathbf{x}_1$, the second particle in a volume element $d\mathbf{x}_2$ around the location $\mathbf{x}_2$, and so on. The term "specific" corresponds to the fact that N is the total number of particles in the system. The specific probability density function satisfies the constraint that its integral over all possible locations of all the particles must yield unity:

$$\int_v G_N(\mathbf{x}_1, \mathbf{x}_2, \ldots, \mathbf{x}_N) \, d\mathbf{x}_1 \, d\mathbf{x}_2 \cdots d\mathbf{x}_N = 1 \,. \tag{3.29}$$

If the integration is over all particles except one, then we obtain the generic one-particle probability density function

$$f_1(\mathbf{x}_1) = N \int_v G_N(\mathbf{x}_1, \mathbf{x}_2, \ldots, \mathbf{x}_N) d\mathbf{x}_2 \cdots d\mathbf{x}_N \,. \tag{3.30}$$

Note that the term "generic" represents the fact that the one particle is one out of N particles within the system and as a consequence f_1 is not the same as G_1. The latter is the specific probability density function when there is only one particle in the entire domain, while the former corresponds to the probability of finding a particle at location $\mathbf{x}_1$ in a system where there are a total of N particles. In the above equation, the prefactor N accounts for the fact that any one of the N indistinguishable particles can be at location $\mathbf{x}_1$. From the generic one-particle probability density function we can obtain other dispersed-phase-related quantities. For example, the ensemble-averaged volume fraction at a point $\mathbf{x}$ can be written as

$$\langle \phi_d \rangle(\mathbf{x}) = \frac{\pi}{6} d_p^3 f_1(\mathbf{x}) \,, \tag{3.31}$$

which uses the fact that f_1 is the probability of finding one particle and thus is equal to the ensemble-averaged number density. The dependence on time has been suppressed, since in this section we are concerned only with the description of an instantaneous distribution of particles.

The generic two-particle probability density function is similarly defined as the integral over all particles except two:

$$f_2(\mathbf{x}_1, \mathbf{x}_2) = N(N-1) \int_v G_N(\mathbf{x}_1, \mathbf{x}_2, \ldots, \mathbf{x}_N) d\mathbf{x}_3 \cdots d\mathbf{x}_N \,, \tag{3.32}$$

where the prefactor accounts for the fact that there are $N(N-1)$ ways to choose two particles out of N indistinguishable particles. The two-particle probability density represents the probability of finding two particles, one in a volume element $d\mathbf{x}_1$ around the location $\mathbf{x}_1$ and the other in a volume element $d\mathbf{x}_2$ around the location $\mathbf{x}_2$. All other particles can be anywhere within the domain. This definition can be extended to the generic three-particle probability density function, and so on.

3.4.1   Some Examples of Dispersed-Phase Distribution

To solidify some of the concepts introduced in the previous sections, a few sample distributions of monodispersed particles are shown in Figure 3.4. For simplicity, the distributions are shown in two dimensions (2D), but extension to three dimensions (3D) is straightforward. For now we will not worry about the dynamics. That is, we will not worry about how the distribution of particles got there or how the distribution of particles will evolve in the future. We will apply the ideas of homogeneity and isotropy to the statistical distribution of particles.

The top row of Figure 3.4 shows two snapshots of random distributions of particles within the box, and they represent two different realizations of the random distribution. The distributions of particles appear to be spatially uniform, or more precisely, the particle distribution is statistically homogeneous as well as isotropic. In the context of particle distribution, statistical homogeneity means one-particle statistics f_1 is independent of the location $\mathbf{x}_1$. Two-particle statistics is dependent only on the relative position $\mathbf{x}_2 - \mathbf{x}_1$ and independent of the absolute location of the particle pair. Similarly, multi-particle statistics is dependent only on the relative configuration of those particles and not on their absolute location. In other words, in case of homogeneity all the particle-related statistics are invariant under all translations of the coordinate system. In simple terms, if we move to the right, left, up, or down, the individual particle location will be different (due to the random distribution), but the statistical information will not change.

For a particle distribution, statistical isotropy means that the two-particle statistics is dependent only on the distance between the two particles (i.e., only on $|\mathbf{x}_2 - \mathbf{x}_1|$), and independent of both the absolute location of the particle pair and their relative orientation. Similarly, multiparticle statistics is dependent only on the relative configuration formed by the particles. In other words, all the particle-related statistics are invariant under all rotations and reflections of the coordinate system in addition to translational invariance.

In general, the ensemble-averaged volume fraction $\langle \phi_d \rangle$ is a function of space and it is perhaps the most important one-particle statistics describing the distribution of particles. Similarly, the most important two-particle statistics is the radial distribution function, also known as the pair distribution function, $g(\mathbf{x}, \mathbf{x}')$, which is the scaled probability of finding one particle at $\mathbf{x}$ and simultaneously a second particle at $\mathbf{x}'$. We shall consider this further in the next section. In what follows, for the distribution of particles shown in Figure 3.4a and for the other examples to be discussed, we will consider the effect of statistical symmetry on ensemble-averaged particle volume fraction $\langle \phi_d \rangle$ and the two-particle probability density.

If the particle distribution is homogeneous, then $\langle \phi_d \rangle$ is independent of $\mathbf{x}$. Thus, in Figure 3.4a the ensemble-averaged volume fraction is the same at all points within the box. Though the distribution of particles in Figure 3.4a is isotropic, since the volume fraction is a scalar one-particle quantity, there is no further simplification needed or possible. However, due to isotropy of the particle distribution, f_2 is only a function of the separation distance between the two particles.

Let us now consider two realizations of a different particle distribution shown in Figure 3.4b. The individual realizations are clustered and spatial uniformity is violated. If we consider an ensemble of such realizations with equal probability of the clusters appearing at any region within the box, then the ensemble of distributions shown in Figure 3.4b is statistically homogeneous. Similarly, if the clusters have no preferred orientation then the ensemble is statistically isotropic as well. Due to homogeneity, $\langle \phi_d \rangle$ of the ensemble is independent of $\mathbf{x}$. Due to isotropy, the corresponding f_2 is only a function of $|\mathbf{x} - \mathbf{x}'|$. We now begin to see challenges associated with the statistical description. In an unknown problem, if we are given (i) that the ensemble is isotropic and (ii) the value of the ensemble-averaged volume fraction, then it is impossible to say if the particles are uniformly distributed in every realization as in Figure 3.4a, or if there are particle clusters as in Figure 3.4b. Only by looking at the two-particle statistics $f_2(r = |\mathbf{x} - \mathbf{x}'|)$ will we be able to distinguish between the two possibilities. Such differences in the nature of individual realizations also appear in higher-order one-point statistics (e.g., volume fraction variance), and in multipoint statistics.

The distribution of particles in Figure 3.4c is qualitatively similar to those seen in a fluidized bed under some operating condition. Here, the particles are clustered in horizontal layers which are separated by horizontal layers of lower particle volume fraction, which almost resemble void regions. Thus, at any instance, the particle distribution has a one-dimensional (1D) wave-like structure and it travels in the vertical direction. Due to the traveling nature of the horizontal layers of particle clusters, in any given realization, a point within the domain will experience the dense cluster for part of the time and the void regions at other instants. If we ignore real-world complications such as side-wall effects, and the influence of top and bottom boundaries of the domain, then the ensemble can be considered homogeneous. Like in the previous problem, the clusters have equal probability of occurring at any vertical location and are homogeneous along the horizontal direction. However, the statistics is not isotropic, since there is a preferred direction, namely the clusters are preferentially oriented along the horizontal direction. But the ensemble is *axisymmetric*, since there is rotational and reflectional symmetry on horizontal planes. Thus, in this problem $\langle \phi_d \rangle$ is a constant that is independent of $\mathbf{x}$ and t. The dependence of f_2 is given by $f_2(\Delta xz, \Delta y)$, where $\Delta xz = \sqrt{(x = x')^2 + (z - z')^2}$ is the distance between the two particles on the horizontal $x - z$ plane and $\Delta y = y - y'$ is the distance between the two particles in the vertical direction. This is a consequence of axisymmetry.

Finally, in Figure 3.4d we show particle distributions that are statistically inhomogeneous. In each realization, a decrease in particle concentration from the bottom to the top can be observed. This effect will also be observed in ensemble-averaged quantities. For example, the ensemble-averaged volume fraction $\langle \phi_d \rangle$ will now be dependent on y. Similarly, the dependence of two-particle statistics becomes $f_2(\Delta xz, y, y')$ – note the dependence on the y location of both particles. Although not shown here, local clustering of particles such as those shown in Figure 3.4b and c can occur in combination with sustained large-scale variation, such as in Figure 3.4d.

Problem 3.1 Shock–particle interaction is an interesting example that leads to anisotropic (but homogeneous) distribution of particles (Stewart et al., 2018). Consider a uniform distribution of particles, such as the ones shown in Figure 3.4a, traveling in the y-direction and crossing a horizontally oriented stationary planar shock. In this shock-attached frame, the gas and the particles travel at supersonic speed and approach the shock. Downstream of the shock, the gas rapidly slows down to subsonic velocity and correspondingly the gas density increases downstream of the shock. The particle velocity also decreases across the shock and the particle volume fraction downstream of the shock increases. But the particle distribution is preferentially compressed in the vertical direction. That is, the average distances between the particles in the horizontal direction remain roughly the same as they were before the shock, but the average distance in the vertical direction decreases, which is the source of the increased volume fraction. Thus, downstream of the shock the ensemble remains homogeneous, but isotropy is broken due to the preferential compression along the vertical direction. The distribution is axisymmetric since there is no preferred direction along the horizontal planes. The resulting anisotropy is strong a short distance downstream of the shock. But far downstream the distribution will relax to isotropy, due to interparticle interaction. Think about the distribution of particles in this problem and express the dependencies of ensemble-averaged particle volume fraction and f_2:

(a) before the shock;

(b) a little downstream of the shock when the statistics is only axisymmetric;

(c) far downstream of the shock when isotropy returns.

Problem 3.2 Consider a distribution of particles similar to those shown in Figure 3.4a, except the particle locations have been compressed by a factor of two in the vertical y-direction, while they are expanded by a factor of two in the horizontal x-direction. The particles are left unaltered in the z-direction. As a result, the ensemble-averaged volume fraction of particles is exactly the same as that for the distributions shown in Figure 3.4a. However, each particle is closer to its neighbors in the vertical direction (due to compression), further apart in the horizontal x-direction, and the separation is intermediate in the z-direction. This difference between the three directions renders the distribution anisotropic and also axisymmetry is violated.

(a) Discuss if the ensemble remains statistically homogeneous. In other words, are the one, two, and multiparticle statistics translationally invariant? They certainly are not rotationally invariant.

(b) Express the dependencies of ensemble-averaged particle volume fraction and f_2.

Problem 3.3 The distributions shown in Figure 3.5 are not only clustered, but the clusters appear to be preferentially oriented. Again, with a suitably chosen reference volume $\mathscr{V}$ that is smaller than the cluster size, the volume-averaged ϕ_d can be made to retain the spatial non-uniformity and preferred orientation of an individual realization. Assume the ensemble of realizations is such that the preferentially oriented clusters are likely to occur with equal probability anywhere within the box, but always oriented as shown in the figure.

(a) Discuss the properties of the volume-averaged quantities and compare them to the statistical properties of the ensemble average $\langle \phi_d \rangle$.

(b) In what ways will the preferred orientation of the particle clusters be reflected in the two-point and multipoint statistics?

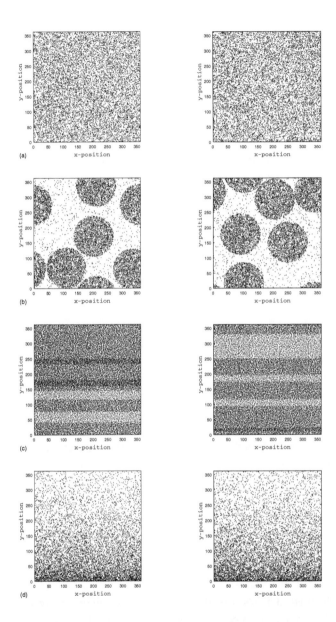

Figure 3.4 (a, first row) Two realizations of the uniform distribution of particles within a two-dimensional periodic domain. (b, second row) Two realizations of a non-uniform distribution with the explicit presence of circular clusters. Since the clusters are likely to occur at any region of the domain, the distribution is statistically homogeneous and isotropic. (c, third row) The same as (b) except the clusters are elliptical and are preferentially oriented with the major axis along the y-axis. Thus the distribution is statistically homogeneous, but not isotropic. (d, last row) The distribution is inhomogeneous along the y-axis and statistically homogeneous along the x-axis.

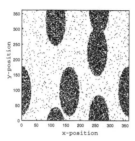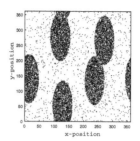

Figure 3.5 Two realizations of a non-uniform distribution with the explicit presence of clusters. Here the clusters are elliptical and are preferentially oriented with the major axis along the y-axis.

The definition of the ensemble average is fundamentally different from the volume average definition given in Eq. (3.7). The volume-averaged ϕ_d of the distributions shown in Figure 3.4a will clearly vary in **x** (will not be perfectly uniform), and the nature of the variation will depend on the size of the reference volume ($\mathcal{V}$) used for averaging. The volume-averaged ϕ_d will show spatial variation mainly on length scales larger than $\mathcal{V}^{1/3}$ and variations on smaller scales are damped by the averaging process. Furthermore, with increasing $\mathcal{V}$, the number of particle within the reference volume increases and as a result, as discussed in Section 3.3, the level of variation in volume fraction will decrease. These differences highlight the difference between the volume average of an individual realization versus the ensemble average of the system.

In Figures 3.4b and c, the individual realizations are clustered and spatial uniformity is violated. This information on spatial clustering is completely erased in the ensemble-averaged volume fraction. Only higher-order one-point and multipoint ensemble-averaged statistics carry the clustering information. Although in principle the ensemble-averaged statistics has all the needed information, it is not straightforward to see the clusters as we do in the individual realizations shown in Figure 3.4. This is where the volume-averaging process has an advantage. By choosing a suitable averaging volume, we can smooth out the microscale details, while retaining the mesoscale structures such as clusters. On the other hand, the volume average will not completely average out the statistical fluctuations described by the Poisson distribution. As discussed in Section 3.3, the relative level of fluctuation will decrease as the number of particles within the averaging volume increases. In contrast, an ensemble average does not suffer this statistical fluctuation, in the theoretical limit of unlimited realizations (averages over a limited number of realizations will suffer from statistical fluctuations). Finally, the volume average of Figure 3.4d will similarly show volume fraction contours that decrease with increasing y. The examples thus illustrate what information is retained in the volume average (with a suitable reference volume for averaging) and what simplifications result from the ensemble average. Table 3.1 summarizes what can be expected from the ensemble and volume-averaged results for the particle distributions shown in Figure 3.4.

Table 3.1 The nature of the spatial and ensemble averages of the particle distributions shown in Figure 3.4. Here the reference volume $\mathscr{V}$ is chosen large enough to average the microscale variations. But the reference volume $\mathscr{V}$ is chosen smaller than the scale of the clusters shown in Figure 3.4b. The volume averages will show a statistical fluctuation whose magnitude will decrease with increasing average volume.

Case	Individual Realization and Local Volume Average	Ensemble Average
(a)	Inhomogeneous	Homogeneous and isotropic
(b)	Inhomogeneous	Homogeneous and isotropic
(c)	Inhomogeneous	Homogeneous
(d)	Inhomogeneous	Inhomogeneous

3.4.2   Radial Distribution Function

From the generic one and two-particle density functions, the radial distribution function (RDF) is defined as

$$g(\mathbf{x}_1, \mathbf{x}_2) = \frac{f_2(\mathbf{x}_1, \mathbf{x}_2)}{f_1(\mathbf{x}_1) f_1(\mathbf{x}_2)} \tag{3.33}$$

and in general it depends on the location of the two particles. The RDF is also known as the pair probability function. For the particle distributions shown in Figures 3.4a and b, isotropy simplifies the RDF. The two-particle probability density function is now only a function of the scalar distance between the two particles (i.e., $f_2(r)$, where $r = |\mathbf{x}_2 - \mathbf{x}_1|$). The corresponding definition of the RDF also simplifies as

$$g(r) = \frac{f_2(r)}{(f_1)^2}. \tag{3.34}$$

The RDF can be interpreted as the probability of finding a particle at a distance of r away from a given reference particle, relative to that for a perfectly uniform distribution. Though both are only functions of r, the RDF of the particle distributions shown in Figure 3.4a will differ from that of Figure 3.4b, and the difference will characterize the level of clustering present in Figure 3.4b.

Given a random distribution of particles as shown in Figure 3.4a or b, the RDF $g(r)$ can be evaluated as follows. For any chosen particle (see Figure 3.6), the space around it is divided into spherical bins. Count the number of particles within the shell of annular thickness dr at radial location r. In a homogeneous system this process can be repeated with all the particles as the reference particle to obtain the average number of particles $dn(r)$ within the shell located at r. The RDF is then given by

$$g(r)\, dr = \frac{1}{4\pi r^2} \frac{dn(r)}{n_d}, \tag{3.35}$$

where $n_d = f_1$ is the number density of the homogeneous distribution. A typical RDF that results from such a calculation is shown in Figure 3.6b, where r has been normalized by the particle diameter d_p. Since the center of no other particle can lie within one diameter of the reference particle, $g(r) = 0$ for $r < d_p$. There is an enhanced probability of finding a particle at near contact distance with the reference particle and this is indicated by the strong peak at $r/d_p = 1$. The RDF decreases to a minimum at a distance of $r/d_p \approx 1.5$ and reaches a second peak at $r/d_p \approx 2.0$. This oscillatory behavior decays and at large distances the RDF reaches the asymptotic value of unity. The value of the RDF at contact plays an important role in many models of interparticle interaction. We shall consider this further when we discuss collisions and coagulations in Chapter 10.

The RDF shown in Figure 3.6b is typical of a nearly uniform distribution, such as those shown in Figure 3.4a. It is important to note that even in this nominally uniform distribution, the probability of finding a neighbor is *not* spatially uniform (i.e., the RDF is not a constant function of unit value). The probability of finding a neighbor at very close range is higher, as indicated by the RDF. This is due to the finite size and finite volume fraction of the particles. It should also be pointed out that the RDF of the distributions shown in Figure 3.4b will have an even stronger peak at $r/d_p = 1$ and show larger-amplitude oscillation. The level of departure from the RDF shown in Figure 3.6b is indicative of the level of clustering. Thus, the RDF can be used as a quantitative measure of clustering. In anisotropic systems, the RDF is a function of the vector quantity **r**. This more complex RDF can be used to quantify the nature of anisotropic clustering.

We close the discussion with a comment on the simple measures of interparticle distance presented in Section 3.3.1 and their relation to the RDF. Measures of mean interparticle distance such as the Wigner–Seitz radius, or cubic lattice spacing, are based on local particle number density, in other words on one-particle statistics. While these simple definitions of interparticle spacing are quite adequate in many circumstances, in order to accurately represent multiphase flow processes that depend on interparticle interaction (e.g., collision between neighbors), it is important to correctly calculate them based on two-particle statistics.

Implicit in the simple definitions of interparticle spacing is the assumption that given the location of any one particle (alternately, from the perspective of any one particle), the probability of finding a second particle is spatially uniform. In other words, a constant RDF of unit value is assumed. As discussed above, this assumption is appropriate in the limit of point particles, which do not occupy volume. This assumption must be relaxed in the realistic case of finite-sized particles.

Problem 3.4 Let us perform a simple computational exercise and calculate the radial distribution function. Consider a cubic box of unit length with N particles within this box (we will call this box the primary cubic box). Let N be a large number – for example, start with $N = 12,500$. Choose all the particles to be of the same diameter and choose this value of d_p such that the average volume fraction of particles within

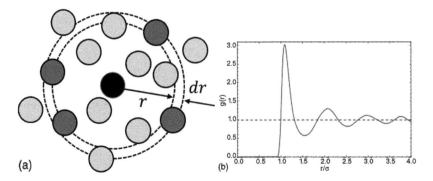

Figure 3.6 (a) Schematic of random particle distribution from which the radial distribution function is calculated. The red sphere is the reference sphere. The blue spheres have a spherical shell of radius between r and $r + dr$ and all other spheres are colored yellow. (b) A typical radial distribution function.

this box is 10%. This will yield a particle diameter of 0.0248. We will assume the primary cubic box with the particles to be periodically repeated in all three directions.

(a) From the above information, verify that the Wigner–Seitz radius of this system is $l_{d,ws} = 0.0534$ (little more than twice the particle diameter).

We will place the particles within the primary cubic box using a process known as *random sequential addition*. The center of the first particle is randomly chosen within the box. The center of the second particle is then chosen randomly as well. This position is accepted if the second particle does not overlap with the first particle. If there is overlap, the position is rejected and a new random position is selected to be tested for overlap. Once the second particle is successfully placed within the box, the process continues to the third particle and so on. Each new particle is randomly placed within the periodic box with uniform probability and with the constraint that it cannot overlap with all the particles that have already been placed within the box. In other words, each new ith particle must satisfy the condition

$$d_{i,j} > \frac{1}{2}(d_i + d_j) \quad \text{for all } j = 1, 2, \ldots, i-1, \tag{3.36}$$

where $d_{i,j}$ is the distance between the centers of the ith and the jth particle and d_j is the diameter of the jth particle. In the present example, all the particles are of the same diameter. In a periodic box, care must be taken to check for overlap that can occur across the periodic boundaries. As pointed out by Torquato (2013), the resulting uniform distribution is not an equilibrium distribution, but it is sufficient for illustration purposes.

(b) Follow the process outlined in Figure 3.6a and count the number of particles within spherical shells around each of the N particles in the primary box. For particles that are close to the edge of the primary box, the spherical shells will extend beyond the primary box. In this case, use periodic repetitions to correctly count the number of particles. Consider many spherical shells, each of thickness $l_{d,ws}/20$, starting from a shell of inner radius equal to one particle diameter. Plot the histogram of the resulting particle number density obtained in each shell as a function of mean shell radius.

(c) Scale the histogram to obtain the RDF (note that $g(r) \to 1$ as $r \to \infty$).

(1) Increase N and observe the convergence of $g(r)$. Since the box size is held fixed, for a fixed ϕ_d, as N increases, the particle diameter will decrease. For comparison of $g(r)$ for the different N, you must first rescale to make the particle diameter the same.

(2) Consider several different ϕ_d and repeat the calculation.

Problem 3.5 Repeat the above random sequential addition process to generate a uniform distribution of particles within a periodic cubic box. For each of the N particles within the box, calculate the distance to its nearest neighbor. Scale this nearest-neighbor distance by the Wigner–Seitz radius $l_{d,ws}$ and divide the scaled distance into 25 equal bins over the range $0 \le r/l_{d,ws} \lesssim 2$.

(a) Use this information to create a histogram of the number of particles vs. distance to the nearest neighbor.

(b) Convert the histogram to a probability of distance to the nearest neighbor and compare the result with the analytic probability given in Eq. (3.27) and shown in Figure 3.3b. The difference between the two is the finite particle size effect.

(1) Increase N and observe the convergence.

(2) Consider different ϕ_d and repeat the calculation.

Problem 3.6 Repeat the random sequential addition process of Problem 3.4 to generate a uniform distribution of monosized particles within a periodic cubic box. In the generation process, enforce a certain extra gap between the particles with the condition

$$d_{i,j} > \frac{\xi}{2}(d_i + d_j) \quad \text{for all } j = 1, 2, \ldots, i - 1, \tag{3.37}$$

where the parameter ξ controls the minimum distance between the particles. If $\xi = 1$ as in Problem 3.4 then particles are allowed to touch, and if $\xi > 1$ then particles cannot touch and there must be a minimum separation distance. Consider $\xi = 1.1$ and $\xi = 1.25$ and redo the problem. With increasing ξ there is increased order and this will be reflected in the RDF. Plot the RDF and compare the results with those obtained in Problem 3.4.

The sequential random addition process described in Problem 3.4 works well at low to modest volume fractions. But it faces difficulty at large particle volume fraction, since finding a non-overlapping position for each new particle becomes increasingly hard. Several algorithmic alternatives have been advanced in the literature in the context of close packing of spheres. One approach uses a *swelling algorithm*, which starts with an initial random distribution of the required number of particles, whose initial size is chosen much smaller, so that the starting volume fraction is low. Then the spheres are allowed to swell (or increase in diameter) slowly, while monitoring contact between neighboring spheres. Upon contact, the touching spheres are nudged with a soft contact force or a small random perturbation away from contact. The process of swelling, contact detection, and contact resolution is carried out in an iterative fashion until the desired final volume fraction is achieved (see Woodcock, 1976 and Zinchenko, 1994).

Another possible approach is to start from an initial structured *face center cubic (FCC)* or *body center cubic (BCC)* arrangement of particles at the desired volume fraction. The particles are then given initial random velocity and allowed to undergo a random collisional process. After some time, with sufficient number of collisions, the state of the particles can be considered to be random. In this case, the nature of interparticle collisions and how they are modeled will determine the precise statistical nature of the resulting random particle distribution. We will see more about this when we discuss modeling of interparticle collision in Chapter 10. However, one thing is clear. There are many ways to generate a random distribution of particles and they are not statistically the same. Their differences mainly appear in the two-particle and higher-order statistics. For example, the RDF of two different random distributions at the same average volume fraction, but generated with two different methodologies, will not be the same.

3.4.3   Voronoi Tessellation and Statistics

An important characterization of a distribution of particles that is gaining wider use in multiphase flow is Voronoi tessellation. For example, several researchers (Monchaux et al., 2010; Saye and Sethian, 2011; Garcia-Villalba et al., 2012) have used Voronoi tessellation to measure the degree of preferential accumulation of particles in turbulent flows and Stewart et al. (2018) used Voronoi tessellation to quantify the nature and amount of particle clustering downstream of a shock.

Voronoi tessellation takes a set of N particles and a containing bounding box, and divides the space into N Voronoi cells, each containing a single particle. The boundaries of a Voronoi cell are defined such that all points interior to the cell are closer to the particle center contained within the cell than any other particle. In other words, each bounding face of a Voronoi cell passes through the midpoint of the vector between the given particle and each of its nearest neighbors. Figure 3.7a shows a sample Voronoi tessellation in 2D where the particles are circles and the Voronoi cells are polygons. In 2D, each face of the Voronoi cell is a bisector of the line connecting the center of adjacent particles. Figure 3.7b shows the Voronoi tessellation in 3D where only the edges of the polyhedral Voronoi cells surrounding each particle are shown.

From the Voronoi tessellation we observe that each particle is caged within a bounding box, and we denote as $\mathcal{V}_V$ the Voronoi volume associated with that particle. If the particles were in an FCC or BCC lattice, then the Voronoi volume of all the particles that form the lattice will be the same. In contrast, in a random distribution of particles, the Voronoi volume associated with the particles will be a random variable. In a statistically uniform distribution of particles, such as the one shown in Figure 3.4a, some particles will be surrounded by close neighbors, resulting in small Voronoi volumes, while some other particles by chance will have their neighbors far away, yielding much larger Voronoi volumes.

A number of researchers have studied the statistical properties of the distribution of Voronoi volumes that naturally result from a random distribution of particles with uniform probability. In the limit of point particles that do not occupy volume, the Voronoi volume associated with the Poisson distribution of particles has been studied as the Poisson Voronoi distribution (Tanemura, 2003; Kumar and Kumaran, 2005; Ferenc and Néda, 2007; Lazar et al., 2013). These investigations explored the statistical properties of Voronoi areas for a random distribution of hard disks in 2D and Voronoi volumes for a random distribution of hard spheres in 3D, with the restriction that the disks and spheres do not overlap.

We define the Voronoi free volume as the volume of the Voronoi cell in excess of the close packing limit:

$$\mathcal{V}_{Vf} = \mathcal{V}_V - \frac{3\sqrt{2}\mathcal{V}_p}{\pi}, \tag{3.38}$$

where the second term on the right is the smallest possible Voronoi volume associated with FCC packing. Here, $\mathcal{V}_p = \pi d_p^3/6$ is the particle volume. The Voronoi volume is a random variable, whose distribution has been empirically fitted with the following 3Γ *distribution* to excellent accuracy (Kumar and Kumaran, 2005):

$$f(\mathcal{V}_{Vf}) = \frac{\delta \, \alpha^{m/\delta^2}}{\Gamma(m/\delta^2)} \mathcal{V}_{Vf}^{(m/\delta-1)} e^{-\alpha \mathcal{V}_{Vf}^{\delta}}, \tag{3.39}$$

where Γ in the denominator is the Gamma function. There are three parameters that define the 3Γ distribution. The parameters δ and m are functions of the mean volume fraction $\langle \phi_d \rangle$, and their empirical relations have been plotted in Figure 3.8 (taken from Kumar and Kumaran, 2005). The final parameter α is then constrained to be a function of the other two parameters, δ and m, so that the mean of the above distribution is equal to the mean Voronoi free volume, $\langle \mathcal{V}_{Vf} \rangle$, which can be expressed as

$$\langle \mathcal{V}_{Vf} \rangle = \frac{\mathcal{V}_p}{\langle \phi_d \rangle} - \frac{3\sqrt{2}\mathcal{V}_p}{\pi}. \tag{3.40}$$

This gives the relation

$$\alpha = \left[\frac{\Gamma((m+\delta)/\delta^2)}{\Gamma(m/\delta^2)\langle \mathcal{V}_{Vf} \rangle} \right]^{\delta}. \tag{3.41}$$

Thus, given $\langle \phi_d \rangle$ and $\mathcal{V}_p$, the three parameters δ, m, and α can be obtained, which are then substituted in Eq. (3.39) for the Voronoi free volume distribution. Higher-order moments can be explicitly calculated, which can then be used to obtain the variance, skewness, and kurtosis. The nth moment can be expressed as

$$\langle \mathcal{V}_{Vf}^n \rangle = \frac{\Gamma((m + n\delta)/\delta^2)}{\Gamma(m/\delta^2)\alpha^{n/\delta}}. \tag{3.42}$$

Example 3.7 The advantage of the 3Γ distribution is that it provides a quantitative measure of departure from the random uniform distribution. As an example, here we show in Figure 3.9 a pseudo color plot of the normalized Voronoi volume, defined as $\mathcal{V}_V/\mathcal{V}_p$, projected onto a plane. In this example the green vertical line is the stationary shock and a random uniform distribution of particles at a mean volume fraction of 0.098 approaches the shock on the left (upstream of the shock). As the particle-laden gas passes through the shock, the gas velocity decreases and correspondingly the gas density increases. In a similar manner, the particle velocity also decreases and the particle volume fraction increases. In this particular example, the particle velocity decreases to 33.4% of its upstream value and correspondingly the particle volume fraction increases to a post-shock value of 0.293. But, in this example, due to particle inertia, it takes a few hundred particle diameters downstream of the shock for particles to reach their post-shock equilibrium state.

While in the pre-shock state the particle distribution was uniform, it can be observed that in the post-shock equilibrium state the particle distribution has a remarkable wave-like clustering. These clusters are frozen and travel at the post-shock equilibrium velocity. There are two mechanisms that were responsible for the formation of these clusters. First, the drag force on the particles that slowed them to their post-shock value was volume fraction dependent (i.e., particles in regions of higher local volume fraction experienced a higher drag). It is well established in the fluidized bed research community that such volume fraction-dependent drag leads to mesoscale instability and wave-like clustering of particles (Murray, 1965; Duru et al., 2002). In addition, in the example shown in Figure 3.9, interparticle collisions were dissipative (i.e., the coefficient of restitution was 0.4), which promotes clustering of particles at the microscale (Goldhirsch and Zanetti, 1993; Luding and Hermann, 1999). This is due to the fact that after each collision the particle pair loses momentum and does not separate as much as its separation before the collision. Thus, repeated collisions result in localized clusters of particles.

Here we will demonstrate the use of the Voronoi volume to quantify such a clustered distribution of particles. We will consider the distribution of the Voronoi volume in the pre-shock state upstream of the shock and also in the post-shock state well downstream of the shock where equilibrium conditions are reached. Fluctuations in the Voronoi volume are expected even upstream of the shock due to the random distribution of particles. A histogram of the normalized Voronoi volume obtained from the particles that are located upstream of the shock is shown in Figure 3.10a, along with the theoretical 3Γ distribution calculated for the upstream average volume fraction of 0.098. The agreement is excellent, indicating that the 3Γ distribution is an excellent approximation. We emphasize that the Voronoi volume varies from less than 40% to more than 200% of the mean value, and this variation is the baseline in a random uniform distribution. A variation smaller than this would indicate structure or order in the particle distribution. For example, in the limit of FCC or BCC, the normalized distribution will be a delta function.

On the other hand, a broader distribution indicates clustering. Figure 3.10b shows the histogram of the normalized Voronoi volume obtained from the particles that are located far downstream of the shock in the post-shock equilibrium region. The corresponding 3Γ distribution corresponding to the mean volume fraction of 0.295 is also shown. Due to clustering, the peak has shifted to the left, indicating that many Voronoi volumes are now substantially smaller than the mean. These Voronoi volumes correspond to particles that are in the clustered regions marked in red in Figure 3.9, where the particle volume fraction is substantially larger. The clusters are separated by regions of much lower volume fraction, where there are fewer particles with substantially larger Voronoi volumes, as indicated by the long tail. Thus, the extent of departure from the 3Γ distribution is a clear measure of the degree of clustering.

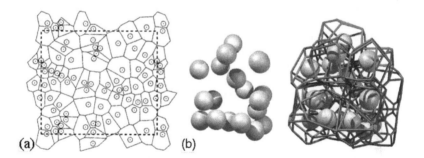

(a)   (b)

Figure 3.7 (a) A random distribution of circular particles in 2D with their Voronoi polygonal cells. The circular particles in this example, taken from Kumar and Kumaran (2005), are arranged in a periodic box whose boundaries are shown as dashed lines. (b) A random distribution of a few spherical particles in 3D with Voronoi tessellation. The edges of the Voronoi polyhedra are shown as a wire frame. Figure reproduced with permission of AIP Publishing.

3.5 Statistical Description of Multiphase Dynamics

The statistical description of the previous section addressed the spatial distribution of the dispersed phase at an instant in time. By restricting attention to a static viewpoint, we were able to obtain a mathematical description. A multiphase flow is far more complicated due to its temporal evolution. The spatial distribution of particles changes over time. A system that started as a uniform distribution can evolve over time and create local particle clusters due to multiphase flow mechanisms such as preferential

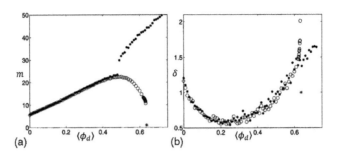

Figure 3.8 (a) A plot of m vs. $\langle\phi_d\rangle$ for a random distribution of hard spheres. (b) A plot of δ vs. $\langle\phi_d\rangle$. In these plots, the solid circles are results for a random distribution generated by the NVE-MC algorithm, the open circles with a swelling algorithm. Both plots, taken from Kumar and Kumaran (2005), are arranged in a periodic box whose boundaries are shown as dashed lines. Figures reproduced with permission of AIP Publishing.

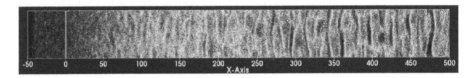

Figure 3.9 A pseudo color plot of the Voronoi volume. A green line at $x = 0$ shows the location of the planar shock. The particles move from left to right in this shock-attached frame of reference. The shades of blue to the left of the shock indicate local fluctuations in volume fraction around the mean value of 0.1. The clustering of particles after passing through the shock can be seen as the red bands to the right of the shock, where the particle volume fraction is substantially larger.

accumulation or turbophoresis. Or an initially stratified system can evolve over time and become well mixed with a uniform distribution of particles.

The complete characterization of the dispersed phase in an individual realization of multiphase flow requires information on the position, velocity, and angular velocity of all the particles within the system. Due to the dynamic nature of the problem, instantaneous location of all the particles within the system is not sufficient to fully characterize the system. Time evolution of the system will additionally depend on the translational and rotational velocities of the particles. This is clear from the governing equations of a realization presented in Section 3.1. As elaborated by Subramaniam (2000, 2001, 2013), one way to represent the statistical variability in the time evolution of the dispersed phase from realization to realization is to consider a *stochastic point process*.

In the stochastic point process, the Lagrangian nature of the dispersed phase is retained. In the limit when the particles reduce to a point and accordingly when each particle within the distribution is completely independent of all other particles, the stochastic point process reduces to a homogeneous Poisson process discussed in Section 3.3 (Stoyan et al., 1995; Daley and Vere-Jones, 2003). For finite-sized particles, the stochastic point process must be represented by the Matérn hard-core

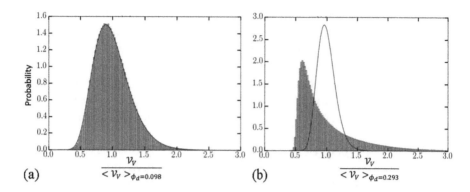

Figure 3.10 Histogram Voronoi volume (green) plotted with the corresponding 3Γ distribution (blue). (a) The upstream state of mean volume fraction 0.098. (b) The downstream equilibrium state of mean volume fraction 0.293.

process. A complete statistical description of the dispersed phase involves a sequence of N-particle specific probability density functions similar to that introduced in Eq. (3.29). For now we ignore the rotational motion of the particles and introduce

$$G_N(\{\mathbf{x}\}^N, \{\mathbf{v}\}^N; t), \tag{3.43}$$

where $\{\mathbf{v}\}^N$ represents the N velocity vectors $\{\mathbf{v}_1, \mathbf{v}_2, \ldots, \mathbf{v}_N\}$.[3] For now, we have also ignored polydispersity by assuming all the particles to be of the same size. Extension to a polydisperse system is straightforward and will be addressed in Section 3.5.1. The above is a more general N-particle specific probability density function (more general than that in Eq. (3.29)) and it represents the probability of finding the first particle in a volume element $d\mathbf{x}_1$ around the locations $\mathbf{x}_1$ with a velocity in the range $[\mathbf{v}_1 - d\mathbf{v}_1/2, \mathbf{v}_1 + d\mathbf{v}_1/2]$, the second particle in a volume element $d\mathbf{x}_2$ around the locations $\mathbf{x}_2$ with a velocity in the range $[\mathbf{v}_2 - (d\mathbf{v}_2/2), \mathbf{v}_2 + (d\mathbf{v}_2/2)]$, and so on. Also note that it is now a function of time. However, since the probability is only in the variables $\{\mathbf{x}\}^N$ and $\{\mathbf{v}\}^N$, the dependence on time is separated with a semicolon (instead of a comma) in this notation. This distinction will be used only if the quantity being considered is a probability.

The temporal dependence introduces an important complication. While in Eq. (3.29) the number of particles N was taken to be a fixed constant, in the present context of a time-evolving multiphase flow we should allow for the possibility of the number of particles N varying over time. Particles can enter or exit from the domain. Or, in the case of droplets and bubbles, their total number can change due to agglomeration and breakup, which will change the droplet size. In Eq. (3.43), G_N represents a sequence of probabilities for $N = 1, 2, \ldots$, where G_1 is the conditional probability density that there is only one particle within the system and G_2 is the conditional probability density that there are only two particles within the system, and so on. This sequence is also known

---

[3] Here we use lowercase $\mathbf{x}$ and $\mathbf{v}$ to denote particle position and velocity, since they are treated as random variables. This convention will be followed in the rest of the book.

as the symmetrized Liouville densities (see Subramaniam, 2013). Thus, the complete description is very complicated and its purpose is to serve as a theoretical framework.

We will restrict our attention to only a one-particle generic density function. For a monodisperse system we follow the definition given in Eq. (3.30) and define a one-particle generic density function, $[f_1]_N(\mathbf{x}, \mathbf{v}; t)$, as the number density of particles in the region $\mathbf{x} - d\mathbf{x}/2$ to $\mathbf{x} + d\mathbf{x}/2$ with velocity $\mathbf{v} - d\mathbf{v}/2$ to $\mathbf{v} + d\mathbf{v}/2$ at time t, conditioned on the fact that there are a total of N particles within the domain:

$$[f_1]_N(\mathbf{x}_1, \mathbf{v}_1; t) = N \int G_N(\{\mathbf{x}\}^N, \{\mathbf{v}\}^N; t) \, d\mathbf{x}_2 \cdots d\mathbf{x}_N \, d\mathbf{v}_2 \cdots d\mathbf{v}_N. \tag{3.44}$$

From this, the *number density function* (NDF) for a monodisperse system can be obtained as

$$f(\mathbf{x}, \mathbf{v}; t) = \sum_{k \geq 1} Pr_k \, [f_1]_k(\mathbf{x}, \mathbf{v}; t), \tag{3.45}$$

where Pr_k is the probability of finding k particles within the system. In the spray community, this quantity is also known as the *droplet density function*. Thus, the NDF is simply the weighted sum of the one-particle generic density functions, for all number of particles within the system. The ensemble-averaged number density of particles can be defined from the NDF as

$$\langle n_d \rangle(\mathbf{x}, t) = \int f(\mathbf{x}, \mathbf{v}; t) \, d\mathbf{v}, \tag{3.46}$$

where the integral is over all possible velocities. The ensemble-averaged number density is related to the ensemble-averaged volume fraction by

$$\langle n_d \rangle(\mathbf{x}, t) = \frac{1}{\mathcal{V}_p} \langle \phi_d \rangle(\mathbf{x}, t). \tag{3.47}$$

We now address the dynamics (i.e., the time evolution) of the statistical description of the dispersed phase. Time evolution of the stochastic point process leads to the *population balance equation* (PBE) for the NDF (Ramkrishna, 2000; Marchisio and Fox, 2007). In the spray literature this equation is known as the *Williams spray equation* (Williams, 1958). The PBE is a transport equation and can be expressed as

$$\frac{\partial f}{\partial t} + \nabla \cdot (\mathbf{v} f) + \nabla_{\mathbf{v}} \cdot (\langle \mathbf{a} \rangle f) = S, \tag{3.48}$$

where ∇ and $\nabla_{\mathbf{v}}$ represent the gradient with respect to position $\mathbf{x}$ and velocity $\mathbf{v}$, respectively, and $\langle \mathbf{a} \rangle$ is the expected value of the particle acceleration conditioned on the particle location $\mathbf{x}$, velocity $\mathbf{v}$, and time t, and should strictly be written as $\langle \mathbf{a} \mid \mathbf{x}, \mathbf{v}, t \rangle$. In the above, the first term on the left corresponds to the rate of change in number density within a unit control volume in the $(\mathbf{x}, \mathbf{v})$ space; the next two terms correspond to the net flux of particles into the control volume along the physical coordinate $\mathbf{x}$ and in the velocity space $\mathbf{v}$ and S is the source/sink term. See Ramkrishna (2000), Marchisio and Fox (2007), and Subramaniam (2013) for further discussion on a complete description of the PBE formulation.

The above is a scalar equation, but its dimensionality is large, since it depends on seven variables: $\mathbf{x}$, $\mathbf{v}$, and t. Despite this complexity, the NDF is complete only as a one-particle statistics. It does not carry information on the relative position of even two particles. Information such as RDF or pair probability function that is crucial for describing the spatial features of the dispersed-phase distribution, such as clustering, is not contained in $f(\mathbf{x}, \mathbf{v}; t)$. Information at the next level will involve two-particle statistics, which is a function of the position and velocity of both particles, in addition to time. One can derive an evolution equation, similar to the PBE, for the two-particle density function. As can be expected, the evolution equation for the two-particle density function will be even more formidable, with a dimensionality of 13, namely, it is dependent on $\mathbf{x}_1$, $\mathbf{v}_1$, $\mathbf{x}_2$, $\mathbf{v}_2$, and t. Thus, the PBE and similar equations for higher-order statistical quantities suffer from the curse of dimensionality. As we will see in the examples below, symmetries such as stationarity, homogeneity, and isotropy will ease the curse of dimensionality greatly.

This sequence of one and two-particle density functions goes on to three-particle statistics and beyond. Clearly, the complete statistical description of an N-particle dispersed-phase assembly is immensely complex. Such a complete statistical framework, although not useful as a practical tool, is of value as a theoretical construct and as a guiding post. We hasten to add that the above discussion has been for a monodisperse case. The statistical formalism can easily be generalized for a polydisperse system, with the particle size included as an additional independent variable.

3.5.1 Some Simple Examples

Here we illustrate the use of the PBE with some examples, which are admittedly chosen to be quite simple.

Example 3.8 Consider a sedimentation example where a large horizontal layer of monodispersed particles initially extends in the region between $y = -h_0$ and $y = 0$. Note that positive y points downward. Let all the particles be settling at their terminal velocity V_s and let the initial average number density of particles within the layer be a constant $\langle n_d \rangle_0$. This problem is homogeneous along the horizontal direction and as a result, the NDF is a function of only the vertical coordinate y, vertical velocity v, and time t. The initial NDF can be expressed as

$$f(y, v, t = 0) = \langle n_d \rangle_0 \left(H(y + h_0) - H(y) \right) \delta(v - V_s), \qquad (3.49)$$

where $H(y)$ is the Heaviside function and δ is the delta function. The corresponding 1D version of the PBE is

$$\frac{\partial f}{\partial t} + \frac{\partial v f}{\partial y} = 0. \qquad (3.50)$$

In this simple example, the sedimentation velocity of the particles is taken to be a constant and acceleration is identically zero. The above PBE can be solved by the method of characteristics to obtain the solution

$$f(y,v,t) = \langle n_d \rangle_0 \left(H(y + h_0 - V_s t) - H(y - V_s t) \right) \delta(v - V_s), \tag{3.51}$$

which simply states that the entire particle-laden layer of height h_0 moves down at the settling velocity V_s.

Example 3.9 For the second problem let us focus only on the lower interface, which was initially at $y = 0$, but consider the particles to have a narrow Gaussian distribution of settling velocities around V_s. This distribution of settling velocities can be thought of as a crude model of polydispersity due to the presence of a range of sediment sizes. The initial NDF will be expressed as

$$f(y,v,t=0) = \langle n_d \rangle_0 \left(1 - H(y) \right) \frac{1}{\sqrt{\pi}\sigma} \exp\left(-\frac{(v - V_s)^2}{\sigma^2} \right). \tag{3.52}$$

In this example, $\langle n_d \rangle_0$ is the average initial number density of particles taking into account every possible settling velocity. Since the focus is on the lower interface, in the above we have ignored the presence of the top interface and assume the layer to extend up to $y \to \infty$. Also, the width of the Gaussian σ will be restricted to be much smaller than V_s.

Though the settling velocity is Gaussian distributed, each particle is assumed to settle at its settling velocity without acceleration and as a result, the conditional average of particle acceleration $\langle \mathbf{a} \rangle = 0$. Thus, the PBE for the present problem is the same as Eq. (3.50). The solution obtained from the method of characteristics is

$$f(y,v,t) = \langle n_d \rangle_0 \left(1 - H(y - vt) \right) \frac{1}{\sqrt{\pi}\sigma} \exp\left(-\frac{(v - V_s)^2}{\sigma^2} \right). \tag{3.53}$$

Again, the interpretation is simple. The interface of each particle size class has moved down to its location $y = vt$ at time t, while maintaining its number density to be $\langle n_d \rangle_0 \exp\left(-(v - V_s)^2/\sigma^2 \right)$.

The average number density of particles, integrated over all the settling velocities, will now be calculated. We recognize that only particles of settling velocity $v > y/t$ could have arrived and be at any location $y > 0$ at time t. Thus, the ensemble-averaged number density is given by the integral

$$n_d(y,t) = \langle n_d \rangle_0 \int_{y/t}^{\infty} \frac{1}{\sqrt{\pi}\sigma} \exp\left(-\frac{(v - V_s)^2}{\sigma^2} \right) dv$$

$$= \frac{1}{2}\langle n_d \rangle_0 \left[1 - \mathrm{erf}\left(\frac{y - V_s t}{\sigma t} \right) \right]. \tag{3.54}$$

As can be expected, the mean interface location evolves as $y = V_s t$, but for $t > 0$ the interface is smoothened to take an error function profile and the thickness of the interface grows as σt. In this simple example, the larger particles of higher settling velocity will be leading the mean interface location, while the smaller particles of lower settling velocity will be lagging.

Even in the case of the monodisperse system considered in Example 1, the leading edge of the interface that is falling down can be expected to smoothen. This smoothening of the interface is due to particle–particle interaction, which at any instant will render some particles to settle faster than V_s, while others to settle slower. However, unlike in Example 2, the settling velocity is not fixed for a particle. Each particle's settling velocity fluctuates in time about the mean settling velocity V_s, depending on the nature of its interaction with its neighbors at the time. One may think of this as the effect of pseudo turbulence. The effect of this particle–particle interaction can be modeled as a diffusion process and the PBE given in Eq. (3.50) can be modified as

$$\frac{\partial f}{\partial t} + \frac{\partial v f}{\partial y} = \mathscr{D}\frac{\partial^2 f}{\partial x^2}, \tag{3.55}$$

where $\mathscr{D}$ is the effective diffusion coefficient. Interestingly, the solution of this equation is the same as Eq. (3.54), but with the denominator of the error function argument replaced by $\sqrt{\mathscr{D}t}$. Thus, in case of diffusion, the thickness of the interface grows as $\sqrt{t}$ as opposed to being linear in time, in the case of polydisperse sedimentation.

Problem 3.10 Consider the monodisperse problem of Example 3.8, but with a time-dependent settling velocity given by $V_s(t)$. This settling velocity is the same for all particles within the system and thus there is a uniform acceleration of $a_y(t) = dV_s/dt$. The solution of this problem is simple – the thickness of the particle-laden layer remains h_0 and its position is given by $\int_0^t V_s(\tau)\,d\tau$. Write the solution in the form of Eq. (3.51) and show that it satisfies the PBE

$$\frac{\partial f}{\partial t} + \frac{\partial v f}{\partial y} + \frac{\partial a_y f}{\partial v} = 0. \tag{3.56}$$

Problem 3.11 As the final example, again consider the monodisperse problem of Example 3.8, but with an initial NDF that decreases linearly from $\langle n_d \rangle_0$ at $y = -h_0$ to zero at $y = 0$.

(a) First show that if the settling velocity were to be a constant, then this initial NDF will simply advect down with velocity V_s. In fact, this result is independent of the initial shape of the NDF.

Now let us complicate the problem by making the settling velocity be a function of the local volume fraction:

$$V_s = V_{s0} - \alpha \mathscr{V}_p \langle \phi_d \rangle = V_{s0} - \alpha \langle n_d \rangle. \tag{3.57}$$

Thus, the drag coefficient increases, and correspondingly the settling velocity decreases linearly, as the local number density increases.

(b) Use a simple numerical approach to solve the PBE for this case. March in time with small time steps and monitor how the NDF changes over time from the initial linear profile.

The above is an interesting example. As can be argued, the top of the layer will be moving slower due to its higher number density, while the bottom of the layer will be moving at a constant velocity of V_{s0}. Thus, the thickness of the particle-laden layer increases over time and correspondingly, the mean number density within the layer decreases, due to conservation of the total number of particles within the layer.

Now consider a different twist, where the initial NDF increases linearly from zero at $y = -h_0$ to $\langle n_d \rangle_0$ at $y = 0$. You may think this is a small change. However, now the bottom of the layer moves slower than the top and as a result, the mean concentration increases. More importantly, particles pile up at the bottom interface to form a shock-like sharp front that intensifies over time. More and more particles from layers above travel faster than those at the bottom front and catch up with the front, thereby making the local number density grow over time. This increasing number density further slows the bottom front. Eventually, the initial layer of thickness h_0 evolves toward an intense layer of infinitesimal thickness. Of course, this mathematical description will not occur in reality, due to particle–particle interaction, which will introduce diffusion-like behavior. The reader is recommended to consult Ramkrishna (2000) for other, more involved examples, solution techniques for the PBE, and many other applications.

We conclude this section by pointing out that in all the above examples, the ensemble-averaged statistics of particle velocity is known. The homogeneous sedimentation problem is unique in this respect. We know the settling velocity of an isolated particle in a quiescent medium. But more importantly, we also have information on the ensemble-averaged settling velocity of a dispersion of particles. As discussed in Section 1.3.2, the average settling velocity is influenced by many different multiphase flow mechanisms, including mesoscale instabilities. Thus, in problems where the ensemble-averaged statistics of v and its evolution (i.e., $\langle \mathbf{a} \rangle$) are known, the PBE can be solved to obtain the NDF.

There are other problems, such as dispersion of droplets in sprays, where the PBE has been used to study the streamwise evolution of the NDF. Here, an externally driven flow, such as an air jet, drives the dispersion of droplets. Thus, one must know the ensemble-averaged velocity statistics of the continuous phase (i.e., one must know the flow), with which one must obtain the required ensemble-averaged particle statistics, since the distribution of v and $\langle \mathbf{a} \rangle$ is needed in the PBE. As we will see in the following chapters, even in the one-way coupled limit, it is not trivial to obtain the dispersed-phase statistics from the continuous phase. The problem gets even more complicated in two-way-coupled applications.

3.6   Micro, Meso, and Joint Ensembles

Several important aspects of multiphase flow must now be clear. (i) Its wide range of scales can best be studied with a divide-and-conquer strategy. (ii) At the microscale, we can focus on pseudo turbulence generated by individual and small collections of particles. (iii) At the mesoscale, we can focus only on intermediate-scale instabilities, particle clusters, and multiphase turbulence. (iv) At the macroscale, the focus will be on the system-level geometric and other details of the multiphase flow application at hand.

Since it is not possible to investigate any realistic multiphase flow problem over the entire range of scales, researchers have devised ingenious ways to study the problem only at the micro, meso, or macroscale, with clever assumptions about the role of scales that have not been included in the study. The multiphase flow literature contains a very large body of work at each of these micro, meso, and macroscales, and the challenge has been to come up with a unified multiscale approach that ties them together. In this section we will revisit the divide-and-conquer strategy by fine-tuning our definitions of ensemble and realization.

The ensemble is a conceptual tool, since it is impractical to repeat any multiphase flow more than a few times. Each realization is unique and differs from any other realization due to randomness in both the distribution of mesoscale structures and the distribution of particles at the microscale within the mesoscale structures. As a result, each realization is unique in its continuous-phase turbulence, as well as pseudo turbulence. We refer to the ensemble of realizations where both these sources of fluctuations are present as the *joint ensemble* and the corresponding realizations as a *joint realization.*

Microcentric View

As argued above, it is sometimes advantageous to look at the dispersed multiphase flow only at the microscale. For example, Beetstra et al. (2007), Tenneti et al. (2011), Zaidi et al. (2014), Bogner et al. (2015), Tang et al. (2015), and Akiki et al. (2016, 2017a) considered a nominally uniform flow over a random array of stationary particles with a focus on the pseudo turbulence generated within the array and its effect on the coupling between the particles and the surrounding flow. It must be emphasized that in order to study a problem only at the microscale, one must choose a suitable mesoscale condition. With such a study then, for that chosen mesoscale condition, the microscale details of the flow are revealed. In the problem of uniform flow through a random array, the mesoscale state is characterized by the mean volume fraction of the array and the mean Reynolds number of the flow. As a result, the microscale studies must be repeated for varying values of these mesoscale parameters. Furthermore, for the chosen value of mesoscale parameters, each investigation of the microscale flow through the random array corresponds to only an individual *micro-realization.* When repeated for many different random arrangements of the particles, for the same mean particle volume fraction and Reynolds number, the collection of micro-realizations will constitute the *micro-ensemble.* The focus of the micro-ensemble is clearly on fluctuations in the random microscale arrangement of the particles and the attendant continuous-phase pseudo turbulence at the microscale.

Mesocentric View

There have been a number of studies whose focus has been on the mesoscale multiphase physics. With an exclusive focus on the intermediate mesoscale, approximations and assumptions have to be made about the micro and macroscales. At this level, the collective effect of the particles on the mesoscale motion of the continuous phase and the mesoscale structure of the particle distribution is of interest. All the fluctuations arising from the random arrangement of the particles at the microscale and the attendant pseudo turbulence at the microscale will be averaged at this mesoscale picture. This averaging of the microscale must be correctly interpreted as an ensemble average over the micro-ensemble. However, stochastic fluctuations in the continuous-phase flow due to mesoscale turbulence are retained. At the other end, in the mesocentric view, the system-level details are not important. However, the larger macroscale state of the flow, in which the mesoscale being studied is embedded, must be specified. Thus, for a chosen value of macroscale parameters, each investigation of the mesoscale turbulent flow is an individual *meso-realization*. An ensemble of these mesoscale realizations can be considered a *meso-ensemble*.

Macrocentric View

Often the interest is to study the multiphase application at the system scale with the inclusion of geometric and other details that are specific to the practical application, such as a fluidized bed or sediment-laden river. With the focus only on the largest scales, the details of the flow at the micro and mesoscales will have to be averaged. Once again, this averaging must be correctly interpreted as an average over the meso-ensemble (which includes an average over the micro-ensemble). Each investigation of the macroscale turbulent flow is an individual *macro-realization*. An ensemble of these realizations can be considered a *macro-ensemble*.

In the above painted picture, the joint realization is a union of the micro, meso, and macro-realizations. That is, the macroscale details of the joint realization come from the chosen macro-realization, the mesoscale details come from a meso-realization that is consistent with the chosen macroscale, and the microscale details come from a micro-realization that is consistent with the chosen macro/mesoscales. It should be stressed that the above partitioning of scales into micro, meso, and macro is arbitrary, though well motivated. But it is sufficient to illustrate the multiscale nature of multiphase flow and to develop a rational approach to solution.

A micro-realization has the advantage that it can be studied from first principles without any averaging or modeling. This gives rise to *fully resolved formulation* at the microscale. However, micro-realizations are far removed from the physical application. Averaging involved in the meso and macro-realizations gives rise to the closure problem. The effect of the unresolved smaller scales must be accounted for through closure models such as drag laws, heat transfer correlations, and Reynolds stress models. Nevertheless, the meso and macro-realizations give rise to the *Euler–Euler formulation* of the multiphase flow – also known as the *two-fluid formulation*. Another popular method for multiphase flow is the *Euler–Lagrange formulation*, which is at the mesoscale as far as the continuous phase, but microscale in terms of the dispersed phase. This formulation can be obtained from a volume average of the joint realization to remove the microscale details of the continuous phase. We will discuss these formulations more in later chapters.

Let us illustrate the above definitions with a practical example. In Figure 3.11 we use the outstanding images of a circulating fluidized bed in a CFB riser obtained at the National Energy Technology Laboratory (Shaffer et al., 2013). Frame (a) shows the schematic of the complete CFB riser arrangement. A macroscale section of the fluidized bed over a length of 1 m is shown in frame (b). This can be considered as one realization of the flow at the macroscale. We focus on a small section of this entire region, where the state of the flow is recorded at many different time instances. Three of these are shown in frame (c). They can be considered meso-realizations, a larger number of which form the meso-ensemble, and they powerfully illustrate the instabilities and clustering that happen at the mesoscale. The flow visualization has gone one step deeper to reveal the microscale physics as well. We have zoomed into the small region shown in frame (c) and in frame (d), three microscale states of the particle distribution are shown. Here, the microscale has been defined such that generally the particle distribution is nearly uniform within it.

An important point illustrated in Figure 3.11 that is relevant to multiphase flow computations must be introduced here. In the macroscale view shown in frame (b), it is clear that there are numerous particles and it is not possible to view individual particles. Thus, at this scale both the dispersed phase as well as the continuous phase are represented as an average over the mesoscale. In the mesoscale view of the present problem shown in frame (c), there are still numerous particles, but individual particles can be identified and followed. Thus, at the mesoscale, as mentioned above, two computational approaches have been pursued. The mesoscale Euler–Lagrange approach where the continuous phase is averaged over the microscale, but the individual particles are tracked. There are also approaches where each computational particle being tracked represents a collection of real particles, which is also an average representation for the particles. The alternative is the mesoscale Euler–Euler approach, where both phases are averaged over the microscale.

The above discussion and the fluidized bed example may have left the reader with the impression that the separation of a problem into micro, meso, and macroscales if straightforward. Unfortunately, in many real applications there may not be a clear separation of scales. Furthermore, what is referred to as mesoscale variation in the figures can itself be superseded by even larger-scale variations. But the example we have considered is sufficient to illustrate the point that in a multiphase flow, the wide range of scales (micro vs. meso) can be separated and analyzed. This divide-and-conquer strategy can be of great help in gaining insight into multiphase-flow physics, or in devising effective multiphase-flow solution strategies, since the problem in its entirety is often too complex to handle.

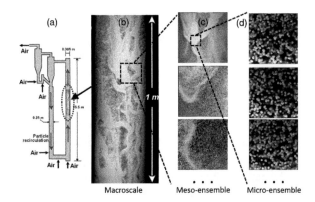

Figure 3.11 Two micro-realizations of a particle distribution with a mesoscale sinusoidal variation in number density. Many such realizations together will form a micro-ensemble. Taken from the experiments of Shaffer and colleagues at NETL (see Shaffer et al., 2013).

3.7   Polydispersity

In this section we briefly generalize the above discussions of monodisperse multiphase flow to the polydisperse situation. The main difference is that instead of a single particle size d_p, the particle diameter is a variable, which we denote ξ. Again we can look at the polydisperse multiphase flow at the level of an individual realization and perform averages over the reference volume to obtain local volume-averaged quantities. We can also extend the statistical framework presented in Section 3.5 to include particle diameter variation. Here we will first discuss at the level of an individual realization before extending the statistical framework.

The generalization of the volume fraction is the function $\phi_{d\xi}(\xi, \mathbf{x}, t)$, which can be termed the size-dependent volume fraction distribution (for lack of a better term). The quantity $\phi_{d\xi}(\xi, \mathbf{x}, t)d\xi$ now corresponds to the volume occupied by all particles of diameter between ξ and $\xi + d\xi$ in a unit volume around $\mathbf{x}$ at time t. From this definition we obtain

$$\phi_d(\mathbf{x}, t) = \int_0^\infty \phi_{d\xi}(\xi, \mathbf{x}, t)\, d\xi,\qquad (3.58)$$

which simply states that when integrated over all the particle sizes we recover the total volume fraction occupied by the particles.

Similarly, for the number density we can define the size-dependent number density distribution as

$$n_{d\xi} = \frac{\phi_{d\xi}(\xi, \mathbf{x}, t)}{\pi \xi^3 / 6}.\qquad (3.59)$$

The above size-dependent number density can also be integrated over all particle sizes to obtain the size-independent number density as

$$n_d(\mathbf{x}, t) = \int_0^\infty n_{d\xi}(\xi, \mathbf{x}, t)\, d\xi.\qquad (3.60)$$

At a practical level, if we suppress the dependence on $\mathbf{x}$ and t,

$$P_\xi(\xi) = \frac{n_{d\xi}(\xi)}{n_d} \tag{3.61}$$

is the probability that a particle within a unit volume around $\mathbf{x}$ at time t is of diameter between ξ and $\xi + d\xi$. This *particle size distribution* is of fundamental importance in describing polydisperse systems. Industrial and naturally occurring polydisperse systems such as droplet spray, rain droplets, or atmospheric aerosol tend to nearly follow well-known size distributions. For example, a lognormal probability distribution can be expressed as

$$P_\xi(\xi) = \frac{1}{\xi}\frac{1}{\sqrt{2\pi}\sigma}\exp\left[-\frac{1}{2}\left(\frac{\ln(\xi) - \mu}{\sigma}\right)^2\right], \tag{3.62}$$

where the average particle size for the distribution is given by

$$d_{p,av} = \int_0^\infty \xi\, P_\xi\, d\xi = \exp\left(\mu + \frac{\sigma^2}{2}\right), \tag{3.63}$$

and the standard deviation is given by

$$d_{p,std} = \left[\int_0^\infty (\xi - d_{p,av})^2 P_\xi\, d\xi\right]^{1/2} = \exp\left(\mu + \frac{\sigma^2}{2}\right)\sqrt{e^{\sigma^2} - 1}. \tag{3.64}$$

In the lognormal distribution, μ has a more direct meaning: it is equal to the log of the mode of the distribution. There are other similar distributions that are well suited to specific applications. For example, the *Rosin–Rammler distribution* is a popular one for droplet size distributions in sprays.

Problem 3.12 Consider a polydisperse spray of water droplets in air with $d_{p,av} = 100$ μm and $\sigma = 20$ μm.

(a) Now use the results of Problem 2.1 to obtain (i) the mean and rms translational time scale of this polydisperse system, (ii) the mean and rms settling velocity of this polydisperse system, and (iii) the mean and rms particle Reynolds number based on settling velocity. Note that you need to calculate (or use Figures 2.2 and 2.4) the translational time scale, settling velocity, and Reynolds number for each droplet size and then average.

(b) Instead, calculate the translational time scale, settling velocity, and Reynolds number for the mean droplet size of 100 μm. You will see that the two different ways of calculating the mean will yield different answers. Clearly the former is the correct approach. Due to the nonlinear relation among these quantities at finite Reynolds number, you must calculate the properties of each droplet size and then weight by its probability and average to arrive at the correct answer.

Statistical Description

We close this chapter by briefly considering the statistical description of an ensemble of polydisperse multiphase flows. The N-particle specific probability density function given in Eq. (3.43) must be further extended to include the particle size as an internal variable

$$G_N(\{\mathbf{x}\}^N, \{\mathbf{v}\}^N, \{\xi\}^N; t), \tag{3.65}$$

where $\{\xi\}^N$ represents the vector of N particle diameters $\{\xi_1, \xi_2, \ldots, \xi_N\}$. Following the steps taken in Section 3.5, we can arrive at the NDF (aka droplet density function) $f(\mathbf{x}, \mathbf{v}, \xi; t)$. Note that the dimensionality of the NDF is now increased to eight.

Correspondingly, the PBE is generalized to

$$\frac{\partial f}{\partial t} + \nabla \cdot (\mathbf{v} f) + \nabla_\mathbf{v} \cdot (\langle \mathbf{a} \rangle f) + \nabla_\xi \cdot (\langle b \rangle f) = \dot{f}_c + S, \tag{3.66}$$

where ∇_ξ represents the gradient with respect to the variable particle diameter ξ. Here, $\langle b \rangle$ is the expected change in particle diameter conditioned on the particle location $\mathbf{x}$, velocity $\mathbf{v}$, size ξ, and time t, and it should strictly be written as $\langle b \mid \mathbf{x}, \mathbf{v}, \xi, t \rangle$. This term thus accounts for the change in number density due to direct changes in the size of the particle (or droplet) due to processes such as evaporation or condensation. On the right-hand side of the above equation, $\dot{f}_c$ corresponds to changes in f due to collisional processes which we shall consider in Chapter 10. If two particles collide and do not coagulate, then the effect of the collision is to change the particle velocities from their pre-collisional values to post-collisional values, and $\dot{f}_c$ accounts for the corresponding changes in the associated particle probabilities in the particle velocity phase only. If two particles collide and coagulate, in addition to changes in the pre- to post-collisional velocity, there is also a change in the particle sizes, which must be correctly accounted for in $\dot{f}_c$. Finally, the breakup process must be considered as well, which also leads to changes in particle size and velocity.

The ensemble-averaged size-dependent number density of particles can be defined from the polydisperse NDF as

$$\langle n_{d\xi} \rangle(\mathbf{x}, \xi, t) = \int f(\mathbf{x}, \mathbf{v}, \xi; t) \, d\mathbf{v}, \tag{3.67}$$

where the integral is over all possible velocities. This can further be integrated over the range of particle diameters to obtain the ensemble-averaged number density

$$\langle n_d \rangle(\mathbf{x}, t) = \int_0^\infty \langle n_{d\xi} \rangle(\mathbf{x}, \xi, t) \, d\xi. \tag{3.68}$$

The ensemble-averaged size-dependent number density is related to the ensemble-averaged size-dependent volume fraction by

$$\langle n_{d\xi} \rangle(\mathbf{x}, \xi, t) = \frac{1}{\pi \xi^3 / 6} \langle \phi_{d\xi} \rangle(\mathbf{x}, \xi, t). \tag{3.69}$$

The local volume fraction occupied by all the particles (independent of size) is given by

$$\langle \phi_d \rangle (\mathbf{x}, t) = \int_0^\infty \frac{\pi \xi^3}{6} \langle n_{d\xi} \rangle (\mathbf{x}, \xi, t) \, d\xi. \tag{3.70}$$

Finally, we can compare the ensemble-averaged number density and volume fraction with those obtained from the volume-averaging process for an individual realization. As discussed for the monodisperse case, the comparison of the two will depend on the reference volume $\mathscr{V}$ used for the volume average and on the temporal and spatial symmetries (stationarity, homogeneity, and isotropy) present in the polydisperse distribution.

Problem 3.13 Consider Example 2, where we arbitrarily assumed a Gaussian velocity distribution.

(a) Repeat the analysis with a lognormal distribution of sediment size with a mean diameter of 100 μm and an rms of 10 μm. You first need to find the corresponding distribution of the terminal velocity and assume this still-fluid settling velocity to still be applicable (unfortunately, this assumption is only good for an exercise).

(b) Define the mean location of the interface to correspond to 100 μm. Evaluate the mean location after 0.1, 1, 10, and 100 s. What is the average number density at the mean location at these times (call this $\langle n_d \rangle_m$)?

(c) Define the interface thickness as the distance from the mean location to the point where the local number density is $\langle n_d \rangle_m / 2$. Obtain the interface thickness at 0.1, 1, 10, and 100 s.

4 Isolated Rigid Particle in an Unbounded Ambient Flow

In this chapter we will consider in detail the interaction of an isolated rigid particle with the surrounding continuous-phase flow. In the low Reynolds number limit, the problem can be solved analytically. At finite Reynolds number, one must resort to numerical simulations. Nevertheless, in both cases, by simultaneously solving the Navier–Stokes equations for the fluid, equations of rigid-body motion for the particle, and coupling them with no-slip and no-penetration boundary conditions, we can obtain complete details of the flow around the particle. At one level, our goal in this chapter is to use such a fully resolved solution to understand the complex flow physics around an individual particle. But in our quest to solve more complex multiphase flow problems, in this chapter we are also interested in making use of the fully resolved information to build accurate closure models that can be used when full resolution of the microscale flow physics is not possible. Most importantly, we want to encapsulate the information on momentum exchange between the particle and the surrounding flow in terms of drag and lift force models. Similarly, mass and energy exchange between the particle and the surrounding flow must be modeled in terms of heat and mass transfer correlations.

These coupling models will form the foundation of Euler–Lagrange (EL) and Euler–Euler (EE) methodologies, to be discussed in later chapters. With the use of these coupling models, there is no need to resolve the flow at the microscale and therefore EL and EE methods focus their attention on mesoscale and macroscale physics. However, their accuracy critically depends on how well the drag, lift, heat and mass transfer coupling models capture the actual physics at the level of individual particles. Thus, the subject matter of this chapter is of fundamental importance. This chapter will mainly be restricted to rigid particles with brief occasional excursions to discuss drops and bubbles. The emphasis of this chapter is on drag force. The following two chapters will respectively focus on lift force and heat transfer of an isolated particle in an unbounded ambient flow.

Consider a particle of diameter $d_p(t)$ and mass $m_p(t) = \pi d_p^3 \rho_p / 6$ translating at velocity $\mathbf{V}(t)$ and rotating at angular velocity $\mathbf{\Omega}(t)$ while immersed in an "undisturbed" ambient flow of velocity $\mathbf{u}_c(\mathbf{x}, t)$ and pressure $p_c(\mathbf{x}, t)$. Here the term "undisturbed" indicates the flow that would exist in the absence of the particle. With the particle located within the flow, continuous-phase velocity will approach $\mathbf{u}_c$ and continuous-phase pressure will approach p_c only sufficiently away from the particle, but close to the particle there will be perturbation flow introduced by the particle. In the presence of the particle, the perturbed velocity and pressure fields will be denoted as $\mathbf{u}_{\text{tot}}(\mathbf{x}, t)$

and $p_{\text{tot}}(\mathbf{x}, t)$, where the subscript "tot" indicates superposition of the undisturbed flow and the perturbation flow induced by the particle. The pressure and velocity gradient of the continuous phase results in a stress field. Let $\sigma_{\text{tot}}(\mathbf{x}, t)$ be the total stress in the continuous phase and it is the sum of stresses due to the undisturbed flow and that due to the perturbation flow. Similarly, let the particle temperature be $T_p(t)$ and the surrounding undisturbed fluid temperature be $T_c(\mathbf{x}, t)$. Let $T_{\text{tot}}(\mathbf{x}, t)$ be the total temperature of the continuous phase and it is the sum of the undisturbed and the perturbation temperatures.

With the above descriptions of the particle and the undisturbed continuous phase, the time evolutions of particle mass, momentum, and energy are given by

$$\frac{dm_p}{dt} = \dot{m} = -\mathcal{D}_\alpha \, \rho_f \oint_{S_p} (\nabla Y_{\alpha,\text{tot}} \cdot \mathbf{n}) \, dA,$$

$$m_p \frac{d\mathbf{V}}{dt} = \mathbf{F} = \oint_{S_p} (\sigma_{\text{tot}} \cdot \mathbf{n}) \, dA + \mathbf{F}_{\text{ext}},$$

$$I_p \frac{d\mathbf{\Omega}}{dt} = \mathbf{T} = \oint_{S_p} \mathbf{x} \times (\sigma_{\text{tot}} \cdot \mathbf{n}) \, dA + \mathbf{T}_{\text{ext}},$$

$$m_p C_p \frac{dT_p}{dt} = \dot{Q} = -k_f \oint_{S_p} (\nabla T_{\text{tot}} \cdot \mathbf{n}) \, dA + \dot{Q}_{\text{ext}},$$

(4.1)

where the integrations are over the surface S_p of the spherical particle, $\mathbf{n}$ is the outward normal unit vector to the surface of the particle, and $\mathbf{x}$ is the position vector from the origin located at the center of the particle. Thus, $\oint (\sigma_{\text{tot}} \cdot \mathbf{n}) \, dA$ represents the integration of the tractional force $\sigma_{\text{tot}} \cdot \mathbf{n}$ exerted by the fluid on the particle over the entire surface of the particle. As a result, the integral yields the fluid force on the particle. The right-hand side $\mathbf{F}$ is the sum of all forces on the particle and it is composed of the surface fluid force and other external body forces $\mathbf{F}_{\text{ext}}$, such as gravity. Similarly, $\mathbf{T}$ is the sum of all torques on the particle and it is composed of the torque exerted by the surrounding fluid (given by the integral), and other external torques $\mathbf{T}_{\text{ext}}$. On the left-hand side, I_p is the moment of inertia of the particle. The total heat source/sink to the particle includes both heat transfer from the surrounding fluid, represented by Fourier's law in the first term on the right-hand side, and any additional external heating represented by $\dot{Q}_{\text{ext}}$. In the first equation, the term on the right-hand side represents the mass source or sink arising from condensation, evaporation, or sublimation processes. On the right-hand side, in Fick's law, $\mathcal{D}_\alpha$ is the mass diffusivity of species α whose diffusion is the source of mass transfer and $Y_{\alpha,\text{tot}}$ is the mass fraction of this species in the continuous phase (for a more detailed discussion on mass transfer, see Bird et al., 1960; Crowe et al., 2011). Over the rest of the chapter we will ignore mass transfer and set m_p, d_p, and I_p as constants.

If the details of the continuous-phase flow at the microscale are known, then the distribution of quantities such as tractional force and thermal gradient can be evaluated on the surface of the particle. Thus, a complete knowledge of the continuous phase at the microscale will allow us to precisely evaluate the integrals on the right-hand side and thereby obtain the fluid force, torque, heat and mass transfer. *Our quest is to model the fluid force, torque, heat and mass transfer in terms of the mesoscale fluid quantities* $\mathbf{u}_c(\mathbf{x},t)$, $p_c(\mathbf{x},t)$, *and* $T_c(\mathbf{x},t)$, *and particle quantities* $\mathbf{V}(t)$, $\mathbf{\Omega}(t)$, *and* $T_p(t)$ *by accurately accounting for the perturbation flow at the microscale.* Such models, when built on sound fundamental theoretical foundations, will encapsulate the essence of the coupling between the phases at the microscale and therefore can be used in EL and EE methods.

The focus of this chapter is to first develop a model of the force on an isolated spherical particle in terms of the mesoscale quantities. The resulting force expression is the elementary building block upon which more complex multiphase flow models will rest. This may seem like an easy task, since it involves flow over only one particle. But, as we will discuss below, an explicit expression for force on a particle is quite complex even under conditions of simple ambient flow. The problem becomes analytically intractable at higher Reynolds number and in the presence of ambient turbulence. This challenge makes multiphase flow very exciting.

This chapter will proceed in a step-by-step manner, with each step including the effect of one additional complexity. We begin Section 4.1 with five related steady Stokes flow problems. The two key features of these five problems that enable analytic solutions are that the flow is steady and the solutions are in the Stokes limit of zero Reynolds number (i.e., inertial effects are ignored). The five problems are: (i) steady uniform Stokes flow over a stationary rigid particle; (ii) steady uniform Stokes flow around a spherical droplet or bubble; (iii) steady torque on a spinning rigid sphere immersed in a quiescent ambient fluid; (iv) evaluation of modified viscosity by considering a steady linear straining flow around a stationary rigid particle; and (v) Faxén's law of steady drag on a rigid particle in an arbitrary non-uniform flow. Together, these five problems can be considered the fundamental solutions of flow around an isolated particle in the steady Stokes flow regime.

In Section 4.2 we will extend the analysis of the uniform ambient flow to include unsteady effects and obtain the classic *Basset–Boussinesq–Oseen (BBO) equation* of motion (Boussinesq, 1885; Basset, 1888; Oseen, 1927). This derivation will be presented with a detailed discussion of the various mechanisms that contribute to the overall force. In Section 4.3 we will combine the effects of spatial non-uniformity of the ambient flow and unsteadiness (i.e., $\mathbf{u}_c$ is now a function of both space and time) to obtain the *Maxey–Riley–Gatignol (MRG) equation* of motion (Gatignol, 1983; Maxey and Riley, 1983). Thus, the MRG equation can be considered as the extension of the Faxén theorem to the BBO equation of motion.

Before we proceed to finite Reynolds number, in Section 4.4 we consider the inviscid limit and present the analytical results of Auton and co-workers (Auton, 1987; Auton et al., 1988). Then, Section 4.5 will address the finite Reynolds number effect. The analytical results of the previous sections apply strictly to the Stokes and the inviscid limits. These results provide the functional form for the force on a particle and in Section 4.5 we seek empirical extensions to finite Reynolds number.

4.1 Five Steady Stokes Flow Problems

Before we discuss more complex situations, in this section we will discuss five canonical problems of steady flow over a sphere. All five problems are of fundamental importance and exact solutions will be obtained in the zero Reynolds number limit. These five problems are as follows.

- We will start with the classic problem of a steady uniform flow over a stationary rigid particle and obtain the *Stokes drag* law. In this process we will obtain Stokeslet and doublet as two fundamental solutions of Stokes flow.
- We will then consider the problem of a steady uniform flow over a spherical droplet or bubble and obtain the classical result by Hadamard and Rybczynski. The important difference is that the Stokes flow must be solved both inside and outside the sphere with proper matching at the interface.
- Next, we will consider the problem of flow over a spinning sphere submerged in a quiescent fluid and obtain an explicit expression for the induced fluid motion and the fluid torque on the particle.
- Then, we will consider the problem of a steady straining flow over a rigid particle, at zero relative velocity. The resulting perturbation flow induced by the particle changes the mean stress within the fluid for an imposed strain field. This modified stress–strain relation due to the presence of the particle is what we earlier referred to as the thermodynamic/transport effect and it gives rise to a particle volume fraction-dependent viscosity.
- Finally, we will discuss the remarkable result of Faxén (1923) that generalizes Stokes drag to any arbitrary steady non-uniform flow over a rigid particle.

4.1.1 Stokes Drag of a Rigid Particle

Consider a stationary spherical particle of radius $R_p = d_p/2$ immersed in a steady uniform ambient flow $\mathbf{u}_c$. The presence of the particle introduces a steady perturbation flow and let that be $\mathbf{u}(\mathbf{x})$. Together, the total continuous-phase velocity around the particle is given by $\mathbf{u}_{\text{tot}}(\mathbf{x}) = \mathbf{u}_c(\mathbf{x}) + \mathbf{u}(\mathbf{x})$. We consider the limit of zero Reynolds number and the perturbation flow satisfies the steady Stokes equation

$$\nabla \cdot \mathbf{u} = 0 \quad \text{and} \quad 0 = -\nabla p + \mu_f \nabla^2 \mathbf{u} \tag{4.2}$$

and the boundary conditions are

$$\text{On the sphere: } \mathbf{u} = -\mathbf{u}_c,$$
$$\text{Far field: } \mathbf{u} \to 0. \tag{4.3}$$

In the above, $p(\mathbf{x})$ is the perturbation pressure field and μ_f is the dynamic viscosity of the fluid. By taking the divergence of the momentum equation and applying the incompressibility condition, we obtain the perturbation pressure to satisfy the Laplace equation

$$\nabla^2 p = 0. \tag{4.4}$$

There are different techniques to solve the above equations and obtain the steady Stokes flow solution. Here we will follow an approach that involves fundamental harmonic solutions of the Laplace equation. To follow the discussion of this and subsequent sections, it is important for the student to have absolute mastery of index notation and vector calculus, which most readers of this book must have. For the benefit of those who want to strengthen their background, we discuss some essentials in Appendices A and B. It is crucial that the reader be familiar with the identities presented in the appendices to follow the rigorous derivations to be presented below.

There are two families of harmonic solutions. The first is the family of growing harmonics that increase and become unbounded as $r \to \infty$ and the second is the family of decaying harmonics that decay and become zero as $r \to \infty$. Note that r is the length of the position vector $\mathbf{x}$. That is, $r = (\mathbf{x} \cdot \mathbf{x})^{1/2}$. On the other hand, the growing harmonic solutions are well behaved at the origin ($r = 0$), while the decaying harmonic solutions are singular at the origin. For the present problem of flow outside a rigid particle that should decay far away from the particle, we choose the decaying family of harmonic solutions. The decaying harmonic solutions will be denoted as h_{-n} for $n = 1, 2, \ldots$ and they are given by the general expression

$$h_{-n} = \frac{\partial^{n-1}}{\partial x_i \partial x_j \cdots} \left(\frac{1}{r} \right). \tag{4.5}$$

The first few of these decaying harmonic solutions can be explicitly expressed as

$$h_{-1} = \frac{1}{r}, \quad h_{-2} = \frac{\partial}{\partial x_i} \left(\frac{1}{r} \right) = \frac{x_i}{r^3}, \quad h_{-3} = \frac{\partial^2}{\partial x_i \partial x_j} \left(\frac{1}{r} \right) = \frac{x_i x_j}{r^5} - \frac{\delta_{ij}}{3r^3},$$

$$h_{-4} = \frac{\partial^3}{\partial x_i \partial x_j x_k} \left(\frac{1}{r} \right) = \frac{x_i x_j x_k}{r^7} - \frac{x_i \delta_{jk} + x_j \delta_{ki} + x_k \delta_{ij}}{5r^5}, \quad \cdots. \tag{4.6}$$

While the derivation of h_{-2} is given in Eqs. (B.4), the derivation of higher-order terms is given as an exercise in Appendix B. The appendix also shows that each harmonic function satisfies the Laplace equation (i.e., $\nabla^2 h_{-1} = 0$, $\nabla^2 h_{-2} = 0$, and so on). From the above, we infer that h_{-1} is a scalar, h_{-2} is a vector, h_{-3} is a second-rank tensor, and so on.

The growing harmonic solutions are related to their decaying counterpart through

$$h_n = r^{2n+1} h_{-(n+1)} \tag{4.7}$$

and the first few can be explicitly written as

$$h_0 = 1, \quad h_1 = x_i, \quad h_2 = x_i x_j - \frac{r^2}{3} \delta_{ij}, \quad \cdots. \tag{4.8}$$

Again, h_0 is a scalar, h_1 is a vector, h_2 is a second-rank tensor, and so on. Also, each solution satisfies the Laplace equation $\nabla^2 h_k = 0$, and thus a linear combination of the above harmonic solutions will be a solution to the Laplace equation.

The perturbation pressure is obtained in the following way. From Eq. (4.4), p is a harmonic function and it can be written as a linear combination of the above listed harmonic solutions. In order to choose which of the harmonic solutions to combine, we make use of the following three requirements:

(1) The solution must decay as $r \to \infty$. This excludes all growing solutions.
(2) Since the solution (perturbation flow) is driven by the ambient flow, it must be a linear function of $\mathbf{u}_c$. In other words, the solution must contain the term $\mathbf{u}_c$.
(3) The harmonic solution must be a scalar quantity.

The only harmonic function that can satisfy the above requirements is h_{-2}. Though it is a vector, its dot product with $\mathbf{u}_c$ will correctly yield the required scalar, thus satisfying requirements (2) and (3). No other harmonic functions can be linearly combined with $\mathbf{u}_c$ to yield a scalar. Thus, the only possible solution is

$$p(\mathbf{x}) = a_1 u_{cj} \frac{x_j}{r^3},$$

(4.9)

where a_1 is a yet-to-be-determined constant.

The perturbation velocity field is sought as a sum of a harmonic component and a particular solution, each of which satisfy the following equations:

$$\nabla^2 \mathbf{u}^{(h)} = 0 \quad \text{and} \quad \mu_f \nabla^2 \mathbf{u}^{(p)} = \nabla p.$$

(4.10)

Problem 4.1 (a) Show that the particular solution can be expressed as

$$\mathbf{u}^{(p)} = \frac{\mathbf{x}p}{2\mu_f}.$$

(4.11)

(b) Also show that in order to satisfy the continuity equation:

$$\nabla \cdot \mathbf{u}^{(h)} = -\nabla \cdot \mathbf{u}^{(p)} = -\frac{1}{2\mu_f} (3p + \mathbf{x} \cdot \nabla p).$$

(4.12)

The harmonic part of the velocity can be expressed as a sum of the harmonic solutions listed earlier. But the solution must satisfy the following facts:

(1) The solution must decay as $r \to \infty$. This excludes all growing solutions.
(2) It must be a linear function of $\mathbf{u}_c$. In other words, the solution must contain the term $\mathbf{u}_c$.
(3) The harmonic solution must be a vector quantity.

There are now two choices. Both $h_{-1}\mathbf{u}_c$ and $h_{-3} \cdot \mathbf{u}_c$ satisfy the above conditions. This yields

$$u_i^{(h)} = \frac{a_2}{2\mu_f} \frac{1}{r} u_{ci} + \frac{a_3}{2\mu_f} \left(\frac{x_i x_j}{r^5} - \frac{\delta_{ij}}{3r^3} \right) u_{cj},$$

(4.13)

where a_2 and a_3 are yet-to-be-determined constants and they have been divided by $2\mu_f$ just for algebraic convenience. For the particular solution, we use the perturbation pressure field given in Eq. (4.9) to obtain

$$u_i^{(p)} = \frac{a_1}{2\mu_f} \frac{x_i x_j}{r^3} u_{cj}$$

(4.14)

and combining the two we obtain the total perturbation velocity to be

$$u_i = \frac{u_{cj}}{2\mu_f} \left(a_1 \frac{x_i x_j}{r^3} + a_2 \frac{\delta_{ij}}{r} \right) + \frac{u_{cj}}{2\mu_f} a_3 \left(\frac{x_i x_j}{r^5} - \frac{\delta_{ij}}{3r^3} \right). \tag{4.15}$$

Problem 4.2 Require that the divergence of the above velocity field is zero. That is, set $\partial u_i / \partial x_i = 0$. Show that $a_2 = a_1$ in order for the velocity to be divergence free. You can also prove this using Eq. (4.12).

The remaining two constants are obtained by satisfying the velocity boundary condition on the surface of a sphere. Setting $r = R_p$ in Eq. (4.15), we obtain

$$u_i\big|_{@r=R_p} = \frac{u_{ci}}{2\mu_f} \left(\frac{a_1}{R_p} - \frac{a_3}{3R_p^3} \right) + \frac{u_{cj} x_i x_j}{2\mu_f} \left(\frac{a_1}{R_p^3} + \frac{a_3}{R_p^5} \right). \tag{4.16}$$

Now, setting the right-hand side equal to the velocity boundary condition $u_i\big|_{@r=R_p} = -u_{ci}$, we obtain two conditions. Requiring the coefficient of the term containing $u_{cj} x_i x_j$ to be zero, we obtain $a_3 = -a_1 R_p^2$. Equating the coefficients of u_{ci}, we obtain the second relation $a_1 = -3\mu_f R_p/2$. Substituting for these constants into Eq. (4.15), we obtain the final perturbation velocity field to be

$$u_i = -\frac{3}{4} R_p u_{cj} \left(\frac{x_i x_j}{r^3} + \frac{\delta_{ij}}{r} \right) + \frac{3}{4} R_p^3 u_{cj} \left(\frac{x_i x_j}{r^5} - \frac{\delta_{ij}}{3r^3} \right). \tag{4.17}$$

The above perturbation flow is of fundamental importance as a lot can be learned from analyzing it. The first term on the right-hand side is the Stokeslet contribution, where

$$\textbf{Stokeslet}: \quad \frac{F_{fj}}{8\pi\mu_f} \left(\frac{x_i x_j}{r^3} + \frac{\delta_{ij}}{r} \right) \tag{4.18}$$

is defined as the axisymmetric flow induced by a point force F_{fj} applied to the fluid at the origin in an unbounded medium. In the present context of a uniform flow over an isolated sphere, the above Stokeslet solution with force vector $F_{fj} = -6\pi\mu_f R_p u_{ci}$ (which we will derive as the negative of the force exerted on the particle by the fluid) gives the first term on the right-hand side of Eq. (4.17). Based on Eq. (4.9), the pressure field due to the Stokeslet is given by $p = F_{fj} x_j/(4\pi r^3)$.

The second term on the right-hand side of Eq. (4.17) is the doublet contribution, where the velocity field of a potential doublet is

$$\textbf{Doublet}: \quad \frac{\mu_i}{4\pi} \left(3\frac{x_i x_j}{r^5} - \frac{\delta_{ij}}{r^3} \right), \tag{4.19}$$

where μ defines the strength and orientation of the doublet. It can be shown that the above doublet solution is the Laplacian of the Stokeslet solution. As a result, the doublet solution decays must faster than the Stokeslet, as $1/r^3$. It can also be seen that the doublet solution is identically the same as that in potential flow. However, in the present Stokes limit, the pressure field of the doublet is identically zero over the entire domain.

Thus, the second term on the right-hand side of Eq. (4.17) corresponds to a doublet of strength $\pi R_p^3 \mathbf{u}_c$. The doublet portion of the perturbation flow, when added to the ambient flow u_{ci}, satisfies the no-penetration boundary condition on the sphere. The doublet is thus the inviscid potential flow perturbation due to a steady uniform flow over a stationary particle. Also, the net force on the particle due to the doublet is zero. Together, the doublet and the Stokeslet ensure that the combined flow

$$(\mathbf{u}_c + \mathbf{u}_{doublet} + \mathbf{u}_{Stokeslet}) \tag{4.20}$$

satisfies both the no-penetration and the no-slip conditions on the particle.

A more transparent way to see these flow components is in terms of axisymmetric stream functions. In an axisymmetric coordinate aligned with the ambient flow, the stream functions can be expressed as

$$
\begin{aligned}
\psi_{uniform} &= \frac{1}{2}|\mathbf{u}_c| r^2 \sin^2 \theta, \\
\psi_{doublet} &= \frac{1}{4}|\mathbf{u}_c| \frac{R_p^3}{r} \sin^2 \theta, \\
\psi_{Stokeslet} &= -\frac{3}{4}|\mathbf{u}_c| r R_p \sin^2 \theta .
\end{aligned}
\tag{4.21}
$$

The corresponding radial and circumferential velocities are given by

$$u_r = \frac{1}{r^2 \sin\theta}\frac{\partial\psi}{\partial\theta} \quad \text{and} \quad u_\theta = -\frac{1}{r\sin\theta}\frac{\partial\psi}{\partial r}. \tag{4.22}$$

Figure 4.1 shows the streamlines around the particle for the following cases: (a) perturbation flow only due to the doublet given by $\psi_{doublet}$; (b) perturbation flow only due to the Stokeslet given by $\psi_{Stokeslet}$; (c) total perturbation flow given by $\psi_{doublet} + \psi_{Stokeslet}$; (d) total flow, which is a superposition of all three contributions.

Of particular importance is the manner in which the doublet and the Stokeslet perturbations decay away from the particle. In the case of a doublet, the perturbation decays faster, as $1/r^3$, while the decay of the Stokeslet is quite slow, at $1/r$. This overall slow decay of the perturbation flow induced by a spherical particle suggests that the Stokes solution is not uniformly convergent. The source of this difficulty can be traced back to our original assumption of negligible fluid inertia. Recall that our analysis is in the limit $\text{Re} = d_p|\mathbf{u}_c|/\nu_f \to 0$, which we used to justify the neglect of the inertia terms in the Navier–Stokes equation. However, the limit of $\text{Re} \to 0$ is a singular perturbation problem. No matter how small the value of Re, there exists a distance $\mathcal{L}$ beyond which the importance of fluid inertia cannot be neglected. The Reynolds number based on this length (known as the Stokes length) is unity (i.e., $\mathcal{L}|\mathbf{u}_c|/\nu_f = 1$). An explicit expression for the Stokes length is

$$\frac{\mathcal{L}}{R_p} = \frac{2}{\text{Re}}. \tag{4.23}$$

Only in the inner region for $r < \mathcal{L}$ is the neglect of inertia and the use of the Stokes equation (4.2) appropriate. For $r > \mathcal{L}$, inertial terms are important and the Stokes flow solution obtained in this section is not appropriate. We shall discuss this aspect later in the chapter, in the unsteady context of viscous history force.

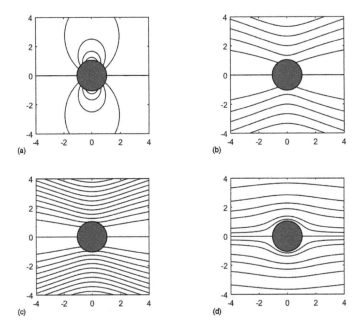

Figure 4.1 Streamlines of flow around a stationary rigid spherical particle. (a) Perturbation flow only due to the doublet. (b) Perturbation flow only due to the Stokeslet. (c) Total perturbation flow. (d) Total flow, which is a superposition of uniform flow plus the doublet and Stokeslet contributions. Note that all flows are axisymmetric. Also, if the ambient flow is from left to right, the direction of the perturbation flows in frames (a), (b), and (c) is right to left.

Fortunately, even though the solution is inaccurate in the outer region, the inner solution is accurate and can be used to calculate the drag force on the particle. In order to calculate the force on the particle, we use the tractional force distributions derived in Problem 4.3. The tractional force distribution due to the Stokeslet can be integrated around the surface of the sphere. This evaluation gives

$$F_i = \oint_{S_p} \frac{9}{2} \mu_f u_{cl} \frac{x_i x_l}{R_p^3} dS = 6\pi \mu_f R_p u_{ci}, \tag{4.24}$$

which is known as the *Stokes drag* on an isolated sphere due to steady uniform flow in the zero Reynolds number limit. In the above integral, S_p indicates the surface of the sphere and we have used the relation $\oint_{S_p} x_i x_l dS = R_p^2 \oint_{S_p} n_i n_l dS = 4\pi R_p^4 \delta_{il}/3$. This force can also be obtained by integrating the total tractional force $3\mu_f \mathbf{u}_c/(2R_p)$ obtained in Problem 4.3 over the surface of the sphere. This clearly indicates that the force on the particle is entirely due to the Stokeslet and the force on the particle due to the doublet is identically zero. Also note that in Figure 4.1 the flows are top–bottom and left–right symmetric. In other words, the departure of the streamlines from horizontal lines is symmetric and is the same both upstream and downstream of the particle. The pressure contribution to the Stokes drag can be obtained as $2\pi\mu_f R_p u_{ci}$ by surface integration of the pressure force. The corresponding viscous stress contribution to the Stokes drag is $4\pi\mu_f R_p u_{ci}$. Thus, in the Stokes limit, one-third of the drag is due to pressure and is known as the form drag and two-thirds is due to surface shear and is known as skin friction drag. As we will see later, this split between pressure and viscous drag changes with increasing Re, with pressure drag becoming the dominant contribution.

The Stokes flow over a particle is qualitatively similar to a potential flow over a particle. For example, the streamlines of the inviscid potential flow over a particle will qualitatively resemble Figure 4.1d. However, the perturbation pressure field associated with the Stokes flow will be left–right antisymmetric and contribute to the drag on the particle. In contrast, the pressure field of the potential flow will be left–right symmetric and result in zero drag (D'Alembert's paradox).

Problem 4.3 (a) Obtain the stress field of the Stokeslet as

$$\left[\sigma_{ij}\right]_{Stokeslet} = \left[-p\delta_{ij} + \mu_f\left(\frac{\partial u_i}{\partial x_j} + \frac{\partial u_j}{\partial x_i}\right)\right]_{Stokeslet} = \frac{9}{2}\mu_f R_p u_{cl}\frac{x_i x_j x_l}{r^5}, \quad (4.25)$$

where the pressure field given in Eq. (4.9) is entirely due to the Stokeslet.

(b) Obtain the corresponding tractional force distribution on the surface of the sphere by evaluating $\left[\sigma_{ij}n_j\right]_{Stokeslet}$ for $r = R_p$ as

$$\frac{9}{2}\mu_f u_{cl}\frac{x_i x_l}{R_p^3}, \quad (4.26)$$

where you need to use the definition of surface normal $n_j = x_j/r$.

(c) Obtain the tractional force distribution of the doublet as

$$\left[\sigma_{ij}n_j\right]_{doublet} = \left[\mu_f\left(\frac{\partial u_i}{\partial x_j} + \frac{\partial u_j}{\partial x_i}\right)n_j\right]_{doublet} = -\frac{9}{2}\mu_f u_{cl}\frac{x_i x_l}{R_p^3} + \frac{3}{2}\mu_f\frac{u_{ci}}{R_p}. \quad (4.27)$$

Note that the pressure distribution of the doublet is a constant and therefore ignored in the above.

(d) By adding the two tractional force distributions, obtain the total force distribution on the surface of the sphere to be given by $3\mu_f\mathbf{u}_c/(2R_p)$. Thus, amazingly, the net force on every part of the sphere is a constant and is oriented in the direction of the ambient flow. We will use this fact later in the reciprocal theorem.

4.1.2 Stokes Drag of a Drop or Bubble

We now proceed to extend the above solution to the problem of a steady uniform flow over a spherical droplet or bubble. The difference is that there is flow inside the sphere and that the no-slip boundary condition does not apply anymore on the surface of the sphere. However, we will assume the droplet or bubble to remain as a sphere of radius R_p and thus the no-penetration boundary condition applies for both the outside and inside fluids. The density and dynamic viscosities of the outside fluid are ρ_f and μ_f, respectively. The corresponding values of the inside fluid will be taken to be ρ_p and μ_p.

We start with the flow outside the sphere. The governing equations of the perturbation flow (i.e., deviation from the undisturbed uniform ambient flow) remain the same as those given in Eq. (4.2). Thus, the functional form of the solution remains the same as for the rigid-particle case considered in the previous section. The outside pressure is given by Eq. (4.9) and with $a_1 = a_2$ the outside velocity from Eq. (4.15) is given by

$$u_i = a_1 \frac{u_{cj}}{2\mu_f} \left(\frac{x_i x_j}{r^3} + \frac{\delta_{ij}}{r} \right) + a_3 \frac{u_{cj}}{2\mu_f} \left(\frac{x_i x_j}{r^5} - \frac{\delta_{ij}}{3r^3} \right) . \tag{4.28}$$

Unlike the rigid particle, in the case of a drop or bubble there is internal flow, which we will now solve. As we did for the outside flow, the flow inside the spherical region of radius R_p will also be solved as perturbation from the far-field uniform velocity of $\mathbf{u}_c$. The inside perturbation flow is also governed by Eq. (4.2) and the inside solution can be written as a superposition of the growing harmonic solutions h_0, h_1, h_2, and so on, given in Eq. (4.8). The decaying solutions are not appropriate, since they become singular at the origin, while the growing solutions do not pose a problem, since the inside solution extends only up to the radius of the sphere. For the pressure, we employ the facts that it is a scalar field and must be a linear function of $\mathbf{u}_c$ to obtain

$$p'(\mathbf{x}) = a_4 u_{cj} x_j . \tag{4.29}$$

Here, the prime denotes flow inside the sphere. The constant a_4 needs to be determined along with all the other constants. There is an additional constant pressure Δp that must be added to the inside solution of the droplet/bubble to balance the surface tension. This is a static component that exists even in the absence of fluid motion. This constant pressure difference is not of importance to the present discussion and therefore has been ignored. Here, the prime is used to denote the inside solution within the droplet/bubble.

Following the steps outlined in the previous section, we can obtain the inside velocity field to be

$$u_i' = a_4 \frac{u_{cj}}{2\mu_p} x_i x_j + a_5 u_{ci} + a_6 u_{cj} \left(x_i x_j - \frac{r^2}{3} \delta_{ij} \right) , \tag{4.30}$$

where the first term on the right is the particular solution. The next two terms are the only harmonic components that can form a vector field linear in the ambient velocity. Simplifying the above with the incompressibility condition (see Problem 4.4), we obtain

$$u_i' = a_5 u_{ci} + a_4 \frac{u_{cj}}{10\mu_p} \left(-x_i x_j + 2r^2 \delta_{ij} \right) . \tag{4.31}$$

Problem 4.4 Use the incompressibility requirement that $\partial u_i'/\partial x_i = 0$ to obtain the relation $a_6 = -3a_4/(5\mu_p)$. Also apply the boundary conditions and obtain the relations given in Eqs. (4.32) to (4.36).

The next major task is to apply the appropriate boundary conditions to evaluate the remaining four constants a_1, a_3, a_4, and a_5. The three boundary conditions to be applied are: (i) velocity continuity (i.e., $u_i = u'_i$) at $r = R_p$; (ii) no penetration through the spherical surface applies for both the outside and inside fluids. Since both flows are defined as perturbation from $\mathbf{u}_c$, the no-penetration condition yields the following relations for the normal component of outside and inside perturbation velocities: $u_i n_i = u'_i n_i = -u_{ci} n_i$; and (iii) continuity of tractional force across the surface of the sphere (ignoring the static pressure difference between the inside and outside that arises due to surface tension). These boundary conditions lead to the following relations between the constants. Velocity continuity leads to the two relations

$$\textbf{Velocity continuity}: \quad -\frac{a_4}{10\mu_p} = \frac{a_1}{2\mu_f R_p^3} + \frac{a_3}{2\mu_f R_p^5},$$

$$a_5 + \frac{a_4 R_p^2}{5\mu_p} = \frac{a_1}{2\mu_f R_p} - \frac{a_3}{6\mu_f R_p^3}, \tag{4.32}$$

where the first equation appears multiplied by $u_{cj} x_i x_j$, while the second appears multiplied by $u_{cj}\delta_{ij}$. The no-penetration condition can be applied either to the outside or the inside velocity field. Both lead to the relation

$$\textbf{No penetration}: \quad \frac{a_1}{\mu_f R_p} + \frac{a_3}{3\mu_f R_p^3} = -1. \tag{4.33}$$

The tractional force on the surface of the sphere is given by the stress tensor dotted with the unit normal vector and can be written as $\sigma_{ij} x_j / R_p$ and $\sigma'_{ij} x_j / R_p$ for the flow fields on the outside and inside of the sphere, respectively, where $n_j = x_j / R_p$ is the unit outward normal vector. The outside tractional force vector on the surface of the sphere can be expressed as

$$
\begin{aligned}
\left[\sigma_{ij} \frac{x_j}{r} \right]_{@r=R_p} &= \left[-p\delta_{ij}\frac{x_j}{r} + \mu_f \left(\frac{\partial u_i}{\partial x_j} + \frac{\partial u_j}{\partial x_i} \right) \frac{x_j}{r} \right]_{@r=R_p} \\
&= \frac{u_{cj}}{R_p} \left[\frac{a_3}{R_p^3}\delta_{ij} - 3\left(\frac{a_1}{R_p^3} + \frac{a_3}{R_p^5} \right) x_i x_j \right].
\end{aligned} \tag{4.34}
$$

Similarly, the tractional force on the surface of the sphere from the inside flow can be expressed as

$$
\begin{aligned}
\left[\sigma'_{ij} \frac{x_j}{r} \right]_{@r=R_p} &= \left[-p'\delta_{ij}\frac{x_j}{r} + \mu_p \left(\frac{\partial u'_i}{\partial x_j} + \frac{\partial u'_j}{\partial x_i} \right) \frac{x_j}{r} \right]_{@r=R_p} \\
&= \frac{u_{cj}\, a_4}{10 R_p} \left[3\delta_{ij}R_p^2 - 9 x_i x_j \right].
\end{aligned} \tag{4.35}
$$

We again note that the above expression ignores the constant static pressure inside the drop/bubble due to surface tension, which only affects the normal component of the tractional force. For tractional force balance, we equate the expressions (4.34) and (4.35). Of the two resulting equations, only the balance of tractional force in the direction of ambient flow matters, which can be expressed as

$$\textbf{Tractional force balance :} \quad \frac{a_3}{R_p^3} = \frac{3a_4}{10}R_p^2 \,. \tag{4.36}$$

From the above four relations presented in Eqs. (4.32), (4.33), and (4.36), we obtain all four constants in terms of the viscosity ratio $\mu = \mu_p/\mu_f$ as

$$
\begin{aligned}
\frac{a_1}{\mu_f R_p} &= -\frac{3\mu + 2}{2(\mu + 1)}, & \frac{a_3}{\mu_f R_p^3} &= \frac{3\mu}{2(\mu + 1)}, \\
\frac{a_4 R_p^2}{\mu_p} &= \frac{5}{\mu + 1}, & a_5 &= -\frac{1}{2(\mu + 1)} \,.
\end{aligned}
\tag{4.37}
$$

As in the case of a rigid particle, only the Stokeslet part of the outer solution contributes to the force on the spherical droplet/bubble. Note that the constant a_1 determines the strength of the Stokeslet. It is equal to $-3\mu_f R_p/2$ in the case of a rigid sphere, while we have obtained its value to be $(3\mu + 2)\mu_f R_p/(2(\mu + 1))$ in the case of a droplet/bubble. The corresponding force on the droplet/bubble can then be written as

$$\textit{Hadamard–Rybczynski drag :} \quad F_i = 2\pi\mu_f R_p u_{ci}\frac{3\mu + 2}{\mu + 1} \,. \tag{4.38}$$

The rigid-particle limit can be recovered by setting the inside-to-outside viscosity ratio $\mu \to \infty$. In this limit, we recover the Stokes drag. In the intermediate limit of $\mu = 1$, we obtain the drag force to be $F_i = 5\pi\mu_f R_p u_{ci}$. In the limit of a bubble, where the inside viscosity can be ignored (i.e., $\mu \to 0$), we obtain the drag on a spherical bubble to be $F_i = 4\pi\mu_f R_p u_{ci}$.

It must, however, be cautioned that the above result applies only for the case of an air bubble in ultra-pure water. In the presence of impurities, the surface of the bubble becomes immobile and the bubble behaves more like a rigid sphere with no internal flow. As a result, the drag on a bubble in typically impure water is better approximated by the Stokes drag of a rigid particle.

The above Stokes drag on a droplet/bubble is known as the *Hadamard–Rybczynski drag*. The drag on a bubble arises entirely due to the pressure distribution around the bubble. Since the inside viscosity is nearly zero, and due to the matching of shear stress, the net frictional force on the bubble is zero. In the case of a rigid particle, it can be shown that one-third of the Stokes drag is due to pressure distribution around the particle, which is lower than that for a bubble, and the remaining two-thirds of the drag force is due to skin friction.

Again, the outside flow streamlines can be calculated as a superposition of uniform flow plus contributions from doublet and Stokeslet portions. These streamlines around a droplet/bubble will qualitatively resemble that shown in Figure 4.1d. The streamlines of flow inside the droplet/bubble can be obtained from the inside solution given in Eq. (4.31), with the constants a_4 and a_5 taken from Eq. (4.37). The resulting flow field when expressed as a stream function simplifies to

$$\psi_{inside} = \frac{|\mathbf{u}_c|}{4(1 + \mu)}r^2\left(1 - \frac{r^2}{R_p^2}\right)\sin^2\theta \,. \tag{4.39}$$

Figure 4.2 shows the streamlines plotted inside the sphere. This toroidal vortex flow inside the sphere is known as *Hill's spherical vortex*. Clearly, the strength of the vortex decreases with increasing viscosity of the inside fluid.

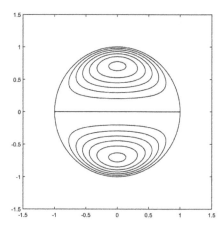

Figure 4.2 Streamlines of flow inside a stationary spherical droplet or bubble. This flow is known as Hill's spherical vortex.

Problem 4.5 For the case of a bubble, show that the outer velocity field will be such that the shear stress on the surface of the bubble is zero. In other words, for a bubble we could have ignored the inner problem and just solved the outer problem with no-stress boundary condition on the surface of the bubble.

Problem 4.6 Consider the case of a spherical bubble of radius R_p rising in water at its terminal velocity. For what value of R_p can the Reynolds number (based on diameter and rise velocity) be considered less than 1, so that Stokes flow analysis applies.

4.1.3   Flow Around a Spinning Sphere

In this subsection, we will use the harmonic solutions to solve the problem of flow around a spinning sphere and calculate the resulting torque. For this, consider a rigid particle of radius R_p spinning at a constant angular velocity of Ω. Since the undisturbed ambient flow is quiescent, the perturbation flow $\mathbf{u}$ induced by the spinning sphere is the total flow. The governing equations of the perturbation flow $\mathbf{u}(\mathbf{x})$ are again the steady Stokes flow equations given in Eq. (4.2). As a result, the solution procedure to be followed is the same as in the previous two problems.

The first step is to obtain a harmonic pressure field $p(\mathbf{x})$ that satisfies the following conditions: (i) must be a linear function of the angular velocity; (ii) must decay away from the sphere; and (iii) must be a "true" scalar field.

Here we have to introduce a new twist. We distinguish "true" tensors from "pseudo" tensors. A true tensor remains equivariant under both rotation and reflection of the coordinate system. Whereas a pseudo tensor remains equivariant under only proper rotation. A pseudo tensor will change sign under reflection (reflection is an improper rotation). An example is provided in Figure 4.3, where in frame (a) the flow field around a few particles is plotted. The vectors indicate the velocity field and the color contours indicate pressure. The velocity vector at one point is highlighted, which is associated with counter-clockwise vortex (or positive vorticity). Frame (b) is the same as the first frame, but rotated 90° about the center. Note that the pressure at the highlighted point remains the same. Though the highlighted velocity vector is not identically the same in both frames, one is the rotational transformation of the other. This is what is meant as equivariance. Also, the fluid circulation at the highlighted point remains the same. Now consider frame (c), which is a reflection of the first frame about the vertical line (y-axis). Again, pressure and the velocity vectors transform as expected. However, fluid circulation (or vorticity) has changed sign. This is why vorticity ($\omega = \nabla \times \mathbf{u}$) is a pseudo vector, while velocity ($\mathbf{u}$) is a true vector.

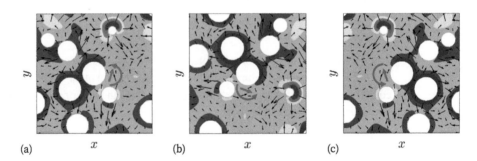

Figure 4.3 Schematic illustrating the effects of rotation and reflection on "true" and "pseudo" vectors. (a) Pressure contours and velocity vector plot around a few particles in a single plane. One point in the flow is highlighted with the in-plane velocity vector and the out-of-plane vorticity denoted as a counter-clockwise arrow. (b) Upon clockwise rotation by 90°, we see that the value of pressure and out-of-plane vorticity at the rotated point remains the same and the velocity vector also transforms properly. (c) Upon reflection about a vertical line passing through the center of the box, pressure correctly remains the same, but vorticity changes sign, indicating its property as a "pseudo" vector.

The rules of the outcome of an outer or dot product between two vectors are summarized in the left side of Table 4.1, while the outcome of a cross product between two vectors is summarized in the right side of Table 4.1. In essence, any cross product or curl operation will change the nature of the vector or tensor it is acting on. Since $\mathbf{u}$ is a true vector, $\nabla \times \mathbf{u}$ is a pseudo vector. On the other hand, $\nabla \times \omega$ will again become a true vector. Thus, the new twist is to make sure that we use the correct number of cross

Table 4.1 Inner outer, and cross-product rules of true and pseudo vectors.

	Outer or Dot Product		Cross Product	
	b = True	**b** = Pseudo	**b** = True	**b** = Pseudo
a = True	$\mathbf{a}\,\mathbf{b}$, $\mathbf{a} \cdot \mathbf{b}$ = True	$\mathbf{a}\,\mathbf{b}$, $\mathbf{a} \cdot \mathbf{b}$ = Pseudo	$\mathbf{a} \times \mathbf{b}$ = Pseudo	$\mathbf{a} \times \mathbf{b}$ = True
a = Pseudo	$\mathbf{a}\,\mathbf{b}$, $\mathbf{a} \cdot \mathbf{b}$ = Pseudo	$\mathbf{a}\,\mathbf{b}$, $\mathbf{a} \cdot \mathbf{b}$ = True	$\mathbf{a} \times \mathbf{b}$ = True	$\mathbf{a} \times \mathbf{b}$ = Pseudo

products along with the harmonic solutions and the input constant angular velocity $\boldsymbol{\Omega}$ to result in the desired true pressure and true velocity field. Note that $\boldsymbol{\Omega}$ is a pseudo vector, since it is angular velocity and similar to vorticity.

Going back to our evaluation of the pressure field for the spinning sphere, we realize that we cannot use the decaying harmonic solution h_2 as we did in Eq. (4.9). This is because $\Omega_j x_j / r^3$ is a pseudo scalar, although it is a decaying scalar which is linear in $\boldsymbol{\Omega}$. It is a pseudo scalar since it is formed by the dot product of a pseudo vector with a true vector. No other harmonic solution will satisfy all three requirements and thus, the perturbation pressure due to a spinning particle is zero.

The particular solution of the velocity field is also zero. So we proceed to solve for $\mathbf{u}^h$, which satisfies the conditions: (i) must be a linear function of the angular velocity; (ii) must decay away from the sphere; and (iii) must be a "true" vector field. Fortunately, the cross product between a pseudo vector and a true vector gives a true vector. Thus, the velocity solution of the Stokes flow around a spinning sphere is given by[1]

$$u_i = u_i^{(h)} = a_1 \epsilon_{ijk} \Omega_j \frac{x_k}{r^3},$$ (4.40)

where ϵ_{ijk} is the third-rank Levi-Civita tensor used to represent $\boldsymbol{\Omega} \times \mathbf{r}$ in index notation. We find the constant a_1 by imposing the boundary condition that $\mathbf{u} = \boldsymbol{\Omega} \times \mathbf{r}$, which in index notation becomes

$$[u_i]_{@r=R_p} = \epsilon_{ijk} \Omega_j x_k .$$ (4.41)

Equating the above two equations, we obtain $a_1 = R_p^3$. Thus, we obtain the final result

$$u_i = \epsilon_{ijk} \Omega_j x_k \frac{R_p^3}{r^3} .$$ (4.42)

The expression for the tractional force on a surface can be obtained as

$$f_i = \sigma_{ij} n_j = -3\mu_f \frac{R_p^3}{r^5} \Omega_l (\epsilon_{ilk} x_j x_k + \epsilon_{jlk} x_i x_k) n_j .$$ (4.43)

We then evaluate the above force at every point on the surface of the rotating sphere, to obtain

$$[f_i]_{@r=R_p} = -\frac{3\mu_f}{R_p} \epsilon_{ijk} \Omega_j x_k .$$ (4.44)

Integration of the above over the surface of the sphere yields zero net force on the particle. The torque on the sphere can be computed as

---

[1] Defining the symbol $\mathbf{t}$ to represent a true vector and $\mathbf{p}$ to represent a pseudo vector, the rule is as follows: $\mathbf{t} \cdot \mathbf{t} \Rightarrow \mathbf{t}$, $\mathbf{t} \cdot \mathbf{p} \Rightarrow \mathbf{p}$, $\mathbf{t} \times \mathbf{t} \Rightarrow \mathbf{p}$, and $\mathbf{t} \times \mathbf{p} \Rightarrow \mathbf{t}$. This rule can be extended to higher-rank tensors as well.

$$T_i = \oint_{S_p} \epsilon_{ijk} x_j f_k \, dA = -\frac{3\mu_f}{R_p} \Omega_l \, \epsilon_{ijk} \, \epsilon_{klm} \oint_{S_p} x_j x_m dA = -8\pi \mu_f R_p^3 \Omega_i. \qquad (4.45)$$

In the above equation, we observe the expected behavior that the torque vector $\mathbf{T}$ on the spinning sphere by the surrounding fluid is directed opposite to the sphere's angular velocity vector $\mathbf{\Omega}$.

We now introduce the *rotlet* as one fundamental solution of the steady Stokes equation and the velocity field, given by

$$\text{\textbf{Rotlet}}: \quad \frac{\mathbf{T}_f \times \mathbf{x}}{8\pi \mu_f \, r^3}. \qquad (4.46)$$

The above is the velocity field that results from the application of a torque $\mathbf{T}_f$ on an unbounded fluid at the origin and can be compared with the Stokeslet solution that results from the application of a force $\mathbf{F}_f$ on an unbounded fluid at the origin. The rotlet joins the Stokeslet and the doublet as the leading-order fundamental solutions of the Stokes equation. In the case of the rotlet, velocity decays more rapidly than for the Stokeslet, as r^{-2}. As shown by Batchelor (1970), the rotlet is the antisymmetric part of the Stokes doublet that can be obtained as the derivative of the Stokeslet. It should be noted that Batchelor refers to the rotlet as a couplet. By comparing with Eq. (4.42), it can readily be seen that the flow induced by a spinning sphere is a rotlet of appropriate strength.

Problem 4.7 Consider the Stokes flow between two concentric spinning spheres of radii R_{p1} and R_{p2}.

(a) Using vector harmonics, obtain the solution for the general case where the spheres are spinning with angular velocity vectors $\mathbf{\Omega}_1$ and $\mathbf{\Omega}_2$.

(b) Specialize the solution for the case where only the inner sphere is rotating about the x-axis.

4.1.4   Strain-Rate Effect and Effective Viscosity

In Section 2.2 we discussed the thermodynamic/transport effect in a multiphase flow that arises from the influence of suspended particles in modifying the effective properties of the mixture. In this subsection, we will consider the modification of the stress–strain-rate constitutive relationship in a multiphase mixture from that of the single-phase flow. We further note that this change in the stress–strain-rate relation can be represented as a modified effective viscosity of the multiphase medium that depends on the local volume fraction of particles. In this section, we will consider this strain-rate effect and derive the famous Einstein's viscosity correction, which he derived during his doctoral studies.

In order to focus attention only on the strain-rate effect, we will set the slip velocity between the particle and the local undisturbed fluid to be zero. We will consider a steady undisturbed ambient flow of constant strain rate, given by

$$u_{ci} = S_{cij} x_j, \tag{4.47}$$

where S_{cij} is the constant symmetric strain-rate tensor. The undisturbed ambient flow is incompressible and as a result the trace of the strain-rate tensor is zero (i.e., $S_{cii} = 0$). Let a stationary rigid particle of radius R_p be located with its center at the origin. Thus, the undisturbed flow velocity at the center of the particle is zero, and as a result there is no mean relative velocity between the particle and the ambient flow. The perturbation flow is entirely due to the interaction between the particle and the ambient strain. We define $|\mathbf{S}_c| d_p^2 / \nu_f$ to be the strain-based Reynolds number and restrict our investigation to the limit Re $\to 0$. In this limit the inertial effects can be ignored in the neighborhood of the particle and the perturbation flow is obtained by solving the Stokes equation (4.2).

As in the previous sections, we seek harmonic solutions for the pressure and velocity fields. The perturbation pressure field: (i) must be a linear function of S_{cij}; (ii) must decay as $r \to \infty$; and (iii) must be a true scalar field. The only harmonic solution that satisfies the above requirements involves h_{-3} and can be written as

$$p(\mathbf{x}) = a_1 S_{cjk} \left(\frac{x_j x_k}{r^5} - \frac{\delta_{jk}}{3r^3} \right) = a_1 \frac{S_{cjk} x_j x_k}{r^5}, \tag{4.48}$$

where we have used $S_{cii} = 0$ and a_1 is a constant that must be determined. From equation (4.11), the particular part of the perturbation velocity is obtained from the above pressure field as

$$u_i^{(p)} = \frac{a_1}{2\mu_f} \frac{S_{cjk} x_j x_k x_i}{r^5}. \tag{4.49}$$

The harmonic solution of the velocity field: (i) must be a linear function of S_{cij}; (ii) must decay as $r \to \infty$; and (iii) must be a true vector. The only harmonic solutions that satisfy the above conditions involve h_{-2} and h_{-4}. Using them, we obtain

$$u_i^{(h)} = a_2 \frac{S_{cij} x_j}{r^3} + a_3 S_{cjk} \left(\frac{x_i x_j x_k}{r^7} - \frac{x_i \delta_{jk} + x_j \delta_{ki} + x_k \delta_{ij}}{5r^5} \right). \tag{4.50}$$

The three constants a_1, a_2, and a_3 can be determined by requiring the incompressibility condition (4.12) and by demanding that the total velocity satisfies no-penetration and no-slip conditions on the surface of the particle, which can be expressed as

$$\left[u_i^{(p)} + u_i^{(h)} \right]_{@r=R_p} = -S_{cij} x_j. \tag{4.51}$$

We will skip the details for the reader to prove. Substituting the constants, the final expressions for the perturbation pressure and velocity are

$$p(\mathbf{x}) = -5\mu_f R_p^3 \frac{S_{cjk} x_j x_k}{r^5},$$

$$u_i(\mathbf{x}) = -\frac{R_p^5}{r^5} S_{cij} x_j + \frac{5}{2r^2} \left(\frac{R_p^5}{r^5} - \frac{R_p^3}{r^3} \right) S_{cjk} x_j x_k x_i. \tag{4.52}$$

It is clear from the above expression that the perturbation velocity decays as $1/r^2$. Though faster than a Stokeslet, the above decay is still slow. The solution is not uniformly convergent over the entire region outside the particle. No matter how small we choose the value of the strain-rate-based Reynolds number $|S_{cij}|d_p^2/\nu_f$, there exists a length $\mathcal{L}$ at which $|S_{cij}|\mathcal{L}^2/\nu_f = 1$ and this length in terms of particle radius is given by

$$\frac{\mathcal{L}}{R_p} = \sqrt{\frac{4}{\text{Re}}}. \tag{4.53}$$

For radial distances $r > \mathcal{L}$, inertial effects must be included, which will help in the faster decay of the perturbation. Fortunately, we are satisfied with the above solution in the region $r < \mathcal{L}$ closer to the sphere, which we will use for further analysis.

We take this opportunity to introduce the *stresslet* as another fundamental singular solution of the steady Stokes equations, whose velocity can be expressed as

$$\textbf{Stresslet}: \quad \frac{-3S_{fjk}x_ix_jx_k}{8\pi\mu_f\,r^5}. \tag{4.54}$$

The above is the velocity field that results from the application of a stress tensor $\mathbf{S}_f$ on the fluid at the origin in an unbounded domain. This solution can be compared with the Stokeslet, doublet, and rotlet solutions introduced earlier. The far-field decay of the velocity of the stresslet goes as r^{-2} and is similar to that of the rotlet, but faster than that of a Stokeslet. Comparing with the above stresslet, it can be seen that the perturbation solution (4.52) includes other terms that decay faster and account for the finite size of the particle. Far from the particle, the flow is similar to that of a point stress applied at the origin.

As discussed in Batchelor (2000), one way to evaluate the strain effect of particles in modifying the effective viscosity of the fluid is by calculating the added dissipation due to the particle. An increase in dissipation with the particle, over that in the absence of the particle, would indicate an increase in effective viscosity. In the absence of the particle, the strain rate in the fluid is a constant S_{cij} and the corresponding dissipation $2\mu_f S_{cij}S_{cij}$ is also a constant over the entire volume of the fluid. With the addition of the particle, the total strain rate in the fluid is given by

$$S_{cij} + S_{ij} = S_{cij} + \frac{1}{2}\left(\frac{\partial u_i}{\partial x_j} + \frac{\partial u_i}{\partial x_j}\right) \tag{4.55}$$

and the added dissipation is given by

$$2\mu_f(S_{cij} + S_{ij})(S_{cij} + S_{ij}) - 2\mu_f S_{cij}S_{cij}, \tag{4.56}$$

which includes a contribution from the perturbation flow as well. Since the perturbation velocity given in Eq. (4.52) goes as $1/r^2$, the corresponding strain rate decays as $1/r^3$. As a result, the added dissipation due to the particle decays as $1/r^3$ as $r \to \infty$ and thus its volume integral over the infinite expense of fluid around the particle does not absolutely converge. Here we will follow the elegant procedure presented in Batchelor (2000) in order to evaluate the modified viscosity.

We start with a large domain of volume D_∞ with a distribution of particles. The first step is to evaluate the added dissipation given in Eq. (4.56) of a single particle.

The detailed steps of this derivation are given in Appendix C. The key result is that the added dissipation of an individual particle can be rigorously obtained to be $(20\pi/3)\mu_f R_p^3 S_{cij} S_{cij}$. We now proceed to calculate the effect of a random distribution of particles within the volume D_∞ with an average particle volume fraction of ϕ. The number of particles within the large domain can be evaluated as $D_\infty \phi / \mathcal{V}_p$. Furthermore, we assume that all the particles are moving at the local fluid velocity and therefore their perturbation flow is only due to the strain effect and has no slip effect. Let the volume fraction be sufficiently low that the perturbation flow due to each particle is uninfluenced by that due to the other particles. As a result of these assumptions, the contribution to the added dissipation from each particle is the same as that given above. So, multiplying by the number of particles within the volume, we obtain the total added dissipation to be $5D_\infty \phi \mu_f S_{cij} S_{cij}$. When included along with the dissipation of the undisturbed flow, we obtain the total dissipation to be

$$2\mu_f D_\infty S_{cij} S_{cij} + 5 D_\infty \phi \mu_f S_{cij} S_{cij} . \tag{4.57}$$

We now propose that the multiphase mixture behaves as a single-phase fluid of effective viscosity μ_m. Since the particles are always moving with the local fluid without any slip, it is reasonable to consider the mixture as an effective fluid. For the externally imposed strain rate of S_{cij}, the corresponding dissipation within the homogeneous mixture is given by

$$2\mu_m D_\infty S_{cij} S_{cij} , \tag{4.58}$$

which, when equated to the total dissipation that we have just computed, yields the final relation for the effective viscosity of the multiphase mixture obtained by Einstein (1906) as

$$\mu_m = \mu_f \left(1 + \frac{5}{2}\phi\right) . \tag{4.59}$$

In summary, we see the important aspect of the thermodynamic/transport effect of the suspended particles. While the viscosity of the fluid is μ_f and that of a rigid particle can be taken to be infinite, the effective viscosity is not a simple weighted sum. Correspondingly, even though the particles are moving with the fluid (i.e., zero slip effect), there is a nonzero perturbation flow in response to the ambient strain rate in order to satisfy the no-slip and no-penetration boundary conditions. This complex perturbation flow modifies the resulting stress field and thereby contributes to the modified viscosity of the mixture. A similar approach can be used to obtain the effective thermal conductivity of the mixture, which we shall consider in Chapter 16.

4.1.5   Faxén's Law

We started this section with the investigation of the perturbation flow induced by placing a particle, droplet, or bubble in a steady uniform ambient flow (Sections 4.1.1 and 4.1.2). Section 4.1.3 considered the case of a spinning sphere immersed in an unbounded stagnant fluid. But these results apply equally to the case of a stationary nonspinning particle subjected to an ambient fluid undergoing rigid-body rotation. In Section 4.1.4, we considered the interaction of a particle with a straining ambient flow. Any linearly varying ambient flow can be expressed as a sum of uniform, rotational, and straining flows. Thus, the interaction of a particle with a linearly varying ambient flow can be expressed as a superposition of the results obtained in the previous subsections. In particular, we draw the important conclusion that only the uniform component of the undisturbed ambient flow contributes to the drag force. Only the rotational component of the ambient flow contributes to torque, and only the straining component of the ambient flow contributes to modified effective viscosity.

In this subsection, we will consider an elegant generalization of Stokes drag law for any arbitrary steady ambient flow. As discussed in the previous paragraph, the drag force on a particle in a linearly varying ambient flow is not affected by the linear variation. Thus, we are interested in more complex variations of $\mathbf{u}_c(\mathbf{x})$ in the neighborhood of the particle and their effect on drag. Based on the complexity of the solution procedure presented in the previous sections, it may seem like this generalization for any spatially varying ambient flow will require an even more difficult and complex solution procedure. However, this is not the case. The solution procedure is greatly simplified with the use of a reciprocal theorem for Stokes flow, to be discussed below. This remarkable result has been credited to Faxén (1923). Here we will only discuss his result for drag, but it should be noted that similar generalizations can be obtained for torque and induced stress as well.

As in Section 4.1.1, we consider a steady ambient flow over a stationary isolated rigid particle. Let the undisturbed flow that exists in the absence of the particle be $\mathbf{u}_c(\mathbf{x})$. We restrict the ambient flow to be steady, but it can vary in space, although we will restrict the ambient flow to be incompressible (i.e., $\nabla \cdot \mathbf{u}_c = 0$). We are interested in obtaining an expression for the force on a particle of radius R_p placed within this ambient flow. The introduction of the particle will induce a steady perturbation flow $\mathbf{u}(\mathbf{x})$. The perturbation flow must decay to zero as $r \to \infty$. On the surface of the spherical particle, the perturbation flow must cancel the undisturbed ambient flow such that the total velocity becomes zero. This requirement of no penetration and no slip on the surface of the rigid stationary particle centered at the origin can be expressed as $\mathbf{u}(|\mathbf{x}| = R_p) = -\mathbf{u}_c(|\mathbf{x}| = R_p)$.

To solve this problem, we first introduce the *reciprocal theorem* due to Lorentz (1896). Consider two different incompressible Stokes flow perturbations, named $\mathbf{u}_1$ and $\mathbf{u}_2$, around the same object whose surface will be denoted as ∂D. The two flows differ in their boundary values on the surface of the object. Both the flows decay as $1/r$ far away from the object, as required by the Stokeslet solution. The corresponding pressure fields are p_1 and p_2 and they too decay as $1/r^2$, as given in Eq. (4.9). The stress tensors of the two flows can then be written as

$$\sigma_1 = -p_1\mathbf{I} + 2\mu_f\mathbf{S}_1 \quad \text{and} \quad \sigma_2 = -p_2\mathbf{I} + 2\mu_f\mathbf{S}_2, \quad (4.60)$$

where $\mathbf{S}_1 = (\nabla\mathbf{u}_1 + (\nabla\mathbf{u}_1)^T)/2$ is the symmetric part of the velocity gradient tensor and $\mathbf{S}_2$ is defined similarly. We can now obtain the following symmetry relation:

$$\sigma_2 : \mathbf{S}_1 = -p_2 \nabla \cdot \mathbf{u}_1 + 2\mu_f \mathbf{S}_2 : \mathbf{S}_1 = 2\mu_f \mathbf{S}_1 : \mathbf{S}_2 = \sigma_1 : \mathbf{S}_2 , \qquad (4.61)$$

where we have used the fact that the flows are divergence free. Furthermore, since they are symmetric, $\mathbf{S}_1 : \mathbf{S}_2 = \mathbf{S}_2 : \mathbf{S}_1$.

From the separation of the velocity gradient tensor into its symmetric and anti-symmetric parts as $\nabla \mathbf{u} = \mathbf{S} + \mathbf{R}$, we obtain

$$\sigma_2 : \mathbf{S}_1 = \sigma_2 : \nabla \mathbf{u}_1 = \nabla \cdot (\sigma_2 \cdot \mathbf{u}_1) . \qquad (4.62)$$

In obtaining the first equality, we use the fact that σ_2 is symmetric. $\mathbf{R}_1$ is antisymmetric and therefore $\sigma_2 : \mathbf{R}_1 = 0$. In obtaining the second relation, we use the fact that Stokes flow satisfies $\nabla \cdot \sigma_2 = 0$. Applying the same steps as in expression (4.62) to the right-hand side of the relation (4.61), we obtain

$$\nabla \cdot (\sigma_2 \cdot \mathbf{u}_1) = \nabla \cdot (\sigma_1 \cdot \mathbf{u}_2) . \qquad (4.63)$$

The final step of obtaining the reciprocal relation is to take the volume integral of the above equation over the entire fluid volume exterior of ∂D. We then use the divergence theorem (see Appendix B) to convert the volume integral to a surface integral. The contribution from the surface integral at infinity is zero. This can be seen through the following argument. As stated above, since $\mathbf{u}$ decays as $1/r$, σ decays as $1/r^2$ and their product decays as $1/r^3$. Since the surface area increases only as r^2, the contribution from the surface at infinity is zero. Retaining only the contribution from the surface integral of the body ∂D, the final *reciprocal relation* in index notation can be expressed as

$$\oint_{\partial D} (\sigma_{2ij} n_j) u_{1i} \, dA = \oint_{\partial D} (\sigma_{1ij} n_j) u_{2i} \, dA . \qquad (4.64)$$

In the above, $f_{1i} = \sigma_{1ij} n_j$ is the tractional force distribution of the first flow field on the surface of the body and correspondingly $f_{2i} = \sigma_{2ij} n_j$ is that due to the second flow field.

Now we are ready to apply the reciprocal relation. Let the first perturbation flow field be that due to a uniform ambient flow $\mathbf{u}_{c0}$, where the subscript 0 has been added to denote the uniform nature of the flow. We let the surface $\partial D = S_p$ be the surface of the spherical particle. From the results of Problem 4.3, the corresponding tractional force distribution is a constant vector given by $f_{1i} = 3\mu_f u_{c0i}/(2R_p)$. The second perturbation field is that due to the non-uniform ambient flow $\mathbf{u}_c$, and let its tractional force distribution be f_{2i}. In both cases, the perturbation fluid velocity on the surface $r = R_p$ is exactly opposite to the local value of the undisturbed ambient flow. In other words, $u_{1i} = -u_{c0i}$ and $u_{2i} = -u_{ci}$. Substituting the above into the reciprocal relation:

$$-u_{c0i} \oint_{S_p} f_{2i} \, dA = -\frac{3\mu_f u_{c0i}}{2R_p} \oint_{S_p} u_{ci} \, dA . \qquad (4.65)$$

In the above, $\oint_{S_p} f_{2i} \, dA = F_i$ is identified as the force on the particle due to the non-uniform ambient flow. On the right-hand side, the surface integral for a spherical particle can be written in terms of the surface-averaged velocity as $4\pi R_p^2 \overline{u_{ci}}^S$. Substituting these, we obtain *Faxén's law* (Faxén, 1923) as

$$\mathbf{F} = 6\pi \mu_f R_p \overline{\mathbf{u}_c}^S, \tag{4.66}$$

which is in precisely the same form as Stokes drag, but written in terms of the undisturbed ambient flow velocity averaged over the surface of the spherical particle. This remarkable result applies for any spatially varying ambient flow, with the only restriction that it has been derived in the Stokes limit of zero Reynolds number. We have thus cleverly used the earlier result of force due to a uniform flow to obtain that drag on the particle due to any steady ambient flow in the Stokes limit. The rationale for the surface average can be presented as follows. We recall that the perturbation flow satisfies the steady Stokes equation and must decay to zero in the far field. The perturbation flow is driven by the requirement that the perturbation flow on the surface of the particle must negate the corresponding local undisturbed ambient flow. Thus, the effective flow that needs to be negated is given by the surface average $\overline{\mathbf{u}_c}^S$.

Problem 4.8 The evaluation of the strain effect in the form of a modified viscosity can be carried out for a dispersion of droplets/bubbles as well. In this case, the analysis will start with a single spherical droplet/bubble placed at the origin in an ambient straining flow.

(a) Follow the steps taken for the rigid particle and carry them out for the droplet/bubble. First, show that the form of the outer flow given in Eqs. (4.48) and (4.50) is still relevant.

(b) Obtain the form of the inner harmonic solution.

(c) Then apply the matching boundary conditions, as we did in the uniform flow over a droplet/bubble in Section 4.1.2. Obtain the following solutions for the outer and inner flows:

$$p(\mathbf{x}) = -\frac{2\mu_f + 5\mu_p}{\mu_f + \mu_p} \mu_f R_p^3 \frac{S_{cjk} x_j x_k}{r^5},$$

$$u_i(\mathbf{x}) = -\frac{\mu_p}{\mu_f + \mu_p} \frac{R_p^5}{r^5} S_{cij} x_j$$

$$+ \left(\frac{5\mu_p}{2(\mu_f + \mu_p)} \frac{R_p^5}{r^7} - \frac{2\mu_f + 5\mu_p}{\mu_f + \mu_p} \frac{R_p^3}{2r^5} \right) S_{cjk} x_j x_k x_i,$$

$$p'(\mathbf{x}) = \frac{21\mu_f}{2(\mu_f + \mu_p)} \frac{\mu_p}{R_p^2} S_{cjk} x_j x_k,$$

$$u_i'(\mathbf{x}) = \left(\frac{5\mu_p}{2(\mu_f + \mu_p)} \frac{r^5}{R_p^2} - \frac{3\mu_f}{2(\mu_f + \mu_p)} \right) S_{cij} x_j$$

$$- \frac{\mu_f}{\mu_f + \mu_p} \frac{1}{R_p^2} S_{cjk} x_j x_k x_i.$$

$$(4.67)$$

(d) Use the above solution to calculate the added dissipation due to the perturbation flow induced by the droplet/bubble to be

$$\frac{4\pi}{3}\frac{2\mu_f + 5\mu_p}{\mu_f + \mu_p}\mu_f R_p^3 S_{cij}S_{cij}\,. \tag{4.68}$$

(e) Continuing along the lines presented for the rigid particle, derive the final expression for the modified viscosity for a suspension of droplets/bubbles to be

$$\mu_m = \mu_f\left(1 + \frac{2\mu_f + 5\mu_p}{2(\mu_f + \mu_p)}\phi\right)\,. \tag{4.69}$$

In the case of rigid particles, $\mu_p \to \infty$ and we recover Einstein's viscosity relation. In the other limit of bubbles, if we take $\mu_p \to 0$, we get $\mu_m = \mu_f(1 + \phi)$.

4.2   Basset–Boussinesq–Oseen Equation of Motion

In this section, we proceed beyond the five steady problems considered in the previous section by introducing time dependence. We consider the classic problem of a spherical rigid particle under arbitrary time-dependent motion in an unsteady spatially uniform ambient flow and obtain an explicit analytical expression for the force on the particle in the zero Reynolds number limit. The restriction that the ambient flow be spatially uniform implies that the particle is much smaller than the spatial scales of the ambient flow. Let the particle be translating at velocity $\mathbf{V}(t)$ while immersed in an "undisturbed" incompressible, spatially uniform ambient flow of velocity $\mathbf{u}_c(t)$ and pressure $p_c(\mathbf{x},t)$. The undisturbed flow satisfies the following momentum equation:

$$\frac{\partial \mathbf{u}_c}{\partial t} = \mathbf{g} - \frac{1}{\rho_f}\nabla p_c\,. \tag{4.70}$$

Note that the advection term does not appear in the above equation, since the flow has been assumed to be spatially uniform and therefore its gradient is zero. To satisfy the above momentum equation, p_c is restricted to be linearly varying. We begin our analysis in a frame of reference that moves with the ambient flow. In this frame of reference, the ambient undisturbed flow is quiescent and the particle moves with velocity $\mathbf{V} - \mathbf{u}_c$. Due to this relative motion of the particle with respect to the ambient, there will be a nonzero perturbation flow around the particle and we will denote this microscale flow as $\mathbf{u}$. Let the corresponding perturbation pressure be $p(\mathbf{x},t)$. In this moving frame, the total velocity and pressure fields are given by

$$\mathbf{u}_{\text{tot}} = \mathbf{u} \quad \text{and} \quad p_{\text{tot}} = p_c + p\,. \tag{4.71}$$

The governing equations in the moving frame of reference are

$$\nabla \cdot \mathbf{u} = 0,$$
$$\frac{\partial \mathbf{u}}{\partial t} + \mathbf{u}\cdot\nabla\mathbf{u} + \frac{\partial \mathbf{u}_c}{\partial t} = \mathbf{g} - \frac{1}{\rho_f}\nabla(p_c + p) + \nu_f\nabla^2\mathbf{u}\,. \tag{4.72}$$

In the momentum equation, the term $\partial \mathbf{u}_c / \partial t$ on the left-hand side arises from the non-inertial frame of reference attached to the undisturbed ambient flow. This term, along with gravity and undisturbed pressure gradient, can be eliminated from the above using Eq. (4.70).

We define the Reynolds number based on relative velocity as

$$\text{Re} = \frac{|\mathbf{u}_c - \mathbf{V}| \, d_p}{\nu_f} \tag{4.73}$$

and consider the Reynolds number to be small, so that the nonlinear inertial term on the left can be ignored. Further, pressure can be eliminated by taking the curl of the momentum equation, which results in the following linear equations:[2]

$$\nabla \cdot \mathbf{u} = 0,$$
$$\frac{\partial (\nabla \times \mathbf{u})}{\partial t} = \nu_f \nabla^2 (\nabla \times \mathbf{u}). \tag{4.74}$$

The above equations must be solved with the following boundary conditions:

$$\begin{aligned} \text{On the sphere:} \quad & \mathbf{u} = \mathbf{V} - \mathbf{u}_c, \\ \text{Far field:} \quad & \mathbf{u} = 0. \end{aligned} \tag{4.75}$$

4.2.1   Solution by Helmholtz Decomposition

We proceed to solve using *Helmholtz decomposition* of the velocity field in terms of a scalar potential ϕ and a vector potential $\boldsymbol{\psi}$:

$$\mathbf{u} = \nabla \phi + \nabla \times \boldsymbol{\psi}, \tag{4.76}$$

where $\nabla \cdot \boldsymbol{\psi} = 0$. Substituting the Helmholtz decomposition in Eqs. (4.74), we obtain the following separate equations for the scalar and the vector potentials:

$$\nabla^2 \phi = 0 \quad \text{and} \quad \nabla^2 \boldsymbol{\psi} - \frac{1}{\nu_f} \frac{\partial \boldsymbol{\psi}}{\partial t} = 0. \tag{4.77}$$

We expect the solution to Eqs. (4.74) and (4.75) to be axisymmetric about the relative velocity $(\mathbf{V} - \mathbf{u}_c)$. This allows simplification of the vector potential as

$$\boldsymbol{\psi} = \nabla \times (\mathbf{e}_r \, \psi), \tag{4.78}$$

where $\mathbf{e}_r$ is the unit vector along the direction of relative velocity and ψ is a scalar field. With this simplification, we obtain

$$\nabla^2 \phi = 0 \quad \text{and} \quad \nabla^2 \psi + k_2^2 \psi = 0, \tag{4.79}$$

where $k_2^2 = -(1/\nu_f)(\partial/\partial t)$. It may seem a bit surprising that the variable k_2 is in fact an operator. It may be easier to think in terms of a Fourier or Laplace transform. If we had taken the Fourier transform of Eq. (4.77), then we would have $k_2^2 = \iota \omega / \nu_f$, where ω is the frequency. Instead, if we had taken the Laplace transform of Eq. (4.77),

---

[2] Here and in what follows we use the following vector identities: $\nabla \cdot \nabla b \equiv \nabla^2 b$; $\nabla \times (\nabla b) \equiv 0$; and $\nabla \cdot (\nabla \times \mathbf{A}) \equiv 0$ for any scalar field b and vector field $\mathbf{A}$.

then $k_2^2 = -s/\nu_f$, where s is the Laplace variable. The advantage of defining k_2 as an operator is that the solution will be obtained in terms of k_2 without taking a Fourier or Laplace transform for now. Later, the solution can be transformed to the time domain by substituting for k_2 and taking an inverse Fourier or Laplace transform.

As expected, ϕ represents the potential flow and it carries no vorticity. It is responsible for enforcing the no-penetration boundary condition on the surface of the particle. All the vorticity generated is captured by the potential ψ and it satisfies the no-slip boundary condition. The solution to the Laplace equation around a moving sphere is given by a doublet:

$$\phi = \frac{A}{r^2} \cos\theta \, f(t).$$ (4.80)

The general solution to the Helmholtz equation for ψ can be written as (see Appendix D)

$$\psi = B \, h_n(k_2 r) \, P_n(\cos\theta) f(t) \quad \text{for} \quad \text{Im}(k_2 r) > 0,$$ (4.81)

where A and B are constants and the superscripts (1) and (0) in the Appendix Eq. (D.9) have been dropped from the Hankel function h_n and the associated Legendre polynomial P_n, respectively. Since we are interested in a solution that decays outside of the spherical particle, we have only chosen the outgoing part of the solution. Here, $f(t)$ is a function of time that will be determined from the boundary condition (4.75) in the following subsection.

4.2.2 Application of Boundary Condition

In order to apply the velocity boundary condition on the sphere, we express the radial and tangential velocity components in terms of velocity potentials as (Guz, 2009; Parmar, 2010)

$$
\begin{aligned}
u_r &= \frac{\partial\phi}{\partial r} - \frac{1}{r\sin\theta}\frac{\partial}{\partial\theta}\sin\theta\frac{\partial\psi}{\partial\theta} \\
&= \left[-\frac{2A}{r^3}\cos\theta - \frac{Bh_n(k_2 r)}{r\sin\theta}\frac{\partial}{\partial\theta}\sin\theta\frac{\partial P_n(\cos\theta)}{\partial\theta} \right] f(t), \\
u_\theta &= \frac{1}{r}\frac{\partial\phi}{\partial\theta} + \frac{1}{r}\frac{\partial}{\partial r}r\frac{\partial\psi}{\partial\theta} \\
&= \left[-\frac{A}{r^3}\sin\theta + \frac{B}{r}\frac{\partial}{\partial r}rh_n(k_2 r)\frac{\partial P_n(\cos\theta)}{\partial\theta} \right] f(t).
\end{aligned}
$$ (4.82)

The velocity boundary condition on the sphere (Eq. (4.75), first line) can be written in terms of radial and tangential components as

$$u_r|_{r=R_p} = (\mathbf{V} - \mathbf{u}_c)\cos\theta \quad \text{and} \quad u_\theta|_{r=R_p} = -(\mathbf{V} - \mathbf{u}_c)\sin\theta,$$ (4.83)

where $R_p = d_p/2$ is the radius of the spherical particle. Comparing the above with Eqs. (4.82), and noting that $P_1(\cos\theta) = \cos\theta$, we observe that the only possible value of n is $n = 1$. Furthermore, we obtain the time dependence to be $f(t) = (\mathbf{V} - \mathbf{u}_c)$. Setting $r = R_p$ in Eqs. (4.82) and equating to (4.83), we obtain

$$-\frac{2A}{R_p^3} + \frac{2Bh_1}{R_p} = 1 \quad \text{and} \quad -\frac{A}{R_p^3} - B\left(k_2 h_0 - \frac{h_1}{R_p}\right) = -1, \tag{4.84}$$

where we have used the following property of the Hankel function: $x(dh_1/dx) = x\,h_0 - 2h_1$. Note that with the arguments included, the Hankel functions are $h_0(k_2 R_p)$ and $h_1(k_2 R_p)$. Solving, we get the constants

$$A = \frac{3}{2}\frac{h_1 R_p^2}{k_2 h_0} - \frac{R_p^3}{2} \quad \text{and} \quad B = \frac{3}{2k_2 h_0}. \tag{4.85}$$

4.2.3   Force on the Spherical Particle

With the expressions for the constants A and B, we now have the perturbation flow field due to the particle from Eqs. (4.76), (4.80), and (4.81). We proceed to calculate the force as the surface integral of the radial and tangential stresses:

$$\mathbf{F}(t) = \oint_{S_p} (-p_c \mathbf{n} + \sigma \cdot \mathbf{n})\, dA, \tag{4.86}$$

where the integral is over the surface of the particle S_p. The first term on the right corresponds to the force on the particle due to the pressure distribution of the undisturbed ambient flow, while the second term arises from the disturbance flow due to the presence of the particle. First, let us evaluate the undisturbed flow force as

$$\mathbf{F}_{un}(t) = -\oint_{S_p} p_c \mathbf{n}\, dA = -\oint_{\mathcal{V}_p} \nabla p_c\, dV = -\mathcal{V}_p \nabla p_c, \tag{4.87}$$

where we use the Gauss theorem to go from the surface to the volume integral. Furthermore, we have employed the fact that according to Eq. (4.70), ∇p_c does not vary over the volume of the particle (it is independent of $\mathbf{x}$).

The integration of the second term in Eq. (4.86) yields the force on the particle due to the perturbation flow. The details of this integration are presented in Appendix E. The results of the appendix, when combined with the pressure gradient force, yield the following:

Basset–Boussinesq–Oseen equation

$$\boxed{\begin{aligned}\mathbf{F}(t) = &-\mathcal{V}_p \nabla p_c + 6\pi\mu_f R_p(\mathbf{u}_c - \mathbf{V}) + \frac{1}{2}m_f\left[\frac{d\mathbf{u}_c}{dt} - \frac{d\mathbf{V}}{dt}\right] \\ &+ 6\pi\mu_f R_p \int_{-\infty}^{t} \frac{R_p}{\sqrt{\pi\nu_f(t-\xi)}}\left[\frac{d\mathbf{u}_c}{dt} - \frac{d\mathbf{V}}{dt}\right]_{@\xi} d\xi.\end{aligned}} \tag{4.88}$$

Note that $\mathbf{V}$ and $\mathbf{u}_c$ are functions of time. In the last term, $[f(t)]_{@\xi}$ represents the value of f evaluated at the instant $t = \xi$. Thus, the force on the particle depends on the past time history of relative acceleration. *This BBO equation is an exact expression for force on a spherical particle in arbitrary motion in a spatially uniform, time-dependent ambient flow, in the limit of zero Reynolds number.* In the following subsection, we will consider each term of the BBO equation in greater detail and explore their physical origin.

The careful reader will have noticed a subtle, but important, point that we glossed over in the above derivation. Note that the boundary condition of Eq. (4.75), first line is applied on the surface of the spherical particle, which in the frame of reference attached to the ambient flow is moving and can be expressed as

$$|\mathbf{x} - \mathbf{x}_p(t)| = R_p \quad \text{where} \quad \mathbf{x}_p(t) = \int_0^t (\mathbf{V}(\xi) - \mathbf{u}_c(\xi))d\xi. \tag{4.89}$$

The application of the velocity boundary condition on a moving sphere is thus more complicated. This complication was not explicitly addressed in the above solution procedure. This was made possible by the following transformation. If we redefine the coordinate variable as $\mathbf{x}' = \mathbf{x} - \mathbf{x}_p(t)$, then Eqs. (4.74) can be rewritten in the new variable as

$$\nabla' \cdot \mathbf{u} = 0,$$
$$\frac{\partial (\nabla' \times \mathbf{u})}{\partial t} - (\mathbf{V} - \mathbf{u}_c) \cdot \nabla' \mathbf{u} = \nu_f \nabla'^2 (\nabla' \times \mathbf{u}). \tag{4.90}$$

In the $\mathbf{x}'$ coordinate, the boundary condition of Eq. (4.75), first line is conveniently applied on a fixed spherical surface given by $|\mathbf{x}'| = R_p$. In the above equation, the second term on the left-hand side is quadratic in velocity and therefore can be ignored in the Stokes limit. Thus, the governing equations revert back to those given in Eqs. (4.74). In essence, the assumption of linearity allows us to apply the velocity boundary condition corresponding to that of a moving particle to a fixed spherical surface located with its center at the origin. This useful trick will be used later in other problems, as needed. The final outcome is that the analysis of this section and the result presented in Eq. (4.88) are exact in the Stokes limit.

The result (4.88) can be derived in other ways, some perhaps even easier. However, as we proceed to consider other effects, the derivation presented here will prove to be useful. In this and the following sections, we will consistently pursue the approach presented here.

4.2.4 Interpretation

Undisturbed Flow Force. The first term of the BBO equation is the force experienced by the particle due to the undisturbed ambient flow, without the flow being disturbed by the presence of the particle. All other force contributions are due to the perturbation flow induced by the presence of the particle. In the absence of the particle, this is the force that acts on the spherical volume of fluid that exits at the location of the particle. In the present context of a spatially uniform flow, this force appears in Eq. (4.88) as the *pressure-gradient force*. In the more general context of a spatially varying flow, the spherical volume of fluid that exits at the location of the particle will experience not only the pressure-gradient force, but also a force due to the viscous stresses. Thus, the more general form of the undisturbed flow force is

$$\mathbf{F}_{un} = \oint_{S_p} \boldsymbol{\sigma}_c \cdot \mathbf{n}\, dA = \oint_{\mathscr{V}_p} \nabla \cdot \boldsymbol{\sigma}_c\, dv = \mathscr{V}_p \overline{\nabla \cdot \boldsymbol{\sigma}_c}^V,\tag{4.91}$$

where $\boldsymbol{\sigma}_c$ is the stress in the undisturbed ambient flow. The total stress, $\boldsymbol{\sigma}_c = -p_c\mathbf{I} + \boldsymbol{\tau}_c$, is the sum of the isotropic pressure and the viscous stress of the undisturbed flow. Thus, the more general undisturbed flow force includes the pressure gradient and the viscous force. In fact, from the Navier–Stokes equation for the undisturbed ambient flow, we have

$$\mathbf{F}_{un}(t) = \mathscr{V}_p\, \nabla \cdot \boldsymbol{\sigma}_c = \underbrace{-m_f\mathbf{g}}_{\substack{\text{hydrostatic}\\ \text{contribution}}} + \underbrace{m_f\frac{D\mathbf{u}_c}{Dt}}_{\substack{\text{hydrodynamic}\\ \text{contribution}}},\tag{4.92}$$

where m_f is the mass of the displaced fluid. The first term corresponds to the upward buoyancy force due to the weight of the displaced fluid and the second term corresponds to the undisturbed flow force that would be experienced by the volume occupied by the particle due to the acceleration of the ambient fluid. Under hydrostatic conditions, the viscous stress is identically zero, and the pressure-gradient force will become the buoyancy force pointing in the upward direction. In this case, the undisturbed flow force is also called the buoyancy or Archimedes force.

The undisturbed flow force takes interesting forms in different problems. Although its definition as the integral of the divergence of the undisturbed stress is quite straightforward, there have been misinterpretations in the past. The following two examples will illustrate some interesting aspects of the undisturbed flow force. These two examples will also illustrate the point that often, the contribution from the pressure gradient is the dominant portion of the undisturbed flow force, hence the undisturbed floor force is also often simply referred to as the pressure-gradient force.

Example 4.9 Consider the problem of settling of a dilute distribution of particles of volume fraction ϕ_d. Let all the particles be settling at the same settling velocity and let the particles be sufficiently apart that they do not directly affect each other. We are interested in evaluating the undisturbed flow force on an individual particle – let us call it the reference particle. In the frame moving with the particle, the undisturbed flow is a steady, spatially uniform up-flow of magnitude the same as the settling velocity of the reference particle. By definition, the undisturbed flow is in the absence of the reference particle, and due to all the neighbors being far away, the undisturbed flow is not affected by them as well. At a scale larger than the individual particles, the undisturbed flow satisfies the following Navier–Stokes equation:

$$\nabla \cdot \boldsymbol{\sigma}_c = \rho_f \frac{D\mathbf{u}_c}{Dt} - \rho_f\mathbf{g} - (\rho_p - \rho_g)\phi_d\,\mathbf{g},\tag{4.93}$$

where the last term on the right accounts for the force exerted by all the other particles within the system. Although this feedback force exerted by the other particles is distributed over the surface of the particles, we have approximated it as a constant volumetric source. This approximation is appropriate in a dilute random distribution of particles. In a steady sedimenting system, there will be a weak up-flow in the frame attached to the falling particle. However, its total derivative, which is the first term on the right-hand side, is zero. The last two terms combine to yield $-\rho_m \mathbf{g}$, where the mixture density is given by $\rho_m = \phi_d \rho_p + (1 - \phi_d)\rho_f$. On the left-hand side, viscous stresses are negligible and we are left with only the pressure gradient.

We finally arrive at the following expression for the undisturbed flow force:

$$\mathbf{F}_{un} = -\mathcal{V}_p \nabla p = -\mathcal{V}_p \rho_m \mathbf{g}. \tag{4.94}$$

Several important observations can be made. The undisturbed flow force is nothing but the upward buoyancy force, but evaluated with the mixture density. This is the case even though the fluid displaced by the reference particle has a density of ρ_f. This is because the collective action of all the particles gives the fluid a higher hydrostatic pressure gradient. The total fluid force on the reference particle must be equal to $-\mathcal{V}_p \rho_p \mathbf{g}$, since it must balance its weight. Thus, in this problem, we explicitly obtain the perturbation force on the particle due to the quasi-steady force as $-\mathcal{V}_p (\rho_p - \rho_m) \mathbf{g}$. Clearly, the assumption of all the particles falling at the same settling velocity is an idealization. In reality, there will be long-range interaction between the particles and each particle's velocity and trajectory will be time dependent and will instantaneously deviate from the average. We have ignored all these complexities.

Example 4.10 In this example, consider a distribution of stationary particles of volume fraction ϕ_d and let there be a uniform flow going over this distribution, which can be taken to be along the x-axis. As in the previous example, we will assume all the particles experience the same force $\mathbf{F}$ along the flow direction. A pressure gradient is required along the x-direction and from control volume analysis we can obtain $-\nabla p = \mathbf{F}\phi_d / \mathcal{V}_p$. Thus, we obtain the pressure gradient force

$$\mathbf{F}_{un} = -\mathcal{V}_p \nabla p = \mathbf{F}\phi_d. \tag{4.95}$$

In this problem, we obtain the perturbation force on the particle due to the quasi-steady force as $\mathbf{F}(1 - \phi_d)$.

Quasi-Steady Force. The second term on the right-hand side of Eq. (4.88) can be identified as the Stokes drag. This and all other forces to follow are due to the perturbation flow created by the imposition of no-slip and no-penetration boundary conditions on the surface of the particle. Even though the particle and the ambient flow velocities are time dependent, here the steady Stokes drag formula has been applied in a quasi-steady fashion based on the instantaneous velocity difference between the particle and the undisturbed ambient flow. Hence, this force is called the *quasi-steady force.*

Although most readers will likely be quite familiar with the Stokes drag, here we must point out some implicit assumptions in its use. First and foremost, the Stokes drag is expressed in terms of the undisturbed flow velocity $\mathbf{u}_c$. Once the flow is disturbed by the imposed no-slip and no-penetration boundary conditions, the fluid velocity very close to the particle must be equal to the particle velocity. Thus, even though the quasi-steady force accounts for the effect of flow perturbation induced by the particle, *it is modeled in terms of the undisturbed flow velocity at the particle*. This is true of all the force components presented in Eq. (4.88), where it should be noted that all the expressions are in terms of undisturbed flow velocity.

Added-Mass Force. While the quasi-steady force is due to the relative velocity between the particle and the undisturbed ambient flow, the third and fourth terms on the right-hand side of Eq. (4.88) are due to the relative acceleration between the particle and the undisturbed ambient flow. These forces are identically zero if both the particle and the ambient fluid motion are steady. The force represented by the third term on the right can be attributed to the inviscid potential part of the perturbation flow generated by the particle. Hence we also call this force the *inviscid-unsteady force*.

In order to understand the physical origin of this force, we reconsider the derivation presented earlier in the inviscid limit of $(\nu_f, \mu_f) \to 0$. Several simplifications occur during the derivation in the inviscid limit: (i) without a viscous effect, the vector potential is identically zero (i.e., we can take $\psi = 0$ and as a result the constant $B = 0$); (ii) only the radial velocity boundary condition given in Eq. (4.83) needs to be satisfied. Solving for the other constant A, the resulting perturbation potential ϕ flow takes the familiar doublet form

$$\phi = -\frac{R_p^3}{2r^2} \cos\theta |\mathbf{V} - \mathbf{u}_c| . \tag{4.96}$$

This perturbation flow decays away from the spherical particle. We can follow the discussion in Batchelor (2000, Section 6.2) and express the total kinetic energy of the perturbation flow as

$$KE = -\frac{1}{2}\rho_f \oint_{S_p} \phi(\mathbf{V} - \mathbf{u}_c) \cdot \mathbf{n}\, dA = \frac{1}{2}\left(\frac{1}{2}m_f\right)|\mathbf{V} - \mathbf{u}_c|^2, \tag{4.97}$$

where the integral is over the surface of the sphere, which when carried out results in an explicit expression for the kinetic energy of the fluid, written in terms of the mass of the displaced fluid (m_f) and the relative velocity. In a frame attached to the ambient fluid, when a particle moves at velocity $(\mathbf{V} - \mathbf{u}_c)$, the kinetic energy of the particle is $m_p|\mathbf{V} - \mathbf{u}_c|^2/2$. But there is additional kinetic energy in the fluid associated with the potential perturbation flow. This additional fluid kinetic energy is given above and it is as if half the displaced volume of fluid is moving with the particle. Thus, an added mass of fluid can be taken to be moving with the particle, and as the particle accelerates with respect to the ambient, this added mass of fluid must accelerate as well.

Taking the time derivative of Eq. (4.97), we obtain

$$\frac{dKE}{dt} = \left[\frac{1}{2}m_f\left(\frac{d\mathbf{V}}{dt} - \frac{\partial\mathbf{u}_c}{\partial t}\right)\right] \cdot (\mathbf{V} - \mathbf{u}_c) . \tag{4.98}$$

The rate of change of kinetic energy is equal to the rate of work done on the fluid by the moving particle. In turn, the rate of work done is equal to the dot product of the force on the fluid by the particle and the relative velocity vector. Therefore, we identify the term within square brackets as the force on the fluid. The corresponding inviscid unsteady force on the particle is

$$\mathbf{F}_{iu} = -\frac{1}{2} m_f \left(\frac{d\mathbf{V}}{dt} - \frac{\partial \mathbf{u}_c}{\partial t} \right). \tag{4.99}$$

From the above discussion it is clear that in order to accelerate the particle relative to the fluid, an additional force $\mathbf{F}_{iu}$ must be applied to the particle, which will go towards changing the kinetic energy of the added mass of fluid that is accelerated with the particle. Hence, this force is also known as the *added-mass force* or the *virtual-mass force*.

Viscous-Unsteady Force. The last term of Eq. (4.88) is the viscous-unsteady force. It is the viscous effect of relative acceleration between the particle and the ambient flow. As we discussed in the previous subsection, the inviscid-unsteady force is due to the irrotational scalar potential ϕ, which satisfies only the boundary condition for the normal velocity component on the surface of the particle. The viscous-unsteady force is then due to the rotational scalar potential ψ, which enforces the no-slip boundary condition given in Eq. (4.83), second line.

When a particle accelerates relative to the surrounding fluid, acoustic waves are generated due to the normal velocity boundary condition (no-penetration boundary condition) at the surface of the particle. In the present context of an incompressible flow, these acoustic waves propagate at infinite speed. This is reflected in the fact that the scalar potential ϕ is obtained by solving the steady Laplace equation (4.77), whose boundary condition is the instantaneous relative velocity. As a result, the inviscid effect of relative acceleration immediately shows up as the instantaneous inviscid-unsteady force.

When a particle accelerates relative to the surrounding fluid, viscous vorticity waves are also generated, due to the requirement of the no-slip boundary condition at the surface of the particle. However, unlike the acoustic perturbation waves, the propagation speed of the viscous waves is finite and is given by $\sqrt{\nu_f/t}$. As a result, the effect of relative acceleration $(\partial \mathbf{u}_c/\partial t - d\mathbf{V}/dt)$ at time ξ is felt at all later times $t > \xi$. In other words, the viscous-unsteady force on the particle at time t depends on all past relative acceleration $(-\infty \leq \xi \leq t)$. This is the source of the convolution history integral in the last term of Eq. (4.88), and the viscous-unsteady force can be expressed as

$$\mathbf{F}_{vu} = 6\pi \mu_f R_p \int_{-\infty}^{t} K_{BH}(t - \xi) \left[\frac{\partial \mathbf{u}_c}{\partial t} - \frac{d\mathbf{V}}{dt} \right]_{@\xi} d\xi. \tag{4.100}$$

In the Stokes limit, the viscous-unsteady kernel that weighs the past acceleration decays as $K_{BH}(\tau) = 1/\sqrt{\tau}$, where $\tau = \pi \nu_f (t - \xi)/R_p^2$ is the nondimensional elapsed time between the instant of relative acceleration ξ and the instant t of its effect on the force. This weight function is known as the *Basset history kernel*.

Two aspects of the Basset history kernel must be pointed out. First, the kernel is singular as $\tau \to 0$, which could have been anticipated from the fact that the speed of propagation of the viscous vorticity wave is also singular as $\tau \to 0$. Fortunately, this singularity is integrable. In other words, the integration in Eq. (4.100) can be performed despite this singularity. Far more problematic is the slow $1/\sqrt{\tau}$ decay of the kernel. For example, if the relative acceleration expressed by the term within the square brackets were to remain a constant, then it can be taken out of the integral and the remaining integral becomes $\int_{-\infty}^{t} K_{BH}(t-\xi)\,d\xi = \int_{0}^{\infty} K_{BH}(\tau)\,d\tau$. This integral does not converge if the kernel decays as $1/\sqrt{\tau}$. As we will see below, this problem is resolved with finite Reynolds number effect. No matter how small the Reynolds number is, provided it is finite, the kernel will decay faster than $1/\sqrt{\tau}$ and the integral will become convergent.

Energy Implication. In this subsection, we investigate the energy implications of the perturbation flow created by the particle. The force on the particle due to the perturbation flow is

$$\mathbf{F}_{\text{pert}} = \mathbf{F}_{qs} + \mathbf{F}_{iu} + \mathbf{F}_{vu}. \tag{4.101}$$

An equal and opposite force acts on the fluid. Thus, the total momentum of the system (particle plus fluid) does not change due to the perturbation flow or the associated force exchange. In terms of energy, $\mathbf{F}_{\text{pert}} \cdot \mathbf{V}$ is the rate of work done on the particle and it contributes to the rate of change of kinetic energy of the particle. Similarly, $-\mathbf{F}_{\text{pert}} \cdot \mathbf{u}_c$ contributes to the rate of change of kinetic energy of the fluid at the macroscale. But there is a difference between these two evaluations. The former is precise due to the rigid-body translational motion of the particle, while the latter is the rate of change of kinetic energy of the fluid only when viewed at the macroscale. At the microscale, the flow around the particle is complex, and the associated change in kinetic energy of the fluid varies around the particle. The problem simplifies at the macroscale, with the macroscale fluid velocity approximated as $\mathbf{u}_c$ and the net force on the fluid taken to be $-\mathbf{F}_{\text{pert}}$.

Since the total energy must be conserved, the difference $\mathbf{F}_{\text{pert}} \cdot (\mathbf{u}_c - \mathbf{V})$ must contribute to the rate of change of kinetic energy of the perturbation flow at the microscale (represented by the scalar potentials ϕ and ψ). This contribution to the perturbation flow kinetic energy at the microscale from each of the force terms in Eq. (4.101) can be investigated. From the definition of the quasi-steady force, it can readily be seen that its contribution to the perturbation flow kinetic energy is guaranteed to be positive definite. Due to the steady viscous nature of this perturbation flow, this steady supply of kinetic energy to the microscale motion must be balanced by an equal amount of energy dissipation. In other words, $\mathbf{F}_{qs} \cdot (\mathbf{u}_c - \mathbf{V})$ is purely dissipative.

In contrast, from the definition of the inviscid-unsteady force in Eq. (4.99), it can be seen that its contribution to the perturbation flow kinetic energy (i.e., $\mathbf{F}_{iu} \cdot (\mathbf{u}_c - \mathbf{V})$) can be positive or negative. As argued in Eq. (4.98), the kinetic energy of the added mass of the perturbation flow can go up or down with increase or decrease in the relative velocity. This kinetic energy in the potential part of the microscale motion will not be dissipated. This transfer of energy to the microscale motion is fully reversible, as can be expected from the inviscid nature of this force.

Finally, $\mathbf{F}_{vu} \cdot (\mathbf{u}_c - \mathbf{V})$ is also not guaranteed to be positive. Due to the viscous nature of this force, it can be expected that at least part of the corresponding microscale kinetic energy will be dissipated. But the energetic consequences of $\mathbf{F}_{vu}$ need further careful investigation.

Example 4.11 Let us now revisit the problem of an isolated particle, released from rest in a quiescent ambient that was earlier considered in Section 2.4.3. We will solve the equation of motion

$$m_p \frac{d\mathbf{V}}{dt} = \mathbf{F} + m_p \mathbf{g} \tag{4.102}$$

to obtain the time evolution of particle velocity from rest to its terminal velocity. We will now evaluate the different components of the force $\mathbf{F}$ using the BBO equation (4.88). Starting with the undisturbed flow force, the hydrostatic pressure gradient in the ambient is $\nabla p_c = \rho_f \mathbf{g}$, and thus the first term on the right-hand side of the BBO equation becomes the buoyancy force given by $-m_f \mathbf{g}$, where m_f is the mass of fluid displaced by the particle. Note that the undisturbed ambient fluid is quiescent and therefore $\mathbf{u}_c = 0$. The quasi-steady force in Eq. (4.88) becomes $-6\pi \mu_f R_p \mathbf{V}$. The added-mass and history forces also simplify, since $d\mathbf{u}_c/dt = 0$. Substituting the different force components and simplifying, we obtain

$$\frac{d\mathbf{V}}{dt} + \frac{\mathbf{V}}{\tau_p} + \frac{1}{\tau_p} \int_{-\infty}^{t} K_{BH}(t - \xi) \left[\frac{d\mathbf{V}}{dt} \right]_{@\xi} d\xi = \frac{V_s \, \mathbf{e}_g}{\tau_p}, \tag{4.103}$$

where the terminal settling velocity V_s and the particle time scale τ_p were discussed in Section 2.4.3 and are given by

$$V_s = \frac{\rho - 1}{\rho + 1/2} \tau_p g \quad \text{and} \quad \tau_p = \frac{(1 + 2\rho)d_p^2}{36\nu_f}, \tag{4.104}$$

where the gravity vector $\mathbf{g} = g \, \mathbf{e}_g$ and ρ is the particle-to-fluid density ratio.

From the above, it can readily be seen that in the final steady state when $d\mathbf{V}/dt \to 0$, we have the result $\mathbf{V} \to V_s \mathbf{e}_g$. If we ignore the Basset history force represented by the integral term, then Eq. (4.103) can be solved to obtain an exponential approach to the terminal settling velocity. This was the basis on which we defined the translational time scale of the particle in Section 2.4.1. It can be seen in the above equation that the history force opposes the gravitational force and thus slows down the approach to the terminal velocity. In other words, the history force tends to increase the translational time scale of the particle. We shall discuss this more later in Problem 4.17, with the inclusion of the finite Reynolds number effect.

Problem 4.12 In this problem, we will consider a particle in a uniform oscillatory flow along the x-direction given by

$$\mathbf{u}_c(t) \cdot \mathbf{e}_x = \cos(\omega t), \tag{4.105}$$

where ω is the frequency of oscillation. As a result of the oscillatory flow, the particle will also be executing an oscillatory motion. The quantities of interest are the amplitude and phase of the periodic variation in the particle velocity, in comparison with the amplitude and phase of the oscillatory flow given above.

In the zero Reynolds number limit, with the force on the particle given by the BBO equation, this problem has an exact analytical solution. Express the oscillatory flow in complex variables as

$$\mathbf{u}_c(t) \cdot \mathbf{e}_x = \mathfrak{R}\left[\exp(\omega t)\right], \qquad (4.106)$$

where $\mathfrak{R}[\ ]$ indicates the real part. The corresponding particle velocity in the linear limit is also sinusoidal and can be expressed as

$$\mathbf{V}(t) \cdot \mathbf{e}_x = \mathfrak{R}\left[\hat{v}\exp(\omega t)\right], \qquad (4.107)$$

where $\hat{v}$ is the complex amplitude of the sinusoidal particle velocity. We seek an analytic expression for this quantity, since its magnitude and phase correspond to the amplitude and phase of the periodic variation in the particle velocity.

In this case, it is easier to work with a Fourier transform. In the Fourier transform space, the equation of motion can be written as

$$\iota\omega m_p \hat{v} = \hat{F}_{un} + \hat{F}_{qs} + \hat{F}_{iu} + \hat{F}_{vu}. \qquad (4.108)$$

Obtain the following expressions for the Fourier transform of the different force components of the BBO equation:

$$\hat{F}_{un} = \iota\omega m_f, \quad \hat{F}_{qs} = 6\pi\mu_f R_p(1 - \hat{v}),$$

$$\hat{F}_{iu} = \frac{1}{2}\iota\omega m_f(1 - \hat{v}), \quad \hat{F}_{qs} = 6\pi\mu_f R_p\sqrt{\frac{R_p^2\omega}{2\nu_f}}(\iota - 1)(1 - \hat{v}). \qquad (4.109)$$

Define the Stokes number as $\mathrm{St} = \omega\tau_p$, where the particle time scale τ_p has been defined in Eq. (4.104). Substituting the above in Eq. (4.108), obtain the following expression

$$\hat{v} = \frac{\left(1 - \dfrac{3}{2}\sqrt{\dfrac{\mathrm{St}}{\rho + 1/2}}\right) + \iota\left(\dfrac{3}{2}\dfrac{\mathrm{St}}{\rho + 1/2} + \dfrac{3}{2}\sqrt{\dfrac{\mathrm{St}}{\rho + 1/2}}\right)}{\left(1 - \dfrac{3}{2}\sqrt{\dfrac{\mathrm{St}}{\rho + 1/2}}\right) + \iota\left(\mathrm{St} + \dfrac{3}{2}\sqrt{\dfrac{\mathrm{St}}{\rho + 1/2}}\right)}. \qquad (4.110)$$

The above complex velocity can be expressed as an amplitude $|\hat{v}|$ and a phase. The amplitude measures the extent of oscillatory motion of the particle compared to that of the surrounding fluid and the phase measures the phase difference between the two oscillations. Figure 4.4 plots the magnitude and phase angle of the normalized particle velocity for different values of particle-to-fluid density ratio. It is clear that the amplitude of oscillation is larger for a lighter-than-fluid bubble, while the amplitude is lower for a heavier-than-fluid particle. The phase shift varies between leading or lagging the fluid oscillation for varying Stokes number. The phase shift is substantial only for the largest density ratio.

In the inviscid limit, we can set $\mu_f = 0$ and obtain the simple result

$$\hat{v} = \frac{3}{1 + 2\rho} .$$ (4.111)

Thus, the particle oscillates in phase with the fluid. As the density of the particle becomes much larger than that of the fluid (i.e., as $\rho \to \infty$), the amplitude of oscillation of the particle velocity goes to zero. In the limit of a neutrally buoyant particle (i.e., $\rho \to 1$), the amplitude of oscillation of the particle matches that of the fluid, as can be expected. The more interesting case is that of a bubble whose density ratio $\rho \to 0$. In this case, the amplitude of bubble oscillation is three times that of the surrounding fluid, independent of the oscillation frequency. Of course, this oscillation will be somewhat damped by viscous effects.

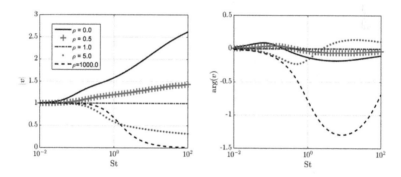

Figure 4.4 Amplitude and phase angle of particle oscillation in an oscillatory ambient flow as a function of particle Stokes number for varying particle-to-fluid density ratios.

Problem 4.13 Instead, consider the problem of a quiescent fluid in which the particle is moving at

$$\mathbf{V}(t) \cdot \mathbf{e}_x = \mathfrak{R}\left[\hat{v} \exp(\omega t)\right] .$$ (4.112)

Since the ambient fluid is quiescent, $\hat{F}_{un} = 0$.

(a) Show that the Fourier transform of the other forces remains the same. [Note that Problems 4.12 and 4.13 have been solved in the *Fluid Mechanics* book of Landau and Lifschitz, 1987.]

(b) What external force must be applied in order to make the particle move in the above fashion?

(c) Evaluate the rate of work done on the particle by this external force.

(d) What can you say about the energetics of this problem in terms of kinetic energy of the fluid (which in this case is all microscale) and dissipation?

4.3   Maxey–Riley–Gatignol Equation of Motion

Here we revisit the classic solution for the equation of motion presented by Maxey and Riley (1983) and Gatignol (1983). Their quest was to extend the BBO equation for ambient flows that vary on the scale of the particle diameter. Note that in the analysis of the previous section, the undisturbed ambient flow $\mathbf{u}_c$ was spatially uniform and varied only in time. There are many situations where the undisturbed ambient flow varies on the scale of the particle. For example, if a spherical particle were to be located in a converging nozzle such as the one shown in Figure 2.6, and if the particle size is comparable to the nozzle size, then the undisturbed nozzle flow that would exist in the absence of the particle would vary on the scale of the particle.

If we decide to use the BBO expression for force on the particle, a question remains as to what values of $\mathbf{u}_c$ and $\partial \mathbf{u}_c / \partial t$ should be used in the force expression (4.88), as these quantities vary over the volume of the particle. A reasonable choice is to take their values at the center of the spherical particle. But a rigorous analysis by Maxey and Riley (1983) and Gatignol (1983) yielded a satisfactory answer to the above question.

This problem in fact has a long history and it starts with the remarkable result by Faxén (1923) that was discussed in Section 4.1.5. Faxén generalized the Stokes drag expression for an arbitrary spatially varying steady flow. The result of Faxén in the present notation can be written as

$$\mathbf{F}_{qs} = 6\pi \mu_f R_p (\overline{\mathbf{u}_c}^s - \mathbf{V}), \qquad (4.113)$$

where $\overline{()}^s$ represents an average over the surface of the spherical particle. Since $\mathbf{u}_c(\mathbf{x})$ varies over the volume of the particle, the dilemma as to where it should be evaluated in the Stokes drag formula was resolved by precisely defining an average over the surface of the particle. The above formula is now known as the *Faxén theorem* and also referred to as the Faxén formula. The remarkable feature of this theorem is that it is exact and it applies to any arbitrary spatially varying steady ambient flow; the only restriction is that the Stokes limit of zero Reynolds number applies. There have been several efforts to generalize the Faxén theorem to unsteady flows (see Tchen, 1947; Corrsin and Lumley, 1956; Soo, 1975), culminating in Gatignol (1983) and Maxey and Riley (1983).

4.3.1   Undisturbed Flow Force

We again consider a spherical rigid particle of diameter d_p and mass $m_p = \pi d_p^3 \rho_p / 6$ translating at velocity $\mathbf{V}(t)$ while immersed in an "undisturbed" incompressible, spatially varying ambient flow of velocity $\mathbf{u}_c(\mathbf{x}, t)$ and pressure $p_c(\mathbf{x}, t)$. The undisturbed flow is now allowed to vary both in space and time, and as a result it satisfies the following more general governing equation:

$$\nabla \cdot \mathbf{u}_c = 0,$$

$$\frac{\partial \mathbf{u}_c}{\partial t} + \mathbf{u}_c \cdot \nabla \mathbf{u}_c = \mathbf{g} - \frac{1}{\rho_f} \nabla p_c + \nu_f \nabla^2 \mathbf{u}_c . \qquad (4.114)$$

The undisturbed flow force on the particle due to this ambient flow can readily be calculated as

$$\mathbf{F}_{un} = \oint_{S_p} \nabla \cdot \boldsymbol{\sigma}_c \, dA = \oint_{\mathcal{V}_p} (-\nabla p_c + \mu_f \nabla^2 \mathbf{u}_c) \, dV . \qquad (4.115)$$

The integrals in the above equation are over the surface or volume of the region occupied by the particle that is in motion. Unlike in a uniform flow, the viscous stress of the undisturbed flow is nonzero and its integral around the volume occupied by the particle now makes an added contribution. If we further use the momentum equation given as Eq. (4.114), then the undisturbed flow force can also be expressed as

$$\oint_{\mathcal{V}_p} \rho_f \left(\frac{D\mathbf{u}_c}{Dt} - \mathbf{g} \right) dV , \qquad (4.116)$$

where $D\mathbf{u}_c/Dt$ is the total acceleration of the fluid. However, the form given in Eq. (4.115) is more fundamental, since the fluid momentum equation may involve forces other than those shown in Eq. (4.114).

4.3.2   Perturbation Flow

The particle in motion will perturb the otherwise undisturbed ambient flow. Since the undisturbed ambient flow is spatially varying, there is no far-field velocity to attach the frame of reference as in the derivation of the BBO equation. Let the total flow velocity and pressure in the presence of the particle be $\mathbf{u}_{tot}$ and p_{tot}, and it can be separated into the undisturbed flow and the perturbation flow as

$$\mathbf{u}_{tot} = \mathbf{u}_c + \mathbf{u}, \quad p_{tot} = p_c + p, \quad \boldsymbol{\sigma}_{tot} = \boldsymbol{\sigma}_c + \boldsymbol{\sigma} . \qquad (4.117)$$

The governing equations of this total flow are the standard Navier–Stokes equations

$$\nabla \cdot \mathbf{u}_{tot} = 0,$$
$$\frac{\partial \mathbf{u}_{tot}}{\partial t} + \mathbf{u}_{tot} \cdot \nabla \mathbf{u}_{tot} = \mathbf{g} - \frac{1}{\rho_f} \nabla p_{tot} + \nu_f \nabla^2 \mathbf{u}_{tot} , \qquad (4.118)$$

which are satisfied in the region outside the moving particle. By subtracting Eq. (4.114) from the above, we obtain the equations for the perturbation flow. We now consider the Reynolds number of the perturbation flow (see Eq. (4.73)) to be small and neglect terms that are nonlinear in the velocity. Furthermore, pressure can be eliminated from the resulting linearized equations by taking the curl of the momentum equation, which yields

$$\nabla \cdot \mathbf{u} = 0,$$
$$\frac{\partial (\nabla \times \mathbf{u})}{\partial t} = \nu_f \nabla^2 (\nabla \times \mathbf{u}) . \qquad (4.119)$$

As pointed out in Maxey and Riley (1983), for the above linearization step to be valid, in addition to zero Reynolds number, we also require the undisturbed flow variation on the scale of the particle to be small. This additional condition can be expressed as

$$\frac{\Delta U_c \, d_p}{\nu_f} \ll 1, \tag{4.120}$$

where ΔU_c is the characteristic velocity variation of the undisturbed ambient flow over a particle diameter, in the neighborhood of the particle. Thus, we are interested in the regime where the ambient flow varies over the size of the particle (i.e., $\Delta U_c \neq 0$), but not too large to violate the condition given above.

The above equations for the perturbation flow must be solved with the following boundary conditions:

$$\begin{array}{ll} \text{On the sphere:} & \mathbf{u} = \mathbf{V} - \mathbf{u}_c \ \text{ at } \ |\mathbf{x} - \mathbf{x}_p(t)| = R_p, \\ \text{Far field:} & \mathbf{u} = 0 \ \text{ at } \ |\mathbf{x}| \to \infty. \end{array} \tag{4.121}$$

Equation (4.118) and the above boundary conditions are precisely the same as those in Eqs. (4.74) and (4.75). However, there is a big difference. In the previous section, $\mathbf{u}_c$ was not a function of $\mathbf{x}$, and only a function of time. As a result, the velocity boundary condition on the sphere could be expressed as in Eq. (4.83), which allowed us to focus only on mode $n = 1$. Such simplification is not possible here. Since the undisturbed ambient flow $\mathbf{u}_c$ is not a constant and varies on the surface of the particle, analysis cannot be restricted to only the $n = 1$ mode. An elegant solution to this complicated problem using the reciprocal theorem is presented in Appendix F. The key outcome of this rigorous derivation is the following expression for the force on the particle due to the perturbation flow:

$$\begin{aligned} \mathbf{F}_{\text{pert}}(t) = {}& 6\pi\mu_f R_p (\overline{\mathbf{u}_c}^s - \mathbf{V}) + \frac{1}{2} m_f \left[\frac{d\overline{\mathbf{u}_c}^v}{dt} - \frac{d\mathbf{V}}{dt} \right] \\ & + 6\pi\mu_f R_p \int_{-\infty}^{t} \frac{R_p}{\sqrt{\pi \nu_f (t - \xi)}} \left[\frac{d\overline{\mathbf{u}_c}^s}{dt} - \frac{d\mathbf{V}}{dt} \right]_{@\xi} d\xi, \end{aligned} \tag{4.122}$$

where $\overline{\mathbf{u}_c}^s$ and $\overline{\mathbf{u}_c}^v$ are the undisturbed flow averaged over the surface and volume of the particle, defined in Eqs. (F.12). To the above, we add the undisturbed flow force from Eq. (4.115) to obtain an expression for the total force in a spatially varying ambient flow.

Maxey–Riley–Gatignol equation

$$\boxed{\begin{aligned} \mathbf{F}(t) = {}& \oint_{\mathcal{V}_p} (-\nabla p_c + \mu_f \nabla^2 \mathbf{u}_c) \, dV + 6\pi\mu_f R_p (\overline{\mathbf{u}_c}^s - \mathbf{V}) + \frac{1}{2} m_f \left[\frac{d\overline{\mathbf{u}_c}^v}{dt} - \frac{d\mathbf{V}}{dt} \right] \\ & + 6\pi\mu_f R_p \int_{-\infty}^{t} \frac{R_p}{\sqrt{\pi \nu_f (t - \xi)}} \left[\frac{d\overline{\mathbf{u}_c}^s}{dt} - \frac{d\mathbf{V}}{dt} \right]_{@\xi} d\xi. \end{aligned}} \tag{4.123}$$

Note that since the surface and volume averages $\overline{\mathbf{u}_c}^v$ and $\overline{\mathbf{u}_c}^s$ are only functions of time, their time derivative has been written as d/dt (i.e., following the particle). Furthermore, in the last term, $[f(t)]_{@\xi}$ represents that f is evaluated at the instant $t = \xi$. *This MRG equation is an exact expression for force on a spherical particle in arbitrary motion in a spatially non-uniform time-dependent ambient flow, in the limit of zero Reynolds number.*

Interpretation of the different terms in the above equation is the same as in Section 4.2.4 for the various terms of the BBO equation (4.88). The differences are in the volume and surface averages. Clearly, these averages arise in a spatially varying undisturbed ambient flow because quantities such as $\mathbf{u}_c$ vary over the volume occupied by the particle. The inviscid part of the perturbation flow represented by the potential ϕ enforces the no-penetration condition. In the context of a spatially varying flow, the normal component of undisturbed ambient velocity along the surface of the particle varies in an arbitrary manner, as specified by the ambient flow $\mathbf{u}_c$. It turns out that it is the volume average of $\mathbf{u}_c$ that dictates the no-penetration boundary condition and thus appears in the inviscid unsteady force (the third term on the right-hand side). The viscous part of the perturbation flow enforces the no-slip boundary condition on the surface of the particle. In the context of a spatially varying flow, it is the surface average of $\mathbf{u}_c$ that dictates the effective no-slip boundary condition and thus appears both in the quasi-steady drag (the second term) and in the viscous unsteady force (the fourth term on the right-hand side). The sample problems to be presented below illustrate the advantage of the MRG equation over the BBO equation in the case of spatially varying flows. In closure it must be pointed out that compressible flow versions of both the BBO and MRG equations have been derived and they can be found in Parmar et al. (2008, 2010, 2011) and Annamalai and Balachandar (2017).

Problem 4.14 Consider a particle of diameter d_p with its center located at the origin $\mathbf{x} = 0$ in a unidirectional linear shear flow

$$\mathbf{u}_c = (U_0 + Gy)\,\mathbf{e}_x, \tag{4.124}$$

where G is the linear shear and U_0 is the undisturbed ambient flow at the center of the particle. For this linear shear flow, show that

$$\overline{\mathbf{u}_c}^v = U_0\,\mathbf{e}_x \quad \text{and} \quad \overline{\mathbf{u}_c}^s = U_0\,\mathbf{e}_x. \tag{4.125}$$

Substituting the above in the MRG equation, show that Eq. (4.123) becomes exactly the same as the BBO equation, except for the undisturbed flow force. In other words, in a linear shear flow, one can safely ignore the velocity variation across the size of the particle and calculate the perturbation force just using the undisturbed flow velocity at the center of the particle. In fact, it can easily be shown that for any arbitrary linear variation in $\mathbf{u}_c$, the surface and volume averages are the same as the velocity at the particle center.

The above statement, that for any linearly varying flow the use of the BBO equation is adequate, is correct only in the Stokes limit. We will soon see in Section 4.5 that at finite Reynolds number, the particle will experience an additional lift force that must be considered.

Problem 4.15 In this example, consider a particle of diameter d_p with its center located at the origin $\mathbf{x} = 0$ in a unidirectional parabolic velocity profile:

$$\mathbf{u}_c = (U_0 - \alpha y^2)\,\mathbf{e}_x,\qquad\qquad (4.126)$$

where U_0 is the undisturbed ambient flow at the center of the particle and α determines how the velocity varies along the y-direction. We restrict $\alpha \ll U_0/d_p^2$ so that the velocity variation across a particle diameter is small, to satisfy the requirements of the MRG equation. For this simple flow, obtain expressions of $\overline{\mathbf{u}_c}^v$ and $\overline{\mathbf{u}_c}^s$ and show that they are lower than U_0. This is to be expected since the maximum velocity is at the particle center.

Problem 4.16 In this problem, we will generalize the results of the previous examples. Again, consider a particle of diameter d_p with its center located at the origin $\mathbf{x} = 0$ in a complex undisturbed ambient flow. Let us expand the undisturbed ambient flow about the particle center as

$$\mathbf{u}_c(\mathbf{x}, t) = \mathbf{U}_0 + \mathbf{x} \cdot [\nabla \mathbf{U}]_0 + \frac{1}{2}\mathbf{x}\mathbf{x} : [\nabla\nabla\mathbf{U}]_0 + \cdots,\qquad (4.127)$$

where $\mathbf{U}_0$ is the undisturbed velocity vector at $\mathbf{x} = 0$, $[\nabla \mathbf{U}]_0$ is the velocity gradient at the origin, and so on. Note that in the above expansion, all the spatial variations are contained in the $\mathbf{x}$ terms, since the coefficients of the expansion, $\mathbf{U}_0$, $[\nabla \mathbf{U}]_0$, and so on, are evaluated at the origin and can only depend on time. This simplified representation of the ambient flow allows explicit integration over the surface and volume of the particle. Derive the following leading-order expressions:

$$\overline{\mathbf{u}_c}^s \approx \mathbf{U}_0 + \frac{d_p^2}{24}\left[\nabla^2 \mathbf{U}\right]_0 \quad \text{and} \quad \overline{\mathbf{u}_c}^v \approx \mathbf{U}_0 + \frac{d_p^2}{40}\left[\nabla^2 \mathbf{U}\right]_0.\qquad (4.128)$$

Thus, the surface and volume-averaged undisturbed flow velocities deviate from the velocity at the center of the particle. To leading order, these deviations depend on the Laplacian of velocity evaluated at the particle center. This approximation for surface and volume averages can be substituted into Eq. (4.123), thereby eliminating the appearance of surface and volume averages. This is the form in which Maxey and Riley (1983) presented their force expression. Since the analysis is restricted to small velocity variation over the size of the particle, the above approximation is sufficient.

Also, given an undisturbed ambient flow, it is far easier to calculate $\overline{\mathbf{u}_c}^s$ and $\overline{\mathbf{u}_c}^v$ in terms of the above approximation rather than to explicitly calculate the surface and volume averages. Finally, we point out that, as stated in Problem 4.14, for a linearly varying ambient velocity, the Laplacian is zero and the second term on the right-hand side in Eq. (4.128) does not contribute to either the surface or the volume average.

4.4   Force on a Particle in the Inviscid Limit

The discussion so far has been limited to zero Reynolds number flows, as the equations we solved are the steady and unsteady Stokes equations. Terms that were nonlinear involving either the particle or the fluid velocity were systematically ignored. Before we proceed to finite Reynolds number, in this section we will obtain a general expression for the force on the particle in the inviscid limit. As with the BBO and MRG equations, we will consider the undisturbed ambient flow to be unsteady and spatially varying. In particular, we will consider the undisturbed flow to be rotational. However, we will consider the perturbation flow due to the presence of the particle to be inviscid. The inviscid perturbation flow will be obtained by solving the Euler equation in the incompressible limit, with the application of only the no-penetration boundary condition on the particle. The relative motion and acceleration of the surrounding fluid as seen by the moving particle contributes to the potential part of the perturbation flow. Although no new vorticity is generated on the surface of the particle due to the inviscid assumption, the solution also includes a vortical component that accounts for the advection of the ambient vorticity around the particle.

The solution of this inviscid problem was obtained by Auton and co-workers (Auton, 1987; Auton et al., 1988) in the limit when the velocity variation across the particle diameter due to inhomogeneity of the undisturbed ambient flow was small compared to relative velocity. The resulting inviscid force on the particle is given by (ignoring the buoyancy effect of gravity)

$$\textbf{Inviscid}: \quad \mathbf{F} = m_f \frac{D\mathbf{u}_c}{Dt} + \frac{1}{2}m_f \left[\frac{D\mathbf{u}_c}{Dt} - \frac{d\mathbf{V}}{dt} \right] + \frac{1}{2}m_f(\mathbf{u}_c - \mathbf{V}) \times \omega_c, \quad (4.129)$$

where ω_c is the undisturbed ambient flow vorticity, which along with $\mathbf{u}_c$ and $D\mathbf{u}_c/Dt$ is evaluated at the location of the particle. By comparing the above with the BBO equation (4.88), we can identify the first term on the right-hand side to be the undisturbed flow force, since $-\nabla p = \rho_f D\mathbf{u}_c/Dt$ in an inviscid flow ignoring the effect of gravity. The second term on the right-hand side is similar to the inviscid unsteady force (i.e., the added-mass force) in the BBO equation, except for the replacement of fluid acceleration following the particle $d\mathbf{u}_c/dt$ with total acceleration following the fluid parcel $D\mathbf{u}_c/Dt$. The difference between the two can be expressed as

$$\frac{D\mathbf{u}_c}{Dt} - \frac{d\mathbf{u}_c}{dt} = (\mathbf{u}_c - \mathbf{V}) \cdot \nabla \mathbf{u}_c. \quad (4.130)$$

This difference is due to the BBO force being obtained by solving the unsteady Stokes equation in the zero Re limit, while the above inviscid force is obtained by solving the nonlinear Euler equation. For example, in obtaining the BBO equation, in Eq. (4.90), the nonlinear interaction represented by the second term on the left was neglected. Such nonlinear effects are retained in the inviscid analysis, although the viscous term has been ignored. As a result, the inviscid unsteady force is modified to account for the nonlinear interaction between the relative velocity vector and the ambient flow velocity gradient. In other words, the effect of convective acceleration of the ambient flow starts to play a role at finite Re, and contributes to the unsteady force in addition to the effect of temporal acceleration. The temporal and convective accelerations are the two terms on the left-hand side of Eq. (4.90), and both terms are important in determining the perturbation force.

The effect of the ambient flow's convective acceleration on particle force can be interpreted in the following manner. In a frame moving with the particle, the impulse vector acting on the fluid surrounding the particle due to the relative rectilinear motion of the particle can be expressed as (Lamb, 1932; Landau and Lifshitz, 1987; Ohl et al., 2003)

$$\mathbf{p} = C_M \, m_f \, (\mathbf{V} - \mathbf{u}_c), \tag{4.131}$$

where the added-mass coefficient of a sphere $C_M = 1/2$ and m_f is the mass of the displaced volume of fluid. The inviscid unsteady reaction force on the surrounding continuous phase, which is the opposite of the inviscid unsteady force $\mathbf{F}_{iu}$ that acts on the particle, is given by

$$-\mathbf{F}_{iu} = \frac{d\mathbf{p}}{dt} + (\mathbf{u}_c - \mathbf{V}) \cdot \nabla \mathbf{p}, \tag{4.132}$$

where d/dt is the time derivative following the particle. This inviscid unsteady force is nothing but the added-mass force. The first term corresponds to the traditional added-mass force due to the time rate of change of impulse. The origin of the second term was brilliantly articulated by G. I. Taylor (1928) in his investigation of inviscid forces on a body in a converging stream of fluid. Under the steady condition given by $(\mathbf{u}_c - \mathbf{V})$ = constant, though $d\mathbf{p}/dt = 0$, the spatial variation in the ambient flow will give rise to spatial variation in impulse and an attendant added-mass force. Thus, the first term arises from the temporal relative acceleration (in the frame moving with the particle), while the second contribution is due to convective acceleration of the ambient flow. From the above relation, the added-mass force can be expressed as

$$\begin{aligned} \mathbf{F}_{iu} &= \frac{1}{2} m_f \left[\frac{d(\mathbf{u}_c - \mathbf{V})}{dt} + (\mathbf{u}_c - \mathbf{V}) \cdot \nabla(\mathbf{u}_c) \right] \\ &= \frac{1}{2} m_f \left[\frac{D\mathbf{u}_c}{Dt} - \frac{d\mathbf{V}}{dt} \right], \end{aligned} \tag{4.133}$$

where D/Dt is the total derivative following the fluid and it includes both temporal and convective accelerations of the undisturbed ambient flow. The reader should consult

Tollmien (1938), Voinov et al. (1973), Landweber and Miloh (1980), and Lhuillier (1982) for further discussion on the inviscid unsteady force and its generalization.

The final term on the right-hand side of Eq. (4.129) corresponds to the vorticity-induced lift force, as it is directed perpendicular to the direction of relative velocity. Similar to convective acceleration, this term is also nonlinear in origin and therefore does not appear in the BBO equation. We shall see a lot more about the lift force in the next chapter, especially about the inertial origin of the lift force and why it is identically zero in the Stokes limit.

We finally note that the inviscid force expression of this section can be considered as the Re $\rightarrow \infty$ result. However, this result is in the absence of viscosity and therefore ignores the formation of the boundary layer and other important physics.

Problem 4.17 Consider a simple inviscid planar stagnation point flow given by

$$\mathbf{u}_c = [U_0 + \alpha x, -\alpha y, 0] \,, \tag{4.134}$$

where α is the strength of the stagnation point flow.

(a) Show that the total acceleration of this flow is entirely from convective acceleration and at the origin obtain

$$\frac{D\mathbf{u}_c}{Dt} = \frac{\partial \mathbf{u}_c}{\partial t} + \mathbf{u}_c \cdot \nabla \mathbf{u}_c = [U_0 \alpha, 0, 0] \,. \tag{4.135}$$

(b) Let us consider a stationary particle of diameter d_p with its center located at the origin. At the location of the particle, the fluid acceleration is $U_0 \alpha \mathbf{e}_x$. Obtain the undisturbed flow force on the particle to be $\mathbf{F}_{un} = m_f U_0 \alpha \mathbf{e}_x$.

(c) Obtain the added-mass force on the particle to be $\mathbf{F}_{iu} = m_f U_0 \alpha \mathbf{e}_x / 2$.

(d) Let us now consider the example investigated by Auton et al. (1988). In the above flow, instead of a stationary particle, consider the massless bubble released at the origin at $t = 0$ with velocity $V_{x0} \mathbf{e}_x$ (i.e., the bubble's initial velocity matches the local fluid velocity). The bubble is allowed to move freely and it is easy to verify that it will move along the x-axis. Since the mass of the bubble is zero, show that the bubble's motion is governed by

$$\frac{1}{2}\frac{d\mathbf{V}}{dt} = \frac{3}{2}\frac{D\mathbf{u}_c}{Dt} \,. \tag{4.136}$$

(e) Solve the x-component of this equation to obtain the result $V_x^2 - V_{x0}^2 = 3((U_0 + \alpha X)^2 - U_0^2)$, where $(U_0 + \alpha X)$ is the local fluid velocity at the bubble's location X. In other words, the bubble's velocity is substantially higher.

Problem 4.18 Let us consider an inviscid $45°$ planar straining flow given by

$$\mathbf{u}_c = [U_0 + \alpha y, \alpha x, 0] \,, \tag{4.137}$$

where α is the strength of the straining flow. Consider a stationary particle of diameter d_p with its center located at the origin. Show that the undisturbed flow force, inviscid unsteady force, and lift force given by the three terms on the right-hand side of Eq. (4.129) are $m_f U_0 \alpha \mathbf{e}_y$, $m_f U_0 \alpha / 2 \mathbf{e}_y$, and 0. Thus, the undisturbed flow and inviscid unsteady forces are in fact oriented perpendicular to the relative velocity vector.

Problem 4.19 Consider an ambient uniform flow superposed on a solid-body rotation given by

$$\mathbf{u}_c = [U_0 + \alpha y, -\alpha x, 0] \,, \tag{4.138}$$

where α is the strength of the vortical flow. Consider a stationary particle of diameter d_p with its center located at the origin. Show that the undisturbed flow force, inviscid unsteady force, and lift force given by the three terms on the right-hand side of Eq. (4.129) are $-m_f U_0 \alpha \mathbf{e}_y$, $-m_f U_0 \alpha / 2 \mathbf{e}_y$, and $m_f U_0 \alpha$.

Problem 4.20 Consider an ambient shear flow given by

$$\mathbf{u}_c = [U_0 + \alpha y, 0, 0] \,, \tag{4.139}$$

where α is the strength of the shear. Consider a stationary particle of diameter d_p with its center located at the origin. Show that the undisturbed flow force, inviscid unsteady force, and lift force given by the three terms on the right-hand side of Eq. (4.129) are 0, 0, and $m_f U_0 \alpha$.

4.5   Extension to Finite Reynolds Number

In this section, we will address the effect of finite Reynolds number and the associated effect of nonlinearity on each of the force components in the BBO and MRG equations. In essence, in this section, we will obtain finite Reynolds number versions of the BBO and MRG equations. Towards this end, we will use the nonlinear version of the inviscid unsteady force that was derived in the previous section with the inviscid analysis. We will first focus attention on ambient flows whose spatial variation on the scale of the particle is small and as a result we will start with the finite-Re BBO equation, before proceeding to the finite-Re MRG equation, which in addition will involve volume and surface averages.

We start with the undisturbed flow force given in Eq. (4.115). It only depends on the undisturbed flow and as a result it is unaffected by the linearization of the perturbation velocity. Thus, the expression given in Eq. (4.115) is appropriate at all Re.

4.5.1   Finite-Re Quasi-Steady Force

At small but finite Reynolds number, the inertial correction to the steady Stokes drag was obtained by Oseen (1927) as

$$\mathbf{F}_{Oseen} = 6\pi\mu_f R_p(\mathbf{u}_c - \mathbf{V})\left(1 + \frac{3}{16}\,\mathrm{Re}\right), \qquad (4.140)$$

where the particle Reynolds number based on relative velocity is defined as $\mathrm{Re} = 2R_p|\mathbf{u}_c - \mathbf{V}|/\nu_f$. Higher-order corrections to the steady Stokes drag have been obtained through singular perturbation expansion (Proudman and Pearson, 1957; Chester et al., 1969). But the Reynolds number range over which these analytic solutions remain accurate is limited.

There are a number of empirical correlations of steady drag force that have been developed based on careful experimental measurements and direct numerical simulations. These steady-state correlations can be used for the quasi-steady drag force. One popular model that is widely used is (Schiller and Naumann, 1933; Clift et al., 1978)

$$\mathbf{F}_{qs} = 6\pi\mu_f R_p(\mathbf{u}_c - \mathbf{V})\Phi(\mathrm{Re}) \quad \text{where} \quad \Phi(\mathrm{Re}) = \left(1 + 0.15\,\mathrm{Re}^{0.687}\right). \qquad (4.141)$$

Here, $\Phi(\mathrm{Re})$ is the finite Reynolds number drag correction that has been observed to be adequate for particle Reynolds number up to about 800. But there are other models and the reader is referred to Crowe et al. (2011).

4.5.2   Finite-Re Inviscid-Unsteady Force

It must be pointed out that the inviscid unsteady force given by the second term on the right-hand side of Eq. (4.129) remains appropriate even at finite particle Reynolds number. Though this term was derived in the inviscid limit, it remains appropriate even at finite Re, where viscous effects are included. This is due to the fact that the inviscid unsteady force is not affected by viscous effects. Only in the zero Re limit does the nonlinear contribution become zero, and the inviscid force can be expressed as that given in the BBO equation.

It should be noted that the added-mass coefficient of a sphere has been established as $C_M = 1/2$ in both the Stokes limit and the potential flow limit. It then seems natural to expect the added-mass coefficient to be independent of Reynolds number and, for a sphere, to remain equal to 1/2 at all Re. However, questions may arise at finite Re. Consider a particle in a steady uniform ambient flow at a finite Reynolds number, where the flow is characterized by the presence of a steady recirculating region in the wake of the particle (see Figure 4.5a). Such a recirculation region does not exist either in the Stokes limit or in the inviscid limit. The presence of the wake together with the spherical particle can be considered to have changed the effective shape of the particle, as far as the ambient flow around it is considered. Thus, if the particle is accelerated, will the added-mass coefficient of this finite-Re flow equal that of a sphere or that of a sphere plus the wake?

Motivated by the above considerations, researchers have proposed an added-mass coefficient to depend on the particle Reynolds number (Odar and Hamilton, 1964; Odar, 1966; Cheng et al., 1978). In addition, they defined an acceleration parameter Ac that measures the relative strength of acceleration and suggested the added-mass coefficient to depend on this parameter. Such a model may appear reasonable. Just like the quasi-steady drag coefficient is a function of Re, the added-mass coefficient can also be a function of Re. However, using numerical simulations, researchers (Rivero et al., 1991; Chang and Maxey, 1994, 1995) have isolated the added-mass force acting on a particle undergoing rapid acceleration at finite Re. The added-mass coefficient was observed to remain 1/2 over the entire range of Reynolds and acceleration numbers considered. There is now additional evidence suggesting that the added-mass coefficient of a sphere remains 1/2 no matter what the ambient flow is at the time of relative acceleration (Bagchi and Balachandar, 2002b, 2003a; Mougin and Magnaudet, 2002). It is, however, not clear why the inviscid result is recovered even at finite Re.

Wakaba and Balachandar (2007) addressed this question with numerical simulations of flow around a rigid particle subjected to a sudden acceleration (or deceleration). It was shown that the immediate effect of a sudden change in relative velocity is to establish a corresponding additive potential flow that superposes over any flow that existed before the introduction of the sudden change. For example, Figure 4.5a is the flow in a frame attached to a moving sphere whose Reynolds number is 50. As shown in Figure 4.5c, Re was rapidly changed to 62.5 over a very short period, rapidly increasing the velocity of the sphere. The corresponding time history of force is shown in Figure 4.5d, where the drag coefficient was small for $t < 0$, corresponding to that of Re = 50. The drag coefficient sudden increases when the particle is undergoing acceleration (i.e., relative velocity is increasing), and this increase is clearly due to added mass. Careful inspection of frame (b), which is at the time marked t_2 in frame (c), shows that the flow around the particle changes and it is dominated by the potential flow due to particle acceleration; the wake bubble can be seen to have shifted downstream and shrunken a bit. For sufficient acceleration, the additive potential flow dominates, and the net effect is to blow away the boundary layer and any pre-existing recirculation region in the wake. In the case of sufficiently strong deceleration, the recirculation region rapidly advances forward and engulfs the sphere to establish a potential flow (not shown here). Thus, a potential flow perturbation establishes instantaneously after the rapid change, while viscous effects develop on a slower time scale. Their carefully constructed numerical simulations isolate the added-mass effect and establish $C_M = 1/2$, independent of Re.

4.5.3   Finite-Re Viscous-Unsteady Force

In Section 4.2.4 we discussed the slow decay of the Basset history kernel and mentioned that at finite Re the kernel will decay faster. Here we will address this finite Reynolds number effect. Even at finite Reynolds number, the Basset history kernel is accurate for small values of τ, and must be modified only for large values of τ. In particular, there exists a time beyond which the kernel must decay faster than $1/\sqrt{\tau}$ (Mei and Adrian, 1992).

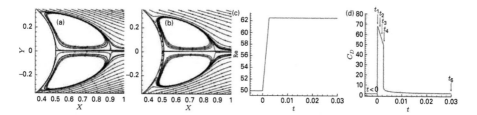

Figure 4.5 Rapidly accelerating flow over an isolated sphere. (a) The flow in a frame attached to a moving sphere whose Reynolds number is 50. (b) The flow around the particle at a time marked t_2 in frame (c). (c) How the particle velocity is increased so that Re changes rapidly to 62.5 over a very short period. The corresponding time history of force is shown in (d), where the drag coefficient was small for $t < 0$ corresponding to that of Re = 50.

To illustrate this, define a Stokes length as $\mathcal{L} = \nu_f / |\mathbf{u}_c - \mathbf{V}|$. This is the length scale whose Reynolds number based on relative velocity is unity. From this definition, it is clear that the local Reynolds number of the perturbation flow at distances larger than the Stokes length away from the particle is greater than unity. Also, we see that the ratio of Stokes length to particle diameter goes as $\mathcal{L}/d_p = 1/\text{Re}$. Thus, no matter how small we take Re to be, there exists a Stokes length beyond which inertial effects become important. Beyond the Stokes length, the neglect of the nonlinear term (inertia effect) in going from Eq. (4.72) to Eq. (4.74) cannot be justified. Mei and Adrian (1992) performed a matched asymptotic analysis where they solved the unsteady Stokes equation in the inner region of $r < \mathcal{L}$ and Oseen equations in the outer region of $r > \mathcal{L}$ and matched the two solutions. The resulting viscous-unsteady force can still be represented by the general convolution integral given in Eq. (4.100), but the revised viscous-unsteady kernel is given by

$$\text{Mei–Adrian:} \quad K_{vu}(\tau; \text{Re}) = \frac{1}{\sqrt{\tau_{vu}}} \frac{1}{\left[(\tau/\tau_{vu})^{1/4} + \tau/\tau_{vu} \right]^2}, \tag{4.142}$$

where the viscous-unsteady time scale is defined as

$$\tau_{vu} = (16\pi)^{2/3} \left(\frac{0.75 + 0.105\,\text{Re}}{\text{Re}} \right)^2. \tag{4.143}$$

In the above, we have included Re as a parameter in $K_{vu}(\tau; \text{Re})$ to emphasize the Reynolds number dependence of the history kernel. For $\tau \ll \tau_{vu}$ the above kernel approaches $1/\sqrt{\tau}$ behavior of the Basset kernel. But for $\tau \gg \tau_{vu}$ the kernel decays faster as $1/\tau^2$. Thus, for finite Re, no matter how small, there exists a $\tau = \tau_{vu}$ beyond which the kernel will decay faster than the Basset kernel.

The advantage of the above kernel over the Basset kernel is that it is applicable for finite Re. Figure 4.6 shows a plot of the nondimensional Basset and Mei–Adrian kernels (i.e., $\sqrt{\tau_{vu}}K_{vu}$) as a function of the nondimensional time τ/τ_{vu}. The change in slope at longer time is clear. It must be stressed that the kernel of Mei and Adrian is not unique. The problem of time-dependent vorticity diffusion is unfortunately very complicated. The precise nature of the viscous-unsteady kernel depends on whether relative acceleration is positive, negative, or cyclic, and so on. For further results and discussion on this topic, see Lovalenti and Brady (1993a,b) and Kim et al (1998).

Evaluation of the convolution integral for the viscous unsteady force is computation-ally expensive, since past history of relative acceleration must be stored and integrated, irrespective of the form of the finite-Re history kernel we choose to use. There have been a few ideas that exploit the properties of the kernel to reduce the computational cost (Brush et al., 1964; Michaelides, 1992; Dorgan and Loth, 2007; Elghannay and Tafti, 2016). More recently Parmar et al. (2018) used the rational theory of Beylkin and Monzón (2005) to approximate any history kernel in the form of exponential sums to reformulate the viscous unsteady force in a differential form to any desired level of accuracy. This removes the need for long-time storage of the acceleration histories of the particle and the fluid, and thereby considerably speeding up the calculation of the viscous unsteady force.

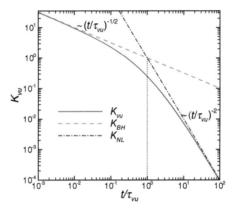

Figure 4.6 Nondimensional viscous-unsteady kernel $\sqrt{\tau_{vu}}K_{vu}$ as a function of t/τ_{vu} plotted as the solid line, compared to the Basset history kernel (K_{BH} shown as dashed line) and the long-time quadratic decay of the kernel (shown as dashed–dotted line).

4.5.4   Finite-Re BBO and MRG equations

As the final step towards writing the finite-Re versions of the BBO and MRG equations, we draw inspiration from the force expression (4.129) obtained in the inviscid limit. In particular, it is noted that nonlinear interaction between the relative velocity and the local flow vorticity gives rise to a lift force at nonzero Re. The presence of the lift force has been revealed by the inviscid analysis. However, we do not expect its precise form to be the same as that given in Eq. (4.129). In particular, at small but nonzero Re, the viscous effect on the lift force can be expected to be significant. We will devote the whole of Chapter 5 to obtaining appropriate lift expressions at finite Re. For now, in the finite-Re BBO and MRG equations, we will include an additional quasi-steady lift term $\mathbf{F}_{qsL}$. Here, "quasi-steady" indicates that the lift force is based on the instantaneous state of relative velocity and spatial gradients of the undisturbed flow at the particle. It should also be noted that there can be a viscous-unsteady contribution to lift that depends on the past history of relative velocity and local flow strain and vorticity. But this effect will be ignored here, since our understanding is currently limited.

We now assemble the finite Reynolds number version of the quasi-steady, inviscid unsteady, and viscous unsteady forces and first write the finite Reynolds number version of the BBO equation as

Finite-Re Basset–Boussinesq–Oseen equation

$$
\begin{aligned}
\mathbf{F}(t) = {} & -\mathcal{V}_p \nabla p_c + 6\pi \mu_f R_p (\mathbf{u}_c - \mathbf{V}) \Phi(\mathrm{Re}) + \mathbf{F}_{\mathrm{qsL}} \\
& + \frac{1}{2} m_f \left[\frac{D\mathbf{u}_c}{Dt} - \frac{d\mathbf{V}}{dt} \right] \\
& + 6\pi \mu_f R_p \int_{-\infty}^{t} K_{\mathrm{vu}}(\tau; \mathrm{Re}) \left[\frac{d\mathbf{u}_c}{dt} - \frac{d\mathbf{V}}{dt} \right]_{@\xi} d\xi .
\end{aligned}
\tag{4.144}
$$

In the above, $\tau = \pi \nu_f (t - \xi)/R_p^2$. Also in the viscous unsteady force we have retained $d\mathbf{u}_c/dt$ instead of $D\mathbf{u}_c/Dt$, unlike the corresponding change in the added-mass force.

The finite Reynolds number version of the MRG equation can be developed similarly. It must be emphasized that unlike the BBO and MRG equations, their finite-Re versions are empirical. Clearly, the finite-Re force expressions are motivated by the form of the analytic force expressions rigorously derived in the previous sections in the zero Reynolds number limit. But they now include finite-Re quasi-steady drag correction $\Phi(\mathrm{Re})$ and finite-Re viscous unsteady kernel K_{vu}. Both of which are empirical fits through experimental and computational data. In any case, the final expression of the finite-Re MRG equation is

Finite-Re Maxey–Riley–Gatignol equation

$$
\begin{aligned}
\mathbf{F}(t) = {} & \oint_{\mathcal{V}_p} (-\nabla p_c + \mu_f \nabla^2 \mathbf{u}_c)\, dV \\
& + 6\pi \mu_f R_p (\overline{\mathbf{u}}_c^{\,s} - \mathbf{V}) \Phi(\mathrm{Re}) + \mathbf{F}_{\mathrm{qsL}} \\
& + \frac{1}{2} m_f \left[\frac{D\overline{\mathbf{u}}_c^{\,v}}{Dt} - \frac{d\mathbf{V}}{dt} \right] \\
& + 6\pi \mu_f R_p \int_{-\infty}^{t} K_{\mathrm{vu}}(\tau; \mathrm{Re}) \left[\frac{d\overline{\mathbf{u}}_c^{\,s}}{dt} - \frac{d\mathbf{V}}{dt} \right]_{@\xi} d\xi .
\end{aligned}
\tag{4.145}
$$

In the above, the total derivative following the fluid can be interpreted as $D/Dt = \partial/\partial t + \overline{\mathbf{u}}_c^{\,v} \cdot \nabla$.

Example 4.21 Let us now revisit the problem of an isolated particle released from rest in a quiescent ambient, but with the finite Reynolds number effect taken into account. We follow the steps outlined in Example 4.11 and obtain

$$
\frac{d\mathbf{V}}{dt} + \frac{\mathbf{V}}{\tau_p} + \frac{1}{\tau_p} \int_{-\infty}^{t} K_{vu}(t - \xi) \left[\frac{d\mathbf{V}}{dt} \right]_{@\xi} d\xi = \frac{V_s\, \mathbf{e}_g}{\tau_p},
\tag{4.146}
$$

starting from the finite-Re BBO equation (4.144). The main difference is that the definition of particle time scale now includes the finite-Re drag correction as

$\tau_p = (1 + 2\rho)d_p^2/(36\nu_f\Phi(\mathrm{Re}))$. Though the above equation is strictly valid only for a rigid particle, investigate the three scenarios of (i) water droplet falling through air, (ii) sand particle falling through water, and (iii) air bubble rising in water considered in Section 2.4. For each scenario, consider different values of d_p and plot the time history of the different forces by integrating the above equation of motion. At each instant, calculate the Reynolds number and use the corresponding finite-Re quasi-steady drag correction and the finite-Re viscous kernel. Integrate long enough that the particle velocity is close to 1% of the terminal value. Also, make a table of the time it took for the particle to reach factor $(e - 1)/e$ of the terminal velocity. Compare this value to the particle time scale. Any difference is due to the viscous-unsteady force, which was ignored in the analysis. Show that inclusion of the viscous-unsteady force slows the approach to the steady state.

Example 4.22 In this problem, we will consider the finite Reynolds number extension of Problem 4.12. We will consider a particle in a uniform oscillatory flow along the x-direction given by Eq. (4.105). As a result of the oscillatory flow, the particle will execute an oscillatory motion. We will again investigate the three scenarios considered in the previous problem. For each scenario, consider different values of d_p and plot the time history of the different forces. Integrate long enough that the particle velocity is close to the terminal periodic state. Note that at finite Reynolds number, the particle motion will be periodic but not a pure sinusoid. In the periodic state, the quantities of interest are the amplitude and phase of the periodic variation in the particle velocity, in comparison to the amplitude and phase of the oscillatory flow.

4.5.5   Particle Motion in Steady Linear Flows

Exact solutions to the BBO equation (without the history term and for small Re) can be obtained for the case of particle motion in steady linearly varying flows. A steady linear ambient flow can be expressed as

$$\mathbf{u}_c = \mathbf{G} \cdot \mathbf{x}, \tag{4.147}$$

where $\mathbf{G}$ is a constant second-rank velocity gradient tensor. Though the flow has no temporal acceleration, its convective acceleration is nonzero and as a result it can be readily derived that

$$\frac{D\mathbf{u}_c}{Dt} = \mathbf{u}_c \cdot \nabla\mathbf{u}_c = \mathbf{G}^2 \cdot \mathbf{x}. \tag{4.148}$$

Now we consider the following equation of particle motion:

$$m_p \frac{d\mathbf{V}}{dt} = \mathbf{F} + m_p \mathbf{g}, \tag{4.149}$$

where the force on the right-hand side is that from the BBO equation (4.144) without the history term. For a linearly varying steady ambient flow, the undisturbed flow force can be expressed as

$$-\mathcal{V}_p \nabla p_c = m_f \left[\frac{D\mathbf{u}_c}{Dt} - \mathbf{g} \right]. \tag{4.150}$$

Substituting this, quasi-steady, and added-mass forces into the BBO equation and rearranging, we can express the equation of particle motion as

$$\frac{d\mathbf{V}}{dt} = \frac{\mathbf{u}_c - \mathbf{V}}{\tau_p} + \beta \frac{D\mathbf{u}_c}{Dt} + (1 - \beta)\,\mathbf{g}, \tag{4.151}$$

where the density parameter $\beta = 3/(2\rho + 1)$, with ρ being the particle-to-fluid-density ratio and τ_p the particle time scale.

The above equation can be written in terms of particle position $\mathbf{X}$ using the definition $\mathbf{V} = d\mathbf{X}/dt$. It is further specialized for the linear shear flow by substituting for both the fluid velocity and fluid acceleration at the particle location as $\mathbf{u}_c = \mathbf{G} \cdot \mathbf{X}$ and $D\mathbf{u}_c/Dt = \mathbf{G}^2 \cdot \mathbf{X}$. With these substitutions, we obtain the following equation of particle position:

$$\frac{d^2\mathbf{X}}{dt^2} + \frac{1}{\tau_p} \frac{d\mathbf{X}}{dt} - \left(\frac{\mathbf{G}}{\tau_p} + \beta \mathbf{G}^2 \right) \cdot \mathbf{X} = (1 - \beta)\,\mathbf{g}. \tag{4.152}$$

The above equation must be solved with initial conditions for the particle position and velocity, which are stated as

$$\mathbf{X}(t = 0) = \mathbf{X}_0 \quad \text{and} \quad \frac{d\mathbf{X}}{dt}(t = 0) = \mathbf{V}_0. \tag{4.153}$$

We will now discuss three types of linear flow: (i) a particle in a planar straining flow; (ii) a particle in a vortical flow; and (iii) a particle falling through a linear shear flow. We have already considered the latter two in the context of Stokes number in Section 2.4.5. Because the above ordinary differential equation is linear and second-order with constant coefficient, it is not hard to solve it in all three situations: see the problems below.

Problem 4.23 Consider a two-dimensional planar straining flow with compression (or convergence) along the x-axis and extension (or divergence) along the y-axis given by

$$u_{cx} = -kx, \quad u_{cy} = ky, \quad u_{cz} = 0, \tag{4.154}$$

where k is the strain-rate magnitude. Its velocity gradient tensor and its square are then

$$\mathbf{G} = \begin{bmatrix} -k & 0 & 0 \\ 0 & k & 0 \\ 0 & 0 & 0 \end{bmatrix} \quad \text{and} \quad \mathbf{G}^2 = \begin{bmatrix} k^2 & 0 & 0 \\ 0 & k^2 & 0 \\ 0 & 0 & 0 \end{bmatrix}. \tag{4.155}$$

The analysis to be discussed below follows that presented in Ferry et al. (2003). We ignore the effect of gravity in this problem and thus the equation is unforced. Furthermore, the particle motion along the x- and y-directions decouples, and there is no motion along the z-direction.

Solution of the y Component

(a) Show that the governing equation along the extensional y-direction becomes

$$\tau_p \frac{d^2 Y}{dt^2} + \frac{dY}{dt} - \gamma_y Y = 0, \tag{4.156}$$

where $\gamma_y = k(1 + k\tau_p\beta)$ is positive, since all its arguments are positive. The roots of this ODE's characteristic equation are

$$r_{y\pm} = \frac{-1 \pm \sqrt{1 + 4\gamma_y \tau_p}}{2\tau_p}. \tag{4.157}$$

The roots are guaranteed to be real.

(b) Show that the solutions for particle position and velocity have the form

$$Y(t) = c_{y+} \exp(r_{y+} t) + c_{y-} \exp(r_{y-} t),$$
$$\frac{dY}{dt}(t) = c_{y+} r_{y+} \exp(r_{y+} t) + c_{y-} r_{y-} \exp(r_{y-} t), \tag{4.158}$$

where c_{y+} and c_{y-} are the integration constants.

(c) Determine the integration constants by applying the boundary conditions (4.153), and simplify the solutions. Note that r_{y-} is the smaller of the two roots and is guaranteed to be negative. Thus, the second terms of Eqs. (4.158) decay away on a faster time scale given by $1/r_{y-}$. Once the transients die away, the particle position and velocity are related by the multiplicative constant r_{y+}.

(d) Show that the y-component of the particle velocity can be related to the y-component of the fluid velocity through the relation

$$V_y = \eta_y u_{cy} \quad \text{where} \quad \eta_y = \frac{-1 + \sqrt{1 + 4\,\mathrm{St}(1 + \beta\,\mathrm{St})}}{2\,\mathrm{St}}. \tag{4.159}$$

In the above, η_y is a nondimensional constant and it only depends on the particle Stokes number $\mathrm{St} = k\tau_p$, with the time scale of the straining flow being $1/k$. It is interesting to note that beyond the transient time of $1/r_{y-}$, the particle velocity is independent of its initial condition. That is, V_y depends only the particle's current y position and is independent of where it was released and at what initial velocity.

Solution of the x Component

(e) Show that the governing equation along the compressional x-direction becomes

$$\tau_p \frac{d^2 X}{dt^2} + \frac{dX}{dt} + \gamma_x X = 0, \tag{4.160}$$

where $\gamma_x = k(1 - k\tau_p\beta)$.

(f) Obtain the roots of this ODE's characteristic equation as

$$r_{x\pm} = \frac{-1 \pm \sqrt{1 - 4\gamma_x\tau_p}}{2\tau_p}. \tag{4.161}$$

The roots are real only when $\gamma_x\tau_p \leq 1/4$, otherwise the roots are complex.

(g) Show that the solutions of the particle's x position and velocity have the form

$$X(t) = c_{x+}\exp(r_{x+}\,t) + c_{x-}\exp(r_{x-}\,t),$$
$$\frac{dX}{dt}(t) = c_{x+}\,r_{x+}\exp(r_{x+}\,t) + c_{x-}\,r_{x-}\exp(r_{x-}\,t), \tag{4.162}$$

where c_{x+} and c_{x-} are the integration constants.

(h) Determine the integration constants by applying the boundary conditions (4.153), and simplify the solutions. When the two roots are real, both are guaranteed to be negative, but r_{x-} is the more negative of the two. Thus, the second terms of Eqs. (4.162) decay away on a faster time scale given by $1/r_{x-}$. Once the faster transients die away, the particle position and velocity are related by the multiplicative constant r_{x+}.

(i) Show that the x-component of the particle velocity can be related to the x-component of the fluid velocity through the relation

$$V_x = \eta_x\,u_{cx} \quad \text{where} \quad \eta_x = \frac{1 - \sqrt{1 - 4\,\mathrm{St}(1 - \beta\,\mathrm{St})}}{2\,\mathrm{St}}. \tag{4.163}$$

In the above, η_x is a nondimensional constant and it only depends on the particle Stokes number. Again, beyond the transient time of $1/r_{x-}$, the particle velocity is independent of its initial condition. It depends only on the particle's current x position and is independent of X_0 or V_{0x}. As considered above, when both roots are real, the x location of the particle and the x-component of velocity continue to decrease in magnitude and the particle monotonically approaches the $x = 0$ plane in an exponentially decaying fashion.

(j) We now consider the case when $\gamma_x\tau_p > 1/4$. The roots are complex conjugate and both must be retained in the analysis. The resulting particle motion is oscillatory. In this case, the particle inertia is too strong, so that the particle velocity cannot decrease rapidly enough to monotonically approach the $x = 0$ plane. So the particle overshoots and crosses the $x = 0$ plane, but its forward velocity eventually decays and the particle turns back and heads toward the $x = 0$ plane. In other words, the particle approaches $x = 0$ in an oscillatory manner. A schematic of the flow and the particle trajectories are shown in Figure 4.7. The dashed lines are the flow streamlines, while the solid lines are the particle pathlines. Figure 4.7a shows the scenario when $\gamma_x\tau_p \leq 1/4$, so the different particle trajectories approach $x = 0$ monotonically. In Figure 4.7b, since $\gamma_x\tau_p > 1/4$, the particle trajectory is oscillatory.

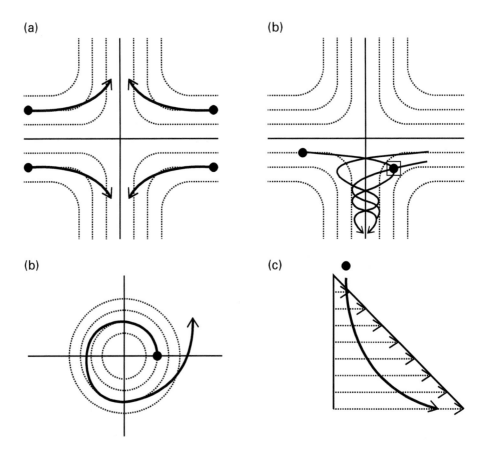

Figure 4.7 Schematic of flow streamlines (dotted lines) and sample particle pathlines (solid lines) for four different cases. (a) Planar straining flow with $\gamma_x \tau_p \le 1/4$ so the particle paths are monotonic. (b) Planar straining flow with $\gamma_x \tau_p > 1/4$ so the particle paths are oscillatory. (c) Vortical flow and (d) particle falling through a linear shear flow.

Problem 4.24 Consider a two-dimensional vortical flow of solid-body rotation given by

$$u_{cx} = -ky, \quad u_{cy} = kx, \quad u_{cz} = 0, \tag{4.164}$$

where k is the rotation-rate magnitude. Its velocity gradient tensor and its square are then

$$\mathbf{G} = \begin{bmatrix} 0 & -k & 0 \\ k & 0 & 0 \\ 0 & 0 & 0 \end{bmatrix} \quad \text{and} \quad \mathbf{G}^2 = \begin{bmatrix} -k^2 & 0 & 0 \\ 0 & -k^2 & 0 \\ 0 & 0 & 0 \end{bmatrix}. \tag{4.165}$$

We again ignore the effect of gravity and thus the equation of particle motion is unforced. Though the x and y motions of the particle do not decouple, the problem can be simplified by introducing a complex variable for the particle position that combines the x and y locations as $\xi = X + \iota Y$ (here $\iota = \sqrt{-1}$).

(a) Show that the x and y equations of particle motion can now be combined as

$$\tau_p \frac{d^2\xi}{dt^2} + \frac{d\xi}{dt} + \gamma_\xi \xi = 0,$$

(4.166)

where $\gamma_\xi = k(-\iota + k\tau_p \beta)$.

(b) The solution procedure follows that of the previous problem (also see Ferry et al., 2003). Obtain the following roots of the characteristic equation:

$$r_{\xi\pm} = \frac{-1 \pm \sqrt{1 - 4\gamma_\xi \tau_p}}{2\tau_p}.$$

(4.167)

Both the roots are now complex.

(c) Obtain the time history of particle position and velocity in the complex ξ plane as

$$\xi(t) = c_{\xi+} \exp(r_{\xi+} t) + c_{\xi-} \exp(r_{\xi-} t),$$
$$\frac{d\xi}{dt}(t) = c_{\xi+} r_{\xi+} \exp(r_{\xi+} t) + c_{\xi-} r_{\xi-} \exp(r_{\xi-} t),$$

(4.168)

where $c_{\xi+}$ and $c_{\xi-}$ are integration constants.

(d) Determine the integration constants by applying the boundary conditions (4.153), and reconcile the present analysis with that presented in Section 2.4.5.

(e) Show that the real part of $r_{\xi-}$ is more negative than that of $r_{\xi+}$ and as a result, the second terms of Eqs. (4.168) decay fast. Using this approximation, obtain the following result:

$$V_x + \iota V_y = \eta_\xi (u_{cx} + \iota u_{cy}),$$

(4.169)

where the nondimensional complex constant $\eta_x i$ depends only on the particle Stokes number and is given by

$$\eta_\xi = -\iota \left(\frac{-1 + \sqrt{1 - 4\,\mathrm{St}(-\iota + \beta\,\mathrm{St})}}{2\,\mathrm{St}} \right).$$

(4.170)

Thus, if we ignore the fast-decaying initial transient, here again the particle velocity can be written entirely in terms of the local fluid velocity at the particle location and the complex transfer coefficient is η_ξ.

Problem 4.25 As the final example, consider the problem of a particle falling down due to gravity in a linear shear flow. For the linear shear flow, the velocity gradient tensor and its square are

$$\mathbf{G} = \begin{bmatrix} 0 & k & 0 \\ 0 & 0 & 0 \\ 0 & 0 & 0 \end{bmatrix} \quad \text{and} \quad \mathbf{G}^2 = \begin{bmatrix} 0 & 0 & 0 \\ 0 & 0 & 0 \\ 0 & 0 & 0 \end{bmatrix},$$

(4.171)

where k is the shear magnitude. Steady unidirectional shear flow has no acceleration and therefore $\mathbf{G}^2 = 0$. In this case, let the vertical motion of the particle be in the y-direction and show that the vertical motion is the same as in still fluid.

(a) Following the solution procedure of the previous two problems, solve for the particle position and velocity along the x-direction.

(b) Identify the fast-decaying component. For longer times after the decay of the transient, express the particle velocity in terms of the local fluid velocity and the still-fluid settling velocity.

(c) Compare the results with those presented in Section 2.4.5 and in Ferry et al. (2003).

5 Lift Force and Torque in Unbounded Ambient Flows

We have completed our discussion of the drag force, where the term "drag" has been used to represent the force on a particle that is in the direction of ambient flow as seen in a frame of reference attached to the particle (i.e., drag is the force component along the direction of relative velocity). But there are many situations where the force on the particle is not only directed along the ambient flow, but also has a component that is perpendicular to the direction of ambient flow. In this case, the particle not only experiences a "drag" force, but also is subjected to a "lift" force.

The lift force on an airfoil is due to its asymmetric shape. When subjected to a uniform ambient flow, the airfoil is so shaped that the flow on the top side is faster than on the bottom side. The resulting differential pressure between the top and the bottom generates the lift force. In the case of a spherical particle subjected to a uniform flow, there is no asymmetry between the top and the bottom, and as a result there will be no lift force on the particle. Only by breaking the top–bottom symmetry can a lift force be induced on a spherical particle. The two main ways of symmetry breaking are (i) by introducing a shear component to the ambient flow and (ii) by spinning the particle with the axis of spin being perpendicular to the direction of ambient flow. In both these cases, the ambient flow velocity on one side of the particle becomes higher than the opposite side and this difference leads to a lift force.

While lift force is oriented perpendicular to the relative velocity vector, it is not appropriate to consider every force on the particle that is normal to relative velocity as a lift force. An example is the upward-directed buoyancy force, $-m_f \mathbf{g}$, which is an undisturbed flow force due to the hydrostatic pressure gradient in the ambient continuous phase. If we consider a particle moving horizontally in a stagnant ambient medium, then the relative velocity $\mathbf{u}_c - \mathbf{V}$ is horizontally oriented opposite to the particle motion. Though the buoyancy force is perpendicular to relative velocity, it should not be considered as a lift force. Similarly, in the BBO equation (4.88), the relative acceleration vector $d(\mathbf{u}_c - \mathbf{V})/dt$ can be oriented in any manner with respect to the relative velocity vector $(\mathbf{u}_c - \mathbf{V})$. As a result, inviscid and viscous unsteady forces may have a component in the direction normal to the relative velocity vector. At finite Re, we have seen that the inviscid unsteady force includes the effect of convective acceleration, which again can have a component normal to the relative velocity vector. For example, in Problems 4.18 and 4.19, we observed the inviscid unsteady force due to ambient strain and rotation to be along $\mathbf{e}_y$, while the relative velocity was along $\mathbf{e}_x$.

In this chapter and henceforth, we reserve the term "lift force" for the normal component of force induced on the particle due to the vortical component of the perturbation flow. The potential component of the perturbation flow results in the inviscid unsteady force. In the inviscid limit, the vortical component of the perturbation flow is due to the advection of ambient vorticity around the particle and the resulting lift force is captured by the third term on the right-hand side of Eq. (4.129). At finite Re, the vortical component of the perturbation flow also includes the contribution from vorticity generated on the particle surface due to the no-slip boundary condition. Lift force due to this viscous vortical component will be the main focus of this chapter.

We first emphasize the fact that the lift force is an inertial phenomenon. In the zero Reynolds number Stokes limit, the lift force on a particle will be identically zero. This can easily be demonstrated with a simple example. Consider a spherical particle subjected to a linear shear flow and let the particle be spinning as shown in the schematic of Figure 5.1. In the zero Reynolds number limit, owing to the linear nature of the governing equations, the above problem can be separated into superposition of three simpler problems. The first is a particle subjected to a uniform flow, the second is a particle subjected to a shear flow with velocity at the particle center being zero, and the third is a spinning particle in a quiescent ambient fluid. In all three subproblems, it can easily be seen that the upper and lower half of the particle are symmetric and as a result there cannot be any lift force. Lift force arises only due to the nonlinear interaction between these components. The combination of uniform flow and linear shear gives rise to *shear lift*, while the combination of uniform flow and particle spin gives rise to *Magnus lift*. Both these lift mechanisms require a nonzero finite Reynolds number.

In this chapter, we will first consider the important problem of lift force on a particle subjected to a linear shear flow. We will obtain the classic result of Saffman (1965) that is appropriate for small, but finite, shear Reynolds number. After obtaining this low Re asymptotic result for the shear-induced lift force, we will present its extension to finite Re. In Section 5.2 we will consider the problem of vortical lift force on a particle subjected to rotational and 45° planar straining undisturbed ambient flows. These results will first be presented in the small Re limit, with further extensions to larger values of Re. Section 5.3 will then present a general representation for the finite-Re quasi-steady force that includes both drag and lift contributions for linearly varying undisturbed ambient flows. Next, in Section 5.4 we will consider the problem of particle rotation and the resulting Magnus lift. This section will also consider the combined effects of particle rotation and ambient shear-induced lift force. Finally, in Section 5.5 we will take a brief look at the complex problem of unsteady vortical lift force under conditions where the ambient shear varies in a strongly time-dependent manner.

5.1   Saffman Shear Lift

Saffman (1965) considered a spherical particle of diameter d_p located at the origin subjected to the following steady linear shear flow:

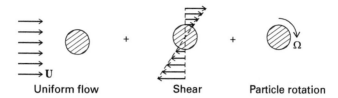

Uniform flow   Shear   Particle rotation

Figure 5.1 Schematic of a shear flow over a spinning particle in the Stokes limit, which can be expressed as a sum of the three problems: (i) uniform flow; (ii) linear shear with zero velocity at the particle center; and (iii) a spinning sphere in a quiescent ambient. By top–bottom symmetry about a horizontal line passing through the center of the particle, it can be seen that the lift is zero in all three flows.

$$\textbf{Shear}: \quad \mathbf{u}_c(\mathbf{x}) = (U_0 - Gy)\,\mathbf{e}_x, \tag{5.1}$$

where the unidirectional flow at the center of the particle is U_0. Here, G is the z-vorticity of the flow, which is a constant both in space and time. The shear G will be taken to be negative, so that the ambient flow is faster above the particle than below the particle (i.e., velocity is higher for $y > 0$). We define the particle Reynolds number based on relative velocity and shear Reynolds number to be

$$\mathrm{Re} = \frac{U_0 d_p}{\nu_f} \quad \text{and} \quad \mathrm{Re}_G = \frac{|G| d_p^2}{\nu_f}. \tag{5.2}$$

In what follows, we will (i) first present the low Reynolds number analysis of Saffman that is valid in the limit $\mathrm{Re} \ll \mathrm{Re}_G^{1/2} \ll 1$, (ii) then present the results of McLaughlin (1991) who relaxed the restriction to $\mathrm{Re}, \mathrm{Re}_G \ll 1$, and (iii) conclude with empirical results of shear-induced lift for larger values of Re and Re_G.

Saffman recognized the singular perturbation nature of the problem. No matter how small the Reynolds number, sufficiently far away from the sphere, there exists a distance beyond which inertial effects become important and will balance the viscous effects. An inner solution close to the sphere and an outer solution away from the sphere are required, with a matching between the two in the intermediate region. We define the Stokes length and the Saffman (shear) length scales as

$$\mathcal{L}_s = \frac{\nu_f}{U_0} \quad \text{and} \quad \mathcal{L}_G = \left(\frac{\nu_f}{|G|}\right)^{1/2}. \tag{5.3}$$

At small radial distances $r \ll (\mathcal{L}_s, \mathcal{L}_G)$, the inertial effects of both the uniform velocity U_0 and the shear G can be neglected. In this inner region, to leading order, we obtain the Stokes flow corresponding to a uniform flow. At radial distances of $O(\mathcal{L}_s)$ the inertial effect of the uniform flow U_0 becomes comparable to the viscous effect and at radial distances of $O(\mathcal{L}_G)$ the inertial effect of shear becomes comparable to the viscous effect. Thus, at leading order, the governing equations for the outer solution that are valid for distances of order $\mathcal{L}_s$ and $\mathcal{L}_G$ are

$$\nabla \cdot \mathbf{u}_{out} = 0,$$

$$\mathbf{u}_c \cdot \nabla \mathbf{u}_{out} + \mathbf{u}_{out} \cdot \nabla \mathbf{u}_c = -\frac{1}{\rho_f} \nabla p_{out} + \nu_f \nabla^2 \mathbf{u}_{out} - \frac{\mathbf{F}}{\rho_f} \delta(\mathbf{r}), \qquad (5.4)$$

where $\mathbf{u}_c$ is the undisturbed ambient shear flow and $\mathbf{u}_{out}$ is the disturbance flow generated by the particle in the outer region. In the outer solution, the details of the particle do not matter and the only matching with the inner solution is through the force $\mathbf{F} = 6\pi \mu_f R_p U_0 \mathbf{e}_x$, which acts at the origin. The subscript *out* is used to remind us that this solution is valid only in the outer region away from the particle. In the above momentum equation, the time derivative has been ignored due to the steady nature of flow. The nonlinear term $(\mathbf{u}_c + \mathbf{u}_{out}) \cdot \nabla (\mathbf{u}_c + \mathbf{u}_{out})$ has been approximated as the left-hand side. The undisturbed flow term $\mathbf{u}_c \cdot \nabla \mathbf{u}_c$ is balanced by the undisturbed flow pressure and viscous stresses and therefore not included. The quadratic term involving the disturbance flow $\mathbf{u}_{out} \cdot \nabla \mathbf{u}_{out}$ has been neglected as being small. For the specific linear shear flow given in Eq. (5.1), the above Oseen approximation of the inertial terms on the left-hand side yields

$$\mathbf{u}_c \cdot \nabla \mathbf{u}_{out} + \mathbf{u}_{out} \cdot \nabla \mathbf{u}_c = (U_0 - Gy) \frac{\partial \mathbf{u}_{out}}{\partial x} - (\mathbf{u}_{out} \cdot \mathbf{e}_y) G \mathbf{e}_x. \qquad (5.5)$$

After substituting the above in Eq. (5.4), the resulting linear equations are solved in an unbounded region, where the far-field boundary condition is to require that the disturbance flow decays (i.e., $\mathbf{u}_{out} \to 0$ as $\mathbf{r} \to \infty$). Since this is the outer solution, it is influenced by the presence of the particle only through the force $\mathbf{F}$.

By imposing the additional restriction that $\mathrm{Re} \ll \mathrm{Re}_G^{1/2}$, Saffman was able to ignore the inertial effect of the uniform part of the undisturbed flow. In other words, while substituting the right-hand side of Eq. (5.5) into the left-hand side of Eq. (5.4), the term $U_0 \partial \mathbf{u}_{out}/\partial x$ was ignored by Saffman. This simplifies the outer solution. The important aspect of the outer solution is the inertial migration velocity. This is the y-component of the outer solution evaluated at the origin. Saffman obtained the following explicit expression for the inertial migration velocity:

$$U_{im} = \mathbf{u}_{out}(\mathbf{x} = 0) \cdot \mathbf{e}_y = -0.343 R_p U_0 \, \mathrm{sgn}(G) \left(|G|/\nu_f \right)^{1/2}. \qquad (5.6)$$

The direction of the inertial migration velocity can be determined as follows. If the ambient velocity relative to the particle is directed in the positive x-direction (i.e., U_0 is positive), and the ambient shear is such that velocity increases with increasing y (i.e., G is negative), then U_{im} is positive, indicating that the inertial migration velocity points in the positive y-direction (i.e., from the low-velocity to the high-velocity side of the shear flow).

At the next level of the inner solution, by matching with the above outer solution, we observe that the particle is subjected to a uniform cross flow given by the inertial migration velocity. In other words, at the next level of the inner solution, the particle sees not only the undisturbed ambient shear flow, but also a uniform cross flow. Again, the inner solution satisfies the Stokes equation for flow around a sphere located at the origin. The Stokes drag corresponding to this cross flow results in the Saffman lift force $6\pi\mu_f R_p U_{im}\mathbf{e}_y$. After substituting for U_{im}, Saffman's shear-induced lift force, in general terms, can be expressed as

$$\text{Shear}: \quad \mathbf{F}_{qsL} = 6.46\,\mu_f R_p^2\,|\mathbf{u}_c|\,\left(|\omega_c|/\nu_f\right)^{1/2}\mathbf{e}_L, \tag{5.7}$$

where $\mathbf{u}_c$ and ω_c are the undisturbed shear flow's velocity and vorticity at the particle and the lift direction is given by $\mathbf{e}_L = \mathbf{u}_c \times \omega_c/|\mathbf{u}_c \times \omega_c|$. The above expression is for a stationary nonrotating particle. If the particle translates with velocity $\mathbf{V}$, then one must use the relative velocity $\mathbf{u}_c - \mathbf{V}$ in place of $\mathbf{u}_c$. In case of a spinning particle with nonzero angular velocity, there is an additional contribution to the lift force, which we shall consider in Section 5.4.

5.1.1 Generalization of Saffman's Analysis and Finite-Re Extension

Saffman's pioneering analysis has been extended in several ways. It was originally restricted by the condition $\text{Re} \ll \text{Re}_G^{1/2}$, but it has been observed that in applications such as particle motion in a pipe or channel flow, $\text{Re}_G^{1/2}$ is often smaller than Re. McLaughlin (1991) relaxed this restriction and his analysis only required a low Reynolds number limit of $\text{Re}, \text{Re}_G^{1/2} \ll 1$, without any additional restriction on the relative value of the two Reynolds numbers. As a result, McLaughlin retained all the inertial terms that were on the right-hand side of Eq. (5.5). This somewhat complicated the solution process, nevertheless McLaughlin was able to modify Saffman's lift prediction with a simple correction function that depended on the ratio

$$\epsilon = \frac{\text{Re}_G^{1/2}}{\text{Re}} = \frac{|\omega_c|^{1/2}\nu_f^{1/2}}{|\mathbf{u}_c|}.$$

The resulting modified lift can be expressed as

$$\mathbf{F}_{qsL} = 6.46\,\mu_f R_p^2\,|\mathbf{u}_c|\,\left(|\omega_c|/\nu_f\right)^{1/2} J(\epsilon)\,\mathbf{e}_L, \tag{5.8}$$

where the dependence of the correction function J on ϵ was tabulated in McLaughlin (1991). In the Saffman limit of $\epsilon = \text{Re}_G^{1/2}/\text{Re} \to \infty$, we recover the Saffman lift force with $J \to 1$. In the other limit of $\epsilon \ll 1$, McLaughlin obtained the limiting solution of $J \approx -140\epsilon^5 \ln(1/\epsilon^2)$. An empirical relation for the correction function that is accurate over a wide range of ϵ values was obtained by Mei (1992) as

$$J(\epsilon) = 0.3\left\{1 + \tanh\left[2.5\left(\log_{10}\epsilon + 0.191\right)\right]\right\}\left\{0.667 + \tanh\left[6\epsilon - 1.92\right]\right\}. \tag{5.9}$$

The above theoretical results obtained from the perturbation analysis do not apply for Reynolds numbers larger than unity. At finite Reynolds numbers, the nondimensional lift force is expected to depend on both Re and Re_G in a complicated way, and one must resort to fully resolved numerical simulations or experiments to extend the theoretical results. To do so, let us first define the nondimensional shear magnitude to be

$$s = \frac{|G|d_p}{U_0} = \frac{|\omega_c|d_p}{|\mathbf{u}_c|}. \tag{5.10}$$

Thus, s measures the ratio of velocity change across a particle diameter due to shear to the average relative velocity at the center of the particle. It is related to the particle and shear Reynolds numbers through $s = \mathrm{Re}_G/\mathrm{Re}$. Direct numerical simulation of linear shear flow over a stationary spherical particle was carried out by Kurose and Komori (1999) and Bagchi and Balachandar (2002a) for varying values of Re and s, and the computed lift force was nondimensionalized to obtain the lift coefficient as

$$C_{L,qs} = \frac{|\mathbf{F}_{qsL}|}{\frac{1}{2}\pi\rho_f R_p^2 |\mathbf{u}_c|^2} . \qquad (5.11)$$

The numerical results from Bagchi and Balachandar (2002a), along with the $C_{L,qs}$ predicted using Saffman's lift formula (5.7) and with McLaughlin's correction function are shown in Table 5.1 for a range of particle Reynolds numbers at two different values of s.

Several important observations can be made. The simulation results are closer to McLaughlin's prediction in the low Re range. At higher Reynolds number, McLaughlin's lift expression decays much faster and quickly becomes zero. The lift prediction using Saffaman's expression significantly overpredicts the lift force, especially at finite Re. It must be noted that at Re $\gtrsim$ 100 the lift force, although small in magnitude, is directed from the high-speed side to the low-speed side, as indicated by the negative lift coefficient. The general conclusion is that the Saffman lift formula with McLaughlin correction is adequate over a range of Reynolds numbers and shear magnitudes, although systematic departures are expected at larger Re.

Finally, by comparing the lift coefficient with the standard drag coefficient of $C_D = (24/\mathrm{Re})(1 + 0.15\,\mathrm{Re}^{0.687})$, it can be seen that the lift force is generally much smaller than the drag force. This means that the particle motion is dominantly aligned with the ambient fluid motion. However, even a small cross-stream motion of the particle induced by the lift force can be quite important, as this is the only way to explain the lateral migration of particles in laminar flows.

5.2   Lift Force in Other Linearly Varying Flows

The important message of the previous section is the following: if the undisturbed ambient flow varies on the scale of the particle, then a lift force is generated due to asymmetric vorticity generation and advection induced by the perturbation flow of the particle. In the case of a linear shear flow, this results in the generation of Saffman shear lift force. But there are many other ambient flow configurations where the undisturbed velocity can vary linearly on the scale of the particle. The linear shear flow given in Eq. (5.1) is an important flow configuration that is often encountered in applications. Three other canonical, linearly varying ambient flow configurations are listed below

Table 5.1 List of lift coefficients for different combinations of particle Reynolds number Re and nondimensional shear magnitude s. The direct numerical simulation (DNS) results are presented along with $C_{L,qs}$, calculated with Saffman's lift formula given in Eq. (5.7) and with McLaughlin's correction given in Eq. (5.8).

Re	s	$C_{L,qs}$ DNS	$C_{L,qs}$ Saffman	$C_{L,qs}$ McLaughlin
5	0.2	0.034	0.823	0.002
20	0.2	0.011	0.411	0.0
50	0.2	0.002	0.260	0.0
100	0.2	−0.016	0.184	0.0
200	0.2	−0.058		0.0
0.5	0.4	2.15	3.678	2.30
1.0	0.4	1.40	2.601	1.238
5.0	0.4	0.186	1.163	0.045
10.0	0.4	0.039	0.822	0.002
20.0	0.4	0.019	0.582	0.0
50.0	0.4	0.003	0.367	0.0
100.0	0.4	−0.03	0.260	0.0

$$\textbf{45}° \textbf{ Plane strain}: \quad \mathbf{u}_c(\mathbf{x}) = \left(U_0 - \frac{G}{2}y\right)\mathbf{e}_x - \frac{G}{2}x\mathbf{e}_y,$$

$$\textbf{Vortex}: \quad \mathbf{u}_c(\mathbf{x}) = \left(U_0 - \frac{G}{2}y\right)\mathbf{e}_x + \frac{G}{2}x\mathbf{e}_y,$$

$$\textbf{Transverse shear}: \quad \mathbf{u}_c(\mathbf{x}) = U_0\mathbf{e}_x + Gx\mathbf{e}_y. \tag{5.12}$$

While the linear shear flow given in Eq. (5.1) is unidirectional, the above three flows are two-dimensional. But in all the cases, the ambient flow velocity at the particle center is the same. If we define the velocity gradient magnitude as the sum of the absolute value of its nine components, then we see that the linear shear flow of the previous section is equivalent to the above three flows in that they all have the same net velocity gradient of G. This allows direct comparison of their effect on the force on the particle. Again, the only two important parameters are the particle Reynolds number based on the relative velocity Re and the velocity gradient Reynolds number $\text{Re}_G = |G|d_p^2/\nu_f$.

The linear shear, plane strain, vortex and transverse shear flow configurations around a particle located at the origin are shown in Figure 5.2. Note that the ambient flow vorticity, defined as $\omega_z = \partial v/\partial x - \partial u/\partial y$, is equal to G in the case of linear shear and vortex flow, while it is identically zero in the plane-strain flow. It can readily be observed that the linear shear flow is nothing but a superposition of plane-strain and vortex flows. In all the flow schematics, the velocity gradient G is taken to be negative, so that the ambient flow is faster above the particle than below the particle (i.e., x-velocity is higher for $y > 0$) in frames (a), (b), and (c). Thus, in the case of linear shear, plane-strain, and vortex flows, the pressure above the particle will be lower than that below the particle, resulting in a lift force in the positive y-direction.

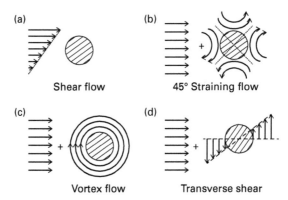

Figure 5.2 Schematic of four linearly varying flows: (a) linear shear flow; (b) 45° straining flow; (c) rotational flow; and (d) transverse shear flow. All are superposed with a uniform component. The equations for the last three are given as Eqs. (5.12).

However, the details of the flow and the resulting force on the particle in these three flows are different. Let us first investigate these differences in the low Reynolds number limit of $\mathrm{Re}, \mathrm{Re}_G \ll 1$. The lift force on the particle in a linear shear flow was obtained as Eq. (5.8). A similar low Reynolds number analysis for a particle subjected to a vortex flow was considered by Herron et al. (1975), who expressed the lift expression as

$$\textbf{Vortex}: \quad \mathbf{F}_{qsL} = 8\,\mu_f R_p^2\,|\mathbf{u}_c|\,\left(|\omega_c|/\nu_f\right)^{1/2}\mathbf{e}_L\,. \qquad (5.13)$$

Thus, in the limit of small Reynolds number, the lift force on a particle subjected to a vortex flow will be 24% larger than in a linear shear flow of the same vorticity. A low Reynolds number analysis for a particle subjected to plane strain was considered by Perez-Madrid et al. (1990), who obtained the lift as

$$\textbf{45° Plane strain}: \quad \mathbf{F}_{qsL} = 7.35\,\mu_f R_p^2\,|\mathbf{u}_c|\,\left(|G|/\nu_f\right)^{1/2}\mathbf{e}_L\,. \qquad (5.14)$$

In the case of 45° strain, the lift direction is given by $\mathbf{e}_L = \mathbf{u}_c \cdot \mathbf{S}_c/|\mathbf{u}_c \cdot \mathbf{S}_c|$, where $\mathbf{S}_c$ is the strain-rate tensor evaluated at the particle. Thus, in the limit of small Reynolds number, the lift forces in the three configurations are quite similar, with the lift force on a particle subjected to a 45° plane strain in between those of linear shear and vortex flow of the same magnitude.

Finite-Re Effect

These differences in the behavior of the different linear flows continue even at finite particle Reynolds numbers. In fact, the finite Reynolds number results for the linear shear flow reported in the previous section apply only to the linear shear flow and do not carry over for the vortex flow or the plane strain. For example, at $\mathrm{Re} = 50$, the lift coefficient of a particle subjected to linear shear flow of $s = 0.1$ is $C_{L,qs} = 0.001$, while that of a particle subjected to a vortex of strength $s = 0.1$ is $C_{L,qs} = 0.347$ (Bagchi and

Balachandar, 2002c). For a particle subjected to a vortex flow as given in Eq. (5.12), Bluemink et al. (2008) observed the lift force to be lower than that presented by Bagchi and Balachandar (2002c), and obtained the following expression for the lift coefficient for $s = 0.2$ (note that their definition of lift coefficient and vorticity parameter are different):

$$C_{L,qs} = \frac{4}{3}s \left(0.51 \log_{10} \text{Re} -0.22\right) . \tag{5.15}$$

This finite-Re solid-body rotation-induced (or vortex-induced) quasi-steady lift force is substantially larger than that due to linear shear flow. This increased lift in a vortex has been observed in the case of bubbles and particles trapped in vortices (Sridhar and Katz, 1995; Bluemink et al., 2008). Thus, for the same ambient vorticity, a particle in vortex flow experiences a far higher lift force, which cannot be predicted using the lift formula of a shear flow. The corresponding lift coefficient for a particle subjected to 45° plane strain is $C_L = -0.21$ (Bagchi and Balachandar, 2002b), which is different from those of linear shear and vortex flows.

Earlier, in Problems 4.18 and 4.19, we saw that the undisturbed flow and inviscid unsteady forces could be directed normal to the relative velocity. These force contributions are oriented in the same direction as the quasi-steady lift force; following our earlier definition of lift force, we will not refer to them as lift force. For the shear flow given in Eq. (5.1), and for the vortex and 45° straining flows given in Eqs. (5.12), the sum of undisturbed flow and inviscid unsteady forces can be expressed as

$$m_f(1 - C_M)\frac{D\mathbf{u}_c}{Dt} = \begin{cases} 0 & \textbf{shear}, \\ -\pi R_p^3 \rho_f U_0 G \, \mathbf{e}_y & \textbf{vortex}, \\ \pi R_p^3 \rho_f U_0 G \, \mathbf{e}_y & \textbf{45° strain}. \end{cases} \tag{5.16}$$

The relative importance of these force contributions compared to the quasi-steady lift can be examined. The undisturbed flow and inviscid unsteady forces scale as $\rho_f R_p^3 U_0 G$, while at small values of Re, the quasi-steady lift scales as $\rho_f R_p^2 U_0 \sqrt{G \nu_f}$. The ratio $|\mathbf{F}_{qsL}|/|\mathbf{F}_{un} + \mathbf{F}_{iu}|$ therefore goes as $\sqrt{\text{Re}_G}$ or $\sqrt{\text{Re}_s}$. Thus, at low Reynolds numbers, quasi-steady lift dominates and this is the reason why the analyses of Saffman, Herron, and Perez-Madrid et al. yielded only the quasi-steady lift. However, as the Reynolds number approaches $O(1)$ and above, the above scaling does not apply and one must consider all the force contributions.

We now recall the vorticity-induced lift force that was obtained in the inviscid limit, which in the case of a stationary particle becomes $m_f(\mathbf{u}_x \times \omega_c)/2$. This force is nothing but $\mathbf{F}_{qsL}$ in the inviscid limit. As we discussed in Section 4.4, this force arises due to the advection of ambient vorticity around the particle by the inviscid perturbation flow. In the finite-Re limit, with the inclusion of viscous effects, new vorticity is generated due to the imposition of the no-slip boundary condition, and furthermore the viscous effect alters the perturbation flow and the advection of vorticity around the particle. The net effect is that $m_f(\mathbf{u}_x \times \omega_c)/2$ is not a good predictor of $\mathbf{F}_{qsL}$ at finite values of Re.

We now conclude this finite-Re discussion by pointing out that there have been a number of research efforts that have focused on direct numerical simulations of shear, vortex, plane, and axisymmetric straining flows over a particle at finite Reynolds number (Magnaudet et al., 1995; Kurose and Komori, 1999; Bagchi and Balachandar, 2002a, 2002b, 2002c, 2003a, Bluemink et al., 2008). These studies clearly illustrate the challenges associated with predicting the drag and lift forces on a particle subjected to a linearly varying flow. As an additional example, consider the transverse shear flow shown in Figure 5.2d, where the x-velocity is a constant and the shear flow is in the y-direction. Though the shear magnitude itself is the same as that considered in Eq. (5.1), it is now oriented differently, with a 90° rotation with respect to the linear shear flow configuration. The difference in force between these two cases has been highlighted by Bluemink et al. (2008) through numerical simulation results. We recommend the reader to consult these and other recent research developments to gain a better understanding of forces on a particle subjected to linearly varying ambient flows of different kinds at finite Re.

5.3   General Representation

In this subsection, we consider the theoretical generalization of the results presented above. Consider a frame of reference where the particle of diameter d_p is located with its center at the origin $\mathbf{x} = 0$. Let the undisturbed ambient flow have a linear variation around the particle location, with the most general representation of this linearly varying flow being

$$\mathbf{u}_c(\mathbf{x}) = \mathbf{U}_0 + \mathbf{x} \cdot [\nabla \mathbf{U}]_0 = \mathbf{U}_0 + \mathbf{x} \cdot [\mathbf{S}_0 + \mathbf{R}_0] \,, \qquad (5.17)$$

where $\mathbf{U}_0$ is the undisturbed ambient flow at the origin and $[\nabla \mathbf{U}]_0$ is the gradient of the undisturbed ambient flow at the origin. Furthermore, the velocity gradient tensor has been separated into a symmetric strain-rate part $\mathbf{S}_0$ and an antisymmetric rotation-rate part $\mathbf{R}_0$.

The linear shear, vortex, 45° plane strain, and transverse shear problems of the previous section are special cases. In these examples, the uniform flow component is chosen as $\mathbf{U}_0 = U_0 \mathbf{e}_x$ along the x-direction. For example, the linear shear given in Eq. (5.1) is obtained when

$$\textbf{Shear}: \quad [\nabla \mathbf{U}]_0 = \begin{bmatrix} 0 & 0 & 0 \\ G & 0 & 0 \\ 0 & 0 & 0 \end{bmatrix}, \qquad (5.18)$$

which can be decomposed into symmetric and antisymmetric components as

$$\mathbf{S}_0 = \begin{bmatrix} 0 & G/2 & 0 \\ G/2 & 0 & 0 \\ 0 & 0 & 0 \end{bmatrix}, \quad \mathbf{R}_0 = \begin{bmatrix} 0 & -G/2 & 0 \\ G/2 & 0 & 0 \\ 0 & 0 & 0 \end{bmatrix}. \qquad (5.19)$$

The velocity gradient tensor of the 45° plane strain is symmetric and is the same as $\mathbf{S}_0$ given above, with zero antisymmetric component. In contrast, the velocity gradient tensor of the vortex flow is antisymmetric and is the same as $\mathbf{R}_0$ given above, with zero symmetric component. The velocity gradient tensor of the transverse shear flow is the negative transpose of that given in Eq. (5.18), whose symmetric and antisymmetric components can readily be obtained as well.

The quest now is to obtain a general expression for the force on a particle subjected to this general linearly varying flow under steady conditions. Before we embark upon this, let us reconsider the simpler case when the velocity gradient is identically zero and the ambient flow is spatially uniform, given by $\mathbf{U}_0$. The only force on the particle is drag, which from Eq. (4.141) can be expressed in nondimensional form as

$$ \mathbf{C}_{D,qs} = \frac{\mathbf{F}_{qs}}{\frac{\pi}{2} R_p^2 \rho_f |\mathbf{U}_0|^2} = \left[\frac{24}{\text{Re}} \right] \Phi(\text{Re})\, \mathbf{e}_U , \qquad (5.20) $$

where the force coefficient $\mathbf{C}_{D,qs}$ is the nondimensional quasi-steady force. Written in the above form, the three factors on the right-hand side have a precise interpretation: (i) $\mathbf{e}_U$ is the unit vector along the ambient flow direction, defined as $\mathbf{e}_U = \mathbf{U}_0/|\mathbf{U}_0|$. This factor shows that irrespective of the value of all other quantities affecting the force, the force vector will be directed along the ambient flow seen by the particle. This arises from the fact that the only vectorial quantity that the force depends on is the ambient flow velocity. (ii) The factor 24/Re is the nondimensional force in the low Reynolds number limit. It depends on the only nondimensional scalar parameter (i.e., the Reynolds number) that can be formed out of the input parameters (i.e., relative velocity, diameter, and kinematic viscosity). In the low Reynolds number limit, we have the requirement that the force be linear in $\mathbf{U}_0$. This yields $\mathbf{C}_{D,qs} \propto 1/\text{Re}$ as the only option. The numerator 24 is, however, problem specific and cannot be obtained from the above dimensional argument. (iii) The factor $\Phi(\text{Re})$ is the finite Reynolds number correction to quasi-steady force. Its functional form is also problem specific and cannot be determined from a dimensional argument.

The above modeling framework will be used to obtain an expression for the force coefficient in the case of the linearly varying ambient flow given in Eq. (5.17). Our quest is as follows: given the ambient flow in terms of the velocity vector $\mathbf{U}_0$, strain-rate tensor $\mathbf{S}_0$, and rotation-rate tensor $\mathbf{R}_0$ (or equivalently, the velocity gradient tensor $[\nabla \mathbf{U}]_0$), obtain the general form of the quasi-steady force coefficient in terms of the relevant nondimensional parameters of the problem. As a first step, we identify all the unique vector quantities that can be formed out of $\mathbf{U}_0$, $\mathbf{S}_0$, and $\mathbf{R}_0$, and define unit vectors along them. Apart from $\mathbf{e}_U$, we can define the following unit vectors:

$$ \mathbf{e}_S = \frac{\mathbf{U}_0 \cdot \mathbf{S}_0}{|\mathbf{U}_0 \cdot \mathbf{S}_0|}, \quad \mathbf{e}_{S^2} = \frac{\mathbf{U}_0 \cdot \mathbf{S}_0^2}{|\mathbf{U}_0 \cdot \mathbf{S}_0^2|}, $$

$$ \mathbf{e}_\Omega = \frac{\mathbf{U}_0 \cdot \mathbf{R}_0}{|\mathbf{U}_0 \cdot \mathbf{R}_0|}, \quad \mathbf{e}_{\Omega^2} = \frac{\mathbf{U}_0 \cdot \mathbf{R}_0^2}{|\mathbf{U}_0 \cdot \mathbf{R}_0^2|}, $$

$$ \mathbf{e}_{S\Omega} = \frac{\mathbf{U}_0 \cdot (\mathbf{S}_0\mathbf{R}_0 - \mathbf{R}_0\mathbf{S}_0)}{|\mathbf{U}_0 \cdot (\mathbf{S}_0\mathbf{R}_0 - \mathbf{R}_0\mathbf{S}_0)|} . \qquad (5.21) $$

According to the representation theorem (Wang, 1970; Smith, 1971; Bagchi and Balachandar, 2003a), there will be a contribution to force along each of the six unit vectors.

Next we identify the low Reynolds number form of the effect of the ambient flow's strain and rotation rates to be

$$C(I_S, I_\Omega) \frac{\sqrt{\mathrm{Re}_S}}{\mathrm{Re}} \quad \text{and} \quad C(I_S, I_\Omega) \frac{\sqrt{\mathrm{Re}_R}}{\mathrm{Re}}, \tag{5.22}$$

where $\mathrm{Re}_S = ||\mathbf{S}_0|| d_p^2 / \nu_f$ and $\mathrm{Re}_R = ||\mathbf{R}_0|| d_p^2 / \nu_f$. Here the matrix norms $||\mathbf{S}_0||$, $||\mathbf{R}_0||$ are suitably defined to extract the magnitudes of the symmetric strain-rate and anti-symmetric rotation-rate tensors. Only with the above dependence on $||\mathbf{S}_0||$, $||\mathbf{R}_0||$, and Re will we recover the functional form of the lift forces given in Eqs. (5.7), (5.13), and (5.14). In the above, I_S represents the list of scalar invariants associated with $\mathbf{S}_0$. This list includes measures of relative orientation of $\mathbf{S}_0$ with respect to the velocity vector $\mathbf{U}_0$. Similarly, I_Ω represents the list of scalar invariants associated with $\mathbf{R}_0$, including its relative orientation with respect to the velocity vector $\mathbf{U}_0$.

Compared with the functional form presented in Eq. (5.20), for the general linearly varying flow, we now have the six unit vectors (which are the terms analogous to $\mathbf{e}_U$), and the low-Re functional forms (which correspond to [24/Re]). This line of argument yields the following general expression for the force coefficient:

$$\mathbf{C}_{F,qs} = \left(\frac{24}{\mathrm{Re}} \Phi(\mathrm{Re}) + C_1 \frac{\sqrt{\mathrm{Re}_S}}{\mathrm{Re}} \Phi_1 \right) \mathbf{e}_U + C_2 \frac{\sqrt{\mathrm{Re}_S}}{\mathrm{Re}} \Phi_2 \, \mathbf{e}_S$$

$$+ C_3 \frac{\sqrt{\mathrm{Re}_S}}{\mathrm{Re}} \Phi_3 \, \mathbf{e}_{S^2} + C_4 \frac{\sqrt{\mathrm{Re}_R}}{\mathrm{Re}} \Phi_4 \, \mathbf{e}_\Omega$$

$$+ C_5 \frac{\sqrt{\mathrm{Re}_R}}{\mathrm{Re}} \Phi_5 \, \mathbf{e}_{\Omega^2} + C_6 \left(\frac{\mathrm{Re}_S}{\mathrm{Re}^2} \frac{\mathrm{Re}_R}{\mathrm{Re}^2} \right)^{1/4} \Phi_6 \, \mathbf{e}_{S\Omega}, \tag{5.23}$$

where the factors C_1 to C_6 are functions of I_S and I_Ω. The finite Reynolds number correction functions Φ_1 to Φ_6 are functions of all the nondimensional scalar invariants: Re, Re_S, Re_R, I_S, and I_Ω. By definition, the correction functions approach unity as Re, Re_S, and Re_R become small.

The above general expression comes from the fact that the relation between the force coefficient and $\mathbf{U}_0$, $\mathbf{S}_0$, and $\mathbf{R}_0$ must be isotropic. However, the representation theorem can only yield the functional form of the force and cannot yield the exact dependence of C_1 to C_6 and Φ_1 to Φ_6 on the scalar invariants. They must be obtained from numerical simulations or experiments.

You may consider the force expression (5.23) to be far too complex. Indeed, you are correct. It is a sobering thought that the above complex expression is just for an ambient flow that is restricted to being linear and steady. In the problems that follow, by limiting our attention to specific flows such as linear shear, plane strain, and vortex flows, we will shed light on the nature of the scalar invariants I_S and I_Ω. You will obtain explicit expressions for factors C_2 to C_5 using the low Reynolds number lift force expressions given in Eqs. (5.7), (5.13), and (5.14). Direct numerical simulation results of shear, vortex, plane, and axisymmetric straining flows over a particle (Magnaudet et al., 1995; Kurose and Komori, 1999, Bagchi and Balachandar, 2002a, 2002b, 2002c, 2003a, Bluemink et al., 2008) have been used to obtain expressions for finite Reynolds number corrections applicable in these restricted classes of linearly varying flows.

Problem 5.1 In this problem, we will consider the list of scalar invariants that characterize the strain-rate tensor. Using elementary linear algebra, show that the 3×3 symmetric matrix $\mathbf{S}_0$ can be diagonalized and the three eigenvalues can be ordered as $\lambda_1 \geq \lambda_2 \geq \lambda_3$. Show that in an incompressible flow, the sum of eigenvalues $\lambda_1 + \lambda_2 + \lambda_3 = 0$. This constraint guarantees λ_1 to be positive and λ_3 to be negative, with the sign of λ_2 being either positive or negative. In other words, if we choose an orthogonal coordinate system given by the eigenvectors, along the first eigenvector there is an extensional strain rate, along the third eigenvector there is a compressional strain rate, and along the intermediate eigenvector the nature of the strain rate (extensional or compressional) depends on the eigenvalue.

(a) Show that the qualitative nature of strain is entirely characterized by the parameter $b = \lambda_2/\lambda_1$.

(b) Show that its value is bounded by $-1/2 \geq b \geq 1$. When $b = -1/2$, the three eigenvalues are $(\lambda_1, -\lambda_1/2, -\lambda_1/2)$ and correspond to axisymmetric compression, where stretching is along the first eigenvector and is compensated by compression along the plane formed by the other two eigenvectors. When $b = 0$, the three eigenvalues are $(\lambda_1, 0, -\lambda_1)$ and correspond to plane strain, where stretching is along the first eigenvector and compression along the third eigenvector. When $b = 1$, the three eigenvalues are $(\lambda_1, \lambda_1, -2\lambda_1)$ and correspond to axisymmetric elongation, where there is compression along the third eigenvector, compensated by elongation along the plane formed by the other two eigenvectors.

The last two nondimensional scalar parameters characterize the relative orientation of the strain-rate tensor with respect to the ambient flow $\mathbf{U}_0$. Since the principal directions of the strain-rate tensor are given by the three orthogonal eigenvectors, show that the relative orientation of the vector $\mathbf{U}_0$ is given by the two angles Θ_U and Φ_U. The strength of the strain-rate tensor is measured in terms of the norm $||\mathbf{S}_0||$. The other three scalar invariants associated with the strain-rate tensor are $I_S = \{b, \Theta_U, \Phi_U\}$.

Problem 5.2 (a) From elementary fluid mechanics, show that the rotation-rate tensor can be written in terms of vorticity components as

$$\mathbf{R}_0 = \frac{1}{2} \begin{bmatrix} 0 & \omega_{z0} & -\omega_{y0} \\ -\omega_{z0} & 0 & \omega_{x0} \\ \omega_{y0} & -\omega_{x0} & 0 \end{bmatrix}, \tag{5.24}$$

where the vorticity vector $\omega_0 = \nabla \times \mathbf{u}_c$ of the linearly varying flow given in Eq. (5.17) is a constant.

(b) Verify the following statements. The relative orientation of the rotation-rate tensor with respect to the ambient flow $\mathbf{U}_0$ is given by the angle between the vorticity vector ω_0 and $\mathbf{U}_0$. If the linear variation of the ambient flow contains both the strain rate and the rotation rate, then the relative orientation of the vorticity vector in the coordinate system formed by the eigenvectors of $\mathbf{S}_0$ is given by the two angles Θ_ω

and Φ_ω. The strength of the rotation-rate tensor is measured in terms of the matrix norm $||\mathbf{R}_0||$, and it is related to the vorticity magnitude. Two other scalar invariants are needed to characterize the rotation-rate tensor, and they are $I_\Omega = \{\Theta_\omega, \Phi_\omega\}$.

(c) Finally, show that

$$\mathbf{U}_0 \cdot \mathbf{R}_0 = -\frac{1}{2}\,\omega_0 \times \mathbf{U}_0. \tag{5.25}$$

Therefore, the unit vector $\mathbf{e}_\Omega$ given in Eq. (5.21) is orthogonal to both the ambient vorticity and relative velocity directions.

Problem 5.3 We now consider the problem of a particle located at the origin in a vortex flow given by Eq. (5.12).

(a) Show that in this example the unit vectors defined in Eq. (5.21) are

$$\mathbf{e}_U = \mathbf{e}_x, \quad \mathbf{e}_\Omega = \mathbf{e}_y, \quad \mathbf{e}_{\Omega^2} = -\mathbf{e}_x. \tag{5.26}$$

(b) Substituting into Eq. (5.23), obtain

$$\mathbf{C}_{F,qs} = \frac{24}{\mathrm{Re}}\Phi(\mathrm{Re})\,\mathbf{e}_x + (C_1\Phi_1 - C_5\Phi_5)\frac{\sqrt{\mathrm{Re}_S}}{\mathrm{Re}}\,\mathbf{e}_x + C_4\frac{\sqrt{\mathrm{Re}_R}}{\mathrm{Re}}\Phi_4\,\mathbf{e}_y, \tag{5.27}$$

where the second term on the right is the drag correction due to the vortex, while the third term is the lift force due to the vortex. Note that in this example we have $||\mathbf{R}_0|| = G/2$.

(c) Compare this lift force with the Herron lift formula given in Eq. (5.13) and obtain (after setting $\Phi_4 = 1$, appropriate for the low Reynolds number limit)

$$C_4 = -\frac{16\sqrt{2}}{\pi}, \tag{5.28}$$

which is the value of the coefficient C_4 for the specific configuration where the ambient vorticity vector is perpendicular to the ambient velocity vector $\mathbf{U}_0$.

Problem 5.4 We now consider the problem of a particle located at the origin in a 45° plane-strain flow given by Eq. (5.12).

(a) Show that in this example the unit vectors defined in Eq. (5.21) are

$$\mathbf{e}_U = \mathbf{e}_x, \quad \mathbf{e}_S = -\mathbf{e}_y, \quad \mathbf{e}_{S^2} = \mathbf{e}_x. \tag{5.29}$$

(b) Substituting into Eq. (5.23), obtain

$$\mathbf{C}_{F,qs} = \frac{24}{\mathrm{Re}}\Phi(\mathrm{Re})\,\mathbf{e}_x + (C_1\Phi_1 + C_3\Phi_3)\frac{\sqrt{\mathrm{Re}_S}}{\mathrm{Re}}\,\mathbf{e}_x - C_2\frac{\sqrt{\mathrm{Re}_R}}{\mathrm{Re}}\Phi_2\,\mathbf{e}_y, \tag{5.30}$$

where the second term on the right is the drag correction, while the third term is the lift force due to the plane strain. Note that in this example, $||\mathbf{S}_0|| = G/2$.

(c) Compare this lift force with the lift formula given in Eq. (5.14) and obtain

$$C_2 = \frac{(14.7)\sqrt{2}}{\pi},$$ (5.31)

which is the value of C_2 for the specific configuration of the ambient velocity vector at 45° to the principal axis of strain.

Problem 5.5 Now consider the problem of a particle located at the origin in a linear shear flow given by Eq. (5.1). Note that linear shear is a superposition of plane strain and rotation, and therefore both S_0 and R_0 are nonzero.
(a) Obtain the unit vectors defined in Eq. (5.21).
(b) Substitute into Eq. (5.23) and obtain a simplified expression for the force coefficient as in the previous problems.
(c) Consider the low Reynolds number limit and compare with the Saffman lift force formula.

5.4   Particle Rotation, Spin-Induced Torque, and Lift

We started this chapter with equations of time evolution of particle mass, momentum, and energy, where the motion of the particle included both translational and rotational motion. So far, our discussion has been limited to different drag and lift contributions to force on the particle. In this section, we will discuss the torque on the particle and the resulting rotational motion of the particle.

5.4.1   Pure Rotation and Torque

The problem of particle rotation in an unbounded ambient fluid medium, without any relative translational motion with respect to the surrounding fluid, was considered in Sections 2.4.6 and 4.1.3 in the context of the rotational time scale of the particle and torque on a spinning particle. If the angular velocity of the particle is different from that of the surrounding fluid, then the particle will experience a torque, which will tend to angularly accelerate/decelerate the particle toward the angular velocity of the surrounding fluid. In the absence of translational relative velocity, the particle will not experience any force. This is the state of *pure rotation* and we are interested in the torque on the particle.

Let us consider the steady state, where a particle is spinning about its center at a constant angular velocity of Ω in a quiescent ambient fluid. As discussed in Section 4.1.3, the torque on the particle can be expressed as

$$\mathbf{T}_{qs} = -8\pi\mu_f R_p^3 \, \mathbf{\Omega} \, \Phi_R \, , \tag{5.32}$$

where the subscript qs stands for the quasi-steady component of the torque. The torque correction factor Φ_R depends on the rotational Reynolds number $\mathrm{Re}_\Omega = |\mathbf{\Omega}| d_p^2/\nu_f$. In the Stokes limit ($\mathrm{Re}_\Omega \to 0$), the finite Reynolds number correction factor $\Phi_R \to 1$ and the quasi-steady torque approaches the classic result of Basset (1888).

Since we have defined many different Reynolds numbers in the above discussions, here we take a short detour to list all the different Reynolds numbers. In all these definitions, the Reynolds number measures the inertial effect versus the viscous effect on the scale of the particle due to free-stream velocity, shear, strain, or rotation.

- $\mathrm{Re} = |U_0| d_p/\nu_f$: particle Reynolds number based on relative velocity U_0.

- $\mathrm{Re}_G = |G| d_p^2/\nu_f$: Reynolds number based on local shear G.

- $\mathrm{Re}_S = |\mathbf{S}| d_p^2/\nu_f$: Reynolds number based on the magnitude of local fluid strain-rate tensor.

- $\mathrm{Re}_R = |\mathbf{R}| d_p^2/\nu_f$: Reynolds number based on the magnitude of local fluid rotation-rate tensor.

- $\mathrm{Re}_\Omega = |\mathbf{\Omega}| d_p^2/\nu_f$: Reynolds number based on the angular velocity $\mathbf{\Omega}$ of the particle.

The solution of the unsteady Stokes equation for the translational motion of a particle in a quiescent fluid in the zero Reynolds number limit led to the derivation of the BBO equation in Section 4.2. A similar solution of the unsteady Stokes equations for the arbitrary rotational motion of a particle in the zero Reynolds number limit will lead to the derivation of the rotational BBO equation. This derivation was presented by Feuillebois and Lasek (1978) and Gatignol (1983), and the resulting rotational BBO equation is given below.

Rotational BBO equation

$$
\boxed{
\begin{aligned}
\mathbf{T}(t) &= I_f \frac{D(\omega_c/2)}{Dt} + 8\pi\mu_f R_p^3((\omega_c/2) - \mathbf{\Omega}) \\
&\quad + \frac{8\pi\mu_f R_p^3}{3} \int_{-\infty}^{t} K_\Omega(t-\xi) \left[\frac{d((\omega_c/2)-\mathbf{\Omega})}{dt} \right]_{@\xi} d\xi \\
&\text{where} \quad K_\Omega(\tau) = \frac{R_p}{\sqrt{\pi\nu_f\tau}} + \exp\left[\frac{\nu_f\tau}{R_p^2}\right] \mathrm{erf}\sqrt{\frac{\nu_f\tau}{R_p^2}} \, .
\end{aligned}
}
\tag{5.33}
$$

In the above, the first term on the right is the undisturbed flow torque, where I_f is the moment of inertia of the fluid that was displaced by the particle. This is the torque that would be experienced by the fluid even in the absence of the particle. All other terms in the equation are torque generated due to the perturbation flow induced by the presence of the particle and the relative rotation rate with respect to the ambient. In these terms, ω_c is the vorticity vector of the undisturbed flow at the particle center and $\mathbf{\Omega}$ is the angular velocity vector of the particle. In the limit $((\omega_c/2) - \mathbf{\Omega}) \to 0$, these additional torque contributions will become zero. The second term on the right is the quasi-steady torque given in Eq. (5.32) in the limit of zero rotational Reynolds number. The last term is the unsteady viscous torque, where K_Ω is the rotational history kernel, which again shows the slow $1/\sqrt{\tau}$ decay.

The finite Reynolds number correction depends on the rotational Reynolds number. Lee and Balachandar (2010) considered direct numerical simulations of a particle spinning adjacent to a flat wall. Based on their results, in the limit of the wall being far away from the particle, we can obtain the finite Reynolds number correction for a particle spinning in an unbounded fluid as

$$\Phi_\Omega(\text{Re}_\Omega) = 1 + 3.52 \times 10^{-5}\,\text{Re}_\Omega^2 \quad \text{for} \quad \text{Re}_\Omega \lessgtr 100. \tag{5.34}$$

They observed that the finite Reynolds number correction is quite weak and that the Stokes limit applies even to Re_Ω as large as 10. The above empirical correction has been developed using simulations up to $\text{Re}_\Omega = 100$, and therefore its applicability is limited to only this range of Re_Ω.

5.4.2 Magnus Lift Force

A spinning sphere in an unbounded quiescent fluid will only involve torque on the particle and the direction of the torque vector will oppose the angular velocity of the particle, just like the drag force will be directed opposite to the direction of particle motion. A spinning sphere cannot induce a force on the particle. However, it is a well-known fact that a spinning sphere in the presence of cross flow will experience a lift force. For example, a tennis ball hit with a top spin will experience a downward force (negative lift), while a ball hit with a bottom spin will experience an upward lift force. This spin-induced lift force is known as the *Magnus lift force*.

In the low Reynolds number limit, Rubinow and Keller (1961) obtained the following analytic expression for the spin-induced Magnus lift force:

$$\mathbf{F}_{L,Mg} = \pi R_p^3 \rho_f \mathbf{U}_0 \times \mathbf{\Omega}, \tag{5.35}$$

where $\mathbf{U}_0$ is the ambient flow velocity in the frame attached to the spinning particle, which therefore can be interpreted as the local uniform fluid velocity relative to the particle. Also, we use the subscript Mg to distinguish the spin-induced lift force from those due to ambient shear, vorticity, or strain. The lift force is perpendicular to both the relative velocity vector and the angular velocity vector. If the axis of particle spin aligns with the velocity vector $\mathbf{U}_0$, then there will be no lift force. The Magnus lift force will be the largest when the axis of spin is perpendicular to $\mathbf{U}_0$.

It is now of interest to consider the effect of finite Reynolds number on the Magnus lift force. Following the standard approach pursued in earlier sections, we introduce a finite Reynolds number correction factor Φ_{Mg} and redefine the Magnus lift force as

$$\mathbf{F}_{L,Mg} = \pi R_p^3 \rho_f (\mathbf{U}_0 \times \mathbf{\Omega}) \, \Phi_{Mg}, \qquad (5.36)$$

where Φ_{Mg} can be expected to be a function of both Re and Re_Ω. Interestingly, the experimental measurements by Tanaka et al. (1990) and Tri et al. (1990) on the lift force on a spinning particle in a uniform cross flow yielded nearly constant values of $\phi_{Mg} = 0.25$ and 0.4, respectively, independent of both the Reynolds numbers. The direct numerical simulation results of Bagchi and Balachandar (2002a) for a spinning sphere in a cross flow yielded a near constant correction factor of $\Phi_{Mg} \approx 0.55$ for Re in the range of zero to few hundred.

We can now compare the above spin-induced lift force against the corresponding lift forces due to ambient shear or vortex. The angular velocity of the fluid in the shear flow (5.1) and the vortex flow (5.12) cases are $G/2$, and therefore for fair comparison we set the angular velocities of the particle and the fluid to be the same (i.e., we set $G/2 = |\mathbf{\Omega}|$). First, we perform the comparison in the low Reynolds number limit, where we compare the Herron lift force (5.13) with the above Magnus lift force to obtain

$$\frac{|\mathbf{F}_{L,Herron}|}{|\mathbf{F}_{L,Mg}|} = \frac{32}{\pi} \frac{1}{\text{Re}_G^{1/2}}. \qquad (5.37)$$

Except for the multiplicative constant, the above result is valid for the Saffman shear lift to the Magnus spin lift ratio as well. Thus, in the low Reynolds number limit (i.e., for $\text{Re}_\Omega, \text{Re}_G \ll 1$), the shear and vortex lifts are far more important than the equivalent spin-induced lift. In fact, in his original derivation, Saffman (1965) showed that a particle subjected to a linear shear flow will rotate and the steady-state rotation rate of the particle is the same as the rotation rate of the shear flow, and that the lift due to shear is much stronger than the lift due to particle rotation.

Of course, the above conclusion that spin-induced lift is less important is relevant only in the low Reynolds number limit. The importance of spin-induced lift must be investigated at higher Reynolds numbers. Furthermore, even in the low Reynolds number limit, spin-induced lift is small only when the particle spin is of the same order of magnitude as the rotation rate of the ambient fluid. In situations such as a tennis ball with top or bottom spin, the rotation rate of the particle will far exceed the rotation rate of the fluid and the spin-induced lift force will be a major contributor to particle motion.

5.4.3 Combined Effect of Shear and Spin

Consider the problem of a spherical particle released in a linear shear flow and let the particle move freely in response to the force and torque acting on it. Understanding and modeling of the force and torque acting on the particle are of great interest due to the many practical applications of this flow configuration. The complexity of this problem arises from the coupling between the translational and rotational particle motion. Due to shear, the particle will start to spin, and the spin along with the cross flow will in turn affect the cross-stream motion of the particle. We want to simplify this problem by considering the following two possibilities:

1. Is the spin-induced lift force on the particle important compared to the shear-induced lift? If not, the rotational motion of the particle can be ignored as far as calculating the translational motion of the particle.
2. If it turns out that the spin-induced lift cannot be ignored, does the particle reach rotational equilibrium faster than the translational motion? If so, then the translational motion of the particle can be considered assuming the particle to be in rotational equilibrium, where the rotational equilibrium will correspond to the torque-free state. In this case, the translational motion of the particle can be calculated without simultaneously integrating the rotational motion of the particle, since the rotational state of the particle is given by the torque-free state.

The purpose of exploring these possibilities is simple. Without these simplifications, the combined problem is quite complex and quickly gets out of control. The issue is not with the integration of the angular motion of the particle, which can be done quite easily with the same methodology being used for integrating the translational motion. The difficulty is with the modeling of the force and torque. Without the above simplifications, for example, the lift force on the particle must be known as a function of the three Reynolds numbers Re, Re_G, and Re_Ω that characterize the translational motion, ambient shear, and particle rotation. Similarly, the torque on the particle must be known in terms of these parameters. Such complete knowledge over a wide range of Reynolds numbers is lacking, thus compelling us to pursue simpler options.

As can be seen from the ratio in Eq. (5.37), in the low Reynolds number limit, the spin-induced lift force is small and thus in this limit, the rotational motion of the particle can be ignored when computing the translational particle motion. Unfortunately, this simplification does not hold at finite Reynolds number. At finite Re, the shear-induced lift force rapidly decreases, while the spin-induced lift force correction reaches a constant value. Thus, it appears that particle rotation can influence and modify the shear-induced lift force at finite Re.

However, Bagchi and Balachandar (2002a) found the following superposition to work even at finite Reynolds numbers. The lift force due to shear predicted ignoring the rotational motion can be added to the lift force due to spin predicted without the shear, in order to obtain the overall lift force due to their combined effect. This is a surprising and fortunate result, since such a superposition is theoretically justified only at low Re. Thus, Bagchi and Balachandar proposed the following simple model:

$$C_L(\text{Re}, \text{Re}_G, \text{Re}_\Omega) = C_{L,nr}(\text{Re}, \text{Re}_G) + C_{L,Mg}(\text{Re}, \text{Re}_\Omega), \qquad (5.38)$$

where the shear and spin contributions to lift coefficients are assumed to simply add up. Bluemink et al. (2008) observed the above superposition to be reasonably valid even in the case of vortex flow. Therefore, in the above expression, $C_{L,nr}$ is the lift coefficient computed for a nonrotating particle.

Table 5.2 Translational and rotational time scales normalized by $|\mathbf{U}_0|/d_p$ and their ratios. The time scale is defined as the time taken to reach 63.2% of ambient velocity for a freely moving sphere in a uniform flow, and the time taken for a rotating particle to reach 63.2% of its terminal torque-free asymptotic rotation rate in a linear shear flow. The Reynolds number of the uniform component and the particle-to-fluid density ratio are varied.

ρ	Re	Translational time scale	Rotational time scale	Time-scale ratio
1.05	0.5	0.172	0.02	0.116
5	0.5	0.295	0.055	0.186
1.05	20	1.73	0.78	0.451
5	20	4.17	2.10	0.504
1.01	200	5.58	4.30	0.771
5	200	15.0	13.8	0.922

Approach to Rotational Equilibrium and Torque-Free State

The time scale on which a particle approaches rotational equilibrium compared to the time scale of translational equilibrium plays an important role. In Table 2.2, it was seen that in the low Reynolds number regime, the particle will approach rotational equilibrium 16.6 times faster for particle-to-fluid density ratio 0.1 and 3.3 times faster for large values of particle-to-fluid density ratio. For large time-scale ratios, the particle can reasonably be assumed to be in rotational equilibrium as far as translational motion is concerned. Unfortunately, the faster convergence of the rotational motion to its equilibrium compared to the translational motion is not valid at higher Re. The time-scale ratio at finite Re was considered by Bagchi and Balachandar (2002a) and their results are summarized in Table 5.2.

The problem considered for translational motion is that of a sphere allowed to move freely in a uniform ambient flow. The sphere is initially held fixed, and the governing equations are solved until a fully developed steady flow is attained. The sphere is then released and allowed to move in response to the force. The sphere accelerates, and its velocity approaches that of the uniform ambient flow. Since the ambient flow is uniform, the sphere experiences no net torque and does not undergo any rotation. For the rotational problem, the sphere is initially held fixed in a linear shear flow until a fully developed steady state is established. The particle is then allowed to rotate freely due to viscous stresses on its surface. However, the particle is not allowed to translate. The particle accelerates angularly and reaches a final torque-free state.

As can be seen in Table 5.2, the two time scales are of comparable value at higher Re. Nevertheless, the terminal torque-free rotation rate of a particle is observed to be of the same order as the rotation rate of the ambient shear. The idea is that, since the rotation rate of the shear flow is known, the torque-free rotation rate of the particle can be calculated as a fraction. From the torque-free rotation rate, the spin-induced lift force can be calculated and added to the shear-induced lift force.

Here we present the results from Bagchi and Balachandar (2002a) for a particle held stationary in a linear shear flow. The translational slip velocity $\mathbf{U}_0$ is held constant but the particle is allowed to rotate freely under the action of torque. Starting from rest, the particle angular velocity accelerates in response to the torque. A steady state is obtained when the torque on the sphere reaches zero. The corresponding torque-free rotation rate Ω_{tf}, normalized by the rotation rate of the ambient shear flow, was investigated as a function of the shear Reynolds number. The results were compared against the low Reynolds number analytical expression of Lin et al. (1970b) and the experimental results of Poe and Acrivos (1975). For small Reynolds number, Lin et al. (1970b) obtained the following behavior:

$$\frac{\Omega_{tf}}{G/2} = 1 - 0.0385\,\mathrm{Re}_G^{3/2}, \tag{5.39}$$

where we have used the fact that for a shear rate of G, the angular velocity of the fluid is $G/2$. As $\mathrm{Re}_G \to 0$, $\Omega_{tf} \to G/2$. With the inclusion of inertia, the torque-free rotation rate decreases. In their experiments, Poe and Acrivos observed the torque-free rotation rate to decrease at a slower rate. The finite-Re simulations of Bagchi and Balachandar (2002a) showed that for $\mathrm{Re}_G \leq 5$, Ω_{tf} decreased faster, however, at higher Re_G, the decay of torque-free rotation rate was slower.

Figure 5.3 shows the results of Bagchi and Balachandar (2002a), where $1 - 2\Omega_{tf}/G$ is plotted as a function of Re for varying values of shear magnitude s. The terminal rotation rate of the particle decreases rapidly with Re. For example, at Re = 200, Bagchi and Balachandar (2002c) observed the torque-free rotation rate to be only 17% of the ambient rotation rate. The most important observation is the collapse of the results for different shear magnitudes. Therefore, at finite Re, $2\Omega_{tf}/G$ depends only on Re and a curve fit of the data yields

$$\frac{\Omega_{tf}}{G/2} = \begin{cases} 1 - 0.0364\,\mathrm{Re}^{0.95}, & \text{for} \quad 0.5 < \mathrm{Re} \leq 5, \\[2mm] 1 - 0.0755\,\mathrm{Re}^{0.455}, & \text{for} \quad 0.5 \leq \mathrm{Re} \leq 200. \end{cases} \tag{5.40}$$

The combined effect of a vortex and a spinning particle was similarly considered by Bluemink et al. (2008), who gave corresponding expressions for torque-free rotation rate and a superposition of vortex-induced and spin-induced lift forces. Interestingly, in the vortex flow they observe the torque-free rotation rate of the particle to be larger than that of an undisturbed vortex flow's rotation rate. For further discussion on this topic, the reader must refer to these papers.

5.5 Unsteady Lift

The discussion of lift force has so far been focused on steady conditions. As a result, time variation has not been a factor. Since lift arises from the nonlinear interaction of a uniform flow with either the ambient shear or particle rotation, if these quantities were to vary over time, then the effect of unsteadiness must be considered. The time scales of variation of ambient shear or particle rotation must be compared with the advection

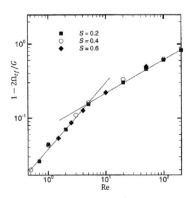

Figure 5.3 Plot of $1 - 2\Omega_{tf}/G$ vs. Re for varying values of shear magnitude s. Taken from Bagchi and Balachandar (2002a), and reproduced with the permission of AIP Publishing.

and diffusion time scales of perturbation vorticity. If the time scale of ambient shear or particle rotation is much longer, then the perturbation flow is able to adjust to the instantaneous flow condition and the results of the previous sections can be applied in a quasi-steady manner. When the time-variation ambient flow or the particle motion is rapid, unsteady effects on lift force must be considered along with unsteady effects on drag. Unsteady lift force will be the focus of this section.

It must be admitted that our understanding of unsteady lift is limited. Although unsteady lift can be significant under some practical scenarios, there have only been a few analyses. These studies have only considered the effect of time variation of relative velocity, or equivalently the time variation of Reynolds number. The motivation for these studies has been to consider situations where a particle accelerates, thus changing its relative velocity while immersed in a region of uniform shear. We will briefly consider these results in this section. However, there are also scenarios where the ambient shear or particle rotation can change with time sufficiently rapidly that their time variation can be the source of unsteadiness. Investigation of these additional unsteadiness mechanisms has not been carried out.

One of the earliest investigations of the lift force on a particle in a time-dependent situation was by Miyazaki et al. (1995). This early work was followed by Asmolov and McLaughlin (1999), who considered the relative velocity between the particle and the ambient fluid to be sinusoidally varying. Thus, their solution can be considered as the lift force in the frequency domain and furthermore it can be written in the same form as the Saffman lift force given in Eq. (5.7). In Section 5.1, we considered a steady uniform flow of velocity U_0 with a steady uniform shear of G as the undisturbed ambient flow. Asmolov and McLaughlin (1999) considered the ambient flow to be sinusoidally varying as $U(t) = U_0 + U_\omega \exp(-\iota\omega t)$, oriented along the x-direction. They obtained the corresponding oscillatory lift force on the particle to be

$$\mathbf{F}_L = -\frac{9}{\pi}\mu_f R_p^2 U_\omega \, \mathrm{sgn}(G) \left(|G|/\nu_f\right)^{1/2} J_\omega \, \exp(-\iota\omega t)\, \mathbf{e}_y, \qquad (5.41)$$

where the function J_ω in the low and high-frequency limits is

$$J_\omega = \begin{cases} 2.254 + 3.894\iota\dfrac{\omega}{G} & \text{for } \dfrac{\omega}{G} \ll 1, \\[3mm] \dfrac{7\pi^2(1+\iota)}{60\sqrt{2\omega/G}} & \text{for } \dfrac{\omega}{G} \gg 1. \end{cases} \tag{5.42}$$

It can be verified that in the limit when $\omega \to 0$, the above lift force expression reduces correctly to the Saffman lift force and thus one could multiply the above with the function J_ϵ given in Eq. (5.9). By making some reasonable assumptions, Candelier and Souhar (2007) were able to transform the above into a time domain to obtain the following expression for the time-dependent lift force:

$$\mathbf{F}_L = \frac{21\sqrt{\pi}}{20}\mu_f R_p^2 \operatorname{sgn}(G)\frac{|G|}{\nu_f^{1/2}}J(\epsilon)\,\mathbf{e}_y \int_{-\infty}^{t} K_L(t-\xi)\,U(\xi)\,d\xi. \tag{5.43}$$

The above unsteady lift force must be interpreted as the sum of the quasi-steady and the viscous unsteady contributions. Thus, if we remove the quasi-steady lift contribution, the difference $\mathbf{F}_L - \mathbf{F}_{L,qs}$ corresponds to the lift version of the viscous unsteady force given in the BBO equation.

The above expressions have been derived with the low-Re number assumption and thus the corresponding low-Re lift force history kernel is given by

$$K_L(\tau) = \frac{\exp(-0.136(G\tau)^{3/2})}{\sqrt{\tau}}. \tag{5.44}$$

When the relative velocity of the undisturbed ambient flow $U(t)$ varies very slowly, it can be taken out and carrying out the integral one can recover the Saffman drag force with McLaughlin's correction. Unfortunately, extensions to finite Re have been very difficult. From K_L we can obtain the time scale of decay of the lift history kernel to be $3.78/G$. If the variation of $U(t)$ is on a slower time scale, then the lift can simply be calculated as $\mathbf{F}_{L,qs}$ using the quasi-steady approximation. When the variation of $U(t)$ is on a faster time scale, the above unsteady force expression must be used in place of $\mathbf{F}_{L,qs}$ in the finite-Re BBO equation.

While the above low-Re lift kernel is monotonic, at finite Reynolds number this is not guaranteed. The time history of the lift force at finite Re has been observed to be quite complex (Legendre and Magnaudet, 1998; Wakaba and Balachandar, 2005). This complexity is due to the fact that the mechanisms contributing to the lift force are different and more complex at finite Re.

6 Heat and Mass Transfer from an Isolated Sphere

In this chapter, we investigate the problem of heat transfer from an isolated rigid sphere subjected to a cross flow of different temperature. This thermal problem is analogous to the flow problem considered earlier, and the interest here is to establish an expression for heat transfer in terms of the undisturbed ambient flow, which must now be characterized both in terms of relative velocity and temperature difference. We will start our investigation with rigorous analytical results in the Stokes and the small Péclet number regime. First, we will consider the problem of steady heat transfer and obtain the appropriate Nusselt number correlation, which is analogous to the steady drag correlation. Then, we will consider the problem of unsteady heat transfer in the zero Péclet number limit and obtain the unsteady heat transfer equation, which is the thermal equivalent of the BBO equation. Then we will consider the problem of heat transfer from a finite-sized particle (i.e., the undisturbed flow and thermal fields vary on the scale of the particle) and obtain the thermal equivalent of the MRG equation. Finally, we will discuss the nonlinear finite Reynolds and Péclet number regime and obtain an empirical extension to the thermal MRG equation. The above outlined progression precisely matches that presented in the context of force modeling. However, the present discussion of heat transfer will be brief. For more information, the reader is referred to a recent review (Balachandar and Michaelides, 2022). Due to heat and mass transfer analogy, the results to be presented for heat transfer apply directly for mass transfer as well.

6.1 Steady Heat Transfer in the Small Péclet Number Limit

Following the discussion of Section 4.1, let us consider a particle of radius R_p immersed in a steady uniform ambient flow of velocity $\mathbf{u}_c$ and temperature T_c. Let the particle be stationary and at a constant uniform temperature of T_p. While the drag on the particle was considered earlier, here we are interested in evaluating the net heat transfer rate to the particle from the surrounding. In order to do so, in addition to the linearized continuity and momentum equations given as Eqs. (4.2), the steady thermal equation

$$\mathbf{u} \cdot \nabla T = \kappa_f \nabla^2 T \tag{6.1}$$

must also be solved, where κ_f is the thermal diffusivity of the fluid. The velocity boundary conditions on the sphere are no-slip and no-penetration. For the above temperature equation, the appropriate boundary conditions are

$$\text{On the sphere}: T = T_p \quad \text{and} \quad \text{Far field}: T \to T_c. \quad (6.2)$$

Here we consider the buoyancy effect due to temperature variation within the fluid to be negligible and as a result the linearized mass and momentum equations given as Eqs. (4.2) remain unaffected by the fluid temperature field. That is, the flow is decoupled from the temperature and the velocity and pressure fields are given by the Stokes flow solution obtained in Section 4.2.1. In the temperature equation, if the advection term on the left-hand side is ignored, then the problem reduces to a pure heat condition from a sphere of radius R_p at temperature T_p in an infinite ambient of far-field temperature T_c. Without the advective effect of the flow, the temperature field is spherically symmetric and the solution to the spherical conduction problem outside the spherical particle is

$$T(r) = T_c + (T_p - T_c)\frac{R_p}{r}. \quad (6.3)$$

In general, the heat transfer rate to the spherical particle from the surrounding fluid can be evaluated as

$$\dot{Q} = -k_f \oint_{S_p} (\nabla T \cdot \mathbf{n}) \, dA = -\pi R_p^2 \, k_f \, \overline{\nabla T \cdot \mathbf{n}}^S, \quad (6.4)$$

where the integral is over the surface of the spherical particle whose outward surface normal is $\mathbf{n}$, k_f is the thermal conductivity of the fluid, and $\overline{\nabla T \cdot \mathbf{n}}^S$ the surface-averaged normal thermal gradient in the fluid. The heat transfer rate is normally expressed in terms of the nondimensional Nusselt number, defined as

$$\text{Nu} = \frac{2|\dot{Q}|}{\pi R_p \, k_f \, |T_p - T_c|}. \quad (6.5)$$

Substituting the temperature variation of Eq. (6.3), we obtain the result that Nu = 2 for steady conduction.

To include the effect of Stokes flow on heat transfer, the advection term on the left-hand side of Eq. (6.1) must be retained. The analytical solution can be obtained as an asymptotic expansion using singular perturbation. The singular nature of the problem arises from the fact that even at very small particle Reynolds number, there exists a distance away from the sphere beyond which the fluid inertia will become important. This singular perturbation problem was solved by Acrivos and Taylor (1962), who obtained

$$\text{Nu} = 2 + \frac{1}{2}\text{Pe} + \frac{1}{4}\text{Pe}^2 \ln(\text{Pe}) + \cdots, \quad (6.6)$$

where the Péclet number Pe = Re Pr. The above expression has been shown to be accurate and useful for nonzero values of Pe in the range 0 to 0.6 (Michaelides, 2003). For larger value of Péclet number, one must resort to empirical measurements of Nu, which will be considered below.

6.2   BBO-Like Equation of Heat Transfer

In this section, we will consider modeling the heat exchange of a particle with the surrounding fluid under *unsteady* conditions. Here the term "unsteady" refers to the fact that as the heat is being exchanged between the particle and the surrounding fluid, the condition under which the exchange is taking place is changing as well on a fast enough time scale. If the unsteady effects are unimportant, then at each instance the heat transfer rate can be taken to be the same as what would exist had the problem been held steady at the instantaneous relative velocity and temperature difference – this heat transfer is termed the *quasi-steady contribution*. If the actual heat transfer rate deviates from the quasi-steady contribution, then the difference is the intrinsic effect of unsteadiness and can be termed the *unsteady contribution*.

At the end of this subsection we will discuss conditions under which the unsteady effects on heat transfer are important and must be accounted for in modeling the heat transfer rate. In this subsection, we will present the result of unsteady heat transfer analysis in the zero Péclet number limit, which yields a heat transfer relation analogous to the BBO equation for force. The derivation follows along the lines pursued earlier in the context of force on the particle and therefore only the results will be presented. For a rigorous derivation and additional interpretation, the reader should consult Michaelides and Feng (1994), Michaelides (2003), and Coimbra et al. (2004).

First, we introduce time dependence into the problem by considering a particle whose temperature varies as $T_p(t)$. However, the entire particle is assumed to be at the same temperature (i.e., the particle is considered to be highly conductive corresponding to a very small Biot number). The far-field temperature of the particle is taken to be spatially uniform, but temporally varying as $T_c(t)$. The analysis has been carried out only in the linear limit, so that the advection effects are neglected. The resulting unsteady heat transfer rate to the particle from the surrounding fluid can be expressed as follows (Michaelides and Feng, 1994; Michaelides, 2003):

BBO-like equation for unsteady heat transfer

$$\boxed{\begin{aligned} \dot{Q} = {} & m_f C_{p,f} \frac{d[T_c]_@}{dt} + 4\pi k_f R_p ([T_c]_@ - T_p) \\ & + 4\pi k_f R_p \int_{-\infty}^{t} \frac{R_p}{\sqrt{\pi \kappa_f (t - \xi)}} \left[\frac{dT_c}{dt} - \frac{dT_p}{dt} \right]_{@\xi} d\xi, \end{aligned}}$$

(6.7)

where m_f is the mass of the fluid displaced by the particle and $C_{p,f}$ the specific heat of the fluid. On the right-hand side, each term can be given a precise explanation. The first term is the undisturbed heat transfer and it is analogous to the undisturbed flow force. This heat exchange will happen even in the absence of the particle in the form of heat transfer between the spherical volume of fluid which occupies the particle position and its surrounding. The second term is the quasi-steady heat transfer. In the present zero Péclet number limit, the quasi-steady heat transfer corresponds to a Nusselt number of 2. The quasi-steady heat transfer is proportional to the temperature difference between the far-field fluid (as seen by the particle) and the particle temperature. The last term is analogous to the viscous-unsteady force and it accounts for the unsteady diffusional effects of heat transfer. The integral in this term accounts for the past history of thermal difference between the particle and ambient, and it is weighted by the thermal history kernel that decays as $1/\sqrt{t-\xi}$.

In many respects, the above equation closely mirrors the BBO equation of motion of a particle. The one significant difference is the absence of the added-mass term. While there was added-mass force, there is no such analogous added-mass heat transfer. This is because the vector momentum needs to satisfy both no-slip and no-penetration conditions and thus the unsteady perturbation flow due to the particle has an inviscid component arising from the no-penetration condition, which gives rise to the added-mass force, and a viscous component arising from the no-slip condition, which contributes to the viscous-unsteady force. In case of the temperature equation, the isothermal condition is analogous to the no-slip boundary condition and gives rise to only unsteady diffusional heat transfer.

Note that in the above heat transfer expression, the temperature T_c and its time variation dT_c/dt are to be interpreted as those of the undisturbed fluid that would exist in the absence of the particle (this is precisely the interpretation that we gave to the fluid velocity and acceleration in the BBO equation of motion). Also, when the above equation is used in the context of complex particle-laden flows, where the continuous-phase temperature T_c varies within the computational volume, the particle size is taken to be much smaller than the flow scales and therefore the ambient fluid temperature seen by a particle is essentially spatially uniform. In this very small particle limit, the undisturbed fluid temperature T_c and its time variation dT_c/dt are simply taken to be their values evaluated at the location of the particle center.

6.2.1 Conditions of Unsteadiness

In this subsection, we address the following question: under what conditions can the heat transfer between a particle and the surrounding be treated as quasi-steady? If these conditions are not satisfied, then the unsteady contribution is significant and must be included in the heat transfer analysis. To answer this question we define the following time scales, and a comparison of these time scales can be used to evaluate the importance of unsteadiness. There are three primary ways by which the conditions of heat transfer change: (i) changes in particle temperature; (ii) changes in particle velocity; and (iii) changes in ambient flow temperature or velocity seen by the particle.

The thermal time scale τ_T of the particle was defined earlier in Section 2.4.2. Provided the ambient fluid temperature and the relative velocity between the particle and the surrounding remain the same, τ_T is the time scale on which the conditions of heat transfer change. Instead, if the ambient fluid velocity and the temperature difference between the particle and the surrounding are held independent of time, then unsteadiness can only be due to the acceleration (or deceleration) of the particle

Table 6.1 The rows represent conditions where τ_T, τ_V, and τ_f are smaller than the other two, while the three columns represent the largest of convection, thermal-diffusion, and momentum-diffusion time scales. The corresponding conditions for considering heat transfer to be quasi-steady are given in the table entries.

	(Pe, Re) < 1	(Pe, Pr) > 1	(Re, 1/Pr) > 1
$\tau_T < (\tau_V, \tau_f)$	$\dfrac{\mathrm{Pe}}{6\gamma\,\mathrm{Nu}} \gg 1$	$\dfrac{1}{6\gamma\,\mathrm{Nu}} \gg 1$	$\dfrac{\mathrm{Pr}}{6\gamma\,\mathrm{Nu}} \gg 1$
$\tau_V < (\tau_T, \tau_f)$	$\dfrac{\rho\,\mathrm{Re}}{18\,\Phi} \gg 1$	$\dfrac{\rho}{18\,\mathrm{Pr}\,\Phi} \gg 1$	$\dfrac{\rho}{18\,\Phi} \gg 1$
$\tau_f < (\tau_T, \tau_V)$	$\dfrac{\tau_f\,U}{d_p} \gg 1$	$\dfrac{\tau_f\,\kappa_f}{d_p^2} \gg 1$	$\dfrac{\tau_f\,\nu_f}{d_p^2} \gg 1$

towards the surrounding fluid velocity. This momentum time scale τ_p of the particle was also defined in Section 2.4.1.

As the third source of unsteadiness, we consider either the velocity or the temperature of the surrounding fluid seen by the particle changing on the time scale τ_f. While τ_T and τ_V are determined by the thermodynamic and transport properties of the fluid and the particle, τ_f is dependent on the nature of the flow in which the particle is immersed and therefore must be carefully evaluated in any given problem. We now define the time scale of unsteadiness to be the smallest of the three:

$$\tau_{un} = \min\{\tau_T, \tau_V, \tau_f\}, \tag{6.8}$$

since the smallest time scale will dictate the unsteady nature of heat transfer.

This time scale of unsteadiness must be compared with both the advection time scale d_p/U and the diffusion time scales d_p^2/κ_f and d_p^2/ν_f of the flow, where U is the relative velocity between the particle and the flow. For Péclet and Reynolds numbers $\mathrm{Pe} = d_p U/\kappa_f, \mathrm{Re} = d_p U/\nu_f > 1$, the diffusion time scales are larger than the advection time scale, while for $\mathrm{Pe}, \mathrm{Re} < 1$, the advection time scale is larger than the other two. In order for heat transfer to be considered quasi-steady (i.e., to ignore unsteady effects), τ_{un} must be larger than the advection and diffusion time scales. That is, the condition for validity of quasi-steady heat transfer is

$$\tau_{un} \gg \max\left\{\frac{d_p}{U}, \frac{d_p^2}{\kappa_f}, \frac{d_p^2}{\nu_f}\right\}. \tag{6.9}$$

Only then can the inertial and diffusional adjustment of the velocity and temperature fields of the fluid be considered faster than the change in the conditions of heat transfer. The above compactly expressed condition can be more explicitly expressed as the nine different conditions given in Table 6.1.

In this work, we will primarily consider unsteadiness in the simpler context of a particle subjected to a steady ambient stream of uniform velocity at a constant temperature different from that of the particle. The relative velocity between the particle and the ambient will be held fixed and the temperature of the particle is allowed to evolve toward that of the ambient stream. Thus, we consider the case $\tau_T \ll (\tau_V, \tau_f)$. In this limit, for a quasi-steady approximation to be valid we require the volumetric heat capacity of the fluid to be significantly smaller than that of the particle (i.e., the ratio γ must be substantially smaller than the smallest of $1/(6\,\mathrm{Nu})$, $\mathrm{Pe}/(6\,\mathrm{Nu})$, and $\mathrm{Pr}/(6\,\mathrm{Nu})$).

6.3   MRG-Like Equation of Heat Transfer

Michaelides and Feng (1994) extended the above BBO-like formulation to consider situations where there is non-negligible spatial variation in the undisturbed thermal field on the scale of the particle. That is, they considered T_c to vary over a particle diameter, but they limited their attention to problems where the length scale of such variation is much longer than the particle diameter. This condition is similar to that considered by Maxey and Riley (1983) in their analysis of force on a particle. In which case, the temperature of the undisturbed flow at the particle cannot be simply taken to be the value at the particle center. The use of Faxén's law allows systematic replacement by volume and surface averages. A process similar to that used by Maxey and Riley (1983) was implement in the context of the temperature equation to obtain the following modified expression for particle heat transfer (Michaelides and Feng, 1994):

MRG-like equation for unsteady heat transfer

$$
\boxed{
\begin{aligned}
\dot{Q} = {} & m_f C_{p,f} \overline{\frac{dT_c}{dt}}^v + 4\pi k_f R_p (\overline{T_c}^s - T_p) \\
& + 4\pi k_f R_p \int_{-\infty}^{t} \frac{R_p}{\sqrt{\pi \kappa_f (t-\xi)}} \left[\frac{d\overline{T_c}^s}{dt} - \frac{dT_p}{dt} \Big|_\xi \right] d\xi,
\end{aligned}
}
\tag{6.10}
$$

where $\overline{(\cdot)}^s$ and $\overline{(\cdot)}^v$ are averages over the surface and volume of the particle, respectively. The three terms on the right have the same interpretation as in Eq. (6.7), but they are now evaluated using the surface and volume averages of the undisturbed fluid temperature-related quantities, and thereby accommodate a spatially varying undisturbed thermal field.

Problem 6.1 This problem is the same as Problem 4.16 but for the temperature field. Let us expand the undisturbed ambient flow temperature about the location of particle center as

$$
T_c(\mathbf{x}, t) = T_0 + \mathbf{x} \cdot [\nabla T]_0 + \mathbf{x}\mathbf{x} : [\nabla \nabla T]_0 + \cdots,
\tag{6.11}
$$

where T_0 is the undisturbed temperature at $\mathbf{x} = 0$, $[\nabla T]_0$ is the temperature gradient at the origin, and so on. Note that in the above expansion, all the spatial variations are contained in the $\mathbf{x}$ terms, since the coefficients of the expansion, T_0, $[\nabla T]_0$, etc., are evaluated at the origin and can only depend on time. Derive the following leading-order expressions:

$$\overline{T_c}^{s} \approx T_0 + \frac{d_p^2}{24}\left[\nabla^2 T\right]_0 \quad \text{and} \quad \overline{T_c}^{v} \approx T_0 + \frac{d_p^2}{40}\left[\nabla^2 T\right]_0 . \tag{6.12}$$

Thus, the deviations in the surface and volume-averaged temperatures from the undisturbed temperature at the center of the particle depend on the Laplacian of temperature evaluated at the particle center. This approximation for surface and volume averages can be substituted into Eq. (6.10) and thereby eliminate the appearance of surface and volume averages. This is the form in which Michaelides and Feng (1994) presented their expression. Since the analysis is restricted to small variation on the scale of the particle, the above approximation is quite good.

6.4   Heat Transfer at Finite Pe

We finish this chapter by empirically extending the MRG-like expression for heat transfer to finite Reynolds number. This extension will be similar to that considered in the context of force. In the context of heat transfer, Péclet number plays the role of Reynolds number. Therefore, the resulting finite-Pe MRG-like equation for heat transfer can be expressed as follows:

Finite-pe MRG-like equation for unsteady heat transfer

$$
\dot{Q} = m_f C_{p,f} \overline{\frac{DT_c}{Dt}}^{v} + 2\pi k_f R_p (\overline{T_c}^{s} - T_p)\, \mathrm{Nu(Re,Pr)}
$$
$$
+ 4\pi k_f R_p \int_{-\infty}^{t} K_T(t-\xi; \mathrm{Re,Pr}) \left[\frac{d\overline{T_c}^{s}}{dt} - \frac{dT_p}{dt}\bigg|_{\xi} \right] d\xi . \tag{6.13}
$$

The differences between this and Eq. (6.10) will now be discussed term by term. In the unsteady heat transfer term, the volume average of the derivative of undisturbed ambient fluid temperature following the particle has been replaced by the volume average of the total temperature derivative following the fluid (i.e., dT_c/dt has been replaced with DT_c/Dt). The difference between the two is $\overline{(\mathbf{u}_c - \mathbf{V}) \cdot \nabla T_c}^{v}$ and it is analogous to the convective acceleration that contributed to undisturbed flow force in the nonlinear regime. This difference is zero in the zero Péclet number limit. This form is exact and it precisely corresponds to the heat exchange between the volume of fluid occupied by the particle and the surrounding in the absence of the particle. In the quasi-steady heat transfer term, in the zero Péclet number limit, the Nusselt number is 2. At finite Péclet number, the Nusselt number is a function of both the Reynolds and the Prandtl numbers. An expression for Nu that is valid in the limit Pe < 1 was presented in Eq. (6.6). While that expression was rigorously derived, we must resort to empirical correlations at larger values of the Péclet number. Two widely used empirical Nusselt number expressions were introduced in Section 2.4.2, which are repeated here for completeness:

$$
\text{Nu(Re, Pr)} = \begin{cases} 2 + (0.4\,\text{Re}^{1/2} + 0.06\,\text{Re}^{2/3})\,\text{Pr}^{0.4} & \text{Whitaker (1972),} \\[2mm] 2 + 0.6\,\text{Re}^{1/2}\,\text{Pr}^{1/3} & \text{Ranz and Marshall (1952)} . \end{cases}
$$

(6.14)

From these expressions it is clear that the finite-Pe Nusselt number can be much larger than the zero-Pe value of 2. Clearly, flow-induced convective transfer of heat is a far more effective way to transfer heat between the particle and the surrounding fluid.

In the third term on the right-hand side, the zero-Péclet number thermal kernel has been replaced by the finite-Pe kernel, which depends on both Re and Pr. The one-over-square-root decay of the thermal kernel is appropriate at short times, even in case of finite Péclet number. But at finite Péclet number, beyond a certain critical time (which will depend on the Péclet number), the decay will be faster. This faster decay is often important for the history integral to converge and yield a finite viscous unsteady contribution to heat transfer. For example, if the argument within square brackets is a constant in Eq. (6.13) (i.e., if the time rate of change of temperature difference between the particle and the surrounding is a constant), then for the integral to converge, the decay must be slower than $1/(t - \xi)$ for short time and faster than $1/(t - \xi)$ for longer time. Thus, the short-time singularity of one-over-square-root decay is integrable. To avoid long-time singularity, a sufficiently faster long-time decay of the kernel is needed.

6.4.1   Thermal Kernel for Small Pe

Feng and Michaelides (2000) extended the unsteady analysis to small but finite Péclet number and obtained results that complement the low Péclet number steady-state results of Acrivos and Taylor (1962). The small-Pe form of the thermal history kernel was obtained by considering the exact solution of the temperature field around a spherical particle subjected to Stokes flow. The particle and the surrounding fluid are initially at the same temperature, but at $t = 0$ the temperature of the particle is suddenly decreased and thereafter the temperature difference is maintained at $T_c - T_p = \Delta T$. From the exact solution of the governing equations, the nondimensional heat transfer to the particle is obtained as

$$
\frac{m_p\,C_p}{\pi d_p k_f \Delta T}\,\frac{dT_p}{dt} = \underbrace{2 + \frac{\text{Pe}}{2} + \frac{1}{4}\,\text{Pe}^2\,\ln(\text{Pe})}_{= \text{Nu}} + \underbrace{\frac{2\sqrt{\text{Pe}}}{\sqrt{\pi \tilde{t}}}\,\exp\left(-\frac{\text{Pe}\,\tilde{t}}{16}\right) - \frac{\text{Pe}}{2}\,\text{erfc}\left(\frac{\sqrt{\text{Pe}\,\tilde{t}}}{4}\right)}_{\text{unsteady contribution}},
$$

(6.15)

where erfc is the complementary error function. In the above, $\tilde{t} = tU/d_p$ is the nondimensional time, where $U = |\mathbf{u}_c - \mathbf{V}|$ is the constant relative velocity magnitude. We compare the above to the three heat transfer contributions given on the right-hand side of the unsteady model in Eq. (6.13). Since the ambient flow is at constant temperature, the undisturbed flow heat transfer is zero. The other two contributions are nonzero. When nondimensionalized by $(\pi d_p k_f \Delta T)$, the quasi-steady contribution to heat transfer is simply the Nusselt number Nu, which in the present problem, for $t > 0$, is time independent. Immediately after the change in particle temperature, unsteady effects will contribute and in nondimensional terms this contribution can be obtained by performing the convolution integral in Eq. (6.13). In evaluating the integral, we can set $\left[d(T_c - T_p)/dt \right]_{\xi} = \Delta T \delta(\xi)$, where δ is the delta function. We then obtain the third term on the right-hand side of Eq. (6.13) to be

$$\sqrt{\frac{\text{Pe}}{\pi}} \tilde{K}_T \,, \tag{6.16}$$

where $\tilde{K}_T = d_p K_T / U$ is the nondimensional kernel. Comparing the above with Eq. (6.15), we obtain the thermal history kernel to be (Balachandar and Ha, 2021)

Small Pe limit : $\qquad \tilde{K}_T(\tilde{t}) = \dfrac{2}{\sqrt{\tilde{t}}} \exp\left(-\dfrac{\text{Pe}\,\tilde{t}}{16} \right) - \dfrac{\sqrt{\pi\,\text{Pe}}}{2} \operatorname{erfc}\left(\dfrac{\sqrt{\text{Pe}\,\tilde{t}}}{4} \right).$ $\qquad$ (6.17)

For a wide range of Pe, it can be verified that the first term on the right is much larger than the second term. Henceforth, we will ignore the second term and simply consider the small-Pe thermal history kernel to be given by the first term.

6.4.2   Thermal Kernel for Large Pe

The behavior of the kernel for Pe > 1 can be expected to be substantially different, as indicated by the difference between the asymptotic quasi-steady Nusselt number relation (6.6) that is valid for small Péclet number and the empirical correlations given in Eq. (6.14). However, extension of the low Péclet number analytical results to finite Péclet numbers is not straightforward. In the case of an unsteady viscous kernel, it was seen earlier in Eq. (4.142) that K_{vu} decayed as $1/\sqrt{t}$ at small time and as $1/t^2$ at longer time. Though a long time, faster decay is expected in case of the thermal kernel as well, differences between the two are likely to exist, since the analogy between momentum and thermal diffusion is imperfect.

Balachandar and Ha (2021) obtained the finite-Pe thermal history kernel by considering the problem of heat transfer from a spherical particle subjected to a steady uniform flow of finite Re. The particle and the surrounding fluid are initially at the same temperature, but at $t = 0$ the temperature of the particle suddenly decreased. However, the temperature of the particle is allowed to evolve freely toward the constant ambient fluid temperature. No analytic solution is possible at finite Reynolds number, therefore they used the particle-resolved simulation results of Balachandar and Ha (2001).

For a wide range of Reynolds number Re and heat capacity ratio γ, Balachandar and Ha (2001) observed the heat transfer coefficient to start at a large value and decrease rapidly to reach a constant value. Correspondingly, the particle temperature initially increases more rapidly, but quickly settles to an exponential approach to the surrounding constant fluid temperature. Motivated by the form of the thermal history kernel in the small-Pe limit, they proposed the following simple form for the finite-Pe thermal history kernel

$$\textbf{Finite Pe}: \quad \tilde{K}_T(\tilde{t}) = \frac{2}{\sqrt{\tilde{t}}} \exp(-b\,\tilde{t}).$$ (6.18)

By comparing the above with Eq. (6.17), it is clear that $b = \text{Pe}/16$ for small Pe. For each combination of Pe and γ, the corresponding values of b were obtained from the particle-resolved simulations of Balachandar and Ha (2001). These results are plotted as symbols in Figure 6.1 for Pe ranging from 7 to 350 and for $\gamma = 0.004$, 0.02, and 0.1 plotted as different lines. Although the value of b evaluated from the simulations shows some variation with γ, the primary dependence seems to be on Pe. Based on the simulation results and on the analytical small-Pe behavior given in Eq. (6.17), Balachandar and Ha (2021) proposed the following curve fit for b, which when applied in Eq. (6.18) yields a thermal history kernel that remains appropriate over a wide range of Péclet numbers:

$$b = (1.63 - 0.92\,\text{erf}(0.017(\text{Pe} - 80))) \left[1 - 0.4\exp\left(-\frac{\text{Pe}}{16}\right) - 0.6\exp\left(-\frac{\text{Pe}^2}{30}\right) \right].$$ (6.19)

For large Péclet number, the term within square brackets becomes 1 and the value of b is simply given by the expression within the first parentheses. On the other hand, for Pe $\ll 1$ the Taylor series expansion of the above correctly approaches the limit $b \to 1/16$.

Problem 6.2 As a simple example, let us consider the problem of oscillatory particle temperature, where the ambient fluid velocity $\mathbf{u}_c$ and temperature T_c are held steady and spatially uniform. The particle is stationary and its temperature is lower than that of the ambient fluid. The temperature of the particle is varied as $T_p(t) = T_f - \Delta T + \Delta T \alpha e^{-\iota \tilde{\omega} \tilde{t}}$, where α is smaller than unity and thus the particle temperature remains smaller than the ambient fluid. This problem was studied by Balachandar and Ha (2001) with particle-resolved simulations for varying Péclet number and for varying frequency of oscillation (also see Balachandar and Ha, 2021).

In this problem, we will predict the time variation of heat transfer using the unsteady heat transfer model given in Eq. (6.13). Note that undisturbed fluid velocity and temperature are spatially uniform, so surface and volume averages are the same as the values at the center of the particle.

(a) Substitute the ambient fluid and particle temperatures into the right-hand side and nondimensionalize by $\pi d_p k_f \Delta T$ to obtain the undisturbed flow heat transfer to be zero.

(b) Obtain the quasi-steady heat transfer as $\text{Nu}(1 - \alpha e^{-\iota \tilde{\omega} \tilde{t}})$.

(c) Obtain the unsteady history heat transfer to be given by

$$\hat{q}_{hi} e^{-\iota \tilde{\omega} \tilde{t}} = \left[\frac{2\alpha\sqrt{\text{Pe}}}{\sqrt{b^2 + \tilde{\omega}^2}} \iota\tilde{\omega}\sqrt{b + \iota\tilde{\omega}} \right] e^{-\iota \tilde{\omega} \tilde{t}}.$$ (6.20)

In performing the convolution integral, take the thermal history kernel to be given by Eq. (6.18).

(d) Obtain the real and imaginary parts of the Fourier coefficients (i.e., the right-hand side within square brackets) and plot them as a function of $\tilde{\omega}$ for a few different Péclet numbers (as shown in Figure 6.2), which can be compared with the particle-resolved simulation results.

(e) Obtain the small and large oscillation frequency solutions as

$$\hat{q}_{hi} = 2\alpha\sqrt{\text{Pe}}\left[-\frac{\tilde{\omega}^2}{2b^{3/2}} + \iota\frac{\tilde{\omega}}{b^{1/2}}\right] \tag{6.21}$$

and

$$\hat{q}_{hi} = \alpha\sqrt{2\,\text{Pe}\,\tilde{\omega}}(-1 + \iota). \tag{6.22}$$

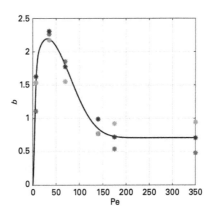

Figure 6.1 The value of the exponent b for varying combinations of Pe and $\gamma = 0.004, 0.02,$ and 0.1 are plotted in red, blue, and green symbols. The black line shows the curve fit presented in Eq. (6.19). Taken from Balachandar and Michaelides (2022).

6.5   Unsteadiness Due to Velocity Variation

The thermal kernel obtained above is expected to be quite accurate in situations where unsteadiness arises mainly from a time-dependent particle or ambient fluid temperature. In such situations, the Reynolds and Péclet numbers of the flow do not change and only $\left[d(\overline{T_f}^S - T_p)/dt\right]$ varies over time. Particle-resolved simulations of this scenario were used in developing the finite-Pe thermal history kernel and as a result the model performance will be good under similar conditions of thermal variation. Consider a different scenario where the temperature difference $(\overline{T_f}^S - T_p)$ remains a constant, but the flow Reynolds and Péclet numbers change due to time variation of relative velocity.

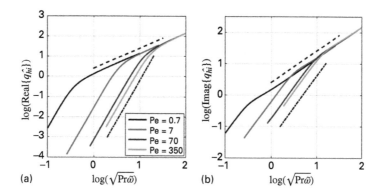

Figure 6.2 (a) Real part of $\hat{q}_{hi}$ as a function of $\sqrt{\mathrm{Pr}\,\tilde{\omega}}$ for four different values of Pe. Also plotted are $(\mathrm{Pr}\,\tilde{\omega})^2$ and $(\mathrm{Pr}\,\tilde{\omega})^{1/2}$ as dashed–dotted and dash, lines, respectively. (b) Imaginary part of $\hat{q}_{hi}$ as a function of $\sqrt{\mathrm{Pr}\,\tilde{\omega}}$. Also plotted are $(\mathrm{Pr}\,\tilde{\omega})$ and $(\mathrm{Pr}\,\tilde{\omega})^{1/2}$ as dashed–dotted and dashed lines. Taken from Balachandar and Michaelides (2022).

This unsteadiness will also result in unsteady heat transfer, which is not addressed by the above model.

As the relative velocity between the particle and the ambient changes, the flow field around the particle varies, which in turn contributes to the time-dependent heat transfer rate. For instance, let the relative velocity between the particle and the surrounding flow be a constant U_1 for $t < 0$, suddenly changed to a new constant value of U_2 for $t > 0$, while the temperature difference ΔT is maintained constant. Before the change in relative velocity, the flow and thermal fields are steady and the nondimensional quasi-steady heat transfer is given by the Nusselt number correlation (6.14). Long after the change in relative velocity, the flow and thermal fields become steady and the nondimensional quasi-steady heat transfer is again given by the correlation (6.14), but for the new Reynolds number. Although the relative velocity changed as a step function, due to the diffusive response of the thermal field we expect the heat transfer rate to change from the pre-jump to the post-jump steady state in a smooth manner. In the small Péclet limit, the flow field will be given by the unsteady Stokes flow as the ambient flow seen by the particle changes from U_1 to U_2. With the unsteady Stokes flow, the corresponding thermal field is obtained by solving the unsteady temperature equation

$$\frac{\partial T}{\partial t} + \mathbf{u} \cdot \nabla T = \kappa_f \nabla^2 T \,. \tag{6.23}$$

A surface integration of the normal gradient of the time-dependent thermal field will then yield the time history of heat transfer to the particle. This interesting problem does not seem to have been addressed in the literature either in the small-Pe limit or at finite Pe using particle-resolved simulations.

It is important to note that according to the model given in Eq. (6.13), the only contribution to heat transfer is from the quasi-steady term, which cannot capture the smooth transition from pre- to post-jump steady state. The only way to account for the unsteady effects of velocity change is through an additional term that is similar to the history term in force modeling. We can conjecture that this additional term must be added to Eq. (6.13) to take the following form:

$$4\pi R_p^2 \rho_f C_{pf}(\overline{T_c}^s - T_p) \int_{-\infty}^t \tilde{K}_{TV}(t - \xi; \text{Re}, \text{Pr}) \left| \frac{d\overline{\mathbf{u}_c}^s}{dt} - \frac{d\mathbf{V}}{dt} \right|_\xi d\xi, \qquad (6.24)$$

where the nondimensional kernel $\tilde{K}_{TV}$ captures the history effect of velocity variation on heat transfer. Unfortunately, its precise form is not known, although it can be conjectured to be similar to the Mei–Adrian kernel given in Eq. (4.142). Clearly, more work is needed to obtain a comprehensive model that encompasses all possible modes of unsteadiness.

7   Particle–Turbulence Interaction in the Dilute Limit

In Chapter 4, we started with a rigorous derivation of force on a spherical particle in the limit of zero Reynolds number in a time-dependent uniform ambient flow, which led to the BBO equation. We then extended the analysis to spatially varying flows in the Stokes limit and obtained the MRG equation. At finite Reynolds number, due to the introduction of fluid inertia, we saw how difficult a complete solution of the hydrodynamic force on a particle can become. In this chapter, we plan to boldly venture into the difficult topic of interaction between a particle and a turbulent flow. The problem of particle–turbulence interaction is a very rich subject, with thousands of worthy papers written on the topic. This complex subject cannot be fully addressed in one chapter, let alone in a book. Thus, our goal in this chapter is to take a first look at some important aspects of particle–turbulence interaction that we are ready to study based on what has been learned in the earlier chapters. Since our understanding so far has been for an isolated particle, we will restrict attention to low particle volume fraction. In Chapter 9, when we study particle–particle interaction, we will reconsider the problem of particle–turbulence interaction in the context of finite volume fraction multiphase flows.

Our discussion will address the two sides of particle–turbulence interaction: (i) how the presence of particles contributes to turbulence generation in the continuous phase, both at the level of an individual particle and at the level of the collective action of a dilute distribution of particles; and (ii) the influence of continuous-phase turbulence, both on the hydrodynamic force and the resulting motion of an individual particle and on the dispersion of a distribution of particles. While the former pertains to the *effect of particles on turbulence*, the latter concerns the *effect of turbulence on the particles*. The various sections of this chapter will be sequenced as follows.

In Section 7.1, we will start with a detailed look at the complex evolution of flow around an isolated stationary particle subjected to a steady uniform flow. While the theoretical studies of the previous chapter generally restricted us to $\mathrm{Re} \lessgtr O(1)$, in this section we will consider particle Reynolds number as large as $O(10^6)$, where the flow around the particle, including the particle's boundary layer, is highly turbulent. It is now well established that the wake behind a particle becomes turbulent when the particle Reynolds number exceeds a few thousand. In other words, at higher Reynolds number, turbulence is self-generated by the particle, even when the oncoming flow is steady, uniform, and laminar. We will refer to this as *self-generated turbulence*. In this section, we will address the role of self-generated turbulence on both the force on the particle and on the turbulent velocity fluctuations in the particle wake.

In Section 7.2, we will consider the complex evolution of flow around a freely moving particle in a quiescent ambient, again for a wide range of particle Reynolds numbers. The focus will be on understanding the difference between the flow around a freely moving particle and that of a corresponding stationary particle. We note that even in the case of a freely falling particle, the wake flow progresses from steady to unsteady to chaotic to turbulent with increasing Re.

Even when the wake of an individual particle is nonturbulent, the overall flow created by the wakes of a random distribution of freely moving particles will be turbulent, which we earlier termed "pseudo turbulence." This turbulence is due to the random superposition of the particle wakes. In Section 7.3, we will consider pseudo turbulence generated by the superposition of particle wakes in the dilute limit, where wake–wake and particle–wake interactions are not important.

The focus of the remaining sections will be on the effect of ambient turbulence on the particles. We will first be concerned with turbulence in the undisturbed ambient flow approaching a particle. The source of this ambient turbulence is not of direct concern, as long as it is generated upstream. This turbulence may be created by other particles that are located upstream of the particle whose hydrodynamic force is under consideration, or may be created by other mechanisms. First, we will consider the limit when the scales of turbulence are larger than the particle size (i.e., the Kolmogorov scale is much larger than the particle diameter). In this limit, the ambient flow seen by the particle can be considered spatially uniform and the effect of ambient turbulence primarily reduces to a time-dependent flow seen by the particle. As a result, the finite-Re BBO equation, or its simplification, is adequate to accurately represent the force on the particle. The effect of turbulence is then seen in the motion of the particles and the resulting mixing, and sometimes de-mixing, of particles. In Section 7.4, we will investigate the effect of turbulence on the effective settling velocity of a particle. Section 7.5 will then investigate the effect of turbulence on diffusing particles from regions of high to low concentration.

We will then consider the limit when the particle size is larger than the Kolmogorov scale. In this limit, the effect of turbulent eddies of size comparable to the particle size can be expected to be even more complicated than the influence of linearly varying flows considered in Section 7.5. In Section 7.6, we will investigate the effect of ambient turbulence on the drag and lift force on particles, and how the BBO or MRG equation must be modified to account for the effect of turbulence. The purpose of this section is to review our current understanding of the effect of turbulence on momentum coupling between the particle and the surrounding fluid.

7.1   Flow Around an Isolated Stationary Particle with Increasing Re

A steady uniform ambient flow approaching a stationary spherical particle will appear to be a simple problem – after all, it involves only a single particle, that stationary too, and the approaching flow is the simplest possible steady uniform flow. As we will see in this section, with increasing particle Reynolds number, the flow around the particle will get more and more complicated, with fully developed turbulence first appearing in the wake region and progressively penetrating into the shear layer and then into the boundary layer around the particle. Thus, even a single stationary particle can contribute to substantial turbulence generation. The results of this section on flow around an isolated stationary particle will be used as the foundation upon which we will consider progressively more complex scenarios: (i) first an isolated freely moving particle in Section 7.2; (ii) then a random dilute distribution of freely moving particles, where particle–particle interactions can be ignored, in Section 7.3; and (iii) finally a random distribution of nondilute particles in Section 9.5. The non-dilute case can be considered only after we gain sufficient insight into particle–particle interaction.

The problem of uniform flow over an isolated spherical particle is a great example of how the flow transitions from a steady laminar state to a progressively more turbulent state. This problem has been studied extensively and the results are well summarized by Tiwari et al. (2020a,b), which will form the basis of our discussion here. As the Reynolds number is systematically increased, the flow around the stationary spherical particle undergoes a definite sequence of instabilities and the different flow regimes are summarized in the following list.

Regime R1: Creeping flow (Re $\lesssim$ 1)
This is the Stokes flow regime whose analytical solution we considered in Section 4.1. The streamlines of the resulting Stokes flow around the particle were shown in Figure 4.1d. An important feature of this flow is that the streamlines have a perfect left–right symmetry across the central plane passing through $x = 0$. The drag force on the particle is given by the Stokes drag (Eq. (4.24)).

Regime R2: Steady, unseparated flow (1 $\lesssim$ Re $\lesssim$ 22)
At finite Reynolds number, inertial effects begin to play a role. But for Re less than about 22, the inertial effects are not strong enough and the flow remains attached to the surface of the sphere. The streamlines computed for Re = 10 are shown in Figure 7.1a (Dennis and Walker, 1971). Even at this modest Reynolds number, unfortunately, an exact close-form solution is not possible. Due to the finite value of Re, it can clearly be seen that the left–right symmetry is broken and the shape of the streamlines approaching the sphere from the left is slightly, but noticeably, different from those leaving the sphere on the right. But topologically, the flow in this regime is the same as the Stokes flow.

Regime R3: Steady, axisymmetric, separated flow (22 $\lesssim$ Re $\lesssim$ 210)
Above a Reynolds number of 22 the flow still remains axisymmetric, but bifurcates to a new topological state that includes a steady recirculating wake behind the sphere. Figure 7.1b shows the streamlines plotted over the upper half-plane at Re = 40 (Dennis and Walker, 1971). The wake of the particle is marked by the recirculating region, which

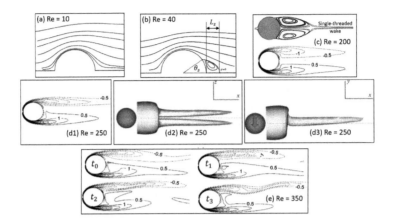

Figure 7.1 (a) Regime R2 of unseparated wake shown as flow streamlines computed for Re = 10 from Dennis and Walker (1971). (b) Flow streamlines of steady axisymmetric separated wake at Re = 40 from Dennis and Walker (1971). (c) Flow streamlines and circumferential vorticity contours of steady axisymmetric wake at Re = 200 from Bagchi and Balachandar (2001). The lengthening of the wake can be seen compared to frame (b). (d) Regime R4 of planar symmetric wake at Re = 250. The three frames show vorticity contours, and the vortical structure extracted as an isosurface of swirling strength observed in two different views from Bagchi and Balachandar (2001, 2002b). The double-threaded wake can clearly be seen and the direction of lift force is marked on the sphere. (e) Vorticity contours at four different time instances in Regime R5 of periodic planar symmetric vortex shedding from Bagchi and Balachandar (2001).

is in fact a symmetric vortex ring due to the axisymmetric nature of the flow. Here, θ_s is the angle at which the attached streamline from the front stagnation point separates from the surface of the sphere. The length of the recirculation region is also marked on the figure as L_s. This qualitative feature of the flow remains the same till Re $\simeq 210$, but quantitatively the size of the separated region (i.e., the recirculating wake) increases with increasing Reynolds number. That is, both θ_s and L_s increase with Re, which can be seen in Figure 7.1c, where streamlines and circumferential vorticity contours are shown for Re = 200 (Bagchi and Balachandar, 2001). As can be seen in the figure, the wake downstream of the vortex ring forms a single filament, known as the *single-threaded wake*.

Regime R4: Steady, planar-symmetric, separated flow (210 $\lesssim$ Re $\lesssim$ 270)
The axisymmetric nature of the flow remains stable until a Reynolds number of about 210. That is, below this Reynolds number, any perturbation to the flow, either as noise in the ambient flow approaching the particle or as imperfections in the shape of the spherical particle, gets damped quickly and the flow maintains its axisymmetric character. Whereas, as the Reynolds number increases above 210, any small noise in the ambient flow or in the shape of the particle will be amplified and the flow loses its axisymmetry and becomes planar symmetric. An example of a planar symmetric flow at Re = 250 taken from Bagchi and Balachandar (2001) is shown in Figure 7.1d.

The circumferential vorticity contours on the x–y symmetry plane passing through the center of the particle and aligned along the flow direction are shown. The flow on either side of this plane is perfectly symmetric about the plane. But the flow is not symmetric on the symmetry plane, which can be noticed in the difference between the positive and negative circumferential vorticity contours (which was not the case in Figure 7.1c).

Due to loss of axisymmetry, the recirculation region in the wake is no more a perfect vortex ring. The region behind the recirculating flow is now characterized by a *double-threaded wake*, which can be seen in the vortex structure shown in Figure 7.1d (Bagchi and Balachandar, 2002b). The view of the vortex structure in frame (d2) is perpendicular to the view shown in the circumferential vorticity contours. As a result, the vortex structure is perfectly planar-symmetric and the two identical vortex threads are on either side of the symmetry plane. However, while the single-threaded wake is aligned along the streamwise axis of the sphere (i.e., along the x-axis), the two threads of the double-threaded wake are tilted away from the x-axis, as shown in frame (d3). Most interestingly, even though the flow is steady, due to the tilted nature of the double-threaded wake, there is a sustained nonzero side (or lift) force on the particle that is marked in Figure 7.1d3. Note that the particle experiences this lift force even though the ambient flow is uniform and the particle is not rotating. This lift force in this sense is self-induced. While the direction of lift in the case of Saffman or Magnus lift depends precisely on the direction of ambient shear or particle rotation, the direction of self-induced lift force depends on the orientation of the one-sided double-threaded wake. The orientation of the one-sided double-threaded wake in turn depends on any slight imperfection that may have existed either in the ambient flow or the shape of the particle. Thus, the orientation of the self-induced lift can be considered to be sensitively dependent on the problem.

Regime R5: Periodic, planar-symmetric, vortex shedding ($270 \lesssim$ Re $\lesssim 360$)

The steady, planar-symmetric, double-threaded wake state becomes unstable and undergoes Hopf bifurcation and the flow becomes unsteady above a Reynolds number of about $Re_{cr1} = 270$ (Natarajan and Arcivos, 1993). So far, we have listed four different bifurcations, where the flow topologically changes its character, and each bifurcation is characterized by a critical Reynolds number. For simplicity, we will only name this critical Reynolds number for *Hopf bifurcation* as Re_{cr1}. Below this Reynolds number the flow around the particle remains steady and any small-amplitude perturbation in the incoming flow will be damped out and the wake flow will return back to the steady state. However, above this Reynolds number even infinitesimally small perturbations in the incoming flow will be amplified and vortex shedding will be initiated in the wake of the particle.

This instability is an *absolute instability*, as opposed to convective instability. In other words, once a small asymmetric perturbation is introduced into the incoming flow, the perturbation is amplified over time and is retained within the recirculating region of the wake and eventually the unsteady vortex-shedding process emerges. The small asymmetric perturbation in the inflow that initiated this instability can cease to exist after the seeding process. Due to the absolute nature of the instability, the perturbation, once introduced into the wake, is retained thereafter. The vortex-shedding process will continue forever, even if the inflow returns back to a perfect uniform state. In this sense, the vortex shedding of a sphere at post-critical Reynolds numbers is self-generated. In fact, in a high-quality simulation of flow over a sphere, to initiate vortex shedding the inflow must be perturbed for a short duration. Typically, a small shear component is introduced, say by having the inflow on one side of the sphere sightly faster than in the other. Once the perturbation enters the recirculating wake of the sphere, the perturbation of the inflow can stop, and the disturbance will continue to grow in the wake and periodic vortex shedding will initiate.

The periodic vortex-shedding state at Re = 350 is shown in Figure 7.1e, where circumferential vorticity contours are plotted at four different times. While the vorticity contours in Figures 7.1c and d remain the same at all times, a cyclical variation is observed in Figure 7.1e. The vortex-shedding process is characterized by the shedding period T_{vs} and the associated shedding frequency $f_{vs} = 2\pi/T_{vs}$.

It is important to note that the vortex shedding in this regime is planar symmetric and not axisymmetric. If the periodic vortex-shedding regime had followed directly after Regime R3, then the shed vortices would be axisymmetric in the form of axisymmetric vortex rings. However, this is not the case. The shed vortices have only a plane of symmetry, and in Figure 7.1e the plane of symmetry is the plane on which the vorticity contours are drawn. On this plane, it can be observed that though the vorticity evolution in both the top and bottom half are cyclic, the two halves are not the same. Thus, shedding of vortices in the wake of a stationary sphere in Regime R5 is *one-sided*.

As a consequence of periodic vortex shedding, the force on the particle oscillates over time. This periodic oscillation is observed in both the drag and lift forces, but the amplitude of oscillation is much more pronounced in the time variation of the lift force. As vortices are shed in the wake, the low pressure associated with these vortex cores results in a strong lift-force fluctuation. This is shown in the phase plot of y- and z-direction force coefficients for Re = 350 in Figure 7.2a, taken from Mittal (1999). The plot shows $C_y(t)$ vs. $C_z(t)$ and it is clear that the symmetry line (marked ξ) is at an angle of $45°$ clockwise from the y-axis in the y–z plane. The symmetry plane is then the x–ξ plane. The effect of one-sided vortex shedding is clear: the lift coefficient along the ξ-direction is always positive, though it oscillates over time. We again emphasize that the orientation of the symmetry line in the y–z plane is in general arbitrary. In the particular simulation of Mittal (1999), the initial perturbation was such that the flow resulted in this particular choice of symmetry plane. With a different choice of initial perturbation, the vortex shedding can be reoriented to any other symmetry plane.

The significance of the critical Reynolds number Re_{crl} is that beyond this Reynolds number, the hydrodynamic force on the particle will fluctuate over time and the fluctuation will be uncorrelated with the flow that approaches the particle from upstream. The force fluctuation is self-induced due to instabilities that dominate as the flow goes around the particle.

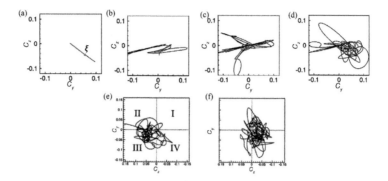

Figure 7.2 Phase portrait of force coefficient along the two transverse directions at six different Reynolds numbers. The time variations of C_y and C_z are plotted against each other. Here $C_y = F_y/(\rho_f \pi d_p^2 U^2/8)$, where F_y is the dimensional force in the y-direction, U is the flow velocity and a similar definition applies for C_z. (a) Re = 350, (b) Re = 375, (c) Re = 400, (d) Re = 425, (e) Re = 500, (f) Re = 650. Taken from Mittal (1999); Mittal et al. (2002).

Regime I6: Chaotic vortex shedding ($360 \lesssim$ Re $\lesssim 3000$)

In the regimes discussed so far, the flow is regular and deterministic and hence these regimes are called R1 to R5. In comparison, the flow in this and the next three regimes will be irregular (chaotic and turbulent) and thus will be denoted I6 to I9. As the Reynolds number increases above a value of about 360, the shedding process that used to be periodic starts to become more complex. This is best illustrated in the phase portrait of C_y vs. C_z shown in Figures 7.2b to f, where Re increased steadily from 375 to 650. At Re = 375 it appears that there is almost a plane of symmetry for the vortex shedding, but there is occasional departure from this plane of symmetry, unlike at the lower Re of 350. These chaotic departures from planar symmetry increase steadily with increasing Reynolds number and for Re $\gtrsim 500$ the wake of the particle can be classified as chaotic. Interestingly, with the onset of chaos, the one-sidedness of the shedding process seems to be lost and the time-averaged mean side force on the particle becomes nearly zero. This random nature of the shedding process is also shown in Figure 7.3a, where an isosurface of vorticity at Re = 400 obtained by Homann et al. (2013) is reproduced. The wake is characterized by hairpin-like vortical loops and the orientation of these shed vortices can be observed to vary over time. In the chaotic regime, these vortical structures have equal probability of appearing on any circumferential side of the sphere in a nondeterministic fashion. The intensity of the chaotic fluctuations in the wake increases as Re increases above 500, and the downstream wake region can be considered fully developed turbulence at a Reynolds number of about 3000.

Regime I7: Subcritical turbulent flow ($3000 \lesssim$ Re $\lesssim 3.4 \times 10^5$)

The far wake of the spherical particle becomes fully turbulent and the classical theory of turbulent free-shear flow applies (Pope, 2001). As examples, the vortical structures in the far wake of a particle at Re = 3700 and 10^4, as obtained by Yun et al. (2006),

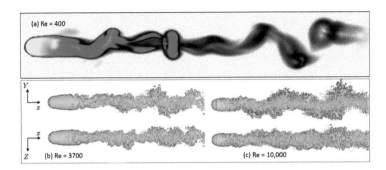

Figure 7.3 (a) Vortex structure in the wake of a sphere in Regime I6 at Re = 400. (b) Turbulent wake of a sphere in Regime I7 at Re = 3700 and 10^4. Taken from Homann et al. (2013); Yun et al. (2006).

are shown in Figures 7.3b and c in two different views. It can be observed that the intensity of turbulence in the wake increases with increasing Reynolds number. In fact, the structure of the far wake depends only on the momentum lost by the fluid in the form of drag on the particle. Details such as the precise shape of the particle do not matter. With increasing Re, turbulence in the wake increases and the region of turbulence moves upstream, closer to the sphere. Turbulence first migrates into the shear layer that forms after the flow separates from the surface of the sphere. That is, the separated shear layer of the sphere becomes turbulent and exhibits properties of a turbulent mixing layer. This scenario is depicted in Figure 7.4a, where we show vortical structures in the near wake region of a sphere at Re = 2×10^4, as obtained by Geier et al. (2017). The turbulent vortical structures in the shear layer are clearly visible.

Regime I8: Critical regime ($3.4 \times 10^5 \lessgtr$ Re $\lessgtr 4.4 \times 10^5$)

Tiwari et al. (2020b) define this Reynolds number range as the critical Reynolds number range when turbulence starts to migrate from the separated shear layer into the attached boundary layer on the sphere. This process of the attached boundary layer becoming turbulent depends on details such as the level of disturbance in the incoming flow and the level of surface finish and smoothness of the sphere. Recall that the ratio of the thickness of an attached boundary layer to its length scales as $Re^{-1/2}$. This means, in the critical regime, that the boundary-layer thickness is only about 0.12% of the particle diameter and thus even a small amount of roughness can influence the transition process by which the attached boundary layer changes from laminar to turbulent. Nevertheless, experimentally it has been observed that in most cases, by Re = 4.4×10^5, the transition to turbulence of the attached boundary layer is complete. As we will see below, this transition is of importance, since it has a dramatic effect on the drag on the sphere. The process of turbulence beginning to encroach into the boundary layer is captured in Figure 7.4b at Re = 10^5 (Geier et al., 2017). The Reynolds number at which the attached boundary layer transitions from the laminar to the turbulent state is often referred to as the second critical Reynolds number, Re_{cr2}.

Regime I9: Supercritical turbulent flow ($4.4 \times 10^5 \lesssim$ Re)

The dramatic effect of the attached boundary layer becoming turbulent is illustrated in Figure 7.4c at Re = 1.14×10^6, as obtained by Geier et al. (2017). A turbulent boundary layer stays attached on the surface of the sphere longer than a laminar boundary layer. As a result of this delayed separation, the wake region is considerably smaller. This reduction in the low-pressure region on the lee side of the sphere is responsible for a dramatic drop in the drag force. This rapid reduction in the drag coefficient is known as the *drag crisis*. Nevertheless, over the entire Reynolds number range, the shedding of coherent vortices persists in the wake, in addition to the wake being turbulent. This self-induced vortex shedding and turbulence results in substantial force fluctuation. Again, these fluctuations are uncorrelated with the flow that approaches the particle from upstream.

The drag crisis is the reason why golf balls are dimpled. Since the Reynolds number of a well-hit golf ball is in the range of the critical value Re_{cr2}, it is important to promote a transition to turbulence in the attached boundary layer of the golf ball, so that the ball will experience a much lower drag and travel farther. The purpose of the dimples is to ensure prompt transition of the attached boundary layer to a turbulent state. Interestingly, in a golf ball, to ensure that it travels straight, it is important to promote this transition all around the golf ball, so there is no side force.

The scenario is quite different in a baseball or a cricket ball, where mastery includes the ability to throw a curve ball or to bowl a reverse swing. The idea is then to promote transition on only one side of the ball, which is rougher than the other side. The attached boundary layer on one side of the ball versus early separation on the other results in a significant side force and a pronounced curved trajectory of the ball. By carefully placing the rough versus the smooth side of the ball, the pitcher or bowler can create magic. Those interested in sports ball fluid mechanics should refer to Mehta (1985) and the many other publications that have followed.

In closing, we make the following important point. Although the very high Reynolds number flow regimes I7 and I8 are of great interest, they are less relevant in many multiphase flow problems involving particles smaller than a millimeter. A millimeter-sized particle must gain a relative velocity of about 1000 m/s in air or 100 m/s in water to reach a Reynolds number of 10^5. Thus, the regimes where the shear layer and the attached boundary layer are turbulent are important only in problems involving substantially larger particles traveling at high speeds. However, there are many applications involving particle Reynolds number $O(10^3)$ where the wake is turbulent.

7.1.1   Wake Properties

In this subsection, we will investigate the effect of the above-listed flow transitions on key wake properties. Again, an excellent summary has been provided by Tiwari et al. (2020b). The first quantity we present is the wake length L_s from the rear stagnation point to the reattachment point along the x-axis (see Figure 7.1b) as a function of Re. This quantity is plotted in Figure 7.5a, where the different regimes are marked. Note

Figure 7.4 (a) Vortex structure in the wake of a sphere in Regime I7 at Re = 2×10^4. (b) Vortex structure around a sphere at Re = 10^5. Turbulence can be seen starting to penetrate into the boundary layer. (c) Regime I9 where the attached boundary layer is turbulent and as a result separation is delayed. Reprinted from Geier et al. (2017), with permission from Elsevier.

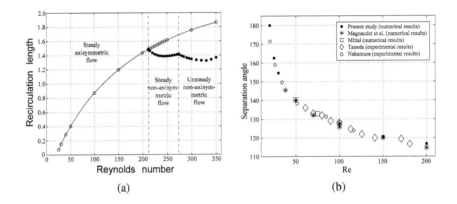

Figure 7.5 (a) Normalized length of the recirculation region as a function of Re; the different flow regimes are marked. (b) Separation angle as a function of Re. Reproduced from Bouchet et al. (2006). Elsevier Masson SAS. All rights reserved.

that the present definitions of the different regimes are slightly different from those introduced therein. The recirculating wake vortex ring starts forming at the transition from Regimes R2 to R3 at Re $\approx$ 22. The wake length increases in Regime R3 to more than one particle diameter (see Figure 7.1c). In the one-sided planar symmetric regime, with increasing Re, the wake length decreases slightly. In the unsteady regime, the wake length can only be defined in the time-averaged sense.

Another quantity of interest is the angle at which the boundary layer separates from the surface of the sphere, measured from the front stagnation point (see Figure 7.1b). Variation of θ_s as a function of Re over a wider range of Reynolds numbers is shown in Figure 7.5b. With increasing Reynolds number, the separation point moves upstream as indicated by the increasing angle of separation. This trend continues even after the wake becomes chaotic and turbulent. Some scatter in the data can be observed at higher Reynolds number. The separation angle is expected to decrease at the second critical Reynolds number Re_{cr2}, when the attached boundary layer becomes turbulent and remains attached to the sphere over a longer extent.

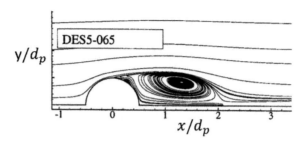

Figure 7.6 Streamlines of the ensemble-averaged mean flow around a stationary particle at Re = 10^4 obtained from detached eddy simulation. Taken from Constantinescu and Squires (2003).

The velocity field of the flow around the particle can be well characterized by (i) the axial and radial components of the mean flow $\langle u_x \rangle(x,r)$ and $\langle u_r \rangle(x,r)$, which are functions of the axial and radial directions and (ii) the nonzero components of the Reynolds stress tensor $\langle u_x'^2 \rangle(x,r)$, $\langle u_r'^2 \rangle(x,r)$, and $\langle u_x' u_r' \rangle(x,r)$. Here the ambient flow direction x is taken to be the axial direction and r is the radial direction from this axis. The prime indicates perturbation away from the mean (i.e., $u_x' = u_x - \langle u_x \rangle$ and $u_r' = u_r - \langle u_r \rangle$). The mean and the Reynolds stress are only the first two of infinitely many single-point statistics that can be defined, but they provide an adequate description of the flow. In Regimes R1 to R3, where the flow is axisymmetric and steady, the ensemble average represented by the angle brackets is not needed and the Reynolds stresses are identically zero. In Regime R4, though the flow is steady, axisymmetry is broken. In this case, the ensemble average is an average over the θ-direction. In other words, the mean corresponds to the axisymmetric part of the planar wake shown in Figure 7.1d1 and the Reynold stress will provide a measure of departure from axisymmetry. In the unsteady regimes of R5 to I9, the ensemble average represents an average over both θ and time. As a result, both the mean and the Reynolds stress fields are only functions of the axial and radial distance from the center of the particle.

As an example, the streamlines of the ensemble-averaged mean flow at Re = 10^4 obtained from a detached eddy simulation by Constantinescu and Squires (2003) are shown in Figure 7.6. Despite the very complex nature of the instantaneous flow seen in Figure 7.3c, the ensemble-averaged flow qualitatively looks much like the laminar separated wake seen in Figure 7.1c. Only the angle at which the flow separates and the wake length have increased, consistent with the trends seen in the vortex structure. Figure 7.7 shows the three nonzero components of the Reynolds stress at the two different Reynolds numbers of Re = 3700 and 10^4, whose wake vortical structures were shown in Figure 7.3 (Yun et al., 2006). The important observation to be made is that the velocity fluctuation measured in terms of Reynolds stress is significant only about two diameters downstream at the lower Reynolds number of 3700. At the higher Reynolds number, the turbulence has advanced close to the particle and penetrated well into the shear layer. This upstream migration of turbulence into the separated shear layer continues with increasing Re, and the turbulence eventually penetrates into the attached boundary layer.

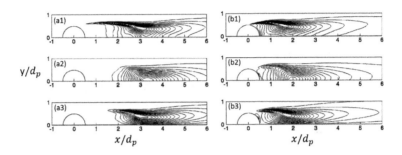

Figure 7.7 Reynolds stress profiles in the wake of a sphere. (a1) $\langle u_x'^2 \rangle / U^2$ at Re = 3700; (a2) $\langle u_y'^2 \rangle / U^2$ at Re = 3700; (a3) $\langle u_x' u_y' \rangle / U^2$ at Re = 3700. Frames (b1), (b2), and (b3) are the same at Re = 10^4. Taken from Yun et al. (2006).

All the above discussion pertains only to the near wake of the particle, which extends to about five diameters downstream of the particle. This is where the recirculation region of the near wake is located. Farther downstream, defined as the far wake, the velocity profile takes on a self-similar shape. That is, the velocity deficit, defined as $U_\infty - u_x(x, r)$, at different downstream x locations, collapses onto a universal profile when scaled by the centerline velocity deficit and by the wake thickness (here U_∞ is the velocity of the uniform flow approaching the stationary particle). In the laminar regime, the boundary layer theory can be used to obtain an analytic solution for the wake velocity profile. The only key input parameter to the theory is the drag on the particle, which is also the momentum lost by the flow. Let Θ be the momentum radius, or the radius of an imaginary disk of oncoming fluid, whose momentum is lost due to drag. An expression for this can be obtained as

$$\frac{1}{2}\pi d_p^2 \, \rho_f \, U_\infty^2 \, C_D = \pi \Theta^2 \, \rho_f \, U_\infty^2 \quad \Rightarrow \quad \Theta^2 = \frac{C_D}{8} d_p^2 \, . \tag{7.1}$$

Thus, Θ serves as a proxy for drag and in terms of Θ, the self-similar axial and radial laminar velocity profiles are expressed as

Laminar wake
$$\begin{cases} 1 - \dfrac{u_x(x,r)}{U_\infty} &= \dfrac{\Theta^2 U_\infty}{4\nu_f x} \exp\left(-\dfrac{r^2 U_\infty}{4\nu_f x}\right), \\[3mm] \dfrac{u_r(x,r)}{U_\infty} &= \dfrac{r\,\Theta^2 U_\infty}{32\nu_f x^2} \exp\left(-\dfrac{r^2 U_\infty}{4\nu_f x}\right). \end{cases} \tag{7.2}$$

Thus, the self-similar wake's centerline velocity deficit decays as x^{-1}, while the thickness of the wake, $2\sqrt{\nu_f x / U_\infty}$, increases slowly as $x^{1/2}$. In the turbulent regime, the boundary layer theory can again be used, with additional assumptions about the Reynolds stress distribution, to obtain the following self-similar wake profiles (Parthasarathy and Faeth, 1990; Tennekes and Lumley, 2018):

Turbulent wake
$$\begin{cases} 1 - \dfrac{\langle u_x \rangle(x,r)}{U_\infty} &= 2.23 \left(\dfrac{\Theta}{x}\right)^{2/3} \exp\left(-\dfrac{r^2}{2l_w^2}\right), \\[3mm] \dfrac{\langle u_r \rangle(x,r)}{U_\infty} &= 0.74 \dfrac{\Theta^{2/3} r}{x^{5/3}} \exp\left(-\dfrac{r^2}{2l_w^2}\right), \end{cases} \tag{7.3}$$

where the wake thickness goes as $l_w = 0.47\Theta^{2/3}x^{1/3}$. The centerline velocity deficit decays as $x^{-2/3}$. Because of the slower decay, the turbulent wake of a particle can extend over a very long distance downstream of the particle. Note that the mean axial velocity deficit has its peak along the wake axis, while by symmetry the radial velocity is zero along the axis.

In the case of the turbulent wake, the boundary layer theory can be used to obtain the distributions of the nonzero components of the Reynolds stress tensor as well. Following Tennekes and Lumley (2018), we obtain these components to be

$$\textbf{Turbulent wake} \begin{cases} \dfrac{\langle u_x'^2 \rangle}{U_\infty^2} & \approx 2.65 \dfrac{\Theta^{2/3}r}{x^{5/3}} \exp\left(-\dfrac{r^2}{2l_w^2}\right), \\[2ex] \dfrac{\langle u_x'u_r' \rangle}{U_\infty^2} & = 1.06 \dfrac{\Theta^{2/3}r}{x^{5/3}} \exp\left(-\dfrac{r^2}{2l_w^2}\right), \end{cases} \tag{7.4}$$

where a prime indicates perturbation from the mean. Turbulence in the far wake is nearly isotropic and therefore $\langle u_x'^2 \rangle = \langle u_r'^2 \rangle = \langle u_\theta'^2 \rangle$. According to the above relations, all the Reynolds stress components become zero along the wake axis and reach their peak at $r = l_w$ (i.e., at the edge of the wake). While this may be a reasonable approximation for $\langle u_x'u_r' \rangle$, it is not so for the mean square velocity fluctuations. Turbulent kinetic energy $\langle u_i'u_i' \rangle/2$ does not become zero along the wake axis – it remains comparable to the off-axis peak value. Hence the expression for $\langle u_x'^2 \rangle$ has been denoted approximate.

7.1.2 Drag and Lift Coefficients and Shedding Frequency

Next we consider the important question of the dependence of the drag coefficient on Re. The compilation of many experimental and numerical results on C_D vs. Re obtained by Tiwari et al. (2020b) is shown in Figure 7.8. In the Stokes regime (Regime R1), C_D decreases as 24/Re, which is a straight line in the log–log plot. The deviation from this relation increases slowly in Regime R2. The standard drag law correction to the Stokes drag given in Eq. (2.3) captures the C_D variation quite well up to a Reynolds number of about 10^4. Note that in Regime R5 and above, since the flow is unsteady, the drag force is time dependent and the C_D presented in the figure corresponds to time-averaged drag force. For Re > 10^4, C_D reaches a nearly constant value up until the second critical Reynolds number. There is a substantial reduction in the drag coefficient at Re_{cr2}, due to the delayed flow separation and the shrinking of the wake. It can be noticed that the estimation of Re_{cr2} varies between the different measurements. Some place the critical Reynolds number closer to 10^5, while others estimate $Re_{cr2} \approx 4 \times 10^5$. This variation is due to the convective nature of boundary layer instability and the transition process. Unlike wake instability, which is absolute in nature, the onset of instability in the attached boundary layer over the front portion of the sphere and transition to turbulence depend largely on the level of perturbation that exists in the oncoming flow and on the smoothness of the sphere. Thus, measurements that indicate a lower value of Re_{cr} are perhaps influenced by a higher level of perturbation that promotes early transition. Nevertheless, nearly all the available results tend to suggest about the same,

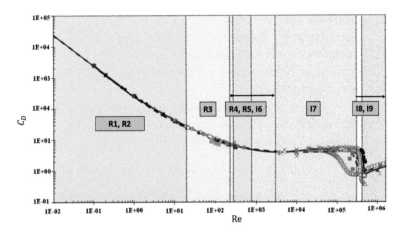

Figure 7.8 C_D vs. Re with the different flow regimes marked. Reprinted from Tiwari et al. (2020b), who compiled the results of many different experiments and simulations; consult the paper for the original data sources. Figure reproduced with permission from Elsevier.

substantially lower value of C_D after transition. With further increase in Re in the supercritical Regime I9, C_D starts to increase again.

Figure 7.9a summarizes the lift coefficient as a function of Re. C_L is identically zero in the axisymmetric regimes. In Regime R4, as Re increases above 210, the magnitude of the lift coefficient increases and reaches a peak value of about 0.068 at Re $\approx$ 300. The corresponding C_D at this Reynolds number is about 0.684, and thus the lift force reaches about 10% of the drag force. With further increase in Re, the lift coefficient decreases in magnitude. Note that the direction of the lift force is arbitrary and thus the negative value of C_L simply indicates that in the measurements, the lift force is taken to be directed along the negative y-direction as shown in Figure 7.1d3. At moderate values of Re, the time-averaged lift force remains nonzero even when the flow becomes unsteady. But eventually, at large enough Re, when the wake becomes sufficiently chaotic, the lift force changes direction in a chaotic manner and the time-averaged lift force becomes zero.

In the unsteady regime, we define the nondimensional shedding frequency as the Strouhal number, Sr $= f_{vs}d_p/U_\infty$. A plot of Sr vs. Re in Regime R5 and above, as compiled by Tiwari et al. (2020b), is shown in Figure 7.9b. Typically, the dominant frequency is extracted from the time variation of the lift force. Only in Regime R5, when the shedding process is periodic, can a single Strouhal number be identified. At higher Re, when the wake flow becomes chaotic and turbulent, the time evolution of the lift force also becomes chaotic and its Fourier transform presents a range of frequencies. Only the Strouhal number corresponding to the dominant frequencies is presented in Figure 7.9b. As indicated in the figure, above a Reynolds number of a few hundred, two different dominant frequencies are observed, which continue their existence even at very high Reynolds numbers. Here, the higher frequency branch is the shedding frequency at which vortices are shed in the wake. The low frequency

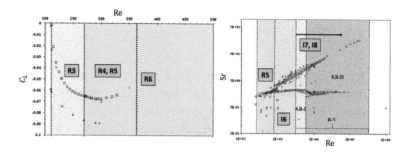

Figure 7.9 (a) Lift coefficient vs. Re and the different flow regimes are marked. (b) Strouhal number plotted as a function of Re. In Regime I6 and above, two different frequencies are observed. The figures are taken from Tiwari et al. (2020b), who compiled the results of many different experiments and simulations; consult the paper for the original data sources. Figure reproduced with the permission of Elsevier.

corresponds to a flapping mode, which has been observed in some wake flows (Najjar and Balachandar, 1998).

7.2 Flow and Dynamics of a Freely Moving Particle

The detailed discussion of the previous section is about flow past a stationary particle. But in most multiphase flow applications, the particles are in free motion. The purpose of this section is to present the different flow regimes that are encountered when the particle is allowed to move freely in response to the hydrodynamic forces acting on it. It will be established that the results of the stationary particle are relevant even in the case when the particle is allowed to move freely. However, there are some differences which will be highlighted.

In the case of a stationary particle, the ambient flow was imposed and parameterized in terms of the particle Reynolds number Re. When the particle is allowed to move freely, a second parameter enters the picture – the particle-to-fluid density ratio $\rho = \rho_p/\rho_f$ plays an important role. In this section, we study the motion of a particle allowed to move freely in response to gravity in a quiescent ambient. A heavier-than-fluid particle (i.e., $\rho > 1$) will fall freely, while a lighter-than-fluid particle will ascend through the ambient fluid. Both these regimes are of interest, since the response differs. With the introduction of the additional parameter, the problem is richer, since the flow depends not only on hydrodynamic instabilities, but also on the dynamics of the particle, which can be quite complex and chaotic.

In the case of the stationary particle, the relative velocity between the ambient and the particle (i.e., U_∞) is imposed. This and the corresponding Re remain time-invariant. In the case of a falling or ascending particle, the velocity of the particle, which is also the relative velocity, is an outcome and furthermore it is often time dependent. Thus, instead of Re, we will consider a related parameter. The Galileo number is defined as

$$\text{Ga} = \frac{\sqrt{(\rho - 1)g d_p^3}}{\nu_f}.$$

(7.5)

The advantage of Ga is that it only depends on the size of the particle and the density ratio. The problem of a freely settling or ascending particle has been studied both experimentally and through particle-resolved simulations for a wide range of the two relevant parameters, ρ and Ga (Jenny et al., 2004; Veldhuis and Biesheuvel, 2007; Horowitz and Williamson, 2010; Zhou and Dušek, 2015). When the particle motion is unsteady, the instantaneous particle settling velocity $V_g = \mathbf{V} \cdot \mathbf{g}$ obtained from an experiment or simulation will vary in time, which can be averaged over the particle trajectory to obtain the mean settling velocity $\langle V_g \rangle_p$, where $\langle \cdot \rangle_p$ indicates an average following the particle. The Reynolds number based on the average settling velocity can be defined as

$$\langle \text{Re} \rangle_p = \frac{\langle V_g \rangle_p \, d_p}{\nu_f} = \text{Ga} \left[\frac{\langle V_g \rangle_p}{\sqrt{(\rho - 1)\, g d_p}} \right],$$

(7.6)

which relates the average Reynolds and Galileo numbers. Note that while Re was a constant in the previous section where a stationary particle was subjected to a steady uniform flow, in the present context of a freely falling particle, the Reynolds number is time varying in the unsteady regime.

Here we will summarize the results of Zhou and Dušek (2015), who considered a density ratio in the range $0 \leq \rho \leq 10$ and a Galileo number in the range $150 \leq \text{Ga} \leq 500$. The comprehensive regime map that they obtained is shown in Figure 7.10. It can be observed that below Ga < 155.8, the particle falls vertically down or ascends vertically up at a constant velocity and thus covers the axisymmetric Regimes R1, R2, and R3 of the previous section. Since there is no side or lift force in these regimes, the particle follows a strict vertical trajectory. Note that the Galileo number Ga $= 155.8$ corresponds to a Reynolds number Re $= 210$, where the flow around a stationary sphere bifurcates from the axisymmetric to the planar symmetric regime. For the freely moving particle, the steady oblique state in Figure 7.10 corresponds to Regime R4 of planar symmetric steady wake. The double-threaded wake is also seen in this regime for the freely moving particle. Due to the nonzero lift force, the particle not only falls down or ascends up, but also drifts along the horizontal direction. Both the vertical and horizontal components of velocity (V_g and V_h) are constant in this steady regime, although the horizontal component is much smaller than the vertical velocity. The precise direction of horizontal motion is arbitrary, and depends on small perturbations in the ambient fluid or in the shape of the particle. The particle thus travels at an angle $\tan^{-1}(V_h/V_g)$ to the vertical at a constant speed of $\sqrt{V_h^2 + V_g^2}$. Note that due to the one-sided nature of the double-threaded wake, the particle experiences a net torque and undergoes steady angular rotation.

In all the above cases, since the particle velocity is a constant, in the frame of reference attached to the particle, the problem is Galilean-invariant. Furthermore,

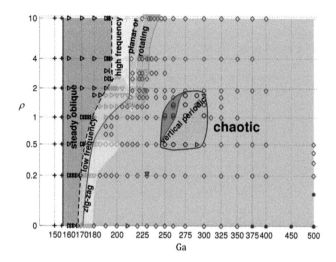

Figure 7.10 Regime map of the different flow dynamics of a falling or ascending spherical particle for varying Ga and ρ. The different regimes are separated by lines and marked. The symbols represent the different cases simulated and the resulting particle dynamics: Regime R3; black right triangle, Regime R4; green right triangle, low-frequency oblique shedding; green down triangle, low-frequency quasi-periodic oblique shedding; yellow right triangle, high-frequency oblique oscillating; cyan circle, zig-zag; cyan down triangle, intermittent zig-zag; blue circle, vertical oscillating planar; blue down triangle, vertical quasi-periodic; blue right triangle, periodic planar slightly inclined; blue circle, helical; pink diamond, chaotic. Reprinted from Zhou and Dušek (2015), with permission from Elsevier.

when the particle motion is strictly vertical, the net torque on the particle is zero and the particle has no rotational motion. In these regimes, the flow around the particle is identical to that of a stationary particle of corresponding Re. The flow behavior is independent of the density ratio and depends only on Ga (or equivalently Re). In the case of steady oblique particle motion, even though the flow is steady in the frame attached to the particle, since the particle has nonzero rotational motion, it is not identically the same as Regime R4 of the stationary particle. The difference between the stationary and freely moving particle becomes relevant only when the flow and the particle motion are unsteady. Above a critical Galileo number, denoted as Ga_{cr1}, the flow undergoes Hopf bifurcation and becomes unsteady with periodic shedding of vortices. The critical Galileo number, however, depends on the density ratio. Table 7.1, taken from Zhou and Dušek (2015), presents Ga_{cr1} along with the vertical, horizontal, and angular velocities of the particle at the critical Galileo number. The table also presents the nondimensional angular frequency. Several points can be observed in the table. For $\rho \gtrsim 3$, the critical Reynolds number for the onset of vortex shedding is about the same as that for a stationary particle. A stationary particle can in fact be considered as the heavy particle limit, since such a heavy particle will not change its velocity easily, and the frame of reference can be attached to the moving particle. As ρ

Table 7.1 Results on the critical Galileo number for the onset of unsteady motion in the case of a freely falling or ascending particle in a quiescent fluid presented as a function of particle-to-fluid density ratio. The table also presents the nondimensional vertical, horizontal, and angular velocities of the particle at the critical Ga. The nondimensional shedding frequency is also presented. In the last column, $\mathrm{Re_{cr1}} = \mathrm{Ga_{cr1}} V_g / \sqrt{(\rho - 1)g d_p}$. Information taken from Zhou and Dušek (2015).

ρ	$\mathrm{Ga_{cr1}}$	$\dfrac{\langle V_g \rangle_p}{\sqrt{(\rho - 1)g d_p}}$	$\dfrac{\langle V_h \rangle_p}{\sqrt{(\rho - 1)g d_p}}$	$\dfrac{\langle \Omega \rangle_p d_p}{\sqrt{(\rho - 1)g d_p}}$	$\dfrac{f\, d_p}{\sqrt{(\rho - 1)g d_p}}$	$\mathrm{Re_{cr1}}$
0.0	167.18	1.3355	0.0969	0.0149	0.0701	223.27
0.2	169.23	1.3388	0.1031	0.0151	0.0667	226.57
0.5	172.52	1.3443	0.1108	0.015	0.0644	231.92
1.0	178.55	1.3544	0.1224	0.0139	0.0672	241.83
1.3	182.50	1.3622	0.1275	0.0129	0.0677	248.60
1.7	187.35	1.3704	0.1327	0.0112	0.0711	256.74
2.0	190.69	1.3763	0.1356	0.0099	0.0729	262.45
2.5	196.08	1.3859	0.1387	0.0077	0.076/0.175	271.74
3.0	195.19	1.3843	0.1383	0.0081	0.1741	270.20
4.0	195.18	1.3842	0.1383	0.0081	0.1751	270.17
10.0	195.06	1.3838	0.1384	0.0082	0.1771	269.92

decreases, the system becomes unstable earlier due to the lower inertia of the particle and $\mathrm{Ga_{cr1}}$ (or equivalently $\mathrm{Re_{cr1}}$) decreases. Thus, the steady oblique shedding regime shrinks at lower particle-to-fluid density ratio.

The horizontal velocity of the particle due to one-sided vortex shedding is more than an order of magnitude lower than the vertical velocity and the ratio approaches 0.1 for a heavy particle. The angular velocity of the particle is also small and decreases with increasing ρ. Two modes of vortex shedding were observed by Zhou and Dušek (2015). At density ratios $\rho \geq 2.5$, the shedding frequency was similar to those observed for a stationary particle, whereas for $\rho \leq 2.5$, vortex shedding was at a much lower frequency, which persisted even for lighter-than-fluid particles. At $\rho = 2.5$, both the higher and lower shedding frequency modes were observed. Figure 7.11, taken from Zhou and Dušek (2015), shows the vortex structure in the wake of (a) a falling sphere of $\rho = 3$ and Ga = 199 (corresponding to Re = 274.6), (b) a stationary sphere of Re = 277.9, and (c) a falling sphere of $\rho = 2$ and Ga = 196. The first two vortex structures are visually identical and their nondimensional shedding frequencies are 0.129 and 0.13, respectively. This shows that the periodic shedding state of a falling particle of $\rho \geq 2.5$ is quite similar to that of a stationary particle. In the case of a lighter particle, the shedding pattern is farther apart, consistent with the lower frequency of shedding.

Zhou and Dušek (2015) analyzed the horizontal velocity of the particle for both the high and low-frequency cases. It was first noted that the time-averaged horizontal velocity is nonzero and about the same for both cases. This indicates that the free-falling particle will not fall directly below the point of release, but will drift to one side and the level of drift is about the same for both the high and low-frequency shedding. There is, however, a substantial difference in the level of time variation – it is much

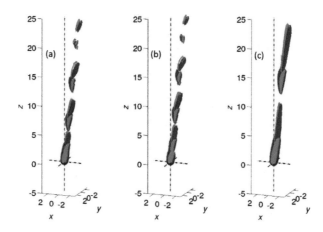

Figure 7.11 Vortex structure in the wake of (a) a falling sphere of $\rho = 3$ and Ga = 199 (corresponding to Re = 274.6), (b) a stationary sphere of Re = 277.9, and (c) a falling sphere of $\rho = 2$ and Ga = 196. Reprinted from Zhou and Dušek (2015), with permission from Elsevier.

stronger in the low-frequency shedding. In both cases, the particle will continue to drift to one side, however, the speed of this drift fluctuates about the mean.

With a further increase in Galileo number, for all values of density ratio, both the flow as well as the particle motion will become quasi-periodic and chaotic, with the appearance of additional frequencies in addition to the primary frequency of vortex shedding. The regime of chaotic motion is marked in Figure 7.10. As with many other dynamical systems, the regime of chaotic motion features many other subregimes of regular and quasi-regular motion, such as zig-zag, helical, and vertical oscillatory motion, that are marked in the figure with different symbols (Zhou and Dušek, 2015). This is the Reynolds number regime where the wake of a stationary particle becomes chaotic. Although the free motion of a particle at higher Galileo numbers (or Reynolds numbers) in excess of 10^4 has not been considered, it can be conjectured that the sequence of events seen for a stationary particle – namely, wake, shear-layer, and boundary-layer turbulence – will occur in case of a freely moving particle as well, albeit the transition Reynolds numbers will be different.

We now address the important question of how different the drag coefficient of a freely moving particle is from the plot of C_D vs. Re presented in Figure 7.8. In the steady vertical regimes of Regimes R1 to R3, the problem of free motion is identical to the stationary problem with a Galilean transformation. The difference between the two is significant only for Ga > Ga_{cr1}, presented in Table 7.1. To evaluate this difference, we present in Figure 7.12a a contour plot of nondimensional fall (or rise) velocity of a freely moving particle. The ratio $\langle V_g \rangle_p / \sqrt{(\rho - 1) g d_p}$ is plotted as a function of Ga and ρ (taken from Zhou and Dušek, 2015). Note that the settling velocity varies with time in the unsteady regimes and therefore a time average of the particle velocity is used. We now establish the relation between Ga and the nondimensional relative velocity

Table 7.2 The relation between Ga and nondimensional relative velocity of a stationary particle whose force is set equal to that of a freely falling particle.

Ga	160	200	240	280	320	360	400
$\tilde{U}$	1.296	1.372	1.434	1.488	1.534	1.576	1.613

of a stationary particle, which is defined as $\tilde{U}_\infty = U_\infty/\sqrt{(\rho - 1)g\delta_p}$, by requiring the force on the particle to be the same as that on a falling particle. This yields the relation

$$3\pi \mu_f d_p U_\infty \Phi(\text{Re}) = \frac{\pi d_p^3}{6}(\rho_p - \rho_f)g, \tag{7.7}$$

where the finite Reynolds number standard drag correction $\Phi(\text{Re}) = 1 + 0.15\,\text{Re}^{0.687}$ provides an excellent fit for the C_D data presented in Figure 7.8. The above equation can be rearranged to obtain

$$\tilde{U}_\infty = \frac{\text{Ga}}{18\Phi(\text{Ga}\,\tilde{U}_\infty)}, \tag{7.8}$$

where we have used the definition $\text{Re} = \text{Ga}\tilde{U}_\infty$. The above implicit equation can easily be solved to obtain $\tilde{U}_\infty$ as a function of Ga for the case of a stationary drag law. The results for a few different values of Ga are presented in Table 7.2 for comparison with Figure 7.12a.

Comparing the nondimensional velocity in the table with that of the contour plot in Figure 7.12a, it is clear that the two are in reasonable agreement for $\rho \gtrsim 2$. Thus, for heavier-than-fluid particles, the standard drag relation can be reliably applied even for a freely moving particle. Whereas, in the case of a lighter-than-fluid or a particle of comparable density, the departure can be substantial. In the extreme case of a bubble (i.e., $\rho = 0$), the average rise velocity in Figure 7.12a is close to that obtained from the standard drag when Ga is close to Ga_{cr1}. But with increasing Ga, the average rise velocity decreases, indicating a higher drag due to the unsteady free motion of the bubble. Also presented in Figure 7.12b is the rms vertical velocity variation as a function of Ga and ρ. The velocity fluctuation is quite small in the case of a heavier-than-fluid particle. The velocity fluctuation is larger for a lighter-than-fluid or a neutrally buoyant particle, but still much smaller than the mean vertical velocity.

7.3 Pseudo Turbulence at the Dilute Limit

A topic of great interest in turbulent multiphase flow is the question of how turbulence in the continuous phase is modified by the presence of the suspended dispersed phase. In other words, how does multiphase turbulence differ from single-phase turbulence? This is a very deep question that we are not yet ready to answer in its entirety. A schematic representation of the energy budget for a multiphase flow was presented by Crowe et al. (2011) and a slightly expanded version is shown in Figure 7.13. According to the schematic, in a single-phase flow, the amount of turbulence in a system changes because of two simple mechanisms:

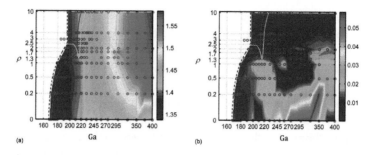

Figure 7.12 (a) Mean vertical velocity normalized by $\sqrt{(\rho - 1)\,g\delta_p}$ plotted over the Ga–ρ plane. (b) The corresponding contour plot of rms vertical velocity. Reprinted from Zhou and Dušek (2015), with permission from Elsevier.

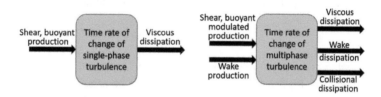

Figure 7.13 Schematic of production and dissipation balance of (a) single-phase and (b) multiphase turbulence.

(1) Shear or buoyant production of turbulence.
(2) Viscous dissipation predominantly by Kolmogorov-scale eddies.

The balance of turbulence gets more complicated in a multiphase flow, with the following additional mechanisms:

(3) Turbulence generation by particle wakes.
(4) Enhancement of viscous dissipation due to the small scales generated by the particles.
(5) Interparticle collisional and frictional dissipation.

The problem is harder than what the schematic displays – the non-uniform distribution of particles within the system will create turbulent motions on scales much larger than interparticle distances. In other words, shear and buoyant turbulent production in a multiphase flow can be substantially modulated by the random distribution and motion of the dispersed phase. This is represented in Figure 7.13 as modulated turbulence production.

   In this section, we will consider the limit when mechanisms (1) and (5) can be ignored. Thus, turbulence generation is only by the wakes of a random distribution of particles and turbulence destruction is by viscous dissipation. The general question of turbulence modulation will be deferred to in later chapters. Consider a simple system of a large number of randomly distributed heavier-than-fluid particles settling down through an otherwise quiescent fluid medium, or equivalently a large number of randomly distributed lighter-than-fluid bubbles ascending through a quiescent fluid. Our analysis will follow Parthasarathy (1990) and Parthasarathy and Faeth (1990). We make the following assumptions: (i) all forms of shear and buoyant production of turbulence are negligible and thus the only source of turbulence is the particle wakes; (ii) the dispersed phase is sufficiently dilute that nonlinear interaction between the wakes of different particles can be ignored – allows linear superposition of the particle wakes; (iii) the particles are randomly distributed over the entire volume with uniform probability (i.e., there is no preferential accumulation of particles); and (iv) particles are sufficiently far apart that collisional/close interactions between them are very rare.

Consider a very large triply periodic system of volume $\mathcal{V}$ containing N particles at a volume fraction of ϕ. In a frame of reference, where the average fluid velocity is zero, let the particles move down (or up) at an average relative velocity of U_∞. Following Batchelor (1972), the ensemble average of a fluid quantity A can be expressed as

$$\langle A \rangle = \frac{1}{N!} \int_\Omega A(\mathbf{x}, t; C_n) P(C_n) \, dC_n \,, \tag{7.9}$$

where the integration is over all possible configurations of the N particles within the volume. Each particular configuration, or the arrangement of the N particles within the box, is denoted as C_n, with the associated probability of that configuration denoted as $P(C_n)$. The property A depends on the spatial point $\mathbf{x}$, time t, and the particle configuration C_n. The factor in front of the integral is for proper normalization of the probability. The above integration simplifies considerably under the assumptions of dilute non-interacting particles and uniform random distribution of particles. Under these assumptions, the above can be replaced by a simple volume integral over the wake of an isolated particle as

$$\langle A \rangle = \frac{6\,\phi}{\pi\,d_{\mathrm{p}}^3} \int_{\mathcal{V}} A(\mathbf{x}, t) dV \,. \tag{7.10}$$

In other words, the above integral exploits the fact that any point chosen within the volume $\mathcal{V}$ has equal probability of lying anywhere in the wake of a particle. Here the pre-factor is simply the number of particles within a unit volume.

As the first example, let the perturbation flow due to each particle be just the potential flow given by the doublet (4.19) and shown in Figure 4.1a. If we take the quantity A to be the Reynolds stress $u_i u_j$ of the perturbation flow, then the potential flow can be substituted in the volume integral and evaluated as (Lance and Bataille, 1991)

$$\langle u_i u_j \rangle = \phi \, U_\infty^2 \begin{pmatrix} \frac{1}{5} & 0 & 0 \\ 0 & \frac{3}{20} & 0 \\ 0 & 0 & \frac{3}{20} \end{pmatrix}. \tag{7.11}$$

The important scaling result we obtain from this simple analysis is that the rms streamwise and cross-stream velocity fluctuations (i.e., $\langle u_\parallel^2 \rangle^{1/2}$ and $\langle u_\perp^2 \rangle^{1/2}$) in the fluid phase due to the wakes of the particles go as $\sqrt{\phi}$ and are proportional to the relative velocity. Furthermore, the wake turbulence is anisotropic, with the rms velocity fluctuation in the direction of relative velocity larger by a factor of 4/3 than the cross-stream rms velocity. However, the above Reynolds stress is homogeneous and stationary (i.e., independent of both **x** and t).

The potential wake assumed in the above analysis is inappropriate for modeling the particle wake at finite Re. The perturbation velocity of the flows shown in Figures 7.1 and 7.3 is substantially different from the potential flow. The laminar and turbulent wake velocity fields defined in Eqs. (7.2) and (7.3) can be used in the integral in Eq. (7.10) instead of the potential flow. Such an analysis was carried by Parthasarathy (1990) and Parthasarathy and Faeth (1990). However, the integral does not converge for both the laminar and the turbulent wake profiles, since the velocity given in Eqs. (7.2) and (7.3) does not decay fast enough along the x-direction. This nonconvergence problem arises in other multiphase flow contexts. Later, in Section 9.4.3, we will see a clever solution by Batchelor (1972) using renormalization in the context of Stokes flow. This elegant approach is not possible in the context of finite Re wake flows. The simple solution offered by Parthasarathy (1990) and Parthasarathy and Faeth (1990) was to limit the integral in Eq. (7.10) to an annular volume between an inner sphere of radius two particle diameters to an outer sphere of radius $\tilde{L}_o d_p$. The rationale is that within the inner diameter, the far wake profiles given in Eqs. (7.2) and (7.3) do not apply and farther than a radius of $\tilde{L}_o d_p$, for a suitably chosen $\tilde{L}_o$, the wake must decay faster and should not contribute to the integral. Their results in terms of the present variables can be expressed as

$$\textbf{Laminar wake} \begin{cases} \langle u_\parallel^2 \rangle &= 0.0117\, C_D^2\, \phi\, U_\infty^2\, \text{Re}\, \ln(\tilde{L}_o/2), \\ \langle u_\perp^2 \rangle &= 8.9 \times 10^{-5}\, C_D^2\, \phi\, U_\infty^2\, \dfrac{\tilde{L}_o - 2}{\tilde{L}_0}, \end{cases} \tag{7.12}$$

and

$$\textbf{Turbulent wake} \begin{cases} \langle u_\parallel^2 \rangle &= 0.313\, C_D^{4/3}\, \phi\, U_\infty^2\, \tilde{L}_o^{1/3}, \\ \langle u_\perp^2 \rangle &= 6.6 \times 10^{-4}\, C_D^2\, \phi\, U_\infty^2\, \dfrac{\tilde{L}_o - 2}{\tilde{L}_0}. \end{cases} \tag{7.13}$$

As in the potential flow, the streamwise and cross-stream rms velocity fluctuations scale as $\sqrt{\phi}\, U_\infty$. Also, the rms velocity fluctuation in the laminar regime scales linearly with the Reynolds number and becomes independent of Re once the wake flow becomes turbulent. We also note that for all the wake models, the Reynolds stress $\langle u_\parallel u_\perp \rangle$ is identically zero. The predicted anisotropy of wake turbulence is quite strong, with the cross-stream velocity fluctuations much weaker than the streamwise component. These last two points are somewhat surprising, since in the case of the turbulent wake of an isolated particle, as given in Eq. (7.4), the Reynolds stress is nonzero and the turbulent fluctuations are nearly isotropic. This is because, in the superposition of wakes, the velocity fluctuations measured at a fixed spatial location are due to sampling of different points in the wake of randomly moving particles. Thus, the contribution to Eq. (7.13) comes from the spatial variation of the mean velocity given in Eq. (7.3). In fact, the small contribution from the turbulent fluctuations of individual wakes was ignored in obtaining Eq. (7.13). Since the mean streamwise velocity deficit was much larger than the mean radial velocity, the turbulence resulting from superposition of random wakes is highly anisotropic.

We finish this section by pointing out that the above approach of evaluating rms velocity fluctuation by linearly superposing the wakes of different particles and assuming the wake of each to be that of an isolated particle is appropriate only at very low volume fraction. Even at modest volume fraction, wake–wake interaction and wake–particle interaction will become important. The problem of nonconvergent integrals is indicative of the limitations of the simple superposition approach. Nevertheless, the results presented above are illustrative of the effect of pseudo turbulence. Later, in Chapter 9, we will consider evaluation of superposition in the context of finite volume fraction, by accounting for the interaction of particles and their wakes. Furthermore, as pointed out by Parthasarathy and Faeth (1990), the above approach is valid only when the rms fluid velocity is much smaller than the particle–fluid relative velocity, since such a limit implies the importance of wakes and pseudo turbulence. In situations where rms fluid velocity fluctuation is comparable to or larger than the particle–fluid relative velocity, the fluid turbulence must have substantial shear and buoyant production and the importance of wake-generated pseudo turbulence decreases. In this latter scenario, one must consider the more complex problem of turbulence modulation by the suspended particles.

7.4   Effect of Turbulence on Particle Settling Velocity

So far we have been concerned with the generation of fluid-phase turbulence by the particles. In this and the next two sections, we will investigate the effect of fluid-phase turbulence back on the particles in terms of their motion and dispersion. This section will consider the effect of turbulence on the mean settling velocity of particles. Let us first examine some existing experimental evidence on the mean settling velocity of particles falling through a field of ambient turbulence. These experiments report a wide variation in the turbulence effect on settling velocity – some report a substantial reduction in the mean settling velocity (Nielsen, 1993; Brucato et al., 1998), some others report a significant increase in the mean settling velocity (Aliseda et al., 2002, Yang and Shy, 2003, 2005), and yet others find a transition from increased to reduced settling velocity (Kawanisi and Shiozaki, 2008; Good et al., 2014). The reason for this diverse behavior is due to the five competing mechanisms that influence the mean settling velocity of particles in a turbulent flow. These mechanisms were discussed in Section 1.3.2 and they are nonlinear drag, loitering, particle trapping, trajectory bias, and two-way coupling effects.

The nonlinear drag, loitering, and particle trapping contribute to a reduction in settling velocity. In the case of nonlinear drag, the higher drag during instances of large than average relative velocity outweights the lower drag during instances of smaller than average relative velocity, thus contributing to an increase in the mean drag and a lower settling velocity. In the case of loitering effect, a particle that is in an updraft region falls slowly and therefore stays in that region far longer than a particle that is in a downdraft region which falls faster and goes out of that region quickly. The net effect of this bias toward a longer duration of slower fall velocity causes the mean settling velocity to be lower than the still-fluid settling velocity. The extreme form of the loitering effect is when a particle is trapped in a strong enough eddy that it remains trapped within it for a long time, thus contributing to a much lower mean settling velocity.

The trajectory bias and two-way coupling effects contribute to an increase in mean settling velocity. In the case of trajectory bias or fast tracking, particles tend to preferentially be located in the downwelling streams rather than in the upwelling streams. This not only avoids the loitering effect, but in fact the particles in the downwelling streams end up settling substantially faster than the still-fluid settling velocity. Two-way coupling is a powerful mechanism, where regions of higher particle concentration tend to be heavier than the surrounding region and the resulting convective instability contributes to a far higher settling velocity than that of an isolated particle in still ambient (Bosse et al., 2006).

Depending on the dominance of the combined effect of the nonlinear drag, loitering, and particle trapping mechanisms versus the combined effect of the trajectory bias and two-way coupling, the overall mean settling velocity can increase or decrease. The three primary parameters that influence these effects are: (i) the particle Stokes number; (ii) the particle Reynolds number; and (iii) the particle volume fraction. As the value of these parameters varies, the relative importance of the five mechanisms varies as well, and as a result past experiments have shown a conflicting influence on the settling velocity, ranging from substantial increase (Maxey, 1987) to modest reduction (Fung, 1997). Direct numerical simulations have been very useful in exploring these mechanisms, and observing the resulting increase and decrease in settling velocity (Wang and Maxey, 1993; Bec et al., 2014; Ireland et al., 2016).

Let us consider the problem of an isolated spherical particle falling through a turbulent field, much like a raindrop or aerosol particle settling down through a turbulent ambient. An important parameter of the problem is the still-fluid settling velocity of the particle, which was obtained in Section 2.4.3 as

$$V_s = \frac{d_{\mathrm{p}}^2}{18\nu_f\,\Phi(\mathrm{Re}_s)}(\rho - 1)\,g, \tag{7.14}$$

where Φ is the finite Reynolds number correction function and $\mathrm{Re}_s = V_s d_p/\nu_f$ is the particle Reynolds number calculated based on the still-fluid settling velocity V_s. Thus, the above is an implicit equation for V_s. Note that as defined above, the still-fluid settling velocity is a constant. Given the material properties of the particle and the fluid (i.e., given $\rho_{\mathrm{p}}, \rho_f, \nu_f$), the still-fluid settling velocity depends only on the particle diameter. Therefore, V_s can simply be considered as a proxy for d_p.

If we consider the particle to be falling through a turbulent field, then the particle velocity will not be a constant over time. If we were to observe different particles, at any instance, their settling velocities will not be the same either. Even in a still ambient, as was discussed in the previous section, at higher Reynolds number, the settling velocity fluctuates due to chaotic vortex shedding. In these situations, the settling velocity (or the vertical component of the particle velocity) will be a random variable. Also, all three components of particle velocity will be nonzero (i.e., the particle trajectory will deviate from a vertical line). One can define the mean velocity component in the gravity direction $\langle V_g \rangle_p$ as the mean settling velocity. Since the vertical component of velocity varies over time, the angle bracket defines the long-time average. Also, an average over many similar particles can be performed. Here again the subscript p has been used to denote that this is a Lagrangian average following the particle.

In general, $\langle V_g \rangle_p = V_s$ only in case of a particle falling through a still ambient. Even here the equality applies only for particle-to-fluid density ratios larger than about 2. As seen in the previous section, for lighter particles, $\langle V_g \rangle_p$ will be smaller than V_s, since the standard drag relation used in calculating V_s underpredicts the actual drag. In the presence of ambient turbulence, in general, $\langle V_g \rangle_p \neq V_s$, and the difference can be attributed to the effect of turbulence. If the mean settling velocity of the particle through a turbulent field is greater than that in still fluid (i.e., $\langle V_g \rangle_p > V_s$), then the effect of turbulence is to decrease the time-averaged drag on the particle. On the other hand, if $\langle V_g \rangle_p < V_s$, then turbulence has increased the effective drag on the particle.

Here we define $\langle V_g \rangle_p$ precisely as the time average of the settling velocity following the particle. However, as pointed out by Davila and Hunt (2001), caution is required in comparing the mean settling velocity reported in prior experiments and simulations. The definition of mean settling velocity varies between the different studies. The mean settling velocity can also be calculated based on the average time particles take to fall down a fixed vertical distance. Instead, the trajectory of a few particles could be followed over time while averaging their instantaneous vertical velocity over time and over all the particles. Yet another option is to observe a large swarm of particles at a few instances and average their instantaneous vertical velocities. For the same system, these definitions of mean settling velocity may not necessarily be identical. Such differences also partially account for the observed and computed differences in settling velocity.

In the following subsections, we will consider two simple analyses: the first will illustrate the increase in mean settling velocity by trajectory bias, or the decrease in mean settling velocity by loitering; the second will illustrate the decrease in mean settling velocity by nonlinear drag effect.

7.4.1   Trajectory Bias or Loitering

In this example, we will obtain an explicit expression for the mean settling velocity in a turbulent field, by assuming the particle to be sufficiently smaller than the scales of turbulence. The surrounding flow will then appear spatially uniform from the perspective of the small particle. However, the ambient flow will be time dependent due to turbulence. Therefore, the finite-Re BBO equation is appropriate to describe the time-dependent motion of the particle. Following the steps of Example 4.11, we rewrite the particle equation of motion (4.102) to obtain

$$\frac{d\mathbf{V}}{dt} = -\frac{2\nabla p_{c@}}{1+2\rho} + \frac{\mathbf{u}_{c@}-\mathbf{V}}{\tau_p} + \frac{1}{1+2\rho}\frac{d\mathbf{u}_{c@}}{dt}$$

$$+ \frac{1}{\tau_p \Phi}\int_{-\infty}^{t} K_{vu}(t-\xi)\left[\frac{d\mathbf{u}_{c@}}{dt}-\frac{d\mathbf{V}}{dt}\right]_{@\xi}d\xi + \frac{V_s\,\Phi(\mathrm{Re}_s)}{\tau_p\,\Phi(\mathrm{Re})}\mathbf{e}_g, \quad (7.15)$$

where $p_{c@} = p_c(\mathbf{X}_p,t)$ and $\mathbf{u}_{c@} = \mathbf{u}(\mathbf{X}_p,t)$ denote the undisturbed fluid pressure and velocity at the particle location, respectively. The terminal settling velocity V_s in the above equation is as defined in Eq. (7.14). The particle time scale τ_p includes the finite-Re drag correction and is given by

$$\tau_p = \frac{(1+2\rho)d_p^2}{36\nu_f\Phi(\mathrm{Re})} \quad \text{where} \quad \mathrm{Re} = \frac{|\mathbf{u}_{c@}-\mathbf{V}|\,d_p}{\nu_f}. \quad (7.16)$$

The last term on the right-hand side of Eq. (7.15) is simply the scaled gravitational vector. By substituting for V_s and τ_p, this term can be varied to be equal to $(\rho - 1)\mathbf{g}/(\rho + 1/2)$. The above equation for a turbulent flow is more complex than that for a quiescent ambient. In order to obtain an explicit solution, we make some simplifying assumptions. Ignoring the pressure gradient, added-mass and Basset history forces represented by the first, third, and fourth terms on the right-hand side, we obtain

$$\frac{d\mathbf{V}}{dt} = \frac{\mathbf{u}_{c@}-\mathbf{V}+V_s(\Phi(\mathrm{Re}_s)/\Phi(\mathrm{Re}))\,\mathbf{e}_g}{\tau_p}. \quad (7.17)$$

In this example, we simplify the problem by considering the limit of very small Stokes number (i.e., the limit $\tau_p \to 0$). In this limit, as the denominator goes to zero the numerator must also go to zero. Therefore, the particle is simply moving with the surrounding fluid, except for an additional vertical velocity due to settling as given by the solution

$$\mathbf{V} = \mathbf{u}_{c@} + V_s\frac{\Phi(\mathrm{Re}_s)}{\Phi(\mathrm{Re})}\mathbf{e}_g. \quad (7.18)$$

We further assume the Reynolds number to be small, so that $\Phi(\mathrm{Re}_s), \Phi(\mathrm{Re}) \approx 1$. Then the time variation of particle velocity $\mathbf{V}(t)$ is simply due to the time variation of the local undisturbed fluid velocity seen by the particle. The time average of the particle velocity along the gravity direction is given by

$$\langle V_g\rangle_p = \langle \mathbf{u}_{c@}\cdot\mathbf{e}_g\rangle_p + V_s. \quad (7.19)$$

It is clear that, depending on the value of $\langle\mathbf{u}_{c@}\cdot\mathbf{e}_g\rangle_p$, the mean settling velocity can be larger or smaller than the still-fluid settling velocity. Note that $\langle\mathbf{u}_{c@}\cdot\mathbf{e}_g\rangle_p$ is the Lagrangian time average of the fluid velocity at the particle location along its trajectory. This Lagrangian average can be different from zero, even if the ensemble average of the fluid velocity is zero. This is due to the fact that a particle may preferentially sample only selective flow regions. For instance, as it falls, if a particle prefers downwelling regions of fluid and avoids upwelling regions, then $\langle\mathbf{u}_{c@}\cdot\mathbf{e}_g\rangle_p$ will be positive and the resulting $\langle V_g\rangle_p$ will be larger than V_s. This is what we referred to in Chapter 1 as trajectory bias or fast tracking. Instead, if the particle gets trapped and spends more time in upwelling regions than in downwelling regions, then $\langle \mathbf{u}_{c@} \cdot \mathbf{e}_g \rangle_p$ will be negative and $\langle V_g \rangle_p$ will be smaller than V_s. This is the loitering effect.

The above results of trajectory bias are rigorous only in the limit of particles being much smaller than the Kolmogorov scale. Only in this limit are the approximations made above appropriate. As will be shown in Figure 13.3 of Chapter 13, for particles smaller than the Kolmogorov size, the particle Reynolds number is smaller than unity and this justifies the earlier assumption of $\Phi(\mathrm{Re}_s), \Phi(\mathrm{Re}) \approx 1$. For such a particle, the undisturbed ambient flow seen by the particle can be well approximated to be spatially uniform. This justifies the use of the finite-Re BBO equation, which relies on fluid properties evaluated at the particle center (otherwise the MRG equation with surface and volume averages must be used). Particle-resolved direct numerical simulations have shown the finite-Re BBO equation to be quite accurate in capturing the motion of sub-Kolmogorov-sized particles. As observed in the simulations of a small particle subjected to isotropic turbulence (Bagchi & Balachandar 2003), or a small particle at the center of a turbulent pipe flow (Merle et al., 2005), provided the time history of the continuous-phase velocity at the particle location is known, the finite-Re version of the BBO equation can be used to accurately compute the time evolution of force. In fact, for such small particles, the dominant contribution to force comes from the quasi-steady term and the unsteady contributions are typically negligible. This justifies the approximation involved in reducing Eq. (7.15) to Eq. (7.17).

7.4.2   Nonlinear Drag

This example is the same as the previous one, but instead considers the limit of a particle of very large Stokes number (i.e., take the limit of τ_p much larger than the fluid time scale). In this limit, the particle is largely unresponsive to the surrounding fluid and the particle settling velocity can be taken to be a constant. Let us still denote this constant settling velocity as $\langle V_g \rangle_p$, as in the previous subsection, although the angle bracket is not needed now due to the constancy of the settling velocity. As we will see below, the constant settling velocity will be different from the still-fluid settling velocity (i.e., $\langle V_g \rangle_p \neq V_s$). This is because, as the particle settles at the constant velocity, its relative velocity with respect to the surrounding fluid will vary over time due to ambient turbulence. The drag on the particle will vary over time as well. The time-averaged drag will determine the value of $\langle V_g \rangle_p$.

The governing equation is again a simplification of Eq. (7.17). Even though the particle velocity changes over time, the time rate of change is small. At any given instance, the particle will have a nonzero acceleration, since the time-varying drag force cannot precisely match the constant gravitational force. But the acceleration of the particle will be negligible due to the large Stokes number of the particle. Hence, we neglect the ensemble average of particle acceleration in the above equation. Correspondingly, the ensemble of the numerator on the right-hand side goes to zero as well. We now define the relative velocity $\mathbf{V}_r = \mathbf{V} - \mathbf{u}_{c@}$ to be the negative of the fluid velocity seen by the moving particle. Rearranging the numerator on the right-hand side of Eq. (7.17), taking the dot product with $\mathbf{e}_g$, and the Lagrangian average over the particle trajectory, we obtain

$$\langle V_{rg}\Phi(\mathrm{Re})\rangle_p = \Phi(\mathrm{Re}_s)V_s\,, \qquad (7.20)$$

where $V_{rg} = \mathbf{V}_r \cdot \mathbf{e}_g$ is the relative velocity along the gravity direction.

The above equation can be interpreted in the following way. Given the particle and ambient fluid properties, the still-fluid settling velocity V_s and the still-fluid finite-Re correction $\Phi(\mathrm{Re}_s)$ are known. So, the right-hand side of the above equation is a time-independent constant. On the left-hand side, we make an assumption that the relative velocity is primarily in the vertical direction, which yields

$$\Phi(\mathrm{Re}) = 1 + 0.15\left(\frac{d_p\,V_{rg}}{\nu_f}\right)^{0.687}. \qquad (7.21)$$

Note that although the particle settling velocity is taken to be a constant due to the large inertia of the particle, the relative velocity V_{rg} with respect to the surrounding fluid in the gravity direction varies over time as the particle passes through different regions of the turbulent flow. Equation (7.20) states that the time average of this relative velocity weighted by Φ must match the right-hand side. This simply implies that the drag on the particle averaged over time (left-hand side) must balance the constant gravitational force (right-hand side).

The time-dependent relative velocity can be separated into a time-averaged mean and a fluctuating part as

$$V_{rg}(t) = \langle V_{rg}\rangle_p(1 + \epsilon(t))\,, \qquad (7.22)$$

where $\epsilon(t)$ is the fluctuating component scaled by the time average. Substituting this decomposition into Eqs. (7.20) and (7.21), and Taylor series expanding for small ϵ, we obtain the following:

$$\langle V_{rg}\rangle_p \left[1 + 0.15\langle \mathrm{Re}_{rg}\rangle_p^{0.687}\left(1 + 0.687\langle \epsilon^2\rangle_p\right)\right] + O(\langle \epsilon^4\rangle_p) = \Phi(\mathrm{Re}_s)\,V_s\,, \qquad (7.23)$$

where the Reynolds number based on the time-averaged relative velocity is $\langle \mathrm{Re}_{rg}\rangle_p = d_p\langle V_{rg}\rangle_p/\nu_f$. Also, in the above equation, we have used the fact that by definition, the time average of the fluctuation $\langle \epsilon\rangle_p = 0$. Here, $\langle \epsilon^2\rangle_p$ is the mean square measure of how much the relative velocity fluctuates about the mean. Since particle velocity is a constant, $\langle \epsilon^2\rangle_p$ is related to the mean square fluid velocity fluctuation seen by the particle normalized by the mean relative velocity.

In the limit $\langle \epsilon^2\rangle_p \to 0$ (i.e., in the limit where the ambient turbulent velocity fluctuation is small compared to the mean relative velocity), $\langle \mathrm{Re}_{rg}\rangle_p \to \mathrm{Re}_s$ and the term within the square brackets on the left-hand side equals $\Phi(\mathrm{Re}_s)$ on the right-hand side. Thus, we recover the result $\langle V_{rg}\rangle_p = V_s$.

In the zero Reynolds number limit of Re_{rg}, $\mathrm{Re}_s \to 0$, we also obtain the result $\langle V_{rg}\rangle_p = V_s$. This is valid even when $\langle \epsilon^2\rangle \neq 0$, since the Reynolds number correction factor $\Phi(\mathrm{Re}_s) = 1$. Thus, at finite Re, when fluctuations in the ambient fluid velocity are significant, the effect of $\langle \epsilon^2\rangle_p > 0$ is to decrease the value of relative velocity and result in $\langle V_{rg}\rangle_p < V_s$. This important result can be generalized. *At finite Re, if a particle sees a time-varying relative velocity, then the time average of the time-dependent standard drag on the particle calculated based on its instantaneous velocity will be larger than the standard drag calculated based on the time-averaged relative velocity.* The difference is due to the nonlinear dependence of standard drag on relative velocity.

As the final step, let us derive the expression for the constant settling velocity. Ensemble averaging the vertical component of the relative velocity definition $\mathbf{V}_r = \mathbf{V} - \mathbf{u}_{c@}$, we obtain

$$\langle V_g \rangle_p = \langle \mathbf{u}_{c@} \cdot \mathbf{e}_g \rangle_p + \langle V_{rg} \rangle_p, \qquad (7.24)$$

where $\langle V_g \rangle_p = \mathbf{V} \cdot \mathbf{e}_g$ is the settling velocity of the particle, which in the present case is a constant. The above equation for the large inertia limit is analogous to Eq. (7.19) for the small inertia limit. In the context of small particle inertia, we discussed the possibility of $\langle \mathbf{u}_{c@} \cdot \mathbf{e}_g \rangle_p$ being positive and contributing to an increase in effective settling velocity due to trajectory bias or fast tracking. In the present case of a particle of large inertia, since the particle is unresponsive to turbulent fluctuations, the likely scenario is that the particle will sample all regions of the turbulent flow equally without any bias toward upwelling or downwelling flow and as a result we expect $\langle \mathbf{u}_{c@} \cdot \mathbf{e}_g \rangle_p \approx 0$.

As a result, we obtain the result $\langle V_g \rangle_p = \langle V_{rg} \rangle_p$, and thus the constant particle settling velocity is the same as the average relative velocity of the particle with respect to the surrounding fluid. Since $\langle V_{rg} \rangle_p < V_s$, we expect the net settling velocity of the inertial particle to be lower than the still-fluid settling velocity. This effect is the nonlinear effect of finite-Re settling. In essence, the higher drag at higher Re contributes more to net drag than the lower drag at lower Re. As a result, the effective drag is somewhat higher and correspondingly the particle settling velocity is lower than in still fluid.

7.4.3   Finite Stokes Number Effect on Settling Velocity

The above two analyses employed small and large particle inertia values, respectively. The inertial behavior of a particle is best described in terms of the particle Stokes number $St_\eta = \tau_p/\tau_\eta$, which is defined as the ratio of particle time scale τ_p to Kolmogorov time scale τ_η. At small particle inertia, we were able to demonstrate the trajectory bias that led to increased settling velocity and at large inertia we demonstrated the nonlinear drag effect that reduced settling velocity. It should be emphasized that in the limit $St_\eta \to 0$, when particle inertia is completely negligible, the particle will simply behave as a fluid tracer and there will be no increase in settling velocity due to trajectory bias. A small but nonzero inertia effect is necessary to obtain enhanced settling. In fact, experiments and simulations show that the biggest increase in settling velocity is realized for $St_\eta \sim 1$, since this is when particles best preferentially accumulate along the downwelling streams. Similarly, in the limit $St_\eta \to \infty$, the nonlinear effect of decreased settling velocity vanishes, since the settling velocity is so much larger than the fluid velocity fluctuations. Therefore, in the limit $St_\eta \to \infty$, the mean settling velocity becomes equal to the still-fluid settling velocity. A large but finite Stokes number is needed to realize the nonlinear effect.

The complex variation of settling velocity as St_η is varied from 0 to ∞ has been studied both experimentally and using point-particle Euler–Lagrange simulations. By

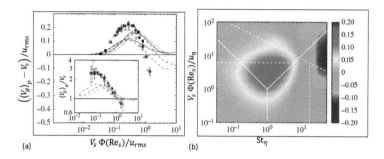

Figure 7.14 Figure taken from Good et al. (2014). (a) A plot of $(\langle V_g \rangle_p - V_s)/u_{\text{rms}}$ as a function of $V_s \Phi(\text{Re}_s)/u_{\text{rms}}$; the inset shows a plot of $\langle V_g \rangle_p / V_s$. The results of both experiments (symbols) and simulations (lines) are shown. (b) A contour plot of $(\langle V_g \rangle_p - V_s)/u_{\text{rms}}$ presented as a function of $\text{St}_R \, \eta$ and $V_s \Phi(\text{Re}_s)/u_\eta$. For additional details, refer to Good et al. (2014).

employing some variant of the finite-Re BBO equation for the point-particle force, the simulations have been successful in demonstrating the nonlinear drag, loitering, trajectory bias, and two-way coupling effects of turbulence. The enhancement and suppression of mean settling velocity has been observed in the simulations in the appropriate parameter regimes. The other two nondimensional parameters of importance are the particle-to-fluid density ratio and the ratio of still-fluid settling velocity to the Kolmogorov velocity scale (i.e., the ratio V_s/u_η), or alternately the ratio of still-fluid settling velocity to the rms turbulent velocity (i.e., the ratio V_s/u_{rms}). Here we present the comprehensive summary provided by Good et al. (2014) for the case of droplets in air, which corresponds to a large particle-to-fluid density ratio of $O(1000)$. The results are presented as mean settling velocity deviation from the still-fluid value nondimensionalized by the rms turbulent velocity fluctuation as $(\langle V_g \rangle_p - V_s)/u_{\text{rms}}$, which is plotted in Figure 7.14a as a function of the nondimensional still-fluid settling velocity. (Note that their definition was without the finite-Re correction, which introduces the additional factor $\Phi(\text{Re}_s)$ along the x-axis.) With increasing nondimensional settling velocity, both the experimental and simulation results clearly show the transition from a regime of increased settling to a regime of decreased settling. Based on results from an extensive set of simulations, their contour plot of nondimensional change in settling velocity as a function of both St_η and $V_s \Phi(\text{Re}_s)/u_\eta$ is presented in Figure 7.14b, where it can clearly be observed that for increasing St_η, $(\langle V_g \rangle_p - V_s)/u_{\text{rms}}$ increases above zero, reaches a peak positive value around $\text{St}_\eta \sim 1$, then decreases to take on negative values. Though not shown in the picture, they argue that with further increase in St_η, $(\langle V_g \rangle_p - V_s)/u_{\text{rms}}$ should go back to zero.

Studies of the settling behavior of particles of size smaller than the Kolmogorov scale offer the benefit of Euler–Lagrange simulations with point-particle models. Extension to particles of size larger than the Kolmogorov scale will require particle-resolved simulations of the sedimentation process. Several groups have pursued this approach very successfully in the past few years (Chouippe and Uhlmann, 2015, 2019; Fornari et al., 2016, 2019; Uhlmann and Chouippe, 2017; Zaidi, 2018; Willen and Prosperetti,

2019). The five mechanisms mentioned in the context of small particles continue to apply even in the context of finite-sized particles. However, due to the increased computational cost of the particle-resolved simulations, only a few combinations of key parameters have been studied in detail. Clearly, more research is needed for a comprehensive understanding.

7.5   Turbulent Diffusion of Particles

The previous section considered the effect of turbulence on the mean settling velocity of particles and we noted that depending on the value of particle inertia, still-fluid settling velocity, and turbulence intensity, the mean settling velocity in a turbulent flow can be higher or lower than that in still ambient. In this section, we will consider the problem of turbulent diffusion. Let us perform a mental experiment of releasing particles at a fixed point in a stationary isotropic turbulence of zero mean velocity. Due to homogeneity and stationarity of turbulence, the point of particle release can be taken to be the origin, and the time instant of release can be set to zero. Let the still-fluid settling velocity of the particle be V_s. We repeatedly release a particle at this point at well-separated times and observe its location at time $t = t_*$ after each release. All the released particles form an ensemble. In a quiescent fluid, assuming Re_s to be sufficiently small that settling is steady, all the particles would have fallen down precisely the same distance $V_s t_*$.

In isotropic turbulence, the trajectory of each particle will cease to be perfectly vertical, and depending on the prevailing turbulent velocity fluctuations, each particle's trajectory will randomly vary about the vertical path. This scenario is depicted in Figure 7.15. The resulting ensemble of particle positions at time t_* after release, denoted in Figure 7.15 as dispersed dots, forms a cloud, whose size is indicative of the level of particle diffusion at time t_*. The mean position of the cloud is approximately at a distance $\langle V_g \rangle_p t_*$ below the point of release, where the mean settling velocity $\langle V_g \rangle_p$ can be different from the still-fluid settling velocity due to the influence of turbulence, as seen in the previous section.[1] In mathematical terms, if the position of a particle in the ensemble at a later time is denoted by $\mathbf{X}(t_*)$, then the ensemble average of the position over all particles (i.e., $\langle \mathbf{X} \rangle$) will yield the centroid of the cloud seen in Figure 7.15. In contrast, the root mean square of the displacement of the ensemble of particles from the mean (i.e., $\langle (\mathbf{X} - \langle \mathbf{X} \rangle)^2 \rangle^{1/2}$) will indicate the radius of the dispersion. The object of this section is to obtain an estimate of the radius of spread of the particle cloud around the mean. In other words, our focus is on the quantification of turbulent diffusion of particles.

In the above discussion, the diffusion of particles is due to externally imposed ambient turbulence. However, even at the level of an isolated particle, as the wake of the particle becomes turbulent, above a Reynolds number of about a few thousand, its trajectory will be chaotic. If the experiment of repeatedly releasing an ensemble of

---

[1] If the particles are released from rest, they need to accelerate to reach the average fall velocity. We assume this transient time to be small compared to the total time of observation.

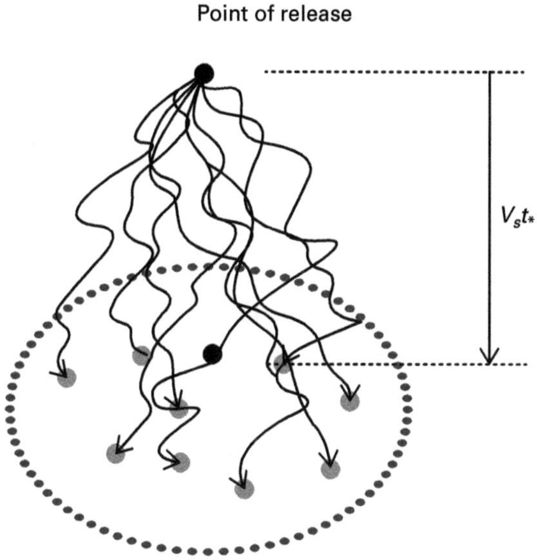

Point of release

Figure 7.15 Particle diffusion in isotropic turbulence. The particles released at the origin at well-separated time instants take a different trajectory and at time t_* their positions are indicated by the gray points. The particles have a mean settling velocity and thus the centroid of the cloud is located at a distance $V_s t_*$ below the point of release. The size of the cloud is indicative of the level of particle diffusion.

particles were to be performed with such large particles, turbulent diffusion of particles would be observed in the position of the particle, even in an otherwise quiescent ambient medium. Clearly, the turbulence in this case is self-generated. But the diffusional effect of this self-generated wake turbulence is somewhat weak. In the context of a dilute distribution of sedimenting particles, a particle may be subjected to wake turbulence of its upstream neighbors. However, as we discussed in Section 7.3, this pseudo turbulence generated by particles is anisotropic in nature. Thus, by considering isotropic ambient turbulence, the rest of this section implicitly assumes that the turbulence was generated by some external mechanism and the role of particle-generated pseudo turbulence is relatively unimportant. But the analysis can be extended to consider a similar effect of particle-generated anisotropic pseudo turbulence on particle diffusion.

7.5.1   Taylor's Example of Discontinuous 1D Movements

As a first step, we will visit the simple example considered by G. I. Taylor (1922) about a century ago. Consider a 1D problem where a particle is constrained to move along a straight line, either forward with velocity $+V$ or backward with velocity $-V$. Each small forward or backward motion lasts for a time interval δt, and thus in each of these steps the particle moves forward by a distance $\delta X = +V\delta t$ or slides backward by a distance $\delta X = -V\delta t$. Although this example seems quite contrived and far from the reality of a turbulent flow, the scaling relations obtained are quite powerful, and therefore we will continue with this simple example.

Since the forward and backward motions are equally likely, the mean distance moved by the particle is identically zero. So, we are interested in the standard deviation or rms of the distance moved by a particle after n successive steps. The mean square distance $\langle X_n^2 \rangle$ can be expressed as

$$\langle X_n^2 \rangle = \langle (\delta X_1 + \delta X_2 + \cdots + \delta X_n)^2 \rangle, \tag{7.25}$$

where the angle bracket indicates an ensemble average over many particles. For each particle of the ensemble, each step of the particle motion is a random variable – either forward or backward. By definition, we have $\langle \delta X_q \delta X_q \rangle = V^2 \delta t^2$. In order to evaluate the right-hand side, we make an assumption that for each particle, two consecutive steps are not completely independent and that they have a correlation coefficient of c. That is, we assume $\langle \delta X_q \delta X_{q+1} \rangle = c V^2 \delta t^2$, where c is a positive constant less than unity. By ignoring higher-order correlations between steps that are separated by more than one, we further assume $\langle \delta X_q \delta X_{q+p} \rangle = c^p V^2 \delta t^2$. With these correlations, the expression for the mean square distance, after expanding out the square of the summation, becomes

$$\langle X_n^2 \rangle = V^2 \left(\frac{1+c}{1-c} \delta t\, t - \frac{2c^2(1-c^n)}{(1-c)^2} \delta t^2 \right), \tag{7.26}$$

where we have set the elapsed time of n steps to be $t = n\delta t$.

Following Taylor (1922), we then approach the continuous limit by taking an infinite number of time steps of infinitesimal magnitude (i.e., $n \to \infty$ and $\delta t \to 0$), but their product $t = n\delta t$ is finite. In this limit, for the mean square distance to converge to a finite value, we require the correlation coefficient to approach unity. In other words, we require $c \to 1$, which makes sense, since as $\delta t \to 0$, each step must be nearly perfectly correlated with the next one. However, as $\delta t \to 0$ and $c \to 1$, their ratio must approach

$$\frac{\delta t}{1-c} \to \tau_{\text{cor}}, \tag{7.27}$$

where τ_{cor} has the interpretation that it is the time scale on which the correlation coefficient decays. In other words, in the continuous limit the correlation $\langle \delta X(t) \delta X(t+\tau) \rangle$ must decay as $\langle \delta X^2 \rangle e^{-t/\tau_{\text{cor}}}$. Substituting this limit into Eq. (7.26), we obtain the final result

$$\langle X_n^2 \rangle = 2V^2 \tau_{\text{cor}} \left(t - \tau_{\text{cor}} \left(1 - e^{-t/\tau_{\text{cor}}} \right) \right). \tag{7.28}$$

We can now interpret the above result to obtain the key scaling relations. In the limit of small time ($t \ll \tau_{\text{cor}}$), we can expand the exponential and obtain the root mean square distance to be $\langle X_n^2 \rangle^{1/2} \simeq Vt$. In other words, for times much smaller than the correlation time, successive steps remain correlated with the very first step and thus the particle continues to move either forward or backward over a distance Vt, which is then the rms distance traveled by the particle. In the limit of large time ($t \gg \tau_{\text{cor}}$), we obtain the scaling $\langle X_n^2 \rangle^{1/2} \simeq V\sqrt{2\tau_{\text{cor}}t}$. In summary, we have obtained the fundamental result that in correlated random motion, the diffusion of particles is initially proportional to t at small times, but diffusion slows down and changes to $\propto \sqrt{t}$ for long times.

7.5.2   Taylor's Theorem on Diffusion by Continuous Movements

The purpose of this theorem is to obtain an explicit expression for the mean square displacement of an ensemble of particles away from their mean position, and the square root of this quantity characterizes the size of the cloud of dispersed particles, such as the one shown in Figure 7.15. In this case, we will consider the motion of particles in 3D, and the position and velocity of the particles will be denoted by the time-dependent vectors $\mathbf{X}(t)$ and $\mathbf{V}(t)$. For simplicity, we will consider the mean velocity of the particles to be zero and as a result the mean particle displacement at all times is zero as well (i.e., $\langle X_i \rangle = 0$).

We start by calculating the time derivative of the covariance of particle displacement:

$$\frac{d\langle X_i X_j \rangle}{dt} = \left\langle X_i \frac{dX_j}{dt} \right\rangle + \left\langle X_j \frac{dX_i}{dt} \right\rangle = \langle X_i(t)V_j(t) \rangle + \langle X_j(t)V_i(t) \rangle, \tag{7.29}$$

where we have explicitly written that the covariance of position and velocity is evaluated at time t. We then express the particle position as the integral of particle velocity, $X_i(t) = \int_0^t V_i(\xi)d\xi$, and commute the integral and the ensemble average to obtain

$$\frac{d\langle X_i X_j \rangle}{dt} = \int_0^t \left(\langle V_i(\xi)V_j(t) \rangle + \langle V_j(\xi)V_i(t) \rangle \right) d\xi. \tag{7.30}$$

We now introduce the two-time particle velocity correlation tensor:

$$R_{ij}^{\text{tvd}}(\xi) = \frac{\langle V_i(\xi)V_j(0) \rangle}{\langle V_i V_j \rangle}, \tag{7.31}$$

where the superscript "tvd" corresponds to temporal correlation of the particle (or the dispersed-phase) velocity following the particle. Since the turbulence is taken to be stationary, the numerator depends only on the time separation. The integrand in Eq. (7.30) can then be written as $2\langle V_i V_j \rangle R_{ij}^{\text{tvd}}(t - \xi)$, whose integral can be expressed as

$$\mathcal{D}_{ij}^d = \frac{1}{2} \frac{d\langle X_i X_j \rangle}{dt} = \langle V_i V_j \rangle \int_0^t R_{ij}^{\text{tvd}}(\xi) \, d\xi. \tag{7.32}$$

The above expression defines the particle diffusivity tensor and like other diffusion coefficients, such as kinematic viscosity or thermal diffusivity, it has dimensions of length^2/time. Thus, a diffusion term of the form $\mathcal{D}_{ij}^d \partial^2 \phi / (\partial x_i \partial x_j)$ should be included in the particle concentration or volume fraction equation in order to account for the turbulent diffusion of particles from regions of high concentration to regions of low concentration. A couple of important points must be noted. First, unlike molecular processes which lead to isotropic kinematic viscosity and thermal diffusivity, the effect of turbulence on particles can be anisotropic and thus the general form of particle diffusion is tensorial. In particular, in the case of sedimenting particles in isotropic turbulence, the vertical direction (denoted by "3") will be different from the other two directions (i.e., $\mathcal{D}_{11}^d = \mathcal{D}_{22}^d \neq \mathcal{D}_{33}^d$), and all other off-diagonal terms of the particle diffusion tensor are identically zero by symmetry. Second, as defined above, the particle diffusivity tensor is time dependent. However, this time dependence is important only for short duration. To see this, a particle integral time-scale tensor can be defined as

$$T_{ij}^{\text{tvd}} = \int_0^\infty R_{ij}^{\text{tvd}}(\xi)\, d\xi. \tag{7.33}$$

In a turbulent flow, all the components of the correlation tensor R_{ij}^{tvd} decrease with increasing ξ and become zero. This is because a large value of ξ corresponds to widely separated time instants and the particle velocity perturbations are completely uncorrelated at such large time separation. The characteristic time scale on which R_{ij}^{tvd} decays to zero and becomes uncorrelated is the particle integral time scale T_{ij}^{tvd}. Thus, for $t \gg T_{ij}^{\text{tvd}}$, the integral on the right-hand side of Eq. (7.32) can be replaced by T_{ij}^{tvd}. This yields the long-time particle diffusion tensor $\mathcal{D}_{ij}^d = \langle V_i V_j \rangle T_{ij}^{\text{tvd}}$, which remains anisotropic, but is independent of time. It should be noted that $\mathcal{D}_{ij}^d$ can spatially vary depending on the inhomogeneous nature of $\langle V_i V_j \rangle$ and T_{ij}^{tvd}.

Integrating Eq. (7.32), we obtain the following final expression for the covariance of particle displacement:

$$\langle X_i X_j \rangle = 2\langle V_i V_j \rangle \int_0^t \int_0^{\xi_*} R_{ij}^{\text{tvd}}(\xi)\, d\xi\, d\xi_*. \tag{7.34}$$

We now consider the asymptotic behavior of the above equation. In the small time limit, when the two-time correlation is almost unity, the above double integral reduces to $t^2/2$ and we obtain $\langle X_1 X_1 \rangle = \langle V_1^2 \rangle t^2$, and thus the short-time rms particle displacement is proportional to t. This is precisely the result we obtained in the previous section. For $t \gg T^{\text{tvd}}$, we go back to Eq. (7.32), where the integral can be evaluated as T_{ij}^{tvd}. Integrating this, we obtain the result $\langle X_1 X_1 \rangle = 2\langle V_1^2 \rangle T_{11}^{\text{tvd}} t$. Again, as in the previous section, the long-time rms particle displacement goes as $\sqrt{t}$.

From the above discussion, it becomes clear that R_{ij}^{tvd} plays a fundamental role in characterizing turbulent diffusion of particles. Several important particle-related quantities can be obtained from this two-time particle velocity correlation. The time scale of particle velocity correlation, particle diffusivity tensor, and radius of the dispersing cloud can be obtained from Eqs. (7.33), (7.32), and (7.34), respectively. Unfortunately, R_{ij}^{tvd} is not an easy quantity to measure, since it requires following the path of an ensemble of particles and correlating the velocity of individual particles at two different times, when the particles are likely to be at different positions at the later time. As we will see in the next section, R_{ij}^{tvd} can be modeled with some judicious approximations and thereby particle diffusion can be evaluated.

7.5.3   Eddy Decay and Crossing Trajectory Effects

We go back to the problem of particles falling through a field of isotropic turbulence, as depicted in Figure 7.15. In this section, our goal is to relate the hard-to-evaluate two-

Table 7.3 Listing of all velocity correlations to be discussed in this section defined in terms of the superscript. The integral time scales associated with the temporal correlations are also denoted in the second column.

Correlation Type	Superscript	Description
Temporal (two-time)	$\mathbf{R}^t, T^t$	Eulerian temporal correlation between fluid velocities at the same point in space (can only be fluid velocity)
	$\mathbf{R}^{\mathrm{tuc}}, T^{\mathrm{tuc}}$	Lagrangian temporal correlation between fluid velocity ($\underline{u}$) at two points following the fluid (continuous phase)
	$\mathbf{R}^{\mathrm{tud}}, T^{\mathrm{tud}}$	Lagrangian temporal correlation between fluid velocity ($\underline{u}$) seen at two points following the particle (dispersed phase)
	$\mathbf{R}^{\mathrm{tvd}}, T^{\mathrm{tvd}}$	Lagrangian temporal correlation between particle velocity ($\underline{v}$) seen at two points following the particle (dispersed phase)
Spatial (two-point)	$\mathbf{R}^x, \mathbf{R}^y$, or $\mathbf{R}^z$	Correlation between fluid velocity at two spatially separated points along x, y, or z-direction
Space-time (two-point, two-time)	$\mathbf{R}^{tx}, \mathbf{R}^{ty}$, or $\mathbf{R}^{tz}$	Correlation between fluid velocity at two spatially separated points at two different times

time particle velocity correlation to the more easily computable Eulerian two-time fluid velocity correlation (also known as the fixed-point fluid velocity correlation). Before proceeding, we note that in the present context, only the diagonal elements of $\mathbf{R}^{\mathrm{tvd}}$ are nonzero, and all the off-diagonal terms are zero by symmetry. Even among the diagonal elements, only the vertical direction is different due to its preferred direction of particle settling and therefore $R_{11}^{\mathrm{tvd}} = R_{22}^{\mathrm{tvd}}$, where direction "3" is chosen to be the vertical direction.

In what follows, we will systematically proceed in a step-by-step manner to relate the particle velocity correlation to the two-time Eulerian velocity correlation and in the process introduce two other velocity correlations. Notational consistency is important in this process. In all cases, the velocity correlation tensor will be denoted $\mathbf{R}^{\square}$, where the characters in the superscript will precisely denote what kind of velocity correlation is being addressed. Individual components of the tensor will be denoted with the subscript ij. Table 7.3 introduces the different superscripts and their meaning.

Consider a fluid velocity probe fixed in position in isotropic turbulence. The two-time velocity correlation tensor, $\mathbf{R}^t$, obtained from such a probe will be diagonal and isotropic (the superscript t corresponds to time correlation in the Eulerian frame of a fixed probe). The two-time Eulerian velocity correlation also decays, since the turbulent eddies that contribute to velocity fluctuation at the probe decay over time.

This process has been termed the *eddy decay effect*, and the associated time scale is defined in a similar manner as[2]

$$T_{ij}^t = \int_0^\infty R_{ij}^t(\xi)\,d\xi \,. \qquad (7.35)$$

Here and henceforth we will generally drop the tensorial subscripts when referring to the time scales and diffusivities, when the indices are not needed for the discussion.

In our quest to relate R_{ij}^{tvd} to R_{ij}^t, as the next step, instead of considering a fixed probe, we follow the trajectory of an individual fluid parcel over time to obtain the two-time Lagrangian fluid velocity correlation, $\mathbf{R}^{\text{tuc}}$, which too will decay over time and the associated time scale can be denoted T^{tuc}. The important difference between the Eulerian and Lagrangian fluid velocity correlation is that in the latter, the velocity at the second time is measured at the location of the fluid parcel (i.e., following the fluid parcel), as opposed to at the location of the fixed probe. Larger turbulent eddies tend to advect both the smaller eddies and the fluid parcel in the same wake, which will tend to make the Lagrangian correlation slower to decay. On the other hand, the fluid parcels may advect out of the larger eddies, which will tend to make the Lagrangian correlation decay faster. In general, it can be expected that $T^{\text{tuc}} \neq T^t$.

Since the interest here is not in the fluid parcels, but the particles that make up the dispersed phase, as a next step we introduce the correlation tensor $\mathbf{R}^{\text{tud}}$, which corresponds to two-time correlation between the fluid velocity seen by the particle (see Table 7.3). This is similar to $\mathbf{R}^{\text{tuc}}$ in the sense that the correlation is between fluid velocity components, but the important difference is that the two fluid velocities being correlated are at the particle location at the two times (i,e., following the particle), instead of following the trajectory of the fluid parcel. The trajectory of the particle will differ from that of the fluid parcel that surrounded it at $t = 0$ due to both the inertial and the settling effects. As we discussed earlier, a heavier-than-fluid inertial particle will spiral out of the fluid vortices, due to its inertia. Furthermore, a heavier-than-fluid particle, acted upon by gravity, will continue to fall relative to the surrounding fluid. As a result of both inertia and gravity, the trajectory of the particle will cut across fluid pathlines (i.e., the particle pathline will deviate from that of the fluid parcel) and this phenomenon was termed the *crossing trajectory effect* (Yudine, 1959; Csanady, 1963). Clearly, the effect of crossing trajectory is to accelerate the rate of decorrelation of fluid velocity and the time scale $T^{\text{tud}} \sim \int_0^\infty \mathbf{R}^{\text{tud}}(\xi)\,d\xi$ can be expected to be smaller (i.e., $T^{\text{tuc}} > T^{\text{tud}}$).

As the final step, we consider the two-time particle velocity correlation $\mathbf{R}^{\text{tvd}}$, which was the quantity of interest in Taylor's theorem. Here too, we follow the trajectory of the particle, but along the trajectory we correlate the particle velocity, while in $\mathbf{R}^{\text{tud}}$ the correlation was between the fluid velocity seen by the particle. The effect of the particle's "inertial memory" will be to slow down the decay of $\mathbf{R}^{\text{tvd}}$ and thereby

---

[2] It must be stressed that the above definition is in the context of isotropic turbulence without any mean velocity. In this case, there is no mean advection velocity and any decorrelation at a fixed probe is due to eddy decay. If there is a mean flow, turbulent eddies will be swept downstream at the mean flow velocity and decorrelation of a fixed problem will be due to both eddy decay and the advection of the turbulent eddies. In case of a nonzero mean flow, one must consider a probe traveling downstream at the mean velocity in order to recover only the eddy decay effect.

increase the associated time scale T^{tvd}. As a result, the *inertial memory effect* will tend to increase T^{tvd} over t^{tud}, while the crossing trajectory effect due to particle inertia will contribute to the lowering of T^{tvd}. We thus see that starting from the familiar two-time Eulerian correlation, which can be measured, as we progress

$$\mathbf{R}^t \Rightarrow \mathbf{R}^{\text{tuc}} \Rightarrow \mathbf{R}^{\text{tud}} \Rightarrow \mathbf{R}^{\text{tvd}}, \tag{7.36}$$

several competing effects come into play, complicating the behavior of two-time particle velocity correlation and the associated time scale.

7.5.4 Lagrangian Velocity Correlation without Inertial Effect

In this subsection, we will follow the work of Csanady (1963) and develop a simple model for the two-time particle velocity correlation and use that to obtain an explicit expression for the particle diffusivity tensor. Like Csanady, we will ignore the inertia effect and only consider the effect of particle settling (i.e., we will consider the limit $\text{St} \to 0$). In the next section, we will include the effect of particle inertia as well. In both these sections, we will present explicit expressions for (i) the ratio of rms particle velocity fluctuation to fluid velocity fluctuation (i.e., $\langle V^2 \rangle^{1/2} / \langle u^2 \rangle^{1/2}$), (ii) the ratio of particle integral time scale to Eulerian integral fluid time scale (i.e., T^{tvd} / T^t), and (iii) the ratio of particle diffusivity to diffusivity of fluid tracer (i.e., $\mathcal{D}^d / \mathcal{D}^c$).

In this section, we will present these ratios as functions of nondimensional settling velocity. In the next section, we include the effect of particle inertia and these ratios will be presented as functions of both nondimensional settling velocity and the inertia parameter (the Stokes number). Given the values of rms fluid velocity $\langle u^2 \rangle^{1/2}$, Eulerian fluid time scale T^t, and diffusivity of fluid tracer $\mathcal{D}^c$, the ratios allow us to obtain the corresponding particle quantities. Although fluid turbulence will be taken to be isotropic, due to settling, particle statistics will be non-isotropic, with the vertical coordinate being the special direction. Therefore, the ratios will be provided separately for the vertical and horizontal directions.

Csanady made analytical progress by assuming the mean particle settling velocity to be larger than the rms turbulent fluid velocity fluctuations. By ignoring the effect of particle inertia, the particle velocity fluctuation can be taken to be that of the surrounding fluid velocity fluctuation at the particle location. In other words, $\mathbf{V}' = \mathbf{u}_@$. Furthermore, anisotropy does not affect the particle velocity statistics in the zero inertia limit. As a result, we have

$$\langle V_1'^2 \rangle = \langle V_2'^2 \rangle = \langle V_2'^2 \rangle = \langle u^2 \rangle, \tag{7.37}$$

where the isotropic fluid velocity fluctuation $\langle u^2 \rangle = \langle u_1^2 \rangle = \langle u_2^2 \rangle = \langle u_3^2 \rangle$. Note that the fluid has no mean velocity, so prime notation is not used to denote fluctuation. However, in the case of particles, there is mean vertical settling velocity and therefore a prime indicates fluctuation from this mean.

The two-time particle velocity correlation tensor can be approximated as

$$R_{ij}^{\text{tvd}}(\xi) \approx \frac{\langle u_i(0,0,0,0) u_j(0,0,-V_s\xi,\xi) \rangle}{\langle u^2 \rangle}, \tag{7.38}$$

where in the denominator, the rms particle velocity fluctuation is taken to be that of the isotropic fluid velocity fluctuation. In the numerator, the particle velocity fluctuation is taken to be that of the local fluid velocity. Furthermore, the spatial location of the particle at the latter time is taken to be $V_s \xi$ vertically below the initial location. In actuality, as seen in Figure 7.15, the particle location will fluctuate about this mean position. Assuming statistical homogeneity, in the limit where settling velocity dominates, the numerator can be approximated as given above.

The numerator in Eq. (7.38) has been approximated by the two-point, two-time Eulerian fluid velocity correlation. The off-diagonal terms are zero by symmetry and due to axisymmetry about the vertical direction $\langle u_1 u_1 \rangle = \langle u_2 u_2 \rangle$. Thus, we are concerned with only two components – the vertical velocity correlation (which will be called the longitudinal correlation, since the separation of the two points is in the vertical direction) and the horizontal or lateral velocity correlation. Nondimensional time and space separation are introduced as

$$\frac{\xi \langle u^2 \rangle^{1/2}}{L^E} \quad \text{and} \quad \frac{V_s \xi}{L^E}, \tag{7.39}$$

where the Eulerian longitudinal length scale is defined in terms of the two-point Eulerian velocity correlation as $L^E = \int_0^\infty \langle u_1(0,0,0,0) u_1(\xi,0,0,0) \rangle / \langle u^2 \rangle$. Owing to isotropy, in the above definition, the two points can instead be separated along the "2" or "3" direction, with the velocity components taken along those directions, and they will yield the same Eulerian longitudinal length scale L^E.

Csanady (1963) advanced the hypothesis that the longitudinal component of the two-point, two-time Eulerian fluid velocity correlation is constant along ellipses and that the correlation decays exponentially both in time and spatial separation. This hypothesis, when applied to the correlation on the right-hand side of Eq. (7.38), yields

$$R_\parallel^{\text{tvd}}(\xi) = R_{33}^{\text{tvd}}(\xi) = \exp \left\{ -\left[\left(\frac{\xi \langle u^2 \rangle^{1/2}}{\beta L^E} \right)^2 + \left(\frac{V_s \xi}{L^E} \right)^2 \right]^{1/2} \right\}, \tag{7.40}$$

where it can be argued that the constant β must take the form (Csanady, 1963)

$$\beta = \frac{\langle u^2 \rangle^{1/2} T^{\text{tuc}}}{L^E}. \tag{7.41}$$

For later use, the above longitudinal correlation has also been denoted as the component parallel to the direction of mean particle motion. A sample contour plot of the longitudinal correlation taken from Csanady (1963) is presented in Figure 7.16a. The longitudinal two-time particle velocity correlation can be substituted into Eq. (7.33) and carrying out the integration, we obtain the following explicit expression for the longitudinal particle time scale:

$$T_\parallel^{\text{tvd}} = T_{33}^{\text{tvd}} = T_{33}^{\text{tud}} = T^{\text{tuc}} \left(1 + \frac{(T^{\text{tuc}})^2 V_s^2}{(L^E)^2} \right)^{-1/2}. \tag{7.42}$$

Because of the neglect of the inertial effect of the particles, in the above equation, the two-time longitudinal particle time scale is exactly the same as the two-time

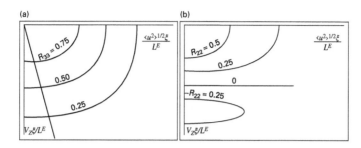

Figure 7.16 Contour plots of longitudinal two-time particle velocity correlation R_{33}^{tvd} plotted as a function of $\xi \langle u^2 \rangle^{1/2}/L^E$ and $V_s \xi/L^E$. Taken from Csanady (1963). Published 1963 by the American Meteorological Society.

longitudinal fluid time scale as seen by the particle. Both are expressed in terms of the two-time Lagrangian fluid velocity correlation (i.e., following a fluid parcel) times a correction factor. The correction factor captures the difference between the particle trajectory and the fluid parcel trajectory. In the present limit of zero particle inertia, this difference is only due to gravitational settling and thus the correction factor is the crossing trajectory effect. Based on results obtained from direct numerical simulations of non-inertial particle dispersion in isotropic turbulence, Wang and Stock (1993) proposed the following relation between the two-time Lagrangian and Eulerian fluid velocity correlation time scales:

$$\frac{T^{\text{tuc}}}{T^t} = 0.356 \,. \tag{7.43}$$

Putting these results together, we obtain the following relation between the time scale of particle velocity correlation and the easily measurable time scale of Eulerian fluid velocity correlation:

$$T_{33}^{\text{tvd}} = \underbrace{\left(\frac{T_{33}^{\text{tvd}}}{T_{33}^{\text{tud}}} \right)}_{\substack{\text{inertia} \\ \text{effect}}} \underbrace{\left(\frac{T_{33}^{\text{tud}}}{T^{\text{tuc}}} \right)}_{\substack{\text{crossing} \\ \text{trajectory}}} \underbrace{\left(\frac{T^{\text{tuc}}}{T^t} \right)}_{\substack{\text{eddy} \\ \text{decay}}} T^t$$

$$= 0.356 \left(1 + \frac{(T^{\text{tuc}})^2 V_s^2}{(L^E)^2} \right)^{-1/2} T^t \,. \tag{7.44}$$

The asymptotic behavior of the particle time scale can now be investigated. With increasing settling velocity, T_{33}^{tvd} decreases and this is the expected decorrelating effect of the crossing trajectory. In the limit of large settling velocity, $T_{33}^{\text{tvd}}/T^t \sim L^E/V_s$, which simply states that for large V_s, the two-time particle velocity correlation remains nonzero for only a small time separation ξ, since the corresponding spatial separation between the two particle positions goes as $V_s \xi$. Thus, T_{33}^{tvd} is essentially related to the length scale associated with the one-time, two-point Eulerian velocity correlation. On the other hand, for small settling velocity we obtain the result $T_{33}^{\text{tvd}} \sim T^{\text{tuc}}$. The particle time scale is the same as the Lagrangian fluid time scale, and this of course ignores the inertial effect of particle motion.

Since the long-time asymptotic value of particle diffusivity is proportional to particle time scale, we also obtain the result

$$\mathcal{D}_{\|}^d = \mathcal{D}_{33}^d = \mathcal{D}^c \left(1 + \frac{(T^{\text{tuc}})^2 V_s^2}{(L^E)^2}\right)^{-1/2}, \tag{7.45}$$

where $\mathcal{D}^c$ is the scalar diffusivity of a fluid parcel (continuous phase) in isotropic turbulence.

We now proceed to investigate the transverse or horizontal components R_{11}^{tvd} and R_{22}^{tvd} of Eq. (7.38). The key result to be employed in the evaluation is the well-known relation between the longitudinal and lateral two-point fluid correlations in isotropic turbulence. If we define $R_{33}^z = \langle u_3(0,0,0,0)u_3(0,0,\xi,0)\rangle\langle u^2\rangle$ and $R_{11}^z = \langle u_1(0,0,0,0)u_1(0,0,\xi,0)\rangle\langle u^2\rangle$ as the longitudinal and lateral two-point fluid velocity correlations, then from continuity they must satisfy the following relation:

$$R_{11}^z(\xi) = R_{33}^z(\xi) + \frac{1}{2}\xi \frac{dR_{33}^z}{d\xi}. \tag{7.46}$$

The assumption of exponential decay in Eq. (7.40) is equivalent to an exponential longitudinal correlation as $R_{33}^z(\xi) = \exp(-\xi/L^E)$, which when substituted into the above continuity constraint yields $R_{11}^z(\xi) = (1 - \xi/(2L^E))\exp(-\xi/L^E)$. The lateral Lagrangian velocity correlation taken from Csanady (1963) is shown in Figure 7.16b, where the negative loop can be observed. The important *continuity effect* is that while R_{33}^z is positive for all values of ξ, $R_{11}^z(\xi)$ becomes negative for $\xi > 2L^E$. As a consequence, the correlation length in the lateral direction is smaller than that in the longitudinal direction. A similar continuity requirement can be placed on the relation between the longitudinal and the lateral two-time correlations to obtain an explicit expression for R_{11}^{tvd} based on the expression given in Eq. (7.40) for R_{33}^{tvd}. Integrating the resulting expression, we obtain the final result

$$\begin{aligned} T_{11}^{\text{tvd}} &= T_{22}^{\text{tvd}} = T^{\text{tuc}} \left(1 + \frac{4(T^{\text{tuc}})^2 V_s^2}{(L^E)^2}\right)^{-1/2}, \\ \mathcal{D}_{11}^d &= \mathcal{D}_{22}^d = \mathcal{D}^c \left(1 + \frac{4(T^{\text{tuc}})^2 V_s^2}{(L^E)^2}\right)^{-1/2}. \end{aligned} \tag{7.47}$$

For small settling velocity, the lateral particle time scale is about the same as the longitudinal particle time scale and isotropy is approached. On the other hand, for larger settling velocity, the lateral particle time scale reduces to about half the longitudinal particle time scale. In terms of particle diffusion, the lateral particle diffusion is lower than the longitudinal particle diffusion, which in turn is lower than the diffusion of fluid parcels.

In summary, we have expressed the ratio of longitudinal and transverse particle time scale to Eulerian fluid time scale and the ratio of longitudinal and transverse particle diffusivity to fluid diffusivity in terms of two quantities that characterize the isotropic turbulence, namely, (i) the Eulerian integral fluid time scale T^t, (ii) the Eulerian integral length scale L^E, and one quantity that characterizes the particle, namely, (iii) the particle settling velocity V_s. In the following subsection, we will

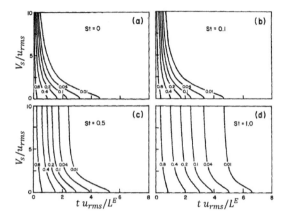

Figure 7.17 Contour plots of longitudinal two-time particle velocity correlation R_{33}^{tvd} plotted as a function of $t u_{\text{rms}}/L^E$ and V_s/u_{rms} for different values of Stokes number. Taken from Wang and Stock (1993). © American Meteorological Society. Used with permission.

include particle inertia in the analysis, which will introduce particle Stokes number as the fourth quantity.

7.5.5   Lagrangian Velocity Correlation with Inertial Effect

The results of the previous subsection capture the essential difference between the velocity correlation of particles versus fluid parcels and highlight the lower diffusivity of particles relative to tracer diffusivity in a turbulent flow. The attractiveness of the approach is the simplicity of the results obtained. However, the analysis ignores the important effect of particle inertia. The analysis of Csanady (1963) has been revisited by others to include the effect of particle inertia (Reeks, 1977; Pismen and Nir, 1978; Wang and Stock, 1993). Explicit algebraic expressions for longitudinal and lateral particle time scales and diffusion coefficients can be obtained as a function of particle Stokes number, particle settling velocity and rms fluid velocity fluctuation.

The results of Wang and Stock (1993) will be discussed briefly below. Here we define the particle Stokes number as $\text{St} = \tau_p/T^t = \tau_p u_{\text{rms}}/L^E$ (here and henceforth in this section we use the notation $u_{\text{rms}} = \langle u^2 \rangle^{1/2}$). This definition of the Stokes number is with respect to the integral scale eddies and it differs from St_η defined earlier with respect to the Kolmogorov scale. This gives rise to an important point regarding the choice of proper Stokes number. In the context of average settling velocity, it is the interaction of particles with the rapidly varying Kolmogorov-scale eddies that is of primary importance. On the other hand, in evaluating long-term dispersion, the coherent motion of integral-scale eddies plays an important role. Again, with the inclusion of the inertial effect parameterized in terms of the modified Stokes number

$$\text{St}_T = \frac{\tau_p}{T_0^{\text{tud}}} = \text{St}\,\frac{T^t}{T_0^{\text{tud}}}, \tag{7.48}$$

we will present the results on (i) the ratio of rms particle velocity fluctuation to fluid velocity fluctuation, (ii) the ratio of particle integral time scale T^{tvd} to Eulerian fluid time scale T^t, and (iii) the ratio of particle diffusivity to diffusivity of fluid tracer. In the above definition of modified Stokes number, Wang and Stock define T_0^{tud} as the Lagrangian time scale of fluid velocity following a particle, but in the absence of settling effect (hence the subscript 0). Thus, T_0^{tud} differs from the Lagrangian fluid time scale T^{tuc} following a fluid parcel only due to the inertial effect. They provide the following relation between T_0^{tud} and the easily measurable Eulerian time scale T^t:

$$T_0^{\text{tud}} = T^t \left(1.0 - \frac{0.644}{(1 + \text{St})^{0.4(1+0.01\,\text{St})}} \right), \qquad (7.49)$$

which reduces to Eq. (7.43) for $\text{St} \to 0$. Note that the difference between T_0^{tud} and T^{tuc} lies in following the particle (without settling) versus following the fluid parcel, and this difference disappears in the zero Stokes number limit. Combining Eqs. (7.43) and (7.49), we obtain the ratio

$$\frac{T_0^{\text{tud}}}{T^{\text{tuc}}} = \frac{1}{0.356} \left(1.0 - \frac{0.644}{(1 + \text{St})^{0.4(1+0.01\,\text{St})}} \right). \qquad (7.50)$$

Instead of Eqs. (7.40) and (7.46) for the vertical (longitudinal) and horizontal (lateral) components of the two-time particle velocity correlations, Wang and Stock obtained more complex expressions for R_{33}^{tvd} and R_{11}^{tvd} to account for the effect of Stokes number. Figure 7.17 shows contour plots of the longitudinal two-time particle velocity correlation R_{33}^{tvd} plotted as a function of $t u_{\text{rms}}/L^E$ and V_s/u_{rms} for different values of Stokes number St. As can be expected, the correlation decays from a value of unity with increasing $t u_{\text{rms}}/L^E$, but remains always positive. With increasing St, the decay rate decreases and this indicates that the correlation time scale of particle velocity T^{tvd} increases with particle inertia. Although the correlation's decay rate increases with increasing settling velocity, at higher Stokes number the decrease is noticeable only at small values of V_s/u_{rms}.

Figure 7.18 shows contour plots of the lateral two-time particle velocity correlation R_{11}^{tvd} plotted as a function of $t u_{\text{rms}}/L^E$ and V_s/u_{rms} for different values of Stokes number. As with the longitudinal correlation, the lateral correlation also decays from its peak value of unity with increasing $t u_{\text{rms}}/L^E$. However, the big difference is that the correlation becomes negative. The negative loop in the correlation decreases with increasing Stokes number and is virtually absent at $\text{St} = 1.0$. Also for Stokes number values larger than about 0.3, there exists a settling velocity above which the correlation does not become negative.

Wang and Stock also obtained the following explicit results. The ratio of particle to fluid rms velocity along the longitudinal and lateral directions can be expressed as

$$\frac{\langle V_3'^2 \rangle^{1/2}}{u_{\text{rms}}} = \left(\frac{1}{1 + \text{St}_T \sqrt{1 + \gamma^2}} \right)^{1/2}, \qquad (7.51)$$

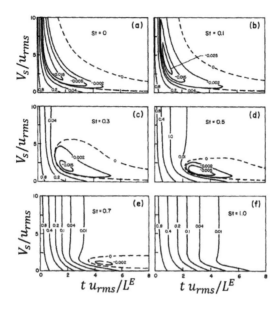

Figure 7.18 Contour plots of lateral two-time particle velocity correlation R_{11}^{tvd} plotted as a function of tu_{rms}/L^E and V_s/u_{rms} for different values of Stokes number. Taken from Wang and Stock (1993). © American Meteorological Society. Used with permission.

$$\frac{\langle V_1'^2 \rangle^{1/2}}{u_{\text{rms}}} = \frac{\langle V_2'^2 \rangle^{1/2}}{u_{\text{rms}}} = \left(\frac{1}{1 + \text{St}_T \sqrt{1 + \gamma^2}} - \frac{0.5\,\text{St}_T\,\gamma}{1 + \text{St}_T \sqrt{1 + \gamma^2}} \right)^{1/2}, \qquad (7.52)$$

where we have introduced the simplifying notation $\gamma = T_0^{\text{tud}} V_s^2 / L^E$. If we set the settling effect to zero (i.e., $\gamma \to 0$), this leads to the isotropic result $\langle V'^2 \rangle^{1/2}/u_{\text{rms}} = 1/(1 + \text{St})$. Thus, we recover the classic result of Tchen (1947). This inertial behavior of particles that their rms velocity goes as $(1 + \text{St})^{-1}$ times the fluid fluid rms velocity has been widely quoted in earlier multiphase literature (Soo, 1956; Hinze, 1959).

In the previous subsection, in the limit of zero Stokes number, the particle-to-fluid rms velocity ratios were unity and we clearly see the effect of particle inertia as the departure of the right-hand sides from unity. Interestingly, the effect of settling velocity affects the rms velocity ratio only in the presence of particle inertia. Contour plots of the ratio of particle to fluid rms velocity along the longitudinal and lateral directions is presented in Figure 7.19. Although the above equations are in terms of St_T, they can be recast in terms of St using the relations (7.48) and (7.49). Accordingly, the plots are in terms of St. For all combinations of St and V_s, the particle velocity fluctuation is smaller than unity. At small values of St, the reduction is nearly independent of settling velocity. Only at larger values of St the reduction of particle rms velocity enhanced by settling velocity. This figure is consistent with the simple result we obtained in Eq. (4.110) in Problem 4.12, but only in the large density ratio limit. This is because the analysis of Wang and Stock uses only the gravitational and quasi-steady forces that act on the particle and neglects the added-mass and Basset history forces – in their title itself they correctly state that the results are for heavy particles. For particles that are lighter than or comparable in density to the surrounding fluid, the above analysis does not apply and the analysis must be repeated with added-mass and Basset history effects taken into account.

We now consider the ratio of particle to fluid time scale and present the following results obtained by Wang and Stock:

$$\frac{T_{33}}{T^{\text{tuc}}} = \frac{1 + \text{St}_T \sqrt{1 + \gamma^2}}{\sqrt{1 + \gamma^2}} \left(\frac{T_0^{\text{tud}}}{T^{\text{tuc}}}\right), \tag{7.53}$$

$$\frac{T_{11}}{T^{\text{tuc}}} = \frac{\left(\sqrt{1 + \gamma^2} - 0.5\,\text{St}_T\,\gamma\right)\left(1 + \text{St}_T\sqrt{1 + \gamma^2}\right)^2}{\left(1 + \gamma^2\right)\left(1 + \text{St}_T\sqrt{1 + \gamma^2} - 0.5\,\text{St}_T\,\gamma\right)}\left(\frac{T_0^{\text{tud}}}{T^{\text{tuc}}}\right), \tag{7.54}$$

where the ratio $T_0^{\text{tud}}/T^{\text{tuc}}$ was given earlier in Eq. (7.50). Note that Eq. (7.53) correctly reduces to Eq. (7.42) in the zero Stokes number limit. But this is not so for the lateral component. Nevertheless, contour plots of the longitudinal and lateral time scale ratios have been plotted in Figure 7.20 for a range of nondimensional settling velocity and Stokes number. The influence of settling velocity is small and the dominant effect is that the particle time scale increases with Stokes number and the difference between the longitudinal and lateral time scales is not very large.

The ratio of longitudinal and lateral particle diffusivities to fluid diffusivity was given by Wang and Stock as

$$\frac{\mathcal{D}_{33}}{T^{\text{tuc}}u_{\text{rms}}^2} = \frac{1}{\sqrt{1 + \gamma^2}}\left(\frac{T_0^{\text{tud}}}{T^{\text{tuc}}}\right), \tag{7.55}$$

$$\frac{\mathcal{D}_{11}}{T^{\text{tuc}}u_{\text{rms}}^2} = \frac{\sqrt{1 + \gamma^2} - 0.5\gamma}{1 + \gamma^2}\left(\frac{T_0^{\text{tud}}}{T^{\text{tuc}}}\right). \tag{7.56}$$

Interestingly, the right-hand sides are independent of the Stokes number and the dependence on Stokes number comes only though the relation between T_0^{tud} and T^t given in Eq. (7.49). Again, the right-hand side of the longitudinal expression is the same as that given in the previous subsection for zero particle inertia. Contour plots of the longitudinal and lateral particle-to-fluid diffusivity ratios have been plotted in Figure 7.21 for a range of nondimensional settling velocity and Stokes number. In this case, the influence of St is small and the dominant effect is that of particle settling velocity. At small settling velocities, particle diffusivity is larger than fluid diffusivity, due to finite particle inertia. But at large settling velocity, particle diffusivity is lower than that of tracer fluid particles, irrespective of St. At small settling velocity, as expected, the diffusivity is nearly isotropic, whereas with increasing settling velocity, the longitudinal diffusivity becomes nearly twice the later diffusivity.

We close our discussion of this topic with the following cautionary notes. (i) The above analysis of Wang and Stock was based on Stokes drag. The results can be extended to finite-Re drag as well by properly adjusting the particle time scale in

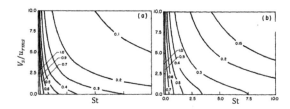

Figure 7.19 Contour plots of ratio of rms particle velocity fluctuation to rms isotropic fluid velocity fluctuation plotted as a function of St and V_s/u_{rms}. (a) $\langle V_3'^2\rangle^{1/2}/u_{\mathrm{rms}}$, (b) $\langle V_1'^2\rangle^{1/2}/u_{\mathrm{rms}} = \langle V_2'^2\rangle^{1/2}/u_{\mathrm{rms}}$. Taken from Wang and Stock (1993). © American Meteorological Society. Used with permission.

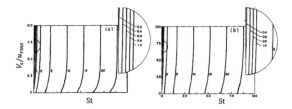

Figure 7.20 Contour plots of ratio of Lagrangian particle time scale to Eulerian fluid time scale plotted as a function of St and V_s/u_{rms}. (a) T_{33}/T^{tuc}, (b) $T_{11}/T^{\mathrm{tuc}} = T_{22}/T^{\mathrm{tuc}}$. Taken from Wang and Stock (1993). © American Meteorological Society. Used with permission.

the definition of St and the settling velocity to account for the nonlinear drag (i.e., by including the finite-Re quasi-steady drag correction factor in the definitions of τ_p and V_s). (ii) The above analysis assumes the settling velocity to be large compared to the velocity fluctuation, and thus the results strictly apply to fast-settling heavy particles. For particles that settle slowly, or for lighter-than-fluid bubbles, not only will the velocity fluctuation be comparable to V_s, but also the effects of added-mass and history forces will become important. A similar simple analytic approach to obtaining rms velocity ratio, time scale ratio, and diffusivity ratio is very much needed.

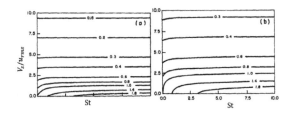

Figure 7.21 Contour plots of ratio of particle diffusivity to fluid diffusivity plotted as a function of St and V_s/u_{rms}. (a) $\mathcal{D}_{33}/(T^{\mathrm{tuc}}u_{\mathrm{rms}}^2)$, (b) $\mathcal{D}_{11}/(T^{\mathrm{tuc}}u_{\mathrm{rms}}^2) = \mathcal{D}_{22}/(T^{\mathrm{tuc}}u_{\mathrm{rms}}^2)$. Taken from Wang and Stock (1993). © American Meteorological Society. Used with permission.

7.5.6   Langevin Model of Particle Motion

In this section, we will study the rudimentary aspects of the *Lagrangian stochastic method* (LSM), where the instantaneous trajectory of a larger number of particles, along with their properties, can be tracked in a time-averaged fluid flow by statistically taking into account the effect of turbulent velocity fluctuations (Haworth and Pope, 1986; Pozorski and Minier, 1998; Minier et al., 2014; Minier, 2015). There is extensive literature on LSM and on Langevin models, and the reader is recommended to read the above references and works cited in them.

Before we consider the random motion of particles due to ambient turbulence, let us consider the simpler problem of *Brownian motion* of particles, where the random motion of a particle is due to its collisions with the fluid molecules. If we ignore any background continuum scale fluid motion, then the equation of motion for a particle can be expressed as

$$dV = K \, dW \quad \text{and} \quad dX = V \, dt, \qquad (7.57)$$

where the Wiener white-noise process dW is a random stochastic process of zero mean, $\langle dW \rangle = 0$, and variance equal to the time step, $\langle |dW|^2 \rangle = dt$. The above equation is in the difference form, where dV is the change in particle velocity over a time step of dt and the corresponding change in particle position is dX. The Brownian constant K is given by

$$K = \sqrt{\frac{2k_b T_c}{m_p \tau_p}}, \qquad (7.58)$$

where k_b is the Boltzmann constant, T_c is the temperature of the continuous phase surrounding the particle, m_p is the mass of the particle, and τ_p is the particle time scale. From the Brownian constant we can also obtain the Einstein diffusion coefficient as

$$\mathcal{D}_B = \frac{k_b T_c}{3\pi \mu_f d_p}, \qquad (7.59)$$

where we have used the definition given in Eq. (7.32). From the definition of the particle time scale given in Eq. (2.2), it can readily be seen that $K \propto d_p^{-5/2}$ and thus Brownian motion is effective only for small particles. Under normal conditions, Brownian diffusion can be ignored for particles larger than a few microns in size.

The key aspect of the above methodology is that at each time step, the velocity includes a random component and the position is advanced accordingly. In order to extend this stochastic approach to turbulence, let us write the equations of particle motion as

$$\frac{dX}{dt} = V \quad \text{and} \quad \frac{dV}{dt} = \frac{(\langle u_@ \rangle + u''_@) - V_s e_g - V}{\tau_p}, \qquad (7.60)$$

where the fluid velocity seen by the particle is separated into a mean component, $\langle u_@ \rangle$, and a stochastic component, $u''_@$. In the previous subsections, we worked toward finding the two-time correlation of particle velocity along its trajectory, without needing to solve the particle equation of motion. In contrast, when solving the above equations of particle motion, particle velocity is part of the solution. From the particle velocity of an ensemble of particles, quantities such as two-time correlation can readily be obtained. It is also assumed here that the mean fluid flow information $\langle \mathbf{u} \rangle$ is available as a 3D vector field, perhaps from a RANS simulation, from which the fluid velocity at the particle position can be interpolated. The real challenge to solving the above equations is the fluctuating fluid velocity at the particle location.

For the reasons discussed in earlier subsections, the random component $\mathbf{u}''_@$ remains correlated over time and its time scales of correlation are $T_\parallel^{\text{tud}}$ for the longitudinal component and $T_\perp^{\text{tud}}$ for the transverse component. Thus, while solving the above equations of particle motion, $\mathbf{u}''_@$ must be randomly selected by satisfying the following two conditions: (i) its temporal correlation must equal $\mathbf{R}^{\text{tud}}$ and (ii) $\langle \mathbf{u}''^2_@ \rangle$ must equal the mean square fluid velocity seen by the particles. Note that $\langle \mathbf{u}''^2_@ \rangle$ equals the mean square fluid velocity fluctuation $\langle u^2 \rangle$ only if the particle ergodically explores all parts of the turbulent flow with equal probability. However, this is not the case for inertial particles, which tend to preferentially accumulate in regions of high strain, if they are heavier than the fluid.

The task at hand thus reduces to appropriate modeling of $\mathbf{u}''_@$. But before we proceed to model $\mathbf{u}''_@$ for an inertial particle, let us first consider a model for $\mathbf{u}''_@$ in the simpler limit of a fluid parcel – this corresponds to the case of a particle with zero inertia and zero settling velocity. In this limit, the fluid parcel explores all parts of the turbulent flow with equal probability and therefore $\langle \mathbf{u}''^2_@ \rangle = \langle u^2 \rangle$.

An elegant model that applies for both homogeneous and inhomogeneous turbulence, but for a fluid parcel, was proposed by Pope (1994) as

$$d\mathbf{u}''_{@f} = -\frac{1}{\rho_f}\nabla\langle p \rangle \, dt - \frac{\mathbf{u}''_{@f}}{T^{\text{tuc}}} dt + \sqrt{C_0 \langle \epsilon \rangle} \, d\mathbf{W}, \qquad (7.61)$$

where $\langle \epsilon \rangle$ is the ensemble-averaged dissipation. In the above equation, the subscript $@f$ corresponds to velocity fluctuation following a fluid parcel. Correspondingly, the second term on the right-hand side of the above equation uses the Lagrangian fluid time scale, as opposed to the Lagrangian particle time scale. The first term on the right accounts for the effect of mean pressure gradient and the last term is the Wiener process, where C_0 is a constant, whose value has been suggested to be 2.1 (Haworth and Pope, 1986). In the second term on the right, the Lagrangian fluid time scale of the fluid parcel is modeled as

$$T^{\text{tuc}} = \frac{1}{\left(\frac{1}{2} + \frac{3}{4}C_0\right)} \frac{k}{\langle \epsilon \rangle}, \qquad (7.62)$$

where the turbulent kinetic energy is given by $k = 3\langle u^2 \rangle/2$. The difference equation (7.61) can be advanced in time to evaluate the stochastic component $\mathbf{u}''_@$, which along with $\langle \mathbf{u}_@ \rangle$ is the total fluid velocity at the particle, which is also the velocity of the tracer particle.

We now proceed to model the fluid velocity fluctuation as seen along the trajectory of an inertial particle. We will essentially use the same procedure as that employed above. From the analysis of Csanady (1963), we recognize the difference between the longitudinal direction versus the transverse direction. In other words, while the model of fluid velocity along the trajectory of a fluid parcel remains isotropic, as presented in Eq. (7.61), the stochastic modeling of fluid velocity along the trajectory of a particle must be treated as only axisymmetric. We generalize the direction of axisymmetry to be along the mean relative velocity vector, instead of the vertical direction considered in the case of particle settling. In the case of particles falling in isotropic turbulence, the mean relative velocity vector was in the vertical direction. Thus, the longitudinal direction, denoted by $\|$, is given by

$$\mathbf{e}_\| = \frac{\langle \mathbf{V} - \mathbf{u}_@ \rangle}{|\langle \mathbf{V} - \mathbf{u}_@ \rangle|} \tag{7.63}$$

and the other two orthogonal transverse directions are statistically equivalent, denoted by $\mathbf{e}_\perp$.

The modeling of the fluid velocity fluctuation along the longitudinal and transverse directions must proceed independently, since as indicated by the crossing trajectory analysis of Section 7.5.3, the correlation time scale along the transverse direction is smaller than that along the longitudinal direction. The Langevin models of fluid velocity perturbation seen by the particle along the longitudinal and transverse directions are expressed as (Minier et al., 2014)

$$d\mathbf{u}''_{@\|} = -\frac{1}{\rho_f}(\nabla\langle p\rangle \cdot \mathbf{e}_\|)\, dt - \frac{\mathbf{u}''_{@\|}}{T^{\text{tud}}_\|}\, dt + \sqrt{\langle\epsilon\rangle\left(C_0\frac{T^{\text{tuc}}}{T^{\text{tud}}_\|} + \frac{2}{3}\left(\frac{T^{\text{tuc}}}{T^{\text{tud}}_\|} - 1\right)\right)}\, d\mathbf{W},$$

$$d\mathbf{u}''_{@\perp} = -\frac{1}{\rho_f}(\nabla\langle p\rangle \cdot \mathbf{e}_\perp)\, dt - \frac{\mathbf{u}''_{@\perp}}{T^{\text{tud}}_\perp}\, dt + \sqrt{\langle\epsilon\rangle\left(C_0\frac{T^{\text{tuc}}}{T^{\text{tud}}_\perp} + \frac{2}{3}\left(\frac{T^{\text{tuc}}}{T^{\text{tud}}_\perp} - 1\right)\right)}\, d\mathbf{W}.$$

$$\tag{7.64}$$

From Eqs. (7.42) and (7.47), we define the longitudinal and transverse time scales of fluid seen by the particle as

$$T^{\text{tud}}_\| = T^{\text{tuc}}\left(1 + \frac{(T^{\text{tuc}})^2|\langle \mathbf{V} - \mathbf{u}_@\rangle|^2}{(L^E)^2}\right)^{-1/2},$$

$$T^{\text{tud}}_\perp = T^{\text{tuc}}\left(1 + \frac{4(T^{\text{tuc}})^2|\langle \mathbf{V} - \mathbf{u}_@\rangle|^2}{(L^E)^2}\right)^{-1/2}.$$

$$\tag{7.65}$$

The only quantity to still be defined is the Eulerian length scale of the two-point velocity correlation, which (based on simple dimensional analysis) can be expressed as $L^E = C_E k^{3/}/\langle\epsilon\rangle$, where the constant C_E is empirically determined.

The above difference equation for $d\mathbf{u}''_@$ can be used to evaluate the perturbation fluid velocity seen by the particle $\mathbf{u}''_@$ at every time step, which can then be used in Eq. (7.60) to move the particle. This is a powerful approach, since one needs only the ensemble-averaged flow $\langle\mathbf{u}\rangle$, which can be obtained quite easily using RANS. With this information, the stochastic trajectory of any number of particles can be tracked within the flow, without the need for actually solving for the time-dependent turbulent flow. The only information needed about turbulence is the relevant correlation length and time scales. However, the parameterization of the various length and time scales and mean square turbulence quantities, needed in LSM, is often unknown in complex turbulent flows, and this remains a major challenge.

7.5.7   Application to Inhomogeneous Turbulent Flows

There have been many applications of the LSM to a variety of turbulent flows, from the simplest isotropic turbulence to much more complicated industrial-scale turbulent flows. Many different variants of the Langevin model presented above have been pursued, and it is not necessary to go over all of them. In this subsection, our objective is to get a flavor of the kind of results that can be obtained with the LSM. One key challenge in practical applications is to account for the fact that turbulence is often inhomogeneous and as a result is anisotropic as well. One of the simplest Lagrangian random walk models, applicable in wall-bounded turbulence, is described in the early, well-cited work of Kallio and Reeks (1989). Its attractiveness arises from its simplicity in accounting for the strong variation in fluid turbulence along the wall-normal direction, with appropriate fit through the experimental data and a suitable implementation of the random walk process. Despite its simplicity, the model was successful in predicting particle deposition. We shall see more about this model in the next chapter. In what follows, as an example of a more complex implementation of the Langevin model in the context of inhomogeneous flows, we will consider LSM's application to a turbulent channel flow and introduce the results of Dehbi (2010).

   Consider a turbulent channel flow between two parallel plates of infinite extent. The flow is statistically stationary and homogeneous along the streamwise and spanwise directions. The statistics is inhomogeneous along the wall-normal direction and the turbulence is anisotropic (has to be due to inhomogeneity). Consider introducing a distribution of inertial particles into the stream, such that particles are randomly distributed across the entire cross-section with uniform probability. As they travel downstream, their distribution will be non-uniform across the channel height (i.e., the mean concentration of particles at any downstream distance will not be uniform in the wall-normal direction). We consider the limit where the particles are only one-way coupled with the flow, and therefore the turbulent flow remains unaffected by the particles. However, the particle statistics is nonstationary because the non-uniform distribution changes over time.

   In the case of turbulent channel flow, the mean velocity is one-dimensional with only the streamwise velocity that varies along the wall-normal direction. Along with this one-dimensional mean flow, Dehbi (2010) used the following variant of the normalized Langevin equation for the fluid velocity fluctuation $\mathbf{u}''_@$ seen by the particle:

$$
d\left(\frac{u''_{1@}}{\langle u''^2_1\rangle^{1/2}}\right) = -\left(\frac{u''_{1@}}{\langle u''^2_1\rangle^{1/2}}\right)\frac{dt}{T^{\text{tud}}} + \sqrt{\frac{2}{T^{\text{tud}}}}dW_1 + \frac{\partial(\langle u''_1 u''_2\rangle/\langle u''^2_1\rangle^{1/2})}{\partial x_2}\frac{dt}{1+\text{St}},
$$

$$
d\left(\frac{u''_{2@}}{\langle u''^2_2\rangle^{1/2}}\right) = -\left(\frac{u''_{2@}}{\langle u''^2_2\rangle^{1/2}}\right)\frac{dt}{T^{\text{tud}}} + \sqrt{\frac{2}{T^{\text{tud}}}}dW_2 + \frac{\partial\langle u''^2_2\rangle^{1/2}}{\partial x_2}\frac{dt}{1+\text{St}},
$$

$$
d\left(\frac{u''_{3@}}{\langle u''^2_3\rangle^{1/2}}\right) = -\left(\frac{u''_{3@}}{\langle u''^2_3\rangle^{1/2}}\right)\frac{dt}{T^{\text{tud}}} + \sqrt{\frac{2}{T^{\text{tud}}}}dW_3 .
$$

$$(7.66)$$

The above equations are written entirely in terms of velocity perturbation and therefore once the perturbation fluid velocity components are obtained by advancing the above equations, they must be added to the mean streamwise velocity. Here, dW_i are the Wiener processes with zero mean and variance equal to the time step dt. Note that in the turbulent channel flow, the mean square velocity fluctuations along the streamwise, wall-normal, and spanwise directions (i.e., $\langle u''^2_1\rangle(x_2)$, $\langle u''^2_2\rangle(x_2)$, and $\langle u''^2_3\rangle(x_2)$) are not the same; they are functions of the wall-normal (x_2) direction. The Stokes number is defined as the ratio of particle time scale to the time scale of fluid velocity correlation seen by the particle (i.e., $\text{St} = \tau_p/T^{\text{tud}}$).

In applying the above normalized Langevin equation, the only quantity to be further defined is the time scale of fluid velocity seen by the particle. Dehbi (2010) used the following model presented by Wang and Stock (1993) for the time scale of fluid velocity seen by the particle:

$$
T^{\text{tud}} = T^t\left(1 - \left(1 - \frac{T^{\text{tuc}}}{T^t}\right)\left(1 + \frac{\tau_p}{T^t}\right)^{-0.4(1+0.01\tau_p/T^t)}\right),
$$

$$(7.67)$$

where T^{tuc} is the Lagrangian time scale of fluid velocity correlation following the fluid parcel, while T^t is the Eulerian time scale of fluid velocity correlation at a fixed probe. Their ratio has been empirically observed to be $T^{\text{tuc}}/T^t = 0.356$ (Wang and Stock, 1993). It can be seen that in the limit of small particle inertia ($\tau_p \to 0$), the above expression yields the correct result of Lagrangian time scale assuming both the particle and the fluid are the same (i.e., $T^{\text{tud}} \to T^{\text{tuc}}$). On the other hand, according to the above model, a very large value of τ_p leads to the result $T^{\text{tud}} \to T^t$. For final closure, Dehbi (2010) used the following approximation for the Lagrangian time scale of fluid velocity given by Kallio and Reeks (1989):

$$
\frac{T^{\text{tuc}}\langle u_*\rangle^2}{\nu_f} = \begin{cases} 10 & y^+ \leq 5, \\ 7.122 + 0.5731y^+ - 0.00129y^{+2} & 5 < y^+ < 200, \end{cases}
$$

$$(7.68)$$

where u_* is the friction velocity of the channel flow and $y^+ = y\,u_*/\nu_f$ is the distance of any point from the nearest wall in wall units. This completes the model for $\mathbf{u}''_@$, which, along with the mean flow, can be used in Eq. (7.60) to track inertial particles through the channel.

A large number of particles were tracked using the above algorithm after releasing them, uniformly distributed within the turbulent channel. The results were compared

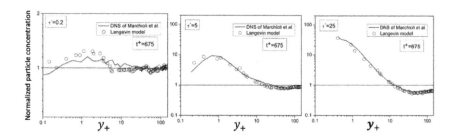

Figure 7.22 A comparison of normalized particle concentration as a function of wall normal distance in a turbulent channel flow as obtained with a Langevin model and from DNS. The results of three different values of nondimensional particle time scales are presented. Taken from Dehbi (2010).

against those obtained by Marchioli et al. (2007) from particle tracking in a direct numerical simulation of turbulent channel flow, which serves as the benchmark. Over time, particles migrate from the center of the channel toward the walls of the channel – a phenomenon we discussed in Section 1.3 as turbophoretic migration. The degree of this wallward migration depends on the Stokes number, which we shall discuss later in Section 12.4.2. The normalized concentration of particles at a nondimensional time of $t\, u_*^2/\nu_f = 675$ after release is shown in Figure 7.22 for three different nondimensional particle time scales (figure taken from Dehbi, 2010). The results of the Langevin model are compared against the direct numerical simulation (DNS) results and the agreement is quite good. While the particles remain nearly uniformly distributed at the lower particle time scale of $\tau_+ = \tau_p\, u_*^2/\nu_f = 0.2$, an accumulation of particles near the channel wall is observed for larger Stokes numbers.

The good agreement seen in Figure 7.22 is due to the careful choice of the various length scales and the choice of the normalized Langevin model. As pointed out by Dehbi (2010), the inertial drift correction represented by the last terms in the x_1 and x_2 components of Eq. (7.66) is quite important in obtaining good answers. The factor $1/(1 + St)$ correctly accounts for the inertial effect of particles. In the limit $St \to 0$ this factor correctly becomes unity, while in the limit $St \to \infty$ this factor correctly becomes zero. Figure 7.23 (taken from Dehbi, 2010) shows the average wall-normal velocity of the fluid seen by the particle ($u_{2@}''$) predicted using the Langevin model against DNS results for $\tau_+ = 25$. The Langevin model results obtained using $\tau_+ = 0$ and $\tau_+ = \infty$ are also plotted. The difference highlights the importance of carefully parameterizing the various quantities.

7.6   Direct Effect of Turbulence on Finite-Size Particle Force

Particle–turbulence interaction becomes complex, and far more interesting, when particle size becomes comparable to or exceeds the continuous-flow scales. An important point related to the results of the previous sections must be emphasized. The finite-Re

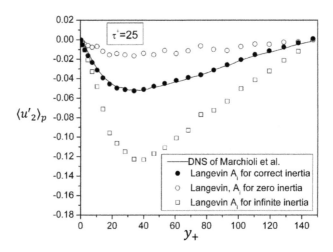

Figure 7.23 A comparison of average wall-normal velocity seen by the particle as a function of wall-normal distance in a turbulent channel flow as obtained with a Langevin model and from DNS. The results of three different values of τ_+ are plotted. Only for the correct value is there a good match between the Langevin model and the DNS. Taken from Dehbi (2010).

BBO equation was assumed to apply and to accurately predict the force on the particle at each instant of its motion through the turbulent flow. Though the BBO equation was derived for a uniform non-turbulent flow, it was applied to calculate particle motion in a turbulent flow. This was possible because the particle was much smaller than the scales of ambient turbulence. The forms of the quasi-steady, added-mass, and history forces in the BBO equation were not changed due to ambient turbulence. In other words, there was no direct influence of turbulence on the different force expressions or the equation of particle motion. The profound effect of turbulence on settling velocity and dispersion was indirect and arose from the temporal fluctuations of the fluid velocity seen by the particle at it moved through the turbulent flow.

In this section, we are concerned with the direct effect of turbulence on particles of size comparable to or larger than the continuous-flow scales. Such an intrinsic or direct effect of turbulence, if it exists and is proven to be significant, must be accounted for by appropriate modification of the BBO or MRG force expressions. For example, the standard drag expression for quasi-steady force will then need to be parameterized not only in terms of particle Reynolds number, but perhaps also in terms of rms turbulent fluctuation.

With a finite-sized particle (defined as one whose size is comparable to the ambient flow scales), the surrounding flow approaching the particle can no longer be considered as a time-varying uniform flow. The turbulent flow can be conceptually separated into three parts: (i) scales much larger than the particle, denoted as $\mathbf{u}_\gg$; (ii) scales much smaller than the particle, denoted as $\mathbf{u}_\ll$; and (iii) scales of the order of the particle size, say from five times smaller to five times larger than the particle diameter, denoted as $\mathbf{u}_{\approx}$. The discussions of the previous sections are in the context where $\mathbf{u}_{\ll}$ and $\mathbf{u}_{\approx}$ are negligibly small due to the small size of the particles.

In the case of finite-sized particles, we will now analyze the effects of $\mathbf{u}_{\gg}$, $\mathbf{u}_{\ll}$, and $\mathbf{u}_{\sim}$. The effect of $\mathbf{u}_{\gg}$ can still be treated as a time-varying, nearly uniform large eddy field, just as we did in the previous sections. The net effect of eddies much smaller than the particles (i.e., the role of $\mathbf{u}_{\ll}$) can be expected to be quite small and can be ignored. If needed, their effect can be treated as Brownian motion due to very small-scale eddy interactions. Typically, eddies of size an order of magnitude smaller than the particle may be considered sufficiently small for heavy particles, while they need to be smaller for the case of bubbles (Balachandar and Eaton, 2010). Finally, we are left with eddies of size comparable to the particle interacting with the particle in a complex way.

Thus, the question to be addressed in this section is what is the role of $\mathbf{u}_{\approx}$ and how best to account for it? To be specific, we are interested in knowing how the presence of $\mathbf{u}_{\approx}$ influences the quasi-steady, added-mass, history, and lift forces, and how the functional forms of these forces must be modified. The implication is that the various force expressions of the BBO equation obtained in Chapter 4 are appropriate only in the limit $\mathbf{u}_{\approx} = 0$, and they may need to be revisited and improved, if necessary, to account for the effect of turbulent scales of the order of particle size. In other words, we are interested in the force expressions of a finite-sized particle. This modification of drag and lift force relations due to turbulence will be referred to as the *direct or instantaneous effect* of turbulence. Effects such as modified settling velocity or diffusion will be referred to as the *indirect or cumulative effect* of turbulence. This latter effect is cumulative, since at any instant the drag law was the same as in laminar flow, and the effect of turbulence manifests only over time due to the correlated action of turbulence.

Since we are interested in extending the point-particle approximation and the corresponding BBO equation of motion to larger particles whose size overlaps the turbulence scales, we recall two such extensions that were discussed in Chapter 4. First, instead of the finite-Re BBO equation, we can consider the finite-Re MRG equation, which allows for ambient undisturbed flow variation on the scale of the particle through surface and volume averages. Second, in Sections 5.3 and 5.4 we considered additional effects of linear variation in the ambient flow on the scale of the particle. These discussions clearly illustrated the complexity of accounting for even the simplest variations in the ambient flow in the models of forces. For larger particles, of size comparable to the turbulent flow scales, leading-order information, such as velocity gradient at the particle location, alone does not provide an adequate description of the undisturbed ambient flow seen by the particle, which can be far more complicated.

In order to assess the direct effect of turbulence, it is not important to consider a freely moving particle. It is sufficient to investigate the effect of turbulence passing over a stationary particle. This allows separation of the direct effect of turbulence from the indirect effect of particle sampling selective regions of turbulence due to its motion. By keeping the particle fixed, it can be subjected to well-controlled turbulent flow, whose effect can be precisely studied. However, in practical applications, the finite-sized particles will be in motion and as a result the particle will be influenced by both the direct and indirect effects. The combined effect is not fully understood. Therefore, here we focus only on the direct effect of some canonical turbulence.

The problem of a finite-sized particle of diameter larger than the Kolmogorov scale subjected to isotropic turbulence was considered by Bagchi and Balachandar (2003b, 2004). Their direct numerical simulations considered particle Reynolds numbers in the range of 50 to 600. The diameter of the particle was varied from 1.5 to 10 times the Kolmogorov scale. Though the mean velocity of isotropic turbulence is zero, a constant free stream of $\langle u_x \rangle$ was added to turbulence and the ratio of rms turbulent velocity fluctuation to mean velocity (i.e., $I = u_{\text{rms}}/\langle u_x \rangle$) was varied from about 10% to 25%. An important outcome of this study is the direct measurement of the effect of free-stream turbulence on particle drag force. The mean drag coefficient obtained as a function of Re from a variety of experimental sources is shown in Figure 7.24. Also plotted in the figure for reference is the standard drag correlation applicable for the case of a stationary particle in a steady uniform ambient flow, shown earlier in Figure 7.8. The scatter in the experimental data clearly illustrates the degree of disagreement as to the effect of turbulence. For example, in the moderate Reynolds number regime, measurements (Uhlherr and Sinclair, 1970; Zarin and Nicholls, 1971; Brucato et al., 1998) indicate a substantial increase in the drag coefficient in a turbulent flow. The numerical study by Mohd-Yusof (1996) also illustrated a drag increase of nearly 40% in a free-stream turbulence of 20% intensity. On the other hand, the results of Rudoff and Bachalo (1988) tend to suggest a reduction in the drag coefficient due to ambient turbulence. In contrast, Warnica et al. (1994) suggest that the drag on a spherical liquid drop is not significantly different from the standard drag. The experiments of J. S. Wu and Faeth (1994a,b) also suggest little influence of turbulence on the mean drag.

The simulation results of Bagchi and Balachandar (2003b, 2004) are also shown in the figure. The scatter in the experimental data is due to the inclusion of the indirect effect of turbulence, since these experiments were not performed for a stationary particle. Thus, the substantial differences in drag are perhaps due to trajectory bias of the particle and nonlinear drag effect. These indirect effects of turbulence have already been addressed in Sections 7.4 and 7.5. In their simulations Bagchi and Balachandar avoid the trajectory bias, though nonlinear effects are somewhat present due to the time variation in the large-scale velocity seen by the particle. The simulation results clearly demonstrate that free-stream turbulence has very little systematic direct effect on the time-averaged mean drag.

While the mean drag seems unaffected by the presence of turbulence in the oncoming flow, the same cannot be said about the time dependence of force on the particle. Figure 7.25 shows the time variation of the streamwise component of the drag coefficient obtained from the particle-resolved simulation. The top frame shows the results for a modest-sized particle of $d_p/\eta = 1.5$, the middle and bottom frames are for larger particles of size $d_p/\eta = 3.8$ and 9.6. The Reynolds numbers based on mean relative velocity for the three cases are 107, 261, and 609, respectively. The turbulence intensity in all three cases was $I = 0.1$. Along with the simulation results, also shown are the drag predicted by (i) only the quasi-steady force given in the finite-Re BBO equation (4.144), (ii) the sum of the quasi-steady, pressure-gradient, and added-mass forces, and

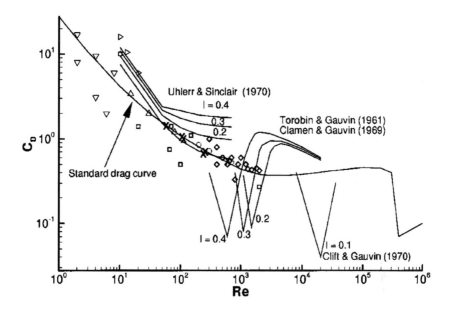

Figure 7.24 A summary of the results on the direct effect of turbulence on the drag coefficient.
× Present results; □ Gore and Crowe (1990); ◇ Sankagiri and Ruff (1997); ○ Zarin and
Nicholls (1971); △ Warnica et al. (1994); ▽ Rudoff and Bachalo (1988); ▷ Brucato et al. (1998).
The standard drag curve (Clift et al., 1978) is also shown. The parameter I is the ratio of the rms
velocity of the free-stream turbulence to the mean relative velocity between the particle and the
fluid. Reprinted from Bagchi and Balachandar (2003b), with the permission of AIP Publishing.

(iii) the sum of the quasi-steady, pressure-gradient, added-mass, and viscous history
forces. In the case of $d_p/\eta = 1.5$ particle, the time variation of force is reasonably
captured, since most of the turbulent scales are much larger than the particle, and
when the velocity seen by the particle increased or decreased due to these much larger
eddies, the finite-Re BBO had no difficulty in capturing these ups and downs. The slight
difference between the computed and the predicted force is the result of Kolmogorov
eddies of the size of the particle. As can be expected, the effect of forces other than the
quasi-steady contribution is not large. As the particle size compared to the Kolmogorov
scale increases, it is clear that the predictions based on the finite-Re BBO equation
depart substantially from the actual force. The mean value is accurately predicted as
shown in Figure 7.24, but the fluctuations are not well predicted, since these are induced
by eddies of size comparable to the particle. The finite-Re BBO equation applies only
when the velocity fluctuations are due to much larger eddies, which appear as time-
dependent uniform flow from the particle's perspective. It is also clear that the direct
effect of turbulence due to the finite particle size cannot be accounted for by the
inclusion of the unsteady added-mass and history force contributions.

Figure 7.26 shows the time variation of the transverse component of the drag
coefficient obtained from the particle-resolved simulation. Again the three frames
show the results for $d_p/\eta = 1.5$, 3.8, and 9.6 particle. For $d_p/\eta = 1.5$ particle, the
instantaneous y-force is well predicted by the BBO equation. This is because, though

the mean flow is along the x-direction, the y-force is not a lift force – it is in fact a component of drag. When turbulent fluctuations are added to the mean streamwise flow, the instantaneous flow direction seen by the particle fluctuates about the x-axis. Therefore, the instantaneous y-force is due to the nonzero y-component of the flow at that time. As a result, the finite-Re BBO equation is able to predict this y-force, just as well as the x-force. For larger-sized particles, since the influence of eddies of comparable size increases, prediction of the finite-Re BBO equation becomes poor.

At the $O(100)$ particle Reynolds number considered by Bagchi and Balachandar (2003b, 2004), the turbulence in the incoming flow can be expected to modify the sequence of flow bifurcation and flow regimes discussed in detail in Section 7.1. Especially in the presence of ambient turbulence, the onset of each transition to a more complex flow will be accelerated. The direct numerical simulations show that the critical Reynolds number $\mathrm{Re_{cr1}}$ for the onset of vortex shedding decreases with increasing level of turbulence in the oncoming flow. Above this critical Reynolds number, due to sustained self-induced vortex shedding, force fluctuations will not be correlated with the oncoming flow – this is true even in the case of a uniform ambient flow. This behavior is most pronounced in the lift component. Thus, at such Reynolds numbers, the inability of the finite-Re BBO equation to predict the instantaneous force on the particle cannot be entirely blamed on ambient turbulence. The presence of turbulence in the ambient flow approaching the particle also affects the second critical Reynolds number for the onset of drag crisis. Torobin and Gauvin (1961) and Clamen and Gauvin (1969) reported the critical Reynolds number $\mathrm{Re_{cr2}}$ for laminar-to-turbulent boundary layer transition on the particle to dramatically decrease with increasing free-stream turbulence. According to their measurements, the drag coefficient dramatically decreased beyond the critical Reynolds number, but with further increase in Re the drag coefficient increased above the standard value.

The problem of an isolated particle in ambient isotropic turbulence was also considered by Burton and Eaton (2005). The corresponding problem of a stationary spherical bubble placed along the centerline of a turbulent pipe flow was considered by Merle et al. (2005). The findings of these studies were confirmed and extended with additional direct numerical simulations of a particle subjected to isotropic turbulence (Naso and Prosperetti, 2010; Botto and Prosperetti, 2012; J. Kim and Balachandar, 2012; Homann et al., 2013) and a particle subjected to channel flow turbulence, but with the particle away from the influence of the channel walls (Zeng et al., 2008, 2010). Together, these studies provide us insight into the direct effect of continuous-phase turbulence on the momentum transfer between the particle and the surrounding turbulent flow, which can be summarized as follows.

- As far as time-averaged mean drag force on the particle is concerned, there is no significant direct effect of ambient turbulence, at least at modest levels of ambient turbulence. The standard drag law can be applied to calculate the force on the particle as a function of time based on the instantaneous relative velocity seen by the particle. The time average of this standard drag adequately captures the actual mean drag force on the particle.

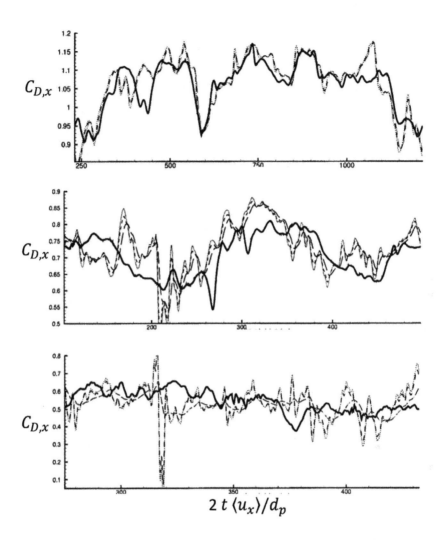

$$2\,t\,\langle u_x \rangle / d_p$$

Figure 7.25 Time variation of the streamwise component of the force coefficient obtained from the particle-resolved simulation (thick solid line). The top frame shows the results for a modest-sized particle of $d_p/\eta = 1.5$, the middle and the bottom frames are for larger particles of size $d_p/\eta = 3.8$ and 9.6. The turbulence intensity in all three cases was $I = 0.1$. Also shown are: long dash line – only the quasi-steady force given in the finite-Re BBO equation (4.144); short dash line – sum of quasi-steady, pressure-gradient, and added-mass forces; and dotted line – sum of quasi-steady, pressure-gradient, added-mass, and viscous history forces. Reprinted from Bagchi and Balachandar (2003b), with the permission of AIP Publishing.

- It must, however, be cautioned that the standard drag law when applied for the time-averaged mean relative velocity will not predict the mean drag. It is important to calculate the drag on an instantaneous basis and then do the averaging. This will allow accurate accounting of the nonlinear effect.

- Only in the case of particles of size smaller than the Kolmogorov scale will the

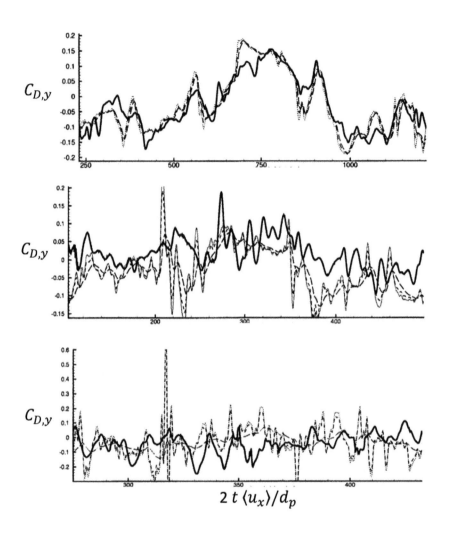

Figure 7.26 Time variation of the transverse component of the force coefficient obtained from the particle-resolved simulation (thick solid line). The top frame shows the results for a modest-sized particle of $d_p/\eta = 1.5$, the middle and bottom frames are for larger particles of size $d_p/\eta = 3.8$ and 9.6. Also shown are the predictions of the finite-Re BBO equation. Reprinted from Bagchi and Balachandar (2003b), with the permission of AIP Publishing.

standard drag relation accurately predict the instantaneous time evolution of the force on the particle.

- For larger particles ($d \gg \eta$), slow variations in particle force arising from scales larger than the particle can be well described by the standard drag law. However, rapid variations in the instantaneous force arising from scales comparable or smaller than the particle cannot be accurately captured by the standard drag.

- For larger particles, the inclusion of added-mass and history forces does not improve the prediction of the instantaneous force.
- At particle Reynolds numbers larger than a few hundred, the lift force fluctuation is chaotic and uncorrelated with the oncoming turbulent flow fluctuations (we shall discuss this further in the next subsection).
- The magnitude of temporal fluctuations in the drag and lift forces scales linearly with both the mean drag and free-stream turbulence intensity.

Based on the above outlined understanding of the turbulence effect on the drag and lift forces on a particle, the following picture emerges. For particles of size smaller than the continuous-flow scales, the finite-Re BBO equation is adequate to predict the instantaneous force on the particle. It must be pointed out that the Reynolds number of such small particles is expected to be small and self-induced vortex shedding will not be present (Balachandar, 2009). For particles of size matching the ambient flow scales, the best strategy will be to consider the force on the particle to comprise a deterministic and a stochastic component. The deterministic component will be based on $\mathbf{u}_\gg$ and will account for the effect of ambient flow scales larger than the particle. Force expressions such as the finite-Re BBO equation can be used for this purpose. The contribution to the force from smaller-scale eddies (i.e., $\mathbf{u}_\ll$ and $\mathbf{u}_\approx$) can best be accounted for in terms of a stochastic component. The Langevin model of the previous section can be used for this purpose. Such modeling of the turbulence effect is fundamentally different from the deterministic approach of Clamen and Gauvin (1969), Uhlherr and Sinclair (1970), and others, where they included relative turbulence intensity as an additional parameter in the drag correlation. If the stochastic component is appropriately modeled, the proposed modeling approach can account for the dispersion of particles that arise from the effect of the smaller scales of ambient motion. However, it should be noted that with the proposed approach, we admit our inability to precisely predict the trajectory of large particles in turbulent flows and resort to a statistical description. This is an appropriate approach, since for large particles Re will be such that self-induced vortex shedding will in any case introduce stochastic fluctuations to the force.

Finally, the above discussion of particle force is equally well applicable to mass and energy exchange between the dispersed and the continuous phases. The question of the turbulence effect on heat transfer has received only limited experimental attention (Raithby and Eckert, 1968; Yearling and Gould, 1995), and these studies report an increase in heat transfer due to ambient turbulence. More recent simulations (Bagchi and Kottam, 2008) have shown the effect of turbulence on heat transfer to be very similar to that of force on the particle (Bagchi and Balachandar, 2003b). The effect of turbulence on evaporation rate has received much attention owing to its importance in the context of fuel droplets (see Birouk and Gokalp, 2006). Here again there is no conclusive understanding of the turbulence effect, nor is there a universally accepted expression for the evaporation rate in a turbulent flow.

8   Particle–Wall Hydrodynamic Interactions

From the range of topics and the depth of physics that were discussed in the previous chapters, it is quite clear that multiphase flow is a challenging subject even at the level of an individual particle. But clearly we need to move forward and begin to consider more complex multiphase-flow physics. Toward this goal, we will progress beyond an isolated particle in an unbounded medium in two different ways. First, in this chapter we will consider the problem of an isolated particle in an ambient flow, but in the presence of a nearby wall. In the framework of what has been considered in the last two chapters, the role of the nearby wall will be twofold: (i) the undisturbed flow that exists in the absence of the particle must be consistent with the presence of the wall (e.g., due to the no-slip boundary condition, a shear flow in the vicinity of the wall is inevitable); (ii) the perturbation flow due to the presence of the particle will also be influenced by the wall, since no-slip and no-penetration boundary conditions must be satisfied not only on the particle, but also along the nearby wall.

These effects significantly alter the hydrodynamic force on the particle and as a result must be studied and modeled. The added effect of a nearby wall can be systematically considered in the context of the various limiting flows studied in the past two chapters. For example, in the zero particle Reynolds number limit, assuming the particle size to be smaller than all the flow scales, the effect of a nearby wall can be included by modifying the BBO equation. The influence of the wall in shaping the undisturbed flow will appear directly as modified undisturbed flow force. The influence of the wall on the perturbation flow will appear as changes to the quasi-steady, added-mass, and viscous unsteady forces. At finite Reynolds number, wall effects must be incorporated into the empirical drag and lift correlations. The importance of the wall in generating cross-stream migration of the particles, due to wall-induced lift force, was first identified in the pioneering work of Segre and Silberberg (1962a,b). In the context of a finite-sized particle subjected to wall turbulence, extensions to the MRG equation that include wall influence should be considered. Unfortunately (or fortunately, depending on one's perspective) our understanding of particle–wall interaction is primarily limited to steady condition. By the steady condition here we imply that the particle is in steady motion in a steady ambient flow at a fixed distance from the wall. In the first part of this chapter we will study the wall effects under steady condition for a range of Reynolds numbers and separation distance between the particle and the wall. Such understanding and corresponding models developed for steady state are still quite useful, as they are generously applied to a wide range of practical problems (that are invariably unsteady) by invoking the quasi-steady assumption.

8.1   Preliminary Discussion

In this chapter, we are interested in understanding the fluid mechanical interaction between a particle and the surrounding flow in the presence of a nearby flat solid boundary, which we will call the "wall." One of the goals is to obtain an expression for the hydrodynamic force on the particle in terms of (i) the particle motion and (ii) the undisturbed flow at the particle location. In an unbounded body of fluid, the force on the particle depended only on the relative velocity between the particle and the surrounding fluid. This symmetry must be carefully considered in the presence of a nearby wall, since it does not always apply.

Figure 8.1a shows a particle moving from left to right, parallel to a solid horizontal wall. Frames (b) and (c) show two different scenarios where the particle is stationary and the laminar flow is right to left. In all these three scenarios, the relative velocity between that of the particle and the undisturbed fluid velocity at the center of the particle is the same. Also, in all three cases the particle diameter d_p, and the distance h between the particle and the wall, are the same. In frame (b) the boundary layer is so thin that the particle is located well outside the boundary layer and the undisturbed flow at the particle can reasonably be approximated as a uniform flow. Whereas in frame (c) the particle is buried well within the boundary layer and the undisturbed ambient velocity is approximately linear around the particle.

It is thus clear that the undisturbed ambient flow near the particle, in the frame attached to the particle, is altered by the wall only in frame (c) – it changes from a uniform flow to a linear profile. But, in all three cases, the perturbation flow induced by the particle will be influenced by the nearby wall, where no-slip and no-penetration boundary conditions are satisfied. This influence of the wall on the perturbation flow depends on the relative distance to the wall in terms of the particle radius (i.e., on the ratio h/R_p). As this ratio approaches the limit $h/R_p \to 1$, the particle is almost in contact with the wall and the wall influence will be the strongest. For large values of h/R_p, particle–wall interaction can be ignored.

The inviscid effect of the wall can be understood by replacing the wall with a mirror particle on the other side of the wall. By considering the potential flow over this pair of

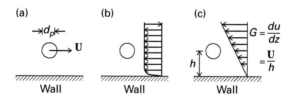

Figure 8.1 Schematic of three configurations, in all of which the relative velocity between the particle and the local undisturbed flow is the same.

particles in an unbounded fluid medium, the inviscid effect of the wall can be explored. As we will see in this chapter, the viscous effect of the wall is far more complex.

In the above paragraphs we only considered the role of particle and fluid velocities in the presence of a nearby wall. Unfortunately, further complications arise, which must be accounted for in the hydrodynamic interaction between the particle, ambient flow, and the wall. For example, the hydrodynamic force will also depend on (iii) whether the undisturbed flow is directed parallel to the wall as in a wall-bounded shear layer or directed normal to the wall as in a stagnation-point flow. Under unsteady conditions, the force will also depend on (iv) the acceleration of the particle and (v) the acceleration of the surrounding flow. Finally, (vi) the rotational motion of the particle will have the impact of the hydrodynamic force and torque on the particle. Thus, the problem quickly becomes quite complex due to the added presence of the wall.

Our strategy in this chapter is as follows. We will first consider the zero Reynolds number limit and restrict our attention to the steady limit. The linear nature of the governing Stokes equations will allow us to decompose the overall problem into five elemental problems, which can be superposed to account for an overall complex flow condition. Then, we will investigate wall-induced lift force as one of the primary effects of the wall. As in the case of Saffman's shear-induced lift, the wall-induced lift force is also due to fluid inertia. Therefore, we will consider the effect of the wall in the low Reynolds number limit and obtain explicit expressions for the drag and lift forces in terms of the controlling parameters. Here again, the overall behavior can be decomposed into contributions from six different mechanisms. We will then consider the full effect of nonlinearity at finite Reynolds number. We will preserve the functional forms of the drag and lift force expressions obtained from the zero and low Reynolds number studies and seek empirical extensions that are valid at finite Re. These finite-Re drag and lift expressions will be based on the results of direct numerical simulations. We will then briefly consider the unsteady effects of particle–wall interaction and close this chapter with an investigation of particle interaction with wall turbulence.

As the title of this chapter suggests, we are only concerned with hydrodynamic interactions between the particle and the wall as mediated by the fluid. Direct interaction between the wall and the particle through collisions, and the resulting momentum exchange, is also very important in many applications. This topic will not be covered in this chapter. Particle–particle collisions are addressed in the next chapter, where we will also address particle–wall collisions.

8.2   Zero Reynolds Number Stokes Limit

Here we consider the problem of a wall-bounded linear shear flow of shear magnitude G with a particle in steady translational and rotational motion within it. We consider the moment when the spherical particle of radius R_p is at a distance h from the wall. The translational velocity of the particle has a component V_x parallel to the wall and a component V_z perpendicular to the wall. Similarly, the rotational motion of the particle has a component Ω_y parallel to the wall and a component Ω_z perpendicular to the wall. The

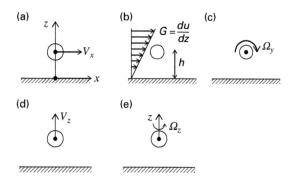

Figure 8.2 Schematic of the five different zero Reynolds number elementary problems of particle–wall interaction to be considered in this section.

flow in this case is governed by the Stokes equation. Due to the linearity of the Stokes equation, the above composite problem can be divided into five elementary problems whose solutions in terms of the force and torque on the particle will be presented below. The solution of the overall problem is then a superposition of the elementary problems. The five elementary problems, shown in Figure 8.2, are as follows:

(I) A particle translating at velocity V_x parallel to a flat wall at a distance h from the wall (frame a).

(II) A stationary particle at a distance h in a wall-bounded linear shear flow of shear magnitude G (frame b).

(III) A particle rotating at an angular velocity Ω_y with the axis of rotation parallel to a flat wall. The particle is located at a distance h from the wall (frame c).

(IV) A particle translating at velocity V_z perpendicular to a flat wall at a distance h from the wall (frame d).

(V) A particle rotating at an angular velocity Ω_z with the axis of rotation perpendicular to a flat wall at a distance h from the wall (frame e).

8.2.1 Problem I: Translation Parallel to the Wall

The flow induced by a spherical particle translating parallel to a flat wall (see Figure 8.2a) is obtained by solving the Stokes equation

$$\nabla \cdot \mathbf{u} = 0 \quad \text{and} \quad 0 = -\nabla p + \mu_f \nabla^2 \mathbf{u} . \tag{8.1}$$

If we consider $\mathbf{u}$ to be the perturbation flow, then the boundary conditions to be satisfied are

$$\begin{aligned}
\text{On the particle:} \quad & \mathbf{u} = V_x \mathbf{e}_x , \\
\text{On the wall:} \quad & \mathbf{u} = 0 , \\
\text{Far field:} \quad & \mathbf{u} \to 0 .
\end{aligned} \tag{8.2}$$

This problem has been solved in several different ways to obtain the net hydrodynamic force and torque on the particle. In the limit where the distance between the particle and the wall becomes large (i.e., for $h/R_p \gg 1$), the hydrodynamic force on the particle should approach the Stokes drag and the torque should be zero. Thus, the interest is to obtain the effect of a nearby wall. As an added bonus, some of the prior investigations have also obtained expressions for the induced force on the wall due to the particle motion.

O'Neill (1964) obtained an exact solution by solving the above equations and boundary conditions in bi-spherical coordinates. This approach placed no restriction on the distance between the particle and the wall. However, the resulting solutions are expressed only as complex series. More transparent expressions of force can be obtained by focusing attention on the limiting cases of a small gap between the particle and the wall $(h/R_p - 1) \ll 1$ or a very large gap $h \gg R_p$. These two limiting cases will be considered below.

Small-Gap Limit

In the limit $(h/R_p - 1) \ll 1$, we briefly present the lubrication analysis of Goldman et al. (1967a), who employed the singular perturbation methodology to solve this narrow-gap problem. Following their analysis, we define $\delta = h - R_p$ as the gap between the bottom of the particle and the wall (see Figure 8.2a) and obtain the solution in cylindrical coordinates (r, θ, z) whose origin is on the wall below the particle. The θ dependence of the solution can be written explicitly as

$$p = \mu_f \, \tilde{p} \cos \theta, \quad u_r = \tilde{u}_r \cos \theta,$$
$$u_\theta = \tilde{u}_\theta \sin \theta, \quad u_z = \tilde{u}_z \cos \theta, \tag{8.3}$$

where all "tilde" quantities are only functions of r and z. Substituting the above into the incompressibility condition and the Stokes equation, the corresponding equations for the "tilde" quantities can be obtained (see Goldman et al., 1967a). The boundary conditions transform to

$$\text{On the particle: } (\tilde{u}_r, \tilde{u}_\theta, \tilde{u}_z) = (V_x, -V_x, 0),$$
$$\text{On the wall: } (\tilde{u}_r, \tilde{u}_\theta, \tilde{u}_z) = (0, 0, 0), \tag{8.4}$$
$$\text{Far field: } (\tilde{u}_r, \tilde{u}_\theta, \tilde{u}_z) \to (0, 0, 0).$$

For small values of $\tilde{\delta} = \delta/R_p$ we seek an asymptotic expansion:

$$\tilde{p} = \tilde{p}_0 + \tilde{\delta} \tilde{p}_1 + \cdots, \quad \tilde{u}_r = \tilde{u}_{r0} + \tilde{\delta} \tilde{u}_{r1} + \cdots,$$
$$\tilde{u}_\theta = \tilde{u}_{\theta 0} + \tilde{\delta} \tilde{u}_{\theta 1} + \cdots, \quad \tilde{u}_z = \tilde{u}_{z0} + \tilde{\delta} \tilde{u}_{z1} + \cdots. \tag{8.5}$$

This expansion is valid everywhere except within the narrow gap and therefore will be termed the "outer solution." Note that the leading-order solution $\tilde{p}_0, \tilde{u}_{r0}, \tilde{u}_{\theta 0}, \tilde{u}_{z0}$ corresponds to the singular limit of the particle in contact with the wall (i.e., to the $\delta/R_p = 0$ limit).

To obtain a solution that is valid within the narrow-gap region, we have to introduce the following coordinate stretching:

$$\rho = \frac{r}{\sqrt{R_p \delta}} \quad \text{and} \quad \zeta = \frac{z}{\delta}. \tag{8.6}$$

The next step is to rewrite the governing incompressibility and Stokes equations in the modified coordinates for $\tilde{p}, \tilde{u}_r, \tilde{u}_\theta, \tilde{u}_z$. Goldman et al. (1967a) proposed the following expansion for the inner solution:

$$\tilde{p} = \tilde{\delta}^{-3/2} \left(\tilde{p}_{I0} + \tilde{\delta}\tilde{p}_{I1} + \cdots \right) , \quad \tilde{u}_r = \tilde{u}_{rI0} + \tilde{\delta}\tilde{u}_{rI1} + \cdots ,$$
$$\tilde{u}_\theta = \tilde{u}_{\theta I0} + \tilde{\delta}\tilde{u}_{\theta I1} + \cdots , \qquad \tilde{u}_z = \tilde{\delta}^{1/2} \left(\tilde{u}_{zI0} + \tilde{\delta}\tilde{u}_{zI1} + \cdots \right) , \tag{8.7}$$

where subscript I indicates the inner solution. Note that terms such as $\tilde{u}_{rI0}$ are now functions of the inner variables ρ and ζ. In the perspective of the inner region of the narrow gap, the bottom surface of the sphere can be approximated as a paraboloid $z = \delta + r^2/(2R_p) + O(r^4)$ and in terms of the stretched coordinates the boundary condition (8.4), first line is applied on the bottom surface of the particle, defined as

$$\text{On the sphere:} \quad \zeta = 1 + \frac{1}{2}\rho^2 . \tag{8.8}$$

In addition, the wall boundary condition (8.4), second line must be applied at $\zeta = 0$. Instead of the far-field boundary condition, the inner solution is required to match the outer solution. We will leave the detailed matching and solution procedure to Goldman et al. (1967a) for the reader to consult.

The leading-order solution of the matched asymptotic analysis correctly captures the logarithmic singularity of the solution as the bottom of the particle approaches the wall. For higher-order correction, Goldman et al. (1967a) used a fitting procedure to match the series solution of O'Neill (1964) and proposed the following analytical expressions for the hydrodynamic drag and torque[1] on the particle:

$$\text{Small-gap limit:} \quad \begin{cases} \mathbf{F}_{V0s} = 6\pi\mu_f R_p V_x \mathbf{e}_x \left(\frac{8}{15} \ln\tilde{\delta} + \frac{64}{375}\tilde{\delta}\ln\tilde{\delta} - 0.952 \right) , \\[2mm] \mathbf{T}_{V0s} = -8\pi\mu_f R_p^2 V_x \mathbf{e}_y \left(\frac{2}{15} \ln\tilde{\delta} + \frac{86}{375}\tilde{\delta}\ln\tilde{\delta} + 0.257 \right) . \end{cases} \tag{8.9}$$

In the above, the subscript $V0s$ indicates hydrodynamic force and torque due to translation parallel to the wall in the zero Reynolds number and small-gap limits. Also, since $\tilde{\delta} \ll 1$, its logarithm is negative, and the drag force is correctly directed opposite to particle motion. The corresponding clockwise torque on the particle can be justified in terms of the strong tangential shear force in the negative x-direction in the gap region. Both the force and the torque become singular as $\tilde{\delta} \to 0$ in a logarithmic fashion, which is the lubrication singularity as the particle approaches the wall. In the limit where the particle is in contact with the wall, it requires infinite force (and torque) to move the particle. The above results and all other small-gap results to follow are taken from Dance and Maxey (2003). They compiled the theoretical results of eight other sources, and reconciled all the errors and inconsistencies in them, to summarize the final results which are reproduced here.

---

[1] We will use the term "torque" for the net moment of all the surface forces that act on the particle (see Eq. (4.1)).

Large-Gap Limit

The limit when the particle is far from the wall was first solved by Faxén (1923) using the *method of reflections*. This iterative methodology starts by considering the motion of the particle in an unbounded fluid, in the absence of the wall. The solution to this problem is well known and we have solved it in detail in Section 4.1.1. At this leading (zeroth) order, the drag on the particle is the Stokes drag with zero associated torque on the particle. If we call this zeroth-order solution $\mathbf{u}_0(\mathbf{x})$, its shortcoming is that it does not satisfy the boundary conditions at the wall. The advantage of limiting the wall to be far from the particle is that the zeroth-order flow at the far-away wall can be approximated to be that due to a Stokeslet. In other words, the particle can be replaced by a point force.

The next step of the iteration is to solve for the Stokes flow due to an imposed velocity of $-\mathbf{u}_0|_{@W}$ at the wall. This imposed velocity will precisely cancel that due to the zeroth-order flow at the wall. If we call this first-order flow due to the imposed wall velocity $\mathbf{u}_1(\mathbf{x})$, then the combined velocity $\mathbf{u}_0 + \mathbf{u}_1$ satisfies the wall boundary condition. Unfortunately, this first reflected flow evaluated at the particle surface (i.e., $\mathbf{u}_1|_{@S}$) will not satisfy the required boundary condition. This gives rise to the second reflection. The second-order solution is then obtained by solving the Stokes equation with the boundary condition at the particle as $-\mathbf{u}_1|_{@S}$ in an unbounded domain. The next step is to reflect off the wall, and then at the particle, and so the process continues. At each stage of reflection, the estimation of force and torque on the particle can be improved.

The method of reflections, in principle, can be applied without any restriction on the value of $\tilde{\delta}$. For small values of $\tilde{\delta}$, at each level of iteration, the flow induced at the wall cannot be approximated as that due to a Stokeslet from a point force at the particle. This approximation simplifies the analysis and can be justified only in the limit $\tilde{\delta} \gg 1$. In the large-gap limit, the expressions of hydrodynamic force and torque obtained from the method of reflections are as follows:

$$\text{Large-gap limit:} \begin{cases} \mathbf{F}_{V0l} = -6\pi\mu_f R_p V_x \mathbf{e}_x \left(1 - \dfrac{9}{16\tilde{h}} + \dfrac{1}{8\tilde{h}^3} - \dfrac{45}{256\tilde{h}^4}\right)^{-1}, \\ \mathbf{T}_{V0l} = 8\pi\mu_f R_p^2 V_x \mathbf{e}_y \left(\dfrac{3}{32\tilde{h}^4}\right)\left(1 - \dfrac{3}{8\tilde{h}}\right), \end{cases} \tag{8.10}$$

where $\tilde{h} = h/R_p = 1 + \tilde{\delta}$. In the above, the subscript $V0l$ indicates hydrodynamic force and torque due to translation parallel to the wall in the zero Reynolds number and large-gap limits. Approximate composite correlations that blend the two asymptotic results can easily be formed in the following way. Define the blending function as

$$B(\tilde{\delta}, c_b, w_b) = \frac{1}{2}\left[1 - \operatorname{erf}\left(\frac{\ln(\tilde{\delta}/c_b)}{\ln(w_b/c_b)}\right)\right], \tag{8.11}$$

where c_b and w_b represent the logarithmic center and half width of the error-function blending. In terms of the blending function, composite correlations that are valid for all gap values are obtained as

$$\text{All gap:} \begin{cases} \mathbf{F}_{V0} = \mathbf{F}_{V0s}\, B(\tilde{\delta}, c_b, w_b) + \mathbf{F}_{V0l}\,(1 - B(\tilde{\delta}, c_b, w_b)), \\ \mathbf{T}_{V0} = \mathbf{T}_{V0s}\, B(\tilde{\delta}, c_b, w_b) + \mathbf{T}_{V0l}\,(1 - B(\tilde{\delta}, c_b, w_b)). \end{cases} \tag{8.12}$$

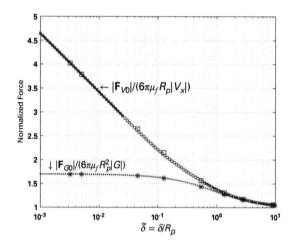

Figure 8.3 The scaled hydrodynamic force on a particle as a function of its nondimensional gap from the nearby wall. The solid line is the small-gap limit given in Eq. (8.9), the dashed–dotted line is the large-gap limit given in Eq. (8.10), and the circles are the composite expression given in Eq. (8.12) that is valid for all values of $\tilde{\delta}$. These lines are for the case of a particle translating parallel to the wall. Also presented as a dotted line is the corresponding composite force expression for a stationary particle in a linear shear flow given in Eq. (8.15). The values given in Table 8.1 are also plotted as asterisks and squares.

The above expressions reduce to the correct small and large-gap limits for small and large values of $\tilde{\delta}$. These composite solutions are neither unique, nor optimized in any way. For example, a different blended correlation for force was presented by Zeng et al. (2009). These functions are sufficient in most applications and can be improved upon if needed. Figures 8.3 and 8.4 plot the normalized force and torque as functions of $\tilde{\delta}$ for both the small-gap and large-gap limits, along with the composite correlations. For the composite force we use $c_b = 0.2$ and $w_b = 0.5$ and for the composite torque we use $c_b = 0.08$ and $w_b = 0.2$.

8.2.2 Problem II: Particle in a Wall-Bounded Linear Shear Flow

In this configuration, the particle is held stationary at a distance h from the wall but in a linear shear flow. The governing equations remain the same as Eq. (8.1) but the boundary conditions change to

$$
\begin{array}{ll}
\text{On the particle:} & \mathbf{u} = 0, \\
\text{On the wall:} & \mathbf{u} = 0, \\
\text{Far field:} & \mathbf{u} \to Gz\mathbf{e}_x .
\end{array} \tag{8.13}
$$

This problem was considered by Goldman et al. (1967b), who provided an elegant solution using the reciprocal relation we discussed in Section 4.1.5.

In order to apply the reciprocal theorem we identify the surface ∂D as the union of the surface of the particle S plus the wall W. The first Stokes flow is identified as the solution to the translating sphere problem that we just considered in the previous subsection. In the frame attached to the particle, the perturbation flow on the surface

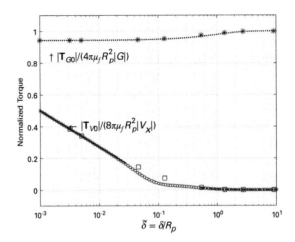

Figure 8.4 The scaled hydrodynamic torque on a particle as a function of its nondimensional gap from the nearby wall. The different lines and symbols are the same as in Figure 8.3.

of the particle is $-\mathbf{e}_x$ (for unit translational velocity), while that on the wall is zero. Thus, in the application of the reciprocal relation (4.64), $\mathbf{u}_1 = \{[-\mathbf{e}_x]_{@S} \cup [0]_{@W}\}$ and the corresponding tractional force distribution on the surface ∂D will be denoted as $\mathbf{f}_1 = \{\mathbf{f}_{1@S} \cup \mathbf{f}_{1@W}\}$. Goldman et al. (1967b) obtained the tractional force distribution on the surface of the sphere ($\mathbf{f}_{1@S}$) from the series solution of Eqs. (8.1) and (8.2) by O'Neill (1964). However, unlike in the case of uniform unbounded flow over an isolated particle, where the tractional force was a constant vector, in the present problem $\mathbf{f}_{1@S}$ varies over the surface of the sphere. As we will see below, the tractional force distribution on the wall $\mathbf{f}_{1@W}$ is not needed, since $\mathbf{u}_2$ will be chosen to be zero on the wall.

The second Stokes flow in the reciprocal theorem is the linear shear flow problem whose solution is the object of interest here. This velocity on the surface ∂D can be expressed as $\mathbf{u}_2 = \{[Gz\mathbf{e}_x]_{@S} \cup [0]_{@W}\}$. The corresponding tractional force distribution will be denoted as $\mathbf{f}_2 = \{\mathbf{f}_{2@S} \cup \mathbf{f}_{2@W}\}$ on the surface of the sphere. Again the tractional force on the wall $\mathbf{f}_{2@W}$ is not needed since $\mathbf{u}_1$ was chosen to be zero on the wall. Substituting these into Eq. (4.64), we can carry out the following simplifications:

$$\int_{\partial D} \mathbf{f}_2 \cdot \mathbf{u}_1 \, dA = \int_{\partial D} \mathbf{f}_1 \cdot \mathbf{u}_2 \, dA,$$

$$\int_S \mathbf{f}_2 \cdot \mathbf{u}_1 \, dA + \int_W \mathbf{f}_2 \cdot \mathbf{u}_1^{\,0} dA = \int_S \mathbf{f}_1 \cdot \mathbf{u}_2 \, dA + \int_W \mathbf{f}_1 \cdot \mathbf{u}_2^{\,0} dA, \qquad (8.14)$$

$$-F_{G0} = -\mathbf{e}_x \cdot \int_S \mathbf{f}_2 \, dA = \left(\int_S \mathbf{f}_1 \cdot (Gz \, \mathbf{e}_x) \, dA \right),$$

where F_{G0} is the shear-induced force on the particle which is directed along the x-direction and this is indicated by the first equality of the last equation above. Most importantly, F_{G0} can be obtained by integrating the surface distribution of the traction force obtained in Problem I weighted by the shear. A similar integral expression can

Table 8.1 Normalized force and torque on a particle subjected to wall-bounded linear shear flow in the Stokes regime. Also presented are results for a translating particle. These values are taken from Goldman et al. (1967a,b).

$\tilde{h}$	$\dfrac{|\mathbf{F}_{G0}|}{(6\pi\mu_f R_p(Gh))}$	$\dfrac{|\mathbf{T}_{G0}|}{(4\pi\mu_f R_p^3 G)}$	$\dfrac{|\mathbf{F}_{V0}|}{(6\pi\mu_f R_p V_x)}$	$\dfrac{|\mathbf{T}_{V0}|}{(8\pi\mu_f R_p^2 V_x)}$
∞	1.0	1.0000	1.0	0.0
10.0677	1.0587	0.99981	1.0591	8.7744×10^{-6}
3.7622	1.1671	0.99711	1.1738	4.2160×10^{-4}
2.3524	1.2780	0.99010	1.3079	2.6423×10^{-3}
1.5431	1.4391	0.97419	1.5675	1.4649×10^{-2}
1.1276	1.6160	0.95374	2.1514	7.3718×10^{-2}
1.0453	1.6682	0.94769	2.6475	0.14552
1.0050	1.6969	0.94442	3.7863	0.34187
1.0032	1.6982	0.94427	4.0223	0.38494
1.0	1.7005	0.94399	∞	∞

also be obtained for the hydrodynamic torque due to the shear flow. Analytical solution of the above integral is not feasible. But numerical solution of the integrals obtained through quadrature was presented in Goldman et al. (1967b), whose tabulated results for drag force and torque are reproduced here in Table 8.1. A composite curve fit for the linear shear-induced drag and torque in the Stokes flow limit can be expressed as (Zeng et al., 2009)

$$\text{All gap:} \begin{cases} \mathbf{F}_{G0} = 6\pi\mu_f R_p(Gh)\mathbf{e}_x \left(1 + 0.138\exp(-\tilde{\delta}) + \frac{9}{16\tilde{h}}\right), \\ \mathbf{T}_{G0} = 4\pi\mu_f R_p^3 G\mathbf{e}_y \left(1 - 0.0541\exp(-\tilde{\delta}) - \frac{0.00191}{\tilde{h}}\right), \end{cases} \tag{8.15}$$

where (Gh) is the shear flow velocity at the particle center. These results are also plotted in Figures 8.3 and 8.4. Again, this expression reduces to the correct asymptotic behavior in the limits of small and large values of the gap.

Several interesting and potentially important features can be observed. Unlike in the translating particle case, there is no logarithmic singularity in the case of a stationary particle subjected to a shear flow. This is quite understandable, since particle motion while in contact with the wall requires squeezing out the liquid layer that is ahead of the particle. Such singularity is not present in the case of a stationary particle that is in contact with the wall.

The limiting value of drag due to a shear flow for a stationary particle in contact with the wall is 70% higher than that for a particle in a uniform flow far away from a bounding wall. Note that in an unbounded medium, linear shear has no effect on the hydrodynamic drag force. This can readily be verified with an application of Faxén's theorem. Thus, we have obtained an important result that the wall effect can substantially increase the Stokes drag. On the other hand, the hydrodynamic torque on the particle is slightly higher in an unbounded shear flow, and the wall effect is to decrease the torque by 5.6%. Furthermore, for the same value of relative velocity (difference between the particle velocity and the undisturbed fluid velocity at the center of the particle), the drag force is consistently higher for the translating particle than for the stationary particle in a shear flow. This observation supports the point made earlier that in the presence of the wall, the drag force is not simply dependent on the relative velocity. Both the particle velocity and the fluid velocity relative to the wall must be considered independently.

8.2.3   Particle III: Particle Spinning About an Axis Parallel to the Wall

In this problem, we will consider a spherical particle spinning about an axis parallel to a wall that is located at a distance h from the particle. Otherwise the fluid is quiescent and therefore the flow is entirely due to the rotational motion of the particle. Thus, this problem is the same as the spinning sphere considered in Section 4.1.3, but with the added presence of the wall. We arbitrarily choose the axis of particle rotation to be along the y-axis with an angular velocity of Ω_y (see Figure 8.2c).

The governing Stokes equations are the same as given in Eq. (8.1), and the appropriate boundary conditions are

$$
\begin{aligned}
\text{On the particle:} \quad & \mathbf{u} = \Omega_y \mathbf{e}_y \times \mathbf{r}, \\
\text{On the wall:} \quad & \mathbf{u} = 0, \\
\text{Far field:} \quad & \mathbf{u} \to 0.
\end{aligned}
\tag{8.16}
$$

This problem was solved using a bi-spherical coordinate system by Dean and O'Neill (1964), who obtained a series solution. Following the process employed for the problem of translating a particle, the two different limiting cases can be considered separately. In the small-gap limit, the lubrication theory leads to a matched asymptotic problem, whose solution was obtained by Goldman et al. (1967a) as

$$
\text{Small-gap limit:} \quad
\begin{cases}
\mathbf{F}_{\Omega 0s} = -6\pi\mu_f R_p^2 \Omega_y \mathbf{e}_x \left(\dfrac{2}{15} \ln \tilde{\delta} + \dfrac{86}{375} \tilde{\delta} \ln \tilde{\delta} + 0.257 \right), \\[2mm]
\mathbf{T}_{\Omega 0s} = 8\pi\mu_f R_p^3 \Omega_y \mathbf{e}_y \left(\dfrac{2}{5} \ln \tilde{\delta} + \dfrac{66}{125} \tilde{\delta} \ln \tilde{\delta} - 0.371 \right),
\end{cases}
\tag{8.17}
$$

where the subscript $\Omega 0s$ indicates the contribution from rotational motion with axis parallel to the wall in the zero Reynolds number and small-gap limits. Here again we observe the logarithmic singularity as the particle approaches the wall. In the limit of the particle touching the wall, infinite torque (and force) will be required to mobilize the lubrication layer and rotate the particle. The leading-order matched asymptotic analysis yielded the logarithmic terms; the constant terms were obtained by matching with the series solutions of Dean and O'Neill (1964).

In the large-gap limit, the method of reflections can be used to obtain the results (Maude, 1961). These results to leading order can be summarized as

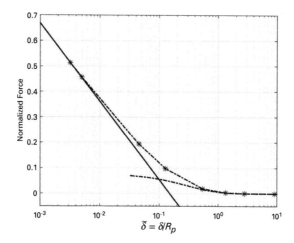

Figure 8.5 The scaled hydrodynamic force on a rotating particle as a function of its nondimensional gap from the nearby wall. The solid line is the small-gap limit given in Eq. (8.17) and the dashed–dotted line is the large-gap limit given in Eq. (8.18). These two limiting cases must be blended in the intermediate values of $\tilde{\delta}$. The tabulated value of Goldman et al. (1967a) is also plotted.

$$\text{Large-gap limit:} \quad \begin{cases} \mathbf{F}_{\Omega 0 l} = 6\pi\mu_f R_p^2 \Omega_y \mathbf{e}_x \left(\dfrac{1}{8\tilde{h}^4}\right)\left(1 - \dfrac{3}{8\tilde{h}}\right), \\[4mm] \mathbf{T}_{\Omega 0 l} = -8\pi\mu_f R_p^3 \Omega_y \mathbf{e}_y \left(1 + \dfrac{5}{16\tilde{h}^3}\right), \end{cases} \tag{8.18}$$

where $\tilde{h} = h/R_p = 1 + \tilde{\delta}$. The rotation-induced force and torque on a particle are presented in Figures 8.5 and 8.6. As can be expected, the normalized torque is much higher in this case than for the translation and shear problems, while the induced force is much lower. A composite curve fit for the rotational drag and torque in the Stokes flow limit can be obtained by blending the two limits. These results are plotted in Figures 8.5 and 8.6, along with the tabulated data of Goldman et al. (1967a).

In the absence of the wall, translational motion of the particle generates only a hydrodynamic force, while particle rotation generates only a hydrodynamic torque. The wall induces cross-coupling. As a result, translational motion also generates a hydrodynamic torque and rotation generates an added hydrodynamic force. In fact, Goldman et al. (1967a) note that these cross-coupling effects are related and obey the following interesting relation:

$$\frac{|\mathbf{F}_{\Omega 0}|/(6\pi\mu_f R_p^2 |\Omega_y|)}{|\mathbf{T}_{V0}|/(8\pi\mu_f R_p^2 |V_x|)} = \frac{4}{3}. \tag{8.19}$$

Problem 8.1 Consider a neutrally buoyant particle of radius R_p in a wall-bounded shear flow of shear rate G. The particle is allowed to translate freely downstream in response to the hydrodynamic force and rotate in response to the hydrodynamic

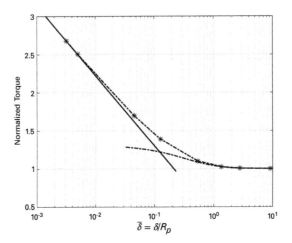

Figure 8.6 The scaled hydrodynamic force on a rotating particle as a function of its nondimensional gap from the nearby wall. The solid line is the small-gap limit given in Eq. (8.17) and the dashed–dotted line is the large-gap limit given in Eq. (8.18). These two limiting cases must be blended in the intermediate values of $\tilde{\delta}$. The tabulated value of Goldman et al. (1967a) is also plotted.

torque. These translational and rotational motions will in turn contribute to additional hydrodynamic force and torque. Since the particle is neutrally buoyant, its motion is strictly horizontal parallel to the wall. In other words, neither the undisturbed shear flow nor the particle has a wall-normal component of velocity. All velocities are in the wall-parallel direction. In this problem, we are interested in the "terminal" steady motion of the particle, where the net hydrodynamic force and torque on the particle are zero. Since there is no net force or torque, the particle will continue to move at the terminal steady state. This problem was discussed by Goldman et al. (1967b). Thus, the equations satisfied at the terminal state are

$$\mathbf{F}_{V0} + \mathbf{F}_{G0} + \mathbf{F}_{\Omega 0} = 0 \quad \text{and} \quad \mathbf{T}_{V0} + \mathbf{T}_{G0} + \mathbf{T}_{\Omega 0} = 0. \qquad (8.20)$$

We now substitute the different force and torque expressions and the above two equations are solved for V_x and Ω_y as a function of the three input parameters of the problem, namely, G, R_p, and h.

(a) First consider the small-gap limit where the translational and rotational forces and torques are given in Eqs. (8.9) and (8.17). Take the corresponding shear-induced force and torque values to be those corresponding to $\tilde{h} = 1.0$ in Table 8.1. Obtain the result

$$\frac{V_x}{Gh} = \frac{0.7431}{0.6376 - 0.2\ln\tilde{\delta}} \quad \text{and} \quad \frac{\Omega_y}{G} = \frac{0.4218}{0.6376 - 0.2\ln\tilde{\delta}}. \qquad (8.21)$$

(b) Next consider the large-gap limit where the translational and rotational forces and torques are given in Eqs. (8.10) and (8.18). To simplify the calculation, restrict the expansions to the leading-order term involving a nonzero power of $1/\tilde{h}$. Take the corresponding shear-induced force and torque values to be those corresponding to $\tilde{h} = \infty$ in Table 8.1. Obtain the result

$$\frac{V_x}{Gh} = 1 - \frac{5}{16\tilde{h}^3} \quad \text{and} \quad \frac{\Omega_y}{G} = \frac{1}{2} - \frac{5}{32\tilde{h}^3}. \tag{8.22}$$

Problem 8.2 Here we will generalize the previous problem by imposing a certain translational velocity on the particle, and seek the corresponding torque-free rotational rate of the particle. Again, we consider a neutrally buoyant particle of radius R_p in a wall-bounded shear flow of shear rate G, translating at velocity V_x. The particle is allowed to rotate freely in response to the hydrodynamic torque and reach the terminal torque-free state. The equation satisfied at the terminal state is

$$\mathbf{T}_{V0} + \mathbf{T}_{G0} + \mathbf{T}_{\Omega 0} = 0. \tag{8.23}$$

We now substitute the different force and torque expressions and the above equation is solved for Ω_y as a function of the four input parameters of the problem, namely, G, V_x, R_p, and h. This torque-free rotational rate of the particle will be termed $\Omega_{y,TF}$.

(a) First consider the small-gap limit where the translational and rotational forces and torques are given in Eqs. (9.9) and (8.17). Take the corresponding shear-induced force and torque values to be those corresponding to $\tilde{h} = 1.0$ in Table 8.1 and obtain an expression for $\Omega_{y,TF}$.

(b) Next consider the large-gap limit where the translational and rotational forces and torques are given in Eqs. (8.10) and (8.18). Again obtain an expression for $\Omega_{y,TF}$. To simplify the calculation, restrict the expansions to the leading-order term involving a nonzero power of $1/\tilde{h}$. Take the corresponding shear-induced force and torque values to be those corresponding to $\tilde{h} = \infty$ in Table 8.1.

8.2.4 Problem IV: Particle Translating Perpendicular to the Wall

This problem is similar to that described by the boundary conditions given in Eq. (8.2), except that the velocity boundary condition on the particle is given by particle velocity in the wall-normal direction (i.e., $\mathbf{u} = V_z \mathbf{e}_z$). As in Problem I, the fluid is quiescent, except for the induced perturbation flow due to the motion of the particle. This problem was solved in bipolar coordinates by Brenner (1961) and reported in the book by Happel and Brenner (1965). In the small-gap limit, the lubrication theory leads to the following solution (see Dance and Maxey, 2003):

$$\text{Small-gap limit:} \quad \begin{cases} \mathbf{F}_{V\perp 0s} = 6\pi\mu_f R_p V_z \mathbf{e}_z \left(-\dfrac{1}{\tilde{\delta}} + \dfrac{1}{5}\ln\tilde{\delta} + \dfrac{1}{21}\tilde{\delta}\ln\tilde{\delta} - 0.848 \right), \\[2mm] \mathbf{T}_{V\perp 0s} = 0. \end{cases} \tag{8.24}$$

Table 8.2 Normalized force on a particle of radius R_p translating at velocity $V_z \mathbf{e}_z$ perpendicular to a flat wall at a distance h from the wall in the Stokes regime.

$\tilde{h} = h/R_p$	$-\mathbf{F}_{V\perp 0}/(6\pi\mu_f R_p V_z \mathbf{e}_z)$
∞	1.0
10.0677	1.1252
6.1323	1.2220
3.7622	1.4129
2.3524	1.8375
1.5431	3.0361
1.1276	9.2518
1.0	∞

Again the large-gap solution is given as an infinite series. The series was evaluated and presented in the form of a table, which is repeated here as Table 8.2. In this problem, the force on the particle is denoted as $\mathbf{F}_{V\perp 0}$ and it is in the wall-normal direction, opposing the direction of particle motion. By symmetry, the particle will not experience any hydrodynamic torque. Again, as the gap between the particle and the wall approaches zero, the singular nature of the problem can be seen as the rapid increase in the hydrodynamic force.

8.2.5   Problem V: Particle Spinning About an Axis Perpendicular to the Wall

Now we complete this section with the final elementary problem of a spinning particle of radius R_p at a distance h from the wall with the axis of rotation perpendicular to the wall. The small-gap solution is given as (Dance and Maxey, 2003)

$$\text{Small-gap limit:} \quad \begin{cases} \mathbf{F}_{\Omega\perp 0s} = 0, \\ \mathbf{T}_{\Omega\perp 0s} = 8\pi\mu_f R_p^3 \omega_z \mathbf{e}_z \left(\frac{1}{2}\tilde{\delta}\ln\tilde{\delta} - 1.202 \right). \end{cases} \tag{8.25}$$

The large-gap solution can also be found in the classic book by Happel and Brenner (1965). If we consider the particle's angular velocity to be $\Omega_z \mathbf{e}_z$, then the torque on the particle to leading order can be expressed as

$$\mathbf{T}_{\Omega\perp 0l} = -\frac{8\pi\mu_f R_p^3 \Omega_z \mathbf{e}_z}{1 - \dfrac{1}{8\tilde{h}^3}}. \tag{8.26}$$

Thus, due to the wall effect, the torque on the particle increases above its value in an unbounded fluid. This increase in torque increases as the distance between the wall and the particle decreases. In the limit where the particle touches the wall, there is a lubrication singularity. Again, by symmetry there is no hydrodynamic force on the particle.

Table 8.3 Summary of composite drag and torque results that are applicable over the entire range of distances from the wall in the zero Reynolds number limit. These composite fits approach the correct asymptotic results in the limits of small and large gaps.

	Problem I: Parallel translation	Problem II: Linear shear	Problem III: Parallel rotation	Problem IV: Perpendicular translation	Problem V: Perpendicular rotation
Drag	Eq. (8.12)	Eq. (8.15)	Eqs. (8.17) and (8.18)	Eqs. (8.24) and Table 8.2	0
Torque	Eq. (8.12)	Eq. (8.15)	Eqs. (8.17) and (8.18)	0	Eqs. (8.24) and (8.26)

8.2.6   Zero Reynolds Number Summary

We now summarize the results of all five cases in Table 8.3. This table presents the zero Reynolds number composite drag and torque formulas that are appropriate for the different problems. Due to the linearity of the Stokes equation, these results are additive. In the more general problem where a particle is both translating and rotating in a wall-bounded shear flow, the solution can be expressed as a superposition of these elementary problems.

Problem 8.3 Reconsider the Stokes flow problem considered in the above section, but without the shear flow. So the fluid is quiescent and the particle has translational and rotational velocities $\mathbf{V}$ and $\mathbf{\Omega}$. In the Stokes limit, the force and torque on the particle can be expressed as linear functions of particle translational and rotational velocity, as (Dance and Maxey, 2003; Kim and Karrila, 2013)

$$\mathbf{F} = 6\pi\mu_f R_p(\mathbf{A} \cdot \mathbf{V} + R_p\mathbf{B} \cdot \mathbf{\Omega}) \quad \text{and} \quad \mathbf{F} = 8\pi\mu_f R_p^2(\mathbf{C} \cdot \mathbf{V} + R_p\mathbf{D} \cdot \mathbf{\Omega}), \quad (8.27)$$

where $\mathbf{A}$, $\mathbf{B}$, $\mathbf{C}$, and $\mathbf{D}$ are second-rank tensors that are only functions of the distance between the particle and the wall.

(a) Use symmetry and argue that $\mathbf{A}$ is diagonal and $A_{22} = A_{33}$.

(b) Also, argue that only B_{23} and B_{32} are nonzero.

(c) Use the Lorentz reciprocal theorem to show that $\mathbf{A}$ and $\mathbf{D}$ are symmetric, while $\mathbf{B}$ and $\mathbf{C}$ are antisymmetric. In addition, $\mathbf{B}^T = \mathbf{C}$.

(d) Show that $\mathbf{D}$ is diagonal and $D_{22} = D_{33}$.

(e) Use the results of Sections 8.2.1, 8.2.3, 8.2.4, and 8.2.5 to obtain the nonzero entries of the matrices $\mathbf{A}$, $\mathbf{B}$, $\mathbf{C}$, and $\mathbf{D}$.

8.3   Small Reynolds Number Results

The five problems considered in the previous section resulted only in hydrodynamic drag (i.e., force along the direction of relative velocity) and hydrodynamic torque in the direction of angular velocity. Because of the zero Reynolds number limit, even though Problems II and III involved ambient shear flow and particle rotation, there was no hydrodynamic lift force (i.e., force perpendicular to the direction of relative velocity). As pointed out in Chapter 5, lift force is an inertial effect and is realized only at non-zero Reynolds number. Therefore, in this section, we will consider the wall effect on lift force at nonzero Reynolds number. However, we will limit our initial attention to small Reynolds number, where the analysis can be simplified and an explicit answer can be obtained.

In the presence of the wall, we expect the lift force on the particle to depend not only on the shear and rotational Reynolds numbers, but also on the distance from the wall to the particle. The effect of the wall is the strongest when the particle is in contact with the wall. The wall effect decays rapidly as the distance between the particle and the wall increases. Based on the results presented in Goldman et al. (1967a,b), for distances of the order of 10 particle diameters or more, the wall effect can reasonably be ignored. For example, the strong effect of the wall can be observed in the case of a stationary particle in contact with a flat wall in a linear shear flow, whose drag is 70% larger than in an unbounded uniform flow. Instead, if we consider a particle translating parallel to a flat wall in a stagnant fluid, in the limit of the particle nearly touching the wall (when the gap between the particle and the wall is 7% of the particle diameter), the drag force is twice that when the wall is far away. Therefore, before we consider the more general problem of particle motion at a distance from a nearby wall, in this section, we start our investigation with lift force on a particle rolling/sliding on a flat wall. Thus, we consider the limit where the particle is in constant contact with the wall (i.e., the $\tilde{\delta} \to 0$ limit).

8.3.1   Particle in Contact with the Wall ($\tilde{\delta} = 0$)

Consider the problem of a wall-bounded linear shear flow of shear magnitude G with a particle of radius R_p in steady translational and rotational motion within it. The translational velocity of the particle, V_x, is parallel to the wall and in the direction of shear flow. The rotational motion of the particle, Ω_y, is due to the ambient shear and therefore is oriented in the y-direction. If the particle is in perfect rolling motion, without any slip, then $V_x = \Omega_y R_p$. In a coordinate system attached to the center of the moving particle, the flow is steady and the governing Navier–Stokes equations are written as

$$\nabla \cdot \mathbf{u} = 0 \quad \text{and} \quad \mathbf{u} \cdot \nabla \mathbf{u} = -\nabla p + \mu_f \nabla^2 \mathbf{u}. \tag{8.28}$$

Here, $\mathbf{u}$ is the perturbation flow and the boundary conditions to be satisfied are

$$\text{On the particle: } \mathbf{u} = \Omega_y \mathbf{e}_y \times \mathbf{r},$$
$$\text{On the wall: } \mathbf{u} = -V_x \mathbf{e}_x, \tag{8.29}$$
$$\text{Far field: } \mathbf{u} \to (Gz - V_x)\mathbf{e}_x.$$

The above scenario is a superposition of the first three problems considered in the previous section: particle translation on a wall (Problem I), particle in a wall-bounded linear shear flow (Problem II), and particle rotation close to a wall (Problem III). In the case of zero Reynolds number, the total drag force is a linear superposition of the three mechanisms: (i) translation, (ii) shear, and (iii) rotation. As we will see below, at small Reynolds number, the lift force is the result of quadratic interaction between the three mechanisms that gives rise to the following six contributions to lift: (i) shear–shear, (ii) translation–translation, (iii) rotation–rotation, (iv) shear–translation, (v) shear–rotation, and (vi) translation–rotation. While the first three are pure quadratic effects of translation, shear, and rotation, respectively, the latter three are cross-coupling mechanisms.

Three different Reynolds numbers can be defined:

$$\text{Re}_V = \frac{|V_x| d_p}{\nu_f}, \quad \text{Re}_G = \frac{|G| d_p^2}{\nu_f}, \quad \text{and} \quad \text{Re}_\Omega = \frac{|\Omega_y| d_p^2}{\nu_f}. \tag{8.30}$$

In the limit $\text{Re}_V, \text{Re}_G, \text{Re}_\Omega \ll 1$, the solution to the above composite problem was presented by Krishnan and Leighton (1995). The particle diameter d_p is chosen as the length scale and one of $\{V_x, G d_p, \Omega_y d_p\}$ is chosen as the velocity scale, U, as appropriate. Correspondingly, the Reynolds number of the problem will be one of $\{\text{Re}_V, \text{Re}_G, \text{Re}_\Omega\}$, which will be chosen as the small parameter $\tilde{\epsilon}$. Krishnan and Leighton obtained a perturbation solution of the form

$$\mathbf{u}(\mathbf{x}) = \mathbf{u}_0(\mathbf{x}) + \tilde{\epsilon} \mathbf{u}_1(\mathbf{x}) + \cdots . \tag{8.31}$$

The pressure scale is chosen to be $\mu_f U/L$ and the governing Navier–Stokes equation in nondimensional form (denoted by tilde) becomes

$$\tilde{\nabla} \cdot \tilde{\mathbf{u}} = 0 \quad \text{and} \quad \tilde{\epsilon} \tilde{\mathbf{u}} \cdot \tilde{\nabla} \tilde{\mathbf{u}} = -\tilde{\nabla} \tilde{p} + \tilde{\nabla}^2 \tilde{\mathbf{u}} . \tag{8.32}$$

Substituting the expansion (8.31) into (8.32), we obtain the leading-order equation

$$\tilde{\nabla} \cdot \tilde{\mathbf{u}}_0 = 0 \quad \text{and} \quad 0 = -\tilde{\nabla} \tilde{p}_0 + \tilde{\nabla}^2 \tilde{\mathbf{u}}_0 \tag{8.33}$$

and the corresponding leading-order boundary conditions are the same as those for the total velocity:

$$\text{On the particle: } \tilde{\mathbf{u}}_0 = \frac{\Omega_y R_p}{U} \mathbf{e}_y \times \tilde{\mathbf{r}},$$
$$\text{On the wall: } \tilde{\mathbf{u}}_0 = -\frac{V_x}{U} \mathbf{e}_x, \tag{8.34}$$
$$\text{Far field: } \tilde{\mathbf{u}}_0 \rightarrow \left(\frac{G R_p}{U} \tilde{z} - \frac{V_x}{U} \right) \mathbf{e}_x .$$

Thus, the leading-order problem satisfies the linear Stokes flow and the boundary conditions can be satisfied by a linear superposition of Problems I, II, and III presented in Section 8.2. This composite leading-order solution can be expressed as

$$\tilde{\mathbf{u}}_0 = \frac{1}{U} \left(\mathbf{u}_I + \mathbf{u}_{II} + \mathbf{u}_{III} - V_x \mathbf{e}_x \right) , \tag{8.35}$$

where $\mathbf{u}_{\mathrm{I}}$ is the solution of Problem I, and so on. Note that the last term within parentheses is due to the change in the frame of reference and the velocity has been suitably nondimensionalized.

However, as discussed earlier, this leading-order solution will not yield the desired lift force on the particle directly. Since the lift force is nonlinear in origin, it is the result of a higher-order solution. It must also be recognized that the leading-order equations (8.33) and the resulting solution are valid only within the inner region, due to the singular nature of the problem. An outer expansion and a matching between the inner and outer solutions is required for a complete asymptotically valid solution. Fortunately, it was shown by Cox and Brenner (1968) and Cox and Hsu (1977) that in the limit when Re $\ll 1$ and $\tilde{\delta} \ll 1$, the lift force can be calculated without the outer solution. If the wall effect is dominantly in the inner region, then the lift force can be calculated only with the inner solution. The present case of a particle in rolling motion on a wall at low Reynolds number satisfies these conditions and the need for an outer solution is avoided.

As pointed out by Krishnan and Leighton (1995), remarkably, the leading-order lift force can be obtained with only the zeroth-order solution, and in nondimensional terms can be expressed as

$$\frac{F_{\mathrm{L0}}}{\mu_f R_p U} = \int_{\mathcal{D}} \tilde{u}_{\mathrm{IV}i} \tilde{u}_{0j} \frac{\partial \tilde{u}_{0i}}{\partial \tilde{x}_j} \, dV, \tag{8.36}$$

where $\mathcal{D}$ is a large domain around the particle, bounded by the wall and a very large sphere centered at the point of contact between the particle and the wall. The subscript L0 stands for lift force in the limit of very small Reynolds number, approaching zero. In the above, $\tilde{u}_{0i}$ is the ith component of the zeroth-order solution given in Eq. (8.35) and $\tilde{u}_{\mathrm{IV}i}$ is the solution obtained in Problem IV for a particle approaching the wall in the perpendicular direction with unit velocity, in the limit when the particle is in contact with the wall (i.e., in the limit $\tilde{\delta} = 0$). The above integral expression for the lift force arises from the application of the reciprocal theorem (see Krishnan and Leighton, 1995 and references cited therein). The perturbation solutions $\tilde{u}_{0i}$ and $\tilde{u}_{\mathrm{IV}i}$ decay sufficiently fast away from the particle and the above integral is convergent. The integral has been evaluated (Krishnan and Leighton, 1995) and the resulting expression for the lift force in dimensional terms is given by

$$F_{\mathrm{L0}} = \rho_f R_p^2 \left(\lambda_1 R_p^2 G^2 + \lambda_2 V_x^2 + \lambda_3 R_p^2 \Omega_y^2 + \lambda_4 R_p G V_x + \lambda_5 R_p^2 G \Omega_y + \lambda_6 R_p \Omega_y V_x \right), \tag{8.37}$$

where $\lambda_1, \ldots, \lambda_6$ are constants that are given as integrals of products of the Stokes solutions of Problems I–IV integrated over the surface of the particle. These integrals were numerically evaluated by Krishnan and Leighton (1995) and their values are

$$\begin{aligned} \lambda_1 &= 9.257, & \lambda_2 &= 1.755, & \lambda_3 &= 0.546, \\ \lambda_4 &= -9.044, & \lambda_5 &= 1.212, & \lambda_6 &= -2.038. \end{aligned} \tag{8.38}$$

As expected, the above lift expression includes contributions from the six quadratic mechanisms. Several other conclusions can be drawn from the above expression. The first three contributions are from (i) shear–shear, (ii) translation–translation, and (iii) rotation–rotation interactions. They are always directed away from the wall in the positive z-direction. In other words, irrespective of whether the shear flow is directed to the right or to the left, as long as the shear flow velocity increases linearly with distance from the wall, the lift force will be directed away from the wall. Similarly, irrespective of the direction of particle translation parallel to the wall, or the direction of rotational axis parallel to the wall, the positive values of λ_2 and λ_3 indicate that the corresponding lift forces are directed away from the wall. Also, among the three, shear-induced lift is the strongest, followed by translation-induced lift, while rotation-induced lift is the weakest.

In the case of shear–translation interaction, the negative value of λ_4 indicates that if the particle's translational velocity is in the direction of shear flow, then the effect of cross interaction between shear and translation will be directed toward the wall and therefore will decrease the lift force contributions by shear and translation. If a particle translates at the shear flow velocity at its center (i.e., $V_x = GR_p$), and is allowed not to rotate, then the combination of shear and translation yields $\lambda_1 + \lambda_2 + \lambda_4 = 1.968$, which is only slightly higher than that due to pure translation (Krishnan and Leighton, 1995).

Interestingly, the positive value of λ_5 suggests that the shear–rotation interaction leads to an additional positive lift force if the particle rotates in the direction of ambient vorticity. On the other hand, in the case of a rolling particle, the nonlinear interaction between translation and corresponding particle rotation contributes negatively to the lift force (λ_6). For the case of a particle rolling perfectly on a wall in a quiescent ambient (i.e., no shear flow), the combined effect of translation and rotation yields an effective coefficient of $\lambda_2 + \lambda_3 + \lambda_6 = 0.263$. Thus, the lift force on a rolling particle is much lower than that on one that is sliding on the wall without any rotation. This completes all the quadratic interactions that contribute to the lift force on a particle.

Experiments on a particle in motion in contact with a plane wall in a linear shear flow were performed by King and Leighton (1997). Their measurements were compared against the theoretical predictions of particle motion. The theoretical predictions were based on the above inertial lift force results. The drag force on the particle in contact with the wall was taken from the results of Goldman et al. (1967a). In order to avoid the logarithmic singularity, in the drag force calculation, the gap between the particle and the wall was set equal to the roughness of the sphere. Good agreement between theory and experiments was observed.

8.3.2   Particle Motion in a Wall-Bounded Shear Flow ($\tilde{\delta} \neq 0$)

This section will consider the particle to be not in contact with the wall. Thus, the results of the previous section will be extended to finite values of $\tilde{\delta}$. This increases the number of parameters by one. Therefore, in order to simplify the analysis, we will seek solutions only in the following two limiting cases of rotational motion of the particle. (i) We will assume the rotational motion of the particle to be negligible and seek a solution for a nonrotating particle. (ii) We will consider the particle to be in steady-state torque-free rotational motion corresponding to a specified ambient shear, translational motion, and distance from the wall. This torque-free rotational motion was considered in Problem 8.2. We will assume this zero Reynolds number solution to remain applicable even at the small Reynolds number being considered here. By focusing attention on the above two limits, we avoid Re_Ω being an active parameter of the problem.

There have been a number of important theoretical analyses of this problem in the small Reynolds number limit (Cox and Hsu, 1977; McLaughlin, 1993; Cherukat and McLaughlin, 1994 and its corrigendum). Two different asymptotic regimes have been considered. The first set of investigations considers the wall to be in the inner region of the perturbation flow induced by the particle. This simply implies that the distance between the particle and the wall is less than the Stokes and Saffman length scales, which were defined in Eq. (5.3). In other words, in the limit

$$h \ll \left\{ L_s = \frac{\nu_f}{|V_x - Gh|}, \ L_G = \left(\frac{\nu_f}{|G|} \right)^{1/2} \right\}, \tag{8.39}$$

the wall effect applies directly as no-slip and no-penetration conditions on the inner solution at all orders. As discussed in the previous section (for more details, see Cox and Brenner, 1968; Ho and Leal, 1974; Cox and Hsu, 1977), in this limit of the wall being in the inner region, we do not need the outer solution in order to evaluate the lift force on the particle. The second set of investigations considers the wall to be in the outer region of the perturbation flow induced by the particle, which leads to a different set of simplifications in the solution of the governing equations. We shall now consider both these limits, one after the other.

Wall in the Inner Region

We will first discuss the limit of the wall being in the inner region. The integral given in Eq. (8.36) still applies and can be used to calculate the lift force on the particle. The various terms in the equation have the same interpretation, with the difference that the particle is now at a nondimensional distance $\tilde{h}$ from the wall and the perturbation velocity fields $\tilde{u}_{IVi}$ and $\tilde{u}_{0j}$ are now for a particle that is not necessarily in contact with the wall. Cherukat and McLaughlin (1994) carried out these integrals numerically and presented the results as both tables and curve fits. The resulting empirical expression for the lift force can be written as

$$F_{L0} = \rho_f R_p^2 (V_x - Gh)^2 \, I(\Lambda_G, \kappa), \tag{8.40}$$

$$I_{NR} = [1.7716 + 0.2160\kappa - 0.7292\kappa^2 + 0.4854\kappa^3]$$
$$- [3.2397 + 1.1450\kappa + 2.0840\kappa^2 - 0.9059\kappa^3]\frac{\Lambda_G}{\kappa} \tag{8.41}$$
$$+ [2.0069 + 1.0575\kappa - 2.4007\kappa^2 + 1.3174\kappa^3]\Lambda_G^2,$$

where the subscript NR stands for the fact that the above expression is valid for a nonrotating particle. In the above, $\kappa = 1/\tilde{h} = R_p/h$ is the inverse nondimensional distance from the wall and

$$\Lambda_G = \frac{GR_p}{(V_x - Gh)} \tag{8.42}$$

is the ratio of velocity change across a particle due to shear versus relative velocity between the particle and the shear flow. Thus, Λ_G measures the intensity of shear compared to relative velocity. In the limit when the wall is far away (i.e., as $\kappa \to 0$), the above result can be shown to approach the theoretical results of Cox and Hsu (1977).

In the case of a particle rotating at the torque-free condition (denoted by the subscript TF), the following empirical curve fit is given in Cherukat and McLaughlin (1994):

$$
\begin{aligned}
I_{\text{TF}} = & \; [1.7669 + 0.2885\kappa - 0.9025\kappa^2 + 0.50763\kappa^3] \\
& - [3.2415 + 2.6729\kappa + 0.8373\kappa^2 - 0.4683\kappa^3]\frac{\Lambda_G}{\kappa} \\
& + [1.8065 + 0.8993\kappa - 1.9610\kappa^2 + 1.0216\kappa^3]\Lambda_G^2 \, .
\end{aligned}
\tag{8.43}
$$

The lift force given in Eq. (8.40) is appropriate only for nonzero values of relative velocity between the particle and the shear flow. Under situations where $V_x - Gh = 0$, the appropriate velocity scale is GR_p and the lift force must be expressed as

$$F_{\text{L0}} = \rho_f R_p^4 G^2 \, I(\kappa) \, . \tag{8.44}$$

Cherukat and McLaughlin (1994) observed the value of I to be nearly independent of the distance from the wall, approximated to be

$$I_{\text{NR}} = \frac{366\pi}{576} \quad \text{and} \quad I_{\text{TF}} = \frac{330\pi}{576} \tag{8.45}$$

for the two cases of a nonrotating particle and torque-free rotation. The lift on a nonrotating particle is slightly higher than that of a freely rotating particle.

In the limit of zero-gap, results are in excellent agreement with those of Krishnan and Leighton (1995). In particular, the results for the case of a nonrotating particle can be recast as an additive superposition of shear, translation, and shear–translation contributions, suggesting the applicability of the superposition presented in their paper even when the particle is away from the wall.

Wall in the Outer Region

The limit of the wall in the outer region of the perturbation flow induced by the particle is defined by $h \gg \max\{L_s, L_G\}$. This problem was originally considered by Vasseur and Cox (1977), who studied the effect of a distant wall on the lift force on a particle in motion in a uniform flow. A comprehensive treatment of this problem was later carried out by McLaughlin (1993), who presented results for the lift force as a wall correction to the shear-induced Saffman lift force on a particle. Thus, following the lift force expression given in Eq. (5.8) for an unbounded shear flow, the corresponding expression that includes the effect of a distant wall that is in the outer region of the perturbation flow is given by

$$F_{\text{L0}} = -6.46 \, \mu_f R_p^2 U_0 \, \text{sgn}(G) \, (G/\nu_f)^{1/2} \left(J(\tilde{\delta}) - \frac{0.833}{(h^2|G|/\nu_f)^{5/6}} \right) \, . \tag{8.46}$$

Blending Between the Inner and Outer Region Results

If we consider the particle to be nonrotating, then the lift force on the particle is a function of the following three nondimensional parameters: $\tilde{h}$, Re_G, and $\text{Re}_r = |V_x - Gh|d_p/v_f$, where Re_r is the Reynolds number based on relative velocity. Note that instead of $\tilde{h}$, we can also equivalently use the nondimensional gap $\tilde{\delta}$ or κ. Similarly, $\text{Re}_r = \text{Re}_V - (\text{Re}_G \tilde{h}/2) = \text{Re}_G/(2\Lambda_G)$, and therefore the dependence on Re_r can be substituted with an equivalent dependence on Re_V or Λ_G. In essence, there are different choices for the three parameters.

In the limit $\tilde{h} = 1$, the lift force will reduce to the expressions obtained in Section 8.3.1 for the particle in contact with the wall. Similarly, in the limit $\tilde{h} \to \infty$, the lift force will reduce to the free-shear expression. Table 8.4 presents a summary of all the lift force expressions that can be used in the small Reynolds number limit. For intermediate values of $\tilde{h}$, interpolation between the inner-wall and outer-wall expressions can be used. A more elaborate approach is presented in Wang et al. (1997), which can be pursued if needed.

In the case of torque-free particle motion, the nondimensional lift force on the particle is again a function of only $\tilde{h}$, Re_G, and Re_r. The corresponding expressions for lift force are also presented in Table 8.4. Note that the lift force expressions for the $\tilde{h} = 1$ and $\tilde{h} \to \infty$ limits are available for any arbitrary rotation of the particle. In the limit of a distant wall, the Magnus contribution to lift due to particle rotation given in Eq. (5.35) has been added to that due to shear and translation.

In the small Reynolds number limit, the composite model presented in Table 8.4 can be used for lift force. For the drag force, a linear superposition of those models presented in Problems I–III of Section 8.2 can be used. Together, these hydrodynamic drag and lift forces can be used to track the motion of small particles in wall-bounded shear flows. This approach has been used successfully not just in laminar shear flows, but also in wall-bounded turbulent channel and pipe flows. However, there are limitations. The first important limitation is that the particle must be much smaller than the scales

Table 8.4 Appropriate lift force expressions in the different regimes of wall position. Results for both nonrotating particle and torque-free particle rotation are presented.

Wall Position	Nonrotating	Torque-Free Rotation
In contact $\tilde{h} = 1$	Eqs. (8.37) and (8.38)	Eqs. (8.37) and (8.38)
Inner $\tilde{h} \ll \{1/\text{Re}_r, 1/\text{Re}_G^{1/2}\}$	Eqs. (8.40) and (8.41)	Eqs. (8.40) and (8.43)
Outer $\tilde{h} \gg \{1/\text{Re}_r, 1/\text{Re}_G^{1/2}\}$	Eq. (8.46)	Eqs. (8.46) and (5.35)
No wall $\tilde{h} \to \infty$	Eq. (5.8)	Eqs. (5.8) and (5.35)

for turbulence, so that the flow around the particle can be locally approximated as a wall-bounded linear shear flow. The second limitation is that the Reynolds numbers based on relative velocity, ambient shear, and particle rotation must all be smaller than unity for the results of this section to be applicable. Despite these limitations, there are a number of applications that have benefited from these drag and lift force expressions. For example, see Wang et al. (1997) for an investigation of the role of lift force in modifying particle deposition rate in a turbulent channel flow.

8.3.3   Particle Translation in a Quiescent Fluid ($\tilde{\delta} \neq 0$)

In this section, we will briefly review our current small Reynolds number theoretical understanding of lift force on a particle translating parallel to a wall in a quiescent fluid medium. This situation is of importance in many practical applications. Consider a spherical particle falling down due to gravity close to a vertical wall. Similarly, consider a bubble or droplet rising close to a vertical wall. In these scenarios, the lift force on the particle is of critical importance, even though its magnitude may be much smaller than the drag force. The sign of the lift force will determine whether the particle will approach the wall or move away, and the magnitude of the lift force will determine the speed of this cross-stream migration. This wall-normal motion will in turn have a strong influence on the drag force due to logarithmic singularity as the particle approaches the wall.

Due to its importance, this problem has been studied analytically and experimentally, resulting in an explicit expression for the lift force that is valid for all distances between the particle and the wall. This is the result that will be presented here. In the low Reynolds number limit, Vasseur and Cox (1977) showed that the lift force is only a function of $\tilde{h}_* = \tilde{h}\mathrm{Re}_V/2$ and obtained an analytical integral expression for the lift coefficient. Takemura and Magnaudet (2003) provided an accurate fit for this expression:

$$F_{L0} = \frac{\pi}{2}\rho_f R_p^2 V_x^2 \begin{cases} \left(\frac{9}{8} + 5.78 \times 10^{-6}\tilde{h}_*^{4.58}\right)e^{-0.292\tilde{h}_*} & \text{for} \quad 0 < \tilde{h}_* < 10, \\ 8.94\tilde{h}_*^{-2.09} & \text{for} \quad 10 < \tilde{h}_* < 300. \end{cases} \tag{8.47}$$

Thus, in the low Reynolds number limit, the lift coefficient decays as $\tilde{h}_*^{-2.09}$ for large $\tilde{h}_*$. We will discuss the finite Re extension to this in Section 8.5.1.

8.4   Finite-Re Results – in Contact with the Wall

As far as the drag force is concerned, the zero-Re results of Section 8.2 remain valid even at small values of Reynolds number. Thus, the focus of the previous section has been on the wall effect on lift force. The reported analytical results were based on matched asymptotic analysis, with the Reynolds number as the small parameter. As we extend our understanding to finite Reynolds numbers, in this section, we must resort to results from laboratory experiments and numerical simulations. One important difference from the previous section is that at finite Re, both drag and lift forces deviate substantially from their Stokes regime values and therefore we must consider both of them.

The nondimensional drag and lift forces, expressed as drag and lift coefficients, depend on $\tilde{h}$, and the translational, shear, and rotational Reynolds numbers. These dependencies at finite Reynolds number can be expected to be more complex than those discussed in the previous section. Our current knowledge is incomplete. It is not possible to present a complete picture with variation of all these parameters. We therefore divide the finite-Re discussion into two sections. In this section, we will discuss drag and lift on a particle translating and rotating in contact with the wall in a linear shear flow. With the restriction of the gap between the particle and the wall being very small, of the order of roughness, we will be able to consider the complete variation of all other parameters. This problem of a particle rolling or sliding on a flat wall at finite Re is important in many applications. We will address the problem when the particle is near a wall, but not in contact, in the next section.

8.4.1   Particle in Motion in Contact with the Wall

This subsection will revisit the problem considered by Krishnan and Leighton (1995) but at finite Reynolds numbers in the range of 0 to 100. The results to be presented here were obtained through direct numerical simulations by Lee and Balachandar (2010) using spectral element methodology. They solved the problem of a spherical particle of radius R_p with translational velocity V_x and rotational velocity Ω_y in contact with the wall in a linear shear flow of shear magnitude G. They assumed the superposition of the six different mechanisms outlined in Eq. (8.37) to be active even at finite Reynolds numbers. Based on this assumption, they proposed the following model for the lift force:

$$F_L = \frac{\pi \rho_f R_p^2}{2} \left[C_{LG,\mathrm{w}} R_p^2 G^2 + C_{LV,\mathrm{w}} V_x^2 + C_{L\Omega,\mathrm{w}} R_p^2 \Omega_y^2 \right. \\ \left. + C_{LGV,\mathrm{w}} R_p G V_x + C_{LG\Omega,\mathrm{w}} R_P^2 G \Omega_y + C_{LV\Omega,\mathrm{w}} R_p \Omega_y V_x \right], \tag{8.48}$$

where $C_{LG,\mathrm{w}}(\mathrm{Re}_s)$ is the lift coefficient of the shear–shear mechanism. The "w" in the subscript stands for the limit of the particle in contact with the wall. Following Lee and Balachandar (2010), we express the lift coefficient $C_{LG,\mathrm{w}}$ as a function of shear velocity at the particle:

$$\mathrm{Re}_s = \frac{(Gh)d_p}{\nu_f} = \mathrm{Re}_G \frac{\tilde{h}}{2}. \tag{8.49}$$

In the present situation of a particle in contact with the wall, $\mathrm{Re}_s = \mathrm{Re}_G/2$. $C_{LV,\mathrm{w}}(\mathrm{Re}_V)$ is the lift coefficient of the translation–translation mechanism and it is only a function of the translation Reynolds number. $C_{L\Omega,\mathrm{w}}(\mathrm{Re}_\Omega)$ is the lift coefficient of the rotation–rotation mechanism and it is only a function of the rotation Reynolds number.

Other than the above pure mechanisms, there are three binary interaction mechanisms. The lift coefficients $C_{LGV,\mathrm{w}}(\mathrm{Re}_G, \mathrm{Re}_V)$, $C_{LG\Omega,\mathrm{w}}(\mathrm{Re}_G, \mathrm{Re}_\Omega)$, and $C_{LV\Omega,\mathrm{w}}(\mathrm{Re}_V, \mathrm{Re}_\Omega)$ are for the shear–translation, shear–rotation, and translation–rotation mechanisms and they depend on the corresponding Reynolds numbers. For simplicity, it is assumed that even when all three mechanisms are active, their nonlinear interaction can be described in terms of the binary cross-couplings. It is thus assumed that simultaneous three-way interaction between shear, translation, and rotation, and its effect on lift force (and drag force), is not important.

When only one lift mechanism is active with the other two mechanisms being absent, the last three binary coupling terms are identically zero. Depending on whether shear $(V_x, \Omega_y = 0)$, translation $(G, \Omega_y = 0)$, or rotation $(G, V_x = 0)$ alone is active, respectively, the first, second, or third term on the right makes the only nonzero contribution. The lift coefficients of these unary mechanisms are plotted in Figure 8.7. Note that the numerical simulations of particle translation and rotation cannot be performed when the particle is in contact with the wall. In these cases, the simulations reported in Lee and Balachandar (2010) considered a particle slightly displaced from the wall at $\tilde{\delta} = 0.01$. From the figure it is clear that shear-induced lift is the strongest of the three and rotation-induced lift is the weakest. For instance, in the limit of small Reynolds number, the shear-induced lift coefficient is 5.87, while that due to translation is more than five times smaller at 1.12, and the rotation-induced lift coefficient is more than sixteen times smaller at 0.348. The following empirical relations for these lift coefficients were proposed by Zeng et al. (2009) and Lee and Balachandar (2010):

$$C_{LG,\mathrm{w}} = \frac{3.663}{\left(\mathrm{Re}_s^2 + 0.1173\right)^{0.22}},$$

$$C_{LV,\mathrm{w}} = 0.313 + 0.812 \exp\left(-0.125\,\mathrm{Re}_V^{0.77}\right),$$

$$C_{L\Omega,\mathrm{w}} = 0.348 - 0.000795\,\mathrm{Re}_\Omega .$$

(8.50)

Note that in the low Reynolds number limit, the above lift coefficients correctly approach their limiting values given in Eq. (8.38). In other words:

$$C_{LG,\mathrm{w}}(\mathrm{Re}_s \to 0) = C_{LG,\mathrm{w0}} = 2\lambda_1/\pi,$$

$$C_{LV,\mathrm{w}}(\mathrm{Re}_V \to 0) = C_{LV,\mathrm{w0}} = 2\lambda_2/\pi,$$

$$C_{L\Omega,\mathrm{w}}(\mathrm{Re}_\Omega \to 0) = C_{L\Omega,\mathrm{w0}} = 2\lambda_3/\pi,$$

(8.51)

where the zero-Re limit of a particle in contact with the wall has been denoted by the subscript w0.

The following expressions of the lift coefficients of the binary mechanisms have been advanced by Lee and Balachandar (2010):

$$C_{LGV,\mathrm{w}} = \frac{2\lambda_4}{\pi} \frac{C_{LG,\mathrm{w}}C_{LV,\mathrm{w}}}{C_{LG,\mathrm{w0}}C_{LV,\mathrm{w0}}} f_{GV},$$

$$C_{LG\Omega,\mathrm{w}} = \frac{2\lambda_5}{\pi} \frac{C_{LG,\mathrm{w}}C_{L\Omega,\mathrm{w}}}{C_{LG,\mathrm{w0}}C_{L\Omega,\mathrm{w0}}} f_{G\Omega},$$

$$C_{LV\Omega,\mathrm{w}} = \frac{2\lambda_6}{\pi} \frac{C_{LV,\mathrm{w}}C_{L\Omega,\mathrm{w}}}{C_{LV,\mathrm{w0}}C_{L\Omega,\mathrm{w0}}} f_{V\Omega}.$$

(8.52)

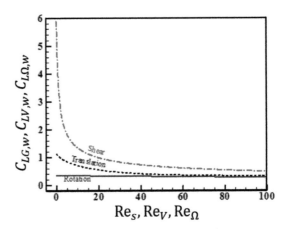

Figure 8.7 Lift coefficients $C_{LG,w}(\mathrm{Re}_G)$, $C_{LV,w}(\mathrm{Re}_V)$, and $C_{L\Omega,w}(\mathrm{Re}_\Omega)$ as a function of their respective Reynolds numbers for a particle in contact with the wall. Taken from Lee and Balachandar (2010).

These expressions are motivated by their low Reynolds number form. For example, the lift contribution from shear–translation coupling is assumed to be proportional to $C_{LG,w}C_{LV,w}GV_x$ and any deviation from this is taken into account with the correction function f_{GV}. Similarly, the functions $f_{G\Omega}$ and $f_{V\Omega}$ account for the finite Reynolds number deviations in the shear–rotation and translation–rotation couplings. In the limit of zero Reynolds number, these correction functions reduce to unity, that is, $f_{GV}(\mathrm{Re}_s \to 0) \to 1$, $f_{G\Omega}(\mathrm{Re}_s \to 0) \to 1$, and $f_{V\Omega}(\mathrm{Re}_V \to 0) \to 1$. With this requirement, the empirical fits for the correction functions obtained from direct numerical simulation results are

$$
\begin{aligned}
f_{GV} &= 1 + \left(2.156\left(\frac{V_x}{GR_p}\right)^2 + 1.789\left(\frac{V_x}{GR_p}\right) + 0.704\right)\tanh(0.02\,\mathrm{Re}_s), \\
f_{G\Omega} &= 1 + \left(0.251\frac{\Omega_y}{G} + 1.018\right)\mathrm{Re}_s^{0.66}, \\
f_{V\Omega} &= 1 + \left(0.0122\frac{\Omega_y R_p}{V_x}\left(\frac{\Omega_y R_p}{V_x} - 2\right) + 0.0548\right)\mathrm{Re}_V^{0.85}.
\end{aligned}
\tag{8.53}
$$

The above form assumes that shear–translation and shear–rotation are driven mainly by the ambient shear flow and that the translational/rotational motion of the particle serves to modify through nonlinear interaction. This is the motivation for the polynomial expansions involving $V_x/(GR_p)$ and Ω_y/G in the first two correction functions. Similarly, in the translation–rotation correction function, translation is taken to be the primary mechanism, with the rotational effect modifying it.

Based on the findings of Lee and Balachandar (2010), the corresponding drag force model can be written as

$$
\begin{aligned}
F_D = \frac{\pi\rho_f R_p^2}{2}\Big[&C_{DG,w}R_p^2 G|G| - C_{DV,w}V_x|V_x| + C_{D\Omega,w}R_p^2\Omega_y|\Omega_y| \\
&- g_{GV}R_p|G|V_x\Big],
\end{aligned}
\tag{8.54}
$$

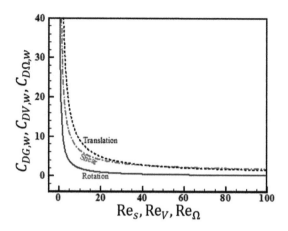

Figure 8.8 Drag coefficients $C_{DG,w}(\text{Re}_s)$, $C_{DV,w}(\text{Re}_V)$, and $C_{D\Omega,w}(\text{Re}_\Omega)$ as a function of their respective Reynolds numbers for a particle in contact with the wall. Taken from Lee and Balachandar (2010).

where $C_{DG,w}(\text{Re}_s)$, $C_{DV,w}(\text{Re}_V)$, and $C_{D\Omega,w}(\text{Re}_\Omega)$ are the drag coefficients of the pure shear, translation, and rotation mechanisms and each are functions of only their respective Reynolds number. Among the binary coupling interaction mechanisms, Lee and Balachandar (2010) observed only the shear–translation to be of importance and this effect is included through the correction function g_{GV}. Explicit expressions for these functions are as follows:

$$
\begin{aligned}
C_{DG,w} &= \frac{40.81}{\text{Re}_s}\left(1 + 0.104\,\text{Re}_s^{0.753}\right), \\
C_{DV,w} &= \frac{81.96}{\text{Re}_V}\left(1 + 0.01(\text{Re}_V)^{0.959}\right), \\
C_{D\Omega,w} &= \frac{18.84}{\text{Re}_\Omega}, \\
g_{GV} &= \left(2.03\frac{V_x}{GR_p} - 8.18\right)\text{Re}_s^{0.3} - \left(2.8\frac{V_x}{GR_p} - 10.73\right)\text{Re}_s^{0.25}.
\end{aligned}
\tag{8.55}
$$

Figure 8.8 shows a comparison of the three drag coefficients $C_{DG,w}$, $C_{DV,w}$, and $C_{D\Omega,w}$ and they monotonically decrease with Re. It is clear that the rotation-induced drag is the weakest of the three. We can see that below a critical value of about 42, translation-induced drag is larger than shear-induced drag. On the other hand, shear-induced drag becomes dominant beyond this Reynolds number. For the cases of a particle either in translation or rotation, the drag force on the particle has a logarithmic singularity and becomes infinite when the particle comes into contact with the wall. But in reality, the drag force on the particle will remain finite, since roughness of the particle and the wall will limit the effective gap to be nonzero. In Figure 8.8, the results for translation and rotation are for a small separation of $\tilde{\delta} = 0.01$.

Particle Translation at Contact in a Linear Shear Flow

Lee and Balachandar (2010) performed simulations of a particle in contact with a flat wall and sliding in the direction parallel to the shear flow. If the direction of sliding is against that of the shear flow, then drag and lift contributions tend to reinforce each other and become large. But a more likely scenario is for the particle to move in the direction of shear flow and in this case the relative velocity as seen by the particle decreases and the contributions to force from shear and translation tend to oppose each other.

Their simulations were performed for a range of Re_s from 0 to 100 and a range of relative translation velocity $V_x/(GR_p)$ between –1 and 0.75. For $V_x/(GR_p) = -1$, the particle speed is the same as the local ambient shear flow but is directed against the shear flow. On the other hand, in the case of a particle translating in the direction of shear flow, the simulations are considered only up to a translation velocity of 75% of local shear flow velocity. Figure 8.9a shows the computed lift coefficient plotted as symbols against Re_s for different values of $V_x/(GR_p)$. In this figure, the nondimensional lift has been calculated as $2F_L/(\pi\rho_f R_p^4 G^2)$ and in this nondimensionalization, the lift coefficient decreases progressively with increasing $V_x/(GR_p)$. Also shown in the figure as lines are the nondimensional lift force evaluated using Eq. (8.48) with the appropriate empirical fits for the lift coefficients and binary correction functions. Since the particle is not rotating, contributions from rotation–rotation, shear–rotation, and translation–rotational mechanisms are absent. In fact, these simulation results were used in developing the shear–translation correction function f_{GV} and thus the agreement is quite good.

The simulation results for the drag coefficient for the different cases are presented as symbols in Figure 8.9b. The drag coefficient decreases rapidly with increasing $V_x/(GR_p)$ and a log scale is used to capture this large variation. In the present limit of the particle being almost in contact with the wall, the effect of translation on drag force dominates over that due to shear at small Reynolds number. This is to be expected owing to the logarithmic increase in translational drag as the wall is approached. In particular, it can be seen that at small Reynolds numbers, the translational drag equals the shear drag when $Re_V \approx Re_s/2$. This in turn leads to $V_x/(GR_p) \approx 0.5$. As a result, for a particle translating in the direction of shear, it can be expected that as nondimensional translational velocity increases above 0.5, the drag force will reverse

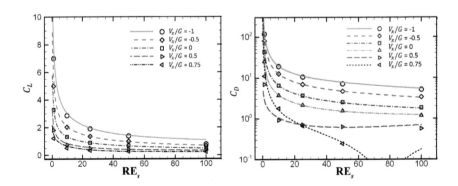

Figure 8.9 Nondimensional (a) lift and (b) drag force as a function of Re_s for a particle in contact with the wall and traveling parallel to the wall in a linear shear flow. Taken from Lee and Balachandar (2010).

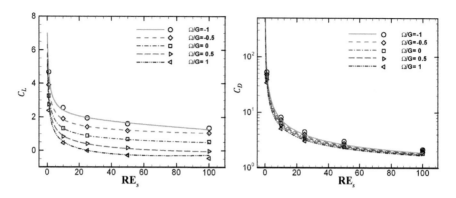

Figure 8.10 Nondimensional (a) lift and (b) drag force as a function of Re_s for a particle in contact with the wall and spinning with its axis parallel to the wall in a linear shear flow. Taken from Lee and Balachandar (2010).

sign and point in the direction opposite to the ambient shear. The present finite-Re simulation results for $V_x/(GR_p) = 0.5$ remain positive, but substantially smaller and closer to zero. The drag force for $V_x/(GR_p) = 0.75$ is observed to be consistently negative at all Reynolds numbers, in agreement with theoretical expectations. Because of the log scale, in the figure, the absolute value of the drag coefficient is shown. Again these simulation data were used in obtaining the drag binary correction function g_{GV} and therefore the agreement between the simulation results and the prediction from the empirical relation is good.

Particle Rotation at Contact in a Linear Shear Flow

Here we report the simulation results of a particle almost in contact with a flat wall ($\tilde{\delta} = 0.01$) and spinning in the linear shear flow. This problem explores the binary interaction between the shear flow and the spinning particle. The particle is fixed in position and only allowed to spin along the y-axis. If the direction of rotation is positive (clockwise as viewed in Figure 8.1) and the shear flow is directed in the positive x-direction, then the drag contribution due to rotation tends to cooperate with that due to shear flow to increase the drag force. Interestingly, as we will see below, the lift superposition behaves in a similar manner.

Simulations were performed for a range of Re_s from 2 to 200 and a range of relative rotation Ω_y/G from -1 to 1. The relative velocity between the particle and the undisturbed ambient flow at the particle center is the same as the shear velocity, since a rotating particle does not have any translational velocity. Figure 8.10a shows the nondimensional lift against Re_s for different relative rotation. In the figure, the results of the numerical simulations are shown as symbols. As discussed above, a substantial increase in lift results in case of positive (clockwise) particle rotation and correspondingly a decrease in lift force can be observed in case of negative (counterclockwise) particle rotation. The effect of rotation persists over the entire range of Reynolds number considered.

In Figure 8.10b, the drag coefficients evaluated from the numerical simulations are plotted as symbols. It can be seen that the effect of particle rotation on drag is not very strong. Thus, in Eq. (8.54) we ignore the effect of the shear–rotation binary coupling correction $g_{G\Omega}$. Then, according to Eq. (8.54), the drag for this case is only a superposition of the shear and rotational contributions, which is also plotted as lines in Figure 8.10b. It is clear that in general, the effect of particle rotation is quite small and linear superposition without considering the coupling term is sufficient.

Particle Translation and Rotation at Contact in a Quiescent Ambient

In this subsection, we report on the results from simulations of a translating and rotating particle in near contact with a flat wall ($\tilde{\delta} = 0.01$) in a quiescent ambient fluid (Lee and Balachandar, 2010). If the particle translates but does not rotate, then the particle is in a pure "sliding" motion. In the limit $\Omega_y R_p / V_x = 1$, the particle rolls perfectly on the wall without any slip. For other nonzero values of Ω_y, the particle is in a partial rolling and partial sliding state. The most likely scenario of rolling motion is one where V_x and Ω_y are of the same sign. But, in the case of $\Omega_y V_x < 0$, the particle is translating with a back spin or a reversed rotation. Theoretical and computational results of a sphere moving down an incline in a stagnant fluid (Cherukat and McLaughlin, 1990; Zeng et al., 2005) suggest that the particle rotation under steady state (zero net drag and zero net torque) will be such that the particle rolls down the incline and the rotation magnitude is weak. However, under conditions of non-Newtonian fluid or in the presence of multiple bounding walls, it has been observed experimentally that a particle can have reversed rotation (Humphrey and Murata, 1992; Liu et al., 1993).

Lee and Balachandar (2010) performed simulations for a range of Re_V between 1 and 100 and a range of scaled rotational velocity $\Omega_y R_p / V_x$ between -1 and 1. In Figure 8.11a the numerically computed nondimensional lift (now defined as $2F_L / (\pi \rho_f R_p^2 V_x^2)$) is plotted as symbols against Re_V for varying values of $\Omega_y R_p / V_x$. It can be seen that the lift coefficient decreases monotonically with increasing Re_V for all values of particle rotation considered. The lift coefficient for the cases of a positive $\Omega_y R_p / V_x$ is larger and that for negative $\Omega_y R_p / V_x$ is lower. In other words, a particle rotating with a back spin will experience higher lift, while a particle rolling on the wall (rotational motion consistent with translation) will experience a lower lift force. From Figure 8.11a it is clear that the effect of particle rotation on lift is significant.

Figure 8.11b shows the drag coefficient plotted against Re_V for different values of $\Omega_y R_p / V_x$. The drag coefficients obtained from the numerical simulations are shown as symbols. As in the case of shear–rotation coupling, the effect of positive (clockwise) particle rotation is to decrease the drag, while the effect of negative rotation is to increase the drag, but the effect of particle rotation on the drag force is quite weak. The lines in Figure 8.11b represent the linear superposition without including the binary correction function $g_{V\Omega}$. From this figure we can see that the effect of translation on drag is dominant and that of rotation is weak and can be ignored.

The flow features of this problem presented in Lee and Balachandar (2010) are shown in Figure 8.12 for $\mathrm{Re}_V = 10, \Omega_y R_p / V_x = 1$ (left column) and $\mathrm{Re}_V = 10, \Omega_y R_p / V_x = -1$ (right column). The streamlines for the two cases are shown in frames (a) and (b). Here,

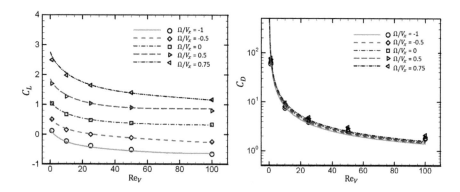

Figure 8.11 Nondimensional (a) lift and (b) drag force as a function of Re_V for a particle translating while in near contact with the wall and also spinning with its axis parallel to the wall in a quiescent ambient. Taken from Lee and Balachandar (2010).

we consider the particle to translate from right to left (i.e., $V_x < 0$). Since the streamlines are plotted in a frame attached to the particle, the main flow is directed from left to right. If the particle's translation and rotation are of the same sign, then the particle's rotation is consistent with a rolling motion and for $\Omega_y R_p / V_x = 1$ the particle is in a perfect rolling motion. In this case, the bottom point of the particle that is in contact with the wall is stationary, while the relative velocity at the top of the particle is twice the translation velocity. The shear around the top surface of the particle is enhanced and as can be observed in Figure 8.12a, the flow behind a particle is dragged from the bottom to the top. In contrast, for $\Omega_y R_p / V_x < 0$, the particle is in back-spin motion and thus the surface motion of the particle is in the same direction as that of the ambient flow due to translation. In frame (b), it can be seen that the flow passes smoothly along the particle surface without separation.

The corresponding pressure contours for the above two cases are shown in frames (c) and (d). As the particle is translating from right to left, the dominant high and low pressures are generated to the left and right of the particle. In frame (c), since the particle rotates in a counter-clockwise direction, relatively higher and lower-magnitude pressures are positioned at the small gap between the particle and the wall. In contrast, in frame (d) we can see that the lower-pressure region is raised somewhat higher behind the particle, since the back spin of the particle contributes to accelerating the flow behind the particle. Vortex formations of a translating and rotating particle are shown in frames (e) and (f) at the higher Reynolds number of 100. For $\Omega_y R_p / V_x > 0$, since the shear above the particle increases, the vortex structure around the particle in the wake region intensifies. At this higher Re, indications of a double-threaded wake can be observed, which is often seen at higher Re even in the absence of particle rotation. In the case of back spin, particle rotation reduces shear above and in the wake of the particle and thus the vortex structure is absent on the symmetry plane, except at the front. A necklace-type vortex structure that is confined to the sides of the sphere is observed. The rotation of the streamwise vortex pair in this case is such that fluid

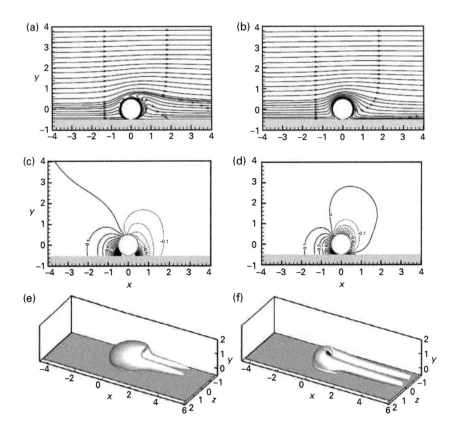

Figure 8.12 Flow features (streamlines, pressure contours, and vortical structures) for a translating–rotating particle in a stagnant ambient fluid for $\mathrm{Re}_V = 10, \Omega_y R_p / V_x = 1$ (left column) and $\mathrm{Re}_V = 10, \Omega_y R_p / V_x = -1$ (right column). Taken from Lee and Balachandar (2010).

is pumped up and away from the wall in the gap between them. Correspondingly, the induced motion is such that the lateral separation between the two legs remains nearly fixed as they extend downstream.

8.5   Finite-Re Results – Not in Contact with the Wall

In this section, we will try to generalize the finite-Re results of the previous section to particles that are located close to the wall, but not necessarily in contact with the wall. Thus, the gap between the particle and the wall becomes an additional parameter and the results of the previous section can be interpreted as the $\tilde{\delta} \to 0$ limit. However, with the introduction of nonzero $\tilde{\delta}$ as an additional parameter, the problem becomes hard and has not been addressed in a complete and comprehensive manner.

Therefore, we restrict ourselves to the five problems that we discussed in Section 8.2, but now extended to finite Re. The five problems that will be described in the following five subsections are: (i) drag and lift on a particle translating parallel to a wall in a quiescent ambient; (ii) drag and lift on a stationary particle in a wall-bounded linear shear flow; (iii) drag, lift, and torque on a particle spinning about an axis parallel to the wall in a quiescent ambient; (iv) drag on a particle translating perpendicular to a flat wall; and (v) torque on a particle spinning about an axis that is perpendicular to a nearby flat wall. In the zero-Re limit, these five problems could be superposed. At finite Re we do not have this luxury. Nonlinear interactions between the different components can be expected to be quite complex. However, the general case of a particle in translational and rotational motion in a wall-bounded flow without being in contact with the wall at finite Re has not been studied in full. Therefore, our discussions will be restricted to the limiting cases.

8.5.1   Problem I: Translating Parallel to the Wall

This subsection on finite Reynolds number effects will focus attention only on the translational effect, where a particle moves at a constant velocity parallel to a flat wall in a quiescent fluid. The flow is entirely induced by the particle motion and is modified due to the presence of the wall. This scenario arises as a rigid particle falls down in a stagnant fluid close to a vertical wall (or equivalently as a droplet or bubble rises in a quiescent fluid). This problem was studied by Takemura and Magnaudet (2003) and Zeng et al. (2005); Zeng et al. (2009).

Drag Force
A smooth particle in contact with a wall presents a lubrication singularity which prevents the particle from translation at finite forces. Zeng et al. (2009) proposed the following power-law Reynolds number dependence of drag:

$$\mathbf{F}_{DV} = \mathbf{F}_{V0} \left(1 + \alpha_t \, \mathrm{Re}_V^{\beta_t} \right), \tag{8.56}$$

where the zero Reynolds number drag force $\mathbf{F}_{V0}$ was given in Eq. (8.12). The prefactor α_t and the exponent β_t depend on $\tilde{\delta}$ and the dependence is given by

$$\alpha_t = 0.15 \left(1 - \exp\left(-\sqrt{\tilde{\delta}/2} \right) \right), \quad \beta_t = 0.687 + 0.313 \exp\left(-\sqrt{2\tilde{\delta}} \right). \tag{8.57}$$

Again, in the limit $\tilde{\delta} \to \infty$, their values correctly approach 0.15 and 0.687, respectively, to match the standard drag law.

Note that in the limit of $\tilde{\delta} \to 0$ we have $\alpha_t \to 0$ and $\beta_t \to 1.0$, and thus the Reynolds number dependence vanishes in the limit of a particle sliding on the wall. Even very close to the wall, for example at $\tilde{\delta} = 0.01$, we have $\alpha_t = 0.01$ and $\beta_t = 0.96$, and thus the Reynolds number effect is non-negligible (see Figure 8.5.1). The above behavior is quite different from that observed for a stationary particle in linear shear flow, where the variation in α_s and β_s was not large. The above empirical correlation is in good agreement with the finite Reynolds number results and reduces to the appropriate lubrication limit for $\mathrm{Re}_V \to 0$. In fact, it has been observed that the drag coefficient at intermediate distances is lower than that for a uniform flow. This can also be seen in Figure 6.16, where drag falls below standard drag at $\tilde{\delta} = 7$ at larger Re_V. This behavior is well captured by the above correlation and the normalized error of the curve fit is typically less than 6%.

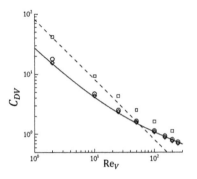

Figure 8.13 Drag coefficient of a particle translating parallel to a wall in stagnant fluid. $\square : \tilde{\delta} = 0.01.$ $\circ \tilde{\delta} = 1.$ $\diamond \tilde{\delta} = 7.$ —— Standard drag correlation. - - - Lubrication theory (Eq. (8.9)) at $\tilde{\delta} = 0.01$.

Lift Force

In this subsection, we will consider the lift force on the translating particle as studied analytically by Vasseur and Cox (1977), experimentally by Takemura and Magnaudet (2003), computationally by Zeng et al. (2005), and summarized in Zeng et al. (2009).

At finite Reynolds number, Takemura and Magnaudet (2003) and Zeng et al. (2005) observed the lift coefficient to be significantly larger than what was predicted in the low Reynolds number limit. The simulations were performed for $2 \leq \mathrm{Re}_V \leq 200$ and $0 \leq \tilde{\delta} \leq 7$. In the low Reynolds number limit, the lift coefficient is only a function of $\tilde{h}_* = \mathrm{Re}_V \, \tilde{h}/2$. In Figure 8.14, the lift coefficient is plotted as a function of $\tilde{h}_*$. Also plotted in the figure is the low Reynolds number limit given by

$$C_{\mathrm{LV0}} = \frac{F_{\mathrm{L0}}}{\frac{\pi}{2} \rho_f R_p^2 V_x^2} , \qquad (8.58)$$

where F_{L0} was given in Eq. (8.47). From Figure 8.14 it can be seen that as long as $\tilde{h}_* < 1$, irrespective of the value of Re_V or $\tilde{h}$, the lift coefficient is well approximated by the asymptotic result C_{LV0}. As pointed out by Zeng et al. (2005), at all distances from the wall, with increasing Reynolds number, the lift coefficient first decreases at a rate slower than predicted by asymptotic theory. For $\tilde{h} \gtrsim 1.5$, after reaching a minimum at around $\mathrm{Re}_V \approx 100$, the lift coefficient increases and the percentile increase is quite dramatic at larger distances from the wall.

In order to understand the mechanisms responsible for such a dramatic increase in lift force, we take a closer look at the wake vortical structure in Figure 8.15, where the surface of constant swirling strength equal to 0.1 is plotted. Swirling strength is defined as the imaginary part of the complex eigenvalue of the velocity gradient tensor and as shown in Zhou et al. (1999) and Chakraborty et al. (2005), it captures compact vortical structures well. For each Reynolds number, both the top view and the side view of the 3D vortex structure are shown. For Re = 50, a weak double thread can be seen in the wake. The double-threaded wake gains strength and extends farther downstream with increasing Reynolds number. Below a Reynolds number of 50, the double-threaded wake is absent.

In an unbounded uniform ambient flow, the flow remains axisymmetric at lower Reynolds numbers. As discussed in Section 7.1, at about Re $\approx$ 210 there is a *perfect bifurcation* to a non-axisymmetric state in which the wake vortex structure takes the double-threaded shape (Natarajan and Arcivos, 1993; Tomboulides et al., 1993; Johnson and Patel, 1999; Bagchi and Balachandar, 2001). With the presence of a nearby wall, the flow around the sphere is non-axisymmetric at all Re. However, based on Figure 8.15, the flow undergoes an *imperfect bifurcation* to a state involving a double-threaded vortical structure in the wake region. At $\tilde{h} = 2$ this bifurcation appears to occur at Re_V below 50. Furthermore, the plane of symmetry is fixed to be wall normal (x–y plane) and as a result the transverse force induced by the double-threaded wake vortex structure is in the wall-normal direction. While the other wall-induced lift mechanisms are weakening with increasing Re_V, the double-threaded wake-induced lift force gains strength after the imperfect bifurcation, resulting in the increase seen in Figure 8.14. For the case of a particle touching or being very close to the wall, the reduction in lift coefficient with increasing Re_V slows down at Reynolds number $O(100)$. However, the turnaround in C_{Lt} is not observed for $Re_V < 200$.

Also, Takemura and Magnaudet (2003) observed the decay rate to increase with Re_V and reach approximately $\tilde{h}^{-4}$ at $Re_V \approx 100$. For the same Reynolds number, based on numerical simulations, Zeng et al. (2009) reported a delay rate of only $\tilde{h}^{-3}$. With further increase in Re_V from 100 to 250, the decay rate appears to fall below

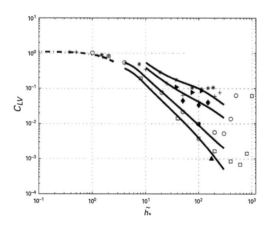

Figure 8.14 Lift coefficient C_L for a particle moving parallel to a nearby wall in an otherwise stagnant fluid. $\Diamond \tilde{\delta} = 0.01$. $*\tilde{\delta} = 0.5$. $+ \tilde{\delta} = 1$. $\circ \tilde{\delta} = 3$. $\square \tilde{\delta} = 7$. $- \cdot -$ The low-Re asymptotic result of Vasseur and Cox (1977). ——— Curve fit by Takemura and Magnaudet (2003) for $\tilde{\delta} = 0.01, 0.5, 1, 3,$ and 7. The solid symbols are data from Kim et al. (1993). $\blacktriangledown \tilde{\delta} = 0.5$. $\blacklozenge \tilde{\delta} = 1$. $\bullet \tilde{\delta} = 3$. $\blacksquare \tilde{\delta} = 6$. $\blacktriangle \tilde{\delta} = 8$. Taken from Zeng et al. (2005).

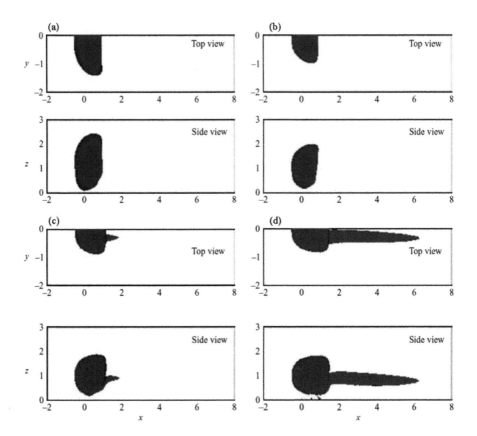

Figure 8.15 Wake vortical structure plotted as the surface of constant swirling strength equal to 0.1. Both the top and side views are shown. (a) $Re_V = 10$; (b) $Re_V = 50$; (c) $Re_V = 100$; (d) $Re_V = 200$. Taken from Zeng et al. (2005).

−3, and this behavior is related to the appearance of a one-sided double-threaded wake behind the particle. At even larger Reynolds numbers, unsteady vortex shedding can be expected and therefore is not pursued. Zeng et al. proposed the following expression for the lift coefficient:

$$C_{LV} = f(\tilde{h}, Re_V) + \left[C_{LV,w} - f(\tilde{h} = 1, Re_V)\right] \exp\left(-11(\tilde{\delta}/\tilde{\delta}_c)^{1.2}\right), \qquad (8.59)$$

where the first term is the approximate expression that provides a good fit away from the wall and is given by

$$f(\tilde{h}, Re_V) = A(Re_V)C_{LV0}(\tilde{h}_*)(\tilde{h}/2)^{g(Re_V)}, \qquad (8.60)$$

where

$$A(Re_V) = 1 + 0.329Re_V + 0.00485Re_V^2,$$
$$g(Re_V) = -0.9\tanh(0.022Re_V). \qquad (8.61)$$

At large Reynolds number the function f itself provides a good fit over the entire range of $\tilde{h}$. However, for $\mathrm{Re}_V < 100$ the power-law behavior holds only for large $\tilde{h}$. For small $\tilde{h}$ there is systematic deviation from the power law. The second term in Eq. (8.59) is a near-wall correction and the distance from the wall over which this correction applies decreases with increasing Re_V. In the correlation term, $\tilde{\delta}_c = 6\exp(-0.17\mathrm{Re}_V{}^{0.7})$ is the gap over which the near-wall correction applies.

Torque
In Figure 8.18 the moment coefficients of a particle translating parallel to a wall are also plotted as filled symbols. At lower Reynolds numbers, the moment coefficient remains very small and changes sign and therefore is not shown in the log–log plot. It is clearly seen that the moment coefficient for a particle translating parallel to a wall is at least an order of magnitude smaller than those in a wall-bounded linear shear flow. Thus, it can be concluded that the torque on the particle is mainly contributed by the shear effect instead of the wall effect.

8.5.2   Problem II: Particle in a Wall-Bounded Linear Shear Flow

The previous subsection was entirely devoted to modeling of lift and drag forces on a particle that is in translational motion parallel to a wall. In this section, we will consider the case of a stationary particle in a wall-bounded linear shear flow and summarize the results of Zeng et al. (2009). In an unbounded linear shear flow, the drag force on the particle is not significantly influenced by ambient shear (Kurose and Komori, 1999; Bagchi and Balachandar, 2002a). Thus, any change in drag force compared to that of a uniform flow is primarily due to the nearby wall. Figure 8.16 shows the drag coefficient over a range of Reynolds numbers and for varying separation distances from the wall. For comparison, also plotted in the figure as a solid line is the standard drag correlation of an isolated particle.

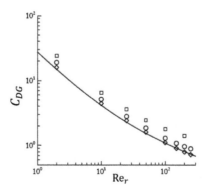

Figure 8.16 Drag coefficient for a stationary particle in a linear shear flow. Drag coefficient vs. Re. □$\tilde{\delta} = 0.01$. ○$\tilde{\delta} = 1$. ◇$\tilde{\delta} = 7$. —— Standard drag correlation. Reprinted from Zeng et al. (2009), with the permission of AIP Publishing.

It is seen that the drag force increases with decreasing distance between the particle and the wall over the entire range of Reynolds numbers considered here. At lower Reynolds numbers, the effect of the wall on the drag force is not entirely negligible, even at separation distance $\tilde{\delta} = 7$. For Re > 100 the effect of the wall is to decrease the drag force below that for a uniform flow for $\tilde{\delta} \geq 3$, however the magnitude of this decrease is quite small. The effect of unbounded shear in the absence of a wall is to slightly increase the drag coefficient at finite Re (Kurose and Komori, 1999) and thus, interestingly, the effect of the nearby wall seems to reverse this trend at larger Reynolds numbers.

Drag Force

Motivated by standard drag law, Zeng et al. (2009) proposed the following Reynolds number dependence of drag in a wall-bounded shear flow:

$$\mathbf{F}_{DG} = \mathbf{F}_{G0} \left(1 + \alpha_s \, \text{Re}_r^{\beta_s} \right), \tag{8.62}$$

where $\text{Re}_r = Ghd_p/\nu_f$ is the Reynolds number based on relative velocity at the particle due to the shear flow. The zero Reynolds number drag force $\mathbf{F}_{G0}$ was given in Eq. (8.15). The prefactor α_s and the exponent β_s are given by

$$\alpha_s = 0.15 - 0.046 \left(1 - 0.04\tilde{\delta}^2 \right) \exp\left(-0.35\tilde{\delta} \right),$$
$$\beta_s = 0.687 + 0.066 \left(1 - 0.19\tilde{\delta}^2 \right) \exp\left(-(\tilde{\delta}/2)^{0.9} \right). \tag{8.63}$$

It can be seen that as $\tilde{\delta} \to \infty$, the drag coefficient goes to the standard drag law. As pointed out by Zeng et al., the above empirical correlation, although not perfect, provides a good fit for the computational data at finite Re and agrees with the theoretical result in the Stokes limit. In fact, it is able to capture the slight decrease in the drag coefficient below the standard drag at Re larger than 100 at intermediate separation distances.

Lift Force

The finite Reynolds number lift force results for a stationary particle in a wall-bounded linear shear flow are from Zeng et al. (2009). Figure 8.17 shows the lift coefficient at different separation distances as a function of Re_r. For larger separation distances, the lift force changes sign at some Re_r and becomes negative. However, for $\tilde{\delta} < 0.5$, the lift coefficient remains positive over the entire range of Re_r considered. Upon further investigation, Zeng et al. observed that with increasing Reynolds number both the pressure and viscous lift coefficients decrease, but the drop in the viscous component is more rapid. Thus, with increasing Re_r, the viscous lift coefficient first becomes negative followed by the pressure lift coefficient, which changes sign at a higher Reynolds number. The total lift coefficient changes sign at a Reynolds number that is in between. Above this Re_r, the particle experiences a force directed toward the wall.

The critical Reynolds number at which $C_L = 0$ (referred to as Re_{cr}) for different $\tilde{\delta}$ is summarized in Table 8.5. The critical Reynolds number in an unbounded linear shear flow is observed to be about 60 (Kurose and Komori, 1999). It is seen that at $\tilde{\delta} = 0.5$, the critical Reynolds number is close to 200 and decreases substantially to $\text{Re}_{cr} \approx 60$ at $\tilde{\delta} = 7$.

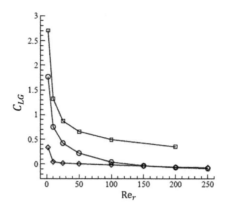

Figure 8.17 Lift coefficient as a function of Reynolds number. $\square\tilde{\delta} = 0.01$. $\circ\tilde{\delta} = 1$. $\diamond\tilde{\delta} = 7$. Reprinted from Zeng et al. (2009), with the permission of AIP Publishing.

Table 8.5 Critical Reynolds number $\mathrm{Re_{cr}}$ for zero lift force in a wall-bounded linear shear flow.

$\tilde{\delta}$	$\mathrm{Re_{cr}}$
0.5	198.19
1	125.52
3	74.70
7	59.11

At low Reynolds numbers the simulation results were satisfactorily compared against the theoretical results of Cox and Hsu (1977), McLaughlin (1993), and Cherukat and McLaughlin (1994). The following empirical correlation for the lift force was proposed that will be valid over a wide range of Re_r and $\tilde{\delta}$:

$$C_{\mathrm{LG}} = \frac{F_{\mathrm{LG}}}{\frac{\pi}{2}\rho_f R_p^2 h^2 G^2} = C_{\mathrm{LG,w}} \exp\left(-0.25\tilde{\delta}(\mathrm{Re}_r/250)^{4/3}\right)$$
$$\times \left[\exp(\alpha_{s\mathrm{L}}(\mathrm{Re}_r)(\tilde{\delta}/2)^{\beta_{s\mathrm{L}}(\mathrm{Re}_r)} - \lambda_{s\mathrm{L}}(\tilde{\delta}, \mathrm{Re}_r)\right], \tag{8.64}$$

where

$$\alpha_{s\mathrm{L}}(\mathrm{Re}_r) = -\exp\left(-0.3 + 0.025\,\mathrm{Re}_r\right),$$
$$\beta_{s\mathrm{L}}(\mathrm{Re}_r) = 0.8 + 0.01\,\mathrm{Re}_r, \tag{8.65}$$
$$\lambda_{s\mathrm{L}}(\tilde{\delta}, \mathrm{Re}_r) = \left(1 - \exp(-\tilde{\delta}/2)\right)(\mathrm{Re}_r/250)^{5/2}.$$

The above expression correctly reduces to $C_{\mathrm{LG,w}}$ in the limit where the particle touches the wall ($\tilde{\delta} \to 0$), whose expression was given in Eq. (8.50). The above correlation captures the variation in lift force over the range $1 < \mathrm{Re}_r < 200$ and $\tilde{\delta} > 0$.

Torque

The symmetry-breaking effects of both the shear flow and the wall induce a net hydrodynamic torque on the sphere about the y-axis. Figure 8.18 shows the moment coefficient versus Reynolds number. It can be observed that for different $\tilde{\delta}$, the moment coefficient nearly shows a power-law decreasing trend with Re_r, except for the very largest Re_r considered, where it can be noticed that C_M is very small.

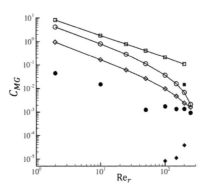

Figure 8.18 Moment coefficient as a function of Reynolds number. $\square \tilde{\delta} = 0.01$. $\circ \tilde{\delta} = 1$. $\diamond \tilde{\delta} = 7$. Filled symbols are the corresponding data for the same $\tilde{h}$ for a particle moving parallel to a wall. The moment coefficient for the shear flow is defined in terms of hydrodynamic torque as $C_{MG} = 2|\mathbf{T}_G|/(\pi \rho_f G^2 R_p^5)$ and for the case of a translating particle the moment coefficient is defined as $C_{MV} = 2|\mathbf{T}_V|/(\pi \rho_f V_x^2 R_p^3)$. Reprinted from Zeng et al. (2009), with the permission of AIP Publishing.

8.5.3   Particle III: Particle Spinning About an Axis Parallel to the Wall

Here we will consider the results of well-resolved simulations reported by Lee and Balachandar (2010), which considered flow induced by a spinning particle located close to a flat wall, with the axis of spinning being parallel to the flat wall. They considered a range of rotational Reynolds numbers (Re_Ω) from 1 to 100 at varying distances from the wall.

Drag Force

The drag coefficient obtained from the simulation results is plotted as symbols in Figure 8.19 as a function of $\tilde{h}$ in log-linear scale for different Reynolds numbers. The logarithmic increase in drag coefficient as the wall is approached is clear. For small and large values of the gap, the results have been compared with the lubrication theory of Goldman et al. (1967a), and with results obtained from the method of reflection. At large gaps the drag force is nearly zero and above a certain Reynolds number the approach seems to be from below. Thus, above a certain Reynolds number the drag force becomes negative at intermediate values of the gap. This behavior is similar to that observed for the cases of a wall-bounded shear flow over a particle and of a translating particle parallel to a wall in a quiescent fluid (Zeng et al., 2009). In both these cases, at intermediate values of the gap, the drag force on the particle was observed to be lower than what it would be in the absence of the wall. Such a reduction was also identified in the theoretical results of Vasseur and Cox (1977), who explained this reduction in terms of a potential contribution induced by an inflow and outflow from the wall boundary layer. A similar mechanism can be expected to be active in the present case of a spinning particle at sufficient wall–particle separation.

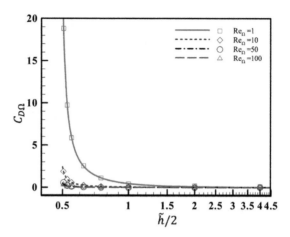

Figure 8.19 Drag coefficient of a spinning particle for different rotation Reynolds numbers and distances between the particle and the wall. The symbols are simulation results and the lines are from the correlation. Taken from Lee and Balachandar (2010).

Lee and Balachandar (2010) presented the following correlation for the drag coefficient for the spinning sphere in the proximity of a wall:

$$C_{D,\Omega} = \frac{|\mathbf{F}_{D\Omega}|}{\frac{1}{2}\rho_f \Omega^2 R_p^4} = \frac{0.2425}{Re_\Omega} \left(0.25\tilde{h}^2 - 0.2126\tilde{h} - 0.02743\right)^{-1}. \qquad (8.66)$$

In Figure 8.19, the drag evaluated using the above correlation is plotted as lines and good agreement with the simulation data can be observed. The above expression provides a good approximation for the numerical results over the parameter range considered. Also, the difference between the above expression and the results of Goldman et al. (1967a) at low Re_Ω is sufficiently small that the above can be taken to be adequate. However, note that in the low Reynolds number limit the above correlation does not capture the logarithmic singularity precisely.

Lift Force

The lift coefficients are presented in Figure 8.20 plotted as a function of $\tilde{h}$ for the different Reynolds numbers. In the limit of a particle touching the wall ($\tilde{h} = 1.0$) and $Re_\Omega \to 0$, the asymptotic theory of Krishnan and Leighton (1995) shows that $C_{L\Omega} \to 0.348$. The results from the low Reynolds number asymptotic analysis (Cherukat and McLaughlin, 1994) is also plotted in Figure 8.20 as symbols, and the agreement with the computational results of $Re_\Omega = 1$ is very good.

The rotation-induced lift force falls off rapidly as the particle position moves away from the wall. The sequence of simulation results for small $\tilde{h}$ can be extrapolated to obtain the lift force on a particle sitting on the wall. The inset in Figure 8.20 shows $C_{L\Omega,w}$ given in Eq. (8.50), which is accurate for $Re_\Omega < 100$. It can readily be seen that at all separations from the wall, for small Reynolds numbers the lift coefficient is nearly a constant, only being weakly dependent on Re_Ω. This is in contrast to the drag coefficient, which has Re_Ω^{-1} behavior at low Reynolds numbers. The following expression for the lift coefficient was found by Lee and Balachandar (2010) to be effective for $\mathrm{Re}_\Omega \leq 100$ in describing the simulation results:

$$C_{\mathrm{L}\Omega} = C_{\mathrm{L}\Omega,\mathrm{w}} \left[1 + \tanh\left((-0.767 - 0.00018\,\mathrm{Re}_\Omega^{1.785})(\tilde{\delta})^{0.563+0.00317\,\mathrm{Re}_\Omega} \right) \right]. \quad (8.67)$$

Note that it reduces to the correct behaviors both in the limit of low Reynolds number and in the limit of the particle touching the wall.

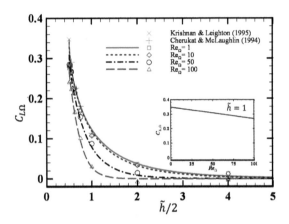

Figure 8.20 Lift coefficient of a spinning particle for different rotation Reynolds numbers and distances. Symbols and lines are the results from numerical simulations and correlation. The inset shows results for the case of a particle touching the wall for varying Reynolds numbers. These results are extrapolated from the numerical simulations to $\tilde{\delta} = 0$. Taken from Lee and Balachandar (2010).

Torque

Figure 8.21 shows the moment coefficient obtained from numerical simulations plotted as a function of $\tilde{h}$ in log–log scale for different values of the rotational Reynolds number. The simulation results are in agreement with the low Reynolds number behavior predicted for small gaps by the lubrication theory and for large gaps by the method of reflections. We observe the computed torque values at small separations to be more sensitive than the force components. Lee and Balachandar (2010) presented the following expression for the moment coefficient of a particle undergoing rotational motion:

$$\begin{aligned}
C_{M\Omega} = {}& \exp(-3\tilde{\delta}) \left[-\frac{64}{\mathrm{Re}_\Omega}(1 + 0.0011\,\mathrm{Re}_\Omega)\left(\frac{2}{5}\ln\tilde{\delta} - 0.3817 \right) \right] \\
& + (1 - \exp(-4\tilde{\delta})) \left[\frac{64}{\mathrm{Re}_\Omega}\left(1 + \frac{5}{16(1+\tilde{\delta})^3} \right) + 0.00225\,\mathrm{Re}_\Omega \right].
\end{aligned} \quad (8.68)$$

The above expression is also shown in Figure 8.21 as lines, and can be seen to well approximate the numerical results over the entire range of Re_Ω and $\tilde{\delta}$ considered in this study.

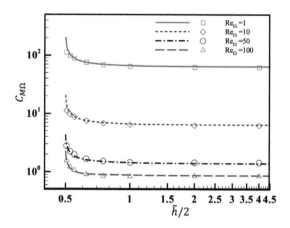

Figure 8.21 Momentum coefficients for a rotating particle for different Reynolds numbers and gap between the particle and the wall. Taken from Lee and Balachandar (2010).

8.5.4 Problem IV: Particle Translating Perpendicular to the Wall

The problem of a particle moving perpendicular to a flat wall at finite Reynolds number results only in a drag force that is also directed normal to the wall. By symmetry, we do not expect a lift force or nonzero torque on the particle, provided the translational Reynolds number of the particle is sufficiently small that the wake does not become one-sided. This problem was numerically solved by Zhou et al. (2017), whose results will be briefly summarized here. Their expression for the force on the particle in present notation can be expressed as

$$\mathbf{F}_{V\perp} = \mathbf{F}_{V\perp 0s} + 6\pi\mu_f R_p V_z\, \mathbf{e}_z \left(6.07 \times 10^{-4} + 0.0351\, \ln(\tilde{\delta})\right) \mathrm{Re}_V, \qquad (8.69)$$

where $\mathbf{F}_{V\perp 0s}$ is the zero-Re force in the small-gap limit. By design, the above expression correctly goes to the right limit at zero Reynolds number and the above expression was developed to be applicable for $\mathrm{Re}_V < 370$. The above expression is, however, valid only for a small gap, since the results were based on simulations performed for $\tilde{\delta} < 0.2$. As a result, the above expression does not go to standard drag when $\tilde{\delta} \to \infty$. By blending the above with the standard drag for the large-gap limit, a composite expression valid for all gaps can be developed.

8.5.5 Problem V: Particle Spinning About an Axis Perpendicular to the Wall

The problem of a particle spinning with its axis perpendicular to a flat wall at finite Reynolds number results only in a torque. By symmetry, we do not expect a force on the particle. This problem was also solved numerically by Zhou et al. (2017), with torque expression

$$\mathbf{T}_{\Omega\perp} = 8\pi\mu_f R_p^3 \Omega_z\, \mathbf{e}_z \left(-1.277 + 0.285\tilde{\delta}^{1/4} - (0.00329 + 0.00141\tilde{\delta})\,(\mathrm{Re}_\Omega/2)\right). \qquad (8.70)$$

The above expression does not tend to the zero Reynolds number expression given in Eq. (8.25), since the authors decided to use the Stokes limit results of Jeffrey (1915). Nevertheless, the above expression was developed to be applicable for $\text{Re}_\Omega < 740$. The above expression is, however, valid only for a small gap, since the results were based on simulations performed for $\tilde{\delta} < 0.4$. As a result, the above expression does not approach the torque expression given in Eqs. (5.32) and (5.34) in the limit $\tilde{\delta} \to \infty$. By blending the above with the torque for the large-gap limit, a composite expression valid for all gaps can be developed.

Problem 8.4 In this problem, we will consider a general representation of wall influence on particle force and torque. We will do this with the use of the general representation theorem discussed in Section 5.3. As the first step, let us simplify the problem by assuming the flow to be quiescent. Thus the force and torque on the problem are only due to the motion of the particle close to a flat wall. We will also assume the conditions to be relatively steady. Note that for a particle with a wall-normal component of velocity, the problem cannot be strictly steady, since the distance between the particle and the wall changes over time. Nevertheless, in this problem, we will assume the conditions to be quasi-steady and ignore the effects of unsteadiness. The independent variables that influence the force and torque are the translational velocity of the particle $\mathbf{V}$, the angular velocity of the particle $\mathbf{\Omega}$, the wall orientation defined by the unit vector $\mathbf{n}$, and the distance from the center of the particle to the plane formed by the wall h. In addition, there are other independent variables such as particle radius and fluid properties.

(a) We first note that force is a true vector, while torque is a pseudo vector, and both are functions of all the above independent variables. The first step is to identify all the (i) scalars, (ii) true vectors, and (iii) pseudo vectors that can be formed out of the independent variables. Show that the three lists are

Scalars: $\quad I_s = \{|\mathbf{V}|, |\mathbf{\Omega}|, h/R_p\ \alpha = \langle\!\langle \mathbf{V}, \mathbf{n} \rangle\!\rangle, \beta = \langle\!\langle \mathbf{\Omega}, \mathbf{n} \rangle\!\rangle, \gamma = \langle\!\langle \mathbf{V}, \mathbf{\Omega} \rangle\!\rangle\}$,
True vectors: $\quad \mathbf{V}, \mathbf{n}, \mathbf{V} \times \mathbf{\Omega}, \mathbf{n} \times \mathbf{\Omega},$ (8.71)
Pseudo vectors: $\quad \mathbf{\Omega}, \mathbf{V} \times \mathbf{n},$

where $\langle\!\langle \mathbf{a}, \mathbf{b} \rangle\!\rangle$ denotes the angle between the vectors $\mathbf{a}$ and $\mathbf{b}$. In the above, I_s denotes the list of all scalar invariants of the independent variables, where $|\mathbf{V}|$ and $|\mathbf{\Omega}|$ can be nondimensionalized to form Re_V and Re_Ω.

(b) With these definitions we can write the general expressions of force and torque coefficients as

$$C_{\mathbf{F}} = C_{f1}(I_s)\mathbf{V} + C_{f2}(I_s)\mathbf{n} + C_{f3}(I_s)\mathbf{V} \times \mathbf{\Omega} + C_{f4}(I_s)\mathbf{n} \times \mathbf{\Omega},$$
$$C_{\mathbf{T}} = C_{t1}(I_s)\mathbf{\Omega} + C_{t2}(I_s)\mathbf{V} \times \mathbf{n},$$ (8.72)

where the coefficients C_{f1}, etc., are functions of all the scalar invariants.

(c) Show that the above forms of the force and torque vectors are consistent with those given in Problem 8.3 in the Stoke limit.

(d) Identify the physical mechanism of each term in the force and torque expressions. For example, the first term in the force expression is the velocity-induced drag, whereas the last term is the drag due to particle rotation in the vicinity of the wall.

8.6 Wall Effect on Added-Mass and History Forces

In this section, we move beyond the wall influence of quasi-steady forces and consider how the presence of a nearby wall influences the unsteady forces on a particle. Much of the discussion will be on the modification of the added-mass force by the nearby wall. Understandably, our understanding of the wall effect on the history force is limited, since its modeling is quite complex – even in the absence of the wall. Therefore, toward the end of this section we will briefly present a simple approach that has been suggested in the literature.

The added-mass force arises from the inviscid mechanism and therefore one needs only to solve the unsteady potential flow around an accelerating particle near a flat boundary and integrate the pressure distribution around the particle to obtain the added-mass force. In fact, as discussed by Batchelor (2000), the added-mass force of a particle can itself be obtained from steady potential flows (see also Brennen, 2005 and the discussion on added mass in Section 4.2.2). The added-mass coefficient of a particle in an unbounded fluid was earlier treated as a constant. However, close to a wall, the added-mass coefficient for particle acceleration parallel to the wall is different than perpendicular to the wall. Thus, in general, the added-mass coefficient must be considered as an added-mass tensor. An explicit expression for the added-mass tensor in terms of the potential flow around the accelerating sphere becomes

$$C_{M,ij} = \frac{\rho_f}{2} \int_{\Omega} u_{ik} u_{jk} \, d\Omega, \tag{8.73}$$

where the integral is over the entire volume of fluid outside the particle and the wall. Here, u_{ij} is the ith component of steady potential flow velocity due to a particle moving in the jth direction, in the presence of the wall. As discussed in Section 4.2.2, in an unbounded domain $C_{M,ij}$ is an isotropic tensor of value $1/2$. With the influence of a nearby wall, the added-mass tensor takes the following form:

$$\mathbf{C}_M \approx \frac{1}{2} \begin{bmatrix} \left(1 + \frac{3}{16\tilde{h}^3}\right) & 0 & 0 \\ 0 & \left(1 + \frac{3}{16\tilde{h}^3}\right) & 0 \\ 0 & 0 & \left(1 + \frac{3}{8\tilde{h}^3}\right) \end{bmatrix}, \tag{8.74}$$

where directions 1 and 2 are parallel to the wall and direction 3 is perpendicular to the wall. The added-mass coefficient becomes isotropic as the influence of the wall decays with increasing $\tilde{h}$. Using the above, the added-mass force expression given in Eq. (4.133) can be generalized to the following:

$$F_{iu,i} = m_f \, C_{M,ij} \left[\frac{Du_{cj}}{Dt} - \frac{dV_j}{dt} \right], \tag{8.75}$$

where it should be noted that we have replaced du_{ci}/dt by Du_{ci}/Dt to account for the nonlinear effect of advective acceleration of the undisturbed ambient flow at the particle location.

The above added-mass tensor includes only the leading-order term in the distance between the wall and the particle, which is adequate for large values of $\tilde{h}$. As pointed out by Ardekani and Rangel (2008) and Yang (2010), the leading-order solution of the wall-normal added-mass coefficient may not be adequate as $\tilde{h}$ becomes small and the particle gets close to contacting the wall. In particular, as $\tilde{h} \to 1$, Eq. (8.74), which accounts for only the leading-order effect, yields a value of $C_{M,33} = 0.6875$. This value is noticeably higher than the 0.606 obtained when all the higher-order effects are taken into account. An improved added-mass coefficient in the wall-normal direction was presented by Yang (2010), whose expression is quite complex. Ardekani and Rangel advanced the following expression, which also accounts for the higher-order effects:

$$C_{M,33} = \frac{1}{2}\left(1 + \frac{1.591}{(1.917 + \tilde{h})^{3.887}}\right).$$ (8.76)

In the wall-parallel direction, as $\tilde{h} \to 1$, Eq. (8.74) gives a value of $C_{M,11} = C_{M,22} = 0.593$. This is quite close to the value of 0.61 predicted by Helfinstine and Dalton (1974). The reader is also referred to Simcik and Ruzicka (2013), where an exhaustive list of past investigations of the added mass of a particle is presented. Their list included not only the effect of a nearby wall, but also the effect of a nearby fluid–fluid interface, and results for nonspherical particles.

We now briefly discuss the influence of the wall on the history force. This problem is hard, since it is likely to depend on the nature of acceleration and deceleration of the particle in the directions parallel and perpendicular to the wall, as well as the nondimensional distance from the wall and the Reynolds number – quite a formidable problem indeed. As a result, there are only a very limited number of investigations that have addressed this challenging problem. Yang (2010) considered the effect of the wall on the potential flow, which in turn affects the growth of the boundary layer around the particle. With this he argued the effect of the wall on the history force to be similar to that on the added-mass force. Based on this argument, he proposed the following expression for the history force. Following the finite-Re BBO equation (4.144), his result can be adapted for the present discussion as

$$F_{vu,i} = 6\pi\mu_f R_p \int_{-\infty}^{t} K_{vu}(\xi; \text{Re})\, C_{M,ij}^{3/2} \left[\frac{du_{cj}}{dt} - \frac{dV_j}{dt}\right]_{@\xi} d\xi.$$ (8.77)

The advantage of this proposal is that it correctly reverts to the unbounded limit when $\tilde{h}$ becomes large. Furthermore, the corrections to wall-normal and wall-parallel relative accelerations are the same as those for the added-mass force. It is very likely that the above proposal can be substantially improved with further research.

8.7   Particle Interaction with Near-Wall Turbulence

We now discuss the hydrodynamic force and particle motion in a turbulent flow close to a boundary. This problem is of great importance in many environmental and industrial applications. Processes such as particle deposition, particle resuspension, and particle–wall impaction depend on the motion of particles within the near-wall region of the flow. Although there are many microscale applications where the flow remains laminar, in many other applications the near-wall flow remains highly turbulent and thus, the effect of the wall must be investigated in the context of turbulent flow. Furthermore, in many practical applications the wall may not be smooth, especially on the scale of the particle. In which case, the effect of wall roughness on particle force and particle motion must be addressed. This is a vast subject, demanding perhaps even an entire book. Here we will only consider this problem briefly, with particular focus on answering the following question: how can the results of the previous sections be used (i) in case of a turbulent flow and (ii) when the wall is rough?

Let us first consider a particle in a turbulent flow over a smooth wall. There is a growing body of information on the effect of wall turbulence on the drag and lift forces of an isolated particle (Hall, 1988; Mollinger and Nieuwstadt, 1996; Zeng et al., 2008, 2010; Van Hout, 2011, 2013; Van Hout et al., 2018; Jebakumar et al., 2019; Peng et al., 2020; Tee et al., 2020). These studies include experimental measurements as well as particle-resolved simulations, where the force on the particle immersed within the turbulent boundary layer is measured directly. There have also been a number of studies where the motion of particles within the turbulent boundary layer has been calculated using the point-particle approach. An important difference is that in these studies, the force and torque on the particle are not internally computed within the simulation, but must be externally given by drag and lift relations, such as those discussed in the previous sections. The reader must be careful, since a wide variety of approximations have been made in these simulations. For example, Saffman lift force and standard drag have often been used even in the near-wall region, although far more accurate lift and drag expressions that account for the wall effect are available.

The reason for continued use of simpler force expressions is due to the fact that the more advanced models that account for the wall effect face a few significant challenges when it comes to their application in a turbulent flow. First, the undisturbed flow surrounding the particle can no longer be considered as a linear shear flow that is parallel to the wall. A turbulent near-wall layer is dominated by occurrences of turbulent burst events where there is strong wall-normal ejection of fluid away from the wall, and sweep events that are marked by a strong inrush of fluid toward the wall. Therefore, the ambient fluid velocity is no longer restricted to be streamwise and parallel to the wall. If we define drag to be force in the direction of relative velocity, then there will be a drag force even in the wall-normal direction. Second, if the particle diameter is comparable to a viscous wall unit, then the particle cannot be taken to be smaller than the flow scales. In case of such complicated ambient flow, in addition to the mean shear of streamwise velocity in the wall-normal direction, there can be shear and vortical regions of flow that are oriented in other directions. Thus, their effect on the drag and lift forces on the particle must be considered. Finally, the flow condition seen by the particle is highly time dependent in a turbulent flow. The effect of each of these complications in terms of force and torque modeling must be considered.

Particle-resolved simulations and careful experimental measurements in a turbulent flow (works that have been referenced above) have provided some insight as to how best to tackle this complex problem of particle force prediction. The following recommendations can be made for the three distinct scenarios.

Particle Located Away from the Wall

If the particle is $O(10)$ diameters away from the wall, then wall effects can be ignored. In fact, the wall influence on drag and lift forces is substantial only when the particle–wall separation is less than about five diameters. If wall effects are unimportant, one can revert to the use of the BBO/MRG equation and follow the discussion of Section 7.6.

Particle Close to a Wall and Smaller than Fluid Scales

We now consider the limit where the particle is smaller than the fluid length scales. This limit presents two important advantages: (i) the particle Reynolds number is then less than unity and (ii) the fluid properties can be taken to be those evaluated at the center of the particle, ignoring any variation on the scale of the particle. Unsteady effects can still be important and therefore one must use the BBO equation. In the following bullet points we consider the best available approach to evaluating all the components of force:

- The *undisturbed flow force* is still given by Eq. (4.91). The effect of the nearby wall is indirect and only through the value of $\nabla \cdot \sigma_c$ which is modified due to the presence off the wall.
- The *quasi-steady drag force* is given as the sum of the five mechanisms which are summarized in Table 8.3. This table is for the limit of zero Reynolds number. If the effects of the small but finite Reynolds number is important, then the formulas given in Table 8.3 must be replaced with the finite-Re counterparts given in Sections 8.5.1 to 8.5.5. It should be noted, however, that the additive nature of the contributions is strictly valid only in the limit of zero Re.
- The *quasi-steady lift force* can be taken to be that given in Table 8.4. These results are applicable only in the limiting cases of no rotation and torque-free rotation. Note that the lift force only accounts for the wall-normal force due to the wall-normal gradient of wall-parallel velocity, the wall-parallel component of relative velocity, and the wall-parallel component of particle rotation. In other words, Table 8.4 only allows for the calculation of the dominant wall-normal component of the lift force.
- The *added-mass force* in the presence of a nearby wall was presented in Eq. (8.75), and the added-mass coefficient is given in Eqs. (8.74) and (8.76).
- The *viscous unsteady force* in the presence of the wall, if important, can be taken to be that given in Eq. (8.77).
- If the rotational motion of the particle is tracked, then one must use the torque induced on the particle due to the five mechanisms that we considered earlier. It must be pointed out that in the near-wall region, particle rotation about the axis perpendicular to the wall is not of great importance.

The above listed approach is necessarily approximate. Nevertheless, the force model given above is likely to capture the actual hydrodynamic force on a particle quite well. This assertion must, however, be confirmed with both high-quality experimental measurements and particle-resolved simulations of a particle in free motion in a near-wall turbulent boundary layer.

Particle Close to a Wall and Comparable in Size to the Flow Scales

As can be expected, this regime, which in Section 7.6 was recognized to be quite complex even in the unbounded case, is a lot harder in the presence of a wall. There are several complicating factors: (i) the nature of wall turbulence is anisotropic and inhomogeneous and therefore quite different from isotropic turbulence considered in the unbounded case; (ii) the particle Reynolds number will now be finite; and (iii) since the particle is of finite size, the fluid properties cannot be evaluated just at the particle center, but must be evaluated as surface and volume averages, as demanded by the MRG equation. The results of Bagchi and Balachandar (2003b, 2004) and Zeng et al. (2008, 2010) indicate that for such large-sized particles, it is hard to accurately predict the instantaneous force variations induced by eddies of the size of the particle. Furthermore, the use of added-mass and history forces does not necessarily improve the accuracy of force prediction. Based on these considerations, we propose the following approximate procedure:

- The *undisturbed flow force* is still given by Eq. (4.91), but one must use the volume integral of $\nabla \cdot \sigma_c$.
- The *quasi-steady drag force* can again be approximated as the sum of the five mechanisms. However, we must now consider the finite-Re versions of the drag formulas given in Sections 8.5.1 to 8.5.5. Of the five, the only three that are of primary importance are the effect of the shear flow and the effect of particle translation parallel and perpendicular to the wall. These drag formulas are given in Eqs. (8.56), (8.62), and (8.69), respectively. Care is required in their superposition, since the additive nature of the contributions must correctly account for the force direction. Furthermore, superposition may need to be adjusted, so that as $\tilde{h}$ becomes large the superposition must correctly go to unbounded quasi-steady force.
- The *quasi-steady lift force* can be taken to be that given in Table 8.4 with appropriate finite-Re corrections. Again, we account for only the dominant wall-normal component of the lift force. An alternate option is to blend the composite lift force on a particle in contact with the wall (i.e., $\tilde{h} = 1$) given in Section 8.4.1 with the unbounded lift force given in Eq. (5.8).
- The *added-mass force* and the *viscous unsteady force* can be ignored.
- As discussed in Section 7.6, a *stochastic contribution* may need to be added in the form of a Langevin model to account for the effect of scales that are of the order of the particle size.

Certainly the above prescription is far less rigorous than that given before for a small particle of small Reynolds number. But we will have to do with this in the absence of more comprehensive understanding and modeling.

8.7.1   Effect of a Rough Wall

Clearly all surfaces are rough at the molecular scale. So we are concerned with wall roughness on the scale of the particle size or on the scale of the turbulent fluid motion. There are different scenarios of wall roughness that are of interest. For example, in the application of dust removal, we consider a surface with some natural roughness with a particle rolling on it or moving close to the surface. On the other hand, in environmental applications one can consider a spherical sand grain moving over a river bed made of many other sand grains. In this latter situation, wall roughness is defined by the particles themselves that are stationary and randomly distributed, forming the bottom boundary.

If wall roughness is $O(10)$ or more in viscous wall units, it is well known that the wall roughness has a strong influence on the nature of near-wall turbulence. This rich topic is not the subject of this book. We consider the undisturbed flow to be known and are interested in evaluating the particle force and particle motion. First and foremost, if wall roughness modifies near-wall turbulence, then the undisturbed flow used in the calculation of the various force components must be based on this roughness-modified turbulence. In addition, the roughness of the wall can affect the perturbation flow induced by the particle and thereby alter the quasi-steady, added-mass, viscous unsteady, and lift force expressions given earlier. Though this is a distinct possibility, we do not have much information. Therefore, the best approach is to accept the models outlined above for a smooth wall and use them even in the context of a rough wall. The one complication that arises in this approach was addressed by Lee and Balachandar (2017) and Li et al. (2019). While the fluid quantities at the particle location and the particle velocity can be used to calculate all the Reynolds numbers that are required in the calculation of the forces, we also require the distance δ from the bottom of the particle to the wall.

This distance δ could be uniquely defined in the case of a smooth wall. But there are different choices for defining this distance in the case of a rough wall. This point is highlighted in Figure 8.22 (Lee and Balachandar, 2017), where three different options of defining δ are presented for the case where the rough wall is in fact a regular distribution of particles of the same size. In the first scenario, the baseline of the wall is defined by preserving the volume under this height to be equal to that of the particles (Ling, 1995). The second option is to find the height where the fluid velocity becomes zero and measure the distance to the particle as shown in the middle frame (Wu and Chou, 2003). Another option is to geometrically obtain the closest distance between the particle and the rough bed. The performance of all these options was tested (Lee and Balachandar, 2017; Li et al., 2019) and the first definition, where $\delta/R_p = 0.3165$ for a roughness made of monodispersed spheres, has been found to be quite accurate in yielding results that compare well with particle-resolved simulation results. This test was made in the context of steady uniform flow (Lee and Balachandar, 2017) and the study was later extended to turbulent flow by Li et al. (2019).

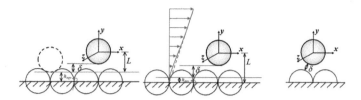

Figure 8.22 Schematic of three different definitions of gap between the particle and a rough bed. Reprinted from Lee and Balachandar (2017), with permission from Elsevier.

8.8   Wall Deposition in Turbulent Flows

One potential application of the wall-normal force that we have discussed lies in the problem of wall deposition. From the quantitative results of the previous sections it must now be clear that, in general, the wall-normal force on a particle is much weaker than the drag force. As a result, again in general, the streamwise velocity of the particle will be much greater than the wall-normal velocity. Although small in magnitude, the wall-normal velocity plays a crucial role in transporting the particles toward the wall in some cases to result in wall deposition, and away from the wall in other cases to result in particle removal.

In a laminar flow, wall deposition is entirely due to the wall-normal (or lift) force, since the flow streamlines are parallel to the wall. If a particle leads the fluid motion (i.e., $V > u$), then the shear lift force is negative and transports the particle toward the wall. If the particle lags the fluid, then the shear lift is positive and the particle is transported away from the wall. As we saw in the previous sections, there are also wall-induced wall-normal force contributions. In the case of heavier-than-fluid particles falling toward a bottom wall, the particle passing through the shear layer will encounter slower-moving fluid and a positive (or negative) lift force will decrease (or increase) the settling rate and deposition. In the Stokes limit, the change in deposition velocity is given by $F_L/(3\pi\mu_f d_p)$.

A far more interesting scenario of deposition arises in the context of wall-bounded turbulent flows. Here the streamlines are not restricted to be parallel to the wall. Near-wall turbulence is characterized by intense turbulence bursts that are characterized by an outrush of fluid away from the wall and strong sweeping motions that bring higher-speed fluid toward the wall. These turbulent sweeping fluid motions are particularly important for wall deposition. Meanwhile, the fluid velocity itself is constrained to go to zero at the wall due to no-slip and no-penetration boundary conditions. The same is not true for inertial particles. Particles that are caught in the wallward sweeping fluid motions, owing to their inertia, will continue to retain their wallward motion and not decelerate as rapidly as the fluid in the immediate vicinity of the wall, and therefore end up in collision with the wall. It is this turbulence-induced wall deposition of particles that we are interested in. While this topic of particle–turbulence interaction fits well with the theme of the previous chapter, we have postponed the discussion to this chapter in order to include the lift force, whose effect on deposition must be ascertained.

Furthermore, to discuss the effect of turbulence and lift forces on deposition, here we will ignore the effect of gravity.

The earliest attempt at quantifying turbulent deposition of particles employed the concept of free flight (Friedlander and Johnstone, 1957; Davies, 1966; Liu and Ilori, 1974). These models assumed the particles to diffuse like fluid parcels away from the near-wall region and only within a stopping distance close to the wall, which depends on particle inertia, will the particles continue to move toward the wall in free flight due to their inertia, resulting in wall deposition. This ignores the role of inertia away from the near-wall region and thus the model prediction of wall deposition has generally been far lower than what is observed experimentally.

Kallio and Reeks (1989) pioneered the use of random walk models in the accurate prediction of wall deposition. They used experimental measurements of mean and rms fluid velocity as a function of wall normal distance in a turbulent wall-layer. They solved the equation of particle motion by including only the Stokes drag and Saffman shear lift forces. The fluid velocity at the particle position was modeled with a simple continuous random walk model, similar to those introduced in the previous chapter. Several important conclusions were drawn from the results, which will be summarized below.

They defined the particle time scale in wall units as $\tau_p^+ = \tau_p u_*^2 / \nu_f$, where u_* is the turbulent friction velocity. τ_p^+ can be interpreted as the Stokes number of the particle. For small particles of $\tau_p^+ \ll 1$, particles behave like tracer particles and their concentration was nearly uniform across the boundary layer. Larger particles of $\tau_p^+ \gg 1$ were not responsive to wall turbulence and therefore their concentration was also nearly uniform. In contrast, particles of $\tau_p^+ \approx 1$ responded the most to wall turbulence and their near-wall concentration peaked at $y^+ \approx 1$, and the peak concentration was observed to be an order of magnitude larger than the average (see Figure 8.23a, reproduced from Kallio and Reeks, 1989). This remarkable increase in near-wall concentration is due to turbophoresis, which we shall see in greater detail in Section 12.4.2. At small values of τ^+ they observed the particle velocity fluctuation to be comparable to the fluid velocity fluctuation. With increasing τ^+ the particle velocity fluctuation was lower, as can be expected from the results presented in Eqs. (7.51) and (7.52).

The deposition velocity V_{dep} can be defined in terms of the average flux of particles $\dot{n}$ depositing on the wall as

$$V_{dep} = \frac{\dot{n}}{n_d}, \tag{8.78}$$

where n_d is the particle number density measured as the number of particles per unit volume, while $\dot{n}$ is measured as the number of particles depositing per unit area of the wall per unit time. Note that the particle concentration and deposition rate depend on both the Stokes number τ_p^+ and the particle-to-fluid density ratio. A plot of nondimensional deposition velocity V_{dep}/u_* is presented in Figure 8.23b, reproduced from Kallio and Reeks (1989). In general, the deposition rate is an order of magnitude larger than what was predicted using the free-flight approach and the difference is in large part due to the enhanced concentration of particles. It is also interesting to note

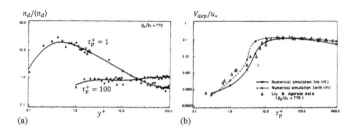

Figure 8.23 (a) Plot of normalized particle number density as a function of y^+ computed with the random walk model. (b) Plot of deposition velocity scaled by u_* as a function of τ_p^+. Reprinted from Kallio and Reeks (1989), with permission from Elsevier.

that the deposition rate decreases for $\tau_p^+ > 10$, mainly due to a reduction in particle velocity fluctuation. Finally, it is observed that the inclusion of shear lift force does increase deposition at intermediate values of τ_p^+.

Although the crux of the problem of wall deposition was well addressed in the original paper by Kallio and Reeks (1989), further improvements in terms of improved Langevin models and an improved equation of particle motion that includes more accurate formulations of wall-induced lift forces was considered by Reeks and co-workers (Jin et al., 2015, 2016). As predicted by most other stochastic models, they observed an underestimation of the deposition rates for small particles with particle time scale normalized in wall units $\tau_p^+ < 5$. The influence on the transport and deposition due to corrections to the drag and lift force in the near-wall region is investigated utilizing the composite correlations proposed by Zeng et al. (2009). As far as deposition rates are concerned, the inclusion of near-wall effects and lift forces is unable to account for the underestimation of deposition for small τ^+ compared to the experimental measurements of Liu and Agarwal (1974). More specifically, the overall effect of near-wall corrections to the drag is found to decrease deposition rates for small particles and increase them for large particles. The inclusion of a lift force based on the formula of Zeng et al. (2009), which accounts for both wall and Reynolds number effects, was somewhat different from that based on Saffman's shear lift. In contrast, the corrected drag does yield significant differences between the particle and carrier flow streamwise velocities for large particles ($\tau_p^+ \approx 20$) in the near-wall region. Given these results, they concluded that whether the wall effects on the hydrodynamic forces should be included or not depends on specific applications.

9 Particle–Particle Interactions

In this chapter, we will consider particle–particle interactions. Here we distinguish two kinds of interactions. The first is direct interaction between particles in the form of collisions. When two particles collide, the time history of force exchange between them is controlled by the solid mechanics of elastic and plastic deformation between the colliding particles. In the context of multiphase flow computations, such collisions are simplified and treated using either a hard-sphere or a soft-sphere collision model, which will be discussed in this chapter. As a special case we will also consider the problem of particle–wall collisions.

We will then proceed to consider particle–particle interaction as mediated by the surrounding fluid. Here again we will address two different classes of particle–particle hydrodynamic interaction. First, we will consider the problem of lubrication flow that dominates the motion of two particles that are in close proximity to each other and are separated only by a thin layer of fluid. The lubrication force can greatly alter their relative motion and thereby influence the manner in which the two particles collide and exchange momentum between them. Therefore, lubrication force is often treated along with interparticle collision. We gain significant analytical advantage by studying this interaction only in the lubrication limit of very small separation between the two particles. In this limit, the hydrodynamic force and torque on the particles are dominated by the pressure and shear stress distributions within the narrow gap between the two interacting particles. Thus, the lubrication force on the particles can be expressed primarily in terms of relative particle position and motion. The details of flow around the two interacting particles on scales larger than the particles is not important in determining the lubrication force. We will consider this lubrication process and treat it much like the zero Reynolds number analysis of particle–wall interaction presented in Section 8.2.

The second class of hydrodynamic interaction between two particles is through the modification of the ambient flow. For example, if a particle is located within the wake of an upstream neighbor, then the fluid velocity the particle sees is lower. As a result, the hydrodynamic drag on the particle will be lower due to sheltering by the upstream neighbor. In addition to the above wake–particle interaction, there are situations of wake–wake interaction between two side-by-side particles.

In summary, the problem of particle–particle interaction is quite complex and has been separated into three classes of interaction. (i) Interaction when particles are physically in contact. Here the dominant force exchange is nonhydrodynamic and is due to the contact mechanics between the particles. (ii) Interaction when the particles are so close to each other that the hydrodynamics of the lubrication layer dominates. Here the dominant force exchange between the particles can be investigated using low Reynolds number theory. Furthermore, the analysis depends only on the motion of the particles and the fluid properties. (iii) Interaction when the interacting particles are sufficiently far away that the modification of the flow by the particle must be considered on the scale of the particle pair, and not just within the gap region. Here, not only the particle motion, but also the nature of the ambient flow and relative position of the interacting particles with respect to the ambient flow matters. Unfortunately, the interaction between the particles at this level cannot be restricted to small Reynolds numbers. Thus, from a hydrodynamic perspective, as we move from class (i) to class (iii) interaction, the difficulty of the problem increases from no hydrodynamics, to low Reynolds number hydrodynamics, and to nonlinear finite Reynolds number hydrodynamics.

Furthermore, interparticle collisional forces can be treated additively. That is, at any time, if a particle simultaneously collides with more than one neighbor, then the collisional forces can be superposed (this of course assumes the contact mechanics and the deformation process within the particle to be linear). Similarly, due to the linearity of the lubrication problem, the lubrication force exchange between the different particle pairs is also additive. However, the hydrodynamic effects of particle–wake and wake–wake interactions are not additive at finite Reynolds numbers due to their nonlinear nature.

The first three sections of this chapter will consider these three classes of particle–particle interactions. These discussions will be limited to the context of two particles. In Section 9.4 we will generalize interparticle interaction from two particles to a distribution of particles. This multiparticle interaction will allow us to investigate the effect of particle volume fraction on quasi-steady drag. We will study this first in the Stokes regime and then extend the analysis to finite Reynolds number. We will also investigate the finite volume fraction effect on added-mass and history forces in the final subsection.

9.1   Particle–Particle Collision

We are interested in the following aspect of particle–particle collision. Given the state of two particles just before collision (pre-collision state), we want to accurately predict the state of the two particles just after collision (post-collision state). Sometimes it is also of importance to accurately account for the duration of collision during which the two colliding particles are in contact. The details of the solid mechanics that occur within the colliding particles and the constitutive relation between the stress distribution inside the particles and their deformation have been studied in great depth in the field of contact mechanics. Here we are only interested in the overall outcome of the collision process and therefore will consider simple models that can be used to describe the behavior of multiphase flows to sufficient accuracy. The simplest is the hard-sphere collision model which will be considered first. The hard-sphere model is approximate in the sense that it assumes infinite force exchange between the two

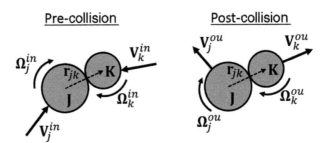

Figure 9.1 Schematic of two spheres going into and out of collision according to the hard-sphere collision model.

particles over an infinitesimal contact period. The resulting impulse is, however, finite and can be used to predict the post-collision state. We will then consider the soft-sphere collision model, which approximates the collision process as a spring dash-pot system.

9.1.1   Hard-Sphere Collision Model

The hard-sphere collision between particle j of radius R_{pj} and particle k of radius R_{pk} is schematically described in Figure 9.1, where the translational velocities of the two particles before collision are denoted as $\mathbf{V}_j^{\text{in}}$ and $\mathbf{V}_k^{\text{in}}$ and their pre-collision angular velocities are $\mathbf{\Omega}_j^{\text{in}}$ and $\mathbf{\Omega}_k^{\text{in}}$. The corresponding post-collision translational and rotational velocities of the two particles are $(\mathbf{V}_j^{\text{out}}, \mathbf{V}_k^{\text{out}})$ and $(\mathbf{\Omega}_j^{\text{out}}, \mathbf{\Omega}_k^{\text{out}})$, respectively. In the hard-sphere model, the entire contact mechanics is abstracted into two parameters: (i) the coefficient of restitution ϵ_c, which controls the amount of energy lost during the collision process; and (ii) the friction coefficient η_c, which controls the ratio of tangential to normal force at the contact point.

We now consider the classic problem in dynamics of evaluating the post-collision state of two particles in terms of their pre-collision state and the two parameters ϵ_c and η_c. We first define the pre- and post-collision relative velocity between the particles to be

$$\mathbf{W}^{\text{in}} = \mathbf{V}_j^{\text{in}} - \mathbf{V}_k^{\text{in}} \quad \text{and} \quad \mathbf{W}^{\text{out}} = \mathbf{V}_j^{\text{out}} - \mathbf{V}_k^{\text{out}}. \tag{9.1}$$

We note that the position vectors of the two particles at the time of their collision are $\mathbf{X}_j$ and $\mathbf{X}_k$, and the unit vector from particle j to particle k is $\mathbf{n}_{jk} = (\mathbf{X}_k - \mathbf{X}_j)/|\mathbf{X}_k - \mathbf{X}_j|$. With this definition, the relative velocity between the particles can be decomposed into normal and tangential components as

$$\begin{aligned}
\mathbf{W}_n^{\text{in}} &= \left(\mathbf{W}^{\text{in}} \cdot \mathbf{n}_{jk}\right) \mathbf{n}_{jk} \quad \text{and} \quad \mathbf{W}_t^{\text{in}} = \mathbf{W}^{\text{in}} - \mathbf{W}_n^{\text{in}}, \\
\mathbf{W}_n^{\text{out}} &= \left(\mathbf{W}^{\text{out}} \cdot \mathbf{n}_{jk}\right) \mathbf{n}_{jk} \quad \text{and} \quad \mathbf{W}_t^{\text{out}} = \mathbf{W}^{\text{out}} - \mathbf{W}_n^{\text{out}}.
\end{aligned} \tag{9.2}$$

For collision to occur, the two particles must be approaching each other prior to collision. In other words, their pre-collision normal component of relative velocity must be positive (i.e., for collision to occur we require $\mathbf{W}^{\text{in}} \cdot \mathbf{n}_{jk} > 0$).

To solve for the post-collisional state of the particles, we appeal to conservation of linear and angular momentum as stated below:

$$\begin{aligned}
\mathbf{J} &= m_j \left(\mathbf{V}_j^{\text{out}} - \mathbf{V}_j^{\text{in}} \right) = -m_k \left(\mathbf{V}_k^{\text{out}} - \mathbf{V}_k^{\text{in}} \right), \\
\mathbf{n}_{jk} \times \mathbf{J} &= \frac{I_j}{R_{pj}} \left(\mathbf{\Omega}_j^{\text{out}} - \mathbf{\Omega}_j^{\text{in}} \right) = -\frac{I_k}{R_{pk}} \left(\mathbf{\Omega}_k^{\text{out}} - \mathbf{\Omega}_k^{\text{in}} \right),
\end{aligned} \tag{9.3}$$

where for a solid particle of mass m_j the moment of inertia is given by $I_j = (2/5) m_j R_{pj}^2$, with a similar definition for particle k. Here, $\mathbf{J}$ is the net impulse imparted on the jth particle during collision, which is the opposite of that imparted on the kth particle. Substituting the above expression for impulse in the definitions of relative relative velocity, it can be shown that

$$\mathbf{J} = \frac{m_j m_k}{m_j + m_k} \left(\mathbf{W}^{\text{out}} - \mathbf{W}^{\text{in}} \right). \tag{9.4}$$

While the normal direction of collision is defined in terms of the particle centers by $\mathbf{n}_{jk}$, the tangential direction of collision can be defined only in terms of their relative motion at the contact point. We evaluate the relative velocity at the contact point to be

$$\mathbf{W}_c = \mathbf{W} + R_{pj} \mathbf{\Omega}_j \times \mathbf{n}_{jk} + R_{pk} \mathbf{\Omega}_k \times \mathbf{n}_{jk}, \tag{9.5}$$

which applies to both pre- and post-collision states. This can then be used to obtain the normal component and by subtraction the tangential component of velocity at the contact point. The tangential components of relative velocity pre- and post-collision can thus be defined as

$$\mathbf{W}_{ct}^{\text{in}} = \mathbf{W}_c^{\text{in}} - \left(\mathbf{W}_c^{\text{in}} \cdot \mathbf{n}_{jk} \right) \mathbf{n}_{jk} \quad \text{and} \quad \mathbf{W}_{ct}^{\text{out}} = \mathbf{W}_c^{\text{out}} - \left(\mathbf{W}_c^{\text{out}} \cdot \mathbf{n}_{jk} \right) \mathbf{n}_{jk}. \tag{9.6}$$

The tangential direction of collision can then be uniquely defined as

$$\mathbf{n}_t = \frac{\mathbf{W}_{ct}^{\text{in}}}{|\mathbf{W}_{ct}^{\text{in}}|}. \tag{9.7}$$

We will require the tangential direction of post-collision motion to be oriented in the same direction as $\mathbf{n}_t$. We can now split the impulse into normal and tangential components as

$$\mathbf{J} = J_n \mathbf{n}_{jk} + J_t \mathbf{n}_t. \tag{9.8}$$

As can be seen from the second equation of (9.3), the torque on the particles imparted by the impulse $\mathbf{J}$ is responsible for the change in their angular momentum. Furthermore, only the tangential component of the impulse J_t contributes to this change in angular momentum.

Apart from the above conservation laws, two other hypotheses are made that essentially model the detailed physics of the contact mechanics:

$$\begin{aligned}
\textbf{Coefficient of restitution}: \quad & \mathbf{W}_n^{\text{out}} = -\epsilon_c \mathbf{W}_n^{\text{in}}, \\
\textbf{Friction factor}: \quad & J_t = \eta_c J_n,
\end{aligned} \tag{9.9}$$

where in the first relation, the post-collision normal velocity is lower than the pre-collision normal velocity, and their ratio is equal to the coefficient of restitution. The second relation is nothing but Coulomb's law of friction between two sliding surfaces that relates the tangential friction force to the normal force through the coefficient of friction. The above two equations are sufficient to solve for the post-collision motion of the particles. Substitution of the definition of the coefficient of restitution in Eq. (9.4) results in

$$J_n = -\frac{m_j m_k}{m_j + m_k}(1 + \epsilon_c)\left(\mathbf{W}^{\text{in}} \cdot \mathbf{n}_{jk}\right) . \qquad (9.10)$$

From the requirement of positivity of the normal-component relative velocity for the collision to occur, we infer that the normal component of the impulse J_n is negative. This guarantees that the velocity of the colliding particles along the normal direction will reverse in the post-collision state. The friction factor definition given in Eq. (9.9) will correspondingly ensure that the tangential impulse $J_t < 0$, so that the frictional force on the jth particle is in the direction of tangential motion of the kth particle with respect to the jth particle. From the normal and tangential impulse expressions, we can obtain the following explicit relations for the post-collision translational and rotational motion of the particles:

$$
\begin{aligned}
\mathbf{V}_j^{\text{out}} &= \mathbf{V}_j^{\text{in}} - (1 + \epsilon_c)\frac{m_k}{m_j + m_k}\left(\mathbf{W}^{\text{in}} \cdot \mathbf{n}_{jk}\right)(\mathbf{n}_{jk} - \eta_c \mathbf{n}_t), \\
\mathbf{V}_k^{\text{out}} &= \mathbf{V}_k^{\text{in}} + (1 + \epsilon_c)\frac{m_j}{m_j + m_k}\left(\mathbf{W}^{\text{in}} \cdot \mathbf{n}_{jk}\right)(\mathbf{n}_{jk} - \eta_c \mathbf{n}_t), \\
\mathbf{\Omega}_j^{\text{out}} &= \mathbf{\Omega}_j^{\text{in}} - \frac{5\eta_c}{2R_{pj}}(1 + \epsilon_c)\frac{m_k}{m_j + m_k}\left(\mathbf{W}^{\text{in}} \cdot \mathbf{n}_{jk}\right)(\mathbf{n}_{jk} \times \mathbf{n}_t), \\
\mathbf{\Omega}_k^{\text{out}} &= \mathbf{\Omega}_k^{\text{in}} - \frac{5\eta_c}{2R_{pj}}(1 + \epsilon_c)\frac{m_j}{m_j + m_k}\left(\mathbf{W}^{\text{in}} \cdot \mathbf{n}_{jk}\right)(\mathbf{n}_{jk} \times \mathbf{n}_t) .
\end{aligned}
\qquad (9.11)
$$

As discussed in Crowe et al. (2011), the above relations assume the final post-collision tangential relative velocity at the contact point between the particles to be nonzero and in the same direction as the pre-collision tangential relative velocity.

The effect of friction at contact can only decrease the tangential velocity (i.e., $|\mathbf{W}_{ct}^{\text{out}}| \leq |\mathbf{W}_{ct}^{\text{in}}|$), but friction cannot reduce the tangential relative velocity to such an extent that it reverses direction. In case of strong frictional effect, the limiting behavior will be such that the post-collision value of relative tangential velocity at the contact point becomes zero. This case is considered in Problem 9.1.

Problem 9.1 Consider the problem where, due to friction, the tangential component of relative velocity at the contact point becomes zero.

(a) Show that this restricts the tangential component of the impulse to

$$J_{t,\text{no-slip}} = -\frac{2}{7}\frac{m_j m_k}{m_j + m_k}\left|\mathbf{W}_{ct}^{\text{in}}\right| . \qquad (9.12)$$

If J_t calculated using Eqs. (9.9) and (9.10) is smaller in magnitude than $|J_{t,\text{no-slip}}|$, then there is a nonzero post-collision tangential slip between the particles and the post-collision motion is described by Eq. (9.11). However, if J_t calculated using Eqs. (9.9) and (9.10) is greater in magnitude than $|J_{t,\text{no-slip}}|$, then we must set $J_t = J_{t,\text{no-slip}}$, and this corresponds to a no-slip post-collision state.

(b) Obtain the corresponding post-collision particle motion given in Crowe et al. (2011) as

$$
\begin{aligned}
\mathbf{V}_j^{\text{out}} &= \mathbf{V}_j^{\text{in}} - \frac{m_k}{m_j + m_k}\left((1 + \epsilon_c)\left(\mathbf{W}^{\text{in}} \cdot \mathbf{n}_{jk}\right)\mathbf{n}_{jk} + \frac{2}{7}\left|\mathbf{W}_{ct}^{\text{in}}\right|\mathbf{n}_t\right), \\
\mathbf{V}_k^{\text{out}} &= \mathbf{V}_k^{\text{in}} + \frac{m_j}{m_j + m_k}\left((1 + \epsilon_c)\left(\mathbf{W}^{\text{in}} \cdot \mathbf{n}_{jk}\right)\mathbf{n}_{jk} + \frac{2}{7}\left|\mathbf{W}_{ct}^{\text{in}}\right|\mathbf{n}_t\right), \\
\Omega_j^{\text{out}} &= \Omega_j^{\text{in}} - \frac{5}{7R_{pj}}\frac{m_k}{m_j + m_k}\left|\mathbf{W}_{ct}^{\text{in}}\right|(\mathbf{n}_{jk} \times \mathbf{n}_t), \\
\Omega_k^{\text{out}} &= \Omega_k^{\text{in}} - \frac{5}{7R_{pk}}\frac{m_j}{m_j + m_k}\left|\mathbf{W}_{ct}^{\text{in}}\right|(\mathbf{n}_{jk} \times \mathbf{n}_t).
\end{aligned}
\tag{9.13}
$$

The advantage of the hard-sphere model is that it does not require the solution of a differential equation to follow the motion of the particles during the collision process. The post-collision motion can be predicted in terms of pre-collision motion through algebraic relations. However, there are other challenges that complicate the implementation of the hard-sphere model in the context of multiphase flow computations. They arise from the fact that the time scale of hard-sphere collision is zero (i.e., the change from pre- to post-collision state happens instantaneously). As a result, during a computational time step, each particle may be subjected to a sequence of collisions. For example, early within the time step let the jth particle first collide with the kth particle, which will redirect the jth particle according to the hard-sphere model. During the remaining portion of the time step, the redirected motion of the jth particle may make it collide with the lth particle, which will again redirect the jth particle. This process will continue and lead to a sequence of collisions. Keeping track of these multiparticle collisions requires a complex algorithm with sub-time scale accuracy. With increasing volume fraction, the time step needed to limit the number of collisions typically is far shorter than that needed for solving the fluid flow. In essence, in the hard-sphere collision model, collisions must be treated sequentially, one after another, with each previous collision influencing the fate of the subsequent collision. This dependency is alleviated in the soft-sphere model to be considered below.

9.1.2   Soft-Sphere Collision Model

In the soft-sphere collision model, the time scale of collision is finite. As a result, a particle could be in collision with more than one neighbor at any given time. However, the force exchange between each particle pair that is in collisional contact is treated separately. Thus, the overall force on a particle due to collisions is the sum of force contributions from all the collisions, each of which will be treated independently.

The important aspect of the soft-sphere model is that the two particles will be allowed to overlap and δ will be used to denote the extent of the overlap (see Figure 9.2). Clearly this is just a mathematical convenience. In reality, the two particles will

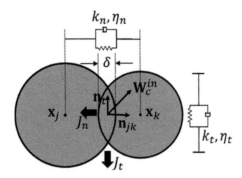

Figure 9.2 Schematic of two spheres undergoing collision according to the soft-sphere model.

undergo elasto-plastic deformation near the point of contact and the distance between the centers of the two particles may decrease below the sum of their radii, by the amount δ. Thus, the overlap distance starts out from a value of zero when the two particles just come into contact, then increases to reach a peak value, when the deformation of the two particles becomes maximum. Subsequently, the shape of the particles elastically rebounds and correspondingly δ decreases from its peak value to become zero again and the two colliding particles separate and pursue their post-collision motion. Thus, the duration of contact in this physical process remains finite.

Like in a linear spring, as δ increases, the repulsive force due to deformation that tends to push the two particles apart increases. The motion of the two particles is then described by Newton's law of motion as a balance between inertia and collisional forces. As a simple first approximation, let us consider the particle deformation upon contact to behave like a simple linear-spring dash-pot system, whose behavior is given by

$$m\frac{d^2\delta}{dt^2} = -k\delta - \eta\frac{d\delta}{dt}, \tag{9.14}$$

where m is the mass, k is the spring constant, and η is the damping coefficient. The above equation is valid only when $\delta \geq 0$ and when δ is negative the particles are out of contact and the above equation does not apply. The appropriate initial conditions are $\delta(t = 0) = 0$ and $d\delta/dt = W_0 > 0$, indicating that at $t = 0$ the particles just came into contact with positive relative velocity W_0. The solution to the above equation is well known in classical physics and the nature of the solution depends on the sign of the discriminant

$$D = \left(\frac{\eta}{2m}\right)^2 - \frac{k}{m}. \tag{9.15}$$

When D is non-negative, δ increases to reach a maximum value and then monotonically decays to zero. Such behavior is physically inappropriate in the present problem, as it implies that the colliding particles will rebound slowly and take infinite time to separate. Thus, we seek a solution only in the limit when $D < 0$ and the solution can be expressed as

$$\delta(t) = \frac{W_0}{\sqrt{|D|}} \exp\left[-\frac{\eta}{2m}t\right] \sin\left(\sqrt{|D|}t\right). \tag{9.16}$$

In particular, the duration of contact can easily be determined as the time when $\delta = 0$, which is given by $t_{\text{cont}} = \pi/\sqrt{|D|}$. As pointed out earlier, Eq. (9.14) and the solution are not valid beyond the point when the particles separate.

Problem 9.2 For the linear-spring dash-pot model, obtain expressions for the maximum overlap δ_{max} and the time at which this maximum deformation occurs in terms of the impact velocity W_0.

Normal Force

There are a variety of models that have been developed for contact physics and are available for use in the context of particle–particle collision in multiphase flows. Here we will only describe the simple *discrete element method (DEM)* pioneered by Cundall and Strack (1979). It has been applied in a number of multiphase flow simulations (Finn et al., 2016). The overlap distance between particles j and k can be defined as

$$\delta = R_{pj} + R_{pk} - |\mathbf{X}_k - \mathbf{X}_j|. \tag{9.17}$$

The normal component of the collisional force depends on both the overlap distance and the normal component of the relative velocity between the particles at the contact point, defined as $\mathbf{W}_{cn} = (\mathbf{W}_c \cdot \mathbf{n}_{jk})\mathbf{n}_{jk}$, where the relative velocity at the contact point $\mathbf{W}_c$ was defined in Eq. (9.5). In the soft-sphere model, both $\mathbf{W}_c$ and its normal component are functions of time. As in the spring dash-pot model, the normal force due to collision on the jth particle can be expressed as

$$\mathbf{F}_{cnj} = -k_n\delta^{3/2}\mathbf{n}_{jk} - \eta_n\mathbf{W}_{cn}, \tag{9.18}$$

where k_n is the normal component of the spring constant and η_n is the normal component of the damping coefficient. The above formula applies only for positive values of δ, otherwise the force on the particles is zero. Note that the above repulsive force is nonlinearly dependent on the overlap distance δ. This nonlinear dependence arises from the classic analysis of Hertz (1882), who considered contact between two elastic bodies of curved surfaces and solved the problem in the small deformation elastic limit. It must be pointed out that in some soft-sphere models the dependence is linearized and the first term on the right is replaced by $-k_n\delta\,\mathbf{n}_{jk}$.

From an analysis of the elastic deformation of the particles, the following model for the normal spring constant is often employed:

$$k_n = \frac{4}{3}\left(\frac{1 - v_j^2}{E_j} + \frac{1 - v_k^2}{E_k}\right)^{-1}\left(\frac{R_{pj} + R_{pk}}{R_{pj}R_{pk}}\right)^{-1/2}, \tag{9.19}$$

where E_j and E_k are the Young's modulus of the materials of the two particles and their Poisson ratios are v_j and v_k, respectively. If the two materials are the same, then the above formula can be simplified. The damping coefficient is typically nonlinear in the overlap distance and a reasonable model for it is given by Tsuji et al. (1992) as

$$\eta_n = g(\epsilon_c)\sqrt{m_{jk}k_n}\delta^{1/4}, \tag{9.20}$$

where the reduced mass m_{jk} is the harmonic average of the mass of the two particles (see Eq. (9.24)). Sierakowski and Prosperetti (2016) proposed the following fit for the function g:

$$g(\epsilon_c) = 2.22 - 2.26\epsilon_c^{0.4}. \tag{9.21}$$

The reader is referred to past implementations of the DEM for appropriate values of the coefficient of restitution ϵ_c and other parameters.

Problem 9.3 Estimates of duration of contact and the maximum overlap as a function of normal velocity at the initiation of contact can be obtained for the above nonlinear Hertzian contact model. Consider two spheres of mass m_j and m_k undergoing normal collision. Ignoring the damping part, the motion of the jth and kth particles is given by the following Newton's law:

$$m_j\frac{d^2 x_j}{dt^2} = -k_n\delta^{3/2} \quad \text{and} \quad m_k\frac{d^2 x_k}{dt^2} = k_n\delta^{3/2}, \tag{9.22}$$

where x_j and x_k are the two particle positions along their line of separation (see Figure 9.2).

(a) Using the definition $\delta = R_{pj} + R_{pk} - |x_j - x_k|$, obtain a differential equation for δ and solve it using the initial conditions that $\delta(t = 0) = 0$ and $d\delta/dt(t = 0) = W_0$. During collision, the particles deform and the overlap distance between the two particles increases to reach a maximum value of $\delta_{\max}$. Then the particles elastically rebound and the two particles go out of contact at time t_{col}.

(b) Obtain the following expressions:

$$\delta_{\max} = \left(\frac{15}{16}\frac{m_{jk}}{E_{jk}R_{jk}^{1/2}}W_0^2\right)^{2/5} \quad \text{and} \quad t_{col} \approx 2.87\left(\frac{m_{jk}^2}{E_{jk}^2 R_{jk}W_0}\right)^{1/5}, \tag{9.23}$$

where the effective radius, reduced mass, and effective Young's modulus are defined as

$$\begin{aligned} \frac{1}{R_{jk}} &= \frac{1}{R_{pj}} + \frac{1}{R_{pk}}, \\ \frac{1}{m_{jk}} &= \frac{1}{m_j} + \frac{1}{m_k}, \\ \frac{1}{E_{jk}} &= \frac{1-v_j^2}{E_j} + \frac{1-v_k^2}{E_k}. \end{aligned} \tag{9.24}$$

Tangential Force

The soft-sphere model can be extended to the tangential component of force as well. In order to do so, we first have to define the tangential overlap or compressional distance δ_t of the tangential spring component. When two particles come into contact, the initial value of δ_t is set to zero. As the two particles move tangentially relative to each other while in contact, δ_t will increase and accordingly the tangential force exchange between the two particles will increase as well. The tangential relative motion will, however, depend on whether the two particles are sliding past each other or whether they are "stuck." Finn et al. (2016) presented the following model for calculating the tangential overlap distance:

$$\delta_t(t) = \begin{cases} 0 & \text{(no contact)}, \\ -(\eta_c |\mathbf{F}_{cnj}| + \eta_t \mathbf{W}_c \cdot \mathbf{n}_t)/k_t & \text{(sliding)}, \\ \delta_{t0} + \int (\mathbf{W}_c \cdot \mathbf{n}_t)\, dt & \text{(sticking)}, \end{cases} \qquad (9.25)$$

where in case of the sticking condition, δ_{t0} is the value of δ_t at the start of the sticking phase of contact. In the above, η_c is the coefficient of friction, k_t is the tangential spring constant, and η_t is the tangential damping coefficient.

The tangential force using the spring dash-pot model is then given by

$$\mathbf{F}_{ctj}(t) = \begin{cases} 0 & \text{(no contact)}, \\ -\eta_c |\mathbf{F}_{cnj}| \mathbf{n}_t & \text{(sliding)}, \\ -(k_t \delta_t + \eta_t (\mathbf{W}_c \cdot \mathbf{n}_t)) \mathbf{n}_t & \text{(sticking)}. \end{cases} \qquad (9.26)$$

In the above relations, during the period of contact, "sliding" versus "sticking" is determined according to the following criterion:

$$\begin{aligned} |k_t \delta_t| > \eta_c |\mathbf{F}_{cnj}| &\Rightarrow \text{(sliding)}, \\ |k_t \delta_t| \leq \eta_c |\mathbf{F}_{cnj}| &\Rightarrow \text{(sticking)}. \end{aligned} \qquad (9.27)$$

Similar to the normal force, an explicit expression for the tangential spring constant can be obtained from considerations of elastic deformation. The resulting expression can be written as

$$k_t = 8\delta^{1/2} \left(\frac{2 - v_j}{G_j} + \frac{2 - v_k}{G_k} \right)^{-1} \left(\frac{R_{pj} + R_{pk}}{R_{pj} R_{pk}} \right)^{-1/2}, \qquad (9.28)$$

where the lateral Young's modulus is $G_j = E_j/(2(1 + v_j))$ and $G_k = E_k/(2(1 + v_k))$, respectively. Thus, according to the above model, the tangential spring constant depends on δ.

9.1.3 Particle–Wall Collision

Particle–wall collision can be treated just as a special case of particle–particle collision, by taking one of the particles to be of infinite radius, which then becomes the wall. The analysis of the previous sections can easily be adapted to consider the case of wall–particle collision.

Consider the colliding particle to be the jth particle and let the wall be the kth particle and as a result, $R_{pk} \to \infty$ and $m_k \to \infty$. Furthermore, in most applications, the wall is considered stationary and therefore $\mathbf{V}_k^{\text{in}} = \mathbf{V}_k^{\text{out}} = 0$. If the wall is moving then the wall's velocity must be left nonzero. In the case of a stationary wall, the normal component of impulse given in Eq. (9.10) can be specialized as

$$J_n = m_j(1 + \epsilon_c)V_j^{\text{in}} \cdot \mathbf{n}_w \,, \tag{9.29}$$

where $\mathbf{n}_w$ is the outward normal from the wall into the fluid and $\mathbf{n}_{jk} = -\mathbf{n}_w$. In the hard-sphere model, the post-collisional motion of the particle after its interaction with the wall becomes

$$
\begin{aligned}
\mathbf{V}_j^{\text{out}} &= \mathbf{V}_j^{\text{in}} - (1 + \epsilon_c)\left(V_j^{\text{in}} \cdot \mathbf{n}_w\right)(\mathbf{n}_w + \eta_c \mathbf{n}_t), \\
\mathbf{\Omega}_j^{\text{out}} &= \mathbf{\Omega}_j^{\text{in}} - \frac{5\eta_c}{2R_{pj}}(1 + \epsilon_c)\left(V_j^{\text{in}} \cdot \mathbf{n}_w\right)(\mathbf{n}_w \times \mathbf{n}_t).
\end{aligned}
\tag{9.30}
$$

Again, the above formula is valid only under slipping condition. In situations of no-slip, the relation given in Eq. (9.13) must be similarly specialized for wall–particle interaction. In both these limits, the effect of the impact on the wall is ignored.

We now consider the application of the soft-sphere model for the case of wall–particle interaction. The normal overlap distance between the particle and the wall is now given by

$$\delta = R_{pj} - L_j \,, \tag{9.31}$$

where L_j is the wall normal distance from the wall to the center of the jth particle. In terms of this overlap distance, the normal force on the particle is given by

$$\mathbf{F}_{cnj} = k_n \delta \mathbf{n}_w + \eta_n(\mathbf{V}_j^{\text{in}} \cdot \mathbf{n}_w) \,. \tag{9.32}$$

The normal spring constant is given by

$$k_n = \frac{4}{3}\left(\frac{1 - \nu_j^2}{E_j} + \frac{1 - \nu_k^2}{E_k}\right)^{-1} R_{pj}^{1/2} \,. \tag{9.33}$$

The expressions for the tangential component of force given in Eqs. (9.25) to (9.28) can similarly be specialized for wall–particle collision.

9.2   Particle–Particle Lubrication Interaction

In this section, we consider the problem of two particles in relative motion when they are very close to each other, but not in contact. In particular, we are interested in estimating the strong hydrodynamic force and torque exerted on the two particles as they approach each other prior to collision. For example, in the case of two solid particles in relative motion heading toward collision in water, for the particles to touch each other, the layer of water (or any other fluid that surrounds the particles) must be drained for the contact to occur. This squeezing of the fluid layer as the gap between the particles becomes very small leads to a lubrication flow due to the vanishing Reynolds number of the flow within the gap. The resulting lubrication force on the particles can be expected to exhibit the logarithmic singularity that we encountered in the zero Reynolds number analysis of particle–wall interaction in Section 8.2. In fact, particle–wall interaction is just a limiting case of particle–particle interaction with the radius of one of the particles tending to very large values.

The importance of particle–particle lubrication force at very small separation distances is primarily in the context of collisional analysis. The pre-collisional velocities of two particles that are rapidly approaching each other toward collision can be greatly altered by a strong lubrication force. This in turn will significantly modify the post-collisional motion of the two particles.

The hydrodynamic interaction between a pair of particles will in general depend on the translational and rotational motions of the two particles, along with the details of the ambient flow that approaches the particle pair. This general problem will be considered in the next section. In this section, by limiting attention to only situations where the particles are in relative motion at very small separation distances between them, we simplify the problem in two significant ways. First, at small separation distances, the hydrodynamic force on the particles is primarily controlled by the inner flow that occurs within the small gap. To leading order this inner flow is dependent only on the translational and rotational motions of the two particles and is independent of the outer ambient flow. Second, even if the outer ambient flow is of finite Reynolds number, the inner flow can be studied in the Stokes limit. These two simplifications allow us to obtain closed-form solutions for the lubrication force and torque on the interacting particles.

The lubrication force between two spherical particles in relative motion was considered by many researchers (Jeffrey, 1915; Goldman et al., 1966; Cooley and O'Neill, 1969; O'Neill and Majumdar, 1970a,b). As in previous chapters, due to the linear nature of the governing Stokes equation, the problem of flow induced by the relative motion between two particles can be separated into four elementary problems, whose solutions can be superposed to obtain the overall solution. These four elementary problems will be briefly described below.

Problem I: Parallel Rotation

Two particles rotating at constant angular velocities about the line joining the particle centers. The angular velocities of the two particles can be different. The flow in this case is axisymmetric and the particles only experience a torque; the direction of the torque is along the line joining the particle centers. An exact solution for this problem was presented by Jeffrey (1915). It was shown that the torque on the particle scales as $\delta \ln \delta$, where $\delta = (d_{jk} - R_{pj} - R_{pk})$ is the nondimensional gap between the particle pair, with d_{jk} being the distance between the particle centers.

Problem II: Parallel Translation

This is the case of two particles in steady translation along their line of separation. For the lubrication analysis within the small gap between the particles, only their relative velocity is of importance. The problem of one sphere approaching another stationary sphere along their line of separation was considered by Cooley and O'Neill (1969). The resulting flow is axisymmetric and a matched asymptotic approach was used to obtain the solution. An inner solution was obtained that is valid within the narrow gap region and an outer solution that is valid on the larger scale containing both the spheres. The analytical approach was quite similar to that presented in Section 8.2.1. They showed that the leading-order force for a small gap between the particles has more than logarithmic singularity and can be expressed as (see Simeonov and Calantoni, 2012)

$$F = 6\pi\mu_f R_p W_0 \left(\frac{1}{4\tilde{\delta}} - \frac{9}{40} \ln(\tilde{\delta}) \right), \qquad (9.34)$$

where $\tilde{\delta} = \delta/R_p$ and for simplicity we have assumed the particles to be of the same radius. Here, W_0 is the relative velocity at which the particles are moving toward each other along the line of separation. In this case the force is repulsive and tends to slow down the approach. Note that the force increases as $\tilde{\delta}^{-1}$ and thus tends to decelerate and slow down the collision. If the particles are moving apart, then the force is attractive and again slows down the speed of separation.

We now present the contribution to lubrication force as discussed in Simeonov and Calantoni (2012) and Sierakowski and Prosperetti (2016). The normal force on the jth particle due to the relative translational motion of the kth particle is given by the following expression:

$$\frac{\mathbf{F}_{j,\mathrm{II}}}{6\pi\mu_f R_{pj}} = \left(\frac{\lambda^2}{(1+\lambda)^2} \left(\frac{1}{\tilde{\delta}} - \frac{1}{\tilde{\delta}_o} \right) - \frac{\lambda(1+7\lambda+\lambda^2)}{5(1+\lambda)^3} \ln\frac{\tilde{\delta}}{\tilde{\delta}_o} \right) (\mathbf{W}_{jk} \cdot \mathbf{n}_{jk})\mathbf{n}_{jk}, \quad (9.35)$$

where $\mathbf{W}_{jk} = \mathbf{V}_k - \mathbf{V}_j$ is the relative velocity and $\lambda = R_{pk}/R_{pj}$ is the radius ratio. Here the notation $\mathbf{F}_{j,\mathrm{II}}$ corresponds to force on the jth particle in the second problem. The above equation is an extension of the analytical result by Cooley and O'Neill (1969) in order to allow smooth numerical implementation of the lubrication force in the context of Euler–Lagrange simulations. The above lubrication force has been tailored such that it is identically zero when the separation distance increases to $\tilde{\delta}_o$, which is a parameter to be suitably chosen such that at distances beyond this, lubrication force is less important. Also, the above force must be applied only for separation distances larger than an inner threshold $\tilde{\delta}_i$, since otherwise the force becomes singular as $\tilde{\delta} \to 0$. Thus, the above expression is applicable only in the range $\tilde{\delta}_i \le \tilde{\delta} \le \tilde{\delta}_o$. The value of the inner limit $\tilde{\delta}_i$ is dictated by the surface roughness of the particles, since the gap between the particles cannot be smaller than this threshold. Similarly, Simeonov and Calantoni (2012) recommend the value of $\tilde{\delta}_0$ to be of the order of two or three grid spacings, since the lubrication force is not resolved on length scales smaller than the grid. Note that there is no torque induced on the particles as a result of their relative normal translational motion.

Problem III: Perpendicular Rotation

The third problem considers two particles at constant angular velocity with the axis of rotation of both the particles being perpendicular to the line of separation. This problem, along with the next one, was considered by O'Neill and Majumdar (1970a), who solved the problem using bi-spherical coordinates. The resulting solution was then simplified for the limit of particles almost in contact (O'Neill and Majumdar, 1970b). The force and torque on the jth particle are given by

$$
\begin{aligned}
\frac{\mathbf{F}_{j,\mathrm{III}}}{6\pi\mu_f R_{pj}^2} &= \frac{2\lambda^2}{15(1+\lambda)^2}\left(\boldsymbol{\Omega}_{jk} + \frac{4}{\lambda}\boldsymbol{\Omega}_j + 4\lambda\boldsymbol{\Omega}_k\right) \times \mathbf{n}_{jk}\ln\frac{\tilde{\delta}}{\tilde{\delta}_o}, \\
\frac{\mathbf{T}_{j,\mathrm{III}}}{8\pi\mu_f R_{pj}^3} &= \frac{2\lambda}{5(1+\lambda)}\left(\boldsymbol{\Omega}_j + \frac{\lambda}{4}\boldsymbol{\Omega}_k - \left(\boldsymbol{\Omega}_j + \frac{\lambda}{4}\boldsymbol{\Omega}_k\right)\cdot\mathbf{n}_{jk}\,\mathbf{n}_{jk}\right)\ln\frac{\tilde{\delta}}{\tilde{\delta}_o},
\end{aligned}
\tag{9.36}
$$

where relative rotation is defined as $\boldsymbol{\Omega}_{jk} = \boldsymbol{\Omega}_j + \boldsymbol{\Omega}_k$. Again the above expression is applicable only for nondimensional separation distances in the range $\tilde{\delta}_i \le \tilde{\delta} \le \tilde{\delta}_o$. Thus, the force and torque go to zero at the outer limit of $\tilde{\delta} = \tilde{\delta}_o$ and increase to large values as defined by the inner roughness limit of $\tilde{\delta} = \tilde{\delta}_i$. As pointed out by Simeonov and Calantoni (2012), both the force and torque have no component along the line of separation. The force and torque components $\mathbf{F}_{k,\mathrm{III}}$ and $\mathbf{T}_{k,\mathrm{III}}$ can be computed using Eq. (9.36) with indices j and k swapped.

Problem IV: Perpendicular Translation

The final problem is to consider two particles in relative translational motion, where the direction of motion of the two particles is perpendicular to the line of separation. Again the solution of this problem was presented in O'Neill and Majumdar (1970a,b). They expressed the force and torque on the jth particle as

$$
\begin{aligned}
\frac{\mathbf{F}_{j,\mathrm{IV}}}{6\pi\mu_f R_{pj}} &= -\frac{4\lambda\left(2+\lambda+2\lambda^2\right)}{15(1+\lambda)^3}\left(\mathbf{W}_{jk} - \left(\mathbf{W}_{jk}\cdot\mathbf{n}_{jk}\right)\mathbf{n}_{jk}\right)\ln\frac{\tilde{\delta}}{\tilde{\delta}_o}, \\
\frac{\mathbf{T}_{j,\mathrm{IV}}}{8\pi\mu_f R_{pj}^2} &= -\frac{\lambda(4+\lambda)}{10(1+\lambda)^2}\left(\mathbf{n}_{jk}\times\mathbf{W}_{jk}\right)\ln\frac{\tilde{\delta}}{\tilde{\delta}_o}.
\end{aligned}
\tag{9.37}
$$

From the definition of relative translational velocity, the force and torque on the kth particle can readily be evaluated as $\mathbf{F}_{k,\mathrm{II}} = -\mathbf{F}_{j,\mathrm{II}}$, $\mathbf{F}_{k,\mathrm{IV}} = -\mathbf{F}_{j,\mathrm{IV}}$, and $\mathbf{T}_{k,\mathrm{IV}} = \mathbf{T}_{j,\mathrm{IV}}$.

The solutions of the above four problems can be analyzed. While the flows in the first two are axisymmetric, the latter two solutions are complex and three-dimensional. Problems I and III that pertain to particle rotation are time independent, while Problems II and IV that involve relative translational motion of the two particles are time dependent, since particle separation $\tilde{\delta}$ changes with time. However, the above lubrication analysis has been based on steady approximation. Furthermore, in all four problems, the translational or rotational velocities of the particles are taken to be time-independent constants. In other words, translational and rotational accelerations of the particles are not considered. The hydrodynamic force and torque on the two particles computed in the four problems will be superposed to obtain the overall solution.

Let the translational and rotational velocities of the two particles be $(\mathbf{V}_j, \boldsymbol{\Omega}_j)$ and $(\mathbf{V}_k, \boldsymbol{\Omega}_k)$. The force and torque on the two particles must be expressed in terms of these quantities and other properties of the fluid and the particles. Following Simeonov and

Calantoni (2012), the overall force and torque on the jth particle are expressed as the sum of individual contributions from the last three problems:

$$\begin{aligned}
\mathbf{F}_{j,\text{lub}} &= \mathbf{F}_{j,\text{II}} + \mathbf{F}_{j,\text{III}} + \mathbf{F}_{j,\text{IV}}, \\
\mathbf{T}_{j,\text{lub}} &= \mathbf{T}_{j,\text{III}} + \mathbf{T}_{j,\text{IV}}.
\end{aligned} \tag{9.38}$$

The torque contribution from the rotational motion of the particles discussed in Problem I has been ignored in the above. This torque is generally weak and more importantly its direct effect is only on the rotational motion of the particles with the axis parallel to the line of separation. As discussed in the hard and soft-sphere collisional models presented in the previous section, this component of rotational motion is not of importance in predicting the post-collisional motion of the particles.

In the case of monodisperse particles, the above equations can be simplified by setting $\lambda = 1$. On the other hand, for the case of particle–wall interaction, the above lubrication force expressions with $\lambda \to \infty$ are consistent with those presented in Problems I to V of Section 8.2.

9.3   Flow-Mediated Particle–Particle Interaction

In this section, we investigate hydrodynamic interaction between two particles that are at a distance from each other. Since the particles are not in contact, the influence of one particle on the other is mediated by the fluid that is present in between. The first of two common scenarios of interaction is the configuration of two particles in tandem, one behind the other, in a cross flow. In this arrangement the downstream particle is in the wake of the upstream particle and as a result can be expected to have a lower drag force, since it is sheltered from the ambient flow. Interestingly, due to the presence of the downstream neighbor, the drag on the upstream particle also decreases, since the wake pressure of the upstream particle increases due to the presence of the downstream neighbor. The second common arrangement that is widely studied is the side-by-side arrangement, where the wakes of the two particles interact and alter the drag force on both the particles and also introduce a nonzero lift force.

The difference between the investigations of this section versus those of the previous section is simple. In the previous section the two particles were considered to be very close to each other and the focus was on the fluid mechanics of the thin layer of fluid within the gap and its consequences on the force and torque on the particle. In contrast, in this chapter, the two particles will not be constrained to be very close to each other. This small difference, however, makes the problems of this section far more complex.

From a geometric point of view, while the lubrication flow problem could be studied in the inner region of the narrow gap, now we must study the flow on the scale of the two particles and their wakes, which can extend over several particle diameters. In addition, the investigations to be pursued in this section cannot be limited to Stokes flow, as the interaction between the two particles and their wakes depends on the Reynolds number of the flow around the particles. Furthermore, the interaction between the particles cannot be described purely in terms of the translational and rotational velocities of the two particles. It also depends on the strength and nature of the ambient flow that approaches the particle pair.

There have been many theoretical, experimental and numerical investigations of the fluid-mediated particle–particle interaction problem over the past few decades. The theoretical investigations have been limited to the Stokes limit, which we will consider first. The finite Reynolds number studies have generally been experimental and computational. Understandably, the finite-Re regime is far more complex and the studies have not been as comprehensive as in the Stokes flow regime. Nevertheless, we have gained significant understanding which will be covered next.

9.3.1 Zero Reynolds Number Limit – Comprehensive Analysis

Here we seek an exact solution for the general problem of a pair of particles subjected to an ambient flow in the Stokes flow regime. Let the jth particle be located at $\mathbf{X}_j$ moving at velocity $\mathbf{V}_j$ and rotating at an angular velocity $\mathbf{\Omega}_j$. Simultaneously let the kth particle be located at $\mathbf{X}_k$ and move at translational velocity $\mathbf{V}_k$ and angular velocity $\mathbf{\Omega}_k$. Let both particles be subjected to an ambient flow of velocity $\mathbf{U}_0$ and angular velocity $\mathbf{\Omega}_0$. In this general context, we are interested in the hydrodynamic force and torque exerted on the two particles. Many aspects of this problem have been studied for many decades using lubrication theory, the method of reflections, and bi-spherical coordinates (Smoluchowski, 1911; Stimson and Jeffery, 1926; Goldman et al., 1966; Lin et al., 1970a; O'Neill and Majumdar, 1970a,b). These works culminated in the comprehensive analysis of Jeffrey and Onishi (1984), who provided the complete solution of the above-stated general problem. They used an elegant twin-multipole expansion approach that yielded results that were cast in a convenient form for subsequent usage. Their analysis and results will be briefly discussed below.

The tractional force distribution on the surface of the two particles can be integrated as defined in Eq. (4.1) to obtain the net force and torque on the two particles. Due to the linearity of the Stokes flow, the dependence of the forces $\mathbf{F}_j$ and $\mathbf{F}_k$ and the corresponding torques $\mathbf{T}_j$ and $\mathbf{T}_k$ on the relative translational and angular velocities of the two particles can be expressed in a compact manner in matrix form as follows:

$$\begin{bmatrix} \mathbf{F}_j \\ \mathbf{F}_k \\ \mathbf{T}_j \\ \mathbf{T}_k \end{bmatrix} = \begin{bmatrix} \mathbf{A}^{jj} & \mathbf{A}^{jk} & \tilde{\mathbf{B}}^{jj} & \tilde{\mathbf{B}}^{jk} \\ \mathbf{A}^{kj} & \mathbf{A}^{kk} & \tilde{\mathbf{B}}^{kj} & \tilde{\mathbf{B}}^{kk} \\ \mathbf{B}^{jj} & \mathbf{B}^{jk} & \mathbf{C}^{jj} & \mathbf{C}^{jk} \\ \mathbf{B}^{kj} & \mathbf{B}^{kk} & \mathbf{C}^{kj} & \mathbf{C}^{kk} \end{bmatrix} \begin{bmatrix} \mathbf{U}_{0j} - \mathbf{V}_j \\ \mathbf{U}_{0k} - \mathbf{V}_k \\ \mathbf{\Omega}_0 - \mathbf{\Omega}_j \\ \mathbf{\Omega}_0 - \mathbf{\Omega}_k \end{bmatrix}, \qquad (9.39)$$

where $\mathbf{U}_{0j}$ is the ambient fluid velocity at the location of the jth particle and similarly $\mathbf{U}_{0k}$ is the ambient fluid velocity at the kth particle, and they include the effects of both ambient translation $\mathbf{U}_0$ and ambient rotation $\mathbf{\Omega}_0$. Here the 3×3 matrices $\mathbf{A}^{jj}$ and $\mathbf{A}^{kk}$ present the linear relation between the hydrodynamic force on a particle to its slip velocity, while matrices $\mathbf{A}^{jk}$ and $\mathbf{A}^{kj}$ account for the influence of the slip velocity of the kth particle on the hydrodynamic force of the jth particles and vice versa. Similarly, the 3×3 matrices $\mathbf{C}^{jj}$ and $\mathbf{C}^{kk}$ present the linear relation between the hydrodynamic torque on a particle to its rotational slip, while the matrices $\mathbf{C}^{jk}$ and $\mathbf{C}^{kj}$ account for the influence of the rotational slip of the kth particle on the hydrodynamic torque of the jth particles and vice versa. The **B** matrices correspond to the cross-coupling between the translation slip and the hydrodynamic torque, and the **B̃** matrices correspond to the cross-coupling between the rotational slip and the force. The overall matrix is called the *resistance matrix* and it provides the force and torque in terms of the particle motion relative to the ambient fluid. In the limit where the two particles are so far away that they stop influencing each other, the resistance matrix becomes strictly diagonal.

As such, the resistance matrix consists of 12 rows and 12 columns and thus contains 144 functions. Each of these functions depends on quantities such as the radius of the two particles and their separation. But symmetries can be exploited to greatly simplify the problem, which will be described below. From the Lorentz reciprocal theorem it can be shown that the resistance matrix is symmetric (Jeffrey and Onishi, 1984), yielding

$$A^{jk}_{lm} = A^{kj}_{ml}, \quad \tilde{B}^{jk}_{lm} = B^{kj}_{ml}, \quad C^{jk}_{lm} = C^{kj}_{ml}, \tag{9.40}$$

where we have used index notation with $\{l, m\} = 1, 2, 3$ and the superscripts j and k indicate the jth and kth particles and they are not tensor indices.

The elements of the resistance matrix are functions of the three variables: the radius of the two particles R_{pj} and R_{pk}, and the separation vector between them $\mathbf{r} = \mathbf{X}_k - \mathbf{X}_j$. Swapping the roles of particles j and k yields the symmetry relations

$$\begin{aligned} \mathbf{A}^{jj}(\mathbf{r}, R_{pj}, R_{pk}) &= \mathbf{A}^{kk}(-\mathbf{r}, R_{pk}, R_{pj}), \\ \mathbf{A}^{jk}(\mathbf{r}, R_{pj}, R_{pk}) &= \mathbf{A}^{kj}(-\mathbf{r}, R_{pk}, R_{pj}), \end{aligned} \tag{9.41}$$

with similar relations for the **B** and **C** matrices.

The next simplification is through the recognition of the fact that each 3×3 tensor of the resistance matrix is only an axisymmetric function of the separation unit vector $\mathbf{n}_{jk}$ that points from the jth particle to the kth particle. Due to axisymmetry, each of the **A**, **B**, **B̃**, and **C** matrices can be written in terms of at most two scalar functions:

$$\begin{aligned} \frac{A^{\alpha\beta}_{lm}}{3\pi(R_{p\alpha} + R_{p\beta})} &= f^{\alpha\beta}_1 n_l n_m + f^{\alpha\beta}_2 (\delta_{lm} - n_l n_m), \\ \frac{B^{\alpha\beta}_{ml}}{\pi(R_{p\alpha} + R_{p\beta})^2} &= \frac{\tilde{B}^{\beta\alpha}_{ml}}{\pi(R_{p\alpha} + R_{p\beta})^2} = f^{\alpha\beta}_3 \, \epsilon_{lmn} n_n, \\ \frac{C^{\alpha\beta}_{lm}}{\pi(R_{p\alpha} + R_{p\beta})^3} &= f^{\alpha\beta}_4 n_l n_m + f^{\alpha\beta}_5 (\delta_{lm} - n_l n_m), \end{aligned} \tag{9.42}$$

where δ_{lm} is the Kronecker delta and ϵ_{lmn} is the Levi-Civita third-rank tensor (see Appendix A). Also, n_l is the lth component of $\mathbf{n}_{jk}$. Again note that $\{\alpha, \beta\} = \{j \text{ or } k\}$ are not tensor indices. They just denote the particle pair.

The functions f_1 to f_5 have been nondimensionalized with appropriate scaling. These nondimensional functions depend only on the following two nondimensional variables:

$$s = \frac{2|\mathbf{r}|}{R_{pj} + R_{pk}}, \quad \lambda = \frac{R_{pk}}{R_{pj}}, \tag{9.43}$$

where the nondimensional distance can take values in the range $2 \leq s \leq \infty$ and the radius ratio can take values in the range $0 \leq \lambda \leq \infty$. Thus, for complete information of the resistance matrix, the following ten functions must be obtained over the domain of the two variables:

$$
\begin{aligned}
&f_1^{jj}(s,\lambda), \quad f_2^{jj}(s,\lambda), \quad f_3^{jj}(s,\lambda), \quad f_4^{jj}(s,\lambda), \quad f_5^{jj}(s,\lambda), \\
&f_1^{jk}(s,\lambda), \quad f_2^{jk}(s,\lambda), \quad f_3^{jk}(s,\lambda), \quad f_4^{jk}(s,\lambda), \quad f_5^{jk}(s,\lambda).
\end{aligned}
\tag{9.44}
$$

In their paper, Jeffrey and Onishi (1984) obtained explicit expressions for the above ten functions by analytically solving the following five pairs of elementary problems. Each pair of elementary problems was solved using twin-multipole expansion and the resulting summation was analytically evaluated for large separation between the particles (i.e., for $s \gg 2$) and for nearly touching spheres (i.e., for $s - 2 \ll 1$). Numerical solutions were obtained for arbitrary values of intermediate distances. The five pairs of elementary problems are as follows (also see the schematic in Figure 9.3):

- In the first problem of the first pair, the j and k particles translate toward each other along their line of separation at equal velocity. In the second problem, the j and k particles translate in the same direction along their line of separation at equal velocity. The particles are allowed only translational motion and they do not rotate. As a result, they experience only hydrodynamic force with no associated torque. This solution pair is used to obtain f_1^{jj} and f_1^{jk}.

- In the second pair, the j and k particles translate along two parallel lines that are perpendicular to the line of separation between the two particles. In the first problem of the pair, the two particles travel in opposite directions at equal velocity along the parallel lines, while in the second problem they travel in the same direction. Again the particles are allowed only translational motion and they do not rotate. The hydrodynamic force on the particle pair is used to obtain f_2^{jj} and f_2^{jk}.

- The third elementary pair is in fact the same as the second pair. Note that in this case when the two particles move perpendicular to their line of separation, they will experience a nonzero torque in addition to the hydrodynamic force. This hydrodynamic torque on the particle pair is used to obtain f_3^{jj} and f_3^{jk}. As can be seen in Eq. (9.42), the f_3 functions represent the cross-coupling between translation (or rotational) motion with torque (or force).

- In the first problem of the fourth pair, the j and k particles rotate in opposite directions at equal angular velocity with their axis of rotation aligned with their line of separation. In the second problem, the j and k particles rotate in the same direction. The particles are allowed only rotational motion and they do not translate. As a result, they experience only hydrodynamic torque with no associated net force on the particles. This solution pair is used to obtain f_4^{jj} and f_4^{jk}.

- In the final pair of elementary problems, the j and k particles rotate with their axis of rotation aligned along two parallel lines that are perpendicular to the line of separation between the two particles. In the first problem of the pair, the two particles rotate in opposite directions at equal angular velocity, while in the second problem, they rotate in the same direction. Again, the particles are allowed only

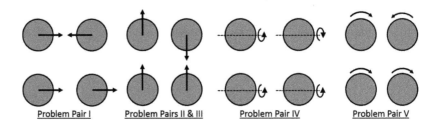

Figure 9.3 Schematic of five pairs of elementary problems where the two spheres undergo specific translational rotational motion.

rotational motion and they do not translate. The resulting hydrodynamic torque on the particle pair is used to obtain f_5^{jj} and f_5^{jk}.

As pointed out by Jeffrey and Onishi (1984), aspects of these elementary problems have been solved by prior researchers. But a composite comprehensive solution was presented in their work. The explicit expressions for the ten functions are quite long and tedious and therefore will not be repeated here. The reader is referred to Jeffrey's paper. The other functions can be obtained from the symmetry conditions as

$$
\begin{aligned}
f_1^{kk}(s,\lambda) &= f_1^{jj}(s,1/\lambda), & f_1^{kj}(s,\lambda) &= f_1^{jk}(s,1/\lambda),\\
f_2^{kk}(s,\lambda) &= f_2^{jj}(s,1/\lambda), & f_2^{kj}(s,\lambda) &= f_2^{jk}(s,1/\lambda),\\
f_3^{kk}(s,\lambda) &= -f_3^{jj}(s,1/\lambda), & f_3^{kj}(s,\lambda) &= -f_3^{jk}(s,1/\lambda),\\
f_4^{kk}(s,\lambda) &= f_4^{jj}(s,1/\lambda), & f_4^{kj}(s,\lambda) &= f_4^{jk}(s,1/\lambda),\\
f_5^{kk}(s,\lambda) &= f_5^{jj}(s,1/\lambda), & f_5^{kj}(s,\lambda) &= f_5^{jk}(s,1/\lambda).
\end{aligned}
\tag{9.45}
$$

We now complete our discussion with the introduction of the related *mobility matrix*, which expresses the translational motion and rotation of the particle pair in terms of their hydrodynamic forces and torque as

$$
\begin{bmatrix}
\mathbf{U}_{0j} - \mathbf{V}_j\\
\mathbf{U}_{0k} - \mathbf{V}_k\\
\boldsymbol{\Omega}_0 - \boldsymbol{\Omega}_j\\
\boldsymbol{\Omega}_0 - \boldsymbol{\Omega}_k
\end{bmatrix}
=
\begin{bmatrix}
\mathbf{a}^{jj} & \mathbf{a}^{jk} & \tilde{\mathbf{b}}^{jj} & \tilde{\mathbf{b}}^{jk}\\
\mathbf{a}^{kj} & \mathbf{a}^{kk} & \tilde{\mathbf{b}}^{kj} & \tilde{\mathbf{b}}^{kk}\\
\mathbf{b}^{jj} & \mathbf{b}^{jk} & \mathbf{c}^{jj} & \mathbf{c}^{jk}\\
\mathbf{b}^{kj} & \mathbf{b}^{kk} & \mathbf{c}^{kj} & \mathbf{c}^{kk}
\end{bmatrix}
\begin{bmatrix}
\mathbf{F}_j\\
\mathbf{F}_k\\
\mathbf{T}_j\\
\mathbf{T}_k
\end{bmatrix}.
\tag{9.46}
$$

While the resistance matrix allows evaluation of the hydrodynamic force and torque on a particle pair resulting from their translational and rotational motion, the mobility matrix allows evaluation of the converse. From their definitions it is quite clear that the mobility matrix is the inverse of the resistance matrix and thus knowing one is equivalent to knowing the other.

Although we have introduced the resistance and mobility matrices in the context of fluid-mediated interaction of a particle pair, their utility extends far beyond this. For example, we can consider the resistance and mobility matrices of an isolated particle in an ambient flow. In this case, given the translational and rotational relative motion of the isolated particle, the resistance matrix can be used to calculate the resulting force and torque on the particle. As explained earlier, in the zero Reynolds number limit, the resistance matrix will simply be diagonal with no coupling between the translational and rotational motion. However, this is not so for a nonspherical particle or for a sphere at finite Re. On the other hand, the resistance and mobility matrices can be extended to include interaction between three particles, four particles, and so on. With increasing number of particles, the matrices will expand in size and a larger number of functions must be computed with many more elementary problems.

9.3.2   Finite Reynolds Number Studies

Except for the above zero Reynolds number theoretical studies, most other finite-Re investigations have been primarily in the limit of the particle pair being stationary relative to each other. With the relative position of the two particles being fixed, they are subjected to varying ambient flows in order to investigate their flow-mediated interaction. If the ambient flow is in the direction of particle separation, then the two particles are in tandem arrangement, while a flow that is perpendicular to the line of separation will constitute the side-by-side arrangement. Thus, unlike in the lubrication limit, flow-mediated finite-Re particle–particle interaction investigations emphasize the nature of the ambient flow part at the expense of not considering the relative translational and rotational motion of the particles during the interaction process. In essence, our understanding of particle pair interaction at finite Re is limited, incomplete, and not as comprehensive as in the Stokes regime.

Due to the large number of experiments and simulations of flow around a pair of spherical particles, only a selected set of investigations will be reviewed here. One of the earliest investigations of interaction between spherical particles in tandem and side-by-side arrangements in cross flow was by Tsuji et al. (1982). They considered both spheres to be of the same size and thus the two key parameters in both these arrangements are the particle Reynolds number (based on the relative velocity between the particle and the ambient flow and the particle diameter) and the separation distance between the sphere centers. They made detailed measurements on flow separation, onset of vortex shedding, and the hydrodynamic force on the particles. This initial study was followed by many additional experimental investigations of both the tandem and side-by-side arrangements (see, e.g., Zhu et al., 1994; Liang et al., 1996). The effect of unequal size of the two particles has also been studied by Zhang et al. (2005).

With the widespread use of numerical simulations, computational investigations of uniform flow past two particles have also yielded valuable information (Kim et al., 1993; Chen and Wu, 2000; Zou et al., 2005). In particular, recent investigations (Prahl et al., 2007; Akiki et al., 2017b) have considered not just tandem or side-by-side arrangements, but all possible close-by arrangements of the two particles. If we consider one of the particles (the jth particle) to be the reference particle with the ambient uniform flow going over it from left to right, then the position of the second particle (the kth particle) can be varied in relation to the reference particle as shown in Figure 9.4. The relative position of the kth particle needs to be varied only over a plane, since the mutual influence of the particle pair is axisymmetric about the ambient flow

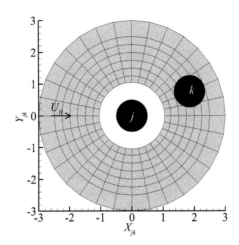

Figure 9.4 Schematic of the two-sphere arrangement. The mesh shows the position of the neighboring sphere for which direct numerical simulation was performed and used to evaluate the hydrodynamic force and torque. Reprinted from Akiki et al. (2017b), with permission from Elsevier.

direction. In Figure 9.4 this relative position is denoted by the pair (X_{jk}, Y_{jk}), which is systematically varied over a region of three diameters in the neighborhood of the reference particle. Also note that in the present case where the particles are of the same size, the center of the kth particle cannot be located very close to the black-shaded reference particle. This excluded region is indicated by the white annular band around the reference particle, where there is no color contour.

The results obtained by Akiki et al. (2017b) are summarized in Figures 9.5, 9.6, and 9.7, respectively. The contours plotted in these figures must be interpreted in the following way. The force or torque on the kth particle whose center is given by the position (X_{jk}, Y_{jk}) is indicated by the contour value. In Figure 9.5, contours of perturbation streamwise force (or drag) are plotted for four different Reynolds numbers. Here, perturbation is from the drag force on an isolated particle at the same Reynolds number and the nondimensionalization is with a force scaling of $4\rho_f R_P^2 |\mathbf{u}_r|^2$, where $\mathbf{u}_r$ is the relative velocity.

Several important observations can be made. The perturbation effect of the reference particle on the drag force of the kth particle seems to vary smoothly over the Reynolds number range considered. Also, white regions indicate that the perturbation effect is locally less than 5%. Thus, except in the wake region of the reference particle, the perturbation effect dies off very fast away from the reference particle. As expected, the perturbation drag on the kth particle is negative (which indicates that the drag decreases compared to that of an isolated particle) when it is located downstream and sheltered by the reference particle. This drag reduction effect decreases both in intensity and extent as the Reynolds number increases. Note that the contours in this figure are truncated at three diameters and it is clear that the wake effect extends farther in the streamwise direction. Interestingly, the drag on the kth particle decreases even

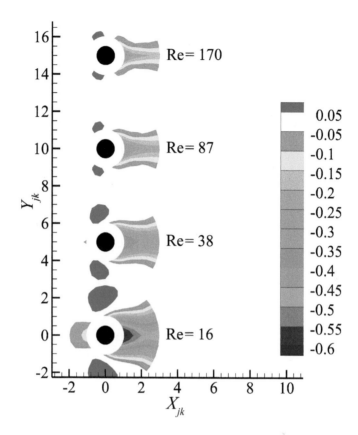

Figure 9.5 Contour plot of the nondimensional streamwise force perturbation on the kth particle whose center is located at (X_{jk}, Y_{jk}) with respect to the reference jth particle located at the origin (marked by the dark circle). Both particles are held fixed in a uniform ambient flow directed along the positive x direction. Results for four different Reynolds numbers are shown. Since the center of the kth particle cannot get closer than one diameter, the contours are absent in the annular region around the center particle. Reprinted from Akiki et al. (2017b), with permission from Elsevier.

when it is located upstream of the reference particle, due to an increase in the wake pressure. But at 5% level this effect can barely be seen at Re = 38 and disappears at even higher Re. The blockage effect of the reference particle is to increase the fluid velocity around it. This contributes to an increase in the drag on the kth particle when it is in the side-by-side arrangement. This increase in drag is weaker than the drag reduction effect of the wake, and again the extent of drag increase shrinks rapidly with increasing Re.

The effect of the reference particle in terms of the lift force on the kth particle is generally negligible in the wake region of the reference particle. However, if the kth particle is located upstream of the reference particle, then the lift (or transverse) force is such that the kth particle is pushed away from the centerline. Again, the effect of the

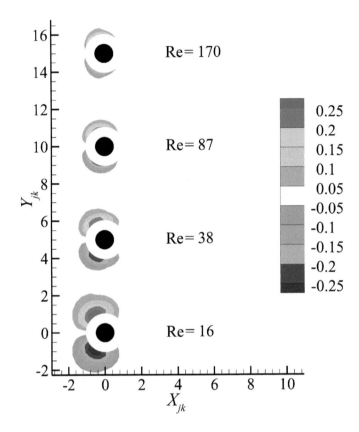

Figure 9.6 Contour plot of the nondimensional transverse y force on the kth particle whose center is located at (X_{jk}, Y_{jk}) with respect to the reference jth particle located at the origin (marked by the dark circle). Both particles are held fixed in a uniform ambient flow directed along the positive x-direction. Results for four different Reynolds numbers are shown. Reprinted from Akiki et al. (2017b), with permission from Elsevier.

reference particle decreases with Re and in all cases to 5% level the influence of the particle pair does not extend beyond little more than two diameters.

Finally, the hydrodynamic torque exerted on the kth particle can be similarly explained with Figure 9.7. Unlike the lift force, the induced torque is primarily limited to the leeward side of the reference particle. The effect of the reference particle is to induce a clockwise torque (as viewed in the picture) on the kth particle located above the centerline and a counter-clockwise torque if located below the centerline. Again the influence of the reference particle decreases with increasing Re and almost goes below 1% for Re $\geq$ 170. Even at low Re, the influence is contained within a region of little more than two diameters.

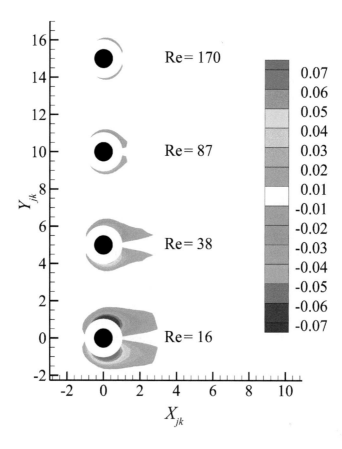

Figure 9.7 Contour plot of the nondimensional z-torque on the kth particle whose center is located at (X_{jk}, Y_{jk}) with respect to the reference jth particle located at the origin (marked by the dark circle). Both particles are held fixed in a uniform ambient flow directed along the positive x-direction. Results for four different Reynolds numbers are shown. Reprinted from Akiki et al. (2017b), with permission from Elsevier.

9.4   Volume Fraction Effect on Drag in the Stokes Flow Limit

So far the focus of this chapter has been interaction between two particles or a particle and a wall. We are now ready to dive into the more complex N-body problem, where there are many simultaneously interacting particles within the fluid. For the most part, our investigation will be restricted to monodisperse systems, since this greatly simplifies the analysis. But the more common scenario of particles of varying size can also be conceptually treated in a similar manner, albeit the analysis will become far more involved. Furthermore, the examples to be considered in this section are in the Stokes limit, where an analytical solution will be obtained. The next section will consider the volume fraction effect at finite Re.

Our interest in this section is to extend the Stokes drag law of an isolated particle to a dispersion of particles. In the zero Reynolds number limit, it is expected that the relation between drag and relative velocity of a particle will change due to the presence and interaction of all the other particles within the dispersion. Similarly, such interparticle interaction will also alter the standard drag relation of a particle in the finite Reynolds number regime. Thus, we seek to obtain drag relations that depend not only on the particle Reynolds number, but also on the particle volume fraction. That is, the correction to the Stokes drag Φ introduced in Eq. (4.141) will now be a function of Re and ϕ_d. In this section, we will obtain analytic expressions for Φ in the Stokes limit of Re = 0, and these results will be extended to finite Re in the following section.

An important distinction must, however, be made. In a random distribution of particles, the force on an individual particle is a random variable, since each particle is surrounded by a unique distribution of neighbors and thus is influenced in a distinct way. In our investigation of the particle–particle interaction effect, we are first interested in the average behavior. Instead of evaluating the force on individual particles within the distribution, we are interested in evaluating the average drag of the distribution as a function of average particle volume fraction within the array. Only later, in Section 9.5.1, will we address particle-to-particle variation in the force.

As in the previous sections, we start our investigation in the Stokes regime. Rigorous analysis and closed-form results are possible only in this zero Reynolds number limit. Even then, much of the early investigations yielded conflicting results that created confusion and controversy. Subsequent seminal studies by Batchelor (1972, 1974) and Saffman (1973), among others, have clarified the complex nature of interparticle interaction, and conclusively resolved the conflicts. Consider the following three related problems:

- First, flow past an infinite structured array of stationary particles arranged in a lattice. In the limit of small-particle volume fraction, the drag increases with volume fraction and the correction function goes as $\Phi(\phi_d) \sim 1/(1 - c\phi_d^{1/3})$, where the constant c depends on the nature of the structured array. In case of a simple cubic array, $c = 1.76$ (Hasimoto, 1959). Note that in this structured arrangement every particle experiences the same drag.

- Second, flow past a random array of stationary particles. The only difference from the previous problem is in the unstructured random particle arrangement. As a result of the random arrangement, for the same macroscale ambient flow, each particle will experience a slightly different drag. In the limit of small volume fraction, the average drag force on the particles again increases and the correction function now goes as $\Phi(\phi_d) \sim 1/(1 - 3\phi_d^{1/2}/\sqrt{2})$ (Childress, 1972; Lundgren, 1972).

- Third, flow past a random array of freely moving particles. This is the scenario of an infinite random distribution of particles settling under gravity that we considered in Section 7.4. In this case, the hydrodynamic force on each particle is the same and balances the excess weight. But at any given instance, the settling velocity of each particle will differ as determined by the spatial location of its neighbors. Again, in the limit of small volume fraction, the average settling velocity of the distribution decreases, which corresponds to an increase in drag above the Stokes drag law with a correction function of $\Phi(\phi_d) = 1/(1 - 6.55\phi_d)$ (Batchelor, 1972; Burgers, 1995).

It is fascinating that the three seemingly similar problems yield results that differ not just in their coefficient, but fundamentally in their functional dependence on ϕ_d. Note that for small ϕ_d the influence of the neighboring particles is strongest for the case of the structured array, followed by the random stationary array. In comparison, the increase in drag in the case of random distribution of freely moving particles is weak. In the following three subsections we will discuss the salient features of these three important problems.

9.4.1   Stokes Drag of a Structured Array of Particles

Different approaches to solving this problem have been presented in the literature. Here we will follow the elegant approach of Saffman (1973). Consider the particles in a cubic lattice of side $b = R_p(4\pi/3\phi_d)^{1/3}$, so that the volume fraction of the array is ϕ_d. Let $\mathbf{x}^\alpha$ for $\alpha = 1, 2, \ldots, \infty$ list all the lattice points, which are the centers of all the particles. Due to the identical flow condition seen by all the particles, they experience the same force $\mathbf{F}^\alpha = \mathbf{F}_0$, which is the average force and in the direction of the ambient flow.

To simplify the analysis, Saffman introduced the *point-particle* assumption, where each particle's effect on the fluid is replaced by a point force $-\mathbf{F}^\alpha$ applied at its center. Mathematically, the governing equation now becomes

$$\nabla \cdot \mathbf{u} = 0 \quad \text{and} \quad 0 = -\nabla p + \mu_f \nabla^2 \mathbf{u} - \sum_\alpha \mathbf{F}^\alpha \, \delta(\mathbf{x} - \mathbf{x}^\alpha). \tag{9.47}$$

By requiring that the average velocity (averaged over the entire volume) be zero, we interpret $\mathbf{u}$ as the perturbation velocity. The summation represents the sum of delta function forces centered at all the particles. Since the particles are replaced with delta function forcing, we do not need to worry about the application of no-slip and no-penetration conditions. This offers the advantage that the above equations can be solved in an unbounded domain. The drawback of this approximation is that the point force is only accurate to leading order and the effect of finite particle size (entering through dipole and higher-order terms) is neglected in the analysis. At low particle volume fraction, interparticle distances are sufficiently large that the Stokeslet solution of the point force is sufficient to accurately capture the average effect of all the neighbors. This leading-order effect of all the neighbors in perturbing the fluid velocity at any given particle will be shown to scale as $\phi_d^{1/3}$ (this can be expected since this is the scaling of interparticle distance b). In comparison, the higher-order effect of finite particle size only scales as ϕ_d and therefore has been sacrificed for the sake of simplicity of the analysis.

Three-dimensional forward and backward Fourier transforms are defined as

$$\hat{\mathbf{u}}(\mathbf{k}) = \frac{1}{(2\pi)^3} \int \mathbf{u}(\mathbf{x}) \exp(-\iota\, \mathbf{k} \cdot \mathbf{x})\, d\mathbf{x},$$

$$\mathbf{u}(\mathbf{x}) = \int \hat{\mathbf{u}}(\mathbf{k}) \exp(\iota\, \mathbf{k} \cdot \mathbf{x}) d\mathbf{k}, \tag{9.48}$$

where $\mathbf{k}$ is the wave vector and $\hat{\mathbf{u}}$ is the Fourier coefficient of velocity. Following Saffman (1973), the momentum equation (9.47) is Fourier transformed to obtain

$$0 = -\iota k \hat{p} - \mu_f |\mathbf{k}|^2 \hat{\mathbf{u}} - \frac{1}{(2\pi)^3} \sum_\alpha \mathbf{F}^\alpha \exp(-\iota\, \mathbf{k} \cdot \mathbf{x}^\alpha). \qquad (9.49)$$

Pressure is then separated into a mean component that linearly decreases along the flow direction, and a perturbation component that is homogeneous. The mean pressure gradient balances the mean drag force applied back on the fluid by the particles and thus

$$- \iota k \hat{p} = -\iota \mathbf{k} \langle \hat{p} \rangle - \iota k \hat{p}' = n_d \mathbf{F}_0 \delta(\mathbf{k}) - \iota k \hat{p}', \qquad (9.50)$$

where $n_d = 1/b^3 = 3\phi_d/(4\pi R_p^3)$ is the number density of particles. The first term on the right is the mean pressure gradient and is equal to the force on the particles per unit volume. Note that $\delta(\mathbf{k})$ indicates that this mean component is spatially uniform as indicated by the $\mathbf{k} = 0$ mode. Substituting, we obtain

$$0 = -\iota k \hat{p}' - \mu_f |\mathbf{k}|^2 \hat{\mathbf{u}} - \frac{1}{(2\pi)^3} \sum_\alpha \mathbf{F}^\alpha \exp(-\iota\, \mathbf{k} \cdot \mathbf{x}^\alpha) + n_d \mathbf{F}_0 \delta(\mathbf{k}). \qquad (9.51)$$

The role of perturbation pressure $\hat{p}'$ is to enforce a divergence-free condition (i.e., $\nabla \cdot \mathbf{u} = 0$), which when Fourier transformed becomes $\mathbf{k} \cdot \hat{\mathbf{u}} = 0$. The effect of pressure in the above equation can be taken into account in the following way. We first compute $\hat{\mathbf{u}}$ without the pressure term. The resulting velocity will not be divergence free and the effect of pressure is to project this velocity onto the divergence-free space. This projection of the velocity field onto the divergence-free space can be achieved using the projection tensor $P_{lm}(\mathbf{k}) = \delta_{lm} - k_l k_m / |\mathbf{k}|^2$. The resulting expression for the perturbation velocity is as follows:

$$\hat{u}_l = P_{lm}(\mathbf{k}) \left[\frac{1}{\mu_f |\mathbf{k}|^2} \left\{ -\frac{1}{(2\pi)^3} \sum_\alpha F_m^\alpha \exp(-\iota\, \mathbf{k} \cdot \mathbf{x}^\alpha) + n_d F_{0m} \delta(\mathbf{k}) \right\} \right], \qquad (9.52)$$

where the term within square brackets is the solution ignoring the pressure term and the projection tensor accounts for the effect of pressure. We have also switched to index notation, and therefore a sum over the index m is implied. We now proceed to evaluate the above by making use of the fact that the force on all the particles is the same and $F_m^\alpha = F_{0m}$ is a constant. Furthermore, from Poisson's summation:

$$\frac{1}{(2\pi)^3} \sum_\alpha \exp(-\iota\, \mathbf{k} \cdot \mathbf{x}^\alpha) = \frac{1}{b^3} \sum_\alpha \delta\left(\mathbf{k} - \frac{2\pi}{b} \mathbf{n}^\alpha \right), \qquad (9.53)$$

where $\mathbf{n}^\alpha$ is the integer triad that denotes the 3D lattice position of the αth particle. When the above summation is substituted into Eq. (9.52), we note that the α value corresponding to the origin (i.e., the α whose $\mathbf{x}^\alpha = 0$ or $\mathbf{n}^\alpha = 0$) cancels the second term in curly brackets to yield

$$\hat{u}_l = -\frac{P_{lm}(\mathbf{k})}{\mu_f |\mathbf{k}|^2} \frac{F_{0m}}{b^3} \sum_{\alpha'} \delta\left(\mathbf{k} - \frac{2\pi}{b} \mathbf{n}^\alpha \right), \qquad (9.54)$$

where the sum over α' indicates a sum over all α except that corresponding to $\mathbf{n}^\alpha = 0$.

The next step is to calculate the perturbation velocity at one of the particle locations, which will be chosen to be the origin. So we inverse Fourier transform the above and evaluate it at $\mathbf{x} = 0$. But, before doing that, we must isolate the perturbation velocity due to all the particles other than the one at the origin. The self-induced perturbation velocity of the point force applied at the origin is included in the above expression. This contribution must be subtracted out, for otherwise the velocity at $\mathbf{x} = 0$ will be singular. Self-induced velocity due to a point force applied at the origin is a Stokeslet, as given in Eq. (4.18). Its three-dimensional Fourier transform is given by

$$\hat{u}_l = -\frac{P_{lm}(\mathbf{k})}{\mu_f |\mathbf{k}|^2} \frac{F_{0m}}{(2\pi)^3}, \tag{9.55}$$

where the negative sign is due to the negative of $\mathbf{F}_0$ applied to the fluid. The above must first be subtracted from the right-hand side of Eq. (9.54) and then Fourier inverse transformed and evaluated at $\mathbf{x} = 0$. This can be expressed as

$$u_l'(\mathbf{x} = 0) = -\frac{F_{0m}}{\mu_f b^3} \int \frac{P_{lm}(\mathbf{k})}{|\mathbf{k}|^2} \left\{ \sum_{\alpha'} \delta\left(\mathbf{k} - \frac{2\pi}{b}\mathbf{n}^\alpha\right) - \left(\frac{b}{2\pi}\right)^3 \right\} d\mathbf{k}, \tag{9.56}$$

where we have used the fact that $\exp(-\iota \mathbf{k} \cdot \mathbf{x}) = 1$ at $\mathbf{x} = 0$. Here the prime in u_l' stands for the perturbation velocity due to all other particles excluding the self-induced perturbation. The above 3D integral over a lattice can be evaluated using techniques originally developed in the theory of ionic crystals. We will simply quote the result presented by Saffman (1973) that the integral is equal to $-\beta b^2/(6\pi^2)$. Substituting, we obtain the final result that

$$\mathbf{u}'(\mathbf{x} = 0) = \frac{\beta \mathbf{F}_0}{6\pi^2 \mu_f b}, \tag{9.57}$$

where the constant β depends only on the lattice arrangement and for a simple cubic lattice $\beta = 8.8$.

The above result must be interpreted in the following way. Suppose the average ambient fluid velocity over the stationary structured array of particles is $\mathbf{U}_0$ and it results in a particle force of $\mathbf{F}_0$, which is experienced by all the particles. The induced perturbation flow at any one of the particles due to the collective action of all the other particles is given by the above expression. This perturbation flow is in the direction of hydrodynamic force on the particle. Thus, the effective relative fluid velocity at a particle becomes $\mathbf{U}_0 + \beta \mathbf{F}_0/(6\pi^2 \mu_f b)$. Applying the Stokes drag law for this effective relative velocity, we obtain

$$6\pi \mu_f R_p \left(\mathbf{U}_0 + \frac{\beta \mathbf{F}_0}{6\pi^2 \mu_f b} \right) = \mathbf{F}_0, \tag{9.58}$$

which when rearranged results in

$$\mathbf{F}_0 = 6\pi \mu_f R_p \mathbf{U}_0 \left(\frac{1}{1 - \frac{\beta R_p}{\pi b}} \right) = 6\pi \mu_f R_p \mathbf{U}_0 \left(\frac{1}{1 - 1.76\phi_d^{1/3}} \right). \tag{9.59}$$

The above relation can be interpreted either as an increase in drag due to the influence of neighbors for a given average ambient flow, or in the context of a falling array of particles as a decrease in fall velocity when the force on each particle is equal to its excess weight over that of the displaced fluid.

9.4.2   Stokes Drag of a Stationary Random Bed of Particles

Here we consider an ambient flow of average velocity U_0 past a random bed of particles within which the particles are distributed with uniform probability. Let the particle volume fraction within the bed be uniform at ϕ_d. In the structured array of particles, the perturbation flow at every particle within the array induced by every other particle is the same and as a result, all the particles experience the same average force. This is not the case in a random array. The perturbation flow at each particle is different and as a result $\mathbf{F}^\alpha$ varies from particle to particle. We define $\mathbf{F}_0$ as the average force, averaged over the entire distribution of particles.

Our quest is to obtain the modified drag law that relates $\mathbf{F}_0$ to $\mathbf{U}_0$. The simplest way to obtaining this relation is to start with Brinkman's equation (Brinkman, 1947) for flow through a porous medium:

$$0 = -\nabla p' + \mu_f \nabla^2 \mathbf{u} - 6\pi \mu_f R_p n_d \mathbf{u} . \qquad (9.60)$$

The key aspect of Brinkman's equation is the last term on the right, which approximates the drag due to a fixed bed of particles to be proportional to the local fluid velocity. This average force per unit volume depends on the number density n_d, which accounts for the average number of particles within the unit volume. As in the previous problem, we use the point-particle approximation within the context of Brinkman's equation to evaluate the average perturbation fluid velocity at a particle location, where the perturbation is due to the collective action of all other particles.

Following the analysis of Saffman (1973), we assume a point-particle to be located at $\mathbf{x}^\alpha$. Consider all possible conditional arrangements of neighbors given the fact that a particle is located at $\mathbf{x}^\alpha$. Each of these realizations is assumed to satisfy Eq. (9.60) and an ensemble average can be written as

$$0 = -\nabla \langle p' \rangle_\alpha + \mu_f \nabla^2 \langle \mathbf{u} \rangle_\alpha - 6\pi \mu_f R_p n_d \langle \mathbf{u} \rangle_\alpha - \mathbf{F}_0 \delta(\mathbf{x} - \mathbf{x}^\alpha) , \qquad (9.61)$$

where $\langle \cdot \rangle_\alpha$ indicates a conditional ensemble average over all realizations in which a particle is located centered at $\mathbf{x}^\alpha$. In the above equation, the delta function in the last term on the right accounts for the point-particle forcing of the fluid due to the particle located at $\mathbf{x}^\alpha$. The effect of all other particles within the random distribution is collectively taken into account through the Dracy drag term. Unlike the heuristic approach presented above, Saffman derived the above averaged Brinkman relation starting from the point-particle equation (9.47) and applying the conditional average in a rigorous manor. His original work must be consulted if one is interested in the rigorous derivation of Eq. (9.60).

We again solve the above linear equation using a three-dimensional Fourier transform. Upon Fourier transformation, we obtain

$$\langle \hat{u}_l(\mathbf{k}) \rangle_\alpha = \frac{P_{lm}(\mathbf{k})}{(|\mathbf{k}|^2 + 6\pi R_p n_d)} \frac{F_{0m}}{(2\pi)^3 \mu_f} \exp(-\iota \mathbf{k} \cdot \mathbf{x}^\alpha). \tag{9.62}$$

As in the previous derivation, the projection tensor P_{lm} accounts for the pressure effect and ensures that the incompressibility condition is satisfied. We now follow the procedure of the previous subsection, by first subtracting the self-induced perturbation flow due to the point force, then inverse Fourier transforming and evaluating the perturbation velocity at the origin to obtain the final relation

$$\mathbf{u}'(\mathbf{x} = 0) = \sqrt{\frac{R_p n_d}{6\pi}} \frac{\mathbf{F}_0}{\mu_f}. \tag{9.63}$$

Following the steps carried out in Eqs. (9.58) and (9.59), we obtain the final result

$$\mathbf{F}_0 = 6\pi \mu_f R_p \mathbf{U}_0 \left(\frac{1}{1 - 3\sqrt{\phi_d}/\sqrt{2}} \right). \tag{9.64}$$

We again obtain the key result that drag force increases with increasing volume fraction. However, for small ϕ_d this increase is slower than that for the structured array of particles.

9.4.3   Stokes Drag of Random Freely Moving Particles

This is the hardest of the three problems, since the leading-order result cannot be obtained with the point-particle approach. The finite size of the particle and the associated higher-order effects must be taken into account. As shown by Saffman, if we pursue the point-particle approach of Section 9.4.1, the resulting perturbation velocity at the αth particle due to the collective action of all the other particles that are statistically uniformly distributed around it is zero. Thus, here we must follow a more elaborate procedure as outlined in Batchelor (1972).

There are important differences between the three problems. The particle arrangements relative to each other are frozen in the former two problems. In the present problem of freely falling particles, the arrangement of where the neighbors are located not only varies from particle to particle, but also varies for any chosen particle over time, since its neighbors are free to move. From an individual particle's perspective, this contributes to an important difference. In a frozen bed of particles, due to the natural Poisson-like random distribution addressed in Section 3.3, some particles will be in regions where the local particle volume fraction is large due to naturally occurring local crowding in a random distribution, while other particles are in regions of lower volume fraction. Correspondingly, their drag will be larger or lower than average. In a frozen system, the fate of each particle is therefore frozen forever. This is not so in a freely moving system. Each particle, over time, may enter into regions of higher-than-average volume fraction and at other times into regions of lower volume fraction.

Most importantly, in a system of freely moving particles, the relative arrangement of particles is time dependent. An initially well-mixed system of particles over time can form micro-structures with clusters of particles randomly distributed over the volume. On the other hand, an initially clustered state can evolve over time to be well mixed. Whether a system stays uniformly distributed or forms micro-structures seems to depend on the particle and flow parameters. Following the analysis of Batchelor (1972), we will assume the distribution of particles to be uniform. Only in this limit are the results to be obtained in this section appropriate. The appearance of structures and self-organization of particles will significantly alter the results.

Since the settling velocity of each particle differs in this problem, let us work in the laboratory frame of reference and let the Stokes settling velocity vector of an isolated particle be $\mathbf{V}_s = V_s \mathbf{e}_g$. As in the previous two problems, in essence, we are interested in the velocity perturbation $\mathbf{u}$ induced at a reference jth particle due to the collective action of all the other particles. We divide this perturbation flow around the reference particle into a series of contributions

$$\mathbf{u}_{j\prime}(\mathbf{x}) = \mathbf{u}_{\mathrm{I}j\prime}(\mathbf{x}) + \mathbf{u}_{\mathrm{III}j\prime}(\mathbf{x}) + \cdots, \tag{9.65}$$

where $j\prime$ indicates that the perturbation flow is induced by particles other than the jth particle. The different terms must be carefully defined. On the right-hand side, $\mathbf{u}_{\mathrm{I}j\prime}$ is the collective perturbation velocity of all the other particles in the absence of the jth particle. In the terminology of Chapter 4, this velocity can be termed the "undisturbed velocity" of the jth particle. That is, this perturbation flow is in the total absence of the jth particle.

Let us now introduce the jth particle back into the flow. The undisturbed flow $\mathbf{u}_{\mathrm{I}j\prime}$ at the jth particle can be used to calculate the force on the jth particle using Faxén's theorem (see Eq. (4.66)) as $6\pi\mu_f R_p \left(\overline{\mathbf{u}_{\mathrm{I}j\prime}}^S - \mathbf{V}_j \right)$, where $\mathbf{V}_j$ is the velocity of the jth particle. But this will not yield the correct force on the jth particle for the following reason. Faxén's theorem will accurately predict the force on an *isolated* particle for any complex flow. However, in the present situation, though we know the undisturbed velocity in the absence of the jth particle, due to the presence of the other particles, Faxén's theorem cannot be applied as given above. We refer back to the method of reflections that was discussed in Section 8.2.1.

Let us consider all the other particles as a single body and call it "particle X." The first term $\mathbf{u}_{\mathrm{I}j\prime}$ on the right-hand side of Eq. (9.65) is only the leading-order perturbation flow that arises from the nonslip and no-penetration conditions of particle X. We shall call this the first perturbation of particle X. The jth particle will reflect this perturbation flow (i.e., the jth particle will create the second perturbation flow in response to $\mathbf{u}_{\mathrm{I}j\prime}$ in order to enforce its nonslip and no-penetration conditions). This second reflected flow will be reflected again by particle X (i.e., a third reflected flow will be created by particle X in order to satisfy its nonslip and no-penetration conditions on the second reflected flow of the jth particle). This third reflected flow of particle X is $\mathbf{u}_{\mathrm{III}j\prime}$ (the second term on the right-hand side). This third reflection will be reflected by the jth particle, the resulting fourth reflection will be reflected by particle X, and so on. This fifth reflection will be the next term on the right-hand side of Eq. (9.65). The sum of these reflections is the "net" undisturbed flow of the jth particle. If we calculate the force on the jth particle with this net undisturbed flow, we get $6\pi\mu_f R_p \left(\overline{\mathbf{u}_{j'}}^S - \mathbf{V}_j \right)$, which is the accurate estimation. Note that $\mathbf{u}_{j'}$ includes the first, third, fifth, and all odd reflections. Now, to calculate the average force and the average settling velocity of all the particles, let us take the ensemble average of the above accurate estimation. After substituting Eq. (9.65) for $\mathbf{u}_{j'}$, we obtain the approximation

$$6\pi\mu_f R_p \left(\langle \overline{\mathbf{u}_{\mathrm{I}j'}}^S \rangle + \langle \overline{\mathbf{u}_{\mathrm{III}j'}}^S \rangle + \cdots - \langle \mathbf{V} \rangle \right). \tag{9.66}$$

We now approximate the surface average of the first term with Faxén's correction in terms of the Laplacian of velocity. The surface average of the second term is simply replaced with its value evaluated at the particle center and thereby we ignore the Faxén correction, since it constitutes a smaller term. Ignoring the higher-order term, we then obtain

$$6\pi\mu_f R_p \left(\langle \mathbf{u}_{\mathrm{I}} \rangle + \frac{R_p^2}{6} \langle \nabla^2 \mathbf{u}_{\mathrm{I}} \rangle + \langle \mathbf{u}_{\mathrm{III}} \rangle - \langle \mathbf{V} \rangle \right), \tag{9.67}$$

where $\langle \mathbf{u}_{\mathrm{I}} \rangle$ is $\mathbf{u}_{\mathrm{I}j'}$ ensemble averaged at the particle center. We now proceed to evaluate the ensemble averages within parentheses.

Let us now follow the approach of Batchelor (1972) and evaluate the first of the perturbation terms. In a system consisting of N neighboring particles, this term can be written as

$$\langle \mathbf{u}_{\mathrm{I}} \rangle = \frac{1}{N!} \int \mathbf{u}_{\mathrm{I}}(\mathbf{x}_j, C_N) \, P(C_N \mid \mathbf{x}_j) \, dC_N. \tag{9.68}$$

Let us define the different variables on the right-hand side. Here, C_N corresponds to the N-particle configuration where the first particle is in a volume element $d\mathbf{x}_1$ around the locations $\mathbf{x}_1$, the second particle is in a volume element $d\mathbf{x}_2$ around the locations $\mathbf{x}_2$, and so on (see the discussion in Section 3.4). The conditional probability $(1/N!)P(C_N \mid \mathbf{x}_j) \, dC_N$ refers to the probability of finding N other particles arranged in the configuration C_N, given that a particle is already located with its center at $\mathbf{x}_j$. Then $\mathbf{u}_{\mathrm{I}}(\mathbf{x}_j, C_N)$ corresponds to the perturbation flow at $\mathbf{x}_j$ due to the neighboring particles defined by their configuration C_N, but in the absence of the jth particle (hence it is the "undisturbed flow" at the jth particle). Here $N!$ is the normalization factor, since $\int P(C_N \mid \mathbf{x}_j) \, dC_N = N!$. Note that $\langle \mathbf{u}_{\mathrm{I}} \rangle$ was defined as the average over all the particles. But as defined in the above equation, the integral is for a fixed particle located at $\mathbf{x}_j$ for all possible arrangements of its N neighbors that are consistent with uniform distribution. These two ensemble averages are identical.

The evaluation of the integral in Eq. (9.68) as it stands is nearly impossible due to the N-body complexity. However, in the limit of a dilute system, we can systematically approximate the integral by considering the influence of only one neighbor to the jth particle. The probability of finding one close neighbor within a chosen volume around the reference particle scales as ϕ_d, while the probability of finding simultaneously two close neighbors within that volume scales as ϕ_d^2 and can be neglected. Thus, the intent is to simplify the integral by replacing $\mathbf{u}_I(\mathbf{x}_j, C_N) P(C_N \mid \mathbf{x}_j) \, dC_N$ with $\mathbf{u}_I(\mathbf{x}_j, \mathbf{x}_k) P(\mathbf{x}_k \mid \mathbf{x}_j) d\mathbf{x}_k$. In this replacement, $P(\mathbf{x}_k \mid \mathbf{x}_j) d\mathbf{x}_k$ represents the probability of finding the kth particle around the location $\mathbf{x}_k$ given a particle at location $\mathbf{x}_j$, and $\mathbf{u}_I(\mathbf{x}_j, \mathbf{x}_k)$ is the velocity perturbation at $\mathbf{x}_j$ due to a second particle at $\mathbf{x}_k$, but in the absence of the particle at $\mathbf{x}_j$. Such replacement has been very useful in other problems of statistical physics, where the argument decays sufficiently fast for the integral to be absolutely convergent. Unfortunately, the Stokeslet portion of $\mathbf{u}_I(\mathbf{x}_j, \mathbf{x}_k)$ decays only as $1/r$ and the conditional probability $P(\mathbf{x}_k \mid \mathbf{x}_j)$ does not decay, thus the replacement is not allowed (or useful).

Batchelor presents a *renormalization* technique that allows for the evaluation of the integral with a two-particle approximation. The simple trick is to replace Eq. (9.68) by its renormalized version

$$\langle \mathbf{u}_I \rangle = \frac{1}{N!} \int \mathbf{u}_I(\mathbf{x}_j, C_N) \left\{ P(C_N \mid \mathbf{x}_j) - P(C_N) \right\} dC_N . \tag{9.69}$$

This replacement is allowed since

$$\int \mathbf{u}_I(\mathbf{x}_j, C_N) \, P(C_N) \, dC_N = 0 . \tag{9.70}$$

The above equation is correct, since the average perturbation velocity at an arbitrary point in a system containing N particles is zero. Since the system is homogeneous, any point within the system is as good as any other and therefore average perturbation (away from the mean) must be zero. With this renormalization, the N-particle integral can be approximated with a two-particle integral as

$$\langle \mathbf{u}_I \rangle \approx \int \mathbf{u}_I(\mathbf{x}_j, \mathbf{x}_k) \left\{ P(\mathbf{x}_k \mid \mathbf{x}_j) - P(\mathbf{x}_k) \right\} d\mathbf{x}_k . \tag{9.71}$$

The above two-particle approximation introduces an error of the order ϕ_d^2.

In a system with uniform distribution of particles, we make the following approximation:

$$P(\mathbf{x}_k) = n_d \quad \text{and} \quad P(\mathbf{x}_k \mid \mathbf{x}_j) \approx \begin{cases} 0 & \text{if } |\mathbf{x}_k - \mathbf{x}_j| < 2R_p, \\ n_d & \text{if } |\mathbf{x}_k - \mathbf{x}_j| \geq 2R_p, \end{cases} \tag{9.72}$$

where the unconditional probability of finding a particle in a unit volume is the same as the number density. For the conditional probability, we have simplified the radial distribution function, so that the probability of finding a neighbor is zero within one diameter of the reference particle and remains uniform outside, equal to the number density. From the above, we readily obtain

$$P(\mathbf{x}_k \mid \mathbf{x}_j) - P(\mathbf{x}_k) \approx \begin{cases} -n_d & \text{if } |\mathbf{x}_k - \mathbf{x}_j| < 2R_p, \\ 0 & \text{if } |\mathbf{x}_k - \mathbf{x}_j| \geq 2R_p. \end{cases} \tag{9.73}$$

As a result of renormalization, the slow decay of $\mathbf{u}_I(\mathbf{x}_j, \mathbf{x}_k)$ is not an issue anymore. We now have the approximation

$$\langle \mathbf{u}_I \rangle \approx -n_d \int_{r < 2R_p} \mathbf{u}_I(\mathbf{x}_j + \mathbf{r}, \mathbf{x}_j) \, d\mathbf{r} . \tag{9.74}$$

Here $\mathbf{u}_{\mathrm{I}}(\mathbf{x}_k + \mathbf{r}, \mathbf{x}_k)$ is nothing but Stokes flow around a sphere located at $\mathbf{x}_k$ falling at $\mathbf{V}_s$ evaluated at a point $\mathbf{x}_k + \mathbf{r}$ that is within a diameter. The Stokes flow solution given in Eq. (4.17) can be used and performing the integration we obtain

$$\langle \mathbf{u}_{\mathrm{I}} \rangle \approx -5.5 \phi_d \mathbf{V}_s, \tag{9.75}$$

where we emphasize that $\mathbf{V}_s$ is the Stokes settling velocity of an isolated particle.

We now go back to Eq. (9.66) to calculate the second term with a similar renormalization procedure. The details are a bit more complex and the reader can consult Batchelor (1972). In the end it can be shown that

$$\frac{R_p^2}{6} \langle \nabla^2 \mathbf{u}_{\mathrm{I}} \rangle \approx 0.5 \phi_d \mathbf{V}_s. \tag{9.76}$$

This brings us to the third term of Eq. (9.66), which can be written as

$$\langle \mathbf{u}_{\mathrm{III}} \rangle = \frac{1}{N!} \int \mathbf{u}_{\mathrm{III}}(\mathbf{x}_j, C_N) \, P(C_N \mid \mathbf{x}_j) \, dC_N. \tag{9.77}$$

As will be shown below, the perturbation flow due to the image system decays fast and therefore a renormalization approach is not needed to make the two-particle approximation. With the two-particle approximation, we obtain

$$\langle \mathbf{u}_{\mathrm{III}} \rangle \approx \int \mathbf{u}_{\mathrm{III}}(\mathbf{x}_j, \mathbf{x}_k) \, P(\mathbf{x}_k \mid \mathbf{x}_j) \, d\mathbf{x}_k. \tag{9.78}$$

We now recognize $\mathbf{u}_{\mathrm{III}}$ as the perturbation velocity of the jth particle due to the image system of the kth particle. Based on this, we write

$$\mathbf{u}_{\mathrm{III}} = \mathbf{V}_j - \left(\mathbf{V}_s - \mathbf{u}_{\mathrm{I}} + \frac{R_p^2}{6} \nabla^2 \mathbf{u}_{\mathrm{I}} \right), \tag{9.79}$$

where $\mathbf{V}_j$ is the velocity of the jth particle in a two-particle system with the kth particle located at $\mathbf{x}_k - \mathbf{x}_j$ relative to the jth particle. We subtract the Stokes settling velocity $\mathbf{V}_s$ and the first reflection in order to accurately account for only the image system.

As defined above, $\mathbf{u}_{\mathrm{III}}$ decays fast with increasing distance $\mathbf{x}_k - \mathbf{x}_j$ and therefore the integral in Eq. (9.78) can be evaluated as

$$\langle \mathbf{u}_{\mathrm{III}} \rangle \approx n_d \int_{r \geq 2R_p} \mathbf{u}_{\mathrm{III}}(\mathbf{x}_j + \mathbf{r}, \mathbf{x}_j) \, d\mathbf{r} \tag{9.80}$$

without the need for renormalization. Batchelor used the two-particle Stokes flow results of Stimson and Jeffery (1926) and Goldman et al. (1966) and evaluated the requisite integral as

$$\langle \mathbf{u}_{\mathrm{III}} \rangle \approx -1.55 \phi_d \mathbf{V}_s. \tag{9.81}$$

We are now ready to put together all the pieces. The actual average settling velocity of the particle is $\langle \mathbf{V} \rangle$ and we are interested in evaluating how this value differs from the Stokes settling velocity $\mathbf{V}_s$. In the frame attached to the particle, the average ambient fluid velocity is $-\langle \mathbf{V} \rangle$. The average undisturbed flow at the particle due to the net effect of all the other particles is the sum of Eqs. (9.75), (9.76), and (9.81), which yields $-6.55 \phi_d \mathbf{V}_s$. Thus, the total relative velocity is $-(\langle \mathbf{V} \rangle + 6.55 \phi_d \mathbf{V}_s)$. The Stokes drag due to this must balance the net weight of the particle. This yields the relation (which can also be obtained from Eq. (9.66))

$$6\pi\mu_f R_p(\langle \mathbf{V} \rangle + 6.55\phi_d \mathbf{V}_s) = (m_p - m_f)\mathbf{g}. \tag{9.82}$$

Since we have the relation $6\pi\mu_f R_p \mathbf{V}_s = (m_p - m_f)\mathbf{g}$, we obtain the final result

$$\langle \mathbf{V} \rangle = (1 - 6.55\phi_d)\mathbf{V}_s. \tag{9.83}$$

This clearly demonstrates that the net settling velocity of a random distribution of freely moving particles is lower than that of an isolated particle. This is the two-way coupling effect on settling velocity that we addressed in Section 7.4, but in the Stokes flow limit. The lower settling velocity indicates enhanced drag in the Stokes limit at finite volume fraction. In a general context, this enhanced drag can be expressed as the following volume fraction corrected drag law:

$$\mathbf{F}_0 = 6\pi\mu_f R_p \mathbf{U}_0 \left(\frac{1}{1 - 6.55\phi_d} \right) \tag{9.84}$$

for some relative velocity $\mathbf{U}_0$ between a random distribution of freely moving particles and a uniform macroscale flow.

The increase in mean drag with increasing volume fraction can also be expressed as a corresponding decrease in the mean settling velocity of a random distribution of particles. According to Eq. (9.84), the mean settling velocity decreases by $6.55\phi_d \mathbf{V}_s$. This effect is known as *hindered settling*. The different contributions to hindered settling can be given the following interpretations (Batchelor, 1972). When a distribution of particles of volume fraction ϕ_d settles at velocity $\mathbf{V}_s$, this downward flux of particles must be compensated with a corresponding flux of fluid in order to satisfy mass balance. This gives rise to a contribution of $-\phi_d \mathbf{V}_s$. As the particles fall down, they drag with them a certain amount of fluid in their vicinity, which must be compensated by an additional counter flow of $-4.5\phi_d \mathbf{V}_s$. These two contributions together account for Eq. (9.75). If the finite size of the particle is taken into account through Faxén's correction, the total counter-flow effect decreases slightly to $-5\phi_d \mathbf{V}_s$. Finally, the estimate of this counter current is improved by fully taking into account two-particle interaction, which gives rise to an additional contribution of $-1.55\phi_d \mathbf{V}_s$.

It should be noted that the collective hindering effect of neighbors is much weaker in the case of random distribution of freely settling particles. When neighbors are organized to form structured arrangement, their collective effect is far stronger. This raises an interesting possibility that when clustering of particles occurs in natural systems, the collective effect of clustering can be much stronger. In the context of instability, this feedback can in turn reinforce the clustering behavior.

9.4.4   Empirical Drag Correction at Finite Volume Fraction

The above three examples yielded results for the leading-order effect of volume fraction. With increasing particle volume fraction, pairwise interaction of neighbors is not sufficient, since simultaneous interaction between more than two particles becomes important. Since a theoretical analysis is out of reach, empirical models of finite volume fraction drag corrections have been sought.

There are many experimental studies against which the theoretical results of the previous subsection have been compared. The experiments have mainly considered sedimentation of a distribution of particles in a large system and the quantity of interest has been the mean settling velocity. In particular, the measured hindered settling velocity is expressed as $\mathbf{V}_s/\Phi$, where $1/\Phi$ is the correction factor to the settling velocity of an isolated particle. The experimental results presented by Kops-Werkhoven and Fijnaut (1981), Buscall et al. (1982), and Tackie et al. (1983) show hindered settling with a correction function of $\Phi(\phi_d) \sim 1/(1 - c\phi_d)$. The experimentally measured constant is slightly lower than the theoretically predicted value of 6.55. On the other hand, based on their experimental measurements, Barnea and Mizrahi (1973) suggest

$$\Phi(\phi_d) = \frac{1 - \phi_d^{1/3}}{(1 - \phi_d)^2} \exp\left(\frac{5\phi_d}{3(1 - \phi_d)}\right), \tag{9.85}$$

which at small volume fraction is consistent with the theoretical results of a structured array. This suggests some degree of organization of particles in their experiments.

A widely used volume fraction correction is (Richardson and Zaki, 1954)

$$\Phi(\phi_d) = \frac{1}{(1 - \phi_d)^n}, \tag{9.86}$$

with the value of the exponent being $n = 5.1$. For small ϕ_d this expression is consistent with that of a random distribution of freely moving particles, but with the constant being 5.1 instead of 6.55. Another popular volume fraction correction model is by Gidaspow (1994), which has found success in the study of fluidized beds. For $\phi_d < 0.2$, the correction follows the same functional form as Eq. (9.86), but with the exponent $n = 1.65$.

9.5 Finite-Re Volume Fraction Effects on Drag Law

Two facts stand out from the previous three sections. First, it is amazing how giants in our field (Batcher and Saffman in particular) have been able to apply the powers of applied mathematics and deep physical insight to obtain rigorous results for complex problems. Second, from the vastly different results of the three problems it is quite clear that the collective effect of a distribution of particles is complex and subtle.

With increasing Re, as we saw in Section 7.1, the flow over even an isolated particle became quite complicated, with vortex shedding and self-induced turbulence. However, the drag force on the particle could reliably be parameterized in terms of a C_D vs. Re curve as shown in Figure 7.8. At small Reynolds numbers and modest volume fractions, the effect of finite Re and finite ϕ_d could be separated and the volume fraction corrections, such as those given in Eq. (9.85) or (9.86), can be multiplied by the finite Reynolds number correction function $\Phi(\text{Re}) = 1 + 0.15\,\text{Re}^{0.687}$ and applied in practical situations. This simple approach will be inaccurate at larger Re and ϕ_d due to nonlinear interactions between the two physics.

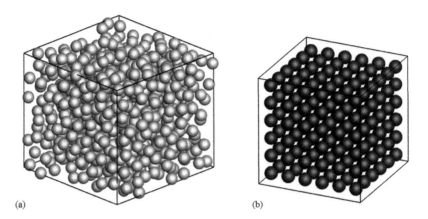

Figure 9.8 (a) Schematic of the geometry for a random distribution of monodisperse spherical particles in a periodic box. The volume fraction of particles within the box shown is about 45%. (b) Schematic of the geometry for a structured array of monodisperse spherical particles.

Recently there have been a number of particle-resolved simulations of flow through a random distribution of spherical particles within a cubic box similar to that shown in Figure 9.8a. By assuming the box to be triply periodic, the random distribution of particles is replicated over space and extended out to infinity. The location of the random distribution of particles is chosen to be of uniform probability and thus the distribution is statistically homogeneous. The results of these simulations have been used to establish the finite (Re, ϕ_d) correction to the Stokes drag. The mean drag experienced by the particles as computed in these simulations has been used to obtain the correction function (Hill et al., 2001; Beetstra et al., 2007; Tenneti et al., 2011; Zaidi et al., 2014; Bogner et al., 2015; Tang et al., 2015; Akiki et al., 2017b). The correction function $\Phi(\mathrm{Re}, \phi_d)$ as a function of Re for three different volume fractions as obtained from the different particle-resolved simulations is shown in Figure 9.9. Also plotted in this figure is the standard drag correlation, which is accurate in the zero volume fraction limit. All the simulations show that the effect of finite volume fraction is to increase drag. However, the different results are in good agreement only at low Re, and the discrepancy increases with increasing Re.

The different force predictions are bounded from below by the correlation of Tenneti et al. (2011) for values up to Re = 300, and by Tang et al. (2015) for Re > 300. The upper bound of the force range is the correlation by Beetstra et al. (2007) for all Re and ϕ_d. As pointed out by Akiki et al. (2016), at the largest Reynolds number considered, the drag force predicted by the different studies varied by about 43%, 43%, and 60%, respectively, for the three volume fractions. Despite the differences, these simulations provide valuable information on the collective effect of neighbors on drag force at finite particle Reynolds number. For instance, the following correction function proposed by Tenneti and Subramaniam (2014) is now widely used in multiphase flow simulations:

$$\Phi(\mathrm{Re}, \phi_d) \qquad\qquad\qquad\qquad\qquad (9.87)$$

$$= \left[\frac{1 + 0.15\,\mathrm{Re}^{0.687}}{(1 - \phi_d)^2} + \underbrace{\frac{5.81\,\phi_d}{(1 - \phi_d)^2} + \frac{0.48\,\phi_d^{1/3}}{(1 - \phi_d)^3}}_{f_1(\phi_d)} + \underbrace{\phi_d^3(1 - \phi_d)\,\mathrm{Re}\left(0.95 + \frac{0.61\,\phi_d^3}{(1 - \phi_d)^2}\right)}_{f_2(\mathrm{Re}, \phi_d)} \right],$$

where Re is based on the mean relative velocity within the random distribution.

The different studies presented in Figure 9.9 use different numerical methodologies (lattice Boltzmann and immersed boundary). Furthermore, different levels of resolution were employed. But it is unlikely that these numerical differences alone could explain the large differences between the plotted results. For example, Figure 9.10 shows the finite (Re, ϕ_d) correction to the Stokes drag for a simple cubic structured array of particles shown in Figure 9.8b for a range of Re at 20% and 40% particle volume fraction. The good agreement between the results obtained from the different methods and resolutions suggests that the cause of disagreement seen in Figure 9.9 is due to other factors.

Based on the Stokes regime results of the previous sections, it can be suggested that the differences in the drag force may be due to hidden structural differences in the random distribution of particles used in the different simulations. The precise source of the wide difference between the direct numerical simulation results has not been identified and needs further investigation. Nevertheless, the following important observations about the strengths and limitations of the finite-Re, finite-ϕ_d drag relation must be made.

(i) Unlike flow over an isolated particle as discussed in Section 7.1, the flow over a random distribution of particles is neither axisymmetric nor planar symmetric. At all Reynolds numbers the flow is three-dimensional with no spatial symmetries. However, a critical Reynolds number Re_{cr1} can be identified below which the flow within the entire domain remains steady. Though the critical Reynolds number for onset of vortex shedding is somewhat dependent on the precise arrangement of particles within the array, any given random array's Re_{cr1} can be expected to be lower than that of an isolated particle (i.e., lower than 270). In a random array, due to the perturbation flow induced by neighbors, the local Reynolds number of the different particles within the array varies substantially from the mean Reynolds number of the entire array (Balachandar, 2020). Thus, even when the mean Reynolds number of the array is less than 270, some particles within the array may experience a faster local fluid flow of higher than critical Reynolds number and start shedding vortices. This early onset of unsteadiness in a random array has been observed in particle-resolved simulations. With further increase in Reynolds number, the flow within the array becomes turbulent in the wake regions of the different particles.

(ii) We can now investigate the effect of particle distribution, whether it is a structured array (Figure 9.8b) or randomly distributed with uniform probability (Figure 9.8a), or randomly distributed with local clustering, on the drag force. A structured array of particles is an extreme case of particle distribution. Statistical quantities such as the radial distribution function are single-valued in a structured array. For example,

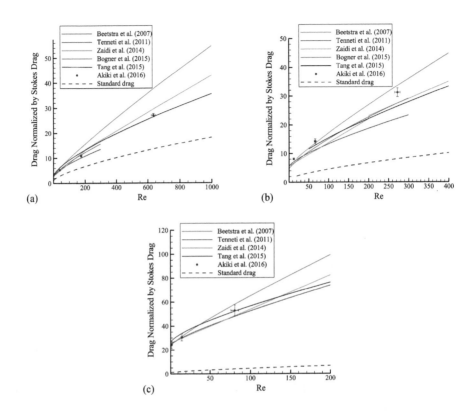

Figure 9.9 Finite Re and volume fraction correction for drag force plotted as a function of Re for (a) $\phi_d = 0.11$, (b) $\phi_d = 0.21$, and (c) $\phi_d = 0.44$ for a random arrangement of spheres. Reprinted with permission from Akiki et al. (2016). Copyright 2016 by the American Physical Society.

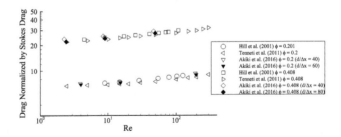

Figure 9.10 The correction function $\Phi(\mathrm{Re}, \phi_d)$ to the Stokes drag vs. particle Reynolds number for the simple cubic arrangement of spherical particles. Results from different simulations that use different numerical methods and resolution are plotted. Reprinted with permission from Akiki et al. (2016). Copyright 2016 by the American Physical Society.

for a simple cubic array, the distribution of Voronoi volume (see Section 3.4.3) is a delta function corresponding to the volume of the cubic lattice. The drag effect of the structured array is the strongest with the largest increase over Stokes drag.

A random distribution of particles with uniform probability may be considered an intermediate case. The Voronoi volumes in a random distribution follow the smooth 3Γ distribution described in Eq. (3.39). Though the mean drag on the particles still increases, the increase is weaker. Stated in a different way, if the distribution of particles becomes more structured than a random distribution of uniform probability, then the corresponding Voronoi volume fraction distribution will be more peaked than the 3Γ distribution. The mean drag on such a more structured distribution of particles can be expected to be higher than in a uniform random distribution. Instead, consider a distribution of particles whose Voronoi volume distribution is more broad than the 3Γ distribution. This would indicate clustering, since the probability of both small Voronoi volume (associated with particles inside clusters) and large Voronoi volume (associated with particles at the edge or in between clusters) increases. Extrapolating the analytical results of the previous section, it can be conjectured that the mean drag on the clustered distribution may decrease.

(iii) The drag corrections, such as those given in Eq. (9.87), are intended for evaluating the average drag of a large number of uniformly distributed particles as a function of their average volume fraction and average Reynolds number. These models are perfect for application in Euler–Euler or two-fluid simulations, where both the continuous and the dispersed phases are averaged over a suitably chosen volume. As a result, only the average drag of a local distribution of particles is required in these simulations. These models are also applicable in a standard implementation of Euler–Lagrange methods, where only the continuous phase is averaged but individual particles are tracked. Consider the scenario depicted in Figure 9.8a to be that of a dense distribution of particles in a finite-volume cell. In this case, the Stokes drag correction function $\Phi(\mathrm{Re}, \phi_d)$, such as that proposed by Tenneti et al. (2011), Eq. (9.87), applies equally to all the particles within the random array, since they are all immersed in a cloud of particles of volume fraction ϕ_d and subjected to the same macroscale flow of Reynolds number Re. Thus, in an implementation of Euler–Lagrange methodology, where particles are much smaller than the grid, the application of Eq. (9.87) for each individual particle is justified. However, this is an accurate picture only in the statistical sense. The resulting estimate of the hydrodynamic force on all the particles within a finite volume will be identical and furthermore, each particle's estimated force will only be in the direction of the macroscale flow, with no transverse force component. The microscale variation is lost in the statistically averaged picture.

(iv) The Eulerian drag correction (9.87) was obtained for a stationary array of particles and therefore it is strictly applicable only in case of frozen particles. But in most applications, the particles are in free motion. In fact, the empirical corrections to Stokes drag obtained from experiments of falling spheres, such as those presented in Eqs. (9.85) and (9.86), have been developed in the context of particles in free motion. Just as in the Stokes limit, the difference between frozen and freely moving particles can be substantial.

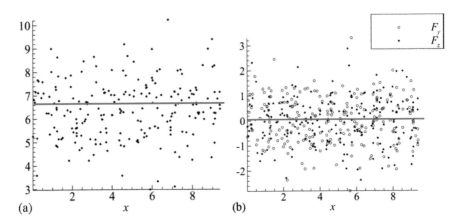

Figure 9.11 (a) The streamwise force on a particle normalized by Stokes drag vs. particle number obtained from a direct numerical simulation of steady flow over an array of 215 particles in a cubic box. The results are for $\phi_d = 20\%$ and Re ≈ 20. The red horizontal line is the average value. (b) The corresponding normalized forces along the transverse y and z-directions.

9.5.1   Statistics of Particle-to-Particle Force Variation

It is expected that each individual particle within the distribution shown in Figure 9.8a will experience a hydrodynamic force that is substantially different from the average. This point is illustrated in Figure 9.11, where in frame (a) the normalized streamwise force computed for a random distribution of 215 particles within a periodic box in a direct numerical simulation is shown. The simulation results are for a volume fraction of $\phi_d = 20\%$ and Re ≈ 20. Also plotted in the figure as a red horizontal line is the average drag of all the particles and this average force is in excellent agreement with Eq. (9.87).

It can be observed that the drag force on individual particles within the array shows substantial variation. Drag on some particles is as low as only half of the mean value, while there are particles whose drag is nearly 50% larger than the mean. Such particle-to-particle variation in the force will be ignored if all the particles within the random distribution are assigned the same mean drag. As discussed in Akiki et al. (2016), the reason for this strong particle-to-particle variation in the drag force is straightforward and can be anticipated. In a random distribution, the neighborhood of no two particles is identical. The drag force on a particle that is sheltered and blocked by one or more neighbors located directly upstream is substantially lowered. This scenario is shown in Figure 9.12a, where the view is such that the flow is directed into the page and the particle shaded black is drafting in the wake of the two upstream neighbors. There are situations where the upstream neighbors are to the side and they serve to channel the flow toward the particle, as opposed to blocking it. Such a scenario is shown in Figure 9.12b and the black shaded particle in this case experiences a substantially enhanced drag force.

Figure 9.11b shows the normalized force on the 215 particles along the transverse y and z-directions, which are orthogonal to the direction of mean flow. When averaged over all the particles, these transverse forces go to zero, since by definition there is no

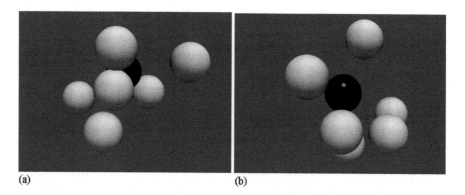

(a)   (b)

Figure 9.12 Upstream view of the scatter of the six closest neighbors for two different particles with the same local volume fraction of particles. For clarity, all other particles within the distribution are removed from this view. (a) The reference particle (shaded black) being sheltered by the two upstream neighbors. (b) The reference particle sees stronger flow directed toward it due to the channeling effect of the upstream neighbors that are shifted off-axis. The reference particle in frame (a) has a substantially lower than average drag, while the reference particle in frame (b) has a substantially higher than average drag. Reprinted with permission from Akiki et al. (2016). Copyright 2016 by the American Physical Society.

mean flow along these transverse directions. But it is striking that these transverse forces are large and for some particles quite comparable to their drag force. The importance of such transverse force can easily be argued when considering cross-stream dispersion of a cloud of particles. Similarly, the strong particle-to-particle variation in the drag force will contribute to streamwise dispersion.

A simple approach to accounting for this particle-to-particle variation in the drag and transverse forces is to introduce a stochastic component to the quasi-steady force model. To do so, Akiki et al. (2016) first investigated the statistical nature of the particle-to-particle force variation. Figure 9.13 shows the histograms of particle drag and lift forces obtained in a direct numerical simulation. Despite being based on only a few hundred particles, it is clear that both the drag and lift variation about their mean values can be satisfactorily approximated as Gaussian.

Based on particle-resolved direct numerical simulations for a range of Re and ϕ_d, Akiki et al. obtained statistical information on the standard deviation, skewness, and kurtosis of force histograms, such as those shown in Figure 9.13. The information on standard deviation of drag and transverse forces for a few cases are given in Table 9.1. They observed the skewness to be nearly zero and the kurtosis to be close to 3, and thus concluded the Gaussian approximation to be reasonable. They proposed the following stochastic model:

$$\mathbf{F}_{qs} = 6\pi\mu_f R_p \mathbf{u}_r \, \Phi(\text{Re}, \phi_d) \left[(1 + N(0, \sigma_D(\text{Re}, \phi_d)))\mathbf{e}_r + N(0, \sigma_L(\text{Re}, \phi_d))\mathbf{e}_n \right],$$
(9.88)

where $\mathbf{u}_r$ is the relative velocity based on mean flow and Re is the Reynolds number based on this relative velocity. The correction to the Stokes drag $\Phi(\text{Re}, \phi_d)$ can be taken to be given by Eq. (9.87). The standard deviations σ_D and σ_L of the normal distributions $N(0, \sigma)$ are functions of both Re and ϕ_d and can be taken from Table 9.1. Finally, $\mathbf{e}_r$ is the unit vector along the direction of relative velocity, and $\mathbf{e}_n$ is a

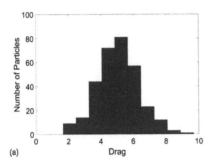
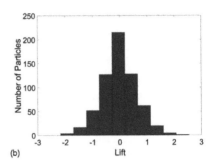

Figure 9.13 Histograms of (a) drag force and (b) transverse force normalized by the corresponding Stokes drag on a random distribution of particles.

uniformly distributed random unit vector that is orthogonal to $\mathbf{e}_r$. The above model is adequate to account for the statistical effect of a distribution of neighbors. Since the assignment of high and low drags in a distribution of particles, such as that shown in Figure 9.8a, is at random, such an assignment will not reflect the reality of the physical cause responsible for such a drag increase and decrease. A physics-based deterministic model is necessary to account for the particle-to-particle variation in the hydrodynamic force in order to better capture close-range interaction between particles. One such deterministic approach based on two-particle interaction approximation has led to the development of the pairwise interaction extended point particle (PIEP) model. This model has been shown to be quite accurate in capturing particle dynamics (Akiki et al., 2017a,b; Moore et al., 2019). More importantly, it has been shown that such an accurate coupling model can make the Euler–Lagrange simulation approach the particle-resolved simulation in accuracy (Balachandar et al., 2020).

Here it must be added that machine learning models are being developed to capture the effect of neighbors in accounting for the particle-to-particle variation in the force. Siddani et al. (2021a,b) attempted to predict the entire flow at the microscale around a random distribution of particles using the Generative Adversarial Network (GAN). There have also been recent attempts to predict the force on a particle by systematically taking into account the location of the neighbors in addition to local flow Reynolds number. These machine learning models have been quite successful in identifying high and low drag particles such as those shown in Figure 9.12 from the location of the neighbors (He et al., 2017; Seyed-Ahmadi and Wachs, 2020; Siddani and Balachandar, 2023). The use of AI tools in multiphase flow modeling is a rich area and we can expect substantial growth in research in the coming years.

9.6   Volume Fraction Dependence of Added-Mass and History Forces

This section will mainly investigate the added-mass force on individual particles immersed in a distribution of particles. In other words, we are interested in knowing the collective effect of all the other neighboring particles on the added-mass force. We will also consider the collective effect of neighboring particles on the history force. So far in this chapter, we were only concerned with the effect of neighboring particles on the quasi-steady force, which was modeled as a correction function to the Stokes drag

Table 9.1 Standard deviation of drag, transverse force, and torque on a random distribution of particles at varying Re and particle volume fraction (Akiki et al., 2017a; Moore et al., 2019). Here, force has been normalized by $4\rho_f R_p^2 U_0^2$ and torque has been normalized by $8\rho_f R_p^3 U_0^2$.

Case	Re	ϕ_d	σ_D	σ_L	σ_T
1	40	0.11	0.309	0.160	0.027
2	70	0.11	0.235	0.160	0.027
3	173	0.11	0.145	0.082	0.006
4	16	0.21	0.730	0.519	0.110
5	89	0.21	0.283	0.171	0.020
6	21	0.45	1.872	1.378	0.139
7	115	0.45	0.600	0.402	0.030

of an isolated particle. In this section, we are interested in establishing the correction to the added-mass force of an isolated particle when surrounded by a distribution of particles. The correction can be expected to depend on the particle volume fraction.

The added-mass force is inviscid in origin and therefore can be investigated using the simpler potential flow. Before considering a distribution of particles, let us start with an isolated particle of radius R_p moving at velocity $V_0 \mathbf{e}_x$ in an unbounded domain. The potential flow around the particle is governed by the potential flow equation $\nabla^2 \phi = 0$, where the velocity potential is defined as $\nabla \phi = \mathbf{u}$. Since the velocity potential satisfies the Poisson equation, the harmonic solution procedure introduced in Section 4.1 can be used to obtain the solution. Since we require the scalar potential to linearly depend on the velocity V_0, the only possible solution is

$$\phi = A V_0 \frac{x}{r^3}, \tag{9.89}$$

which used the decaying harmonic solution h_{-2}. Here, A is a constant to be determined from the boundary condition. On the surface of the particle we have the no-penetration boundary condition, which yields $\partial \phi / \partial r = -V_0 \cos \theta$. Applying the boundary condition we get the constant $A = R_p^3/2$ and substituting:

$$\phi = \frac{V_0}{2} \frac{R_p^3}{r^2} \cos \theta. \tag{9.90}$$

The kinetic energy of the fluid set in motion by the moving particle is

$$T = -\frac{\rho_f}{2} \int \phi \frac{\partial \phi}{\partial r} \, dA = \frac{1}{2} \left(\frac{1}{2} m_f \right) V_0^2, \tag{9.91}$$

which is precisely the same as the result that was presented in Eq. (4.97). There, the term within parentheses on the right was identified as the added mass, which for an isolated particle is equal to one-half the mass of the displaced fluid.

We now want to follow the above derivation to obtain an expression for the added mass of a dispersion. Our first approach will be heuristic and then we will present the results of a more rigorous analysis. In the simple approach, we will approximate each particle of radius R_p to be surrounded by a spherical volume of fluid of radius

$R_o = R_p/\phi_d^{1/3}$. Thus, the volume of the particle compared to the volume of the larger sphere of radius R_o is ϕ_d. Now we consider the potential flow induced by the particle moving at velocity $V_0 \mathbf{e}_x$ inside this spherical shell. Again we seek a harmonic solution to this problem. Since the flow is bounded between the inner and outer spheres, both the decaying and growing solutions are admissible and by combining the harmonic solutions h_{-2} and h_1 (see Section 4.2), we obtain

$$\phi = V_0 \, x \left(\frac{A}{r^3} + B \right) . \tag{9.92}$$

The two constants A and B will be determined from the boundary conditions: $\partial\phi/\partial r = -V_0 \cos\theta$ on the spherical particle and the no-penetration boundary condition $\partial\phi/\partial r = 0$ at the outer sphere. Applying these boundary conditions, we obtain

$$A = \frac{1}{2} \frac{R_p^3 R_o^3}{R_o^3 - R_p^3} \quad \text{and} \quad B = \frac{R_p^3}{R_o^3 - R_p^3} . \tag{9.93}$$

Again the kinetic energy of the fluid set in motion within the outer sphere by the motion of the particle can be calculated as

$$T = \frac{1}{2} \left(\frac{1}{2} m_f \frac{2R_p^3 + R_o^3}{R_o^3 - R_p^3} \right) U_0^2 . \tag{9.94}$$

The term within parentheses is the added mass, now for the case of a dispersion of particles of volume fraction ϕ_d. Using the definition of R_o we obtain

$$C_M(\phi_d) = \frac{1}{2} \left(\frac{1 + 2\phi_d}{1 - \phi_d} \right) \tag{9.95}$$

as the volume-fraction-corrected added-mass coefficient.

The above analysis was originally presented by Zuber (1964), who quoted the work of Lamb (1932). Admittedly the above derivation is not rigorous. A more rigorous derivation of the added-mass force for a distribution of particles was undertaken by Biesheuvel and Spoelstra (1989) and Sangani et al. (1991). The latter considered a distribution of N particles in an oscillatory flow in the zero Reynolds number limit. Although their analysis is quite involved, it is built upon the oscillatory solution of the BBO equation given in Problem 4.12 applied to each particle within the distribution. This oscillatory solution of the BBO equation replaces the steady Stokes flow solution in the Fourier analysis of Saffman (1973) for a periodic array of particles or in the renormalization analysis of Batchelor (1972) in the case of a random distribution of particles. The key result of interest is their volume-fraction-corrected added-mass coefficient

$$C_M(\phi_d) = \begin{cases} \dfrac{1}{2} \left(\dfrac{1 + 2\phi_d}{1 - \phi_d} \right) & \text{simple cubic,} \\ \dfrac{1}{2} (1 + 3.32\phi_d) & \text{random.} \end{cases} \tag{9.96}$$

Thus, the result of the simple analysis given in Eq. (9.95) is recovered in the exact analysis of a simple cubic periodic array of particles. Thus, for small ϕ_d, the rate of increase of the added-mass coefficient in a random array is about 10% more than that of a period lattice of particles.

Interestingly, the analysis of Sangani et al. (1991) also yielded a volume fraction correction to the Stokes drag for a random array of freely moving particles. This result was consistent, but not exactly the same as the results of Batchelor (1972). Their analysis also yielded modified expressions for the Basset history kernel that are applicable for a distribution of particles. Their result of the modified Basset history kernel can be summarized as

$$K_{BH}(\tau, \phi_d) = \frac{1}{\sqrt{\tau}} \begin{cases} \dfrac{1}{(1 - \phi_d)^2} & \text{simple cubic,} \\ \dfrac{1}{2} (1 + 2.28\phi_d) & \text{random.} \end{cases} \tag{9.97}$$

Thus, increasing volume fraction will tend to increase the quasi-steady, added-mass and history force contributions. While the added-mass coefficient remains unaffected by Re, the long-time decay of the history kernel will increase with Reynolds number. In which case, our best possible approach is to apply the above zero Reynolds number volume fraction correction along with the finite Reynolds number history kernel of Mei (1992). Further research is needed to firmly establish the volume fraction effect at finite Reynolds number for the history kernel.

9.7   Pseudo Turbulence at Finite Volume Fraction

So far, the focus of the last few sections of this chapter has been on the effect of flow-mediated particle–particle interaction on the various components of the hydrodynamic force on a particle. Here we will address the effect of a distribution of particles in contributing to flow turbulence in the form of pseudo turbulence. Similar to how the previous sections of this chapter extended the drag correlation to finite volume fraction, in this section we will extend the result of Section 7.3 to include the finite volume fraction effect. An important distinction must be drawn. In the dilute limit, when interaction between the particles can be ignored, the force on an isolated particle is independent of volume fraction. Thus, only at finite volume fraction does ϕ_d enter the drag expression as a correction. This is not so for pseudo turbulence.

Pseudo turbulence, to leading order, was observed in Section 7.3 to be proportional to ϕ_d. This is because velocity fluctuation at any point in the flow is due to the random influence of a nearby particle's wake, and the probability of a neighbor being close to the point of velocity measurement goes as ϕ_d. Thus, even though the particles are non-interacting with each other in the dilute limit, their interaction with the velocity probe cannot be ignored.

We expect the description of pseudo turbulence in the finite volume fraction limit to be somewhat complex. We will first present the semi-analytic results of Risso (2016) to provide a better sense of the role of interacting particle wakes. Then we will present empirical results on pseudo turbulence obtained from particle-resolved simulations.

We will first pursue the approach presented in Section 7.3, where we discussed pseudo turbulence in the dilute limit. Two important changes will be made. (i) In the dilute limit, Parthasarathy and Faeth (1990) used the turbulent wake profile of an isolated particle given in Eq. (7.4) to evaluate the rms pseudo turbulent velocity fluctuation. There it was mentioned that the turbulent wake profile does not decay fast enough for the integral to converge. As a result, the outer limit of the integral was artificially truncated. Instead, Risso (2016) considered an exponentially decaying wake velocity profile of the form (instead of those given in Eq. (7.3))

$$\textbf{Turbulent wake} \begin{cases} 1 - \dfrac{\langle u_x \rangle(x,r)}{U_\infty} = \exp\left(-\dfrac{x}{L_x}\right)\exp\left(-\dfrac{r^2}{L_r^2}\right), \\ \dfrac{\langle u_r \rangle(x,r)}{U_\infty} = 0. \end{cases} \tag{9.98}$$

The longitudinal and transverse length scales L_x and L_r are taken to be independent of particle volume fraction. Also, since the transverse velocity fluctuation in the turbulent wake contributes less to rms pseudo turbulence, it is taken to be zero in the above model. Due to the exponential decay of the wake velocity profile in the axial direction, an integral of the above velocity perturbation over the entire wake converges. The physical reason for the faster than algebraic decay along the axial direction is twofold. First, the assumption of linear superposition of the wakes of all the particles in the evaluation of the perturbation fluid velocity at any point in the domain may not be accurate. Nonlinear interactions between the wakes of neighboring particles may contribute to modify the superposition. Second, the wake flow profile of an isolated particle given in Eq. (7.4) may not be accurate, since it assumes a uniform flow approaching the particle. In the presence of other particles, each particle is subjected to wake turbulence generated by all other upstream particles, which will modify the wake of an individual particle. The net effect of these mechanisms is not fully known and the functional form given in Eq. (9.98) is a model that is consistent with empirical experimental observation.

(ii) The second important difference in the analysis of Risso (2016) is to account for the finite volume fraction effect. In obtaining Eq. (7.10) from Eq. (7.9), it was assumed in Section 7.3 that because the volume fraction was low, the wakes of different particles do not interact. This assumption breaks down with increasing volume fraction. The integral in Eq. (7.9) must be evaluated differently. Risso obtained this integral through Monte Carlo numerical simulations. He considered a large cubic domain consisting of many particles randomly distributed such that the particle volume fraction within the volume is equal to ϕ_d. The velocity at a point induced by all the particles within the system was computed by summing the perturbation contribution of each individual particle. This process was repeated to construct an ensemble of perturbation velocities within the homogeneous system. The only three parameters of the numerical simulations are the streamwise length scale of the individual wake L_x, the transverse length scale L_r, and the volume fraction ϕ_d.

The results on the probability density function of streamwise velocity fluctuation for varying values of the three controlling parameters are shown in Figure 9.14. It is important to note that they considered lighter-than-fluid bubbles and as result, in a frame attached to the mean fluid velocity, the perturbation velocity within the wake of a bubble is positive. Instead, if one were to consider wakes of falling particles, the sign of the perturbation velocity must be changed. At low volume fraction, the probability

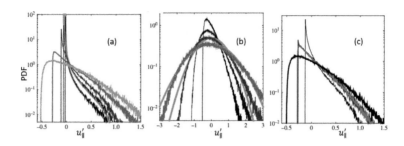

Figure 9.14 PDF of longitudinal velocity fluctuation. (a) For five different volume fractions of $\phi_d = 0.0034, 0.0091, 0.017, 0.044,$ and 0.082. The wake is characterized by $L_x/d_p = 3$ and $L_r/d_p = 0.6$. (b) For four different larger volume fractions of $\phi_d = 0.082, 0.2, 0.4,$ and 0.8. The wake is characterized by $L_x/d_p = 3$ and $L_r/d_p = 0.6$. (c) For $\phi_d = 0.044$. Gray $L_x/d_p = 3$ and $L_r/d_p = 0.4$. Blue: $L_x/d_p = 6.75$ and $L_r/d_p = 0.4$. Red: $L_x/d_p = 1.69$ and $L_r/d_p = 0.8$. Black: $L_x/d_p = 3$ and $L_r/d_p = 0.8$. Taken from Risso (2016).

distribution function (PDF) is highly skewed with a sharp negative peak and a long positive tail, which simply indicates that the large positive velocity perturbations within the finite wake are balanced by a very large region of weak negative counter flow. Frames (a) and (b) show the change in PDF with increasing ϕ_d. The PDFs remain skewed even at a modest volume fraction of 20%. Only at very large volume fractions does the PDF approach a symmetric Gaussian-like shape. Frame (c) shows the PDFs for a fixed volume fraction of $\phi_d = 4.4\%$ for varying wake size. An important scaling property that emerges from these results is that the PDF depends only on the value of the composite parameter $\phi_d L_x L_r^2$, which is proportional to the fractional volume of the system occupied by the particle wakes.

As discussed in Risso (2016), the above wake contribution to velocity fluctuation must be augmented by two other contributions. The wake profile of an individual particle is appropriate only sufficiently downstream of the particle. To account for the perturbation caused by the flow in the immediate vicinity of the particle (or bubble), he considered a potential velocity perturbation. Also, since the particles are randomly distributed, the naturally occurring number density variation will induce large-scale instability. In other words, the sedimentation of particles (or rising of bubbles) is naturally unstable and large-scale turbulence is generated, which contributes to velocity fluctuations. Risso presented a simple model for the turbulent agitation. The PDF from only the wake component was presented in Figure 9.14. The corresponding PDFs of the near-particle potential flow and large-scale turbulent agitation are denoted by $P_{\text{pot}}(u'_x)$ and $P_{\text{turb}}(u'_x)$, and are presented in Risso (2016). Figure 9.15 presents the mean-square longitudinal (vertical) velocity fluctuations computed from the PDF as

$$\langle u_\parallel'^2 \rangle = \int_{-\infty}^{\infty} P_{\text{tot}}(u'_x) \, du'_x, \tag{9.99}$$

where $P_{\text{tot}}(u'_x)$ is the PDF of velocity fluctuation. In the above, $P_{\text{tot}}(u'_x)$ corresponds to the PDF taking into account all three contributions and a similar expression applies

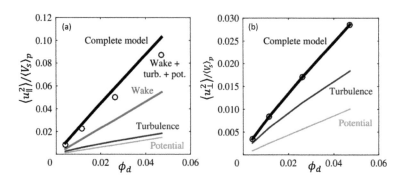

Figure 9.15 Plots of (a) $\langle u_\parallel'^2 \rangle / \langle V_s \rangle_p$ and (b) $\langle u_\perp'^2 \rangle / \langle V_s \rangle_p$ vs. ϕ_d. Taken from Risso (2016).

for the mean-square transverse velocity fluctuation. The mean-square longitudinal and transverse velocity fluctuations as a function of particle volume fraction are presented in Figure 9.15. The particular case shown in the figure is for $d_p = 1.6$ mm bubbles, rising with a mean velocity of 0.335 m/s, which corresponds to Re = 540. As for the wake, its nondimensional length and width are chosen as $L_x/d_p = 4.8$ and $L_r/d_p = 0.4$. In the turbulence model, the turbulent agitation intensity was chosen to be 0.073. Few other similar cases were presented in Risso (2016). In the normalized longitudinal mean-square velocity fluctuation, the largest contribution is from the wake component (red line). The potential and turbulence contributions are relatively small. In the transverse component, as assumed in the wake model of Eq. (9.98), the wake contribution is zero, and the primary contribution is from large-scale turbulent agitation. The composite prediction of both $\langle u_\parallel'^2 \rangle / \langle V_s \rangle_p$ and $\langle u_\perp'^2 \rangle / \langle V_s \rangle_p$, shown in the figure as black lines, compares well with the corresponding experimental measurements of Riboux et al. (2010). It can also be noted that the longitudinal fluctuation is about four times larger than the transverse component, emphasizing the anisotropic nature of particle (or bubble)-induced turbulence. It must also be added that the above turbulence may not precisely classify as pseudo turbulence, since the large-scale turbulent agitation is turbulence generated by shear and buoyancy effects.

9.7.1   Other Empirical Models of Pseudo Turbulence

Particle-resolved simulations of flow over an array of particles have become widespread in recent years. Their benefit in terms of obtaining accurate empirical models of average force on the particle as a function of average Reynolds number Re and average volume fraction ϕ_d was addressed in Section 9.5. In this subsection we will make use of these particle-resolved simulations to obtain empirical models of pseudo turbulence within the random array of particles.

Eulerian Approach

Of particular significance to the present discussion are the particle-resolved simulations of Mehrabadi et al. (2015). They considered triply periodic domains containing a random distribution of particles similar to that shown in Figure 9.8a. A constant pressure gradient along the x-direction drives a steady flow over the random distribution of particles. A volume-averaged mean fluid velocity within the periodic domain is defined, which is also the mean relative velocity since the particles are held stationary. Deviation from this mean defines the perturbation fluid velocity, whose longitudinal (streamwise) and transverse (cross-stream) components can be identified. The Reynolds stress tensor of pseudo turbulence can be calculated by averaging the velocity perturbation over the fluid volume due to the homogeneity of the flow.

The resulting pseudo-turbulent Reynolds stress tensor $\tau_{pt,axi}$ is axisymmetric with the axis aligned in the direction of the mean flow. By performing a suite of simulations at varying mean particle Reynolds number Re and mean volume fraction ϕ_d, Mehrabadi et al. were able to obtain an empirical expression for pseudo turbulence as

$$\tau_{pt,axi} = \frac{\langle u_i' u_j' \rangle}{\langle u_\| \rangle^2} = k_{pt} \phi_d \begin{bmatrix} b_\| + \frac{1}{3} & 0 & 0 \\ 0 & b_\perp + \frac{1}{3} & 0 \\ 0 & 0 & b_\perp + \frac{1}{3} \end{bmatrix}, \qquad (9.100)$$

where the coordinate system is such that the first unit vector is in the direction of mean flow $\langle u_\| \rangle$, while the other two orthogonal unit vectors are on the plane perpendicular to the relative velocity. Due to the axisymmetric nature of the tensor, the orientation of the two orthogonal vectors is arbitrary and the Reynolds stress tensor is diagonal. In the above equation, $k_{pt}\phi_d\langle u_\| \rangle^2/2 = \langle u_i' u_i' \rangle/2$ is the subgrid kinetic energy. From this definition we see that $b_\| + 2b_\perp = 0$. Thus, $b_\|$ and $b_\perp$ measure the degree of anisotropy (if $\langle u_i' u_j' \rangle$ was perfectly isotropic, $b_\| = b_\perp = 0$).

Mehrabadi et al. performed particle-resolved simulations of flow over a random array of stationary particles at four different volume fractions of $\phi_d = 0.1, 0.2, 0.3$, and 0.4, and at nine different Reynolds numbers over the range $0.01 \leq \mathrm{Re}_{su} \lessgtr 300$, where $\mathrm{Re}_{su} = \mathrm{Re}(1 - \phi_d)$ is the Reynolds number based on superficial velocity. Based on these simulation results, they proposed the following curve fit for the pseudo-turbulent subgrid kinetic energy:

$$k_{pt} = 2 + 2.5(1 - \phi_d)^3 \exp\left(-\phi_d(1 - \phi_d)^{1/2}\,\mathrm{Re}_{su}^{1/2}\right), \qquad (9.101)$$

where k_{pt} and the Reynolds stress tensor $\tau_{pt,axi}$ have been normalized by the local kinetic energy of the fluid evaluated based on the relative velocity between the particles and the surrounding fluid. The corresponding curve fit of the anisotropy coefficient is

$$b_\| = \frac{a}{1 + b\exp(-c(1 - \phi_d)\,\mathrm{Re}_{su})} \exp\left(\frac{-d\,\phi_d}{1 + e\exp(-f(1 - \phi_d)\,\mathrm{Re}_{su})}\right), \qquad (9.102)$$

where the constants $a = 0.523$, $b = 0.305$, $c = 0.114$, $d = 3.511$, $e = 1.801$, and $f = 0.005$. The value of $b_\perp$ can then be obtained easily. The plots of $k_{pt}\phi_d$, which stands for the ratio of kinetic energy of turbulent fluctuation to kinetic energy of mean velocity, vs. ϕ_d and Re_{su} are shown in Figure 9.16a and b. As can be expected, the ratio increases steadily with volume fraction and reaches 100% at about $\phi_d = 0.5$. The variation with Re_{su} is somewhat noisy, but we can observe the general trend that the ratio decreases somewhat with increasing Re_{su} at small values of Re_{su} and then remains nearly constant. The anisotropy coefficients $b_\|$ and $b_\perp$ from Mehrabadi et al. (2015) are presented in Figure 9.17. At all volume fractions, anisotropy increases with increasing Reynolds number, reaches a maximum at around $\mathrm{Re}_{su} \approx 30$, and

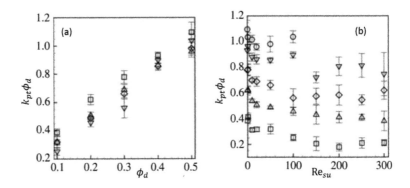

Figure 9.16 Plots of (a) $k_{pt}\phi_d$ vs. ϕ_d: ⊕: $\mathrm{Re}_{su} = 0.01$, ⬫: $\mathrm{Re}_{su} = 20$, ◈: $\mathrm{Re}_{su} = 50$, and ⬙: $\mathrm{Re}_{su} = 100$ and (b) $k_{pt}\phi_d$ vs. Re_{su}: ⊕: $\phi_d = 0.1$, ⬫: $\phi_d = 0.2$, ◈: $\phi_d = 0.3$, ⬙: $\phi_d = 0.4$, and ◈: $\phi_d = 0.5$. Taken from Mehrabadi et al. (2015).

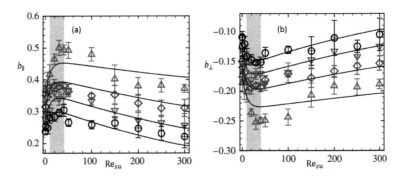

Figure 9.17 Plots of anisotropy parameters (a) $b_{\parallel}$ and (b) $b_{\perp}$ vs. Re_{su}: ⬫: $\phi_d = 0.1$, ⬙: $\phi_d = 0.2$, ⬥: $\phi_d = 0.3$, and ⊕: $\phi_d = 0.4$. Solid lines are the curve fit given by Eq. (9.102). Taken from Mehrabadi et al. (2015).

decreases with further increase in Re_{su}. At any given Reynolds number, the anisotropy is the largest at $\phi_d = 0.1$ and decreases steadily with increasing volume fraction. This is consistent with the observation of Moore et al. (2019) that the prototypical wake (or what they refer to as the superposable wake) of a particle becomes smaller and the aspect ratio becomes closer to unity with increasing volume fraction, since the otherwise elongated wakes of particles are truncated short by the increasing number density of neighbors in the wake. It can also be observed that there is considerable noise in the simulated data and thus the fits presented above must be employed carefully.

Lagrangian Approach

A different approach to modeling the pseudo-turbulent stress tensor was provided by Moore and Balanchandar (2019). In their Lagrangian approach, each particle contributes a certain amount of Reynolds stress based on its Reynolds number and local particle volume fraction. The Reynolds number of the lth particle and the average volume fraction of particles surrounding the lth particle are denoted as Re_l and $\phi_{d@l}$. The perturbation flow around the particle contributes to pseudo turbulence, and the Reynolds stress from the lth particle depends only on Re_l and $\phi_{d@l}$. Moore and Balanchandar also considered particle-resolved simulations of flow over a random array of particles for a range of volume fractions and particle Reynolds numbers. Based on these results, they proposed the following model for the Reynolds stress of a particle:

$$\mathscr{R}(\mathrm{Re}_l, \phi_{d@l}) = |\mathbf{u}_{c@l} - \mathbf{V}_l|^2 \begin{bmatrix} A(\mathrm{Re}_l, \phi_{d@l}) & 0 & 0 \\ 0 & B(\mathrm{Re}_l, \phi_{d@l}) & 0 \\ 0 & 0 & B(\mathrm{Re}_l, \phi_{d@l}) \end{bmatrix} . \quad (9.103)$$

Again, the above axisymmetric form of the tensor is appropriate only in the coordinate system whose first unit vector is aligned along the direction of relative velocity $\mathbf{u}_{c@l} - \mathbf{V}_l$. Here, A and B are the scaled streamwise and transverse components of subgrid stress and the curve fits of them obtained from the simulation are

$$A = 0.943 + 0.00135\,\mathrm{Re}_l - 1.9 \times 10^{-6}\,\mathrm{Re}_l^2 - 4.61\phi_{d@l} + 5.82\phi_{d@l}^2$$
$$- 0.00179\,\mathrm{Re}_l\,\phi_{d@l}, \quad (9.104)$$

$$B = 0.0898 - 1.25 \times 10^{-5}\,\mathrm{Re}_l + 1.09 \times 10^{-6}\,\mathrm{Re}_l^2 - 0.208\phi_{d@l} + 0.103\phi_{d@l}^2$$
$$- 1.18 \times 10^{-4}\,\mathrm{Re}_l\,\phi_{d@l}. \quad (9.105)$$

A plot of A and B as a function of Re for different values of ϕ_d is presented in Figure 9.18. As in the Eulerian modeling approach, both A and B decrease with increasing volume fraction and the dependence on Re is weak.

The subgrid pseudo-turbulent kinetic energy contribution of a particle is then given by

$$\frac{1}{2}|\mathbf{u}_{c@l} - \mathbf{V}_l|^2 \left(A(\mathrm{Re}_l, \phi_{d@l}) + 2B(\mathrm{Re}_l, \phi_{d@l})\right). \quad (9.106)$$

Note that the Reynolds stress $\mathscr{R}$ is that of an individual particle. An Eulerian field of pseudo-turbulent subgrid stress is then obtained by summing the contributions of all the N particles within the system. We shall see more about creating Eulerian fields out of Lagrangian particle data later, in the description of the Euler–Lagrange methodology in Chapter 15. An interesting observation can be made. In the Eulerian approach, the Reynolds stress in Eq. (9.100) to leading order scales as ϕ_d. This scaling was observed in the models of both Parthasarathy and Faeth (1990) and Risso (2016) discussed in the previous subsection and in Section 7.3. However, in the Lagrangian approach, the Reynolds stress $\mathscr{R}$ is of an individual particle and therefore does not include the volume fraction scaling. Only when summed over all the particles within a certain volume does the dependence on ϕ_d appear. In the limit $\phi_d \to 0$, this summation will go to zero due to lack of particles.

The Eulerian way of modeling pseudo-turbulent Reynolds stress given in Eq. (9.100) has the advantage that it is appropriate even in the Euler–Euler (or two-fluid) methodology, since it is based on an Eulerian representation of the dispersed phase. In contrast, the Lagrangian way of modeling Reynolds stress given in Eq. (9.103) can only be applied in the Euler–Lagrange methodology, since it relies on the position and velocity information of individual particles. However, in the case of Euler–Lagrange methodology, Eq. (9.103) presents a few advantages. First, in Eq. (9.103), the Reynolds

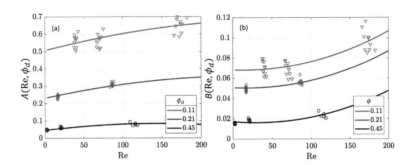

Figure 9.18 Plots of (a) longitudinal and (b) transverse Lagrangian Reynolds stress as a function of Re for different volume fraction. The symbols are simulation data and the lines are curve fit. Reprinted with permission from Moore and Balanchandar (2019). Copyright 2019 by the American Physical Society.

stress contribution of each particle is first computed and then summed over all the particles. Thus, the magnitude and orientation of the microscale flow around each particle is accurately accounted for in the calculation of Reynolds stress. In contrast, in Eq. (9.100) the dispersed-phase velocity is first calculated by averaging over all the particles before evaluating the pseudo-turbulent stress. Due to the quadratic dependence of Reynolds stress on perturbation velocity, the averaging of the particle velocity before the evaluation of Reynolds stress results in a commutation error. Furthermore, it should be noted that the perturbation flow induced by the relative motion of the lth particle is responsible for both the interphase momentum coupling force and the pseudo turbulence. Thus, the summation over all the particles presents a consistent treatment of pseudo turbulence that is similar to the application of the feedback force.

Turbulence modulation due to the presence of suspended particles can also be viewed from the perspective of energy balance depicted in Figure 7.13b. Based on the energy flux balance, Balachandar et al. (2023) proposed a theoretical model to predict the turbulent kinetic energy modulation in isotropic turbulence due to the dispersed phase. Conceptually, they distinguished two different mechanisms of turbulence modulation. At the microscale, the slip velocity between the particles and the fluid due to particle inertia, finite size, and gravity results in pseudo turbulence. At the mesoscale, turbulence is modulated by the gravitational influence on a nonuniform distribution of particulates. They limited attention to only turbulence modulation at the subgrid scale, and assumed the dispersed phase to be uniformly distributed. Their physics-based theoretical model predicted turbulence modulation for a wide range of parameters: (i) ratio of particle diameter to Kolmogorov scale, (ii) Reynolds number of local turbulence, (iii) particle-to-fluid density ratio, (iv) particle volume fraction, and (v) relative mean slip velocity. They tested their model against particle-resolved simulation and experimental results for isotropic turbulence and turbulent channel flow. The validated model is then used to illustrate the physics of turbulence modulation for variations of the five controlling parameters.

9.8   Volume Fraction Effect on Heat Transfer

Let us again consider a random distribution of particles of mean volume faction ϕ_d such as shown in Figure 9.8. Our interest is in the heat transfer between the particles and the surrounding fluid flow. Let the flow of mean Reynolds number Re be driven by a constant streamwise pressure gradient. Let all the particles be maintained at a constant temperature T_d at all times. If the fluid enters the domain at an average temperature of $\langle T \rangle_{c,in}$, it will exit the domain at a higher (or lower) temperature $\langle T \rangle_{c,out}$ depending on whether the fluid temperature is lower (or higher) than the particle temperature, where the angle brackets indicate an ensemble average. It should be noted that while the velocity field can be assumed to be periodic in the streamwise direction across the inlet and outlet planes of the box, periodicity of the thermal field cannot be assumed due to the increase (or decrease) of the fluid temperature along the flow direction through heat transfer. In particular, the difference between the fluid and the particle temperature will decrease along the flow direction, which in turn will affect the local heat transfer along the flow direction. Thus, while the flow can be assumed statistically homogeneous along all three directions, the thermal field is inhomogeneous along the mean flow direction. The challenges associated with this inhomogeneity have been addressed by Sun et al. (2015). The total heat transfer to the fluid from the particles can be evaluated as

$$\dot{Q} = \rho_f \, C_{pf} \langle u \rangle A_{cs} \left(\langle T \rangle_{c,out} - \langle T \rangle_{c,in} \right), \tag{9.107}$$

where C_{pf} is the constant specific heat of the fluid, $\langle u \rangle$ is the mean streamwise velocity of the fluid within the box, and A_{cs} is the average cross-sectional area of the flow, which can be taken to be equal to $1 - \phi_d$. The total heat transfer as measured in an experiment or in a simulation can be nondimensionalized to obtain the Nusselt number, much like the net drag force on the particles can be nondimensionalized to obtain the drag coefficient. The Nusselt number can be obtained as

$$\mathrm{Nu} = \frac{2}{3} \frac{d_p^2}{\mathcal{V} \phi_d} \frac{\dot{Q}}{k_f \left(\langle T \rangle_{c,out} - \langle T \rangle_{c,in} \right)}, \tag{9.108}$$

where k_f is the constant thermal conductivity of the fluid and $\mathcal{V}$ is the volume of the box. In the above, Nu must be interpreted as the average Nusselt number of all the particles that are contained within the box. The average Nusselt number is a function of the mean Reynolds number, the mean volume fraction, and the Prandtl number of the fluid.

The empirical Nusselt number correlations of Ranz and Marshall (1952) and Whitaker (1972) for an isolated particle as a function of Reynolds and Prandtl numbers were introduced in Eq. (2.6) and discussed in Chapter 6. Those classic expressions are valid for an isolated particle subjected to a uniform isothermal ambient flow. In a multiphase flow, the effect of finite volume fraction will influence the heat transfer, just as it greatly increased the average drag coefficient (see Figure 9.10). The enhancement of mean heat transfer between a random array of particles and the surrounding ambient flow has been the subject of great interest in gas–solid flows. The often referred to

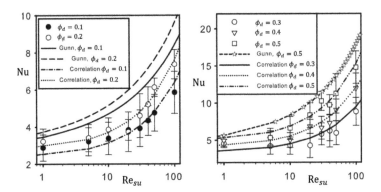

Figure 9.19 Plots of Nusselt number vs. superficial Reynolds number: (a) low volume fraction and (b) large volume fraction. The symbols are simulation data and the lines are curve fit from Gunn (1978) and Sun et al. (2015). Reprinted from Sun et al. (2015), with permission from Elsevier.

finite volume fraction Nusselt number correlation is due to Gunn (1978), who obtained the fit to a collection of experimental measurements from various sources. The Gunn correlation can be expressed as

$$
\begin{aligned}
\mathrm{Nu} = {} & \left(7 - 10(1 - \phi_d) + 5(1 - \phi_d)^2\right)\left(1 + 0.7\,\mathrm{Re}_{su}^{0.2}\,\mathrm{Pr}^{1/3}\right) \\
& + \left(1.33 - 2.4(1 - \phi_d) + 1.2(1 - \phi_d)^2\right)\mathrm{Re}_{su}^{0.7}\,\mathrm{Pr}^{1/3},
\end{aligned}
\tag{9.109}
$$

where $\mathrm{Re}_{su} = \mathrm{Re}(1 - \phi_d)$ is the Reynolds number based on superficial velocity. There are recent particle-resolved simulations of flow through a random array of stationary particles with heat transfer between the particles and the surrounding flow. The results have yielded Nusselt number correlations that are appropriate for a wide range of particle Reynolds number and volume fraction (Tavassoli et al., 2013; Deen et al., 2014; Sun et al., 2015). As an example, the Nusselt number correlation of Sun et al. (2015) is given as

$$
\begin{aligned}
\mathrm{Nu} = {} & \frac{1}{(1 - \phi_d)^3}\left(-0.46 + 1.77(1 - \phi_d) + 0.69(1 - \phi_d)^2\right) \\
& + \left(1.37 - 2.4(1 - \phi_d) + 1.2(1 - \phi_d)^2\right)\mathrm{Re}_{su}^{0.7}\,\mathrm{Pr}^{1/3}.
\end{aligned}
\tag{9.110}
$$

A keen observer will note that the above finite volume fraction Nusselt number correlation does not quite reduce to either the Ranz–Marshall or the Whitaker correlation in the limit $\phi_{d@l} \to 0$. This indicates lack of agreement among the different Nusselt number correlations. There is some uncertainty associated with these correlations. Furthermore, while the experimental results are for freely moving particles that are distributed in a non-uniform way, the simulation results have been developed based on stationary particles that are uniformly distributed. Figure 9.19 shows plots of the above Nusselt number correlations by Gunn and Sun et al., along with the latter's simulation results. In the Stokes limit of $\mathrm{Re} \to 0$ and $\phi_d \to 0$, the Nusselt number $\mathrm{Nu} \to 2$. The Nusselt number increases with volume fraction and also with Reynolds number.

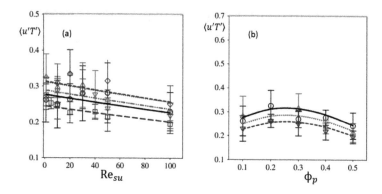

Figure 9.20 Plots of pseudo-turbulent heat flux: (a) as a function of Re_{su} and (b) as a function of volume fraction. Taken from Sun et al. (2015).

9.8.1 Pseudo-Turbulent Heat Flux

Just as we did for momentum, we will now consider the back-effect of particles on the thermal field of the flow. Even if the flow approaching the particles is perfectly isothermal, once the flow goes around the particles, the heat exchange with the particles will create thermal wakes around the particles and the temperature will cease to be uniform. Since the thermal field is statistically homogeneous along the transverse direction, the ensemble average of the fluid temperature will be a function of only the streamwise coordinate, which we denote as $\langle T \rangle(x)$. Thermal fluctuation about the mean is due to the random arrangement of particles within the volume and therefore is part of pseudo turbulence.

Just as the Reynolds stress plays an important role in the ensemble-averaged momentum balance, the Reynolds heat flux plays an important role in the ensemble-averaged energy equation. Here we present the results of Sun et al. (2016) on the pseudo-turbulent heat flux obtained from particle-resolved simulations. Figure 9.20 shows $\langle u_x' \theta' \rangle / \langle u_x \rangle$ plotted as a function of Re_{su} for varying volume fraction (a) and as a function of ϕ_d for varying Re_{su} (b). Here, nondimensional temperature is defined as $\theta' = (T - T_d)/(\langle T \rangle - T_d)$. Only correlation of the thermal fluctuation with those of the streamwise velocity fluctuation is shown, since by symmetry, correlation with the transverse velocity fluctuation is zero. The pseudo-turbulent heat flux decreases slowly with increasing Re_{su} and in terms of volume fraction, it seems to reach a maximum around $\phi_d \approx 0.3$. The following fit for the data is from Sun et al. (2016):

$$\langle u_x' \theta' \rangle = (1 - \phi_d)\left(0.2 + 1.2\phi_d - 1.24\phi_d^2\right)\exp(-0.002\,\mathrm{Re}_{su})\,\langle u_x \rangle . \qquad (9.111)$$

These models of pseudo-turbulent Reynolds stress and heat flux will play a role in the Euler–Lagrange and Euler–Euler approaches of multiphase flow to be discussed in later chapters.

10 Collisions, Coagulation, and Breakup

Collisions among particles, droplets, and bubbles and their growth through coagulation is vital in the understanding of many multiphase problems. Similarly, particles, droplets, and bubbles can also breakup into smaller fragments and daughter droplets and bubbles. For example, it is now well established that collisions and coagulation of droplets play a central role in the formation of precipitation-size raindrops in a cloud (Mason, 1969; Yau and Rogers, 1979; Sundaram and Collins, 1997; Shaw, 2003; Grabowski and Wang, 2013). The droplets that initially nucleate are far too small to fall down as precipitation. Furthermore, droplet growth by condensation is effective only to grow the droplets to a size of about 10–20 μm. Therefore, estimations of droplet growth by condensation alone cannot predict precipitation-size droplets in the lengths of time normally observed in nature. Intense turbulence within the clouds, along with differential settling of different-sized droplets, plays an important role in explaining the observed precipitation process. Another example is soot formation in flames, where tiny nanometer-sized particles (called monomers) collide and agglomerate together to form complex fractal-shaped soot particles. Collisions and agglomerations also play an important role in the cases of cohesive sediments in rivers and estuaries. Here again the monomer particles are very small (micron sized), and collide and agglomerate (also known as flocculate) to form larger-sized flocs. Since the flocs are much larger than the monomers, their settling velocity is substantially larger than the monomer settling velocity. This difference in settling velocity has a strong influence on the suspended sediment concentration. In terms of terminology, we will refer to the process of two particles coming into contact with each other as "collisions." While collisions are driven by particle motion in response to hydrodynamic forces acting on them, what follows after collision is often determined by other physics. If the particles do not bounce off each other after collision, and remain as a combined particle, we will call the process "agglomeration," "coagulation," or "flocculation."

There is a fundamental difference between the coagulation process of rain formation and the soot or sediment floc formation by agglomeration or flocculation. In the case of rain formation, the collision between two droplets results in their coagulation to become a larger droplet. Provided the Weber number is small (i.e., the surface tension effect is large), the colliding droplets as well as the agglomerated droplet are all nearly spherical in shape. In contrast, in the case of soot as well as sediment floc, the individual monomers remain distinctly identifiable even when the floc grows to a very large size. The flocs, as they form, take complex fractal-like structures and do not correspond to the spherical shape of the monomer particles. Thus, in the case of rain droplets, as the droplets grow in size, continued growth of the droplet still depends on collision between two nearly spherically shaped droplets. On the other hand, in the flocculation process, if the monomers are spheres, even a dimer (consisting of two agglomerated monomers) takes the shape of a dumbbell. Larger flocs can take complex fractal-like structure. So the continued growth of the floc cannot be treated as collision between two spheres.

In this chapter, we will limit attention to particles that remain spherical even after a collision that results in coagulation of the colliding particles. In this sense, much of the discussion to follow in this chapter pertains directly to droplet coagulation. However, the basic concepts to be discussed will remain relevant to the agglomeration and flocculation processes as well, but appropriate modifications must be made. It should also be pointed out that collisions between particles occur in many other scientific and industrial applications, such as collisions among suspended particles in pneumatic transport systems, slurry flows, in gas cleaning chambers, and control of industrial emissions. In these examples, the colliding particles bounce off after collision and do not coagulate or agglomerate. Nevertheless, collisions play an important role in generating particle-phase stress. Also, if collisions are inelastic, then interparticle collisions provide an additional mechanism for energy dissipation.

It is conceptually convenient to separate the physics into a *collisional process* followed by a *coagulation* or *agglomeration process*. In this viewpoint, a collision occurs when two particles come into contact. Following the collision, the two colliding particles may or may not end up coagulating or agglomerating, which depends on various close-range forces such as the Van der Waals force, forces due to electrical or magnetic fields, and other adhesive forces that act between the colliding particles. In fact, the rate of collision itself could be augmented or suppressed by the presence of these attractive or repulsive forces among particles.

The earliest work on the theoretical description of the collision process was by Smoluchowski (1918), who obtained a kinetic equation which describes the time rate of change of mean number density of particles of a certain size. His derivation was phenomenological and treats collisions as a deterministic process. In other words, all particles with the same initial volume grow at the same rate in a uniformly distributed cloud. In reality, the collision process is probabilistic in nature. Even in a homogeneous cloud with uniform distribution of particles, any two initially identical particles will not grow at the same rate. Among these identical particles, a few will grow faster than others and a few will grow slower. Gillespie (1972) constructed a stochastically complete model that represents the collision process. Starting with the fundamental definition of the probability that there are n particles of size m in the cloud at time t, he derived an evolution equation for this probability.

Not all close encounters between particles necessarily lead to collisions. Very often the particles, especially small ones, are deflected away from collision by local small-scale aerodynamic forces produced by their relative motion. Each particle perturbs the flow around it and this perturbation flow affects the motion of the other particle as it approaches the first particle. Thus, a collision between two particles that would have occurred had they followed their trajectory as if the other particle did not exist, may not occur in practice, due to the perturbing influence of one on the other. Moreover, not every collision results in particle coagulation, since particles can bounce off one another, or break up after collision. One defines a *geometric collision kernel* to be the probability that a collision can occur while neglecting the local small-scale interactions, *collision efficiency* as the fraction of the geometric collisions that lead to real collisions after local aerodynamic interactions, and *coagulation efficiency* to be the fraction of the real collisions that end up as coagulations. If coagulation efficiency is 100%, then each colliding particle ends up coagulating, whereas a 0% coagulation efficiency corresponds to every collision resulting in the two particles bouncing off. The overall collision kernel is given by a product of the geometric collision kernel and the collision efficiency as

$$\text{Collision kernel} = \text{Geometric collision kernel} \times \text{Collision efficiency}. \quad (10.1)$$

The overall coagulation kernel is given by a product of the overall collision kernel and the coagulation efficiency as

$$\text{Coagulation kernel} = \text{Collision kernel} \times \text{Coagulation efficiency}. \quad (10.2)$$

In this chapter, we shall see several models of geometric collision kernel for the different collisional mechanisms, such as differential settling, particle inertia, fluid shear, and turbulence, that are responsible for bringing the particles to collision. Common to all these collisional mechanisms is the ability to generate negative relative velocity between two particles along their line of separation. This negative relative velocity causes particles to approach each other and result in a collision. This ability of the relative particle motion to induce collisions is measured in terms of the geometric collision kernel. On the other hand, in spite of the numerous theoretical and experimental estimates, modeling of collision and coagulation efficiencies, even for the case of spherical particles, remains complex (Jonas and Goldsmith, 1972; Klett and Davis, 1973; Lin and Lee, 1975; Beard and Ochs, 1983; Rosa et al., 2011).

As pointed out earlier, collisions between droplets or particles can also lead to their breakup. In fact, even without collisions, droplets and particle agglomerates break up due to internal stresses exerted on them by ambient shear and turbulence. In the case of a droplet, under ambient shear, the drop starts to deform. While this deformation is promoted by the inertial effects of ambient shear, surface tension and viscous forces try to slow down the instability. The problem can be characterized by four independent dimensionless parameters: the Weber number, the Reynolds number, and the gas-to-droplet density and viscosity ratios. A large body of knowledge has been gained on droplet breakup based on experiments (Guildenbecher et al., 2009). The experimental results have led to a regime classification of *vibrational, bag, multimodal,* and *sheet-thinning* mechanisms of droplet breakup. The breakup of particle agglomerates, such as soot particles, and sediment flocs is less well studied due to their complex structure. Nevertheless, it is ambient fluid shear on the scale of the agglomerate or floc that is considered to be responsible for the breakup.

The rest of the chapter is organized as follows. In Section 10.1 we will consider the stochastically complete model of collision process described by Gillespie (1972) and from it derive the Smoluchowski equation (Smoluchowski, 1918) describing the time evolution of the number of particles of a certain size as a result of the coagulation process. Section 10.2 will consider different collision mechanisms and their associated geometric collision kernels. In Section 10.3 we will then consider the combined effects of turbulence, particle inertia, and particle settling. In this section we will also consider the close interaction between two approaching particles and obtain expressions for collision efficiency. In Section 10.4 we will consider self-similar solutions to the Smoluchowski equation for different collisional mechanisms. In this section, we will also consider some interesting results on the development of the particle size spectrum as a result of coagulation. Finally, in Section 10.5 we will briefly consider the mechanisms of droplet breakup and other processes.

10.1   Stochastic Collision Model

In deriving the stochastic collision model, we first consider a discrete system where particle sizes are quantized in terms of the smallest particle volume V_{mono}, which defines the monomer volume. All other larger particles are generated by agglomeration of the monomers and therefore their volumes are multiples of V_{mono}. Henceforth, we will refer to a particle agglomerate of volume mV_{mono} as an m-particle. This discrete description is an idealization. In reality, if we consider rain formation, for example, the droplet size will initially increase continuously due to condensation and therefore the droplet sizes in a cloud will not be quantized. Nevertheless, provided we choose V_{mono} to be sufficiently small, the resulting discrete system will be a good approximation to the real continuous system. In this approximation, V_{mono} will be a discretization parameter, just like a time step or grid spacing.

In the discrete stochastic collision model, the appropriate quantity to describe the state of particles in a cloud is the probability function $p(n, m; \mathbf{x}, t)$, which is the probability that there are n particles of size m in a unit volume surrounding $\mathbf{x}$ at time t. Any arbitrary volume satisfying the following two conditions can be chosen as the reference unit volume. The unit volume must be much larger than the particle volume, so that the probability of a few particles being inside this unit volume is meaningful. We also require the unit volume to be much smaller than the cloud volume in order to accurately represent the spatial inhomogeneities. For example, in the case of water droplets in an atmospheric cloud, the size of the cloud is of the order of kilometers and the water droplets are typically smaller than a millimeter in radius. Therefore, the unit volume can be chosen to be a sphere of radius a few centimeters to a few meters. The overall results will be independent of the size of this unit volume, in the spirit of the standard continuum approximation. Gillespie (1975) used the same definition of the probability function in the homogeneous limit without including the space variable $\mathbf{x}$. According to the above definition, $p(n, m; \mathbf{x}, t)$ is a probability function of the variable n. The variables m, $\mathbf{x}$, and t are other parameters. Therefore, the sum of $p(n, m; \mathbf{x}, t)$ over all n should add up to unity and also satisfy the condition $p(n, m; \mathbf{x}, t) > 0$.

The probability function changes due to two basic mechanisms. First, when particles move around in the cloud volume, $p(n, m; \mathbf{x}, t)$ can increase or decrease due to a net inflow or outflow of particles of size m into the unit volume surrounding $\mathbf{x}$. This process merely redistributes the particles inside the cloud and the total number of m-particles inside the cloud does not change. Second, a collision between an m'-particle and an m''-particle changes the probability function $p(n, m; \mathbf{x}, t)$ for the three different values of $m = m', m''$, and $(m' + m'')$. Here a collision is treated purely as a local process where there is no interaction between neighboring volume elements. Moreover, collisions can occur at different rates in different parts of the cloud volume. In an infinitesimal time step, we can assume that these two mechanisms happen independently. In the limit of $\delta t \to 0$ it is valid to decouple the two effects and the overall time rate of change can be expressed as

$$\frac{\partial p(n, m; \mathbf{x}, t)}{\partial t} = \frac{\partial p(n, m; \mathbf{x}, t)}{\partial t}\bigg|_M + \frac{\partial p(n, m; \mathbf{x}, t)}{\partial t}\bigg|_C, \tag{10.3}$$

where the subscript M stands for the contribution due to the movement of the particles and the subscript C represents the contribution due to collisions.

The advective part of the probability can be obtained following standard steps used in the derivation of the Navier–Stokes equation. Here we will focus on the local rate of change of $p(n, m; \mathbf{x}, t)$ due to collisions, following the approach of Gillespie (1975). The rate of change will involve two conditional probabilities, namely $p(n, m | n', m'; \mathbf{x}, t)$, the probability that at time t there are n m-particles in a unit volume surrounding $\mathbf{x}$ given that there are n' m'-particles in the same volume, and $p(n, m \mid n', m'; n'', m''; \mathbf{x}, t)$, the probability that at time t there are n m-particles in a unit volume surrounding $\mathbf{x}$ given that there are n' m'-particles and n'' m''-particles in the same volume. Assuming that the probability of having n m-particles in a unit volume surrounding $\mathbf{x}$ is independent of the number of particles of all other sizes, the following substitutions can be made:

$$\begin{aligned} p(n, m | n', m'; \mathbf{x}, t) &\approx p(n, m; \mathbf{x}, t), \\ p(n, m \mid n', m'; n'', m''; \mathbf{x}, t) &\approx p(n, m; \mathbf{x}, t). \end{aligned} \tag{10.4}$$

This assumption closes the otherwise infinite sequence of conditional probabilities. Gillespie identified four contributions to the collisional rate of change of the probability function, which will be discussed one at a time. The first contribution accounts for the change in the probability function due to all collisions between non-identical particles that lead to the formation of an m-particle. This contribution to the rate of change is given by

$$\dot{p}_{C1} = \frac{1}{2}\left[p(n-1, m) - p(n, m)\right] \sum_{m'=1}^{m-1\dagger} \sum_{n'=1}^{\infty} \sum_{n''=1}^{\infty} n' n'' \, p(n', m') p(n'', m'') \, c_{m', m''}, \tag{10.5}$$

where we have dropped the dependence on $(\mathbf{x}, t)$ in the probability functions. The collision kernel $c_{i,j}$ is defined as the probability that a particle of size i will collide with a particle of size j in unit time in a unit volume around $\mathbf{x}$. The collision kernel is a function of $(\mathbf{x}, t)$ and this dependence is also suppressed. In the first summation, the constraint $m = m' + m''$ ensures that the product of collision (which is assumed to result in a coagulation) is a particle of size m, whose probability is being monitored.

The summations over n' and n'' ensure that the production of an m-particle changes its probability irrespective of the local number of m' and m'' particles. Note that the production of one m-particle will add to the probability $p(n, m)$ if there were $n - 1$ m-particles prior to collision. Meanwhile the production of one m-particle will decrease $p(n, m)$ if there were already n m-particles locally. If the number of m-particles locally was different from n or $n - 1$ prior to collision, then the production of one more particle has no effect of $p(n, m)$. The prefactor $1/2$ accounts for the double counting of all possible collisions by the summations.

The second contribution is due to collisions between 2-particles of equal size and this contribution is zero if m is odd. This contribution can be expressed as

$$\dot{p}_{C2} = \frac{\epsilon(m)}{2} \left[p(n-1, m) - p(n, m) \right] \sum_{n'=1}^{\infty} n' (n' - 1) p(n', m/2) \, c_{m/2, m/2}, \qquad (10.6)$$

where $\epsilon(m)$ is one if m is even and zero if m is odd. Note that this term has the same format as the first term, except that it accounts for collision between two particles of size $m/2$ resulting in an m-particle, thus only the probability $p(n', m/2)$ determines the frequency of collisions. Furthermore, with the inclusion of this term, the first summation in the definition of $\dot{p}_{C1}$ should exclude the possibility $m' = m'' = m/2$, which is indicated by the superscript † in Eq. (10.5).

The third contribution accounts for collisions between m-particles and particles of all other sizes and is given by

$$\dot{p}_{C3} = \left[(n + 1) p(n + 1, m) - n\, p(n, m) \right] \sum_{m'=1}^{\infty \ddagger} \sum_{n'=1}^{\infty} n' \, p(n', m') \, c_{m, m'}. \qquad (10.7)$$

In this contribution, we note that one m-particle will be lost as a result of collision. Therefore, $p(n, m)$ will increase if there were $(n + 1)$ m-particles before collision, whereas $p(n, m)$ will decrease if there were n m-particles before collision. In the above summation, $\ddagger$ in the superscript indicates that the summation excludes $m' = m$. The last contribution to the rate of change of the probability function is from collisions among m-particles. With each such collision, two m-particles will be lost to produce one $2m$-particle. Thus, $p(n, m)$ will increase if there were $(n+2)$ m-particles and $p(n, m)$ will decrease if there were n m-particles before collision. These two contributions can be expressed as

$$\dot{p}_{C4} = \frac{1}{2}(n + 2)(n + 1)\, p(n + 2, m)\, c_{m,m} - \frac{1}{2} n(n - 1)\, p(n, m)\, c_{m,m}. \qquad (10.8)$$

Putting together all four contributions, the total rate of change of the probability function due to collisions can be expressed as

$$\left. \frac{\partial p(n, m; \mathbf{x}, t)}{\partial t} \right|_C = \dot{p}_{C1} + \dot{p}_{C2} + \dot{p}_{C3} + \dot{p}_{C4}. \qquad (10.9)$$

Information of the particles within the cloud can also be studied in terms of the moments of the probability function. The kth moment of the probability distribution function is defined as

$$n_k(m; \mathbf{x}, t) = \sum_{i=0}^{\infty} i^k \, p(i, m; \mathbf{x}, t) \, . \tag{10.10}$$

Since the sum of $p(i, m; \mathbf{x}, t)$ over all i adds up to unity, $n_0(m; \mathbf{x}, t) = 1$ always. The first moment $n_1(m; \mathbf{x}, t)$ of the distribution is of particular importance and represents the mean number of m-particles in a unit volume surrounding $\mathbf{x}$ at time t. For convenience, we shall omit the subscript and refer to the first moment as $n(m; \mathbf{x}, t)$. The variance of the probability distribution is given as

$$\Delta(m; \mathbf{x}, t) = n_2(m; \mathbf{x}, t) - n^2(m; \mathbf{x}, t) \, , \tag{10.11}$$

where $n_2(m; \mathbf{x}, t)$ is the second moment of $p(n, m; \mathbf{x}, t)$. Thus, the variance measures the spread of the probability function. The third and fourth moments measure the skewness and flatness of the probability function. Similar to $p(n, m; \mathbf{x}, t)$, the rate of change of the kth moment can also be split into two parts, one due to the movement of the particles and the other due to collisions between particles. Substituting the definitions of the moments, we get

$$\frac{\partial n_k(m; \mathbf{x}, t)}{\partial t} = \frac{\partial n_k(m; \mathbf{x}, t)}{\partial t}\bigg|_M + \frac{\partial n_k(m; \mathbf{x}, t)}{\partial t}\bigg|_C \, . \tag{10.12}$$

Substituting the above definitions, we obtain an expression for the rate of change of the mean number of m-particles due to the collisional process as (see Gillespie, 1975)

$$\frac{\partial n(m; \mathbf{x}, t)}{\partial t}\bigg|_C = \underbrace{\frac{1}{2} \sum_{m'=1}^{\infty} n(m') \, n(m'') \, c_{m',m''}}_{\text{source}} - \underbrace{n(m) \sum_{m'=1}^{\infty} n(m') \, c_{m,m'}}_{\text{sink}}$$
$$+ \underbrace{\frac{\epsilon(m)}{2} c_{m/2,m/2} \left(\Delta(m/2) - n(m/2) \right)}_{\text{source correction}} - \underbrace{c_{m,m} \left(\Delta(m) - n(m) \right)}_{\text{sink correction}} \, . \tag{10.13}$$

The first term on the right-hand side gives the number of m-particles created by collisions among smaller particles. The second term evaluates the number of m-particles lost due to collisions with particles of all sizes. The third and fourth terms are corrections to the first and second terms to take into account collisions between identical particles. The third term is the correction term when two particles of size $m/2$ collide to form an m-particle and is zero when m is odd. The fourth term is the correction term for the collision of two m-particles. The above equation for the rate of change of the mean particle number density involves the second moment (or variance) and if we derive an equation for the second moment (or variance), it will involve the third moment. Thus, the above equation is the first of an infinite open-ended sequence of equations. We refer to the above equation as the *statistically complete collision equation*.

We note that in the above equation, the coupling to the higher-order term occurs only through collisions between identical particles. If collisions between identical particles are not significant, this coupling can be avoided and the equation for $n(m; \mathbf{x}, t)$ will not involve higher-order statistical terms. Gillespie presented two arguments as to why these higher-order terms can be ignored. (i) As we will see in the next section, the collisional kernel $c_{m', m''}$ depends on the relative motion between two particles. Particles of the same size move similarly and as a result, their relative velocity is small. Since $c_{m,m}$ is small, the neglect of the correction terms can be justified. However, it should be noted that collision between identical particles is not identically zero due to the random motion of particles, either due to Brownian motion or the influence of ambient turbulence. (ii) The source term on the right corresponds to many possible combinations of collisions that will generate a particle of size m, whereas the correction term accounts for only one such possibility where the colliding particles are of the same size. The same applies for the sink terms. Thus, in a large system containing a wide spectrum of particle sizes, the correction terms arising from the collision of similar particles can be considered small and we obtain the equation

$$\frac{\partial n(m; \mathbf{x}, t)}{\partial t} = -\nabla \cdot (\mathbf{v}_m n(m)) + \underbrace{\frac{1}{2} \sum_{m'=1}^{m-1} n(m') n(m - m') c_{m', m-m'}}_{\text{source}} \underbrace{- n(m) \sum_{m'=1}^{\infty} n(m') c_{m, m'},}_{\text{sink}}$$

(10.14)

where the first term on the right accounts for the advective change in the mean number of m-particles, where $\mathbf{v}_m$ is the velocity field of the m particles (see Chapter 12 for a description of the particle velocity field). We have also rewritten the source term in an equivalent form which recognizes the fact that only agglomeration of two smaller particles results in the formation of an m-particle.

We now define $\bar{n}(m; t) = \int_\Omega n(m; \mathbf{x}, t) \mathbf{x} / \Omega$ to be the average number density of particles of size m within the cloud of volume Ω. We average the above equation over the entire volume of the cloud. Advection effects integrate to zero and do not affect the average number of particles. We then make the following approximations:

$$\frac{1}{\Omega} \int_\Omega n(m; \mathbf{x}, t) \, n(m'; \mathbf{x}, t) \, dx \approx \bar{n}(m; t) \bar{n}(m'; t) \quad \text{for all } m, m', \tag{10.15}$$

which implies that the correlation between variations in m and m'-particles and the spatial variance in the number density of the different-sized particles are small. We also consider the collision kernel to be spatially uniform within the cloud. With these approximations, we obtain the following kinetic equation for the average number density of particles of size m:

$$\frac{d\bar{n}(m; t)}{dt} = \frac{1}{2} \sum_{m'=1}^{m-1} \bar{n}(m'; t) \bar{n}(m - m'; t) \, c_{m', m-m'} - \bar{n}(m; t) \sum_{m'=1}^{\infty} \bar{n}(m'; t) \, c_{m, m'}. \tag{10.16}$$

The above kinetic equation was first derived by Smoluchowski in an intuitive manner and we shall refer to it as the *Smoluchowski kinetic equation*. As our derivation shows, it is an excellent equation which only ignores the higher-order statistical effects.

Problem 10.1 *System with only monomers.* Consider a large system consisting of only monomers. Assume that any particle of size two resulting from collision is removed from the system immediately. Assume that the system is initiated with only monomers of average number density n_0. Also, once the system is started, no more monomers are added into the system.

(a) From the Smoluchowski equation, obtain the following ordinary differential equation:

$$\frac{d\bar{n}(1;t)}{dt} = -\bar{n}(m;t)^2 \, c_{1,1} \, . \tag{10.17}$$

(b) If we assume the collision kernel $c_{1,1}$ to be time independent, obtain the following solution:

$$\bar{n}(1;t) = \frac{n_0}{1 - n_0 c_{1,1} t} \, . \tag{10.18}$$

The above solution starts at $\bar{n}(1;t = 0) = n_0$, and correctly evolves for small times. But at larger times the behavior of the solution is incorrect and leads to negative values. This illustrates the limitations of the approximation involved in Eq. (10.16) for collisions among particles of similar size. The effect of the approximation is magnified in the present problem where there are only collisions between particles of like size. As we will see in the following problem, the correct kinetic equation (ignoring the higher-order variance) that must be considered is

$$\frac{d\bar{n}(1;t)}{dt} = -\bar{n}(1;t)(\bar{n}(1;t) - 1) \, c_{1,1} \, . \tag{10.19}$$

The right-hand side of the above equation correctly recognizes the fact that in a unit volume containing $\bar{n}(1;t)$ monomers, there are $\bar{n}(1;t)(\bar{n}(1;t) - 1)/2$ independent pairs with collision probability $c_{1,1}$, and each collision results in the disappearance of two particles. For large $\bar{n}(1;t)$, the difference between Eq. (10.17) and Eq. (10.19) is very small, but as $\bar{n}(1;t)$ decreases the difference becomes large.

(c) Obtain the exact solution of Eq. (10.19) as

$$\bar{n}(1;t) = \frac{n_0}{n_0 - (n_0 - 1)\exp(-c_{1,1}t)} \, . \tag{10.20}$$

This solution has the correct asymptotic behavior $\bar{n}(1, t \to \infty) \to 1$, since there can be no more collisions once the system reduces to only one particle.

Problem 10.2 *System with only monomers – stochastically complete collision equation.* Again consider a large system consisting of only monomers. Let the conditions be identically the same as in the previous problem. In this problem we will go back to the stochastically complete Eq. (10.13). Obtain the following average number density equation from the stochastically complete collision equation:

$$\frac{d\overline{n}(1;t)}{dt} = -\left(\overline{n}(1;t)^2 - \overline{n}(1;t) + \overline{\Delta}(1;t)\right) c_{1,1}. \tag{10.21}$$

It should be noted that the stochastically complete Eq. (10.21) has the right form. If we ignore the variance $\Delta(1,t)$, the rest of the equation is of the form of Eq. (10.19) and therefore will yield an asymptotically correct result. This is one of the advantages of the stochastically complete form in comparison to the standard Smoluchowski kinetic equation. The above equation cannot be explicitly solved without a closure assumption for the variance. Nevertheless, due to the positive nature of the variance, it can be concluded that the true average monomer concentration will decrease faster than predicted by the kinetic equation. This more rapid decrease in the number of monomers is due to the nonlinear nature of the collision process. In the probabilistic picture, the number of collisions will be more frequent in instances of higher-than-average particle number density, and the number of collisions will be less frequent during instances of lower-than-average particle number density. But, when averaged, the increase in the number of collisions during instances of higher-than-average particle number density will outweigh the corresponding reductions in instances of lower-than-average number density. Thus explaining the faster rate of collisions with the inclusion of statistical variation.

This simple observation can be used to at least partially explain the faster formation of precipitation-sized rain droplets in a cloud. Actual clouds are highly inhomogeneous – there are regions of higher-than-average concentration and regions of lower-than-average droplet concentration. This spatial variation in droplet number density is different from the statistical variation discussed above. In a homogeneous cloud, by definition, all the statistical quantities, such as the moments n_k, are independent of spatial location. And the variance Δ represents variation in the number density of particles due to the inherent random nature of particle distribution. On the other hand, in an inhomogeneous cloud quantities such as n and Δ will themselves vary from one part of the cloud to another. However, the effect of spatial variation in droplet number density is expected to be similar to that of statistical variation encoded by Δ. Thus, the effect of number density variation over the volume of the cloud can be a significant factor contributing to further increase in collision among droplets. In Problem 10.2, the system was limited to only monomers. However, if the monomers were allowed to grow, the same increased collision rate will persist among droplets of all sizes and the end result is the faster production of large precipitation-sized droplets.

Bayewitz et al. (1974) obtained analytical solutions of the statistically complete collision equation with constant collision kernel. Their results suggest that in homogeneous systems of large population, the particle size distribution is adequately described by the simpler Smoluchowski kinetic equation. Whereas in inhomogeneous systems or in systems with a small number of particles, the statistical fluctuations will play a significant role and the kinetic equation will not reproduce the particle size distribution satisfactorily. In fact, they also concluded that in homogeneous systems with large number of particles, the probability distribution $p(n, m; t)$ approaches the Poisson distribution as time proceeds. Gillespie inferred the same behavior with a variable collision kernel but in the absence of correlations. Valioulis and List (1984) numerically evaluated the stochastic completeness of the kinetic equation based on a Monte Carlo method and obtained results consistent with the findings of Gillespie and Bayewitz et al.

Problem 10.3 *Monomers and dimers.* Consider a large system consisting of only monomers and dimers. Assume that any particles of size three or four produced from the collisions are removed and do not interact with the monomers and dimers. Also, assume that the system is initiated with only monomers of average number density N_0, without any dimers. Also, once the system is started, no more monomers are added into the system.

(a) From the Smoluchowski coagulation equation, obtain the following ordinary differential equations:

$$\frac{d\overline{n}(1;t)}{dt} = -\overline{n}(1;t)^2 c_{1,1} - \overline{n}(1;t)\,\overline{n}(2;t)\,c_{1,2},$$

$$\frac{d\overline{n}(2;t)}{dt} = \frac{1}{2}\overline{n}(1;t)^2 c_{1,1} - \overline{n}(1;t)\,\overline{n}(2;t)\,c_{1,2} - \overline{n}(2;t)^2 c_{2,2}. \qquad (10.22)$$

(b) Also consider the stochastically complete collision equation for this problem and obtain the corresponding equations for the monomer and dimer number density.

(c) Explore the differences from those obtained above for the simplified kinetic equation.

Problem 10.4 *Constant collision kernel.* In this problem, we will make a rather bold assumption that the collision kernel $c_{m,m'}$ is a constant c that is independent of the size of the two particles colliding. As we will see in the next section, this is not a good assumption. But we will use it for illustration purposes in this problem. The collision kernel can then be taken out of the summations in the Smoluchowski equation (10.16). In this problem, you will obtain an equation for the size-independent number density of a particle. Define

$$\overline{N}(t) = \sum_{m=1}^{\infty} \overline{n}(m,t) \qquad (10.23)$$

to be the overall number density of particles (i.e., the number of particles within a unit volume, irrespective of the particle size).

(a) Obtain the following equation for $\overline{N}(t)$ by summing Eq. (10.16) over all m:

$$\frac{d\overline{N}}{dt} = -\frac{1}{2}\overline{N}^2 c. \qquad (10.24)$$

(b) Assume the collision kernel c to be time independent as well, and obtain the following solution for the size-independent particle number density:

$$\overline{N}(t) = \frac{N_0}{1 - N_0\, ct/2}.$$

(10.25)

Here, N_0 is the initial average number density of particles within the system per unit volume, taking into account all particle sizes. As we saw in Problem 10.1, at larger times the behavior of the solution becomes negative and is therefore incorrect. Again, this incorrect behavior is due to improper accounting of collisions between like particles. But this problem is significant only when the total number density of particles becomes quite small.

(c) The correct approach is to resort to the statistically complete collision equation. Ignore the variance (or higher-order statistics) and sum Eq. (10.13) over all m to obtain

$$\frac{d\overline{N}}{dt} = -\frac{1}{2}\overline{N}(t)(\overline{N}(t) - 1)\,c.$$

(10.26)

For large $\overline{N}$, the difference between Eqs. (10.24) and (10.26) is small.

(d) Obtain the following solution:

$$\overline{N}(t) = \frac{N_0}{N_0 - (N_0 - 1)\exp(-ct/2)}.$$

(10.27)

This solution has the correct asymptotic behavior $\overline{N}(t \to \infty) \to 1$.

10.2   Collision Kernel

In the previous section, we derived a stochastically complete collision equation and simplified it to the classic kinetic equation. In order to complete the model, in this section, we will develop explicit expressions for the collision kernel. First, as an example, we will consider the collision kernel due to gravitational settling, linear shear flow, and Brownian motion. Then, we will derive the collision kernel for two other mechanisms: turbulence-induced collisions and when collisions are induced by the combined action of turbulent shear, particle inertia, and gravitational settling.

Referring to Eq. (10.1), in the following derivation we will consider collision efficiency to be equal to one, which implies that the presence of a particle has negligible effect on the motion of neighboring particles. This is probably not a good assumption when the particles are close to one another (Hocking and Jonas, 1970; Klett and Davis, 1973; Beard and Ochs, 1983). Thus, first we will obtain the geometric collision kernel taking into account the effects of turbulence, particle inertia, and gravity on collision. Once the geometric collision kernels are established, we will consider the local small-scale fluid mechanical perturbing effects of neighboring particles on the collision rate and thereby obtain a better understanding of collisional efficiency.

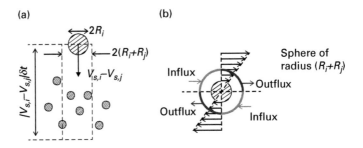

Figure 10.1 (a) Schematic of gravitational collision. (b) Schematic of shear collision.

10.2.1 Gravitational Collision Kernel

We start our discussion with the case of particles falling down due to gravity in an otherwise still ambient fluid. We take each particle to be falling at its still fluid settling velocity, which depends on the size of the particle. In this context, it is easy to visualize the collision mechanism. A larger particle that is falling down sweeps out a cylindrical volume over time and any slower-moving smaller particle that lies within this volume will geometrically collide with the larger particles. In this model, any particle has the potential to collide with a slower-moving particle that is located below it, but cannot catch up with the larger particles that are located below. By the same token, any particle could be swept away through collision by a larger, faster-moving particle falling from above.

Now consider collision of a particle of radius R_i colliding with a particle of radius R_j. Let their volumes be $\mathscr{V}_i = 4\pi R_i^3/3$ and $\mathscr{V}_j = 4\pi R_j^3/3$. Assuming the particles to be small so that their Reynolds number is small, their still fluid settling velocities can be expressed as

$$V_{s,i} = \frac{2(\rho-1)R_i^2}{9\nu_f} \quad \text{and} \quad V_{s,j} = \frac{2(\rho-1)R_j^2}{9\nu_f},\qquad(10.28)$$

where ρ is particle-to-fluid density ratio and ν_f is the kinematic viscosity of the fluid. If the i-particle is larger than the j-particle, $V_{s,i} > V_{s,j}$, and in the frame of reference attached to the falling j-particles, each i-particle sweeps out a cylindrical volume of radius $R = R_i + R_j$ and length $(V_{s,i} - V_{s,j})\delta t$, over a time interval δt. The number of j-particles within this swept volume is $n(j)\pi R^2(V_{s,i} - V_{s,j})\delta t$, where $n(j)$ is the number of j-particles per unit volume of the cloud (see Figure 10.1a). For the present discussion, we assume the cloud to be spatially homogeneous. Further, the number of i-particles within a unit volume is $n(i)$ and each of them sweeps out a volume to collide with the j-particles within its swept volume. Therefore, the total number of collisions per unit volume can be estimated as

$$n(i)\,n(j)\,\pi R^2(V_{s,i} - V_{s,j})\delta t\,.\qquad(10.29)$$

On the other hand, if the i-particle is smaller than the j-particle, $V_{s,j} > V_{s,i}$, and in the frame of reference attached to the falling i-particles, each j-particle sweeps out a cylindrical volume of radius R and length $(V_{s,j} - V_{s,i})\delta t$. Again, the total number of collisions per unit volume is given by Eq. (10.29) with a negative sign added. In either case, the number of collisions within a unit volume of the cloud over unit time becomes

$$n(i)\,n(j)\,\pi R^2 |V_{s,i} - V_{s,j}|\,. \tag{10.30}$$

Substituting for the radius R and the settling velocities in terms of particle volume, we obtain the collision kernel (without the number densities of the colliding particles) as

$$\text{Gravitational kernel:}\quad c_{i,j} = \left(\frac{6}{\pi}\right)^{4/3} \frac{\pi(\rho-1)g}{72\nu_f} \left| \mathscr{V}_i^{2/3} - \mathscr{V}_j^{2/3} \right| \left(\mathscr{V}_i^{1/3} + \mathscr{V}_j^{1/3} \right)^2 . \tag{10.31}$$

Note that the gravitational collision kernel does not have explicit dependence on either $\mathbf{x}$ or t. However, the number of collisions between i and j-particles within the cloud will depend on space and time. The space-time dependence comes from multiplication of $c_{i,j}$ by the number densities of i and j-particles, which are functions of $\mathbf{x}$ and t. It can also be seen that as i, j, and their difference $|i - j|$ increases, the collision kernel increases, indicating the enhanced probability of collisions. Thus, the assumption of constant collision kernel for all particle pairs is not a particularly good one. We also note that the collision kernel is zero between identical particles (i.e., for $i = j$), which validates the claim made in the previous section that in systems containing a large number of particles of varying size, collisions between like particles are of lesser importance. In the next subsections, we will see that this is true only in case of gravitational collision.

10.2.2   Shear Collision Kernel

The collision kernel $c_{i,j}(\mathbf{x}, t)$ is defined as the probability that a particle of size i collides with a particle of size j in unit time in a unit volume surrounding $\mathbf{x}$. This probability is in turn equal to the volume swept by the two particles of size i and j relative to each other. This relative volume swept by two particles of radii R_i and R_j can be evaluated as follows: fix the j-particle at the origin of a moving coordinate system. The radial component of the relative velocity w_r between the two particles on the surface Ω_R of a sphere of radius $R = R_i + R_j$ determines whether the particles will collide or not. A negative radial component of relative velocity indicates that the two particles are moving closer and a positive relative velocity indicates that the particles are moving apart. The common volume swept by the two particles in unit time can be obtained by an integration of the negative part of the radial relative velocity w_r^- over the surface Ω_R. Therefore, the collision kernel can be expressed as

$$c_{i,j}(\mathbf{x}, t) = -\oint_{\Omega_R} w_r^-\, d\Omega, \tag{10.32}$$

where the negative part of the radial relative velocity is defined as

$$w_r^- = \begin{cases} \mathbf{w}\cdot\mathbf{n} & \text{if } \mathbf{w}\cdot\mathbf{n} \leq 0, \\ 0 & \text{if } \mathbf{w}\cdot\mathbf{n} > 0, \end{cases} \tag{10.33}$$

where $\mathbf{w} = \mathbf{V}_i - \mathbf{V}_j$ is the relative velocity between the two particles. The integral is over the surface of the sphere of radius R and $\mathbf{n}$ is the unit normal vector at any point on the surface of the sphere.

In this subsection, we consider a simple linear shear flow where the x-component of velocity increases in the y-direction. We also assume the particles to move simply with the local fluid velocity. In a frame attached to any of the j-particles, the motion of all the i-particles is given by (see Figure 10.1b).

$$\mathbf{w} = Gy = GR\cos(\theta), \tag{10.34}$$

where G is the shear gradient. In the figure, both the particle of radius R_j centered at the origin and the bigger sphere of radius R are shown. It can be seen that the relative velocity of the i-particles is negative in the upper-left and lower-right quadrants, which is where collisions will occur. Particles that are above the $y = 0$ plane move faster and therefore will collide with the j-particle that is located at the origin coming from the left. Similarly, particles that are below the $y = 0$ plane move in the negative x-direction and therefore will collide with the j-particle from the right. There will be no collisions along the top-right and lower-left quadrants. The normal component of the relative velocity can be obtained as

$$w_r = GR\cos(\theta)\sin(\theta)\cos(\phi). \tag{10.35}$$

The integral of only the negative contribution can be calculated as

$$-\oint_{\Omega_R} w_r^- \, d\Omega = 4GR^3 \left(\int_0^{\pi/2} \sin^2(\theta)\cos(\theta)d\theta \right) \left(\int_0^{\pi/2} \cos(\phi)d\phi \right) = \frac{4}{3}GR^3. \tag{10.36}$$

Now, substituting the definition of R in terms of the volume of the colliding particles, we obtain

$$\text{Shear kernel:} \quad c_{i,j} = \frac{G}{\pi} \left(\mathcal{V}_i^{1/3} + \mathcal{V}_j^{1/3} \right)^3. \tag{10.37}$$

Unlike the gravitational collision kernel, the above shear kernel is nonzero even for particles of the same size. Furthermore, as can be expected, the probability of collision, or collision kernel, increases with increasing size of the colliding particles. The spatial homogeneity of the kernel now depends on the spatial homogeneity of shear.

10.2.3   Brownian Collision Kernel

In this subsection, we will discuss the case of particles undergoing Brownian motion in otherwise still fluid. Due to random motion, the particles occasionally come close enough and collide, and we are interested in obtaining the probability that a particle of size i will collide with a particle of size j in unit time in a unit volume, which we will call the collision kernel. Now, consider a coordinate system attached to one of the j-particles, which is undergoing random Brownian motion, and our interest is to find the rate at which the center of the i-particles approaches the spherical surface: $R = R_i + R_j$, which is the sum of the two radii. If the center of an i-particle, due to its

Brownian motion, in the frame attached to the j-particle, reaches the sphere of radius R, it is considered to have collided with the j-particle.

Brownian motion is a diffusion process. The collisional problem can be restated in the following way. In the frame attached to the j-particle, we are interested in the diffusive flux of i-particles toward the surface of the sphere of radius R. This is controlled by the diffusion equation, which under steady state is given by the Laplace equation $\nabla^2 n(i;\mathbf{x}) = 0$. Under dilute conditions, ignoring interparticle interaction effects, this diffusion toward the central j-particle is spherically symmetric and the Laplace equation becomes

$$\frac{1}{r^2}\frac{\partial}{\partial r}\left(r^2\frac{\partial n(i;r)}{\partial r}\right) = 0.\tag{10.38}$$

The appropriate far-field boundary condition is that as the radial distance r goes to infinity, the number density of the i-particle must go to its uniform value (i.e., $n(i, r \to \infty) = n(i)$). On the other hand, since any i-particle that approaches the sphere $r = R$ merges with the j-particle and disappears, $n(i, r = R) = 0$. Solving the Laplace equation with the above two boundary conditions, we obtain the following solution for the number density of i-particles in the immediate vicinity of a j-particle:

$$n(i;r) = n(i)\left(1 - \frac{R}{r}\right).\tag{10.39}$$

The above solution applies equally for the number density of j-particles in the immediate vicinity of an i-particle. In other words, the number densities of i and j-particles are $n(i)$ and $n(j)$, respectively, away from other particles. The number density of each decreases very close to the other particles, due to Brownian collision.

The net diffusional flux of i-particles across the spherical surface of radius R, whose surface area is $4\pi R^2$, is then given in terms of the gradient of the i-particle's number density. This can be expressed as

$$4\pi R^2 \mathcal{D}_{ij}\left(\frac{\partial n(i;r)}{\partial r}\right)_{r=R} = 4\pi R\, \mathcal{D}_{ij}n(i),\tag{10.40}$$

where $\mathcal{D}_{ij}$ is the relative diffusion of i-particles toward j-particles. Since the above flux of i-particles is for each j-particle, the above, when multiplied by the number density $n(j)$, will yield the number of collisions per unit time per unit volume. The final step is to obtain an expression for the diffusion coefficient $\mathcal{D}_{ij}$. We refer back to Einstein's diffusion coefficient derived in Eq. (7.59). That was for the diffusion of a particle from its initial position due to random Brownian motion. Here we are interested in the relative diffusion of the i-particle with respect to the j-particle. A similar approach yields the following expression:

$$\mathcal{D}_{ij} = \frac{k_b T_c}{6\pi\mu_f}\left(\frac{1}{R_i} + \frac{1}{R_j}\right).\tag{10.41}$$

Substituting, we obtain the final expression of the Brownian collision kernel in terms of particle volume as

$$\text{Brownian kernel:} \quad c_{i,j} = \frac{2k_b T_c}{3\mu_f} \frac{\left(\mathcal{V}_i^{1/3} + \mathcal{V}_j^{1/3}\right)^2}{\mathcal{V}_i^{1/3} \mathcal{V}_j^{1/3}}. \tag{10.42}$$

It is clear that each of the above three collisional mechanisms yielded analytical results for the geometric collision rate where the dependence on particle size (or volume) is distinct. We now proceed to obtain the geometric collision kernel for more complex mechanisms.

10.2.4 Turbulence Collision Kernel

Saffman and Turner (1956) derived a collision kernel for shear-induced coagulation in a homogeneous isotropic turbulent cloud. By making simplifying assumptions, Saffman and Turner also included the effects of coagulation due to particle inertia and differential settling velocity in their estimation of the collision kernel. In the past several decades others (Delichatsios and Probstein, 1975; Manton, 1977; de Almeida, 1976, 1979a,b; Grover and Pruppacher, 1985; Sundaram and Collins, 1997; Shaw, 2003; Grabowski and Wang, 2013) have studied the problem of turbulent collisions and arrived at different methods of estimation of the collision rate. For example, unlike Saffman and Turner, de Almeida included the effect of local small-scale aerodynamic interactions on geometric collisions but did not obtain an explicit expression for the collision kernel in terms of the turbulence parameters. There are a few experimental results (Birkner and Morgan, 1968; Okuyama et al., 1978; Lee et al., 1981; Duru et al., 2007; Bordas et al., 2013) to compare with the theoretical estimations of the collision kernel, and even in these experiments it is difficult to isolate the effects of shear, particle inertia, and gravity. Pearson et al. (1984) approximated turbulence as a three-dimensional stochastic motion and performed a Monte Carlo simulation of the coagulation process. Steady-state particle size distributions obtained by Pearson et al. agree well with those obtained from dimensional analysis using Saffman and Turner's collision kernel, which we shall consider in Section 10.4.

We will now summarize the pioneering analysis of Saffman and Turner in establishing the geometric collision kernel for collisions among particles in a turbulent flow. The analysis will be limited to homogeneous isotropic turbulence, so that the average collision kernel to be obtained is independent of spatial variation. Once again we ignore the effects of particle inertia and gravitational settling and thereby assume the particle velocity to be the same as that of local turbulent fluid velocity. In other words, the particles are behaving like fluid tracers. The effects of particle inertia and gravitational settling will be added in the next section. The relative velocity between the two particles can be written in terms of the Eulerian fluid velocity $u(x, t)$ as

$$\mathbf{w} = \mathbf{u}(\mathbf{X}_i(t), t) - \mathbf{u}(\mathbf{X}_j(t), t), \tag{10.43}$$

where $\mathbf{X}_i$ and $\mathbf{X}_j$ are the instantaneous positions of the i and j-particles. In the present scenario, the relative velocity between a particle located at $\mathbf{X}$ and another separated by a vector $\mathbf{r}$ can be expressed as a field variable $\mathbf{w}(\mathbf{r}, t; \mathbf{x})$. If the fluid flow is incompressible, then the relative particle velocity field satisfies $\nabla_r \cdot \mathbf{w}(\mathbf{r}, t; \mathbf{x}) = 0$, where the gradient is with respect to $\mathbf{r}$. The radial velocity w_r that appears in Eq. (10.32) is nothing but the radial component of $\mathbf{w}(|\mathbf{r}| = R, t; \mathbf{x})$. The divergence-free condition stated above leads to

$$\oint_{\Omega_R} w_r^+ \, d\Omega = - \oint_{\Omega_R} w_r^- \, d\Omega, \qquad (10.44)$$

where the positive part of the radial relative velocity is defined similar to the negative part given in Eq. (10.33). Substituting in Eq. (10.32), we get

$$c_{i,j}(\mathbf{x}, t) = \frac{1}{2} \oint_{\Omega_R} |w_r| \, d\Omega. \qquad (10.45)$$

The above-defined collision kernel is a local and instantaneous kernel. It offers an explicit expression for the instantaneous collision probability at location $\mathbf{x}$ at time t only in case we know the relative velocity field. Such precise knowledge of the flow is possible only in case of laminar flows, and we did derive the collision kernel in this manner in the case of linear shear flow. In a turbulent flow, an explicit expression for $\mathbf{u}(\mathbf{x}, t)$ is not available to allow explicit calculation of $c_{i,j}(\mathbf{x}, t)$. A more appropriate quantity in case of turbulent flow is the ensemble average of $c_{i,j}(\mathbf{x}, t)$. Taking an ensemble average of the above equation over all turbulence realizations, we get

$$c_{i,j} = \langle c_{i,j}(\mathbf{x}, t) \rangle = \frac{1}{2} \oint_{\Omega_R} \langle |w_r| \rangle \, d\Omega, \qquad (10.46)$$

where the angle brackets represent an ensemble average. Since the relative velocity between particles has a field representation, in the absence of inertia effects, the collision kernel $c_{i,j}$ is also a field variable and is independent of the initial particle positions and velocities. From its definition, it is clear that the collision kernel is independent of particle number density and any inhomogeneity in particle distribution will not affect the homogeneity of the collision kernel. Therefore, in a homogeneous isotropic turbulent cloud, the collision kernel is homogeneous and isotropic.

In isotropic turbulence, the relative velocity w_r will depend only on the distance between the two particles and not on their relative orientation, so the integration over the surface of a sphere of radius R yields

$$c_{i,j} = 2\pi R^2 \langle |w_x| \rangle, \qquad (10.47)$$

where w_x is the x-component of relative velocity between two particles separated by a distance R along the x-direction. Even in non-isotropic turbulence, the above approximation will be valid for sufficiently small particles, since the small-scale structure of turbulence is nearly isotropic. If the particle radii are smaller than the length scale of the small eddies, which is usually the case in many applications:

$$\langle |w_x| \rangle = R \left\langle \left\| \frac{\partial u}{\partial x} \right\| \right\rangle, \qquad (10.48)$$

where u is the x-component of the fluid velocity. Saffman and Turner assumed $\partial u / \partial x$ to be normally distributed with zero mean as

$$Pr\{\partial u/\partial x = \xi\} = \frac{1}{\sigma\sqrt{2\pi}}\exp\left(-\frac{1}{2}\frac{\xi^2}{\sigma^2}\right), \tag{10.49}$$

where the variance is related to the local dissipation rate ϵ through the relation

$$\sigma^2 = \frac{\epsilon}{15\nu_f}. \tag{10.50}$$

From this he obtained

$$\left\langle\left|\frac{\partial u}{\partial x}\right|\right\rangle = \int_{-\infty}^{\infty}\frac{|\xi|}{\sigma\sqrt{2\pi}}\exp\left(-\frac{1}{2}\frac{\xi^2}{\sigma^2}\right)d\xi = \sqrt{\frac{2\epsilon}{15\pi\nu_f}}. \tag{10.51}$$

Substituting this into Eq. (10.47), we obtain the final expression for the collision kernel as

$$\text{Turbulence kernel:}\quad c_{i,j} = \sqrt{\frac{3\epsilon}{10\pi\nu_f}}\left(\mathcal{V}_i^{1/3} + \mathcal{V}_j^{1/3}\right)^3. \tag{10.52}$$

As can be expected, the turbulence kernel has the same functional form as the shear collision kernel, since in turbulence, local shear is the mechanism that brings particles to collide. By comparing the above with the shear collision kernel in Eq. (10.37), it can be seen that for isotropic turbulence the effective shear $G = \sqrt{3\pi\epsilon/(10\nu_f)}$.

Problem 10.5 *Lognormal distribution.* The estimation of the turbulence collision kernel depends on what we assume about the statistical distribution of $\partial u/\partial x$. In other words, the statistical distribution of relative velocity between the two particles. In this problem you will assume $|\partial u/\partial x|$ to be lognormally distributed (Gurvich and Yaglom, 1967) and infer the effect of this change on the collision kernel. Start with the following distribution instead of Eq. (10.49):

$$Pr\left\{\left|\frac{\partial u}{\partial x}\right| = \xi\right\} = \frac{1}{\sigma\xi\sqrt{2\pi}}\exp\left(-\frac{1}{2}\frac{(\ln\xi - m)^2}{\sigma^2}\right), \tag{10.53}$$

where $m = \langle\ln|\partial u/\partial x|\rangle$ is the logarithmic mean and $\sigma^2 = \langle(\ln|\partial u/\partial x| - m)^2\rangle$ is the logarithmic variance.

(a) First prove the following property of the lognormal distribution:

$$\left\langle\left|\frac{\partial u}{\partial x}\right|^n\right\rangle = \exp\left(nm + \frac{1}{2}n^2\sigma^2\right). \tag{10.54}$$

We now define

$$\left\langle\left|\frac{\partial u}{\partial x}\right|^2\right\rangle = \frac{\epsilon}{15\nu_f}\quad\text{and}\quad S_k = \frac{\langle|\partial u/\partial x|^4\rangle}{\langle|\partial u/\partial x|^2\rangle^2}, \tag{10.55}$$

where the first expression is the same as that used in obtaining the Saffman–Turner collision kernel and the second is the definition of flatness of absolute value of velocity gradient.

(b) Using the above relations, you should obtain $|\partial u/\partial x|$ as

$$\left\langle \left\| \frac{\partial u}{\partial x} \right\| \right\rangle = \exp\left(m + \frac{1}{2}\sigma^2 \right) = \sqrt{\frac{\epsilon}{15\nu_f}} \left(\frac{1}{S_k} \right)^{1/8}. \tag{10.56}$$

(c) Substituting in its definition, obtain the turbulent collision kernel as

$$c_{i,j} = \sqrt{\frac{3\epsilon}{20\nu_f}} \left(\frac{1}{S_k} \right)^{1/8} \left(\mathcal{V}_i^{1/3} + \mathcal{V}_j^{1/3} \right)^3. \tag{10.57}$$

It is clear that the functional form remains the same, but the prefactor changes as the statistical nature of turbulence is differently approximated.

10.3 Combined Effects of Turbulence, Inertia, and Gravity

In general, the collision mechanisms that we considered in the previous section as examples act together in concert with each influencing the other. Of particular importance are the combined effects of particle inertia and gravitational settling in the presence of fluid turbulence. As we saw in the previous section, gravitational settling and the effect of particles faithfully following flow turbulence each independently result in their respective collisional kernels. It can easily be shown that particle inertia also results in interparticle collisions. Inertia-induced collisions are similar to gravitational collision in the sense that they occur only between particles of different sizes. Consider a region of fluid undergoing solid-body rotation, where the circumferential fluid velocity increases linearly with distance from the axis of rotation. If particles within this solid-body rotation were to move perfectly with the fluid, there will be no interparticle collision, since all the particles are rotating about the axis at the same angular velocity (i.e., any two particles move around the axis as if they are part of a solid object with their relative position remaining the same). If we now introduce particle inertia, heavier-than-fluid particles, owing to their inertia, will spiral out of the vortex, while lighter-than-fluid particle will spiral inward toward the center. In both cases, the rate of outward or inward motion will depend on particle inertia (or particle time scale), which in turn depends on particle diameter. Since a particle of larger τ_p will spiral out more than a particle of smaller τ_p, the two can collide due to inertia-induced relative velocity. This mechanism is not active when the two particles are of the same τ_p and thus spiral out the same way.

In the turbulent flow analysis of the previous section, the radial component of relative velocity between the colliding particles was entirely due to the difference in the fluid velocity. That analysis was without the inclusion of the inertia effect on particle motion. With the inclusion of inertia, we expect the average relative velocity between particles of different inertia (or different diameter) to increase and as a result, the rate of collisions between dissimilar particles will increase. Since particle motion is simultaneously influenced by gravitational motion along the vertical direction, the statistics of relative velocity between the colliding particles must be evaluated with the combined effects of turbulence, particle inertia, and particle settling velocity. The latter two mechanisms will add to the effect of turbulence, but for particles of similar size, only turbulence will be active, since the effect of inertia and gravity will be absent. It should also be noted that in isotropic turbulence, the probability of relative velocity between the two particles depends only on their separation length and not on their orientation. The effect of inertia is also isotropic in isotropic turbulence. In contrast, the relative velocity due to gravitational settling is directionally dependent, as we discussed in the previous section. Thus, combining these mechanisms must be carefully handled.

The original attempt at obtaining the collision kernel taking into consideration all three effects was by Saffman and Turner. Some of the limitations of their analysis were addressed and improved upon by Wang et al. (2000). The goal is to get an improved estimate of the surface integral of radial relative velocity given in Eq. (10.46). To achieve this goal, we start with the following expression for the particle velocity:

$$\mathbf{V} = \mathbf{u} - (1 - \beta)\tau_p \left(\frac{D\mathbf{u}}{Dt} - \mathbf{g} \right) + O\left(\mathrm{St}^2 \right), \tag{10.58}$$

which is valid in the limit of small-particle Stokes number (or small-particle time scale compared to the local fluid time scale). This equation will be derived rigorously later in Chapter 12, and here we will accept it as a small Stokes number solution of particle motion. In the above, we recall the density parameter $\beta = 3/(2\rho + 1)$. To leading order, the velocity $\mathbf{V}$ of a particle of small inertia is the same as the local fluid velocity $\mathbf{u}$ and the difference is due to both the total fluid acceleration $D\mathbf{u}/Dt$ at the particle location and gravity. Here the effect of fluid turbulence on particle motion comes through the complex space-time dependence of $\mathbf{u}$ in a turbulent flow. The effect of particle inertia comes from $(1-\beta)\tau_p D\mathbf{u}/Dt$ and the effect of gravitational settling comes from $(1 - \beta)\tau_p \mathbf{g}$, which is equal to the still fluid settling velocity of the particle.

Let us begin with only the effects of turbulence and particle inertia (i.e., without including the effect of gravity, for now). In this limit, the statistics of relative velocity between a pair of particles is isotropic. In other words, the relative velocity between two particles depends only on their separation distance and not on the orientation of their separation vector. To calculate the surface integral of radial relative velocity given in Eq. (10.46), we could simply consider the two particles to be separated along the x-direction by the distance R and use their relative velocity along the x-direction, as was done in obtaining Eq. (10.47). By isotropy, the statistics of radial relative velocity at all other positions of the second particle on the surface of the sphere of radius R will be the same. Again, if we take the particle radii to be smaller than the Kolmogorov length scale, the difference in the x-component of fluid velocity at the location of the two colliding particles that are separated along the x-direction can be taken to be $R(\partial u/\partial x)$ and with the added effect of inertia (without the anisotropic effect of gravity), we can write

$$w_x = R \frac{\partial u}{\partial x} + (1 - \beta)(\tau_{p2} - \tau_{p1}) \frac{Du}{Dt}, \tag{10.59}$$

where we have assumed the fluid acceleration to be the same at the two particles, since their separation R is small. According to the above equation, w_x is a random variable and its randomness comes from both fluctuating $\partial u/\partial x$ and Du/Dt. If we assume the fluctuations of $\partial u/\partial x$ and Du/Dt to be Gaussian with zero mean and independent of each other, then we obtain

$$\langle |w_r| \rangle = \langle |w_x| \rangle = (2/\pi)^{1/2} \left[R^2 \frac{\epsilon}{15\nu_f} + (1-\beta)^2 (\tau_{p2} - \tau_{p1})^2 \left\langle \left(\frac{Du}{Dt} \right)^2 \right\rangle \right]^{1/2}, \quad (10.60)$$

where we have set $\langle |w_x| \rangle$ for two particles separated along the x-axis equal to the required $\langle |w_r| \rangle$, owing to isotropy. The first term within the square brackets will yield the same turbulence collision kernel of Saffman and Turner. The second term accounts for the added effect of inertia. It always increases the ensemble-averaged radial relative velocity and as a result the collision rate. The added effect of particle inertia increases with the difference $\tau_{p2} - \tau_{p1}$. Since $(1-\beta) = (\rho - 1)/(\rho + 1/2)$, the inertia effect is zero if the particles are neutrally buoyant (i.e., if $\rho = \rho_p/\rho_f = 1$).

With the inclusion of gravity, isotropy is broken and the statistics of $|w_r|$ is only axisymmetric. The evaluation of the integral becomes more complicated. For details, we recommend the reader to Hu and Mei (1998), Ayala et al. (2008), and Wang et al. (2000). The collision kernel with all the effects included can be expressed as

$$
\begin{aligned}
\text{Turbulence + Inertia} \\
\text{+ Gravity kernel}
\end{aligned} \Big\} : \quad
\begin{aligned}
c_{i,j} = 2\sqrt{2\pi}R^2 \Big[& R^2 \frac{\epsilon}{15\nu_f} + (1-\beta)^2 (\tau_{p2} - \tau_{p1})^2 \left\langle \left(\frac{Du}{Dt} \right)^2 \right\rangle \\
& + 2(1-\beta)^2 \tau_{p1} \tau_{p2} \left\langle \left(\frac{Du}{Dt} \right)^2 \right\rangle \frac{R^2}{\lambda^2} \\
& + \frac{\pi}{8}(1-\beta)^2 (\tau_{p2} - \tau_{p1})^2 g^2 \Big]^{1/2},
\end{aligned}
$$
$$(10.61)$$

where $R = (3/(4\pi))^{1/3} \left(\mathcal{V}_i^{1/3} + \mathcal{V}_j^{1/3} \right)$ and λ is the longitudinal Taylor microscale of fluid acceleration. There are four terms on the right-hand side. The first two are the same as those discussed above. The third term is a coupling term that accounts for the fact that Du/Dt varies between the two particles separated by distance R, which was assumed to be the same when we derived Eq. (10.60). The last term accounts for the anisotropic effect of relative velocity between the two particles due to gravitational settling. It should also be noted that the factor $(1-\beta)^2 = (\rho-1)^2/(\rho+1/2)^2$ in the above equation appears only as $(1-\rho)^2$ in the references. The difference between them is small for large values of ρ, such as collisions between water droplets in air. But the difference can be large in case of lighter-than-fluid particles, such as collisions between bubbles. The extra factor $\rho + 1/2$ in the denominator accounts for the added-mass contribution.

To complete the above formulation, parameterizations are needed for both the ensemble average of fluid acceleration and the longitudinal Taylor microscale of acceleration. For the former, different proposals have been made (Hill, 2002; Zaichik et al., 2003; Ayala et al., 2008) in terms of properties of Kolmogorov-scale eddies and for the latter, Hu and Mei (1998) provide an appropriate expression.

10.3.1 Effect of Particle Accumulation

In all the above definitions, the rate of collisions between particles of size i and j is obtained by multiplying the collision kernel with $n_i n_j$, which represents the number of particles of size i and j in a unit volume around the point of interest. This assumes that given an i-particle whose center is located at $\mathbf{X}$, the probability of finding a j-particle anywhere within the unit volume is uniform. But this is not so. The fact that we have placed the i-particle at $\mathbf{X}$ changes the probability of finding another particle in its neighborhood. We have seen this earlier in the context of a monodisperse distribution of particles as the radial distribution function. In Section 3.4, the radial distribution function $g(r)$ was defined as the probability of finding a neighbor at a distance r away from a reference particle, relative to a perfectly uniform distribution. Of particular importance in interparticle collision is the value of $g(R)$, which indicates the probability of finding a neighbor at the collisional separation of $r = R$ compared to a perfectly uniform distribution. Thus, for $g(R) > 1$, the actual number of collisions will be greater than that predicted only based on assuming the particles to be perfectly uniformly distributed.

Several additional points must be discussed. First, this effect of pair probability is over and above that of any non-uniformity in the distribution of particles. In other words, in a system that is non-uniformly distributed, at any point within the system the local number density of particles n_i and n_j will be different from their system-wise averaged values. In regions of higher particle concentration, n_i and n_j will be larger than the system average and the enhanced probability of collision in such a local region of higher particle concentration is already taken care of by multiplication of collision efficiency with the product $n_i n_j$. The factor $g(R)$ accounts for the additional effect of close-range inhomogeneity that partly arises from the fact that once the i-particle is placed at $\mathbf{X}$, the neighbors are excluded from this volume due to the finite size effect of the particles and partly due to the action of turbulence and particle inertia. Though $g(r)$ quickly approaches far-field value a few diameters away from the reference particle, it can be substantially different from unity for $r = R$.

In the case of collisions between particles of size i and j, the definition of the radial distribution function of a monodisperse system must be extended. We define $g_{ij}(r)$ as the probability of finding a neighboring j-particle at a distance r away from a reference i-particle, relative to a perfectly uniform distribution of j-particles. In the monodisperse and polydisperse collision cases, we need to multiply the collision kernels defined in the previous sections with $g(R)$ or $g_{ij}(R)$ in order to obtain the correct collision kernel that properly accounts for the close-range inhomogeneity of the pair distribution. This increase in collision rate due to the higher probability of a close-range particle pair is termed the particle accumulation effect.

Due to active research in the context of rain formation, much of the modeling effort of the radial distribution function at particle contact has been primarily in the limit of low particle volume fraction. In this limit, there have been a number of models of $g(R)$ and $g_{ij}(R)$ in terms of the ratio R/η and particle Stokes numbers (η is the Kolmogorov length scale). One such model by Falkovich et al. (2002) is

$$g_{ij}(R) = \left[\frac{R}{\eta} + \left(1 + \frac{g\tau_k^2}{\eta} \right) \sqrt{1 + \left(\frac{\tau_p g}{v_k} \right)^2} \ | \ St_2 - St_1 \ | \right]^{-\alpha}, \qquad (10.62)$$

where τ_k and v_k are Kolmogorov time and velocity scales. An expression for α is given in Falkovich et al. (2002). There are many other models of $g(R)$ and $g_{ij}(R)$ and the reader is referred to the recent review by Wang (2022).

10.3.2 Collision Efficiency

The geometric collision kernel is defined based on the assumption that the paths taken by the i and j-particles are based on the local turbulent flow, the inertial response of the particle to this turbulent flow, and the added effect of gravity. In calculating the collision efficiency, we ignored the crucial fact that as the two particles approach each other, the flow seen by each of them is not simply the ambient turbulent flow. Each particle will also be subjected to the perturbation flow induced by the other.

The perturbative influence of each particle on the other can best be explained by considering the simplest example of a larger particle falling and colliding with a smaller particle, whose fall velocity is smaller. In the frame attached to the larger particle, the smaller particle is moving up. In the evaluation of the gravitational collision kernel given in Eq. (10.31), it is assumed that the relative motion of the smaller particle with respect to the larger one is always directed upward. In this case, the collisional cross-section is a circle of radius $R_i + R_j$. That is, all smaller particles within a circular tube of cross-sectional radius $R = R_i + R_j$ will end up colliding with the larger one. While this vertical motion of a smaller particle is true when the two particles are far away, when the smaller particle approaches closer to the larger one, the motion of the smaller particle is influenced by the perturbed streamlines of the flow around the larger one. Also of importance is the perturbation flow around the smaller particle. The effect of the perturbation flow is to decrease the radius of the collisional cross-section. In other words, only those j-particles whose center lies within a cylinder of radius R' end up colliding with the i-particle. The j-particles whose center falls within the annular region of radius $R \leq r \leq R'$, which would have collided had their trajectory been unaffected by the local flow, have now avoided collision by dodging away from the i-particle. If we define the collision efficiency as the fraction of geometric collisions that end up being real collisions, in this context of falling particles, the collision efficiency becomes R'^2/R^2.

From this example it can be expected that particle inertia will play an important role in determining the collision efficiency. While the flow around each particle as they fall in a quiescent medium may only depend on the Reynolds number, how the particles respond to the perturbation flow of the other particle will also depend on the Stokes number. If the Stokes numbers of both the i and j-particles are large, despite the influence of the perturbation flow around the other particle, each particle will be slugging to deviate from its original vertical motion. As a result, the collision efficiency will be very nearly unity, and the geometric collision kernel will remain appropriate to calculate the actual collision rate. In the other extreme, when both particles are of small Stokes number, both particles will tend to move away from the path of the other, in response to the perturbation flow, and the resulting collision efficiency will be much lower than unity. In this limit, the collision kernel without collision efficiency will yield a substantial overprediction of the actual collision rate.

While general results are hard to obtain, collision efficiency has been obtained numerically by solving the motion of two particles in close proximity. It must be stressed that numerical simulations must resolve the flow around the two approaching particles accurately, especially when their separation is small, in order to correctly predict their future motion and the possibility of collision. Fortunately, we have already presented the solution of this two-particle problem as the comprehensive solution of Jeffrey and Onishi (1984), in the Stokes flow limit. For the case of particles smaller than the Kolmogorov scale, their Reynolds number is small and the results of Jeffrey and Onishi are appropriate. One still needs to track the translational and rotational motion of the i and j-particles over time numerically, while detecting the possibility of collision between the two particles by monitoring the time evolution of the distance between the particle centers. In the case of gravitational settling, the particle motion at each time is influenced by both the gravitational force acting on the particle (minus the buoyancy force) plus the hydrodynamic interaction force between the two particles arising from their perturbation flow, which was expressed in terms of their relative position and relative translational and rotational velocity in Section 9.3.1. By releasing the smaller particle sufficiently below the larger particle and integrating their motion, it can be determined whether they collide or not. The density ratio between the particle and the surrounding fluid is an important parameter and, along with the particle size, it determines the particle time scale and the Stokes number.

Instead of the sedimentation problem, one can consider the motion of particles in isotropic turbulence. Here again, particles that are smaller than the Kolmogorov scale are moved according to the BBO equation of motion. Only when two particles come close enough can the additional forces acting on them due to their perturbation flow be calculated using the formulation of Section 9.3.1. Over a long period of time, the total number of collisions occurring within the system can be calculated with and without the close-range force between the two particles being taken into account. The ratio of number of collisions with the close-range interaction force versus the number of collisions without the close-range force will then be the collision efficiency.

From the above discussions of the sedimentation and isotropic turbulence problems, it must be clear that collision efficiency also depends on the nature of the flow and the problem at hand. Collision efficiency will depend on the relative importance of turbulence, particle inertia, and gravitational settling. In fact, the motion of particles when they are very close to each other is influenced by factors other than hydro or aerodynamics. Interparticle electrostatic and other attractive and repulsive forces can influence their motion when two particles get close to each other, and therefore can greatly modify the collision rate. The effect of these nonhydrodynamic interparticle forces must also be included in the collision efficiency.

10.4   Particle Number Density Distribution

In this section we will consider solutions of the Smoluchowski equation (10.16). We will consider the following continuous version of the Smoluchowski equation in order to obtain analytic solutions:

$$\frac{d\bar{n}(\mathcal{V};t)}{dt} = \frac{1}{2}\int_0^{\mathcal{V}} \bar{n}(\mathcal{V}';t)\,\bar{n}(\mathcal{V}-\mathcal{V}';t)\,c(\mathcal{V}',\mathcal{V}-\mathcal{V}')\,d\mathcal{V}'$$

$$- \bar{n}(\mathcal{V};t)\int_0^{\infty} \bar{n}(\mathcal{V}';t)\,c(\mathcal{V},\mathcal{V}')\,d\mathcal{V}' + S(\mathcal{V}). \qquad (10.63)$$

The continuous version is written in terms of particle volume and the collision kernels have already been written in terms of particle volume, although now particle volume is not restricted to be integer multiples of the monomer volume. Again, the first term on the right-hand side is the rate of production of particles of volume $\mathcal{V}$ due to collisions of two smaller particles and the second term on the right-hand side corresponds to the loss of a particle of volume $\mathcal{V}$ due to collision with another particle of any size. We have also added the third term on the right, which corresponds to an external source or sink of particles into the system.

The exact closed-form solutions of the above integro-differential equation are hard to obtain. In this section, following the work of Hunt (1982), we will obtain self-similar solutions using simple dimensional arguments. Four different conditions are stated for the existence of such steady self-similar solutions. (i) We assume that there is a steady source of particles at small size $\mathcal{V}_{min}$ and there is a steady sink of particles at large size $\mathcal{V}_{max}$. In other words, small particles of volume $\mathcal{V}_{min}$ are randomly nucleated within the domain at a constant rate and at the same time all particles of size larger than $\mathcal{V}_{max}$ that form from collision are immediately removed from the system. In the context of rain formation, one can think of small droplets being created by condensation and larger droplets falling out of the cloud as rain. This source and sink will establish a steady self-similar number distribution between these two particle sizes. Note that at steady state, the rate of volume of small particles introduced into the system will be balanced by the rate of volume of large particles that leave the system. Thus, the total volume of particles within the system will become a constant. (ii) The analysis will assume only one collisional mechanism to be active at a time, which will allow simple scaling arguments. In reality, multiple mechanisms will be simultaneously active, but even then, each mechanism may be dominant only in a certain size range. In which case, the theory to be discussed below will apply within each size range. (iii) As the third assumption, Hunt (1982) considered collision efficiency to be a constant and thus the self-similar number spectrum depends only on the geometric collision kernel. (iv) Finally, each mechanism discussed in the previous section is characterized by a single parameter, which can be listed as

$$\text{Gravity:} \quad \frac{g(\rho-1)}{\nu_f} \quad \left[L^3/l^4 t\right],$$

$$\text{Shear:} \quad G \quad \left[L^3/l^3 t\right], \qquad (10.64)$$

$$\text{Brownian:} \quad \frac{k_b T_c}{\mu_f} \quad \left[L^3/t\right],$$

where the dimension of each key parameter combination is listed as the right column. In the dimension, L stands for the length scale of the fluid (e.g., the size of the box within which the collisions are taking place), l is the length scale of the droplet. How these dimensions are obtained will be discussed below.

We note that $\bar{n}(\mathcal{V})d\mathcal{V}$ stands for the number of particles per unit volume of the cloud, per unit width of particle volume. Therefore, the dimension of $\bar{n}$ is $[1/(L^3 l^3)]$. From this definition and by balancing the different terms of Eq. (10.63), it can be seen that the dimension of the collision kernel is $[L^3/t]$. Substituting this in the definition of the gravitational kernel given in Eq. (10.31), we obtain the dimension of $g(\rho - 1)/\nu_f$ as given above. It is important to note that even though the dimension of $g(\rho - 1)/\nu_f$ can be taken to be one over length-time, it should be properly denoted as $\left[L^3/l^4 t\right]$ in order to obtain the correct result. The other dimensions can be obtained similarly.

We now follow a process similar to that used in obtaining the inertial energy spectrum of turbulence. We define E to be the volume flux of particles that passes through any given particle size. That is, the rate at which the small particles are introduced into the system is E. It then passes through all the intermediate sizes and finally the rate at which the particles are removed from the system is also E. The dimension of E is $[l^3/(L^3 t)]$, which corresponds to the volume of particles per unit volume of the cloud per unit time. In the case of gravitational collision, we hypothesize that $\bar{n}(\mathcal{V})$ must depend only on the variables $\mathcal{V}$, E, and the gravitational collision parameter $g(\rho - 1)/\nu_f$. From dimensional arguments, the only possible self-similar form is

$$\text{Gravitational collision:} \quad \bar{n}(\mathcal{V}) = A_g \left(\frac{E \nu_f}{g(\rho - 1)} \right)^{1/2} \mathcal{V}^{-13/6}, \tag{10.65}$$

where A_g is a constant that cannot be determined from the dimensional analysis. The corresponding results for shear (and turbulent) and Brownian collisions are

$$\text{Shear/turbulent collision:} \quad \bar{n}(\mathcal{V}) = A_s \left(\frac{E}{G} \right)^{1/2} \mathcal{V}^{-2}, \tag{10.66}$$

$$\text{Brownian collision:} \quad \bar{n}(\mathcal{V}) = A_b \left(\frac{E \mu_f}{k_b T_c} \right)^{1/2} \mathcal{V}^{-3/2}. \tag{10.67}$$

These elegant results for gravitational and Brownian collisions were also obtained earlier by Friedlander (1960a,b). For further insightful analysis of Hunt's self-similar results, the reader is referred to Pushkin and Aref (2002).

One of the advantages of the above self-similar solutions is that they are independent of initial conditions and they can be viewed as the intrinsic effect of a collision mechanism. However, in practical situations, the assumptions that went into the derivation of the self-similar results are not valid. For example, the steady source of small-sized particles and the sink at large particles is not the norm in every application. Nevertheless, in a certain range of particle size, if it is known that a particular collision mechanism is the dominant mode of particle growth, then the local value of $\bar{n}$ will follow the self-similar spectra, as long as there is a constant flux of particle volume

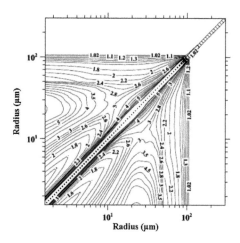

Figure 10.2 Contour plot of the ratio of turbulence-informed collision kernel to the corresponding Hall collision kernel for varying combinations of radii of the colliding particles. Taken from Wang and Grabowski (2009).

through this range of particle size (i.e., there should be no additional source or sink within this range of particle size).

There have also been many attempts to solve the Smoluchowski coagulation equation numerically when conditions of self-similarity are not met. One classic problem is the formation of rain (i.e., larger precipitation-sized droplets) from a cloud that is initially seeded with many droplets of much smaller size. Here the key question is how long does it take for raindrops to form from the initial smaller droplets? Wang and Grabowski (2009) considered an initial number density of

$$\bar{n}(\mathcal{V}, t = 0) = \frac{L_0}{\mathcal{V}_0^2} \exp\left(-\frac{\mathcal{V}}{\mathcal{V}_0}\right), \qquad (10.68)$$

where the constant L_0 was chosen such that one cubic meter of the cloud contains a total volume of $10^{-6}\,\mathrm{m}^3$ of droplets. That is, the volume fraction of droplets within the cloud is 10^{-6}. The initial mean volume of a single droplet is chosen to be $\mathcal{V}_0 = 3.3 \times 10^{-15}\,\mathrm{m}^3$, corresponding to a droplet diameter of 18.6 μm. With this initial condition, Eq. (10.63) was integrated in time to observe the development of $\bar{n}(\mathcal{V})$. The only important quantity yet to be determined is the collision kernel $c(\mathcal{V}, \mathcal{V}')$. First, they used only the gravitational collision kernel (10.31) multiplied by an appropriate collision efficiency function, as was used by Hall, which they called the Hall kernel. This kernel ignored the effect of turbulence and turbulence-induced inertial effects.

For a typical atmospheric turbulence characterized by rms velocity fluctuation of about 2 m/s and a dissipation rate of $0.4\,\mathrm{m}^2/\mathrm{s}^3$, they calculated the collision kernel taking into account the combined effects of turbulence, inertia, gravity, and collision efficiency. The ratio of the turbulence-informed collision kernel of Wang and Grabowski (2009) versus the collision kernel of Hall is shown in Figure 10.2 as contour plots, plotted against the radii of the two colliding droplets. Note that the Hall collision

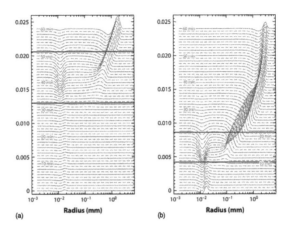

Figure 10.3 The time evolution of the time derivative of the number of drops weighted by their volume plotted as a function of droplet radius at varying times. (a) The results for the Hall kernel; (b) the results for the turbulence kernel. Taken from Wang and Grabowski (2009).

kernel is identically zero when the two colliding droplets are of the same size. Thus, the ratio is undefined along the diagonal. One of the important effects of turbulence is to enhance the rate of collisions between like-sized droplets. Enhancement of the collision kernel is particularly large for the smallest droplets colliding with droplets of radius 20–30 μm. For very large droplets, the turbulence-informed collision kernel is not much higher than the Hall kernel.

The time evolution of the droplet number density was presented in terms of $d(3\mathcal{V}^2\bar{n})/dt$ plotted as a function of $\mathcal{V}$ at varying times. These plots for the Hall kernel are shown in Figure 10.3a and for the turbulence-informed kernel in Figure 10.3b. The difference is striking. Wang and Grabowski (2009) and Grabowski and Wang (2013) identified three phases: (i) the initial auto-conversion phase, where the initial spectrum of droplets slowly shifts to large size due to collision among themselves; (ii) the accretion phase, where the initial smaller droplets are quickly transferred to a secondary peak of larger drizzle-sized droplets; and (iii) the self-collection phase, where the drizzle-sized droplets grow to become larger rain droplets. In both figures, the three phases are demarcated by horizontal red lines. The key effect of turbulence is that the formation of rain-sized droplets (the end of the auto-conversion phase) shifted from 32.5 to 10.5 min with use of the turbulence-informed collision kernel. This provides a mechanism for the observed early onset of precipitation in atmospheric clouds.

The numerical aspects of accurately computing the integral form of the Smoluchowski coagulation equation are often challenging. First and foremost, unlike the discrete form where the volume of the individual droplets within the system is discretized to be multiples of the monomer, in the integral form, the droplet size varies continuously. Thus, a numerical implementation must discretize the droplet volume $\mathcal{V}$ into bins and count the number of droplets within each size bin. As collisions remove droplets from the colliding droplet bins and add the resulting larger droplet into the bin of appropriate size, the removal and additions must be done carefully and consistently in order to conserve volume (Grabowski and Wang, 2013).

10.5   Other Considerations

Due to the complex multiphysics nature of the collision and coagulation process, there are a number of other considerations of practical importance that have not been discussed above. For example, in case of flocculation of sediment particles or soot particles into large-sized structures, even if the monomers were perfectly spherical in shape, the dimers, trimers, and so on that form as a result of collision and coagulation are not spherically shaped. In this sense, rain formation, where collisions are between droplets, is an easier problem for the description of collision and coagulation. Since all the droplets, irrespective of their volume, are taken to be spherical in shape, collisions between particles due to gravity, shear, and Brownian motion can be well described. The collision kernels obtained in the previous section are appropriate only for collisions between spherically shaped particles. Similarly, an evaluation of collision efficiency based on the theoretical results of Jeffrey and Onishi (1984) is appropriate only for close interaction between two spherical particles. Thus, all the theoretical models discussed so far are for problems where the colliding particles remain spherical in shape even as they grow in size.

The Smoluchowski equation, under the different names of population balance equation or Williams spray equation, has been used in describing the flocculation process as well. An important step in the application of the Smoluchowski equation is the definition of the collision kernel for the collision between two flocs into a larger floc. Assuming the growing flocs to be of fractal shape, and defining the characteristic floc diameter, heuristic models of the collision kernel have been proposed (Winterwerp, 1998; Son and Hsu, 2008). As in droplet collision, it has been well established that the effect of turbulence is important in the flocculation process as well.

In addition to agglomeration, we also have the possibility of an existing floc breaking up into smaller particles, both as a result of collisions and due to turbulent shear. Two kinds of breakup have been distinguished. First, the larger floc can break into two sub-pieces of nearly equal size. Second, the larger floc can reduce in size, by steady removal of monomers and small pieces of agglomerates that are located on the outer surface as appendages. In either case, particle breakup is modeled as a breakup kernel, which predicts both the probability of breakup and the likely outcome in terms of the size of the daughter flocs. It should be recognized that breakup is a very complex multiphase flow phenomenon with a lot to still be explored. And the same goes for breakup of larger droplets into smaller ones. Thus, in order to describe the size distribution of flocs within a larger system, one must solve the Smoluchowski equation with additional terms added on the right-hand side to represent floc breakup (see Winterwerp, 1998).

Since there are significant approximations in coming up with collision and breakup kernels for nonspherical particle agglomerates, the solutions of the Smoluchowski equation will be subject to large uncertainty. So an alternate approach is to perform numerical simulations where all the particles are tracked within the turbulent flow by solving their equation of motion. Typically, the particles are treated as points as far as the fluid flow is concerned. That is, the fluid governing equations are solved over the entire computational domain. This is not a problem, since the particles occupy a negligible volume of the entire domain. The force on the particles is computed using the force formulations given in Chapter 4. If all the length scales of turbulence are resolved, then this will be called the point-particle direct numerical simulation or particle-unresolved direct numerical simulation. However, when it comes to particle–particle interaction, the finite size of the particles is recognized. Each time step, by comparing the trajectory of every particle against every other particle in its neighborhood, the possibility of their separation distance becoming smaller than the collision radius R is evaluated. If the separation distance becomes smaller than R, then collision occurs between the two particles. The instance of collision, the position and velocity of the colliding particles just before collision are then established. The soft-sphere collision model of Section 9.1.2 is employed to evaluate the contact force between the colliding particles. This method of detecting collisions among particles and simulating their post-collisional dynamics is termed the *discrete element method*.

In the case of cohesive sediments, when the particles are close to each other, in addition to hydrodynamic forces, the particles are subject to short-range double-layer interaction forces. Once the particles are in contact, there are additional adhesive forces. These added forces are typically modeled using DLVO theory (Subrahmanyam and Forssberg, 1990; Zhang and Zhang, 2011) and the Johnson–Kendell–Robert (JKR) contact model (Kendall, 1971). If these attractive/adhesive forces are strong, then the colliding particles remain in contact and stay together after collision. If the post-collision rebound force is stronger, then the colliding particles or flocs may bounce back. Also, each particle within the floc is subject to hydrodynamic forces and in high-shear regions the force on the different portions of the floc may tend to tear them apart. This is opposed by the attractive/adhesive forces and depending on their relative strength, the floc may remain together or break apart. In fact, the larger the floc, the more likely that different portions of the floc will see different parts of the turbulent flow, which will tend to break the floc into smaller pieces. In general, it is observed that flocs grow in size up to the Kolmogorov scale and beyond that, larger-sized flocs tend to break up.

The point-particle direct numerical simulations have been a great tool to study the flocculation process (see Yu et al., 2021). New insights have been gained with these simulations where the interparticle interactions are simulated with the use of appropriate DLVO and JKR forces. One interesting finding has been the ability of turbulence to reshape the flocs to more stable configurations. If the colliding particles are allowed to stay fixed in their relative position after collision, then the growth of the floc or particle cluster will take a more fractal-like noncompact geometry. Instead, a proper implementation of the soft-sphere collision model will allow contacting particles to stay in touch but roll relative to one another, thus changing the shape of the floc over time. This floc shape rearrangement is in response to forces acting on

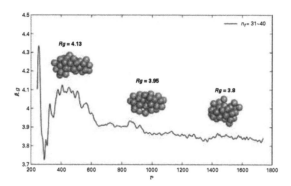

Figure 10.4 The time evolution of the radius of gyration of a particle floc in $Re_\lambda = 33$ isotropic turbulence. The shape of the floc at three different time instances is shown. Taken from Yu et al. (2022).

the different portions of the floc, while the attractive/adhesive forces keep the different portions in contact.

This ability of turbulence to reshape the particle flocs in isotropic turbulence of Taylor microscale Reynolds number $Re_\lambda = 33$ is shown in Figure 10.4 for one sample floc for a short duration of time (Yu et al., 2022). Here, the smallest spherical particle (or monomer) is about half the Kolmogorov length scale. The figure shows the time history of the radius of gyration of this floc as a function of time. During this period, the floc does not increase in size due to collision or decrease in size due to breakup. The change in radius of gyration is only due to reshaping of the floc due to turbulence-induced shear stress on the different portions of the floc. The shape of the floc at three different times is shown. In general, it is observed that turbulence mercilessly breaks away any odd appendages (not shown here). Only stable additions of new particles or flocs are retained. Even these additions are constantly moved from less stable to more stable configurations. This explains the more rounded shape of cohesive sediments often observed in nature.

The point-particle discrete element simulations also provide an excellent tool for studying the dynamics of bigger systems containing a large number of flocs of varying size. The results from two such simulations are shown in Figure 10.5. Both these simulations started with the same initial condition with a random distribution of monomers. The intensity of turbulence in both simulations remains the same, with the only difference between them that the strength of the adhesive force is three times larger in the second simulation than the first. In both cases, the number of flocs of size n_f is counted within the computational domain and plotted as a function of n_f at three different times. In both cases, with increasing time, the number of smaller flocs decreases at the expense of the number of flocs of larger size. In the case where the adhesive force is weak (left frame), the spectrum develops a characteristic peak around the Kolmogorov scale. The case with stronger adhesive force does not undergo breakup easily, and thus exhibits a power-law self-similar spectrum.

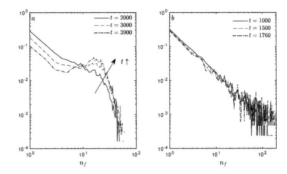

Figure 10.5 The number of flocs of size n_f plotted as a function of n_f at three different times. These simulations were started with a random uniform distribution of monomers (i.e., $n_f = 1$ particle). The adhesive force in the right frame is three times that in the left frame, with both cases having turbulence. Taken from Yu et al. (2021).

11 Filtered Multiphase Flow Equations

From Chapter 4 to Chapter 10 we have studied extensively the interaction of an ambient flow with (i) an isolated particle, (ii) an isolated particle in the presence of a nearby wall, (iii) a pair of particles, and (iv) a large collection of particles. These investigations were at the microscale and we paid great attention to solving for the complete details of the flow around the particles. These studies can be classified as "particle-resolved" or "fully resolved," as they included all the relevant physics. As a result, these studies have yielded reliable results on the hydrodynamic force, torque, and heat transfer on the particles under varying flow conditions.

The challenge with this fundamental approach is that it is not scalable and cannot be applied in the investigation of large-scale multiphase flows that involve billions of particles. As discussed in previous chapters, as the complexity of the multiphase system increases, either in terms of Reynolds number or many random interacting particles, we need to resort to a computational approach, since such problems are beyond current analytical ability. Even with the most powerful computers and sophisticated numerical methodologies we can study the detailed flow around only a limited number of particles, far fewer than what is needed in real problems of practical interest. In this and subsequent chapters we will consider multiphase flow methodologies that will allow us to scale up and consider more complex problems.

Consider a multiphase system that is about a meter in size, consisting of particles of 1 mm diameter. If the average volume fraction is 10%, then there are approximately 100 million particles in the system. The following sequence of length scales is of relevance to this problem: (i) the steady boundary layer on the particle is about one-tenth of the particle diameter (i.e., 0.1 mm, assuming the particle Reynolds number to be $O(10)$ to $O(100)$);[1] (ii) particle diameter of 1 mm; (iii) mean interparticle spacing of about 1.74 mm; and (iv) the system or container size of 1 m. Now, let us assume that with the available computational resources we can only afford a simulation involving 100 grid points along each of the three directions. Clearly, the entire range of flow scales from 0.1 mm to 1 m cannot be resolved with the available computational resources.

In other words, particle-resolved simulations that span the entire volume of interest are not possible. We hasten to note that this is only an idealized example of modest dimensions. Many other particle-laden multiphase flow examples, from fluidized beds

---

[1] The unsteady boundary layer of a suddenly accelerating particle can be much thinner.

to volcanic eruption, present a far wider range of length and time scales. They therefore pose an even greater challenge to understanding by simulation and experiment. To proceed further, we must develop methodologies where we do not need to address the entire range of scales simultaneously. Two approaches have been popularly pursued. The first is a *deterministic approach*, where the entire range of scales is divided into manageable subranges and investigations are pursued within those subranges. Within each subrange the dynamics of the multiphase flow is faithfully pursued – of course with appropriate approximations on how this subrange of scales that are being studied interact with other larger and smaller scales that are not part of the focus. The second is a *statistical approach*, where interest is in the description of the statistical properties of the multiphase flow and not individual realizations. The advantage of this approach is that statistical quantities such as mean and rms are far smoother, with only a limited range of length and time scales which can be studied together without needing to partition them into subranges.

Both these approaches are well developed in the context of single-phase turbulence. In problems where *fully resolved direct numerical simulations* are not possible, one approach that has been developed is to perform *large eddy simulations*, where only the largest scales of motion are resolved and therefore they correspond to a subrange of the entire range of turbulence. The alternate approach is to perform Reynolds-averaged Navier–Stokes (RANS) simulations, where the dynamics of only the mean flow is considered, whose range of length and time scales is often limited. Both these approaches have their strengths and weaknesses, and both have been immensely useful in our understanding of single-phase turbulence. The deterministic approach of large eddy simulation has provided deeper insight into the dynamics of large scales in individual realizations. In contrast, the statistical approach of RANS is quite powerful in extracting the mean and other statistical information in an efficient manner.

In order to focus attention only on a certain subrange of the entire range of scales, one must filter out all scales that are smaller or larger than the chosen subrange. For example, in the large eddy simulation, all the smaller scales are filtered out and only the dynamics of the large scales are considered. This requires the definition of a spatial or temporal low-pass filter that only retains the desired range of scales. From the definition of the filter, appropriate governing equations for the filtered quantities must be derived. On the other hand, in the statistical approach, appropriate definitions of averages must be advanced in order to define statistical quantities such as mean and rms. Although time or space averages have been pursued in the past, the proper approach is to define the ensemble average and obtain ensemble-averaged governing equations that define the evolution of quantities such as the mean and higher-order statistics. In this chapter, we will first define the *filter operation* and obtain filtered governing equations in a rigorous manner. Then we will define the *ensemble average* and obtain the ensemble-averaged governing equations.

Let us go back to the idealized multiphase example, and consider the deterministic approach to investigating subranges of flow scales. As we will see, there are different options in choosing the subranges. One option is to use the 100 grid points in each direction to resolve and study length scales that range from 0.1 to 10 mm. This will allow the simulation to fully resolve each particle and its boundary layer. However, the entire computational domain will extend only $1000 \, \text{mm}^3$ and thus will cover only one-millionth of the entire volume of the system. With this option, the full computational power is deployed at the microscale, allowing complete resolution of all the details of the flow within the smaller computational domain. Such a simulation will, however, be appropriate to study neither the dynamics of the mesoscale instabilities nor the evolution of flow within the system as a whole.

The alternate options are to study the system either at the intermediate mesoscales or at the macroscale by focusing on the system-level scales. At the macroscale, the 100 grid points in each direction will be used to resolve and study length scales that range from 10 mm to 1 m. While this gives us access to the system scale of 1 m, the smallest length scale that can be resolved is only 10 mm, which is ten times bigger than the particle diameter. In this option, at 10% volume fraction each computational cell of volume $1000 \, \text{mm}^3$ will contain about 191 particles on average. The compromise for studying the problem at the system level is that the details of the flow at length scales smaller than 10 mm (which includes the particle and its boundary layer) cannot be resolved.

In the microscale option, the governing equations are the standard Navier–Stokes equation in the region occupied by the fluid and rigid-body equations of motion for the particles. In this option, the challenge is in the specification of the macroscale flow in terms of appropriate boundary conditions at the outer boundaries of the small computational domain. In contrast, in the macroscale approach, the computational boundaries extend over the entire system and appropriate boundary conditions are easy to specify from the statement of the overall problem. In this option, the challenge is that the standard Navier–Stokes equations for the fluid cannot be applied directly at the grid scale of 10 mm, which is far too coarse. Each computational cell will contain both fluid and particles within it.

The quest of this chapter is to obtain appropriate equations for the continuous phase that are applicable at scales much larger than the particle diameter. This will first be achieved through a formal spatial filter operation, where all the small-scale details of the flow will be filtered out and only scales larger than a chosen length scale will be retained. In this chapter, we will first formalize the filter operation and apply it to the continuous-phase equations to obtain the appropriate spatially filtered continuous-phase mass, momentum, and energy equations. The filter operation can also be thought of as a weighted-average operation. As will be discussed below, with the choice of a top-hat filter, the filter operation becomes a running local volume average. The derivation of the continuous and dispersed-phase governing equations using volume, time, or ensemble average has a long and illustrious history in multiphase flow, with many pioneering efforts (Anderson and Jackson, 1967; Ishii, 1975; Drew, 1983; Joseph et al., 1990; Gidaspow, 1994; Zhang and Prosperetti, 1994, 1997; Drew and Passman, 2006). In this chapter, we will follow the spatial filter approach, similar to that pursued in the derivation of single-phase large eddy equations (see Capecelatro and Desjardins, 2013). An important point that will be discussed during this presentation is that such filtered equations will include unknown terms that must be closed with appropriate closure models. We then carry out the filter operation for the dispersed phase and obtain the filtered particulate phase-governing equations.

11.1   Spatial Filter Operation

In this section, we revisit the definition of local volume average introduced in Section 3.2. Our approach is to first formalize the spatial filter operation by precisely defining a filter function with a well-characterized length scale $\mathscr{L}$. We then carry out rigorous filtering of the continuous-phase governing equations. The resulting filtered governing equations then describe the mass, momentum, and energy balance of the continuous phase, but whose spatial variations are only at length scales larger than the length scale $\mathscr{L}$ of the filter. Thus, the key advantage of the spatially filtered governing equations is that it limits the range of length scales of the resulting continuous-phase variables to be larger than $\mathscr{L}$.

11.1.1   Preliminaries

We define a continuous-phase indicator function as

$$I_c(\mathbf{x},t) = \begin{cases} 0 & \text{if } \mathbf{x} \text{ is inside a particle,} \\ 1 & \text{if } \mathbf{x} \text{ is inside the fluid,} \end{cases} \tag{11.1}$$

which complements the dispersed-phase indicator function $I_d(\mathbf{x},t)$ defined in Eq. (3.4). At any time t, every point $\mathbf{x}$ within the system should be occupied by either the particle or the fluid, and therefore by definition $I_c(\mathbf{x},t) + I_d(\mathbf{x},t) = 1$ for all $(\mathbf{x},t)$ within the multiphase flow.

We then introduce a three-dimensional filter function $G(\boldsymbol{\xi})$ as the product of three one-dimensional filter functions that operate along the three Cartesian spatial directions:

$$G(\boldsymbol{\xi}) = G_1(\xi_x)\,G_1(\xi_y)\,G_1(\xi_z), \tag{11.2}$$

where the one-dimensional filter function G_1 is chosen to be the same along all three directions. The two commonly used filters are the *box filter* and the *Gaussian filter*. The one-dimensional box filter is defined as

$$\text{Box filter:} \quad G_1(\xi_x) = \begin{cases} 1/\mathscr{L} & \text{for } |\xi_x| \leq \mathscr{L}/2, \\ 0 & \text{for } |\xi_x| > \mathscr{L}/2, \end{cases} \tag{11.3}$$

where the length scale of the filter is $\mathscr{L}$. When multiplied along the three Cartesian directions we obtain the 3D box filter, which is unity inside a cube of volume $\mathscr{L}^3$, and zero outside. This filter is also known as the top-hat filter and it sharply transitions from one to zero in the physical space.

The one-dimensional Gaussian filter is defined as

$$\text{Gaussian filter:} \quad G_1(\xi_x) = \frac{1}{\sqrt{2\pi}\sigma} \exp\left[-\frac{\xi_x^2}{2\sigma^2}\right], \tag{11.4}$$

where the length scale of the filter can be defined as $\mathscr{L} = 2\sqrt{(2 \ln 2)}\sigma$. This definition is not unique, but has the property

$$G_1(|\xi_x| = \mathscr{L}/2) = \frac{1}{2}G_1(0) . \tag{11.5}$$

Again, when multiplied along the three Cartesian directions, we obtain the 3D Gaussian filter. Unlike the box filter, the decay to zero is not sharp in the case of a Gaussian filter. A number of other filter functions, including triangular, Wendland, cubic spline, and quintic spline, have been employed (Evrard et al., 2019; Capecelatro and Desjardins, 2022). Each have their strengths and weaknesses. The discussion to follow applies for all the filter functions.

The spatial filter of a continuous-phase property ψ can be expressed in one dimension as

$$\overline{\psi}(x,t) = \int \psi(x',t)\, I_c(x',t)\, G_1(x - x')\, dx' , \tag{11.6}$$

where the integral is over the entire domain. Here the overbar indicates a filtered variable and note that multiplication with I_c renders the integration effectively only over the fluid region. Due to the filter function, the value of the filtered quantity $\overline{\psi}$ at a point x depends only on the value of the unfiltered raw variable ψ in the neighborhood of the point x. In the case of a box filter, only the neighborhood of width $\mathscr{L}/2$ on either side of x contributes. In the case of a Gaussian filter, due to the exponential decay, only the value of ψ in the neighborhood of x contributes. In the case of a box filter, the entire region of width $\mathscr{L}$ around x contributes equally, while in the Gaussian filter the value of ψ at x itself contributes the most and the contribution from points farther away decays exponentially. Nevertheless, both filters satisfy the crucial property

$$\int G_1(\xi_x)\, d\xi_x = 1 , \tag{11.7}$$

and this ensures that the mean value of the filtered function $\overline{\psi}$ is precisely the same as the original unfiltered function.

The integral given in Eq. (11.6) is a *convolution integral* that convolves the argument $\psi\, I_c$ defined at (x',t) with the filter function $G(x - x')$ to yield the filtered variable $\overline{\psi}(x,t)$. The level of filtering depends on the filter function employed in the convolution. A very narrow G will return the original argument nearly unmodified, while a broad filter will filter out a wide range of scales resulting in a smooth $\overline{\psi}$.

The effect of the above averaging process can be illustrated with a simple example. Consider a 1D function $\psi(x')$ defined over a periodic interval $0 \le x' \le 1$, such as the one shown in Figure 11.1 as the black line. From periodicity, the function can be extended to the entire range $-\infty \le x' \le \infty$, so that the integral (11.6) can be carried out for all values of x. Since the function ψ is periodic, it can be shown that the filtered function $\overline{\psi}(x)$ is also periodic, and thus needs to be evaluated only over the interval $0 \le x \le 1$.

The original function $\psi(x')$ has been chosen to contain a wide range of scales as seen by the long and short-wavelength oscillations. Also, there are regions over which

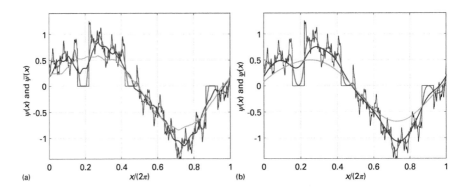

Figure 11.1 The effect of filter operation on a function that fluctuates with a wide range of scales. The function indicated by the black line also has holes, where it is identically zero. These may be considered as regions occupied by the particle in this 1D context. (a) Box filter; (b) Gaussian filter. In both cases, three different filter widths are presented: $\mathscr{L} = 1/30$ (red), $1/10$ (blue), and $1/3$ (green).

$\psi(x')$ is identically zero, and these are the regions where the indicator function $I_c = 0$, indicating that these regions are occupied by the particle, and not by the fluid. This data can be thought of as the streamwise velocity or pressure of the fluid, sampled along a line passing through a periodic box of random distribution of particles. The fluctuations are due to turbulence, such as the illustration shown in Figure 9.8a. Also shown in Figure 11.1 are the filtered velocity for three different filter sizes: $\mathscr{L} = 1/30$, $1/10$, and $1/3$.

As can be expected, in the limit $\mathscr{L} \to 0$, the filter function becomes a delta function (i.e., $G_1(\xi_x) \to \delta(\xi_x)$). None of the length scales are removed (or filtered) and we recover $\overline{\psi}(\xi_x) = \psi(\xi_x)$. With increasing $\mathscr{L}$ we see that more of the small-wavelength, rapid variations are removed by the filter operation and as a result $\overline{\psi}(\xi_x)$ becomes smoother than the original function. In the limit $\mathscr{L} \to 1$ (in the case of a Gaussian filter as $\mathscr{L} \to \infty$), the filter is over the entire periodic domain and thus $\overline{\psi}(\xi_x)$ becomes perfectly flat, equal to the mean value of $\psi(\xi_x)$.

The results for both the box filter and the Gaussian filter are shown in Figure 11.1. From the figure it is clear that the discontinuous jumps in ψ are made continuous by the filter operation. However, $\overline{\psi}$ obtained with the box filter are still discontinuous in the first derivative. In contrast, the Gaussian filter results in smooth functions, where the discontinuities of ψ have been smoothened out even for small filter width.

This is an important feature, for the following two reasons. (i) While the original function $\psi(x)$ was defined only in regions occupied by the fluid, the filtered function $\overline{\psi}(x)$ is defined everywhere and therefore its governing equations can be solved everywhere without distinguishing the inside/outside of the particles. (ii) Since $\overline{\psi}(x)$ is continuous and continuously differentiable, we can seek appropriate differential equations governing $\overline{\psi}$. Although the forms of the mass, momentum, and energy conservation equations to be derived in this chapter are invariant to which filter is used in the filter operation, for the reasons stated above we will interpret these equations to be obtained with the Gaussian filter (Capecelatro and Desjardins, 2013).

The filters given in Eqs. (11.3) and (11.4) are homogeneous filters. They only depend on the separation $\xi_x = x - x'$ and not on the absolute location x. Such homogeneous filters and the corresponding definition of filter operation apply only to regions that are free from boundary effects. That is, they apply strictly in the context of unbounded domains that extend to infinity or to bounded domains with periodic boundary conditions. In the practical case of bounded domains, these filters can be used away from the domain boundaries. If point x in Eq. (11.6) were to lie close to a nonperiodic domain boundary, since there is no fluid on the other side of the boundary, the integral given in Eq. (11.6) cannot be carried out. Inhomogeneous filter functions that are asymmetric and defined only in the fluid region bounded by the domain boundaries must be specified. We will not pursue this further here. For further discussion on filters and their properties, the reader should consult Ghosal and Moin (1995).

In the context of a homogeneous filter, the filter operation defined in Eq. (11.6) can be further analyzed in the Fourier space. We denote the Fourier transform of the product ψI_c symbolically as $\mathcal{F}[\psi I_c]$. The corresponding Fourier transforms of the 1D filter function and the filtered quantity are denoted as $\mathcal{F}[G_1]$ and $\mathcal{F}[\overline{\psi}]$. The convolution integral (11.6) is in the physical space and it becomes a simple multiplication in the Fourier space:

$$\mathcal{F}[\overline{\psi}] = \mathcal{F}[\psi I_c] \, \mathcal{F}[G_1] \,. \tag{11.8}$$

The filtered quantity $\overline{\psi}$ can then be obtained with a Fourier-inverse transform of the left-hand side. For the Gaussian filter, its Fourier transform is also a Gaussian. That is,

$$\mathcal{F}[G_1] = \frac{1}{\sqrt{2\pi}} \exp\left(-\frac{1}{2} k^2 \sigma^2\right) \,, \tag{11.9}$$

where k is the wavenumber. Thus, $\mathcal{F}[G_1]$ that multiplies the Fourier transform of the original data $\mathcal{F}[\psi I_c]$ decays exponentially. Therefore, fluctuations in the original signal with wavenumber k greater than $1/\sigma$ are effectively damped by the Gaussian filter. This is the spectral interpretation of the filter operation. The Fourier transform of the box filter $\mathcal{F}[G_1]$ is the sinc function, which again decays for large k, but in a nonmonotonic fashion.

Problem 11.1 Consider the following simple double-step function in an infinite domain $-\infty \leq x' \leq -\infty$:

$$\psi(x')I_c(x') = \begin{cases} 1 & \text{for } |x'| > 1/2, \\ 0 & \text{for } |x'| < 1/2 \,. \end{cases} \tag{11.10}$$

The particle can be thought of as being located in the region $-1/2 \leq x' \leq 1/2$ with the fluid outside. In this problem, we will apply the box filter to this simple function and investigate its effect on smoothing the jumps in the original function.

Apply the box filter given in Eq. (11.3) to the above function and obtain the result

$$\overline{\psi}(x) = \begin{cases} 1 & \text{for } |x| > (1 + \mathscr{L})/2, \\ \dfrac{2x + \mathscr{L} - 1}{2\mathscr{L}} & \text{for } (1 - \mathscr{L})/2 \leq |x| \leq (1 + \mathscr{L})/2, \\ 0 & \text{for } |x| < (1 - \mathscr{L})/2, \end{cases} \qquad (11.11)$$

for filter values in the range $0 \leq \mathscr{L} \leq 1/2$. By plotting this filtered function it can be observed that the steps have been smoothened to linear ramps. However, the slopes are discontinuous at $x = (1 \pm \mathscr{L})/2$.

Problem 11.2 Apply the Gaussian filter given in Eq. (11.4) to the double-step function presented in the previous problem and obtain the result

$$\overline{\psi}(x) = \frac{1}{2}\left(\text{erf}\left[\frac{1}{\sqrt{2}\sigma}\left(\frac{1}{2} - t\right)\right] + \text{erf}\left[\frac{1}{\sqrt{2}\sigma}\left(\frac{1}{2} + t\right)\right]\right). \qquad (11.12)$$

By plotting this filtered function, it can be observed that the steps have been smoothened to error functions for any nonzero value of σ. For small values of σ, the variation of $\overline{\psi}$ near the steps, though continuous, remains sharp. With increasing σ, higher wavenumbers are more damped and $\overline{\psi}$ becomes very smooth. This example illustrates the power of the Gaussian filter to take discontinuous functions (that arise due to the multiplication of the indicator function) and convert them into smooth functions suitable for differential equations.

Problem 11.3 *Other properties of spatial filters.* The standard definition of the spatial average is over the entire domain and has a uniform weight. In contrast to the filter operation defined in Eq. (11.6), the global spatial average of a continuous-phase variable is defined as

$$\langle \psi \rangle(t) = \frac{1}{\Omega}\int \psi(x',t)\, I_c(x',t)\, dx', \qquad (11.13)$$

where the integral is over the entire domain of size Ω and the spatial average has been denoted by angle brackets. We reserve the angle brackets to denote the ensemble average; since the spatial average of a homogeneous ergodic system is the same as the ensemble average, here we use it in the above expression. You are likely to be familiar with the following properties of this volume average:

(i) The resulting average is a global constant (i.e., $\langle \psi \rangle$ is not a function of x).
(ii) As a result, an average of the average is the same as the average (i.e., $\langle \langle \psi \rangle \rangle = \langle \psi \rangle$). In other words, repeated averages do not alter the average value.
(iii) If we define the fluctuation about the average as $\psi''(x,t) = \psi(x,t) - \langle \psi \rangle(t)$, then the spatial average of the fluctuation is identically zero (i.e., $\langle \psi'' \rangle = 0$).

It can easily be demonstrated that the above useful properties hold for the classical time average that is often used in the Reynolds-averaged approach and for the ensemble average, which we shall discuss later in the book.

These properties do not hold for the box and Gaussian filter operations. The filter operation differs from the above-defined average operation. For example, the box filter has a compact support and does not extend over the entire domain, while the Gaussian filter is non-uniform. We have already seen in Eqs. (11.11) and (11.12) that the filtered quantity $\bar{\psi}$ is still a function of x, and not a constant.

(a) Using the double-step function of Problem 11.1, show that for both the box and Gaussian filters:

$$\bar{\bar{\psi}}(x,t) \neq \bar{\psi}(x,t) . \tag{11.14}$$

(b) Furthermore, if we define fluctuation about the filtered quantity as $\psi'(x,t) = \psi(x,t) - \bar{\psi}(x,t)$, obtain the result

$$\overline{\psi'}(x,t) = \overline{(\psi(x,t) - \bar{\psi}(x,t))} = \bar{\psi}(x,t) - \bar{\bar{\psi}}(x,t) \neq 0 . \tag{11.15}$$

(c) For the double-step function of Problem 11.1, evaluate $\overline{\psi'}(x,t)$ and show that it is not identically zero.

Because of this property, we refer to Eq. (11.6) as the "filter operation," and terms with an overbar as the "filtered variables." The resulting governing equations will be referred to as the "filtered equations." We reserve the term "average" only for those operations that obey the three properties listed above. This difference is well recognized in the context of single-phase turbulent flows, where the Reynolds-averaged Navier–Stokes approach uses a temporal or ensemble average, while large eddy simulation uses a spatial filter.

The 3D extension of the filter operation given in Eq. (11.6) becomes

$$\bar{\psi}(\mathbf{x},t) = \int \psi(\mathbf{x}',t) \, I_c(\mathbf{x}',t) \, G(\mathbf{x} - \mathbf{x}') \, d\mathbf{x}' , \tag{11.16}$$

where the integral is over the entire 3D spatial domain. Setting $\psi = 1$ we obtain the definition of the continuous-phase volume fraction as

$$\phi_c(\mathbf{x},t) = \bar{1}(\mathbf{x},t) = \int I_c(\mathbf{x}',t) \, G(\mathbf{x} - \mathbf{x}') \, d\mathbf{x}' . \tag{11.17}$$

Setting $\psi = \rho_f$, we obtain the filtered continuous-phase density ρ_c as

$$\phi_c(\mathbf{x},t) \, \rho_c(\mathbf{x},t) = \overline{(\rho_f)}(\mathbf{x},t) = \int \rho_f(\mathbf{x}',t) \, I_c(\mathbf{x}',t) \, G(\mathbf{x} - \mathbf{x}') \, d\mathbf{x}' , \tag{11.18}$$

where multiplication by ϕ_c on the left accounts for the fact that the fluid occupies only ϕ_c fraction of the averaging volume. Settling $\phi = p$, we obtain the filtered continuous-phase pressure p_c as

$$\phi_c(\mathbf{x},t) \, p_c(\mathbf{x},t) = \bar{p}(\mathbf{x},t) = \int p(\mathbf{x}',t) \, I_c(\mathbf{x}',t) \, G(\mathbf{x} - \mathbf{x}') \, d\mathbf{x}' . \tag{11.19}$$

For quantities weighted by density, we define the Favre filter in the following way:

$$\overline{(\rho_f \psi)}(\mathbf{x},t) = \int \rho_f(\mathbf{x}',t)\,\psi(\mathbf{x}',t)\,I_c(\mathbf{x}',t)\,G(\mathbf{x}-\mathbf{x}')\,d\mathbf{x}'. \tag{11.20}$$

Setting $\psi = \mathbf{u}$, we obtain the Favre-filtered continuous-phase velocity $\tilde{\mathbf{u}}_c$ as

$$\begin{aligned} \phi_c\,\rho_c\,\tilde{\mathbf{u}}_c &= \overline{(\rho_f \mathbf{u})}(\mathbf{x},t) \\ &= \int \rho_f(\mathbf{x}',t)\,\mathbf{u}(\mathbf{x}',t)\,I_c(\mathbf{x}',t)\,G(\mathbf{x}-\mathbf{x}')\,d\mathbf{x}', \end{aligned} \tag{11.21}$$

where all the quantities on the left-hand side are functions of $(\mathbf{x},t)$. We will use a tilde to denote Favre-filtered quantities.

11.1.2   Gauss Rule and Leibniz Rule

Toward our quest to apply the filter operations to the continuous-phase governing equations, in this subsection we will derive two rules that will allow us to commute the filter integral with time and space derivatives.

First, let us consider filtering the time derivative $\partial\psi/\partial t$ as

$$\overline{\left(\frac{\partial\psi}{\partial t}\right)}(\mathbf{x},t) = \int \frac{\partial\psi}{\partial t}(\mathbf{x}',t)\,I_c(\mathbf{x}',t)\,G(\mathbf{x}-\mathbf{x}')\,d\mathbf{x}'. \tag{11.22}$$

We want to write the above in terms of the time derivative of $\overline{\psi}$. In order to do that, we will carry out the following manipulations of the right-hand side. Using the product rule, we write the right-hand side as

$$\int \frac{\partial\psi\,I_c\,G}{\partial t}\,d\mathbf{x}' - \int \psi\,G\frac{\partial I_c}{\partial t}\,d\mathbf{x}', \tag{11.23}$$

where we have dropped the $\mathbf{x}'$ and t dependence of the different variables and also used the fact that the filter function G is independent of time.

The next step is to write the kinematic equation for the evolution of the indicator function. The indicator function simply moves with the material (i.e., fluid or particles, whichever occupied the point). At the interface between the continuous and the dispersed phases, the indicator function moves with the interface. Thus, the total derivative of the indicator function can be expressed as

$$\frac{dI_c}{dt} = \frac{\partial I_c}{\partial t} + \mathbf{u}_I \cdot \nabla' I_c = 0, \tag{11.24}$$

where $\mathbf{u}_I$ is the velocity of the interface between the particles and the surrounding fluid. We then recognize that the gradient of the indicator function is nonzero only at the interface between the phases. Thus, the gradient of the indicator function can be expressed as $\nabla' I_c = \mathbf{n}\,\delta(\mathbf{x}' - \mathbf{x}'_I)$, where the delta function is nonzero only when the point $\mathbf{x}'$ falls on the particle–fluid interface denoted by $\mathbf{x}'_I$, and $\mathbf{n}$ is the outward normal to the interface that points into the fluid. Substituting the above into Eq. (11.23), we obtain

$$\frac{\partial\overline{\psi}(\mathbf{x},t)}{\partial t} + \int \psi(\mathbf{x}',t)\,G(\mathbf{x}-\mathbf{x}')\,(\mathbf{u}_I \cdot \mathbf{n})\,\delta(\mathbf{x}'-\mathbf{x}'_i)\,d\mathbf{x}', \tag{11.25}$$

where the time derivative has been moved out of the spatial integral in the first term, since the volume integral and the time derivative commute. Furthermore, Eq. (11.16) has been used for $\overline{\psi}$. For compactness, the dependence of $\mathbf{u}_I$ and $\mathbf{n}$ on $(\mathbf{x}', t)$ has been suppressed. In the second integral, we recognize that the argument is nonzero only at the interface between the dispersed and the continuous phases, and therefore the volume integral can be replaced with an integral over the surface of all the particles. With this replacement, we obtain

$$\overline{\left(\frac{\partial \psi}{\partial t}\right)}(\mathbf{x}, t) = \frac{\partial \overline{\psi}(\mathbf{x}, t)}{\partial t} + \int_S \psi(\mathbf{x}', t) G(\mathbf{x} - \mathbf{x}') (\mathbf{u}_I \cdot \mathbf{n}) \, dA', \qquad (11.26)$$

where $\int_S dA'$ stands for the surface integral over all the particles. The above equation is the *Leibniz rule*. As per the above equation, the value of $\overline{(\partial \psi / \partial t)}$ at a point $\mathbf{x}$ depends on the two terms on the right. Due to the rapid decay of the Gaussian filter or the compact support of the box filter, the volume integral in the first term on the right (i.e., $\overline{\psi}$) has a contribution only from the fluid region in the neighborhood of $\mathbf{x}$ and similarly the surface integral in the second term derives a contribution only from the surface of particles that are located in the neighborhood of the point $\mathbf{x}$. In the case of the Gaussian filter, the second term accounts for the fact that as the particles move, their contributions change due to changes in Gaussian weighting at their surface. In the case of the box filter, the second term accounts for the fact that particles cross the volume of compact support and thus contribute to the left-hand side.

We now consider filtering of the spatial derivative $\nabla \psi$ by defining

$$\overline{(\nabla \psi)}(\mathbf{x}, t) = \int \nabla' \psi(\mathbf{x}', t) I_c(\mathbf{x}', t) G(\mathbf{x} - \mathbf{x}') \, d\mathbf{x}', \qquad (11.27)$$

where ∇' denotes the gradient with respect to $\mathbf{x}'$. We want to rewrite the above in terms of the gradient of $\overline{\psi}$. We will again carry out a few manipulations of the right-hand side in order to achieve this goal. We use the product rule to obtain

$$\int \nabla'(\psi \, I_c \, G) \, d\mathbf{x}' - \int \psi \, I_c \, \nabla' G \, d\mathbf{x}' - \int \psi \, G \, \nabla' I_c \, d\mathbf{x}', \qquad (11.28)$$

where we have again dropped the $\mathbf{x}'$ and t dependence of the variables.

The exact differential in the first term can be integrated and it is zero since the filter function G goes to zero as $\mathbf{x}'$ moves farther away from $\mathbf{x}$. In the second term, we note that $\nabla' G(\mathbf{x} - \mathbf{x}') = -\nabla G(\mathbf{x} - \mathbf{x}')$, where in index notation $\nabla' \equiv \partial / \partial x'_i$ and $\nabla \equiv \partial / \partial x_i$, respectively. Then, ∇ can be taken out of the integral, since ψ, I_c, and the integral are functions of $\mathbf{x}'$. In the third term, we again use the identity $\nabla' I_c = \mathbf{n} \delta(\mathbf{x}' - \mathbf{x}'_I)$ to rewrite Eq. (11.28) as

$$\nabla \overline{\psi}(\mathbf{x}, t) - \int \psi(\mathbf{x}', t) G(\mathbf{x} - \mathbf{x}') \mathbf{n} \, \delta(\mathbf{x}' - \mathbf{x}'_i) \, d\mathbf{x}'. \qquad (11.29)$$

As in the derivation of the Leibniz rule, using the property of the delta function, we replace the second volume integral in the above equation with an integral over the surface of all the particles. This results in the following *Gauss rule*:

$$\overline{(\nabla \psi)}(\mathbf{x}, t) = \nabla \overline{\psi}(\mathbf{x}, t) - \int_S \psi(\mathbf{x}', t) \mathbf{n} \, G(\mathbf{x} - \mathbf{x}') \, dA'. \qquad (11.30)$$

As discussed in the Leibniz rule, due to the rapid decay of the filter G, the volume integral in the first term and the surface integral over the particle surfaces derive a contribution only from the neighborhood of the point $\mathbf{x}$. Here the second term on the right accounts for the fact that ψ varies across the interface within the filter volume appropriately weighted by the filter.

Problem 11.4 The Gauss rule presented in Eq. (11.30) has been derived for the gradient of a scalar. Other variants of the Gauss rule can be derived. For example, obtain the following Gauss rule for the divergence of a vector:

$$\overline{(\nabla \cdot \boldsymbol{\psi})}(\mathbf{x},t) = \nabla \cdot \overline{\boldsymbol{\psi}}(\mathbf{x},t) - \int_S \boldsymbol{\psi}(\mathbf{x}',t) \cdot \mathbf{n}\, G(\mathbf{x} - \mathbf{x}')\, dA'. \qquad (11.31)$$

11.2   Filtered Continuous-Phase Equations

11.2.1   Mass Balance

We now apply the filter operation to the continuous-phase mass balance. We multiply the continuous-phase mass balance with the product $I_c(\mathbf{x}',t)\, G(\mathbf{x} - \mathbf{x}')$ and integrate over the entire flow domain to obtain

$$\int \left(\frac{\partial \rho_f}{\partial t} + \nabla' \cdot (\rho(\mathbf{x}',t)\, \mathbf{u}(\mathbf{x}',t)) \right) I_c(\mathbf{x}',t)\, G(\mathbf{x} - \mathbf{x}')\, d\mathbf{x}' = 0. \qquad (11.32)$$

We now use the Leibniz rule for the first term (by setting $\psi = \rho_f$) and the Gauss rule as given in Eq. (11.31) for the second term (by setting $\psi = \rho_f \mathbf{u}$). With these substitutions, the above equation becomes

$$\frac{\partial \overline{\rho_f}}{\partial t} + \nabla \cdot \overline{\rho_f \mathbf{u}} = \int_S \rho_f\, G\, (\mathbf{u} - \mathbf{u}_I) \cdot \mathbf{n}\, dA'. \qquad (11.33)$$

We now recall the definitions of volume fraction, filtered density, and Favre filtered velocity given in Eqs. (11.18) and (11.21) and substitute for them on the left-hand side. In addition, we recognize $\mathbf{u} - \mathbf{u}_I$ as the difference between the fluid velocity and the interface velocity on the surface of the particles. This difference is nonzero only in case there is mass exchange between the dispersed and continuous phases. When there is no mass exchange, from the kinematic condition we require $\mathbf{u} = \mathbf{u}_I$ at the particle surface. If we define

$$\dot{m}(\mathbf{x},t) = \int_S \rho_f(\mathbf{x}',t)\, G(\mathbf{x} - \mathbf{x}')\, (\mathbf{u}_I(\mathbf{x}',t) - \mathbf{u}(\mathbf{x}',t)) \cdot \mathbf{n}\, dA', \qquad (11.34)$$

then $-\dot{m}(\mathbf{x},t)$ accounts for the net mass flux into the continuous phase from the particles weighted by the Gaussian filter. In the evaluation of $\dot{m}$ at the point $\mathbf{x}$, due to the exponential decay of the Gaussian filter, the primary contribution is only from those particles whose surface falls close to the point $\mathbf{x}$ and whose local mass flux given by $\rho_f(\mathbf{u}_I - \mathbf{u}) \cdot \mathbf{n}$ is significant. Particles that are far away do not contribute to the filtered mass balance at $\mathbf{x}$. With these substitutions we arrive at the final filtered continuous-phase mass balance as

$$\frac{\partial \phi_c \, \rho_c}{\partial t} + \nabla \cdot (\phi_c \rho_c \tilde{\mathbf{u}}_c) = -\dot{m} \,. \tag{11.35}$$

If we restrict attention to incompressible flows, then the filtered density ρ_c is a constant equal to the fluid density ρ_f. Further, if we consider zero mass transfer between the phases, the above equation can be simplified to

$$\frac{\partial \phi_c}{\partial t} + \nabla \cdot (\phi_c \tilde{\mathbf{u}}_c) = 0 \,. \tag{11.36}$$

Thus, we have arrived at the important conclusion that even though $\nabla \cdot \mathbf{u} = 0$ at the microscale at every point within the fluid flow, when filtered over a macroscale volume of $\mathscr{L}^3$, the filtered velocity acquires the characteristic of a compressible flow. Rearranging the above equation, we obtain

$$\nabla \cdot \tilde{\mathbf{u}}_c = -\frac{1}{\phi_c} \left(\frac{\partial \phi_c}{\partial t} + \tilde{\mathbf{u}}_c \cdot \nabla \phi_c \right) \,. \tag{11.37}$$

Recall that the particle volume fraction $\phi_d = 1 - \phi_c$. Thus, the divergence of the filtered fluid velocity is related to the net change in the local particle volume fraction both due to the time derivative and due to the advective flux from the surrounding. In essence, if the fractional volume occupied by the particles within a unit volume centered around $\mathbf{x}$ increases, this must be compensated by an outflux of fluid. This contributes to a positive divergence of fluid velocity. Thus, particle motion and rearrangement gives multiphase flow an effective compressible flow-like character.

11.2.2 Momentum Balance

We follow the procedure of the previous subsection and derive the filtered continuous-phase momentum equation. Multiplying the momentum equation with the filter and integrating over the fluid domain, we obtain

$$\int \left(\frac{\partial \rho_f \mathbf{u}}{\partial t} + \nabla' \cdot (\rho_f \, \mathbf{u}\mathbf{u}) - \nabla' \cdot \boldsymbol{\sigma} - \rho_f \mathbf{g} \right) I_c \, G \, d\mathbf{x}' = 0, \tag{11.38}$$

where $\boldsymbol{\sigma}$ is the total fluid stress (which includes both pressure and viscous stress). Using the Leibniz and Gauss rules, the first three terms on the left can be expressed as

$$\frac{\partial \overline{(\rho_f \mathbf{u})}}{\partial t} + \int_S \rho_f \, \mathbf{u} \, (\mathbf{u}_I \cdot \mathbf{n}) \, G \, dA',$$

$$\nabla \cdot \overline{(\rho_f \mathbf{u}\mathbf{u})} - \int_S \rho_f \, \mathbf{u} \, (\mathbf{u} \cdot \mathbf{n}) \, G \, dA', \tag{11.39}$$

$$-\nabla \cdot \overline{\boldsymbol{\sigma}} + \int_S (\boldsymbol{\sigma} \cdot \mathbf{n}) \, G \, dA',$$

and the last term becomes $-\phi_c \rho_c \mathbf{g}$. Before we substitute all the above into Eq. (11.38), we make a few observations.

First, if we consider $\phi_c, \rho_c, p_c,$ and $\tilde{\mathbf{u}}_c$ to be the primary filtered fields that are being solved with the filtered equations, then the filtered nonlinear term $\overline{(\rho_f \mathbf{u}\mathbf{u})}$ falls outside these primary variables, which makes the governing equations unclosed. This is the classic *closure problem* of the filter operation. Following the tradition established in turbulence research, we will write this term as

$$\overline{(\rho_f \mathbf{uu})} = \phi_c \rho_c \tilde{\mathbf{u}}_c \tilde{\mathbf{u}}_c - \phi_c \boldsymbol{\tau}_{\text{sg}}, \tag{11.40}$$

where $\phi_c \boldsymbol{\tau}_{\text{sg}}$ is the subgrid residual stress tensor and accounts for the nonlinear effect of scales that have been filtered out, on the dynamics of the large scales.

We can identify the integrals in the first two equations of (11.39) as being related to mass exchange between the particles and the surrounding fluid. Combining these two terms

$$\dot{\mathbf{M}}(\mathbf{x},t) = \int_S \rho_f \mathbf{u} (\mathbf{u}_I - \mathbf{u}) \cdot \mathbf{n} \, G \, dA', \tag{11.41}$$

where $\dot{\mathbf{M}}$ is the momentum flux associated with the mass exchange. It can be written as $\dot{\mathbf{M}} = \dot{m}\mathbf{u}_{ef}$, with $\mathbf{u}_{ef}$ being the effective velocity of the mass exchange. Since positive (or negative) $\dot{m}$ indicates mass addition (or removal) to (from) the particles (condensation or evaporation), the corresponding $\dot{\mathbf{M}}$ represents the momentum associated with this exchange.

In the last equation of (11.39) we use the definition of the overbar and write $\overline{\sigma} = \phi_c \sigma_c$, where σ_c is the filtered macroscale total stress, which will be expressed in terms of the filtered macroscale pressure and filtered macroscale strain-rate tensor. We will discuss this constitutive relation later in Section 15.3.5 as part of closure relations, and for now leave σ_c as it is. For the integral in the last equation of (11.39) we follow the approach outlined by Anderson and Jackson (1967), Crowe et al. (2011), and Capecelatro and Desjardins (2013) to obtain the final result. We express the total stress as a sum of the filtered macroscale component and a perturbation component: $\sigma = \sigma_c + \sigma'$. Substituting this into the integral, we obtain

$$\int_S (\sigma_c \cdot \mathbf{n}) \, G \, dA' + \int_S (\sigma' \cdot \mathbf{n}) \, G \, dA'. \tag{11.42}$$

Using Gauss's theorem, the first term in the above expression can be converted to a volume integral over the region covered by all the particles. We further manipulate this term in the following manner, which can best be described in index notation as

$$\int_V \frac{\partial}{\partial x_j'} (\sigma_{c,ij}(\mathbf{x}',t) \, G(\mathbf{x} - \mathbf{x}')) d\mathbf{x}'$$

$$= \int_V G(\mathbf{x} - \mathbf{x}') \frac{\partial \sigma_{c,ij}(\mathbf{x}',t)}{\partial x_j'} d\mathbf{x}' + \int_V \sigma_{c,ij}(\mathbf{x}',t) \frac{\partial G(\mathbf{x} - \mathbf{x}')}{\partial x_j'} d\mathbf{x}'$$

$$= \int_V G(\mathbf{x} - \mathbf{x}') \frac{\partial \sigma_{c,ij}(\mathbf{x}',t)}{\partial x_j'} d\mathbf{x}' - \frac{\partial}{\partial x_j} \int_V \sigma_{c,ij}(\mathbf{x}',t) G(\mathbf{x} - \mathbf{x}') \, d\mathbf{x}', \tag{11.43}$$

where, in obtaining the last line, we have used the fact that $\partial G(\mathbf{x} - \mathbf{x}')/\partial x_j' = -\partial G(\mathbf{x} - \mathbf{x}')/\partial x_j$ and the gradient in $\mathbf{x}$ can be taken out of the integral, which is in $\mathbf{x}'$. In the second integral, we make the substitution

$$\sigma_c(\mathbf{x}',t) = \sigma_c(\mathbf{x},t) + \tilde{\sigma}_c(\mathbf{x}',t), \tag{11.44}$$

where we recognize the fact that the filtered stress tensor σ_c is a smooth function that varies slowly only on scales larger than $\mathscr{L}$. Thus, the dominant contribution to the integral comes from $\sigma_c(\mathbf{x},t)$, while the contribution from the filtered stress variation represented by $\tilde{\sigma}_c(\mathbf{x}',t)$ is small. Similarly, in the first term we make the substitution

$$(\nabla \cdot \sigma_c)(\mathbf{x}',t) = (\nabla \cdot \sigma_c)(\mathbf{x},t) + (\widetilde{\nabla \cdot \sigma_c})(\mathbf{x}',t). \tag{11.45}$$

In making these substitutions, terms that are functions of $\mathbf{x}$ can be taken out of the integrals to obtain

$$\int_S (\sigma_c \cdot \mathbf{n}) \, G \, dA' = (\nabla \cdot \sigma_c)(\mathbf{x},t) \int_V G(\mathbf{x} - \mathbf{x}') \, d\mathbf{x}' - \nabla \cdot (\sigma_c(\mathbf{x},t) \int_V G(\mathbf{x} - \mathbf{x}') \, d\mathbf{x}')$$
$$+ \int_V (\widetilde{\nabla \cdot \sigma_c})(\mathbf{x}',t) G(\mathbf{x} - \mathbf{x}') \, d\mathbf{x}' - \nabla \cdot (\int_V \tilde{\sigma}_c(\mathbf{x}',t) G(\mathbf{x} - \mathbf{x}') \, d\mathbf{x}'). \tag{11.46}$$

We can recognize $\int_V G(\mathbf{x} - \mathbf{x}') \, d\mathbf{x}'$, where the integral is over only the regions occupied by the particles, as the particle volume fraction $\phi_d(\mathbf{x},t)$. With this, the first term on the right becomes $\phi_d \nabla \cdot \sigma_c$, and the second term becomes $-\nabla \cdot (\phi_d \sigma_c)$. In comparison, the contribution of the last two terms is small and we simply denote their sum as $\mathcal{N}_m$. This yields the final expression

$$\int_S (\sigma_c \cdot \mathbf{n}) \, G \, dA' = \sigma_c \cdot \nabla \phi_c + \mathcal{N}_m, \tag{11.47}$$

where we have used the fact that $\nabla \phi_d = -\nabla \phi_c$. The above force per unit volume of the multiphase flow is sometimes called the *nozzling effect*, which arises due to the spatial variation in volume fraction that acts like a nozzle, as far as the surrounding flow is concerned, and thus $\mathcal{N}_m$ is the nozzling correction. As we will show in Section 11.2.4, as long as the filter width is much wider than the particle size, this nozzling correction is negligible and the nozzling effect can be well approximated by $\sigma_c \cdot \nabla \phi_c$. So henceforth we will ignore $\mathcal{N}_m$.

As discussed in Chapter 4, the perturbation hydrodynamic force on a particle is given by an integral of the tractional force over the surface of the particle. Thus, the integral $\int_S (\sigma' \cdot \mathbf{n}) \, dA'$ corresponds to the sum of hydrodynamic force on all the particles in the system due to the perturbation flow, since S corresponds to all the particle–fluid interfaces. However, the role of the filter function G is to limit the contribution only to the integral from particles that are close to the point $\mathbf{x}$. We will denote

$$\mathbf{F}'_{\text{hyd}}(\mathbf{x},t) = \int_S (\sigma'(\mathbf{x}',t) \cdot \mathbf{n}) \, G(\mathbf{x} - \mathbf{x}') \, dA' \tag{11.48}$$

as the net perturbation hydrodynamic force on the particles that are located in the neighborhood of the point $\mathbf{x}$ weighted by the filter function. The negative of this force will be applied as the source term in the filtered momentum equation of the continuous phase. From the discussion and definitions of the past three paragraphs, the last equation of (11.39) reduces to

$$- \nabla \cdot (\phi_c \sigma_c) + \sigma_c \cdot \nabla \phi_c + \mathbf{F}'_{\text{hyd}}(\mathbf{x}, t). \tag{11.49}$$

Combining all the above steps for the different terms of Eq. (11.39), and substituting them into Eq. (11.38), we obtain the final equation:

$$\frac{\partial \phi_c \rho_c \tilde{\mathbf{u}}_c}{\partial t} + \nabla \cdot (\phi_c \rho_c \tilde{\mathbf{u}}_c \tilde{\mathbf{u}}_c) - \nabla \cdot (\phi_c \tau_{\text{sg}}) = \phi_c \rho_c \mathbf{g} + \phi_c \nabla \cdot \sigma_c - \mathbf{F}'_{\text{hyd}} - \dot{\mathbf{M}}. \tag{11.50}$$

The physical meaning of each term of the above filtered momentum balance can be examined. The left-hand side corresponds to mass times acceleration of the continuous phase contained within a unit volume at $(\mathbf{x}, t)$. Note that when expressed in terms of the filtered quantities, the effect of velocity fluctuation appears as the subgrid residual stress term. On the right-hand side, other than the gravity term, the interaction between the phases is the source of the other three terms. The nozzling effect due to the gradient of the particle volume fraction is included in the second term on the right-hand side. Momentum exchange between the phases due to their relative motion appears as the perturbation hydrodynamic feedback force and the last term accounts for the momentum exchange associated with mass transfer between the phases.

It is important to note that $\mathbf{F}'_{\text{hyd}}(\mathbf{x}, t)$ does not include the undisturbed flow force due to σ_c, since $\mathbf{F}'_{\text{hyd}}(\mathbf{x}, t)$ accounts for only the effect of σ'. In the terminology of Chapter 4, it includes only those force components that arise due to the perturbation flow induced by the particle's presence. Thus, $\mathbf{F}'_{\text{hyd}}(\mathbf{x}, t)$ includes the quasi-steady force, added-mass force, Basset history force, lift force, and so on, but not the undisturbed flow force. This is the reason we have included the "prime" in its notation. The effect of the filtered macroscale stress field σ_c is to exert an additional force of $\int_{V_l} \nabla \cdot \sigma_c \, d\mathbf{x}$ on each particle, where the integral is over the volume of the lth particle. This is the undisturbed flow force on the particle due to the filtered flow. The back-effect of this undisturbed flow force on the fluid can be calculated as

$$- \mathbf{F}_{un}(\mathbf{x}, t) = - \sum_{l=1}^{N_p} \int G(\mathbf{x} - \mathbf{x}')(\nabla \cdot \sigma_c) d\mathbf{x}' \approx -\phi_d \nabla \cdot \sigma, \tag{11.51}$$

where it is assumed that $\nabla \cdot \sigma_c$ is approximately uniform over the integration volume. If we now define the total hydrodynamic force fed back to the fluid as $-\mathbf{F}_{\text{hyd}} = -\mathbf{F}'_{\text{hyd}} - \mathbf{F}_{un}$, then we can rewrite Eq. (11.50) in the following form as well (Anderson and Jackson, 1967; Capecelatro and Desjardins, 2013):

$$\frac{\partial \phi_c \rho_c \tilde{\mathbf{u}}_c}{\partial t} + \nabla \cdot (\phi_c \rho_c \tilde{\mathbf{u}}_c \tilde{\mathbf{u}}_c) - \nabla \cdot (\phi_c \tau_{\text{sg}}) = \phi_c \rho_c \mathbf{g} + \nabla \cdot \sigma_c - \mathbf{F}_{\text{hyd}} - \dot{\mathbf{M}}. \tag{11.52}$$

Problem 11.5 The subgrid stress tensor is often written in terms of the subgrid velocity fluctuation, which is defined as $\mathbf{u}' = \mathbf{u} - \tilde{\mathbf{u}}_c$.

(a) Starting from the definition of residual stress given in Eq. (11.40), derive the following relation:

$$\phi_c \tau_{\text{sg}} = \underbrace{\phi_c \rho_c \tilde{\mathbf{u}}_c \tilde{\mathbf{u}}_c - \phi_c \rho_c \widetilde{\tilde{\mathbf{u}}_c \tilde{\mathbf{u}}_c}}_{\text{Leonard stress}} \underbrace{- \phi_c \rho_c \widetilde{\mathbf{u}' \mathbf{u}'}}_{\text{Reynolds stress}} \underbrace{- \phi_c \rho_c \left(\widetilde{\tilde{\mathbf{u}}_c \mathbf{u}'} - \widetilde{\mathbf{u}' \tilde{\mathbf{u}}_c} \right)}_{\text{Cross stress}}. \tag{11.53}$$

Except for the appearance of the continuous-phase volume fraction ϕ_c, the above equation is identical to what one gets for residual stress in a large eddy simulation. In the above, the Leonard stress arises from the nonlinear interaction of the filtered large scales, the Reynolds stress arises from the interaction of the small scales, and the cross stress arises from the interaction of the filtered large and small scales. Note that the Leonard stress involves only the filtered variables and therefore can be explicitly calculated, while the other two stresses involve fluctuating quantities and therefore must be evaluated using closure models.

(b) The above expression of the residual stress in terms of filtered and fluctuating components is exact and makes no assumption. Assuming $\tilde{\mathbf{u}}' \approx 0$ (which in general is not true, as discussed in Problem 11.3), obtain the following simplification that the residual stress is the same as the Reynolds stress (Crowe et al., 1996; Capecelatro and Desjardins, 2013):

$$\phi_c \tau_{\mathrm{sg}} = -\phi_c \rho_c \widetilde{\mathbf{u}'\mathbf{u}'}. \tag{11.54}$$

No matter how it is expressed, τ_{sg} must be modeled. Therefore, we prefer its basic definition given in Eq. (11.40). Also, since in general $\tau_{\mathrm{sg}} \neq -\rho_c \widetilde{\mathbf{u}'\mathbf{u}'}$, we refer to τ_{sg} as the residual stress in order to distinguish it from the traditional definition of the Reynolds stress as the correlation of velocity fluctuation.

11.2.3   Energy Balance

In this section, we will repeat the steps of the previous section and obtain the filtered energy equation for the continuous phase. Multiplying the energy equation by $I_c(\mathbf{x}')G(\mathbf{x} - \mathbf{x}')$ and integrating over the fluid domain,

$$\int \left(\frac{\partial \rho_f E}{\partial t} + \nabla' \cdot (\rho_f \mathbf{u} E - \boldsymbol{\sigma} \cdot \mathbf{u} + \mathbf{q}) - \rho_f \mathbf{u} \cdot \mathbf{g} \right) I_c \, G \, d\mathbf{x}' = 0, \tag{11.55}$$

where $E(\mathbf{x}', t)$ is the total energy of the fluid and $\mathbf{q}(\mathbf{x}', t)$ is the heat flux vector. Using the Leibniz and Gauss rules, the first four terms on the left can be expressed as

$$\frac{\partial \overline{(\rho_f E)}}{\partial t} + \int_S \rho_f E \, (\mathbf{u}_I \cdot \mathbf{n}) \, G \, dA',$$

$$\nabla \cdot \overline{(\rho_f \mathbf{u} E)} - \int_S \rho_f E \, (\mathbf{u} \cdot \mathbf{n}) \, G \, dA',$$

$$-\nabla \cdot \overline{(\boldsymbol{\sigma} \cdot \mathbf{u})} + \int_S (\boldsymbol{\sigma} \cdot \mathbf{u}) \cdot \mathbf{n} \, G \, dA', \tag{11.56}$$

$$\nabla \cdot \overline{\mathbf{q}} - \int_S (\mathbf{q} \cdot \mathbf{n}) \, G \, dA',$$

and the last term becomes $-\phi_c \rho_c \tilde{\mathbf{u}}_c \cdot \mathbf{g}$. Again we make the following observations about the different terms.

First, we introduce the following definition of Favre-filtered energy: $\overline{(\rho_f E)} = \phi_c \rho_c \tilde{E}_c$. The nonlinear terms gives rise to the closure problem. We define the subgrid residual energy flux vector and the residual stress-work to be

$$-\phi_c\, Q_{E,\mathrm{sg}} = \phi_c\rho_c\tilde{\mathbf{u}}_c\tilde{E}_c - \overline{\rho_f\mathbf{u}E},$$

$$-\phi_c\, Q_{\sigma,\mathrm{sg}} = \overline{\boldsymbol{\sigma}\cdot\mathbf{u}} - \phi_c\boldsymbol{\sigma}_c\cdot\tilde{\mathbf{u}}_c, \tag{11.57}$$

and they account for the nonlinear effect of the scales that have been filtered out in the filter operation. For notational convenience, we introduce $Q_{\mathrm{sg}} = Q_{E,\mathrm{sg}} + Q_{\sigma,\mathrm{sg}}$ as the sum of the two residual terms. We will refer to Q_{sg} as the residual total energy flux.

Problem 11.6 Follow the steps of Problem 11.5 and express $\phi_c\, Q_{E,\mathrm{sg}}$ and $\phi_c\, Q_{\sigma,\mathrm{sg}}$ in terms of Leonard, Reynolds, and cross-term contributions.

We can identify the integrals in the first two equations of (11.56) as being related to mass exchange between the particles and the surrounding fluid. Combining these two terms, we define

$$\dot{E}(\mathbf{x},t) = \int_S \rho_f\, E\,(\mathbf{u}_I - \mathbf{u})\cdot\mathbf{n}\, G\, dA' \tag{11.58}$$

as the total energy flux associated with the mass exchange, which can be written as $\dot{E} = \dot{m}E_{ef}$, with E_{ef} being the effective total energy carried by the mass exchange. Positive (or negative) $\dot{m}$ and $\dot{E}$ correspond to mass and total energy addition to (or removal from) the particles due to condensation and reverse sublimation (or evaporation and sublimation).

The integral in the third equation of (11.56) can be separated into the following two contributions:

$$\int_S (\boldsymbol{\sigma}\cdot\mathbf{u})\cdot\mathbf{n}\, G\, dA' = \int_S (\boldsymbol{\sigma}_c\cdot\mathbf{u})\cdot\mathbf{n}\, G\, dA' + \int_S (\boldsymbol{\sigma}'\cdot\mathbf{u})\cdot\mathbf{n}\, G\, dA'. \tag{11.59}$$

The first term on the right is the energy analog of the *nozzling effect*. It is due to the spatial variation in particle volume fraction that acts like a nozzle and can be written as

$$\int_S (\boldsymbol{\sigma}_c\cdot\mathbf{u})\,\mathbf{n}\, G\, dA' = -\boldsymbol{\sigma}_c : \nabla(\phi_d\mathbf{u}_d) + \mathcal{N}_e, \tag{11.60}$$

where $\mathbf{u}_d$ is the filtered dispersed-phase velocity that will be defined in Section 11.2 and $\mathcal{N}_e$ is the nozzling energy correction. As we will show in Section 11.2.4, as long as the filter width is much wider than the particle size, this nozzling energy correction is negligible and the nozzling energy effect can be well approximated by the first term on the right.

The second term of Eq. (11.59) is related to work associated with the perturbation hydrodynamic force exchange between the particles and the surrounding fluid. Thus,

$$W'_{\mathrm{hyd}}(\mathbf{x},t) = \int_S (\boldsymbol{\sigma}'\cdot\mathbf{u})\cdot\mathbf{n}\, G\, dA' \tag{11.61}$$

denotes the net perturbation hydrodynamic work on particles that are located in the neighborhood of the point $\mathbf{x}$ weighted by the filter function. Again this term does not include the work associated with the undisturbed flow force on the particles. The negative of W'_{hyd} will be applied as the source term in the filtered total energy equation of the continuous phase.

In the final equation of (11.56) we use the following definition of filtered heat flux: $\overline{\mathbf{q}} = \phi_c \mathbf{q}_c$. We then substitute the separation of the total heat flux as the sum of the filtered and fluctuation components (i.e., $\mathbf{q} = \mathbf{q}_c + \mathbf{q}'$) into the integral to obtain

$$\int_S (\mathbf{q} \cdot \mathbf{n}) \, G \, dA' = \int_S (\mathbf{q}_c \cdot \mathbf{n}) \, G \, dA' + \int_S (\mathbf{q}' \cdot \mathbf{n}) \, G \, dA'. \tag{11.62}$$

The first term on the right is the nozzling effect on heat transfer and can be approximated as

$$\mathbf{q}_c \cdot \nabla \phi_c + \mathcal{N}_q, \tag{11.63}$$

where $\mathcal{N}_q$ is the correction, which will later be shown to be quite small in the limit of particles much smaller than the filter width. Finally, we recognize $-\mathbf{q}' \cdot \mathbf{n}$ as the perturbation heat flux into the particle from the surrounding fluid. The integral $-\int_S (\mathbf{q}' \cdot \mathbf{n}) \, dA'$, which is over the entire interface between the phases, corresponds to the total perturbation heat flux to all the particles from the continuous phase. It does not include the undisturbed flow heat transfer. However, due to multiplication by the filter function,

$$Q'_{\text{hyd}}(\mathbf{x}, t) = -\int_S (\mathbf{q}'(\mathbf{x}', t) \cdot \mathbf{n}) \, G(\mathbf{x} - \mathbf{x}') \, dA' \tag{11.64}$$

accounts for the net heat transfer due to the perturbation flow to the particles that are located in the immediate neighborhood of the point $\mathbf{x}$. The negative of this will be applied as the source term in the filtered total energy equation of the continuous phase.

Combining all the above steps for the different terms of Eq. (11.39) and substituting them into Eq. (11.55), we obtain the final filtered energy equation as

$$\frac{\partial \phi_c \rho_c \tilde{E}_c}{\partial t} + \nabla \cdot (\phi_c \rho_c \tilde{\mathbf{u}}_c \tilde{E}_c) + \nabla \cdot (\phi_c Q_{\text{sg}})$$
$$= \phi_c \rho_c \tilde{\mathbf{u}}_c \cdot \mathbf{g} + \nabla \cdot (\phi_c \boldsymbol{\sigma}_c \cdot \tilde{\mathbf{u}}_c), \tag{11.65}$$
$$+ \boldsymbol{\sigma}_c : \nabla(\phi_d \tilde{\mathbf{u}}_d) - \phi_c \nabla \cdot \mathbf{q}_c - Q'_{\text{hyd}} - W'_{\text{hyd}} - \dot{E}.$$

The left-hand side is the rate of change of total energy of the continuous phase within a unit volume and it includes the residual flux term.[2] The first term on the right-hand side is the work done by gravity and the second term includes viscous heating of the fluid in the bulk and pressure work. The remaining terms account for all the interphase effects on the filtered continuous-phase energy equation. The third term on the right-hand side

---

[2] It should be noted that the sign of the residual flux is the opposite of the residual stress term in the momentum equation. This is simply to match the standard convention used in the single-phase Navier–Stokes equation that the divergence of stress appears with a positive sign on the right-hand side of the momentum equation and the divergence of the heat flux vector appears with a negative sign on the right-hand side of the energy equation. You may find in the literature definitions of residual stress and heat flux with their signs reversed.

accounts for the energy effect of the nozzling term included in the momentum equation. The fourth term accounts for heat conduction, but includes the nozzling-like effect of volume fraction variation on heat flux. The last three terms are as follows: (i) direct heat flux between the dispersed phase and the continuous phase due to temperature difference; (ii) work exchange between the dispersed phase and the continuous phase as a result of the momentum exchange $\mathbf{F}'_{\mathrm{hyd}}$; and (iii) energy exchange between the phases associated with the mass exchange $\dot{m}$.

Thus, we see that a relative difference in species concentration between the phases drives (a-i) mass transfer $\dot{m}$, (a-ii) associated momentum transfer $\dot{\mathbf{M}}$, and (a-iii) energy transfer $\dot{E}$. A relative motion between the two phases (i.e., difference in velocity or acceleration) will drive (b-i) momentum transfer that is modeled as interphase force $\mathbf{F}'_{\mathrm{hyd}}$ and (b-ii) associated work exchange between the phases W'_{hyd}. A difference in temperature between the dispersed and the continuous phases drives (c-i) interphase heat exchange Q'_{hyd}. It should be noted that all these are perturbation effects that are driven by differences in species concentration, motion, and temperature between the phases. Undisturbed flow effects are not included.

Problem 11.7 The derivation of the filtered mass, momentum, and energy balances of the previous three subsections is mathematically rigorous. In the process, we may perhaps have sacrificed physical intuition and transparency. So it is instructive to re-derive these equations using a control volume approach. Follow the approach presented in Crowe et al. (1996) and consider a quasi one-dimensional tube of multiphase flow. Apply the control volume approach and derive the mass, momentum, and energy balances of the continuous phases. Compare the resulting equations with those of the previous sections.

Problem 11.8 *Filtered internal energy and temperature equation.* It is sometimes convenient to write an equation for the internal energy, which can then be converted to an equation for the continuous-phase temperature. The internal energy of the continuous phase is obtained from the total energy by subtracting the kinetic component as $e = E - |\mathbf{u}|^2/2$, which is defined at every point within the fluid phase.

(a) Multiply this with ρ_f and perform the filter operation to obtain the relation

$$\phi_c\,\rho_c\tilde{E}_c = \phi_c\,\rho_c\tilde{e}_c + \frac{1}{2}\phi_c\,\rho_c\tilde{\mathbf{u}}_c\cdot\tilde{\mathbf{u}}_c - \frac{1}{2}\phi_c\,\mathrm{tr}[\tau_{\mathrm{sg}}], \tag{11.66}$$

where $\mathrm{tr}[\cdot]$ indicates the trace of the argument (i.e., the sum of the diagonal terms).

(b) As the next step, derive the following relation:

$$\tilde{\mathbf{u}}_c\cdot\left[\frac{\partial\phi_c\rho_c\tilde{\mathbf{u}}_c}{\partial t} + \nabla\cdot(\phi_c\rho_c\tilde{\mathbf{u}}_c\tilde{\mathbf{u}}_c)\right]$$
$$= \frac{1}{2}\frac{\partial\phi_c\rho_c\tilde{\mathbf{u}}_c\cdot\tilde{\mathbf{u}}_c}{\partial t} + \frac{1}{2}\nabla\cdot(\phi_c\rho_c\tilde{\mathbf{u}}_c(\tilde{\mathbf{u}}_c\cdot\tilde{\mathbf{u}}_c)) - \dot{m}\frac{\tilde{\mathbf{u}}_c\cdot\tilde{\mathbf{u}}_c}{2}. \tag{11.67}$$

(c) Now take the dot product of $\tilde{\mathbf{u}}_c$ with the filtered momentum equation (11.50), and subtract the dot product from the filtered total energy equation given in (11.65). The resulting equation can be combined with Eqs. (11.66) and (11.67) to obtain the final filtered internal energy equation. Follow these steps and obtain

$$\frac{\partial \phi_c \rho_c \tilde{e}_c}{\partial t} + \nabla \cdot (\phi_c \rho_c \tilde{\mathbf{u}}_c \tilde{e}_c) + \nabla \cdot (\phi_c \mathbf{q}_{\mathrm{sg}})$$

$$= \boldsymbol{\sigma}_c : \nabla(\phi_c \tilde{\mathbf{u}}_c + \phi_d \tilde{\mathbf{u}}_d) - \phi_c \nabla \cdot \mathbf{q}_c$$

$$- Q'_{\mathrm{hyd}} - \left[W'_{\mathrm{hyd}} - \tilde{\mathbf{u}}_c \cdot \mathbf{F}'_{\mathrm{hyd}} \right] - \left[\dot{E} - \tilde{\mathbf{u}}_c \cdot \left(\dot{\mathbf{M}} - \frac{\dot{m}\tilde{\mathbf{u}}_c}{2} \right) \right]. \qquad (11.68)$$

(d) Show that the third term on the left-hand side arises from the residual stress and residual total energy flux terms in the momentum and total energy equations and obtain the following relation for the residual contribution to internal energy balance:

$$-\nabla \cdot (\phi_c \mathbf{q}_{\mathrm{sg}}) = \frac{1}{2} \left(\frac{\partial \phi_c \mathrm{tr}[\boldsymbol{\tau}_{\mathrm{sg}}]}{\partial t} + \nabla \cdot (\phi_c \tilde{\mathbf{u}}_c \mathrm{tr}[\boldsymbol{\tau}_{\mathrm{sg}}]) \right) - \nabla \cdot (\phi_c \mathcal{Q}_{\mathrm{sg}})$$

$$- \tilde{\mathbf{u}}_c \cdot \nabla \cdot (\phi_c \boldsymbol{\tau}_{\mathrm{sg}}). \qquad (11.69)$$

In essence, $\mathbf{q}_{\mathrm{sg}}$ is the residual heat flux term that accounts for the nonlinear effect of filtering on internal energy. We note that the filtered internal energy equation can also be derived starting from the internal energy equation of the continuous phase and filtering it. In this manner, it can be shown that the filter of the nonlinear term in the internal energy equation can be expressed as divergence of residual heat flux. The above equation connects the residual heat flux $\mathbf{q}_{\mathrm{sg}}$ in the internal energy equation with the residual total energy flux $\mathcal{Q}_{\mathrm{sg}}$ in the total energy equation.

The third line of Eq. (11.68) can now be interpreted. As to be expected, the heat transfer Q'_{hyd} between the continuous and the dispersed phases contributes entirely to the internal energy of the fluid phase. The terms within the first set of square brackets show that out of the total work $-W'_{\mathrm{hyd}}$ done on the fluid phase, $\tilde{\mathbf{u}}_c \cdot \mathbf{F}'_{\mathrm{hyd}}$ goes toward the kinetic energy and only the difference (if nonzero) contributes to the internal energy of the continuous phase. In the second set of square brackets $-\dot{E}$ is the total energy exchange to the continuous phase and part of it goes to kinetic energy and the balance contributes to internal energy.

(e) Finally, assume an ideal fluid with the following relation between internal energy and temperature: $e_c = C_{vf} T_c$, where C_{vf} is the specific heat of the fluid at constant volume. Obtain the following nonconservative form of the equation for the filtered fluid temperature:

$$\phi_c \rho_f C_{vf} \frac{D\tilde{T}_c}{Dt} = -\nabla \cdot (\phi_c \mathbf{q}_{\mathrm{sg}}) - \phi_c \nabla \cdot \mathbf{q}_c + \boldsymbol{\sigma}_c : \nabla(\phi_c \tilde{\mathbf{u}}_c + \phi_d \tilde{\mathbf{u}}_d) - Q'_{\mathrm{hyd}}$$

$$- \left(W'_{\mathrm{hyd}} - \tilde{\mathbf{u}}_c \cdot \mathbf{F}'_{\mathrm{hyd}} \right) - (\dot{E} - \dot{m}\tilde{E}_c) + \tilde{\mathbf{u}}_c \cdot (\dot{\mathbf{M}} - \dot{m}\tilde{\mathbf{u}}_c). \qquad (11.70)$$

Note that the nozzling corrections are ignored.

Problem 11.9 *Filtered scalar concentration equation.* In this problem, you will practice the spatial filter operation for a scalar transport equation. Start with the following advection–diffusion passive scalar transport equation:

$$\frac{\partial C}{\partial t} + \nabla \cdot (\mathbf{u}C) - \nabla \cdot (\mathbf{q}_C) = 0, \tag{11.71}$$

where C is the volumetric concentration of the passive scalar in the continuous phase. Here, $\mathbf{q}_C$ is the diffusive flux of the passive scalar, which, using Fick's law, can be expressed as $\mathbf{q}_C = -D\nabla C$, where D is the concentration diffusivity.

(a) Apply the filter operation on the above equation and obtain the following filtered scalar concentration equation (Capecelatro and Desjardins, 2013):

$$\frac{\partial \phi_c C_c}{\partial t} + \nabla \cdot (\phi_c \tilde{\mathbf{u}}_c C_c) - \nabla \cdot (\phi_c \mathbf{R}_{C,\mathrm{sg}}) = -\phi_c \nabla \cdot \mathbf{q}_C - \omega'_{\mathrm{hyd}} - \dot{S}, \tag{11.72}$$

where the first term on the right-hand side includes the nozzling effect. The nozzling correction is negligibly small in the limit of particle size much smaller than the filter width, and therefore neglected.

(b) Show that the expressions of (i) the residual concentration flux $\mathbf{R}_{C,\mathrm{sg}}$, (ii) scalar transport between the phases due to their concentration difference, and (iii) scalar transport associated with the mass transport are as follows:

$$\phi_c \mathbf{R}_{C,\mathrm{sg}} = \phi_c \tilde{\mathbf{u}}_c C_c - \overline{\mathbf{u}C},$$

$$\omega'_{\mathrm{hyd}} = -\int_S (\mathbf{q}'_C \cdot \mathbf{n})\, G\, dA', \tag{11.73}$$

$$\dot{S} = \int_S C(\mathbf{u}_I - \mathbf{u}) \cdot \mathbf{n}\, G\, dA'.$$

(c) Simplify the expressions for ω'_{hyd} and $\dot{S}$ in terms of contributions from individual particles by assuming the particle size to be much smaller than the filter width.

Problem 11.10 *Filtered mass fraction equation.* This problem is similar to the previous one, but appropriate for chemically reacting flows. Start with the following advection–diffusion species mass fraction equation:

$$\frac{\partial \rho_f Y}{\partial t} + \nabla \cdot (\rho_f \mathbf{u}Y) + \nabla \cdot (\mathbf{q}_Y) = 0, \tag{11.74}$$

where Y is the mass fraction of the species being considered, whose diffusive flux is $\mathbf{q}_Y$.

Carry out the filter operation and obtain the filtered species mass fraction equation. Show that this equation is similar to that obtained in the previous problem. One important difference is that you need to introduce the Favre-filtered mass fraction, $\tilde{Y}_c$.

> **Problem 11.11** *Filtered equation of state.* In compressible multiphase flows, one must consider the equation of state (EOS) in addition to mass, momentum, and energy balances. The simplest equation of state is the ideal gas relation $p = \rho RT$, where R is the specific gas constant. If we take the continuous phase to obey this relation, apply the filter and obtain the following filtered equation of state:
>
> $$p_c = \rho_c R \tilde{T}_c. \tag{11.75}$$
>
> Thus, we see that the equation of state remains the same, except that the filtered pressure, density, and Favre-filtered continuous-phase temperature are related.

11.2.4 Simplified Filtered Equations

At the microscale, mass, momentum, and energy balances, along with an equation of state that relates pressure, density, and internal energy, are sufficient to uniquely determine the continuous-phase variables $\rho_f(\mathbf{x},t)$, $\mathbf{u}(\mathbf{x},t)$, $p(\mathbf{x},t)$, and $E(\mathbf{x},t)$. At the macroscale, however, the filtered mass, momentum, and energy equations (11.35), (11.50), and (11.65) appear far more complex. They, along with the filtered equation of state, are to be interpreted as the governing equations of the primary filtered variables $\rho_c(\mathbf{x},t)$, $\tilde{\mathbf{u}}_c(\mathbf{x},t)$, $p_c(\mathbf{x},t)$, and $\tilde{E}_c(\mathbf{x},t)$. Note that $\phi_c(\mathbf{x},t)$ is also a primary filtered variable, which can readily be obtained from the instantaneous spatial distribution of the particles using Eq. (11.17). Given this viewpoint, all other variables that appear in Eqs. (11.35), (11.50), and (11.65) are secondary variables and they must be expressed in terms of the primary variables. That is, the secondary variables must be closed by writing them in terms of the primary variables ϕ_c, ρ_c, $\tilde{\mathbf{u}}_c$, p_c, and $\tilde{E}_c$.

The filtered total stress tensor $\boldsymbol{\sigma}_c$ of the continuous phase will be separated into an isotropic pressure part and a deviatoric viscous stress tensor as

$$\boldsymbol{\sigma}_c = -p_c \mathbf{I} + \boldsymbol{\tau}_c, \tag{11.76}$$

where $\mathbf{I}$ is the identity tensor. The filtered continuous-phase pressure p_c will be expressed in terms of other primary variables using the equation of state relation. The filtered viscous stress tensor $\boldsymbol{\tau}_c$ will later be closed in terms of the strain-rate tensor of the filtered velocity through a constitutive relation.

Table 11.1 presents all the secondary variables that appear in the filtered continuous-phase mass, momentum, and energy equations, listed under different categories. We will address the closure models of all the secondary variables listed in Table 11.1 in greater detail in Chapter 15 on the Euler–Lagrange methodology. Before that, in this section we will simplify the governing equations by first assuming the particles to be small and then assuming the fluid to be incompressible.

Table 11.1 List of all the secondary variables that arise from the filter operation and appear in the filtered mass, momentum, and energy equations (11.35), (11.50), and (11.65).

Category	Variables to be modeled
Interphase mass exchange	$\dot{m}$, $\dot{\mathbf{M}}$, $\dot{E}$
Interphase momentum exchange	$\mathbf{F}'_{\text{hyd}}$, W'_{hyd}
Interphase thermal exchange	Q'_{hyd}
Constitutive relation	$\boldsymbol{\tau}_c$, $\mathbf{q}_c$
Subgrid closure	$\boldsymbol{\tau}_{\text{sg}}$ (Q_{sg} or $\mathbf{q}_{\text{sg}}$)

Problem 11.12 *An expression for the filtered macroscale strain rate.* In this problem, we will derive the following important result on the filtered strain rate:

$$\overline{\mathbf{S}} = \frac{1}{2}\overline{\left(\nabla\mathbf{u} + (\nabla\mathbf{u})^{\mathrm{T}}\right)} = \frac{1}{2}\left(\nabla\mathbf{u}_m + (\nabla\mathbf{u}_m)^{\mathrm{T}}\right), \tag{11.77}$$

where the volume-weighted mixture velocity is defined as $\mathbf{u}_m = \phi_c\mathbf{u}_c + \phi_d\mathbf{u}_d$, and superscript T indicates transpose. This is a fundamental result which will be used in Section 15.3.5 in the definition of $\boldsymbol{\tau}_c$.

(a) From the definition of the overbar given in Eq. (11.16), the left-hand side is the filtered strain rate of the continuous phase. Show that it can be expressed as

$$\frac{1}{2}\overline{\left(\nabla\mathbf{u} + (\nabla\mathbf{u})^{\mathrm{T}}\right)} = \int \frac{1}{2}\left(\nabla'\mathbf{u} + (\nabla'\mathbf{u})^{\mathrm{T}}\right) I_c(\mathbf{x}',t)\,G(\mathbf{x}-\mathbf{x}')\,d\mathbf{x}'. \tag{11.78}$$

In the above integral, $\mathbf{u}$ is a function of $(\mathbf{x}',t)$ and the gradient operator $\nabla' = \partial/\partial\mathbf{x}'$. Therefore, the fluid strain rate inside the integral is to be interpreted as a function of $(\mathbf{x}',t)$, similar to the indicator function.

(b) To prove the relation given in Eq. (11.77), we begin by defining a unified velocity field $\mathbf{w}(\mathbf{x}',t)$ that is equal to the fluid velocity $\mathbf{u}(\mathbf{x}',t)$ in regions occupied by the fluid and equal to $\mathbf{v}(\mathbf{x}',t)$ in regions occupied by the particles (the particle velocity field will be precisely defined later in Eq. (11.107)). With this, confirm the following expression:

$$\overline{\left(\nabla\mathbf{u} + (\nabla\mathbf{u})^{\mathrm{T}}\right)} = \int \left(\nabla'\mathbf{w} + (\nabla'\mathbf{w})^{\mathrm{T}}\right) G(\mathbf{x}-\mathbf{x}')\,d\mathbf{x}'. \tag{11.79}$$

In writing the above expression, we use the fact that the strain rate within the rigid particles is identically zero (i.e., $\left(\nabla'\mathbf{w} + (\nabla'\mathbf{w})^{\mathrm{T}}\right) = 0$ inside the particles) and as a result the integral can be taken over the entire volume, instead of being restricted to the fluid with the indicator function. From the right-hand side we subtract $\int \nabla'(\mathbf{w}\,G) + (\nabla'(\mathbf{w}\,G))^{\mathrm{T}}\,d\mathbf{x}'$, since this subtracted term is identically zero – with the Gauss theorem the volume integral can be converted to a surface integral over a large sphere, which goes to zero, since the filter function $G \to 0$ on this surface.

(c) After subtraction, using the product rule, obtain the right-hand side to be

$$-\int \mathbf{w}(\mathbf{x}',t)\left(\nabla'(G(\mathbf{x}-\mathbf{x}')) + (\nabla'(G(\mathbf{x}-\mathbf{x}')))^{\mathrm{T}}\right) d\mathbf{x}'. \tag{11.80}$$

(d) Now we use the property $\nabla G(\mathbf{x} - \mathbf{x}') = -\nabla' G(\mathbf{x} - \mathbf{x}')$ and take the gradient with respect to $\mathbf{x}$ outside the integral, which is in $\mathbf{x}'$, to obtain

$$\nabla \left(\int \mathbf{w}(\mathbf{x}',t)\, G(\mathbf{x} - \mathbf{x}')d\mathbf{x}' \right) + \left(\nabla \left(\int \mathbf{w}(\mathbf{x}',t)\, G(\mathbf{x} - \mathbf{x}')d\mathbf{x}' \right) \right)^{\mathrm{T}}. \qquad (11.81)$$

(e) Now return from the unified velocity to individual velocities, to obtain

$$\int \mathbf{w}(\mathbf{x}',t)\, G(\mathbf{x} - \mathbf{x}')d\mathbf{x}' = \int \mathbf{u}(\mathbf{x}',t)\, I_c(\mathbf{x}',t)\, G(\mathbf{x} - \mathbf{x}')d\mathbf{x}'$$
$$+ \int \mathbf{v}(\mathbf{x}',t)\, I_d(\mathbf{x}',t)\, G(\mathbf{x} - \mathbf{x}')d\mathbf{x}'. \qquad (11.82)$$

We recognize the right-hand side as the volume-weighted mixture velocity $\mathbf{u}_m = \phi_c \mathbf{u}_c + \phi_d \mathbf{u}_d$.

(f) Substituting, and introducing the factor $1/2$, prove the relation (11.77). Thus, we have obtained the fundamental result that the filtered strain rate of the continuous phase is the strain rate of the volume-averaged mixture velocity.

Problem 11.13 *An expression for the filtered macroscale thermal gradient.* In this problem, you will derive the following important result on the filtered thermal gradient:

$$\overline{\nabla T} = \nabla \mathbf{T}_m. \qquad (11.83)$$

Follow the steps of the previous problem to obtain the above result. In the derivation you will assume the particle to be isothermal and thus the thermal gradient inside the particle is zero. With this derivation, we have the fundamental result that the filtered thermal gradient of the continuous phase is equal to the gradient of the volume-averaged mixture temperature. This is a fundamental result which will be used in Section 15.3.5 in the definition of $\mathbf{q}_c$.

Particles Much Smaller Than Filter Length Scale

The filtered mass, momentum, and energy balance equations (11.35), (11.50), and (11.65) are exact and do not involve any approximation (except for the neglect of the nozzling corrections). However, a long list of to-be-closed secondary quantities appear in these equations, as enumerated in Table 11.1. For the filtered equations to be useful, these secondary quantities must be expressed in terms of the primary flow and particle quantities being solved, and this closure presents a formidable challenge. To make further progress, it is necessary to make judicious assumptions that can simplify the governing equations. First, we will consider the situation where the particles are much smaller than the length scale of the filter (i.e., the limit when $d_p \ll \mathcal{L}$). Several simplifications result with this assumption.

Let us start with the net mass flux term given in Eq. (11.34). We will make use of the fact that the surface integral is over the surface of all the particles. If we introduce the notation that the surface S is the union of surfaces S_l of the lth particle for $l = 1, 2, \ldots, N_p$, then Eq. (11.34) can be rewritten as

$$\dot{m}(\mathbf{x}, t) = \sum_{l=1}^{N_p} \left\{ \int_{S_l} \rho_f(\mathbf{x}', t) \, G(\mathbf{x} - \mathbf{x}') \, (\mathbf{u}_I(\mathbf{x}', t) - \mathbf{u}(\mathbf{x}', t)) \cdot \mathbf{n} \, dA' \right\}, \qquad (11.84)$$

where the integral is over the surface of the lth particle. According to the above equation, the mass flux $\dot{m}$ at location $\mathbf{x}$ depends on the contribution from every point $\mathbf{x}'$ that lies on the surface of the particle. The contribution is weighted by the filter function, whose value depends on the distance from point $\mathbf{x}'$ on the particle to the point of evaluation $\mathbf{x}$. If the size of the particle is smaller than the length scale of the filter function, then the value of $G(\mathbf{x} - \mathbf{x}')$ is nearly a constant as $\mathbf{x}'$ varies over the surface of any single particle. However, the value of G will vary significantly from particle to particle. This allows the filter function to be moved out of the integral:

$$\begin{aligned} \dot{m}(\mathbf{x}, t) &\approx \sum_{l=1}^{N_p} G(\mathbf{x} - \mathbf{X}_l) \left\{ \int_{S_l} \rho_f(\mathbf{x}', t) \, (\mathbf{u}_I(\mathbf{x}', t) - \mathbf{u}(\mathbf{x}', t)) \cdot \mathbf{n} \, dA' \right\} \\ &= \sum_{l=1}^{N_p} G(\mathbf{x} - \mathbf{X}_l) \, \dot{m}_l, \end{aligned} \qquad (11.85)$$

where the term within curly brackets is precisely the total mass transfer of the lth particle. It is denoted by $\dot{m}_l$ and the entire mass exchange of the lth particle has been assigned to the particle center $\mathbf{X}_l$. This term is positive if the lth particle gains mass from the surrounding fluid (condensation) and is negative if the lth particle loses mass to the surrounding fluid (evaporation). Note that this term therefore appears in the mass balance of the continuous phase with a negative sign (opposite to that for the lth particle).

Similar approximations for the momentum and energy exchanges associated with the mass exchange yield

$$\dot{\mathbf{M}}(\mathbf{x}, t) = \sum_{l=1}^{N_p} G(\mathbf{x} - \mathbf{X}_l) \, \dot{\mathbf{M}}_l \quad \text{and} \quad \dot{E}(\mathbf{x}, t) = \sum_{l=1}^{N_p} G(\mathbf{x} - \mathbf{X}_l) \, \dot{E}_l. \qquad (11.86)$$

In the above form, it is quite clear that only particles whose $\mathbf{X}_l$ is close to $\mathbf{x}$ will contribute to $\dot{m}$, $\dot{\mathbf{M}}$, and $\dot{E}$. The exponential decay of the filter function will render the contribution from far-away particles to be negligibly small. The fundamental definitions of momentum and energy exchanges of an individual particle are

$$\begin{aligned} \dot{\mathbf{M}}_l(\mathbf{x}, t) &= \int_{S_l} \rho_f(\mathbf{x}', t) \, \mathbf{u}(\mathbf{x}', t) \, (\mathbf{u}_I(\mathbf{x}', t) - \mathbf{u}(\mathbf{x}', t)) \cdot \mathbf{n} \, dA', \\ \dot{E}_l(\mathbf{x}, t) &= \int_{S_l} \rho_f(\mathbf{x}', t) \, E(\mathbf{x}', t) \, (\mathbf{u}_I(\mathbf{x}', t) - \mathbf{u}(\mathbf{x}', t)) \cdot \mathbf{n} \, dA', \end{aligned} \qquad (11.87)$$

where the integrals are over the surface of the lth particle. The exchanges are positive when mass, momentum, and energy go from the continuous to the dispersed phase. In the above expressions, the exchange is measured in the continuous-phase with the surface distribution of continuous phase density $\rho_f(\mathbf{x}', t)$, velocity $\mathbf{u}(\mathbf{x}', t)$, and total energy $E(\mathbf{x}', t)$. Furthermore, it should be noted that $\dot{m}_l$, $\dot{\mathbf{M}}_l$, and $\dot{E}_l$ apply with opposite sign to the particle and the surrounding fluid. The total energy exchange of the lth particle can be split into internal energy and the kinetic energy exchange as $\dot{E}_l = \dot{e}_l + \dot{k}_l$, where the two contributions can be expressed as

$$\dot{e}_l(\mathbf{x},t) = \int_{S_l} \rho_f(\mathbf{x}',t)\, e(\mathbf{x}',t)\, (\mathbf{u}_l(\mathbf{x}',t) - \mathbf{u}(\mathbf{x}',t)) \cdot \mathbf{n}\, dA',$$
$$\dot{k}_l(\mathbf{x},t) = \int_{S_l} \rho_f(\mathbf{x}',t)\, \frac{\mathbf{u}(\mathbf{x}',t) \cdot \mathbf{u}(\mathbf{x}',t)}{2} (\mathbf{u}_l(\mathbf{x}',t) - \mathbf{u}(\mathbf{x}',t)) \cdot \mathbf{n}\, dA'. \tag{11.88}$$

The same trick applies for interphase momentum exchange and the filter function can be taken out of the integral in Eq. (11.48) to obtain

$$\mathbf{F}'_{\text{hyd}}(\mathbf{x},t) = \sum_{l=1}^{N_p} G(\mathbf{x} - \mathbf{X}_l) \left\{ \int_S (\boldsymbol{\sigma}'(\mathbf{x}',t) \cdot \mathbf{n})\, dA' \right\} = \sum_{l=1}^{N_p} G(\mathbf{x} - \mathbf{X}_l)\mathbf{F}'_l, \tag{11.89}$$

where $\mathbf{F}'_l$ is the hydrodynamic force on the lth particle due to the perturbation flow and it includes quasi-steady, inviscid unsteady, viscous unsteady, and lift force contributions, but not the undisturbed flow force. In a similar manner, we can express

$$Q'_{\text{hyd}}(\mathbf{x},t) = \sum_{l=1}^{N_p} G(\mathbf{x} - \mathbf{X}_l)Q'_l, \tag{11.90}$$

where Q'_l is the net heat transferred to the lth particle due to the perturbation flow and it does not include the heat transfer contribution due to the undisturbed flow. The work associated with the above momentum transfer can be approximated in the following way:

$$W'_{\text{hyd}}(\mathbf{x},t) = \sum_{l=1}^{N_p} G(\mathbf{x} - \mathbf{X}_l)W'_l, \tag{11.91}$$

where W'_l is the work associated with the perturbation force of the lth particle.

We now consider the nozzling corrections and show them to be small in the limit of small particles. Consider the expression (11.47) in the limit of a small particle, where the macroscale total stress field $\boldsymbol{\sigma}_c$ can be taken to be nearly uniform over the size of the particle. Therefore we have

$$\int_S (\boldsymbol{\sigma}_c \cdot \mathbf{n})\, G\, dA' \approx \boldsymbol{\sigma}_c \cdot \int_S \mathbf{n}\, G\, dA' = \boldsymbol{\sigma}_c \cdot \nabla \phi_c. \tag{11.92}$$

With this approximation, we arrive at the conclusion that the nozzling correction $\mathcal{N}_m \approx 0$. A similar simplification can be made for Eq. (11.60) in the limit where the particle size is much smaller than the filter width:

$$\int_S (\boldsymbol{\sigma}_c \cdot \mathbf{u}) \cdot \mathbf{n}\, G\, dA' \approx \boldsymbol{\sigma}_c : \int_S \mathbf{u}\, \mathbf{n}\, G\, dA' = -\boldsymbol{\sigma}_c : \nabla(\phi_d \mathbf{u}_d), \tag{11.93}$$

where we have used the fact that $\phi_c = 1 - \phi_d$ and the fluid velocity at the particle surface is the particle velocity (ignoring the mass transfer effect). This leads to the conclusion that $\mathcal{N}_e \approx 0$. Finally, we simplify the nozzling heat flux term as well. The integral is approximated as

$$\int_S (\mathbf{q}_c \cdot \mathbf{n})\, G\, dA' \approx \mathbf{q}_c \cdot \int_S \mathbf{n}\, G\, dA' = \mathbf{q}_c \cdot \nabla \phi_c, \tag{11.94}$$

which again leads to the simplification $\mathcal{N}_q \approx 0$.

With these simplifications the resulting filtered mass, momentum, and energy equations can be written as follows.

When particle size is much smaller than filter width:

$$\frac{\partial \phi_c \rho_c}{\partial t} + \nabla \cdot (\phi_c \rho_c \tilde{\mathbf{u}}_c) = -\sum_{l=1}^{N_p} G(\mathbf{x} - \mathbf{X}_l)\dot{m}_l, \tag{11.95}$$

$$\frac{\partial \phi_c \rho_c \tilde{\mathbf{u}}_c}{\partial t} + \nabla \cdot (\phi_c \rho_c \tilde{\mathbf{u}}_c \tilde{\mathbf{u}}_c) - \nabla \cdot (\phi_c \boldsymbol{\tau}_{\text{sg}})$$
$$= \phi_c \rho_c \mathbf{g} + \phi_c \nabla \cdot \boldsymbol{\sigma}_c - \sum_{l=1}^{N_p} G(\mathbf{x} - \mathbf{X}_l)(\mathbf{F}'_l + \dot{\mathbf{M}}_l), \tag{11.96}$$

$$\frac{\partial \phi_c \rho_c \tilde{E}_c}{\partial t} + \nabla \cdot (\phi_c \rho_c \tilde{\mathbf{u}}_c \tilde{E}_c) + \nabla \cdot (\phi_c Q_{\text{sg}})$$
$$= \phi_c \rho_c \tilde{\mathbf{u}}_c \cdot \mathbf{g} + \nabla \cdot (\phi_c \boldsymbol{\sigma}_c \cdot \tilde{\mathbf{u}}_c)$$
$$+ \boldsymbol{\sigma}_c : \nabla(\phi_d \tilde{\mathbf{u}}_d) - \phi_c \nabla \cdot \mathbf{q}_c - \sum_{l=1}^{N_p} G(\mathbf{x} - \mathbf{X}_l)(Q'_l + W'_l + \dot{E}_l). \tag{11.97}$$

Problem 11.14 This problem is in preparation for the following discussion, where we will derive the governing equations in nonconservative form. To facilitate this derivation, obtain the following equation:

$$\frac{\partial \phi_c \rho_c \psi}{\partial t} + \nabla \cdot (\phi_c \rho_c \mathbf{u}_c \psi) = \phi_c \rho_c \frac{D\psi}{Dt} - \dot{m}\psi, \tag{11.98}$$

where we have used the continuous-phase mass balance to introduce $\dot{m}$. In the above, ψ can be any continuous-phase property, and $D/Dt = \partial/(\partial t) + \mathbf{u}_c \cdot \nabla$ is the total derivative following the filtered continuous-phase velocity. The second term on the right-hand side corresponds to the amount of ψ that resides in the exchanged mass.

Incompressible Flow of Constant Fluid Density

We now consider the next level of simplification by assuming the flow to be incompressible. Then ρ_f is a constant everywhere in the flow and as a result, the filtered continuous-phase density $\rho_c = \rho_f$. With this assumption, the mass balance equation becomes as follows.

Incompressible continuous phase:

$$\frac{\partial \phi_c}{\partial t} + \nabla \cdot (\phi_c \mathbf{u}_c) = -\sum_{l=1}^{N_p} G(\mathbf{x} - \mathbf{X}_l)\frac{\dot{m}_l}{\rho_f}, \tag{11.99}$$

where the constant fluid density has been taken out of the derivatives on the left-hand side and the term on the right can now be interpreted as the volume exchange of material from the dispersed to the continuous phase though evaporation/condensation processes. Another important difference is the absence of the density-weighted Favre filter of the velocity. From the definition of the Favre filter operation in Eq. (11.21), it can readily be seen that with constant fluid density, the Favre filter becomes the same as the regular filter operation.

Fluid density can similarly be extracted from the derivatives on the left-hand side of the momentum and energy equations, leading to similar simplifications. These steps and the resulting equations are straightforward and will not be repeated here. The resulting momentum and energy equations can be expressed as follows.

Incompressible continuous phase:

$$
\begin{aligned}
\phi_c \rho_f \frac{D\mathbf{u}_c}{Dt} = {} & \nabla \cdot (\phi_c \boldsymbol{\tau}_{\text{sg}}) + \phi_c \rho_f \mathbf{g} + \phi_c \, \nabla \cdot \boldsymbol{\sigma}_c \\
& - \sum_{l=1}^{N_p} G(\mathbf{x} - \mathbf{X}_l)(\mathbf{F}'_l + \dot{\mathbf{M}}_l) + \mathbf{u}_c \left(\sum_{l=1}^{N_p} G(\mathbf{x} - \mathbf{X}_l)\dot{m}_l \right),
\end{aligned}
\tag{11.100}
$$

$$
\begin{aligned}
\phi_c \rho_f & \frac{DE_c}{Dt} + \nabla \cdot (\phi_c \mathbf{Q}_{\text{sg}}) \\
& = \phi_c \rho_f \mathbf{u}_c \cdot \mathbf{g} + \nabla \cdot (\phi_c \boldsymbol{\sigma}_c \cdot \mathbf{u}_c) + \boldsymbol{\sigma}_c : \nabla(\phi_d \mathbf{u}_d) - \phi_c \nabla \cdot \mathbf{q}_c \\
& - \sum_{l=1}^{N_p} G(\mathbf{x} - \mathbf{X}_l)(Q'_l + W'_l + \dot{E}_l) + E_c \left(\sum_{l=1}^{N_p} G(\mathbf{x} - \mathbf{X}_l) \right),
\end{aligned}
\tag{11.101}
$$

$$
\begin{aligned}
\phi_c \rho_f C_{vf} & \frac{DT_c}{Dt} \\
& = -\nabla \cdot (\phi_c \mathbf{q}_{\text{sg}}) - \phi_c \nabla \cdot \mathbf{q}_c + \boldsymbol{\sigma}_c : \nabla(\phi_c \mathbf{u}_c + \phi_d \mathbf{u}_d) \\
& - Q'_{\text{hyd}} - (W'_{\text{hyd}} - \mathbf{u}_c \cdot \mathbf{F}'_{\text{hyd}}) - (\dot{E} - \dot{m}E_c) + \mathbf{u}_c \cdot (\dot{\mathbf{M}} - \dot{m}\mathbf{u}_c),
\end{aligned}
\tag{11.102}
$$

where in the third equation, Q'_{hyd}, W'_{hyd}, $\mathbf{F}'_{\text{hyd}}$, $\dot{m}$, $\dot{\mathbf{M}}$, and $\dot{E}_l$ must be replaced with the sum over contributions from the N_p particles as given in Eqs. (11.90), (11.91), (11.89), (11.85), and (11.86). An important point to note in the above equations is that the left-hand sides are written in nonconservative form, by making use of the identity given in Problem 11.14. Correspondingly, the source terms in momentum and energy are modified. Similarly, the compressible versions of the momentum and energy equations presented before can also be written in nonconservative form. However, due to the hyperbolic nature of the dominant advection term, in a compressible multiphase flow it is customary to solve the conservative form of the momentum and energy equations, whereas in the incompressible limit there are many numerical approaches that are based on the nonconservative form. Nevertheless, using the mass balance equation, the left-hand sides of both the compressible and incompressible momentum and energy balance equations can be written in either conservative or nonconservative form.

11.3   Filtered Dispersed Phase Equations

In this section, we will filter the dispersed-phase quantities and work toward filtered dispersed-phase governing equations. The dispersed-phase indicator function is defined as $I_d(\mathbf{x}',t) = 1 - I_c(\mathbf{x}',t)$, which by definition is unity only in regions occupied by the particles. In terms of the indicator function, a filtered dispersed-phase property can be expressed as

$$\overline{\psi}^d(\mathbf{x},t) = \int \psi(\mathbf{x}',t)\, I_d(\mathbf{x}',t)\, G(\mathbf{x} - \mathbf{x}')\, d\mathbf{x}', \tag{11.103}$$

where the integral is over the entire 3D spatial domain and the superscript d indicates that the filter is over the dispersed phase, as opposed to the continuous phase. Setting $\psi = 1$, we obtain the definition of the particle volume fraction as

$$\phi_d(\mathbf{x},t) = \overline{1}^d(\mathbf{x},t) = \int I_d(\mathbf{x}',t)\, G(\mathbf{x} - \mathbf{x}')\, d\mathbf{x}'. \tag{11.104}$$

Setting $\psi = \rho_p$, we obtain the filtered dispersed-phase density ρ_d as

$$\phi_d(\mathbf{x},t)\, \rho_d(\mathbf{x},t) = \overline{(\rho_p)}^d(\mathbf{x},t) = \int \rho_p(\mathbf{x}',t)\, I_d(\mathbf{x}',t)\, G(\mathbf{x} - \mathbf{x}')\, d\mathbf{x}', \tag{11.105}$$

where multiplication by ϕ_d on the left accounts for the fact that the particles occupy only ϕ_d fraction of the filter volume. For quantities weighted by density, we define the Favre filter in the following way:

$$\overline{(\rho_p \psi)}^d(\mathbf{x},t) = \int \rho_p(\mathbf{x}',t)\, \psi(\mathbf{x}',t)\, I_d(\mathbf{x}',t)\, G(\mathbf{x} - \mathbf{x}')\, d\mathbf{x}'. \tag{11.106}$$

In order to calculate the filtered particle velocity, we first define a particle velocity field as

$$\mathbf{v}(\mathbf{x},t) = \begin{cases} \mathbf{V}_l + \mathbf{\Omega} \times (\mathbf{x} - \mathbf{X}_l) & \text{if inside } l\text{th particle,} \\ 0 & \text{if inside fluid,} \end{cases} \tag{11.107}$$

where $\mathbf{V}_l$ and $\mathbf{\Omega}_l$ are the translational and rotational velocity of the lth particle that is centered at $\mathbf{X}_l$. Now, setting $\psi = \mathbf{v}$ we obtain the Favre-filtered dispersed-phase velocity $\tilde{\mathbf{u}}_d$ as

$$\phi_d\, \rho_d\, \tilde{\mathbf{u}}_d = \overline{(\rho_p \mathbf{u})}^d(\mathbf{x},t) = \int \rho_p(\mathbf{x}',t)\, \mathbf{v}(\mathbf{x}',t)\, I_d(\mathbf{x}',t)\, G(\mathbf{x} - \mathbf{x}')\, d\mathbf{x}', \tag{11.108}$$

where all the quantities on the left-hand side are functions of $(\mathbf{x},t)$. Note that while $\mathbf{v}$ is discontinuous, after the filter operation $\tilde{\mathbf{u}}_d$ is smooth and well defined over the entire volume. Here and henceforth we will consider the particles to be of constant density and therefore $\rho_d = \rho_p$. Furthermore, we will drop the tilde notation for the filtered particle quantities, since the density-weighted Favre filter is the same as the regular filter.

Problem 11.15 *Small particle limit* ($d_p \ll \mathcal{L}$). In this problem, you will make use of the assumption that the particle size is much smaller than the filter width $\mathcal{L}$. First, the above volume integrals, in which the argument includes the dispersed-phase indicator function I_d, can be recognized as the sum of integrals over all the particle volumes. With the assumption that the Gaussian does not vary over the volume of each particle, the Gaussian can be taken out of the integral over each particle, as was done in Eq. (11.85).

(a) Following these steps, derive the following simplified definition of particle volume fraction:

$$\phi_d(\mathbf{x},t) = \sum_{l=1}^{N_p} G(\mathbf{x} - \mathbf{X}'(t))\mathcal{V}_l(t), \qquad (11.109)$$

where $\mathcal{V}_l$ is the volume of the lth particle, which can be a function of time. Note that the filter function goes as the inverse filter volume and as a result, each particle in the above summation contributes appropriately to the particle volume fraction field. Due to the decay of the filter function, only particles that are located close to $\mathbf{x}$ contribute to the summation.

(b) Also, obtain the following expression for the particle velocity field:

$$\mathbf{u}_d(\mathbf{x},t) = \frac{1}{\phi_d(\mathbf{x},t)} \sum_{l=1}^{N_p} G(\mathbf{x} - \mathbf{X}'(t))\mathcal{V}_l(t)\mathbf{V}_l(t). \qquad (11.110)$$

(c) In the above summation for small particle limit, show that only the translational velocity of the center of mass of the particle (or droplet/bubble) contributes to the definition of the particle velocity field.

(d) You can obtain similar expressions for particle temperature and other particle-related fields.

Following the steps of Section 11.1.2 we can derive Gauss and Leibniz rules for the dispersed-phase filter operation of the spatial and temporal derivatives. Using these rules, we can follow the procedure of Sections 11.2.1 to 11.2.3 to obtain the filtered dispersed-phase governing equations and we leave it to the reader to carry this out. The filtered dispersed-phase mass balance can be expressed as

$$\rho_p \left(\frac{\partial \phi_d}{\partial t} + \nabla \cdot (\phi_d \mathbf{u}_d) \right) = \dot{m}, \qquad (11.111)$$

where we note that on the right-hand side, $\dot{m}$ accounts for the net mass flux into the dispersed phase from the continuous phase, hence this appears as a source term with a positive sign. The right-hand side is exactly the opposite of that in the continuous-phase mass balance equation (11.35), and thus the total mass of the system is conserved. Furthermore, when there is no mass transfer between the phases, the above can be rewritten in the following way:

$$\frac{1}{\phi_d} \frac{d\phi_d}{dt} = -\nabla \cdot \mathbf{u}_d, \qquad (11.112)$$

where d/dt is the total derivative following the filtered particle velocity $\mathbf{u}_d$. Thus, the change in particle volume fraction following the particle motion is due to the negative of the divergence of the particle velocity field. When the particle paths diverge locally (i.e., when $\nabla \cdot \mathbf{u}_d > 0$), the particle volume fraction will decrease, since the particles are moving farther apart. On the other hand, in regions where the particle paths converge (i.e., in regions where $\nabla \cdot \mathbf{u}_d < 0$), the particle volume fraction increases. If $\dot{m} \neq 0$, then there is an additional change in the particle volume fraction due to nonzero mass exchange with the surrounding continuous phase.

Next we consider the dispersed-phase momentum equation. We start with an equation analogous to Eq. (11.50) for the filtered dispersed-phase momentum balance, then make the following assumptions: (i) the particle density ρ_p is a constant; (ii) the particles are rigid with no internal motion, and as a result the only stress inside the particle is that imposed by the ambient flow; and (iii) the particle is isothermal. With these assumptions you can obtain

$$\rho_p \left(\frac{\partial \phi_d \mathbf{u}_d}{\partial t} + \nabla \cdot (\phi_d \mathbf{u}_d \mathbf{u}_d) \right) - \nabla \cdot (\phi_d \tau_{d,\mathrm{sg}}) = \phi_d \rho_p \mathbf{g} + \phi_d \nabla \cdot \sigma_c + \mathbf{F}'_{\mathrm{hyd}} + \dot{\mathbf{M}} .$$

$$(11.113)$$

Several observations must be made. First, the particle residual stress is defined in an analogous way as

$$\phi_d \tau_{d,\mathrm{sg}} = \phi_d \rho_p \mathbf{u}_d \mathbf{u}_d - \overline{(\rho_p \mathbf{v} \mathbf{v})}^d ,$$

$$(11.114)$$

where $\mathbf{v}$ is the dispersed-phase velocity field defined earlier in Eq. (11.107). The particle residual stress accounts for the deviation of individual particle velocity from the filtered value of $\mathbf{u}_d$ within the filter volume.

The second term on the right-hand side is the undisturbed flow force on the particles and it depends on the divergence of the filtered total fluid stress σ_c. This term combines with the term $\phi_c \nabla \cdot \sigma_c$ that appears in the continuous-phase momentum equation to yield $\nabla \cdot \sigma_c$, which is the total force per unit volume of the mixture. In the above filtered particle momentum equation, the undisturbed flow force combined with $\mathbf{F}'_{\mathrm{hyd}}$ accounts for the entire hydrodynamic interaction between the particle and the surrounding fluid. Finally, the momentum exchange associated with mass exchange is just the opposite of that in the continuous-phase momentum equation. Since the nozzling correction term $\mathcal{N}_m$ is ignored in the filtered continuous-phase momentum equation (11.50), to be consistent a similar term is ignored in the above.

For the energy balance of the dispersed phase, through filter operation we can arrive at an equation analogous to Eq. (11.65). You can then make the same assumptions as for the momentum equation, to obtain

$$\rho_p \left(\frac{\partial \phi_d E_d}{\partial t} + \nabla \cdot (\phi_d \mathbf{u}_d E_d) \right) + \nabla \cdot (\phi_d Q_{d,\mathrm{sg}})$$

$$= \phi_d \rho_p \mathbf{u}_d \cdot \mathbf{g} \qquad (11.115)$$

$$+ \phi_d (\nabla \cdot \sigma_c) \cdot \mathbf{u}_d - \phi_d \nabla \cdot \mathbf{q}_c + Q'_{\mathrm{hyd}} + W'_{\mathrm{hyd}} + \dot{E} ,$$

where the residual energy flux of the dispersed phase is defined as

$$- \phi_d Q_{d,\mathrm{sg}} = \phi_d \rho_p \mathbf{u}_d E_d - \overline{(\rho_p \mathbf{V} E_p)}^d + \overline{(\nabla \cdot \boldsymbol{\sigma}) \cdot V} - \phi_d (\nabla \cdot \boldsymbol{\sigma}_c) \cdot \mathbf{u}_d, \quad (11.116)$$

where $E_p(\mathbf{x}, t)$ is the unfiltered dispersed-phase total energy, which is nonzero only inside the particle. The second term on the right-hand side of Eq. (11.115) accounts for the work done on the particle by the undisturbed flow force and the third term accounts for the heat flux to the particle due to the ambient macroscale flow undisturbed by the particle. The next two terms are opposite to those that appear in the filtered energy equation of the continuous phase and they correspond to heat transfer to the particle and work done on the particle due to the perturbation force $\mathbf{F}'_{\mathrm{hyd}}$. The last term corresponds to the energy flux to the particles associated with the mass transfer.

As with the filtered continuous-phase equations, the filtered dispersed-phase equations also contain a number of terms that must be closed using appropriate closure models. Table 11.2 lists these to-be-modeled secondary particle-related variables that appear in Eqs. (11.111), (11.113), and (11.115). By comparing with Table 11.1, it can readily be observed that the first three rows that account for interphase exchanges are the same. Also, the continuous-phase filtered stress and heat flux (i.e., $\boldsymbol{\tau}_c$ and $\mathbf{q}_c$) act on the dispersed phase as well. The subgrid residual stress $\boldsymbol{\tau}_{d,\mathrm{sg}}$ and subgrid residual total energy flux $Q_{d,\mathrm{sg}}$ depend on particle-to-particle velocity and total energy fluctuations within the filter volume and must be modeled independently, as they are different from their continuous-phase counterparts $\boldsymbol{\tau}_{\mathrm{sg}}$ and Q_{sg}. As we discussed in the context of the continuous phase, if we replace the total energy equation (11.115) of the dispersed phase with an equation for internal energy or particle temperature, then the closure term $Q_{d,\mathrm{sg}}$ will be replaced by the subgrid residual heat flux $\mathbf{q}_{d,\mathrm{sg}}$. We shall consider the derivation of the dispersed-phase temperature equation below. Interparticle collisions introduce additional secondary terms which are also included in the table. We shall discuss these in Section 11.3.1.

Problem 11.16 *Filtered dispersed-phase internal energy equation.* In this exercise you will follow the steps of Problem 11.8 and obtain the internal energy and temperature equations for the dispersed phase. You will first obtain the equation of internal energy, which will then be converted to an equation for the dispersed-phase temperature. As a first step, define the filtered internal energy of the dispersed phase as the total energy minus the kinetic component.

(a) Multiply this with ρ_p and perform the filter operation to obtain the relation

$$\phi_d \rho_p E_d = \phi_d \rho_p e_d + \frac{1}{2} \phi_d \rho_p \mathbf{u}_d \cdot \mathbf{u}_d - \frac{1}{2} \phi_d \, \mathrm{tr}[\boldsymbol{\tau}_{d,\mathrm{sg}}], \quad (11.117)$$

where $\mathrm{tr}[\cdot]$ indicates the trace of the argument (i.e., the sum of the diagonal terms).

(b) As the next step, derive the following relation:

$$\mathbf{u}_d \cdot \left[\frac{\partial \phi_d \rho_p \mathbf{u}_d}{\partial t} + \nabla \cdot (\phi_d \rho_p \mathbf{u}_d \mathbf{u}_d) \right]$$

$$= \frac{1}{2} \frac{\partial \phi_d \rho_p \mathbf{u}_d \cdot \mathbf{u}_d}{\partial t} + \frac{1}{2} \nabla \cdot (\phi_d \rho_p \mathbf{u}_d (\mathbf{u}_d \cdot \mathbf{u}_d)) + \dot{m} \frac{\mathbf{u}_d \cdot \mathbf{u}_d}{2}. \tag{11.118}$$

(c) Now take the dot product of $\mathbf{u}_d$ with the filtered momentum equation (11.113), and subtract the dot product from the filtered total energy equation (11.115). The resulting equation can be combined with Eqs. (11.117) and (11.118) to obtain the filtered internal energy equation. Follow these steps and obtain

$$\frac{\partial \phi_d \rho_p e_d}{\partial t} + \nabla \cdot (\phi_d \rho_p \mathbf{u}_d e_d) + \nabla \cdot (\phi_d \mathbf{q}_{d,\mathrm{sg}})$$

$$= -\phi_d \nabla \cdot \mathbf{q}_c + Q'_{\mathrm{hyd}} + \left[W'_{\mathrm{hyd}} - \mathbf{u}_d \cdot \mathbf{F}'_{\mathrm{hyd}} \right] + \left[\dot{E} - \mathbf{u}_d \cdot \left(\dot{\mathbf{M}} - \frac{\dot{m}\mathbf{u}_d}{2} \right) \right]. \tag{11.119}$$

(d) Show that the third term on the left-hand side arises from the residual stress and residual total energy flux terms in the momentum and total energy equations and obtain the following relation for the residual contribution to internal energy balance:

$$-\nabla \cdot (\phi_d \mathbf{q}_{d,\mathrm{sg}}) = \frac{1}{2} \left(\frac{\partial \phi_d \mathrm{tr}[\boldsymbol{\tau}_{d,\mathrm{sg}}]}{\partial t} + \nabla \cdot (\phi_d \mathbf{u}_d \mathrm{tr}[\boldsymbol{\tau}_{d,\mathrm{sg}}]) \right) - \nabla \cdot (\phi_d Q_{d,\mathrm{sg}})$$

$$- \mathbf{u}_d \cdot \nabla \cdot (\phi_d \boldsymbol{\tau}_{d,\mathrm{sg}}). \tag{11.120}$$

In essence, $\mathbf{q}_{d,\mathrm{sg}}$ is the residual heat flux term that accounts for the nonlinear effect of filtering on internal energy. The above equation connects the dispersed-phase residual heat flux $\mathbf{q}_{d,\mathrm{sg}}$ with the residual total energy flux $Q_{d,\mathrm{sg}}$.

(e) Obtain the following equation for dispersed-phase temperature by substituting $e_d = C_p T_d$, where C_p is the specific heat of the particulate matter and T_d is the dispersed-phase filtered temperature field:

$$\frac{\partial \phi_d \rho_p C_p T_d}{\partial t} + \nabla \cdot (\phi_d \rho_p \mathbf{u}_d C_p T_d) + \nabla \cdot (\phi_d \mathbf{q}_{d,\mathrm{sg}})$$

$$= -\phi_d \nabla \cdot \mathbf{q}_c + Q'_{\mathrm{hyd}} + \left[W'_{\mathrm{hyd}} - \mathbf{u}_d \cdot \mathbf{F}'_{\mathrm{hyd}} \right] + \left[\dot{E} - \mathbf{u}_d \cdot \left(\dot{\mathbf{M}} - \frac{\dot{m}\mathbf{u}_d}{2} \right) \right]. \tag{11.121}$$

The averaged or filtered dispersed-phase governing equations are of fundamental importance in the Euler–Euler approach. They have been rigorously derived by many researchers (Anderson and Jackson, 1967; Zhang and Prosperetti, 1994, 1997; Houim and Oran, 2016; Balakrishnan and Bellan, 2018; Fox, 2019; Fox et al., 2020). For the dispersed-phase governing equations, particular emphasis has been placed on establishing that the governing equations are hyperbolic in the inviscid limit. Ill-posed formulations that are not hyperbolic are known to give rise to spurious numerical instabilities (Lhuillier et al., 2013; Theofanous et al., 2018; Fox, 2019; Fox et al., 2020). When continuous and dispersed-phase equations of motion are faithfully filtered, the resulting filtered equation preserves the well-behaved nature of the underlying equa-

Table 11.2 List of all the secondary variables that arise from the filter operation and appear in the filtered momentum and energy equations of the dispersed phase given in Eqs. (11.111), (11.113), and (11.115).

Category	Variables to be modeled
Interphase mass exchange	$\dot{m}, \dot{\mathbf{M}}, \dot{E}$
Interphase momentum exchange	$\mathbf{F}'_{\text{hyd}}, W'_{\text{hyd}}$
Interphase thermal exchange	Q'_{hyd}
Constitutive relation	$\tau_c, \mathbf{q}_c$
Collisional pressure, stress and heat flux	$p_d, \tau_d, \mathbf{q}_d$
Subgrid closure	$\tau_{d,\text{sg}} \ (Q_{d,\text{sg}} \text{ or } \mathbf{q}_{d,\text{sg}})$

tions. The real issue of hyperbolicity arises when the secondary terms are modeled. Not all models, especially when they are applied in conjunction with each other, guarantee hyperbolicity. If simple ad-hoc closure models are found to be nonhyperbolic, attempts have been made to introduce numerical corrections that will ensure hyperbolicity (Stuhmiller, 1977). More recent rigorous derivations and closure models have presented a self-consistent approach to achieving hyperbolicity (Fox, 2019; Balakrishnan and Bellan, 2021). In this section, we are only concerned with the derivation of the filtered equations. The discussion for closure models of the secondary terms is deferred to Chapter 16. One must consult recent developments in the above references to preserve hyperbolicity.

11.3.1 Summary of Dispersed-Phase Equations Including Interparticle Collisions

The above derivation of the dispersed phase closely followed the steps taken in the derivation of the continuous-phase governing equations. As a result, we have ignored the important effect of interparticle collisions. The above equations are sufficient in dilute systems, where interparticle collisions are rare and contribute little to particulate phase momentum and energy balance. However, in many practical problems, even at moderate values of average particle volume fraction, interparticle collisions can play an important role and therefore collisional effects may need to be included in the governing equations. For example, consider the case of particle-laden pipe flow. As we will see later in Section 12.4.2, over time, particle concentration will build up close to the pipe wall due to turbophoretic migration, and soon the particle volume fraction close to the pipe wall will increase far higher than the average value for the entire pipe. In this example, interparticle collisions, along with particle–wall collisions, play an important role in stabilizing the near-wall particle concentration.

There are few basic facts related to interparticle collisions that play a vital role in determining how collisions modify momentum and energy balances of the particulate phase. First, as discussed in Section 9.1, each collision preserves the total momentum of the colliding particles from before to after collision. Second, in case of inelastic collisions, the total kinetic energy is not preserved. During collision, some of the kinetic energy of the colliding particles is lost and goes to the thermal energy of the particles. Third, in the frame attached to a reference particle, the probability of collision with neighboring particles is isotropic only when both the number density of colliding particles and the probability of radial relative velocity are uniform around the particle. In cases where the relative radial particle velocity is isotropic, collisional probability is larger along directions of higher particle concentration due to an increased chance of encounter with a neighbor. In the presence of mean particle velocity gradient (i.e., $\nabla \mathbf{u}_d \neq 0$), as in the shear flow example of Section 10.2.2, the probability of collisions along a direction $\mathbf{n}$ scales as $-\nabla \mathbf{u}_d : \mathbf{nn}$. This quantity measures the mean influx of neighboring particles along that direction. Even in systems where there is no mean velocity gradient, particle rms velocity fluctuation may vary over space. In this case, collisions are more probable along directions where a neighbor's velocity fluctuation is higher. In other words, there are a number of factors that govern the rate and nature of interparticle collisions.

A detailed investigation of the effects of interparticle collisions can be performed using the kinetic theory of granular material (Jenkins and Savage, 1983; Gidaspow, 1994), which we will consider in greater detail in Chapter 16. Here we will only summarize the key outcomes of kinetic granular theory in terms of their effect on particle-phase governing equations. First and foremost, collisions do not affect the equation of mass balance. In the momentum balance, the net effect of collisions is to introduce a collisional particle stress $\sigma_{d,\text{co}}$, which is analogous to the continuous-phase stress tensor σ_c. The added contribution to the time rate of change of particle momentum from the pressure and frictional effects of collisions can be taken into account with the addition of the term $\nabla \cdot (\phi_d \sigma_{d,\text{co}})$ to the right-hand side of Eq. (11.113). In addition, each collision contributes to particle velocity deviation from the filtered average. Thus, collisions contribute to particle-phase residual stress as defined in Eq. (11.114). The residual stress contribution from collisions is often referred to as *kinetic stress*, which will be distinguished from both the collisional stress $\sigma_{d,\text{co}}$, as well as from particle velocity fluctuations arising from ambient fluid turbulence. More on this will be discussed in Chapter 16.

The direct effect of collisions on the particle temperature and energy equation is not as significant as in the case of particle momentum. Clearly, collisions change the velocity of individual particles quite profoundly, while having only a small influence on particle temperature. The contact time and contact area are typically too small for substantial heat transfer between the particles to occur. In case of inelastic collisions, the kinetic energy lost goes towards heating of the particles. This inelastic dissipative heating of particles is denoted as Γ and appears as an additional source term in the energy equation. In typical applications, this contribution to increasing particle-phase temperature is usually small.

Now we will arrive at the final version of the filtered mass, momentum, and temperature equations of the particulate phase and write them in a form similar to those for the continuous phase given in Eqs. (11.99), (11.100), and (11.101). Recall that we have already made the assumption that the dispersed phase is incompressible. Thus, the equations to be presented below are applicable to rigid particles and droplets. Modifications are required in case of compressible bubbly flows.

In writing the final governing equations, we will need to make the following manipulations. (i) The left-hand sides of the governing equations (11.113) and (11.119) will be written in nonconservative form by making use of the mass conservation equation. (ii) We will assume the thermodynamic relation $e_d = C_p T_d$ to be applicable, where e_d is the filtered particle internal energy field, C_p the specific heat of the particulate material, and T_d the filtered particle temperature field. With these manipulations we arrive at the following filtered mass, momentum, and temperature equations for the dispersed phase.

Incompressible dispersed phase:

$$\rho_p \left(\frac{\partial \phi_d}{\partial t} + \nabla \cdot (\phi_d \mathbf{u}_d) \right) = \dot{m}, \tag{11.122}$$

$$\phi_d \rho_p \frac{d\mathbf{u}_d}{dt} = \nabla \cdot (\phi_d \boldsymbol{\tau}_{d,\text{sg}}) + \underbrace{\nabla \cdot (\phi_d \boldsymbol{\sigma}_{d,\text{co}})}_{\substack{\text{collisional} \\ \text{effect}}} + \phi_d \rho_p \mathbf{g} + \phi_d \nabla \cdot \boldsymbol{\sigma}_c + \mathbf{F}'_{\text{hyd}} + \left(\dot{\mathbf{M}} - \dot{m}\,\mathbf{u}_d \right), \tag{11.123}$$

$$\phi_d \rho_p C_p \frac{dT_d}{dt} = -\nabla \cdot (\phi_d \mathbf{q}_{d,\text{sg}}) - \phi_d \nabla \cdot \mathbf{q}_c + \underbrace{\Gamma}_{\substack{\text{collisional} \\ \text{effect}}} + Q'_{\text{hyd}} + (W'_{\text{hyd}} - \mathbf{u}_d \cdot \mathbf{F}'_{\text{hyd}})$$

$$+ (\dot{E} - \dot{m}E_d) - \mathbf{u}_d \cdot \left(\dot{\mathbf{M}} - \dot{m}\mathbf{u}_d \right). \tag{11.124}$$

The above momentum and energy equations include collisional effects that are clearly marked. In the momentum equation, the residual stress $\boldsymbol{\tau}_{d,\text{sg}}$ is due to particle-to-particle velocity variation about the filtered value of $\mathbf{u}_d$ (see the definition given in Eq. (11.114)), where the particle velocity variation is both due to the action of ambient turbulence as well as interparticle collisions (i.e., kinetic contribution).

The seven terms on the right-hand side of the momentum equation can be interpreted. The first term is the residual stress that accounts for the effects of filtering nonlinear quantities. The second term accounts for the collisional effects on particle momentum. The third term is the filtered buoyancy force, the fourth term accounts for the undisturbed flow force acting on the particles, and the fifth term accounts for the quasi-steady, added-mass, and other perturbation forces on the particles. In the last two terms, $\dot{\mathbf{M}}$ is the momentum exchanged to the dispersed phase due to mass exchange. Of this, $\dot{m}\mathbf{u}_d$ resides with the exchanged mass itself. Only the difference contributes to acceleration or deceleration of the local fluid.

We now interpret the terms on the right-hand side of the energy equation. The first of these terms accounts for all residual heat fluxes arising from filtering of nonlinear terms. The second term corresponds to the undisturbed heat flux to the particles due to the thermal gradient for the surrounding fluid. The third term is dissipative heating of the particles due to kinetic energy lost in the collisional process. In Eq. (11.124), Q'_{hyd} is the heat exchange between the continuous and dispersed phases and it contributes entirely to altering the temperature of the dispersed phase. From the next two terms it can be seen that out of the total work W'_{hyd} done on the dispersed phase, $\tilde{\mathbf{u}}_d \cdot \mathbf{F}'_{\text{hyd}}$ goes toward the kinetic energy and only the difference (if nonzero) contributes to the internal energy. In the next two terms, $\dot{E}$ is the total energy exchanged to the dispersed phase, out of which $\dot{m}E_d$ is the total energy that resides in the mass that was exchanged to the dispersed phase. So the difference must go toward altering the temperature of the dispersed phase. Similarly, the last two terms account for the portion that contributed to the dispersed-phase temperature. In any case, the precise value of these contributions depends on the models that we employ for $\dot{\mathbf{M}}$, $\dot{E}$, and W'_{hyd}. We shall discuss this further in Chapters 15 and 16.

An important effect of collisions is to introduce a particle velocity fluctuation, which in turn becomes the source of additional collisions, and so on. Thus, in granular physics, the mean square particle velocity fluctuation is quantified in terms of the *granular temperature* Θ, which must be distinguished from the thermodynamic temperature T_d of the dispersed phase. As we will see in Chapter 16, Θ plays an important role in the closure models of kinetic and collisional stresses ($\tau_{d,\text{sg}}$, $\sigma_{d,\text{co}}$) and also in the modeling of dissipation due to inelastic collision Γ. Despite its importance, Θ, which measures the particle velocity fluctuation due to collisions, cannot be expressed in terms of the primary particle fields being computed in the governing equations (11.122), (11.123), and (11.124). A separate governing equation for Θ based on collisional physics is needed, which will be derived in Chapter 16.

12   Equilibrium Particle Fields

The Eulerian representation of the continuous phase is natural, where quantities such as fluid velocity $\mathbf{u}(\mathbf{x}, t)$ represent the average velocity of all the fluid molecules within a suitably chosen volume for continuum description. In the previous section, we considered filtering of these flow quantities over a suitably chosen length scale that is much larger than the size of the individual particles. Thus, at any spatial location $\mathbf{x}$, the filtered velocity $\mathbf{u}_c(\mathbf{x}, t)$ accounts for the collective motion of the different parcels of fluid within the filtering volume, with proper weighting by the filter function.

We then proceeded to carry out a similar filter of the dispersed-phase quantities and developed filtered dispersed-phase governing equations. But there is a difference due to the relative inertia of the particles with respect to the surrounding fluid. This can best be explained with an example, shown in Figure 12.1. Here, particle-laden fluid is injected from the two walls of a long channel whose one end is closed and the other end is open. The cylindrical analog of this problem is a very good model of a solid rocket, where the walls of the cylindrical tube are made of solid rocket propellant and the product of combustion along with metal droplets/particles is injected into the flow. In the case of the rocket, the open end of the tube will be connected to a nozzle.

In any case, in our planar example, consider a small volume along the centerline that has been zoomed in the figure. In a single-phase laminar flow, the distribution of streamwise fluid velocity within this small volume will be narrow and centered around the mean value. Similarly, the distribution of wall-normal fluid velocity will be narrow and centered around zero value. Now consider particles being injected with the fluid. The size of the particles, or more precisely the Stokes number of the particles, matters in dictating the nature of their motion.

If particles are very small (i.e., Stokes number much smaller than unity), they essentially follow the fluid and the distribution of streamwise and wall-normal velocities of particles within the small volume shown in Figure 12.1a will closely follow those of the continuous phase. With increasing Stokes number, the velocity of the particles will increasingly deviate from that of the fluid. If the particle inertia is significant, then the particle velocity within the small volume will have a bimodal distribution as, shown in Figure 12.1b. In this case, the injected particles have not forgotten their initial condition (i.e., whether they were injected from the top or the bottom wall) by the time they reach the small volume. Therefore, those that are injected at the top wall continue to have a negative y-velocity, while those injected at the bottom wall have a positive y-velocity.

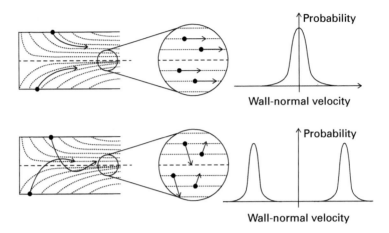

Figure 12.1 Schematic of an injection-driven flow between two parallel planes with one end of the channel closed. The laminar flow streamlines are shown as dotted lines and some sample particle trajectories are also shown. (a) Particles of small enough Stokes number that their velocity deviates only slightly from the local fluid velocity. (b) Larger particles which have not forgotten their initial condition even as they approach the channel centerline. In each frame a zoom-up of a small region along the centerline is shown, along with the probability distribution function (PDF) of the particle-wall normal velocity of all particles passing through this small volume. The PDF is narrow and centered about zero in the case of small particles, while it is bimodal in the large particle limit.

In other words, for sufficiently small particles, their local velocity distribution is similar to that of the fluid and thus can be given a unique Eulerian field representation. This applies not only to the velocity field, but also to other particle-related quantities such as particle temperature or angular velocity. Thus, in the limit of particles of small Stokes number, filtered dispersed-phase quantities, such as $\mathbf{u}_d(\mathbf{x}, t)$ and $T_d(\mathbf{x}, t)$, provide an accurate representation of the actual value of particle velocity and temperature. In contrast, as illustrated in Figure 12.1b, for larger-sized particles the velocity distribution is not single valued. Though the filtered particle velocity field $\mathbf{u}_d(\mathbf{x}, t)$ can still be defined, it now serves as a well-defined mathematical entity. However, none of the particle's velocity is in fact close to $\mathbf{u}_d(\mathbf{x}, t)$. For example, the filtered y-velocity is zero at the sample volume located along the centerline, while every one of the particles will be either moving up or down, depending on where it was injected from. This phenomenon arises due to the inertial memory of the particle.

In this section, we will first establish a precise critical Stokes number. If the particle Stokes number is smaller than the critical value, then the importance of the initial condition is rapidly forgotten and an Eulerian field representation of particle velocity is meaningful. If the particle Stokes number is larger than the critical value, then the local distribution of particle velocity is non-unique and the situation is more like that depicted in Figure 12.1b. There are two options that can be pursued in this later case. (i) Track individual particle trajectories with a Lagrangian approach. By following the inherent Lagrangian nature of particle motion, this approach allows for vastly different velocities for different particles in the same neighborhood of $\mathbf{x}$ and t, depending on the history of their motion. (ii) The alternate option is to adopt a stochastic framework by defining the particle number density function (NDF), $f(\mathbf{x}, \mathbf{v}, t)$, in phase space, as described in Section 3.6, to account for the multivaluedness of particle velocity. Both these approaches account for the crossing trajectory effect, which can be important in problems where close interaction between particles (e.g., collisions) is an important physics.

12.1   Criterion for Uniqueness

For simplicity, let us consider a one-way coupled particle-laden flow. For a specified fluid velocity $\mathbf{u}$, the corresponding particle velocity depends on both $\mathbf{u}$ and the initial condition, given in terms of the initial particle position and velocity. We will investigate uniqueness by considering two different particle velocity solutions and establishing conditions under which differences between any two solutions will die away rapidly to become a single unique solution. Let there be two different particle velocity fields $\mathbf{v}_1$ and $\mathbf{v}_2$ that evolve differently in the same fluid field $\mathbf{u}(\mathbf{x}, t)$, since they started with different initial conditions.[1] This dependence on initial condition is likely to persist for some period of time. Thus, the question of unique particle velocity applies only beyond this initial transient period. It can be anticipated that for particles whose time scale τ_p is smaller than a certain characteristic time scale of the fluid, any two different particle velocity fields will converge exponentially fast toward a unique field. For such particles we may speak of a single velocity field $\mathbf{v}$ once the rapid transients arising from the initial condition decay.

We follow the discussion of Ferry and Balachandar (2001) and let $\Omega(t)$ be the volume of fluid over which the particle velocity field $\mathbf{v}_1(\mathbf{x}, t)$ is defined. Following the discussion of Section 4.5.5, the particle velocity satisfies the following governing equation and initial condition (see Eq. (4.151)):

$$\frac{d\mathbf{v}_1}{dt} = \frac{1}{\tau_p}(\mathbf{u} - \mathbf{v}_1) + \beta \frac{D\mathbf{u}}{Dt} + (1 - \beta)\mathbf{g} \quad \text{and} \quad \mathbf{v}_1(\mathbf{x}, t = 0) = \mathbf{v}_{1_0}(\mathbf{x}). \tag{12.1}$$

The particle velocity field $\mathbf{v}_2$ satisfies the same governing equation, but differs only in the initial condition $\mathbf{v}_2(\mathbf{x}, t = 0) = \mathbf{v}_{2_0}(\mathbf{x})$. The governing equation is the BBO equation and as written above, ignores only the Basset history term in order to simplify the subsequent analysis. In the above, the density parameter $\beta = 3/(2\rho + 1)$, where ρ is the ratio of particle to fluid density. If we define the difference between the two fields as $\delta = \mathbf{v}_1 - \mathbf{v}_2$, then the governing equation of the difference is given by

---

[1] The discussion to follow is not specific to spatial filter operation introduced in the previous chapter. It applies to other forms of time and ensemble averaging as well. Therefore, in the discussion to follow, we refer to fluid and particle velocity fields with the notation $\mathbf{u}$ and $\mathbf{v}$.

$$\frac{\partial \delta}{\partial t} + \mathbf{v}_1 \cdot \nabla \delta = -\frac{1}{\tau_p} \delta - \delta \cdot \nabla \mathbf{v}_2 . \tag{12.2}$$

Consider the trajectory of a particle following the velocity field $\mathbf{v}_1$ and along this trajectory

$$\frac{d\delta}{dt} = -\mathbf{A}\delta \quad \text{where} \quad \mathbf{A} = \frac{\mathbf{I}}{\tau_p} + (\nabla \mathbf{v}_2)^{\mathrm{T}} , \tag{12.3}$$

where $\mathbf{I}$ is the identity tensor and T denotes transpose. Taking the dot product of the above with δ, we obtain

$$\frac{d}{dt}\left(\frac{1}{2}|\delta|^2\right) = -\delta^{\mathrm{T}} \mathbf{B} \delta , \tag{12.4}$$

where

$$\mathbf{B} = \frac{1}{2}\left(\mathbf{A} + \mathbf{A}^{\mathrm{T}}\right) = \frac{\mathbf{I}}{\tau_p} + \mathbf{S}_2 , \tag{12.5}$$

where $\mathbf{S}_2 = \left(\nabla \mathbf{v}_2 + (\nabla \mathbf{v}_2)^{\mathrm{T}}\right)/2$ is the symmetric strain-rate tensor. The symmetric operator $\mathbf{B}$ can be eigen-decomposed to obtain the three eigenvalues $(\lambda_1', \lambda_2', \lambda_3')$ and the corresponding orthonormal set of eigenvectors $(\mathbf{e}_1, \mathbf{e}_2, \mathbf{e}_3)$. In terms of these eigenvalues and eigenvectors, the following bound on the time evolution of the difference between the two particle velocity fields can be obtained:

$$\frac{d}{dt}\left(\frac{1}{2}|\delta|^2\right) = -2\left(\lambda_1'|\mathbf{e}_1 \cdot \delta| + \lambda_2'|\mathbf{e}_2 \cdot \delta| + \lambda_3'|\mathbf{e}_3 \cdot \delta|\right)$$
$$\leq -2\lambda_{\min}'|\delta|^2 . \tag{12.6}$$

Here, $\lambda_{\min}'$ is the smallest eigenvalue of $\mathbf{B}$, which can be expressed as

$$\lambda_{\min}' = \frac{1}{\tau_p} + \lambda_{\min} , \tag{12.7}$$

where $\lambda_{\min}$ is the smallest eigenvalue of the strain-rate tensor $\mathbf{S}_2$, which is the most negative eigenvalue that corresponds to the most compressional strain. Integrating the inequality (12.6) from time $t = 0$ to $t = t_1$, we obtain

$$|\delta^2(t)| \leq |\delta^2(0)| \exp\left(-2\left[\frac{t}{\tau_p} + \int_0^t \lambda_{\min}\, dt\right]\right)$$
$$\leq |\delta^2(0)| \exp\left(-2t\left[\frac{1}{\tau_p} + \tilde{\lambda}_{\min}(t)\right]\right) , \tag{12.8}$$

where $\tilde{\lambda}_{\min}(t)$ is the smallest value of $\lambda_{\min}$ along the particle trajectory from time $t = 0$ to $t = t$

We are now ready to obtain the final criterion under which the two different particle velocity fields will converge to a single unique velocity field. If we define the following as the scalar measure of the maximal difference between the two particle velocity fields:

$$E(t) = \sup_{\mathbf{x} \in \Omega} |\delta(\mathbf{x},t)|^2 , \tag{12.9}$$

where Ω is the entire domain occupied by the particles, then the time evolution of $E(t)$ can be constrained by defining the maximal compressional strain λ_2 as

$$\lambda_2 = \inf_{\mathbf{x} \in \Omega} \tilde{\lambda}_{\min} = \inf_{\substack{\mathbf{x} \in \Omega \\ 0 \le t' \le t}} \lambda_{\min} . \tag{12.10}$$

Note that the first equality corresponds to the lowest value of $\tilde{\lambda}_{\min}$ among all particle trajectories within Ω, which, according to the second equality, is equal to the most negative eigenvalue of the strain-rate tensor within the domain Ω and over the time interval $0 \le t \le t_1$. We then have the global bound

$$E(t_1) \le \exp\left(-2t_1 \left[\frac{1}{\tau_p} + \lambda_2\right]\right) E(0) . \tag{12.11}$$

Thus, the difference between the two particle velocity fields is guaranteed to decay exponentially provided the term within square brackets is positive. This condition can be restated as

$$(-\lambda_2)\tau_p < 1 \tag{12.12}$$

and provided this condition is satisfied, a unique equilibrium particle velocity field can be assumed to exist at the macroscale. Any deviation from equilibrium is guaranteed to decay exponentially and entrain to the unique velocity field. The most compressional strain rate is negative and therefore $-\lambda_2$ can be interpreted as the inverse of the "smallest" relevant fluid time scale. Thus, the left-hand side of the above condition is the ratio of particle to fluid time scale, and when this Stokes number is smaller than unity a unique particle velocity field is meaningful.

Note that by definition, λ_2 is the maximal compressional strain rate taken over all the points within the computational domain and over the entire time range of interest. Thus, $1/(-\lambda_2)$ as the global measure of fluid time scale is quite conservative. A unique particle velocity field is quite possible even when the above criterion is violated.

Two important additional considerations must be discussed. First, the uniqueness of the particulate velocity field pertains only to non-interacting particles. Even in the limit of very small particles, whose Stokes number is much smaller than unity (i.e., criterion (12.12) is satisfied), individual particle velocities can deviate from the equilibrium value due to particle–particle interaction. Since the particles are randomly distributed, the perturbation flow induced by neighbors will vary from particle to particle and due to this random influence of neighbors, the motion of individual particles will fluctuate about the equilibrium value. For example, within the sample volume shown in Figure 12.1a, criterion (12.12) guarantees that the particle velocity is unique and has a field representation only at the macroscale. At the microscale, each particle is uniquely influenced by the distinct random location of its neighbors. Nevertheless, if criterion (12.12) is satisfied, then the particle velocity is dictated entirely by the local fluid flow. As argued above, the local fluid flow consists of a macroscale component and a microscale pseudo-turbulence component, and both determine the particle motion. The initial velocity at which the particles were injected into the flow and the history of their past motion are quickly forgotten when the criterion is satisfied. Thus, provided $(-\lambda_2)\tau_p < 1$, the particle dynamics is dictated by the local fluid dynamics.

Second, the convergence of particle velocity to a unique field should not be confused with the convergence of individual particle trajectories. Suppose at some time t_0 there exist two different particle fields $\mathbf{v}_1(\mathbf{x}, t_0)$ and $\mathbf{v}_2(\mathbf{x}, t_0)$ that differ slightly. Then, condition (12.12) guarantees that the two fields will converge exponentially to a unique field. Instead, if we place two particles at $\mathbf{x}_0$, one with initial velocity $\mathbf{v}_1(\mathbf{x}_0, t_0)$ and the other with initial velocity $\mathbf{v}_2(\mathbf{x}_0, t_0)$, although the velocity fields quickly entrain to a unique field, at all later times the two particles will be at different locations. It is well established that the particle paths can be chaotic even in simple flows, and small initial differences can quickly grow to large values and become permanent. The results of this section concern the convergence of particle velocity fields at the macroscale, and not of individual particle trajectories.

12.2   Other Examples of Non-unique Particle Velocity

Let us revisit Section 4.5.5, where we considered the exact solution of particle motion in linearly varying ambient flows. Three different two-dimensional problems were considered: (i) particle motion in a planar straining flow, where the flow converges along the x-direction and diverges along the y-direction; (ii) particle motion in a vortical flow, where the fluid undergoes solid-body rotation on the x–y plane; and (iii) a particle falling down along the y-direction due to gravity in a linear shear flow that is oriented along the x-direction. In each of these three problems, exact solutions were obtained for particle motion along both the x and y-directions.

We will revisit the unique equilibrium versions of these solutions later in Section 12.3.1. We observe that in all three examples, there exists a rapidly decaying transient component; when this decays away, the particle velocity becomes independent of the initial condition and depends on the current location of the particle and can be expressed in terms of the local fluid velocity (the only exception is the x-component of the particle motion in the planar straining flow). These examples are thus consistent with the existence of a unique particle velocity field that only depends on the particle location. Particle trajectories of these unique particle velocity fields have been depicted in Figure 4.7a, c, and d. Due to the unique particle velocity at every point on the x–y plane, we observe the particle pathlines to be non-intersecting.

The unique equilibrium solution applies even for the x-component of the particle motion in the planar straining flow, provided $\mathrm{St}(1 - \beta\,\mathrm{St}) \leq 1/4$, when the particle Stokes number is defined as $\mathrm{St} = k\tau_p$. Here, k is the magnitude of the planar strain and β is the density parameter. As discussed in Problem 4.23, for larger-sized particles with $\mathrm{St}(1 - \beta\,\mathrm{St}) > 1/4$, the dependence on the initial condition cannot be ignored. Of particular relevance to the present section is the fact that the particle velocity is not unique anymore. This can clearly be seen in Figure 4.7b, where due to the oscillatory approach of the particle trajectories to $x = 0$, the particle velocity is not a unique function of the particle's current position. For example, at the point marked **P**, a particle that arrives from the left half-plane has a positive x-velocity, while that of the particle which arrives from the right is negative. Thus, in this example, the condition for uniqueness is $\mathrm{St}(1 - \beta\,\mathrm{St}) \leq 1/4$, consistent with the condition (12.12) of the previous section.

12.2.1 Condition for Existence of Unique Settling Velocity

In this section, we will consider a special case studied by Meiburg et al. (2000) for the existence of a unique settling velocity for particles falling in a two-way coupled, unidirectional flow. A simplified version of this example will be presented here, as restated by Ferry and Balachandar (2001). Consider the case of particles falling in still fluid under the action of gravity. For simplicity, we only consider particle inertia, Stokes drag, and gravity. The equation of particle motion in the vertical direction can be written as

$$\frac{dV_y}{dt} = \frac{V_s - V_y}{\tau_p}, \tag{12.13}$$

where y is directed down and V_s is the still-fluid settling velocity of the particle. Integrating the above equation twice, we obtain the following explicit solutions for the particle velocity and particle position:

$$V_y(t) = V_{y0}e^{-t/\tau_p} + V_s\left(1 - e^{-t/\tau_p}\right),$$
$$Y(t) = Y_0 + V_s t + \tau_p(V_{y0} - V_s)\left(1 - e^{-t/\tau_p}\right), \tag{12.14}$$

where Y_0 is the initial particle location and V_{y0} is the initial particle velocity.

Our interest lies in obtaining a unique settling velocity that can be expressed as an Eulerian field. If we consider the particle velocity to be an Eulerian field that is a function of both y and t, then the condition of uniqueness can be stated as: $V_y(y,t)$ must be a single-valued function of y for all $t > 0$. We will consider the initial velocity of the particle to be a single-valued function of the y location of release (i.e., $V_{y0}(y_0)$ is a unique function of y_0). For $V_y(y,t)$ to become a multivalued function of y, $\partial V_y/\partial y$ must become infinite at some y and t. Provided $\partial V_y/\partial y$ remains finite for all y and t, single-valuedness (or uniqueness) of the particle velocity is guaranteed. From the first equation of (12.14) we can obtain

$$\frac{\partial V_y}{\partial y} = \frac{dV_{y0}}{dy_0}\left(\frac{\partial y}{\partial y_0}\right)^{-1} e^{-t/\tau_p}. \tag{12.15}$$

The condition for single-valuedness becomes $\partial y/\partial y_0 > 0$. From the second equation of (12.14) we obtain

$$\frac{\partial y}{\partial y_0} = 1 + \frac{dV_{y0}}{dy_0}\tau_p\left(1 - e^{-t/\tau_p}\right). \tag{12.16}$$

From the above relation, Meiburg et al. (2000) obtained the condition for a unique particle velocity field to be

$$-1 < \frac{dV_{y0}}{dy_0}\tau_p. \tag{12.17}$$

If the above criterion is violated, then a multivalued particle velocity will develop in finite time. In other words, particles that started at different initial elevations will have the chance to be at the same y location at time t with different vertical velocities. In which case, one needs a particle NDF $f(\mathbf{x}, \mathbf{v}, t)$, whose evolution in phase space will be governed by William's spray equation or population balance equation.

The above criterion is similar to the more general condition derived in the previous section. Certain distinctions are, however, addressed by Ferry and Balachandar (2001). Both Eqs. (12.12) and (12.17) guarantee that an initially single-valued particle velocity field will remain single-valued at all later times. For the case of settling particles, Meiburg et al. (2000) showed that it is sufficient to set a bound on the initial velocity gradient for uniqueness. In a more general case, however, a bound on initial particle velocity alone will not guarantee single-valuedness, owing to the possibility of a complex underlying fluid velocity field.

The analysis of the previous section implies that even in the case of complex turbulent fluid motion, provided $(-\lambda_2)\tau_p < 1$, a unique particle velocity field can be taken to exist. Following Ferry and Balachandar (2001), we will refer to this as the *equilibrium* particle velocity field, implying that it is asymptotically valid after the initial transients have decayed. Though in general particle velocity depends on both the initial condition and the fluid velocity, an equilibrium particle velocity field is determined solely by the fluid velocity field. However, during the initial transient, the particle velocity can differ from the equilibrium velocity and the difference is due to the the initial particle velocity being different from the equilibrium value.

Even when the condition $(-\lambda_2)\tau_p < 1$ is globally satisfied, individual particle velocity can differ from the unique equilibrium velocity. As discussed at the end of Section 12.1, interaction between neighboring particles, either through direct collisions or due to fluid-mediated interaction, will result in microscale perturbation and the particle velocity will deviate from the unique macroscale velocity. Particles can also collide with the domain boundaries and suffer a sudden change in their velocity. Depending on the specular or diffuse nature of particle–wall collision, the particle NDF $f(\mathbf{x}, \mathbf{v}, t)$ can have a dominant bimodal or dispersed character in the particle velocity $\mathbf{v}$. This non-uniqueness in particle velocity can be expected to remain significant in the near-wall region, and particles need to readjust to the equilibrium velocity away from the domain boundaries.

12.3   Equilibrium Eulerian Approximation

In this section, we will explore the following provocative suggestion. *Under conditions where an equilibrium solution exists for the dispersed phase, the velocity, temperature, and other properties of the dispersed phase can be expressed explicitly in terms of the corresponding local continuous-phase properties without the need for solving the momentum and energy balance equations of the dispersed phase.* This statement must be clarified. In fact, the equilibrium solution that expresses the dispersed-phase quantities in terms of the local continuous-phase quantities was obtained by solving the dispersed-phase momentum and energy equations. The point being made is that the equilibrium solution, while satisfying the governing equations, yields a much simpler algebraic relation between the local dispersed and continuous-phase quantities. Thus, with the equilibrium solution one can avoid solving a system of ODEs for each particle in the Lagrangian framework, or solving a system of PDEs for the filtered dispersed-phase velocity and temperature in the Eulerian framework.

The usefulness of the equilibrium approximation arises from the fact that the only requirement is that it exists. Based on the analysis of the previous section, for the equilibrium solution to be valid at any point within a multiphase flow, the particle Stokes number as defined on the left-hand side of Eq. (12.12) must be smaller than one in the neighborhood of that point. For global validity of the equilibrium solution, Eq. (12.12) must be satisfied at all points and at all times. No further restrictions are placed on the nature of the continuous-phase flow or the particulate phase. In particular, as will be demonstrated below, the equilibrium solution applies even in complex turbulent flows and can be applied irrespective of what hydrodynamic forces govern particle motion.

Our derivation of the equilibrium solution will follow along the investigations of Maxey (1987), Druzhinin (1995), and Ferry and Balachandar (2001). For demonstration purposes we start with the following Lagrangian equation of motion of a particle, derived in Section 4.5.5:

$$\frac{d\mathbf{v}}{dt} = \frac{1}{\tau_p}(\mathbf{u} - \mathbf{v}) + \beta\frac{D\mathbf{u}}{Dt} + (1 - \beta)\mathbf{g}, \tag{12.18}$$

where recall that $\mathbf{u}$ and $D\mathbf{u}/Dt$ are evaluated at the center of the particle. We introduce L as some suitable length scale and τ_f as the time scale, and the nondimensional equation of particle motion can be expressed as

$$\frac{d\tilde{\mathbf{v}}}{d\tilde{t}} = \frac{1}{\text{St}}(\tilde{\mathbf{u}} - \tilde{\mathbf{v}}) + \beta\frac{D\tilde{\mathbf{u}}}{D\tilde{t}} + (1 - \beta)\tilde{\mathbf{g}}, \tag{12.19}$$

where a tilde denotes nondimensional quantities. If we choose the fluid time scale τ_f to be the inverse of the most compressional strain rate $(1/|\sigma_2|)$, then the condition (12.12) for the existence of a unique equilibrium solution becomes $\text{St} < 1$. It is in this limit that we will seek a solution to the above equation. As argued in Section 12.1, this equilibrium solution is independent of the initial condition for times longer than a short initial transient period. We choose the length scale to be such that the acceleration scale L/τ_f^2 provides an excellent measure of the actual fluid acceleration $D\mathbf{u}/Dt$, As a result, the nondimensional fluid acceleration $D\tilde{\mathbf{u}}/D\tilde{t}$ is an $O(1)$ quantity. The value of $\tilde{g} = g/(L/\tau_f^2)$ depends on the relative magnitude of acceleration due to gravity compared to fluid acceleration.

In the limit of neutrally buoyant particles, we expect the equilibrium particle solution to be the same as the local fluid (i.e., as $\beta \to 1$, we expect $\mathbf{v} \to \mathbf{u}$). For non-neutrally buoyant particles, the two velocities will differ by $O(\text{St})$ and therefore to leading order we write

$$\tilde{\mathbf{v}} = \tilde{\mathbf{u}} - (1 - \beta)\,\text{St}\,\tilde{\mathbf{w}}, \tag{12.20}$$

where the vector $\tilde{\mathbf{w}}$ needs to be evaluated from the asymptotic solution. We may now express the time derivative following the particle in terms of the total derivative following the fluid as

$$\frac{d}{d\tilde{t}} = \frac{D}{D\tilde{t}} - (1 - \beta)\,\mathrm{St}\,\tilde{\mathbf{w}} \cdot \tilde{\nabla}\,. \tag{12.21}$$

Substituting into Eq. (12.19), we obtain the following equation for $\tilde{\mathbf{w}}$:

$$\left(1 + \mathrm{St}\,\frac{d}{d\tilde{t}}\right)\tilde{\mathbf{w}} = \left(\frac{D\tilde{\mathbf{u}}}{D\tilde{t}} - \tilde{\mathbf{g}} - \mathrm{St}\,\tilde{\mathbf{w}} \cdot \tilde{\nabla}\tilde{\mathbf{u}}\right)\,. \tag{12.22}$$

For small values of the Stokes number, the operator on the left-hand side can be inverted using Taylor series and to leading order can be written as

$$\left(1 + \mathrm{St}\,\frac{d}{d\tilde{t}}\right)^{-1} = 1 - \mathrm{St}\,\frac{d}{d\tilde{t}} + O(\mathrm{St}^2) = 1 - \mathrm{St}\,\frac{D}{D\tilde{t}} + O(\mathrm{St}^2), \tag{12.23}$$

where the second expression was obtained from Eq. (12.21). Using this inverse operator in Eq. (12.22) and iteratively substituting for $\tilde{\mathbf{w}}$ on the right-hand side, we obtain the leading-order solution

$$\tilde{\mathbf{w}} = \frac{D\tilde{\mathbf{u}}}{D\tilde{t}} - \tilde{\mathbf{g}} - \mathrm{St}\left[\left(\frac{D\tilde{\mathbf{u}}}{D\tilde{t}} - \tilde{\mathbf{g}}\right) \cdot \tilde{\nabla}\tilde{\mathbf{u}} + \frac{D^2\tilde{\mathbf{u}}}{D\tilde{t}^2}\right] + O(\mathrm{St}^2)\,. \tag{12.24}$$

Substituting into Eq. (12.20), we obtain

$$\tilde{\mathbf{v}} \approx \tilde{\mathbf{u}} - (1 - \beta)\,\mathrm{St}\left(\frac{D\tilde{\mathbf{u}}}{D\tilde{t}} - \tilde{\mathbf{g}} - \mathrm{St}\left[\left(\frac{D\tilde{\mathbf{u}}}{D\tilde{t}} - \tilde{\mathbf{g}}\right) \cdot \tilde{\nabla}\tilde{\mathbf{u}} + \frac{D^2\tilde{\mathbf{u}}}{D\tilde{t}^2}\right]\right)\,. \tag{12.25}$$

We can recognize the second term on the right-hand side to be the nondimensional settling velocity of the particle:

$$\tilde{\mathbf{V}}_s = \frac{\mathbf{V}_s}{L/\tau_f} = (1 - \beta)\,\mathrm{St}\,\tilde{\mathbf{g}}\,. \tag{12.26}$$

As pointed out earlier, $\tilde{g} = |\tilde{\mathbf{g}}|$ is an independent parameter and its magnitude depends on the relative magnitude of gravitational acceleration compared to fluid acceleration. Thus, even when $\mathrm{St} \ll 1$, depending on the value of $\tilde{g}$, $|\tilde{\mathbf{V}}_s|$ can be small or large.

Problem 12.1 The derivation presented above started from the equation of motion (12.18), which neglected the Basset history and Saffman-like lift forces on the particle. As shown by Ferry and Balachandar (2001), the derivation of the equilibrium velocity is possible even with the inclusion of these additional forces. Repeat the steps outlined above with these additional forces and obtain an equation corresponding to Eq. (12.25) and compare the result with that given in the aforementioned paper.

We finally arrive at different possible approximations of particle velocity in terms of local fluid velocity. These approximations depend on the relative magnitude of the two nondimensional parameters: Stokes number St and nondimensional settling velocity

$|\tilde{\mathbf{V}}_s|$. Reverting back to dimensional terms, explicit expressions for the particle velocity of the different approximations can be written as

$$
\mathbf{v} = \begin{cases}
\mathbf{u} & \text{(1)} \quad \text{St}, |\tilde{\mathbf{V}}_s| \to 0, \\[2ex]
\mathbf{u} + \mathbf{V}_s & \text{(2)} \quad \text{St} \ll |\tilde{\mathbf{V}}_s| \ll 1, \\[2ex]
\mathbf{u} - (1-\beta)\tau_p \dfrac{D\mathbf{u}}{Dt} & \text{(3)} \quad |\tilde{\mathbf{V}}_s| \ll \text{St} \ll 1, \\[2ex]
\mathbf{u} + \mathbf{V}_s - (1-\beta)\tau_p \dfrac{D\mathbf{u}}{Dt} & \text{(4)} \quad |\tilde{\mathbf{V}}_s|, \text{St} \ll 1, \\[2ex]
\mathbf{u} + \mathbf{V}_s - (1-\beta)\tau_p \dfrac{D\mathbf{u}}{Dt} - \tau_p \mathbf{V}_s \cdot \nabla\mathbf{u} & \text{(5)} \quad \text{St} \ll |\tilde{\mathbf{V}}_s| \sim O(1).
\end{cases}
\tag{12.27}
$$

There are five different approximations that are presented above and they will be discussed now. (1) Corresponds to the limit when both St and $|\tilde{\mathbf{V}}_s|$ are negligibly small. In other words, in this limit, both particle settling and particle inertia are unimportant. This is the dusty gas limit where the particle velocity is simply equal to the local fluid velocity. (2) Corresponds to the limit when the effect of particle settling is important, while the inertial effect is much smaller and therefore can be ignored. This is a very useful limit in many geophysical sediment-laden flows, where the particle velocity can be well approximated as the sum of the local fluid velocity and the still-fluid settling velocity.

The regime (3) is important. It corresponds to the limit where fluid acceleration is much stronger than gravitational acceleration, and as a result the effect of settling can be ignored, while the inertial effect of particles not following curved fluid streamlines is important. The inertial correction appears as the term $-(1-\beta)\,\text{St}\,D\tilde{\mathbf{u}}/D\tilde{t}$ and accounts for the fact that when a local fluid parcel accelerates at $D\tilde{\mathbf{u}}/D\tilde{t}$, the corresponding acceleration of the particle located within the fluid parcel will be different. If the fluid is accelerating (i.e., $D\tilde{\mathbf{u}}/D\tilde{t} > 0$) and the particle is heavier than the fluid (i.e., $(1-\beta) > 0$), then the particle will lag the fluid by an amount given by the inertial correction term. In contrast, a lighter-than-fluid particle ($(1-\beta) < 0$) will lead the fluid, again by an amount given by the inertial correction. It must be stressed that the inertial correction term given in (3) is only the $O(\text{St})$ leading-order term. The solution given in Eq. (12.25) can be used to obtain an $O(\text{St}^2)$ contribution and higher-order terms can be obtained as well. Thus, the expressions given in (12.27) are only the leading-order terms.

In case (4), both the settling and inertial effects of the particle are important and both contribute to the difference between the particle velocity and the local fluid velocity. However, in this case, since both St and $|\tilde{\mathbf{V}}_s|$ are small, their combined quadratic effect is small and therefore has been ignored. In case (5), the effect of settling becomes $O(1)$, then at $O(\text{St})$ there is an additional term that contributes to the difference between particle and local fluid velocity. We shall see the physical interpretation of this term in an example to be discussed below. As a final note, we do not consider the limit St $\sim O(1)$, since the asymptotic expansion in Eq. (12.25) becomes invalid. No such restriction exists for settling velocity, and therefore we have considered the limit of $|\hat{\mathbf{V}}_s| \sim O(1)$.

A significant feature of all the above forms is that the particle velocity is simply related to the local fluid velocity and its spatial and temporal derivatives. Thus, in any computational procedure, once the fluid velocity field $\mathbf{u}(\mathbf{x}, t)$ has been evaluated, the corresponding particle velocity, as given by the equilibrium field $\mathbf{v}(\mathbf{x}, t)$, can be obtained easily from one of the algebraic expressions in Eq. (12.27), without solving additional PDEs or ODEs. Furthermore, it is easy to obtain particle velocity fields for particles of different sizes, since their velocities differ only through their Stokes number. In other words, polydispersity does not pose any additional difficulty in the evaluation of equilibrium velocity.

Before proceeding further, we must clarify the use of the term *equilibrium* in this chapter, since it has been used differently by different authors. In this book, we will consistently call the limit of particles perfectly following the local fluid (i.e., the case of $\mathbf{v} = \mathbf{u}$) the "dusty gas approximation." Our use of the term equilibrium *does not mean* that the particle and fluid quantities are the same. Clearly, the equilibrium particle velocity approximations given in Eq. (12.27) are not identically the same as $\mathbf{u}$. Thus, the term "equilibrium" is used in the sense that the particles tend to move as dictated by the local fluid. The particle velocity is only a function of the local fluid velocity and its derivatives (note that $\mathbf{u}$ and $D\mathbf{u}/Dt$ in Eq. (12.27) are evaluated at the particle location). The equilibrium particle velocity does not depend on the initial location or velocity of the particle, nor does it depend on the past history of particle motion. It is in this sense that we call it the equilibrium approximation.

Problem 12.2 *Thermal equilibrium approximation.* Starting from the temperature equation of a particle, an equilibrium solution for the particle temperature T_p in terms of the local fluid temperature T_f and its gradients can be calculated. Follow the work of Ferry and Balachandar (2005) and obtain the expression

$$T_p \approx T_f - \tau_T \left(\frac{\partial T_f}{\partial t} + \mathbf{v} \cdot \nabla T_f \right), \qquad (12.28)$$

where $\mathbf{v}$ is the equilibrium particle velocity given in Eq. (12.27). Here, τ_T is the thermal time scale defined in Eq. (2.7).

Problem 12.3 *Equilibrium approximation for particle rotation.* Start from the rotational equation of motion of a particle given in Eq. (5.33) and set the hydrodynamic torque on the particle given by the left-hand side equal to $I_p D\mathbf{\Omega}_p / Dt$, where I_p is the moment of inertia of the particle and $\mathbf{\Omega}_p$ is the angular velocity of the particle. For now, ignore the third term involving the history integral (although the result can be obtained even with this term included).

(a) Obtain the nondimensional equation

$$\frac{d\Omega_p}{dt} = \frac{\omega_c/2 - \Omega_p}{\tau_R} + \frac{1}{2\rho}\frac{D\omega_c}{Dt}, \qquad (12.29)$$

where the rotational time scale $\tau_R = \rho d_p^2/(60\nu_f)$, with ρ being the particle-to-fluid density ratio. The rotational time scale is the same as that defined in Eq. (2.34). Here we have assumed $\Phi_R = 1$, corresponding to the Stokes limit.

(b) Follow the steps taken in the previous problems and obtain the equilibrium approximation for the angular velocity of the particle. Here the two parameters are (i) the ratio of particle rotational time scale to fluid time scale $St_R = \tau_R/\tau_f$, which can be termed the rotational Stokes number, and (ii) the nondimensional settling velocity $|\tilde{\mathbf{V}}_s|$ given in Eq. (12.26). The different leading-order asymptotic expressions can be written as

$$\Omega_p = \begin{cases} \dfrac{\omega_c}{2} & \text{①} \ St_R, |\tilde{\mathbf{V}}_s| \to 0, \\[2ex] \dfrac{\omega_c}{2} - \dfrac{\tau_R}{2}\dfrac{\rho-1}{\rho}\dfrac{D\omega_c}{Dt} & \text{②} \ |\tilde{\mathbf{V}}_s|, St_R \ll 1, \\[2ex] \dfrac{\omega_c}{2} - \dfrac{\tau_R}{2}\left(\dfrac{\rho-1}{\rho}\dfrac{D\omega_c}{Dt} + \mathbf{V}_s \cdot \nabla\omega_c\right) & \text{③} \ St_R \ll |\tilde{\mathbf{V}}_s| \sim O(1). \end{cases} \qquad (12.30)$$

Thus, we see that the particle rotation rate under equilibrium can be expressed entirely in terms of local macroscale fluid vorticity ω_c and its total derivative.

(c) Show that in version ③ the effect of strong settling will give rise to an additional contribution due to the particle falling through regions of vertically varying ambient vorticity, which will be the case for a particle falling through a non-uniform shear layer.

(d) Compare this with the behavior to be discussed in Problem 12.6.

12.3.1   Particle Motion in Linear Flows – Revisited

In this section, we will revisit the simple problems considered in Section 4.5.5 and obtain the corresponding equilibrium particle velocities. By comparing them to the exact velocities obtained in Section 4.5.5, we can judge the usefulness and limitations of the equilibrium approximation. Again the discussion will be in the form of three problems whose results the reader should derive and compare with those obtained in Problems 4.23 to 4.25.

Problem 12.4 Consider a two-dimensional planar strain flow with compression (or convergence) along the x-axis and extension (or divergence) along the y-axis given in Eq. (4.154). Here the inverse of the strain-rate magnitude (i.e., $1/k$) will be taken as the fluid time scale, and thus the particle Stokes number is defined as $St = k\tau_p$.

(a) Apply the equilibrium approximation ③ of Eq. (12.27) and obtain

$$V_x = \eta_{ex} u_{cx} \quad \text{where} \quad \eta_{ex} = 1 + (1 - \beta)\,\text{St},$$

$$V_y = \eta_{ey} u_{cy} \quad \text{where} \quad \eta_{ey} = 1 - (1 - \beta)\,\text{St} . \tag{12.31}$$

Here, u_{cx} and u_{cy} are the continuous-phase x and y-velocities at the particle location. Thus, to leading order, the equilibrium approximation predicts the particle velocity along the compressional x-direction to lead the local fluid velocity by $(1 - \beta)\,\text{St}\,u_{cx}$. Owing to its inertia, the particle is not able to slow down as fast as the local fluid as it approaches the $x = 0$ plane (assuming the particle to be heavier than the fluid). The above equilibrium approximation predicts the y-velocity of the particle to lag behind the local fluid velocity by $(1 - \beta)\,\text{St}\,u_{cy}$. In the equilibrium approximation, this difference between the particle and the local fluid velocity is driven by the local fluid acceleration $D\mathbf{u}/Dt$ and the particle's inability to respond synchronously.

(b) Compare the exact solutions given in Eqs. (4.163) and (4.159) with the above equilibrium approximation. For small St, Taylor expand η_x and η_y given in the exact solution and show that to $O(\text{St})$ the equilibrium approximation is in perfect agreement with the exact solution. In other words, for small values of Stokes number the equilibrium particle velocity provides a very good approximation of the actual velocity. In Problem 4.23, however, we noted that for large Stokes numbers above a certain critical value, the particle velocity will be non-unique. Clearly, the equilibrium approximation is invalid in this regime.

Problem 12.5 In this example, we revisit particle motion in a two-dimensional solid-body-like rotational vortex flow considered in Problem 4.24. Here, k is the rotational rate of the vortical flow and again $\text{St} = k\tau_p$.

(a) Apply the equilibrium approximation ③ of Eq. (12.27) and obtain

$$V_x + \iota V_y = \eta_{e\xi}(u_{cx} + \iota u_{cy}) \quad \text{where} \quad \eta_{e\xi} = 1 - \iota(1 - \beta)\,\text{St} . \tag{12.32}$$

(b) Compare the exact solutions given in Eqs. (4.169) and (4.170) with the above equilibrium approximation. For small St, Taylor expand η_ξ of the exact solution and show that to $O(\text{St})$ the equilibrium approximation is in perfect agreement. In other words, for small values of Stokes number the equilibrium particle velocity provides a very good approximation of particle motion in a vortical flow.

(c) Convert the Cartesian velocities to cylindrical coordinates and obtain the following equilibrium radial and circumferential velocities of the particle:

$$V_r = (1 - \beta)\,\text{St}\,u_{c\theta} \quad \text{and} \quad V_\theta = u_{c\theta}, \tag{12.33}$$

where you need to use the fact that the fluid has only circumferential velocity and no radial velocity. It can be seen that to order $O(\text{St})$ accuracy the particle follows the local fluid in the circumferential direction, but with a positive radial velocity drift in the case of a heavier-than-fluid particle $((1 - \beta) > 0)$ or with a negative (inward) velocity in case of a lighter-than-fluid particle. This radial drift arises from the fact that the particles, owing to their inertia, cannot follow curved fluid streamlines. In other words, the nonzero radial component of fluid acceleration $D\mathbf{u}/Dt$ in this case gives rise to the radial slip.

We now go back to the analysis of a particle in a vortex considered in Section 2.4.5. There it was pointed out that the circumferential velocity V_θ of a heavier-than-fluid particle will lag that of the fluid, since the particle as it moves radially outward will always arrive at its current location from its previous radial location of lower circumferential velocity. For a similar reason, the circumferential velocity of a lighter-than-fluid particle will lead that of the local fluid as it travels inward toward the vortex center. As can be seen in Eq. (2.30), this circumferential velocity slip is an $O(\mathrm{St}^2)$ quantity and thus is not captured by the $O(\mathrm{St})$ expansion given in Eq. (12.33). The next term of the asymptotic expansion will include this effect. This can be inferred from the last term of the ⑤ version of Eq. (12.27). In the present problem, this term will appear as $-\tau_p V_r (du_{c\theta}/dr)$. This term accounts for the memory effect of how the circumferential velocity of the fluid changes in the direction of a particle's radial motion. Obtain the circumferential velocity slip using a higher-order equilibrium approximation.

Problem 12.6 As the final application of equilibrium particle velocity, consider the case of a particle falling down due to gravity in a linear shear flow studied in Problem 4.25. This time, you should use the ⑤ version of Eq. (12.27), which accounts for the $O(1)$ effect of settling velocity along with the $O(\mathrm{St})$ inertial effect. In a shear flow, $D\mathbf{u}/Dt = 0$.

(a) Show that the term $-\tau_p \mathbf{V}_s \cdot \nabla \mathbf{u}$ accounts for the streamwise slip as a particle falls down through the shear flow.

(b) Show that the equilibrium approximation in this case exactly recovers the result you obtained in Problem 4.25.

12.3.2 Test of Equilibrium Approximation in Turbulent Flows

The accuracy of the equilibrium approximation has been evaluated in a number of canonical turbulent flows such as isotropic turbulence, homogeneous shear turbulence, and turbulent flow in a wall-bounded channel (Ferry and Balachandar, 2001, 2002; Rani and Balachandar, 2003; Shotorban and Balachandar, 2006, 2007, 2009). In all these cases, two kinds of test were done. (i) An a-priori test in which a large number of particles were tracked in the turbulent flow using their Lagrangian equation of motion. At each instant along the trajectory of each particle, in addition to the exact Lagrangian velocity, the equilibrium particle velocity was also calculated. In the a-priori test, the equilibrium particle velocity was not used to advance the particle position. A comparison between the exact particle velocity and the equilibrium velocity was used to establish the a-priori error. (ii) An a-posteriori test in which the results of a one-way coupled Euler–Lagrange simulation using the exact particle velocities, as integrated from their equations of motion, are compared against one-way coupled Euler–Euler simulations where the particle velocity field was obtained from the equilibrium assumption.

Here we will briefly discuss the results of these two tests for the case of isotropic turbulence. For additional details the reader should consult Rani and Balachandar (2003). In the context of isotropic turbulence in a cubic box, the exact particle velocity is defined to be that obtained from Lagrangian tracking using a quasi-steady force as the only force acting on the particle. We denote this exact velocity of the ith particle to be $\mathbf{V}_{ex,@i}$. In the dusty gas (or zeroth-order) approximation, the particle velocity will be approximated as the fluid velocity interpolated to the center of the particle, $\mathbf{u}_{@i}$. The first-order accurate equilibrium solution given in Eq. (12.27) will be denoted as $\mathbf{V}_{eq,@i}$. Then the rms error in the zeroth and first-order approximations can be defined as

$$
\begin{aligned}
E_0 &= \left[\frac{1}{N_P} \sum_{i=1}^{N_P} |\mathbf{V}_{ex,@i} - \mathbf{u}_{@i}|^2 \right]^{1/2}, \\
E_1 &= \left[\frac{1}{N_P} \sum_{i=1}^{N_P} |\mathbf{V}_{ex,@i} - \mathbf{V}_{eq,@i}|^2 \right]^{1/2},
\end{aligned}
\tag{12.34}
$$

where the average is over all the N_P particles within the system. Figure 12.2 shows the above rms errors plotted as a function of St. All the quantities are made dimensionless by the fluid Kolmogorov length, velocity, and time scales. It is clear that for small particles of St < 1, the first-order equilibrium expansion provides a better approximation to the actual particle velocity and with decreasing particle size, the error decreases more rapidly when the $O(\text{St})$ correction is applied to the particle velocity. The above errors apply equally well at all points within the box of isotropic turbulence, due to spatial homogeneity.

Next, we will consider the rms a-priori errors in a turbulent channel flow (Ferry and Balachandar, 2001). Here the results depend on the distance from the channel walls. The behavior at the center of the channel is likely to be comparable to that seen in isotropic turbulence. So we will present results for a distance of 20 wall units from the wall, which is in the buffer layer. The rms errors are calculated based on those particles that are located around $y_+ \approx 20$. While the isotropic results are only for heavy particles, here the results for heavy particles ($\beta = 0$) and bubbles ($\beta = 3$) are presented.

Figure 12.3 shows the rms values of E_0 and E_1 as a function of nondimensional particle time scale for $\beta = 0$. Both the error in the streamwise velocity as well as the error in the wall normal velocity are shown. Also plotted are the errors in the second-order approximation E_2 (see Ferry and Balachandar, 2001 for a definition). As expected, E_1 decreases more rapidly than E_0 as the particle time scale decreases. The behavior is similar for both the streamwise and wall-normal components. Based on the figure, the extra work involved in using the second-order approximation cannot be justified. The $O(\text{St})$ approximation is more than adequate. Figure 12.4 shows the rms error for

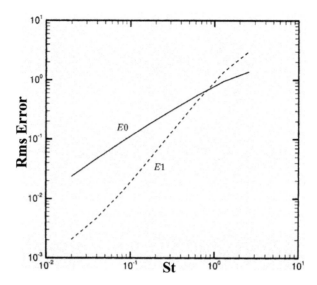

Figure 12.2 Comparison of rms zeroth and first-order errors for heavier-than-fluid particles in isotropic turbulence. Reprinted from Rani and Balachandar (2003), with permission from Elsevier.

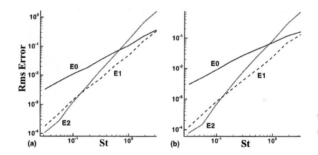

Figure 12.3 Root mean square errors for heavy particles ($\beta = 0$) at a distance of 20 wall units from the channel walls: (a) streamwise errors; (b) wall-normal errors. Reprinted from Ferry and Balachandar (2001), with permission from Elsevier.

a bubble ($\beta = 3$). The results are quite similar to those for heavy particles. In general, there is no significant qualitative difference as β is varied, though the mean errors are of opposite sign since the behavior of bubbles is opposite to that of heavy particles.

The purpose of a-posteriori testing is to evaluate the fidelity of the equilibrium approximation when used to evolve the particle concentration field. We will compare the Eulerian particle concentration field with the corresponding Lagrangian distribution of particles, which serves as the benchmark (see Rani and Balachandar, 2003). Figure 12.5 shows the comparison between the Lagrangian distribution and the Eulerian particle concentration field evolved with the equilibrium velocity field. The particular results shown in the figure are for St = 0.1 and $\tilde{V}_s = 4$. Frame (a) shows the contours of the Eulerian particle concentration field on a vertical plane passing through the cubic box of isotropic turbulence at one instant in time and frame (b) shows the corresponding

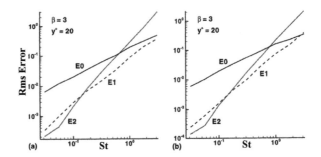

Figure 12.4 Root mean square errors for bubbles ($\beta = 3$) at a distance of 20 wall units from the channel walls: (a) streamwise errors; (b) wall-normal errors. Reprinted from Ferry and Balachandar (2001), with permission from Elsevier.

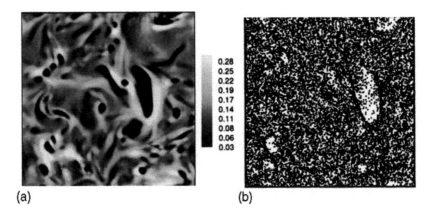

Figure 12.5 (a) Contours of Eulerian particle concentration field on a vertical plane passing through the cubic box of isotropic turbulence. (b) Scatterplot of Lagrangian distribution of particles on a thin slice of volume surrounding this plane. For the case considered, St = 0.1 and $\tilde{V}_s = 4$. Reprinted from Rani and Balachandar (2003), with permission from Elsevier.

Lagrangian distribution of particles on a thin slice of volume surrounding this plane. A satisfactory qualitative agreement between the two frames can be observed, with both showing a non-uniform concentration of particles with preferential accumulation in selected regions of the flow and other locations devoid of particles.

Similar a-posteriori comparisons have been made in turbulent channel flow (Ferry et al., 2003). Figure 12.6 shows the distribution of particles within a thin horizontal slab centered around $y_+ = 30$ at a particular time during the simulation. In the top frame, the Lagrangian particles were evolved in time using the equilibrium particle velocity. As a result, only the position equation $d\mathbf{X}/dt = \mathbf{V}$ was solved for each particle. In the bottom frame, the Lagrangian particles were evolved in time using the standard Lagrangian equations of motion.

The qualitative agreement is rather impressive when one compares it to the alternative of using the dusty gas approximation $\mathbf{V} = \mathbf{u}$, which would give a statistically uniform distribution of particles for all time. The particle Stokes number calculated with the Kolmogorov time scale is 3. Despite this somewhat larger Stokes number,

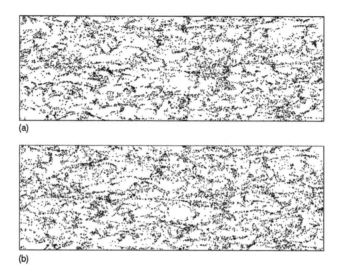

Figure 12.6 Distribution of particles in a turbulent channel flow near the $y_+ = 30$ plane at a particular instant in time for a particle Stokes number based on a Kolmogorov time scale of 3. (a) Particles evolved in time using the equilibrium particle velocity. (b) Particles evolved in time using the full Lagrangian equations of motion. Reprinted from Ferry et al. (2003), with permission from Elsevier.

the comparison is quite good. Clustering of particles and void regions in both cases matches quite well. It must be stressed that we cannot expect a precise match of individual particle position. Since particle evolution in a turbulent flow is chaotic, even small differences will result in a large permanent difference in the actual particle position. However, the reason for good qualitative comparison is that particle behavior depends mainly on the local flow. Regions where particles congregate are easily identified by the equilibrium approximation. In essence, for particles of small Stokes number, the particle velocity can be well approximated by the equilibrium model.

12.4　Multiphase Physics Explored with the Equilibrium Model

Due to the chaotic nature of a particle trajectory, comparison of individual particle position and velocity computed with the different models is not meaningful. In this section, we will present statistics of fluid and particle quantities as evaluated at the distribution of Lagrangian particles and as evaluated from the Eulerian particle concentration field and compare them to assess the accuracy of the equilibrium approximation.

More importantly, we want to explore the multiphase flow mechanisms that contribute to non-uniform distribution of the dispersed phase and explain them with the use of equilibrium approximation. It is natural to expect a well-mixed turbulent multiphase flow with an initial uniform distribution of particles to remain well-mixed and uniform at all later times. In fact, even if the particles were initially maldistributed, turbulent mixing can be expected to take over and render the distribution of particles uniform after a short period of time. This expectation is motivated by our observation of how a passive scalar quickly mixes in a turbulent flow. For example, if red dye is injected in a corner of a turbulent stirred tank, it will quickly mix and be uniformly distributed over the entire fluid volume. Very small particles, whose Stokes number is much smaller than unity, will behave similarly. Based on the pioneering works of John Eaton and Martin Maxey (Maxey, 1987; Eaton and Fessler, 1994; Squires and Eaton, 1991), it is now well accepted that the above picture does not apply for particles whose Stokes number is $O(1)$. Most interestingly, particles can be demixed by turbulence, as we will discuss below.

In this section, we will discuss two important multiphase flow mechanisms that result in sustained non-uniform distribution of the dispersed phase. (i) The first mechanism is preferential accumulation of particles due to their inertia. In simple terms, heavier-than-fluid particles will spin out of vortex cores, while lighter-than-fluid bubbles will congregate at the vortex centers. This mechanism is active even in a laminar flow. But, in a turbulent flow, this mechanism takes special importance, since the flow now includes many intense coherent vortical structures whose dynamics is chaotic. (ii) The second mechanism that we will discuss is turbophoretic migration of particles. As the name suggests, this mechanism is active only in a turbulent flow. In inhomogeneous turbulence, heavier-than-fluid particles will on average migrate up the gradient from regions of low turbulence to regions of large turbulence. Both mechanisms are capable of generating non-uniformities in the spatial distribution of particles, even when started from an initial well-mixed state. Unlike the preferential accumulation mechanism, turbophoretic migration is a statistical behavior that is active only on average in a turbulent flow.

Non-uniformities in macroscale particle distribution, irrespective of whether they are generated by the above demixing mechanisms or initiated by other means, are counteracted by turbulent diffusion of particles. The diffusional flux of particles from regions of high concentration to regions of low concentration will tend to lessen the non-uniformities. This turbulent mixing process is driven by correlations of velocity fluctuation. In a multiphase flow, velocity fluctuations arise both at the microscale due to pseudo turbulence and at the macroscale due to classical turbulence. Diffusion due to pseudo turbulence is present even in macroscopically laminar flows, while diffusion due to macroscale fluctuations requires the multiphase flow to be turbulent. These diffusional mechanisms work to decrease non-uniformity in the dispersed-phase distribution and will not create or enhance non-uniformities.

Finally, we care very much about how the dispersed phase is distributed. If the dispersed phase is non-uniformly distributed, then the mean volume fraction (averaged over the entire domain) and the associated Poisson distribution about the mean are not sufficient to characterize the actual state of the dispersed phase. In many applications, non-uniformity can be sufficiently strong that the local volume fraction of particles can be more than an order of magnitude larger than the mean. In case of droplets, such local clustering can profoundly alter the rate at which the droplets collide, agglomerate, and grow in size. In the case of burning particles, with increasing local volume fraction of particles, locally available oxygen must be shared by a greater number of particles, thus possibly reducing the average burn rate. Shaw (2003) describes how a non-uniform distribution of droplet nuclei may affect the subsequent growth rate of raindrops by both condensation and collision/coalescence in atmospheric clouds. Similarly, Cuzzi et al. (2001) discuss the importance of non-uniform distribution of matter in protoplanetary nebula.

12.4.1 Preferential Particle Accumulation – Revisited

Early in the book, in Section 1.3.1, we introduced the idea of preferential accumulation. In this section, we will elaborate on this phenomenon with mathematical support from the equilibrium particle velocity field.

An example of preferential accumulation behavior is presented in Figure 12.7, which shows the spatial distribution of particles around one vertical plane within a periodic cubic box of $Re_\lambda = 85$ isotropic turbulence (results from Park et al., 2017). The one-way coupled simulation employed a uniform mesh of $(256)^3$ grid points, within which 84 million particles were tracked by solving their Lagrangian equations of motion. Figure 12.7a shows the distribution of heavy particles of Stokes number ($St = \tau_p/\tau_k$, where τ_k is Kolmogorov time scale) equal to 0.5. This must be compared with the expected behavior of very small particles of $St \ll 1$, which will be a uniform distribution of particles over the entire square cross-section. The difference is striking. Particles of $O(1)$ Stokes number respond so well to the turbulent vortices that they are demixed and form highly concentrated regions. As we will see below, these concentrated regions are on the edges of turbulent vortices and they change over time as the vortices advect within the computational domain. Also shown in the figure is the corresponding distribution of $St = 8$ particles. Note that gravity is turned off in these simulations. Though regions of increased and decreased particle concentrations can be seen, clearly the degree of preferential accumulation is far less pronounced. Similar behavior of preferential accumulation was seen earlier in Figures 12.5 and 12.6 as well, but the behavior is most clear in Figure 12.7. Preferential response of $O(1)$ Stokes number particles has been observed in many other turbulent flows.

To explain preferential accumulation, let us take the divergence of the equilibrium particle velocity field given in version ③ of Eq. (12.27) to obtain

$$\nabla \cdot \mathbf{v} = \nabla \cdot \mathbf{u} - (1-\beta)\tau_p \frac{\partial}{\partial t} \nabla \cdot \mathbf{u} - (1-\beta)\tau_p \frac{\partial u_i}{\partial x_j} \frac{\partial u_j}{\partial x_i}. \tag{12.35}$$

In a compressible flow, fluid divergence is passed on to particle divergence, which is the first term on the right-hand side. The other two terms are an $O(St)$ correction. Just as fluid density decreases in regions of positive fluid velocity divergence, particle volume fraction will decrease in regions of positive particle velocity divergence. But a more interesting physics arises from the third term on the right-hand side, and it is active even in incompressible flows. In an incompressible flow we have

$$\nabla \cdot \mathbf{v} = -(1-\beta)\tau_p \frac{\partial u_i}{\partial x_j} \frac{\partial u_j}{\partial x_i} = (1-\beta)\tau_p \left(|\mathbf{S}|^2 - |\mathbf{\Omega}|^2 \right), \tag{12.36}$$

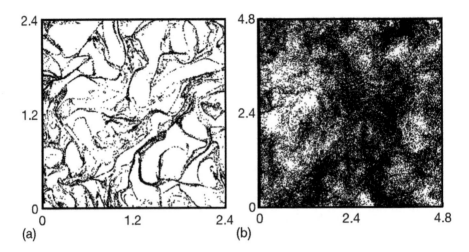

Figure 12.7 Instantaneous spatial distribution of particles (dots) contained in a slice of thickness equal to Kolmogorov length scale for (a) St = 0.5 and (b) St = 8 particles. Reprinted with permission from Park et al. (2017). Copyright 2017 by the American Physical Society.

where $\mathbf{S}$ and $\boldsymbol{\Omega}$ are the symmetric and antisymmetric parts of the velocity gradient tensor $\nabla\mathbf{u}$.

From the dispersed-phase mass balance given in Eq. (11.112), it can be seen that a positive value of particle velocity divergence (i.e., $\nabla \cdot \mathbf{u}_d > 0$) tends to reduce the local volume fraction of the dispersed phase. Thus, heavier-than-fluid particles ($0 < \beta < 1$) tend to accumulate in regions where the strain rate dominates over vorticity, whereas lighter-than-fluid bubbles ($1 < \beta < 3$) tend to accumulate in regions of vorticity. The validity of the above expression, and the linear dependence of preferential accumulation on St, applies only for small St. Large Stokes number particles are sluggish and they only show a small response to turbulence. As suggested by Balachandar and Eaton (2010), in the case of large Stokes number particles, the effect of turbulence can be modeled better as providing small random impulses to particle motion.

In Figure 12.8 we compare the average value of $(|\mathbf{S}|^2 - |\boldsymbol{\Omega}|^2)$ computed for the "exact" particles and the "approximate" particles that are evolved with the equilibrium velocity. These statistics were obtained in a turbulent channel flow and only particles in the buffer region around $y_+ = 20$ and along the channel mid-plane at $y_+ = 180$ are shown. The preferential accumulation of heavy particles ($\beta = 0$) in regions of large strain rate, and bubbles ($\beta = 3$) in regions of large vorticity, is clear from the respective positive and negative mean values of $(|\mathbf{S}|^2 - |\boldsymbol{\Omega}|^2)$. The exact and equilibrium approximation results are in good agreement in the buffer region for St < 1. Near the channel mid-plane, the equilibrium approximation appears to be accurate even for particles of larger Stokes number.

A similar investigation was also carried out in isotropic turbulence and the results are shown in Figure 12.9. Here again, the PDF of $(|\mathbf{S}|^2 - |\boldsymbol{\Omega}|^2)$ is computed three different ways for varying Stokes number: (i) based on Lagrangian particles that were

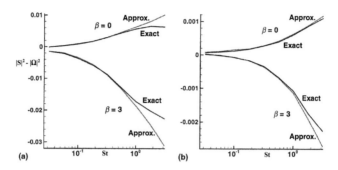

Figure 12.8 Mean values of $(|\mathbf{S}|^2 - |\mathbf{\Omega}|^2)$ evaluated at particle and fluid locations in a turbulent channel flow: (a) $y_+ = 20$ in the buffer region; (b) $y_+ = 180$ at the channel center. Reprinted from Ferry and Balachandar (2001), with permission from Elsevier.

evolved using the exact equations of motion; (ii) based on Lagrangian particles evolved using the equilibrium approximation; (iii) based on the entire fluid volume. Here, the Stokes number is defined as the ratio of particle to Kolmogorov time scale. The average value of $(|\mathbf{S}|^2 - |\mathbf{\Omega}|^2)$ for the equilibrium approximation is in good agreement with the exact particles for $St \leq 0.3$. Note that the average value of $(|\mathbf{S}|^2 - |\mathbf{\Omega}|^2)$, averaged over the entire fluid volume, is identically zero. Above a Stokes number of 0.3, while the average value of $(|\mathbf{S}|^2 - |\mathbf{\Omega}|^2)$ begins to turn around and decrease, the predicted behavior using the equilibrium approximation continues to increase. This illustrates an important point. In the present case of isotropic turbulence, preferential accumulation of particles in regions of excess strain over rotation peaks at a Stokes number around 0.7. For increasing value of Stokes number beyond 0.7, particles do not respond well to turbulent eddies and thus their preferential accumulation decreases. The first-order equilibrium approximation does not capture the existence of this optimal St at which preferential accumulation reaches its maximum. For more advanced analysis of preferential accumulation, the reader is recommended the recent investigations by Esmaily and Mani (2016) and Esmaily and Mani (2020).

For small particles, the equilibrium approximation is able to accurately predict not only the mean value, but even higher-order statistics. This is easily demonstrated by the insets in Figure 12.9, which show the PDF of $(|\mathbf{S}|^2 - |\mathbf{\Omega}|^2)$ for two different values of Stokes number. Comparing the particle PDFs with those of the fluid, we see a clear preference for positive $(|\mathbf{S}|^2 - |\mathbf{\Omega}|^2)$ (in case of bubbles, their preference will be regions of negative $(|\mathbf{S}|^2 - |\mathbf{\Omega}|^2)$). Despite the large deviation in the mean value for large St, the PDFs of the exact and approximate particles are in reasonable agreement, especially when compared to that of the fluid. This shows that equilibrium approximation captures the correct trend even when $St > 1$. For smaller particles, the PDFs of exact and equilibrium particles are in very good agreement.

There is a large body of experimental evidence for preferential accumulation of heavier-than-fluid particles in the strain-dominated regions. Some of the earliest observations are in the planar mixing layer, where particles of large Stokes number

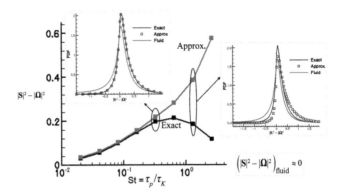

Figure 12.9 Mean values of $(|\mathbf{S}|^2 - |\mathbf{\Omega}|^2)$ evaluated at the particle locations evolved using exact and equilibrium approximation in isotropic turbulence. PDFs of $(|\mathbf{S}|^2 - |\mathbf{\Omega}|^2)$ for two different Stokes number particles and for the fluid volume. All the results are for heavy particles.

were observed to spin out of the dominant Kelvin–Helmholtz vortices (Lazaro and Lasheras, 1989; Wen et al., 1992; Glawe and Samimy, 1993). Medium-sized particles were observed to become trapped in rings around vortices, while small particles simply followed the flow. Studies of preferential accumulation also included axisymmetric jet flows (Chung and Troutt, 1988). They observed that St $\approx$ 1 particles dispersed the most when St was defined based on the vortex passing frequency. Larger St = 10 particles created particle fingers that were expelled from the jet core by the vortices. Pioneering experiments of particle distribution and preferential concentration were conducted by Eaton and co-workers (Longmire and Eaton, 1992; Wicker and Eaton, 2001) whose measurements showed the now familiar results of particles being clustered in high-strain regions between vortex structures.

Before closing, we must stress an important aspect of preferential accumulation. As we discussed early in Section 3.3, even a random distribution of particles with uniform probability will involve volume fraction (or number density) variation. That is, if you look locally there will be regions of particle volume fraction larger than the mean, mixed in with regions of lower volume fraction. However, by choosing the sampling volume to be sufficiently large, the volume fraction variation of a uniform random distribution can be made small. By preferential accumulation we refer to particle volume fraction variation over and above that due to simple random distribution. Preferential accumulation-induced variations in particle concentration persist at the macroscale. A random distribution of particles obeys Poisson statistics in the limit of small volume fraction. Therefore, Fessler et al. (1994) introduced the parameter $(\sigma - \sigma_p)/\mu$ to indicate the level of preferential accumulation as a deviation from the Poisson distribution. Here, σ and σ_p are the standard deviation of the measured and the Poisson distribution with the same mean. This parameter was observed to reach a maximum for St = 1 particles (also see Aliseda et al., 2002).

12.4.2   Turbophoretic Migration

It has been observed in experiments and simulations that in a turbulent pipe or channel flow, the heavier-than-fluid particles migrate toward the channel walls with the consequence of near-wall particle concentration far in excess of the mean concentration (Rani et al., 2004; Winkler et al., 2004). A typical scenario is shown in Figure 12.10, where particle distribution along the length of a pipe is shown projected on a cross-sectional plane. The results were obtained from simulations (Rani et al., 2004), where periodic boundary conditions were used along the pipe axis and thus the streamwise direction was homogeneous. The simulation was initialized with a random distribution of particles that was preferentially distributed toward the center of the pipe, as shown in Figure 12.10a. Over time, the particles evolve into the non-uniform distribution as shown in Figure 12.10b. In both cases, the number of particles is the same, but most of the particles have migrated very close to the pipe wall.

In a laboratory experiment, this behavior will be seen as streamwise evolution of particle distribution. If we start with a uniform distribution of particles at some streamwise location, as the flow and particles evolve downstream, a non-uniform distribution will develop with particle concentration increasing close to the pipe wall. It should be emphasized that Figure 12.10 depicts an exaggerated picture, as the simulation does not capture all the physics of relevance. Most importantly, interparticle and wall–particle collisions were ignored in the simulation, and this is why the particle concentration at the wall has grown unchecked. As shown by subsequent studies, even with the inclusion of collisions, the particles will migrate from the pipe centerline toward the pipe wall. However, as the near-wall particle concentration increases, collisions become dominant. Once near-wall concentration builds up to a certain value, turbulence-induced wallward migration is counterbalanced by collision-induced migration toward the pipe centerline, and a stationary state of non-uniform radial distribution establishes.

Turbulence was responsible for the migration of heavier-than-fluid particles from near the pipe centerline toward the pipe wall. More precisely, it is the variation in the level of turbulence along the wall-normal direction in a pipe that is responsible for particle migration. We now use the equilibrium approximation to explain the turbophoretic migration of particles. We consider particle motion in a turbulent channel flow, where the only inhomogeneous direction is the wall-normal y-coordinate. Version ③ of the equilibrium approximation is chosen, since gravitational settling is ignored in this problem. If we average the y-component of this equation, the resulting equation can be expressed as

$$\langle v_y \rangle = \langle u_y \rangle - (1 - \beta)\tau_p \left\langle \frac{\partial u_y}{\partial t} + \frac{\partial u_x u_y}{\partial x} + \frac{\partial u_y^2}{\partial y} + \frac{\partial u_z u_y}{\partial z} \right\rangle, \qquad (12.37)$$

where the angle brackets represent the ensemble average, which can be approximated as an average over the horizontal x–z planes. Due to stationarity and horizontal periodicity, only the y-gradient term is nonzero. Furthermore, the average wall-normal

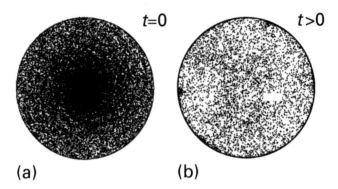

Figure 12.10 Particle distribution within a streamwise periodic pipe flow. (a) The initial distribution of heavy particles. (b) The distribution of the same particles at a later time. Most of the particles have migrated close to the pipe wall.

fluid velocity is identically zero. This yields the following final expression for the mean wall-normal velocity of the particle:

$$\langle v_y \rangle = -(1 - \beta)\tau_p \frac{d\langle u_y^2 \rangle}{dy}. \tag{12.38}$$

Note that since $\langle u_y^2 \rangle$ is only a function of y, the partial derivative has been replaced with the total derivative. This expression was originally obtained by Caporaloni et al. (1975) and Reeks (1983). Even though the average wall-normal velocity of the fluid is zero, the average wall-normal velocity of the particles is not zero and it is proportional to the wall-normal gradient of the rms wall-normal turbulent fluid velocity fluctuation. Hence it is called turbophoresis (a term coined by Reeks, 1983) in analogy with thermophoresis, where particles migrate due to temperature gradient.

In a turbulent pipe flow, $\langle u_y^2 \rangle$ is zero at the wall, has its peak in the buffer region around $y_+ \approx 20$, and then decreases to reach a minimum at the channel center. Thus, in the bottom half of the channel, $d\langle u_y^2 \rangle/dy$ is negative above $y_+ = 20$. For heavier-than-fluid fluid particles $(1 - \beta > 0)$, this means that particles will migrate away from the channel center toward the channel walls. For lighter-than-fluid particles, the effect of turbophoresis is to move the particles toward the channel center. This wallward migration of heavier-than-fluid particles in a turbulent flow is in addition to any effect of wall-normal lift force acting on the particle.

The accuracy of the equilibrium turbophoretic migration prediction given in Eq. (12.38) is tested in Figure 12.11. The figure shows $\langle u_y - v_y \rangle$ averaged over all the particles that are located close to the $y_+ = 20$ horizontal plane in the bottom half of the channel. Thus, the average shown is a Lagrangian average. Lagrangian particles evolved using both the exact velocity and the equilibrium approximation are considered. Fluid velocity averaged at the particle locations is almost zero, and the results reflect wallward motion of the heavier-than-fluid particles while the wallward migration velocity increases with Stokes number. On the other hand, bubbles tend to move toward the channel center. In both cases, the turbophoretic velocity of the exact

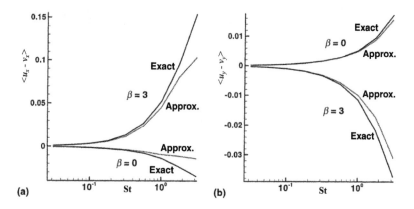

Figure 12.11 (a) Mean streamwise velocity difference between the local fluid and the particle averaged over all particles that are located close to the $y_+ = 20$ horizontal plane. (b) The corresponding mean value of the wall-normal velocity difference between the local fluid and the particles. Reprinted from Ferry and Balachandar (2001), with permission from Elsevier.

particles is reasonably captured by the equilibrium approximation over the entire range of St considered.

Also shown in Figure 12.11b is $\langle u_x - v_x \rangle$ averaged over all the particles that are located close to the $y_+ = 20$ horizontal plane. We expect the same physics as in Problem 12.6. Due to the wall-normal migration of the heavier-than-fluid particles from the channel center plane, their streamwise velocity will be larger than that of the fluid, while the streamwise velocity of bubbles will lag that of the local fluid, since the bubbles' mean wall-normal velocity is from layers of lower streamwise velocity. This expectation is well captured in the figure. Again we observe that the equilibrium approximation captures the actual behavior quite well for St $\lessgtr$ 0.5.

13 Multiphase Flow Approaches

We now have all the background information needed to explore the various computational approaches that are available for solving the wide range of multiphase flows we encounter. In fact, you may feel like you are at the cereal aisle in a grocery store wondering which one cereal among the shelf-full to pick. Fortunately, the process of picking the correct computational approach for a particular multiphase flow problem can be simplified through a rational analysis of the strengths and weaknesses of the different approaches and their suitability to the multiphase flow problem at hand. Each multiphase flow problem is unique, with its list of multiphase physics, input parameters that can be controlled, and outcome quantities of ultimate interest. Thus, there cannot be a universal criterion that can be used to decide upon the best computational approach to any given multiphase flow problem. Good understanding of the unique nature of the multiphase flow problem at hand, along with the general strengths and weaknesses of the different approaches, is essential in choosing the best computational options (yes, there can be multiple good computational options for a multiphase flow).

Though the final decision on the optimal method is problem dependent, broad guidelines can be developed as to which computational approach is best suited for what class of multiphase flow problems. Establishing such guidelines is the focus of this chapter. There are four important parameters to consider in this regard:

(i) The particle Stokes number (St $= \tau_p/\tau_c$) or the ratio of particle to the smallest fluid time scale being computed. In a fully resolved DNS, the smallest fluid time scale τ_c is the Kolmogorov time scale, while in a LES, τ_c will be the time scale of the smallest resolved eddy, which will be larger than the Kolmogorov scale.

(ii) The ratio of particle size to the smallest fluid length scale (i.e., d_p/l_c). Again, in DNS, the smallest fluid length scale is the Kolmogorov length scale $l_c = \eta$, while in LES, l_c is the length scale of the smallest resolved eddy, which will be larger than the Kolmogorov scale. In fact, as will be discussed below, the length scale ratio is related to the time scale ratio, through the particle-to-fluid density ratio.

(iii) The particle volume fraction ϕ_d. Here it should be emphasized that the local particle volume fraction will vary from the domain average even in a well-mixed distribution of particles. But the variation in particle volume fraction away from the domain average can be substantially larger in many multiphase flow problems, due to preferential accumulation and other multiphase flow mechanisms. Thus, decisions cannot be made only based on the domain-average volume fraction of particles in a system. One must also pay attention to regions of larger volume fraction.

(iv) The final parameter concerns the mass of particles compared to the mass of fluid within a unit volume. We define the following ratio:

$$\psi = \frac{\phi_d(\rho_d - \rho_c)}{\rho_c} = \phi_d(\rho - 1), \tag{13.1}$$

where $\rho = \rho_d/\rho_c$ is the particle-to-fluid density ratio. In the above definition, the numerator can be interpreted as the excess mass due to the dispersed phase per unit volume and the denominator is the mass if the unit volume were to be occupied entirely by the continuous phase. Thus, ψ parameterizes the inertia (or mass) of the suspended dispersed phase relative to the suspending continuous phase. There are other ways to define this dispersed-to-continuous-phase mass ratio and they are equivalent. As the particle volume fraction varies over the multiphase flow, so will this inertia parameter. Again, decisions have to be made not on the domain-average value of ψ, but based on large local values.

A regime map outlining the different computational approaches in terms of the four parameters can be drawn. The original delineation of the different multiphase flow regimes was pioneered by Elghobashi (1994), which was subsequently elaborated by others, including Balachandar (2009) and Balachandar and Eaton (2010). Figure 13.1 shows the different computational approaches and their applicability in the context of this map. On the vertical axis we have the Stokes number and length scale ratio plotted, which are related through the particle-to-fluid density ratio. These ratios play the primary role in deciding the suitability of particle-resolved versus Euler–Lagrange versus Euler–Euler methodology. On the horizontal axis the particle volume fraction and the inertia parameter ψ are plotted, which are again related by the particle-to-fluid density ratio. The general recommendation is that for mass loading parameter ψ greater than about 0.1%, two-way coupling effects can become important. Similarly, for particle volume fractions greater than about 0.1%, particle–particle interaction effects through collisions and fluid-mediated interactions can become important. These recommendations are based on domain-average values with the understanding that there can be regions of an order of magnitude larger volume fraction and mass loading within the flow. These recommendations can be relaxed or tightened as required by the specific details of the problem. In the following paragraphs we will briefly introduce and discuss the pros and cons of the different approaches. Later chapters will discuss each of them in much greater detail.

13.1 Introduction to Different Computational Approaches

13.1.1 Particle-Resolved (PR) Approach

By definition, in this approach, each particle is numerically resolved with the grid size being one to two orders of magnitude smaller than the diameter of the particle. In all the other methods, either explicitly or implicitly, the particle is assumed to be smaller

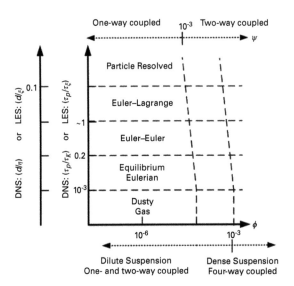

Figure 13.1 A guideline of the different computational approaches to turbulent multiphase flow. Their applicability is separated in terms of Stokes number or the length scale ratio, which are related to each other.

than the computational grid. If the particle is much larger than the grid, there is no reason not to pursue the particle-resolved approach. If the particle size is comparable or smaller than the grid, then particle-resolved simulation is not appropriate.

For a spherical particle, even with second-order spatial discretization, about 20 grid points along each of the three directions is sufficient to geometrically resolve the particle. Thus, by "particle-resolved" we only refer to adequate resolution of the shape of the particle. As the shape of the particle deviates from a sphere and becomes more complex, the resolution requirement for particle-resolved simulation will increase.

When all the scales of ambient turbulence, and the flow scales introduced by the particles (the boundary layers and the wakes), are completely resolved, then the approach qualifies as "fully resolved" direct numerical simulation. When both the particle and the flow scales are fully resolved, such a simulation can be termed PR-DNS – there are many examples of PR-DNS in the literature, from simulations of a single particle to a collection of up to $O(10^5)$ particles in turbulent flows (Pan and Banerjee, 1997; Kajishima et al., 2001; Ten Cate et al., 2004; Uhlmann, 2005, 2008; J. C. Lu and Tryggvason, 2006; Costa et al., 2018; J. Lu et al., 2018; Chouippe and Uhlmann, 2019; Mazzuoli et al., 2020). In PR-DNS, the need for adequate resolution must be based not only on the shape of the particle, but also on the thickness of the attached portion of the boundary layer, the thickness of the separated shear layers, and the range of scales that are generated in the wake, all of which can be more turbulent with increasing Reynolds number, as discussed in Section 7.1 and shown in Figure 7.4. The thickness of the attached boundary layer scales as $d_p/\sqrt{\text{Re}}$, and thus in 3D the number of grid points that are needed to resolve the volume of each particles increases as $\text{Re}^{3/2}$, severely restricting the number of particles that can be considered for the Reynolds number.

At particle Reynolds numbers less than 200, typically about 25 grid points along each of the three directions are sufficient to resolve both the particle and the flow around the particle. With increasing particle Reynolds number, a lot finer grid is needed to resolve the boundary layer and the particle-generated turbulence in the wake. If the grid can be chosen sufficiently fine, then PR-DNS can be performed even at higher Re, however, it may not be possible to consider many such particles in a simulation. An alternate option is to resolve only the particle shape with sufficient number of grid points, but not the flow. In this approach the oncoming turbulence and that generated in the particle boundary layer or wake are not fully resolved. This partially resolved approach is appropriate when the particle Reynolds number becomes so large that the flow around individual particles cannot be fully resolved. In this case one must pursue a large eddy simulation of the flow around the particles.

The advantage of the PR approach is that it is closer to the complete physics of the problem. In the case of DNS, no modeling is needed. In the case of partially resolved LES, the only closure is in terms of classical LES subgrid closure and additional multiphase flow closures are not required. However, the important constraint is that with the available computing power, the number of resolved particles that could be considered in a simulation is limited. Most applications typically involve far more particles in the flow than even a few million, and PR-DNS or LES of such systems is out of the question in the foreseeable future. The particle-resolved approach will be discussed further in Chapter 15.

13.1.2 Euler–Lagrange (EL) Approach

In this approach, the details of the particle surface and the flow in the vicinity of the particle are not resolved. Nevertheless, we retain the true Lagrangian nature of the particles by tracking their position, momentum, and energy with their equations of motion. Since the particles and the microscale flow around them are not resolved, in this approach, one must solve the filtered equations for the continuous phase that were obtained in Section 11.1.

The biggest advantage over the PR approach is that instead of the grid size being one to two orders of magnitude smaller than the particle diameter, it can now be larger than the particle diameter. This offers orders of magnitude computational saving. Thus, with the EL approach, it is possible to solve large-scale multiphase problems involving billions of particles or more. This computational advantage is, however, at the cost of accuracy. In the PR approach, the mass, momentum, and energy exchange between the continuous and the dispersed phases naturally emerge as part of the solution process. In contrast, in the EL approach, these interphase exchanges must be modeled as appropriate drag, lift, and heat transfer laws. Furthermore, as outlined in Table 11.1, there are other quantities, such as residual stresses and fluxes, that must be closed with appropriate models. Thus, the accuracy of the EL approach depends on our ability to model these effects of unresolved microscale processes.

The biggest advantage of the Lagrangian approach over the Euler–Euler approach is that there is no fundamental limitation on the particle Stokes number, since there is no requirement of uniqueness for the particle properties to be described as Eulerian fields. The crossing trajectory effect of different particles having different velocity at the same location is accommodated only in the Lagrangian approach. Furthermore, in the Lagrangian approach, the size of each particle is independent and thus polydisperse systems can be handled easily, while in the Eulerian approach, the particle size spectrum must be partitioned into a finite number of bins and the number of particles within each bin and their velocity and temperature must be treated with a set of field variables.

The coupling of Lagrangian particles with the Eulerian continuous phase poses interesting challenges. Typically, the hydrodynamic force on all the particles within a cell is added and distributed to the neighboring grid points. There needs to be a sufficient number of particles within each cell in order to have a smooth Eulerian representation of the feedback force from the particles. Since cell-to-cell variation in the number of particles goes as the inverse square root of the mean number of particles, even an average of 100 particles per cell gives rise to 10% variation. With fewer number of particles per cell, the Gaussian smoothing must be sufficiently wide for the feedback force to be smooth. The computational cost of the Euler–Lagrange approach increases with increasing number of particles.

13.1.3 Euler–Euler (EE) Approach

In this two-fluid approach, both the continuous and the dispersed phases are treated as interpenetrating fluid media (Druzhinin and Elghobashi, 1999; Fevrier et al., 2005). The particulate phase is also treated as a continuum, and particle properties such as velocity and temperature are given field representation. As in the EL approach, the surface of individual particles and the microscale flow consisting of the boundary layers and wakes around the particles are not resolved in this approach. Thus, in the EE approach one must solve the filtered governing equations for the continuous phase given in Section 11.1. In addition, since the individual particles are not being tracked, the field representation of the dispersed phase requires spatial filtering. Thus, the EE approach requires the solution of the filtered dispersed-phase equations given in Section 11.2.

The main advantage of the EE approach is that it can scale to very large systems. Since both the continuous and dispersed phases are filtered, the filter size can be so chosen that the multiphase problem can be solved at an affordable computational resolution. While there is limitation as to how many particles can be resolved in the PR approach or how many particles can be tracked in the EL approach, there are no such restrictions on the dispersed phase. The clear downside is that the accuracy of the EE approach depends on the closure models that are now needed both in the continuous and dispersed-phase filtered equations (see Tables 11.1 and 11.2).

Another important challenge arises in polydisperse multiphase flows. The entire spectrum of particle sizes cannot be treated with one set of mass, momentum, and energy equations. The wide range of particle sizes must be divided into bins of smaller size range and the evolution of particle mass, momentum, and energy within each bin must be solved separately. Thus, as the number of particle size bins increases, in order to accurately account for the entire spectrum of particle sizes, the number of partial differential equations that must be solved for the dispersed phase also increases.

In the EE approach, we use field representations of particle quantities. As discussed in Section 12.1, when the particle Stokes number is small we can assume the existence of unique values of these quantities. In other words, all particles within a tiny volume of fluid at any given point in space and time can be taken to have the same velocity, temperature, and so on. If the condition of uniqueness is violated then the particle-related quantities of an EE simulation must be considered as averages, which impacts the modeling of the closure terms in the particle-phase filtered mass, momentum, and energy equations.

13.1.4   Simplified Euler–Euler Approaches

As we discussed in Section 12.3, in the limit of small particle Stokes number (or equivalently small particle sizes), the particle velocity is nearly the same as the surrounding fluid or deviates from it by a small amount that can be accurately predicted with an algebraic expression. Similarly, in the limit of small thermal Stokes number, the particle temperature can be algebraically expressed in terms of the local fluid temperature and its derivatives. Under these conditions, the standard EE approach can be considerably simplified. Since the particle velocity and temperature fields can be explicitly obtained in terms of the fluid velocity and temperature fields, we do not need to solve the momentum and energy equations of the dispersed phase either in the Lagrangian frame or in the Eulerian frame. Only the volume fraction of the dispersed phase needs to be solved. However, as in the standard EE approach, the wide range of particle sizes must be divided into bins of smaller size range and the particle mass balance must be solved within each bin.

The most simplified EE approach is the *dusty gas approach* formulation, which was originally proposed by Saffman (1962) and Marble (1970) in the context of particle-laden gas flows. In this approach, it is assumed that the particles are so small (i.e., St $\ll$ 1) that they perfectly follow the local continuous phase. In other words, in this limit, the particle velocity is just the same as the surrounding fluid. This allows the particle-laden flow to be considered as a single fluid, whose density depends on the local mass fraction of suspended particles. The biggest advantage of this approach is its simplicity. In addition to the mass, momentum, and energy equations of the mixture, only the volume fraction equation for the particulate phase needs to be solved. This approach is, however, applicable for only very small particles that closely follow the fluid.

At the next level of EE simplification is the equilibrium EE approach, which can be thought of as an extension of the dusty gas approach, in the sense that it retains the computational simplicity and advantage of the dusty gas approach, but allows for particle velocity and temperature to be different from those of the surrounding continuous phase. In this approach, it is assumed that the particle Stokes number is sufficiently small that the particle motion is dictated only by the surrounding fluid. Particle velocity is explicitly given by the expansion (12.27) and the temperature is similarly given by Eq. (12.28).

The biggest advantage over the dusty gas approximation is that the equilibrium Eulerian approach captures the relative particle motion much more accurately and thereby enables important phenomena, such as preferential particle accumulation and turbophoresis, to be accounted for in the particle motion. A disadvantage over the dusty gas approach is that the particle velocity cannot simply be taken to be that of the local fluid, but is given by an explicit algebraic equation involving the local fluid velocity and its gradients. Since particle velocity depends on particle size, the volume fraction of different-sized particles must be computed independently. Meanwhile in the dusty gas approximation the entire particle size range can be considered together, and it is sufficient to calculate the time evolution of the total particle volume fraction.

The advantage of the EE approach over the simplified EE approaches is that the restriction on particle Stokes number (i.e., $St < 1$) can be relaxed. Thus, the standard EE approach is applicable for particles that are larger than what could be accurately considered using the equilibrium approximation. Furthermore, there are situations where the equilibrium assumption is clearly violated. For example, in cases where the particles are injected into the flow, even for particles of small Stokes number, there may be a region around the injector where the particle velocity is controlled by the injection process. Only sufficiently away from the injector will equilibrium be applicable, and the particle velocity can be accurately described in terms of the local fluid velocity. Another instance of non-equilibrium is particle–shock interaction. In this case, there exists a region of relaxation downstream of the shock where the particle velocity adjusts to the post-shock gas velocity (Rudinger, 2012). Here again, the equilibrium approximation is valid only sufficiently far away from the shock.

13.2   How to Choose the Appropriate Computational Approach

As stated at the beginning of this chapter and presented in Figure 13.1, there are four main parameters that dictate the character of the multiphase flow and the best suited computational methodology to handle it. Of the four, the volume fraction and mass loading are important in determining whether the flow can be considered one-way, two-way, or four-way coupled. This is important information based on which the governing equations can be substantially simplified. For example, one can avoid the expensive process of detecting interparticle collisions. Of the other two parameters, the particle Stokes number plays a pivotal role in deciding which computational method is best suited for a problem. While domain-average values of volume fraction and mass loading are parameters that are readily accessible from the statement of the problem, particle Stokes number is often a derived quantity whose value is not readily apparent. In this section, we want to present a simple analysis which will allow us to judiciously decide upon the expected value of the Stokes number based on which the best computational approach for a turbulent multiphase flow can be chosen. We will follow the discussion presented in Balachandar (2003, 2009) and Ling et al. (2013), and use the equilibrium approximation to obtain the relative velocity between the particle and the surrounding fluid, which will then allow us to obtain estimates of particle Reynolds number and Stokes number. Based on the Stokes number, we will then be able to decide upon the appropriateness of particle-resolved (PR), Euler–Lagrange (EL), Euler–Euler (EE), and simplified Euler–Euler approaches.

13.2.1   Estimate of Relative Velocity and Acceleration

Here we obtain a simple estimate of the relative velocity between the particle and the surrounding fluid in a turbulent multiphase flow. We start with a brief introduction to turbulence scaling. Let L and η denote the largest (integral) and the smallest (Kolmogorov) length scales of the turbulent flow. We are interested in obtaining a general expression for how a particle of diameter d_p will respond to a turbulent eddy of size l, where the size of the eddy can be over the range $\eta \leq l \leq L$ from Kolmogorov to integral scale.

Table 13.1 lists the corresponding time, velocity, and acceleration scales of the Kolmogorov, l-size, and integral eddies. These scales depend only on the local dissipation rate ϵ. By noting that the dimension of ϵ is $(\text{length})^2/(\text{time})^3$, these scales can be obtained simply by a dimensional argument. From turbulence theory we recall that the Kolmogorov length scale is given by $\eta = v_f^{3/4}/\epsilon^{1/4}$, and the integral length is dictated by the geometry of the problem. By definition, the Reynolds number of the Kolmogorov eddy is unity and the Reynolds number of the integral eddy is given by

$$\text{Re}_\eta = \frac{\eta u_\eta}{v_f} = 1 \quad \text{and} \quad \text{Re}_L = \frac{L u_L}{v_f} = \left(\frac{L}{\eta}\right)^{4/3}, \tag{13.2}$$

where Re_L is typically denoted as the macroscale Reynolds number of the turbulence flow. As the Reynolds number of the turbulent flow increases, the length scale ratio of the largest to the smallest Kolmogorov eddy increases. Also, any eddy smaller than the Kolmogorov eddy will have a Reynolds number smaller than one, and as a result gets dissipated quickly. Hence, a Kolmogorov eddy of size η is the smallest relevant length scale of the turbulent flow.

Of particular importance to the discussion that follows is the fact that as the size of the turbulent eddy increases from the Kolmogorov to the integral scale, the associated time scale of the eddy also increases, albeit at a slower rate of two-thirds power. The associated velocity scale also increases, but at an even slower rate of one-third power. However, the acceleration of Kolmogorov eddies is much stronger than that of the larger integral eddies. Therefore, in turbulent flows, the maximum fluid acceleration is dictated by the Kolmogorov eddy, while the maximum fluid velocity scale is represented by integral-scale eddies. As we consider how particles respond to the different turbulent eddies, their time and acceleration scales play an important role.

In a dispersed multiphase flow, a particle will interact with eddies of different sizes. A simple hypothesis can be made that the particle will respond to an eddy of size l only when the time scale of the particle is smaller than that of the eddy (i.e., only when $\tau_p < \tau_l$). This implies that the particle Stokes number with respect to this l-size eddy is < 1 and as a result, the relative velocity due to the l-size eddy can be obtained from the equilibrium approximation given in Eq. (12.27) as

Table 13.1 Estimates of length, time, velocity, and acceleration scales of Kolmogorov, l-size, and integral eddies in a turbulent flow. All estimates are in terms of the eddy size and the dissipation rate ϵ.

Category	Kolmogorov eddy	l-size eddy	Integral eddy
Length scale	η	l	L
Time scale	$\tau_\eta = \eta^{2/3}/\epsilon^{1/3}$	$\tau_l = l^{2/3}/\epsilon^{1/3}$	$\tau_L = L^{2/3}/\epsilon^{1/3}$
Velocity scale	$u_\eta = (\epsilon\eta)^{1/3}$	$u_l = (\epsilon l)^{1/3}$	$u_L = (\epsilon L)^{1/3}$
Acceleration scale	$a_\eta = \epsilon^{2/3}/\eta^{1/3}$	$a_l = \epsilon^{2/3}/l^{1/3}$	$a_L = \epsilon^{2/3}/L^{1/3}$

$$\mathbf{u} - \mathbf{v} \approx (1-\beta)\tau_p \frac{D\mathbf{u}}{Dt}, \tag{13.3}$$

where for now the effect of gravity (particle settling velocity) has been ignored. The above expression is an asymptotic solution of the particle equation of motion with the Stokes number as the small parameter, where the Stokes number corresponding to the l-size eddy is defined as

$$\mathrm{St}_l = \frac{\tau_p}{\tau_l}. \tag{13.4}$$

Under equilibrium particle motion, the relative velocity is dictated by a particle's inability to respond to local fluid acceleration. Using the fluid acceleration of the l-size eddy given in Table 13.1, the equilibrium approximation gives the following estimate of the relative velocity and relative acceleration for $\mathrm{St}_l < 1$:

$$\left. \begin{cases} |\mathbf{u} - \mathbf{v}|_l \approx \tau_p |1 - \beta| \dfrac{u_l}{\tau_l} \\[2mm] \left| \dfrac{D\mathbf{u}}{Dt} - \dfrac{d\mathbf{v}}{dt} \right|_l \approx 0 \end{cases} \right\} \quad \text{if } \mathrm{St}_l < 1. \tag{13.5}$$

Here, subscript l indicates that all the estimates are for l-size eddies. Thus, to leading order, the particle will accelerate as the l-eddy, but with a nonzero slip velocity. In contrast, when $\mathrm{St}_l > 1$, the particle is too sluggish to respond to the l-size eddy motion. Then the contribution to relative velocity from the l-size eddy will be dictated by eddy velocity instead of eddy acceleration. Also, since the particle's response to such an l-size eddy is sluggish, relative acceleration is dictated by l-size eddy acceleration. Therefore, relative velocity and relative acceleration can be approximated as

$$\left. \begin{cases} |\mathbf{u} - \mathbf{v}|_l \approx |1 - \beta| u_l \\[2mm] \left| \dfrac{D\mathbf{u}}{Dt} - \dfrac{d\mathbf{v}}{dt} \right|_l \approx \dfrac{u_l}{\tau_l} \end{cases} \right\} \quad \text{if } \mathrm{St}_l > 1. \tag{13.6}$$

The above estimates are appropriate only for long-time particle behavior for times much larger than τ_p. In this long-time limit, as discussed in Section 12.3, particle motion is dictated by the surrounding turbulent flow and not by the particle initial conditions.

The above estimates of relative velocity can be used to evaluate the particle Reynolds number due to the l-size eddy, which is defined as

$$\mathrm{Re}_l = \frac{|\mathbf{u} - \mathbf{v}|_l d_p}{\nu_f} . \qquad (13.7)$$

In addition, we write the particle time scale in terms of the density parameter β as $\tau_p = d_p^2/(12\beta\nu_f\Phi(\mathrm{Re}_l))$. Substituting this along with the expressions for relative velocity given in Eqs. (13.5) and (13.6), we obtain

$$\begin{cases} \mathrm{Re}_l\,\Phi(\mathrm{Re}_l) \approx \dfrac{|1-\beta|}{12\beta}\left(\dfrac{d_p}{\eta}\right)^3\left(\dfrac{\eta}{l}\right)^{1/3}, & \text{if } \mathrm{St}_l < 1, \\[3mm] \mathrm{Re}_l \approx |1-\beta|\left(\dfrac{d_p}{\eta}\right)\left(\dfrac{l}{\eta}\right)^{1/3}, & \text{if } \mathrm{St}_l > 1, \end{cases} \qquad (13.8)$$

where we have used the turbulent scales given in Table 13.1.

13.2.2   Three Regimes of Particle Response

In this section, we will use the results of the previous section to obtain estimates of maximum relative velocity and particle Reynolds number for a particle subjected to a range of turbulent eddies. Based on the value of particle time scale τ_p in relation to the range of fluid time scales from the Kolmogorov scale τ_η to the integral time scale τ_L, we can identify three different regimes of particle response (Balachandar, 2009; Ling et al., 2013).

Regime-I Particles: $\tau_p < \tau_\eta$
In this regime, $\mathrm{St}_L < \mathrm{St}_l < \mathrm{St}_\eta < 1$, where $\mathrm{St}_L = \tau_p/\tau_L$ and $\mathrm{St}_\eta = \tau_p/\tau_\eta$. Since τ_η is the smallest time scale in the fluid flow, the particle will respond to every eddy in the flow. Therefore, the relative velocity and relative acceleration corresponding to every eddy can be estimated by Eq. (13.5). Since the relative velocity is dictated by eddy acceleration, $|\mathbf{u}_f - \mathbf{u}_p|$ decreases with the eddy time scale, as shown in Figure 13.2a. Since the fluid acceleration of the Kolmogorov eddies is the largest, the maximum velocity difference in this regime is

$$|\mathbf{u} - \mathbf{v}|_{\mathrm{max,I}} \approx \tau_p|1-\beta|\frac{u_\eta}{\tau_\eta} . \qquad (13.9)$$

Similarly, the particle Reynolds number can be estimated by Eq. (13.8). The particle Reynolds number based on maximum relative velocity becomes

$$\mathrm{Re}_{\mathrm{I}}\,\Phi(\mathrm{Re}_{\mathrm{I}}) \approx \frac{|1-\beta|}{12\beta}\left(\frac{d_p}{\eta}\right)^3 . \qquad (13.10)$$

Regime-II Particles: $\tau_\eta < \tau_p < \tau_L$
In this regime, $\mathrm{St}_L < 1 < \mathrm{St}_\eta$. The particle responds to the large eddies in the flow but not to the small ones. There exists an eddy of size l^* ($\eta < l^* < L$), whose time scale matches the particle time scale (i.e., $\tau_{l^*} = \tau_p$). The particle will respond to larger

eddies with $\tau_l > \tau_{l^*}$, and the relative velocity and relative acceleration are represented by Eq. (13.5). On the other hand, the particle does not respond to smaller eddies with $\tau_l < \tau_{l^*}$, and the relative velocity and relative acceleration are given by Eq. (13.6). It can be shown from Eqs. (13.5) and (13.6) that the relative velocity increases with τ_l for $\tau_l < \tau_{l^*}$ and decreases for $\tau_l > \tau_{l^*}$ (see Figure 13.2a). The relative velocity reaches its maximum at τ_{l*}, whose value is given by

$$|\mathbf{u} - \mathbf{v}|_{\max,\mathrm{II}} \approx \tau_p |1 - \beta| \frac{u_{l^*}}{\tau_{l^*}} = |1 - \beta| u_{l^*} \,. \tag{13.11}$$

In this regime, the particle Reynolds number based on the maximum relative velocity is

$$\mathrm{Re}_{\mathrm{II}} \sqrt{\Phi(\mathrm{Re}_{\mathrm{II}})} \approx \frac{|1 - \beta|}{\sqrt{12\beta}} \left(\frac{d_p}{\eta} \right)^2 \,. \tag{13.12}$$

Regime-III Particles: $\tau_p > \tau_L$

In this regime, $1 < \mathrm{St}_L < \mathrm{St}_\eta$. The particle time scale exceeds all the fluid time scales and thus the particle does not respond to any of the turbulent eddies. The relative velocity and relative acceleration are then given by Eq. (13.6). Since the largest fluid velocity scale is dictated by the integral velocity u_L, the maximum relative velocity in this regime can be approximated as

$$|\mathbf{u} - \mathbf{v}|_{\max,\mathrm{III}} \approx |1 - \beta| u_L \,, \tag{13.13}$$

and the particle Reynolds number in this regime is

$$\mathrm{Re}_{\mathrm{III}} \approx |1 - \beta| \left(\frac{d_p}{\eta} \right) \left(\frac{L}{\eta} \right)^{1/3} \,. \tag{13.14}$$

Figure 13.2a shows a schematic of the contribution to relative velocity from the Kolmogorov to the integral scale eddies in the three different regimes. In each regime, the largest contribution to relative velocity is marked as an open circle. The corresponding schematic for relative acceleration for the three different regimes is shown in Figure 13.2b.

The domain of validity of the three regimes can now be recast entirely in terms of the density parameter β and the length scale ratio of particle diameter to Kolmogorov scale. First, at the Reynolds number $\mathrm{Re}_{\mathrm{I,II}}$ that corresponds to the boundary between Regime I and Regime II, both Eqs. (13.10) and (13.12) must be satisfied, which results in the relation

$$\Phi(\mathrm{Re}_{\mathrm{I,II}}) = \frac{1}{12\beta} \left(\frac{d_p}{\eta} \right)^2 \,. \tag{13.15}$$

Similarly, at the Reynolds number $\mathrm{Re}_{\mathrm{II,III}}$ that corresponds to the boundary between Regime II and Regime III, both Eqs. (13.12) and (13.14) must be satisfied, to obtain

$$\Phi(\mathrm{Re}_{\mathrm{II,III}}) = \frac{1}{12\beta} \left(\frac{d_p}{\eta} \right)^2 \mathrm{Re}_L^{-1/2} \,, \tag{13.16}$$

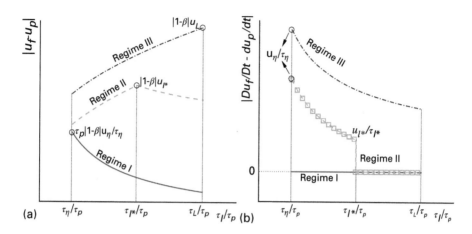

Figure 13.2 Estimates of the relative (a) velocity and (b) acceleration between fluid and particle as functions of τ_l/τ_p for the three different regimes (Regime I: $\tau_p < \tau_\eta$; Regime II: $\tau_\eta < \tau_p < \tau_L$; Regime III: $\tau_p > \tau_L$). Reprinted from Ling et al. (2013), with permission from Elsevier.

where we have used Eq. (13.2) to replace L/η with Re_L. Substituting these into the respective equations, we obtain the Reynolds number corresponding to the boundary between the different regimes as

$$\mathrm{Re}_{\mathrm{I,II}} = |1 - \beta| \frac{d_p}{\eta} \quad \text{and} \quad \mathrm{Re}_{\mathrm{II,III}} = \sqrt{|1 - \beta|} \frac{d_p}{\eta} \mathrm{Re}_L^{1/4} . \tag{13.17}$$

Using the above relations, the three different regimes can be uniquely defined in terms of ρ and d/η as

$$\text{Regime I}: \frac{1}{12\beta} \left(\frac{d}{\eta}\right)^2 < \Phi\left(\mathrm{Re}_{\mathrm{I,II}}\right), \tag{13.18}$$

$$\text{Regime II}: \Phi\left(\mathrm{Re}_{\mathrm{I,II}}\right) < \frac{1}{12\beta} \left(\frac{d}{\eta}\right)^2 < \Phi\left(\mathrm{Re}_{\mathrm{II,III}}\right) \mathrm{Re}_L^{1/2}, \tag{13.19}$$

$$\text{Regime III}: \Phi\left(\mathrm{Re}_{\mathrm{II,III}}\right) \mathrm{Re}_L^{1/2} < \frac{1}{12\beta} \left(\frac{d}{\eta}\right)^2 . \tag{13.20}$$

Problem 13.1 Apply the turbulence scalings presented in Table 13.1 and obtain the results presented in Eqs. (13.15) to (13.20).

Figure 13.3 presents the particle Reynolds number plotted as a function of d_p/η for different values of β or density ratio and Reynolds number of the turbulent multiphase flow. Under equilibrium assumption, neutrally buoyant particles ($\beta = 0$) move with the fluid and as a result their relative velocity and Reynolds number are zero. From Eq. (13.10), it can be seen that in Regime I the Reynolds number of $\beta = 3/5$ particles and $\beta = 3$ bubbles is the same. In Regime II, Eq. (13.12) shows that the Reynolds number of $\beta = 1/3$ particles is the same as that of bubbles. In Regime III, according to Eq.

(13.14), the Reynolds number of bubbles will be the largest and larger than even the heaviest particles. These behaviors can be verified in Figure 13.3. In Regime I, the Reynolds number of $\beta = 0.5$ particles is slightly higher than that of bubbles, while in Regime II, the trend reverses. In both these regimes, with further increase in density ratio, the particle Reynolds number increases substantially.

If we restrict to $d_p/\eta < 0.1$, then the particle Reynolds number remains quite small, even for very heavy particles. In case of large density ratio, as d_p approaches the Kolmogorov scale, the particle Reynolds number increases above 1. In the case of bubbles, only as the bubble diameter increases substantially above the Kolmogorov scale will Re increase above 1. Bubbles and $\beta = 0.5$ particles, even as large as $d_p \approx 100\eta$, are still in Regime II. Over the range of d_p/η shown in the figure, only the heavy particles have entered Regime III. In Regimes I and II, the particle Reynolds number is independent of turbulence intensity (or L/η). Only in Regime III does Re depend on L/η, and as can be expected with increasing turbulence intensity, the particle Reynolds number increases.

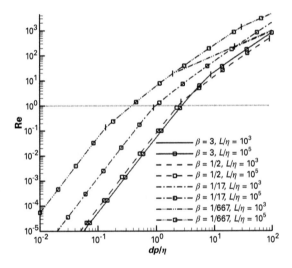

Figure 13.3 Plot of particle Reynolds number vs. nondimensional particle size, for varying density parameter β and turbulence intensity given in terms of ratio of integral to Kolmogorov length scale. Reprinted from Balachandar (2009), with permission from Elsevier.

13.2.3   Criterion for Mesoscale DNS

We are now ready to discuss the criterion which can be used to decide upon the best computational option for a turbulent multiphase flow. In order to do so, we will distinguish five different possible scenarios as depicted in Figure 13.4. These figures show the turbulent energy spectra of continuous-phase velocity as a function of wavenumber and are similar to those discussed in Figure 2.1 early in the book. In the case depicted in frame (a), the particles are an order of magnitude or more smaller than the Kolmogorov scale of continuous-phase turbulence at the mesoscale. In this case, there is scale separation between the primary turbulence and pseudo turbulence, as indicated by the two separate spectra. In contrast, in frame (b), there is no scale separation and the particles are of size larger than the Kolmogorov scale. In the latter case, the only fundamental computational approach that will not require any modeling is to fully resolve all the scales of fluid motion; this approach will be termed *particle-resolved DNS* (PR-DNS).

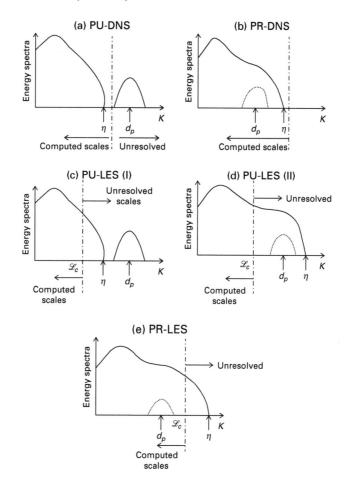

Figure 13.4 Schematic of five different scenarios of multiphase energy spectra. DNS is defined as the computational approach whose cutoff length scale is the same as the Kolmogorov scale (i.e., $\mathcal{L}_c = \eta$) whereas when the cutoff length scale is smaller than the Kolmogorov length scale (i.e., $\mathcal{L}_c < \eta$), only a portion of the turbulence length scales are resolved and give rise to large eddy simulation (LES). Typically, the cutoff length scale is a few times larger than the grid resolution. Depending on the particle diameter compared to the cutoff length scale (or the grid resolution), we have the particle-resolved (PR) or particle-unresolved (PU) approach.

PR-DNS can be pursued even in the case depicted in Figure 13.4a. But a far more efficient approach is not to resolve the particles and the pseudo turbulence by limiting the grid resolution to only mesoscale turbulence. The particles can be treated as points and an Euler–Lagrange approach can be reliably pursued. In this limit of particles being much smaller than the flow scales, even without resolving the microscale flow

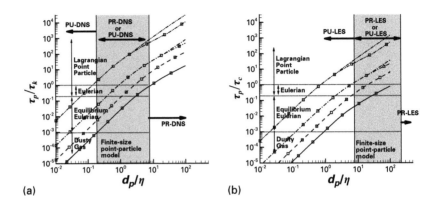

Figure 13.5 (a) Plot of particle Stokes number St_η vs. nondimensional particle size, for varying β and turbulence intensity given in terms of ratio of integral to Kolmogorov length scale. (b) Plot of particle Stokes number St_c vs. nondimensional particle size, for varying β and turbulence intensity given in terms of ratio of integral to Kolmogorov length scale. The results are for LES where the cutoff length scale is 100 times the Kolmogorov scale. The legends for the different lines and symbols are the same as in Figure 13.3. Reprinted from Balachandar (2009), with permission from Elsevier.

around the particles, the mass, momentum, and energy coupling between the phases can be very accurately modeled using the results of Chapters 4 and 5. Thus, there will not be any serious loss of accuracy in pursuing the cheaper EL approach. The scenario depicted in Figure 13.4a will be denoted as *particle-unresolved DNS* (PU-DNS).

Note the nomenclature being employed here. "Particle-resolved" simply means that the grid is more than an order of magnitude smaller than the particle, so that the shape of the particle is well represented by the grid. The term "DNS" is used when all the scales of mesoscale turbulence are fully resolved, which does not include any small-scale flow feature or pseudo turbulence generated by the particle. This terminology will be used consistently in the rest of the book and has generally been used in a growing body of multiphase flow literature. Thus, even though all the scales of fluid motion are not resolved in Figure 13.4a, we have called it DNS, but it is fully resolved only in the sense of mesoscale turbulence. There are three other LES scenarios where only a range of turbulent scales are resolved in the simulation and these will be addressed in the next section. Only the DNS cases will be addressed in this section.

From the definition of particle time scale, we obtain the following expression for the Stokes number:

$$\mathrm{St}_\eta = \frac{\tau_p}{\tau_k} = \frac{1}{12\beta} \frac{1}{\phi(\mathrm{Re})} \left(\frac{d_p}{\eta} \right)^2 . \qquad (13.21)$$

The particle Reynolds number dependence presented in Figure 13.3 can now be used to express St_η in terms of the length-scale ratio d_p/η and the density parameter β. In Figure 13.5, St_η is plotted against d_p/η on log–log scale for different combinations of β and L/η. Problems where the particles are smaller than the Kolmogorov scale of mesoscale turbulence are shown on the left-hand side of Figure 13.5, where

$d_p/\eta \lesssim 0.1$. For these problems it is sufficient to perform particle-unresolved (PU) simulations, which only resolve mesoscale turbulence. The precise choice of PU-DNS computational methodology is primarily based on the Stokes number St_η. The EL approach is appropriate when particle Stokes number $\mathrm{St}_\eta > 1$. The EL approach is applicable even when $\mathrm{St}_\eta < 1$, however, in this regime the EE approach and its simplified versions may offer a computational advantage. Based on detailed examination of the accuracy of the equilibrium approximation discussed in Section 12.3, it may be computationally very efficient to use this approach in cases where $\mathrm{St}_\eta \lesssim 0.2$. In the limit $\mathrm{St}_\eta \ll 1$, one may even simply use the dusty gas approximation. For cases where $0.2 < \mathrm{St}_\eta \sim O(1)$, both the EE and EL approaches are viable options. For larger particles of $\mathrm{St}_\eta > 1$, since uniqueness of particle velocity cannot be guaranteed, the Lagrangian approach may offer an advantage by accounting for the crossing trajectory effect. We note that even for the case of heavy particles in gas ($\beta \ll 1$), the time-scale ratio is generally small and the EL approach is barely the method of choice. However, if we relax the particle size restriction and consider $d_p \sim O(\eta)$, then for heavier-than-fluid particles the time-scale ratio can increase above unity and the EL approach becomes more appropriate. From Figure 13.3 it can be seen that for small values of d_p/η, the particle Reynolds number is generally small. Accordingly, $\Phi(\mathrm{Re}) \sim 1$ and as a result, the time-scale ratio simply shows a quadratic dependence on d/η.

On the other hand, DNS of problems where particles are substantially bigger than the Kolmogorov scale of mesoscale turbulence (i.e., for $d_p/\eta \gtrsim 10$) require a grid that is much finer than the particle and thus fall within PR-DNS. These problems fall on the right-hand side of Figure 13.5. The alternate LES approach where both the particles and turbulence are unresolved will be discussed in the next section.

In the intermediate regime, where the particle size is of the order of the Kolmogorov scale (i.e., $0.1 \lesssim d_p/\eta \lesssim 10$), the choice between PR-DNS and PU-DNS is not clear cut. A lot depends on the balance between what the user needs in terms of resolution based on multiphase-flow physics and what the user can afford in terms of numerical resolution. If it can be afforded, then the grid can be chosen to be smaller than the Kolmogorov scale and an order of magnitude finer than the particle diameter to pursue PR-DNS. As mentioned before, in a system containing many particles, this may not be possible. In which case, the only viable DNS option is PU-DNS.[1]

An important distinction between PU-DNS in the limit $d_p/\eta \lesssim 0.1$ and $d_p/\eta \gtrsim 0.1$ must be made. In both cases, since the particles are not resolved, the coupling between the phases will be based on point-particle models. The point-particle models are on sound theoretical footing in the limit $d_p \ll \eta$. In this limit, the particle Reynolds number is generally small, as indicated in Figure 13.5, and due to the smallness of the particles, the ambient flow seen by the particle can be taken to be slowly varying. This allows rigorous and accurate accounting of the interphase mass, momentum, and energy exchanges through reliable drag, lift, heat, and mass transfer correlations. In recent years, the application of point-particle models has been pushed beyond its original boundaries to address multiphase flow problems where $d_p \sim O(\eta)$, which will

---

[1] Note that in this subsection we decided to focus on DNS. In fact, in the intermediate regime LES may be a better choice and will be discussed in the next subsection.

be the case for PU-DNS in the intermediate shaded region of Figure 13.5. As will be seen in Chapter 15, modeling of interphase mass, momentum, and energy coupling becomes complicated as the particle size increases and approaches the Kolmogorov scale. Unavoidable empiricism needs to be introduced in the point-particle closure models. Therefore, such PU-DNS simulations should be distinguished, since they employ "finite-size point-particle models."

13.2.4   Criterion for Mesoscale LES

We now investigate the computational options that are appropriate when we cannot resolve the entire range of turbulent scales. We define a cutoff length scale $\mathcal{L}_c$ which is larger than the Kolmogorov length scale, and the spectrum of scales is separated into resolved scales and unresolved scales as shown in the last three frames of Figure 13.4. Based on the relative values of d_p, η, and $\mathcal{L}_c$, we can envision three different subcases of LES, which are depicted in these frames. In frame (c) we have the scenario where the particles are much smaller than the Kolmogorov scale and the cutoff length is larger than the Kolmogorov scale. The resulting *particle-unresolved LES* (PU-LES(I)) neither resolves the particles and the associated pseudo turbulence, nor a range of small scales associated with the mesoscale turbulence.

Frame (d) is similar to frame (c), except that the particles are larger than the Kolmogorov scale, but smaller than the cutoff length. The resulting PU-LES(II) does not resolve the pseudo turbulence as well as a range of small scales associated with the mesoscale turbulence. However, there is an important difference between the scenarios depicted in PU-LES(I) and PU-LES(II). In the former, the particles are smaller than the Kolmogorov scale and as a result, the particle Reynolds number is small and the microscale flow around the particles is in the Stokes regime. Whereas, in PU-LES(II), the particle Reynolds number can be larger than one and the direct interaction of particles with the ambient turbulence (as depicted in Figure 13.4d) substantially complicates the modeling of mass, momentum, and energy coupling between the phases.

Finally, frame (e) shows a scenario where the particles are of size larger than the cutoff length, which in turn is larger than the Kolmogorov scale. In this case, the surface of the particle is resolved to quality as a particle-resolved approach. Therefore, the contribution to mass, momentum, and energy exchange between the particles and the resolved scales of the surrounding flow need not be modeled, as they can be obtained as part of the computed solution. However, the unresolved scales also interact with the particles, making additional contributions to interphase mass, momentum, and energy exchange, which must be modeled. This exchange is typically accounted for as part of the wall-model closures in the context of LES. Thus, *particle-resolved LES* is an appropriate computational approach in case of a few large particles in a turbulent flow.

We now investigate the appropriateness of EL and EE approaches in the context of PU-LES, as depicted in frames (c) and (d). The arguments presented above in the context of PU-DNS can be extended to the case of PU-LES as well. Only eddies of size larger than the cutoff are computed; scales below the cutoff are considered subgrid and are modeled. Thus, the smallest resolved length scale ($\mathcal{L}_c$) now plays a role similar to the role the Kolmogorov scale (η) played in the PU-DNS analysis. We can identify Regime I$'$ where the particle time scale is less than the time scale of the smallest resolved eddy ($\tau_c = \mathcal{L}_c^{2/3}/\epsilon^{1/3}$). In this regime, the relative velocity is dictated by the smallest resolved eddy and is given by

$$\text{LES Regime: I}' \ (\tau_p < \tau_c): \quad |\mathbf{u} - \mathbf{v}|_{\max,\text{I}'} \approx \tau_p |1 - \beta| \frac{u_\eta}{\tau_\eta} \left(\frac{\eta}{\mathcal{L}_c}\right)^{1/3} . \tag{13.22}$$

This regime, which is relevant to LES, covers all of DNS Regime I and part of Regime II. In this LES Regime I$'$ the relative velocity induced by the resolved eddies is smaller than what can be expected with the presence of the subgrid scales. For particles whose time scale is larger than τ_c, the dominant eddy which controls the relative velocity is being computed as part of LES and as a result the relative velocity scaling remains the same as those for DNS given in Eqs. (13.11) and (13.13).

In LES, in computing the Lagrangian motion of particles, in addition to the resolved scale flow velocity seen by the particle, a stochastic component that accounts for the effect of the unresolved subgrid scales must also be included (Minier et al., 2004; Shotorban and Mashayek, 2006; Berrouk et al., 2007). If appropriately accounted for, the effect of the subgrid should bring the relative velocity scaling back to the DNS level and as a result, the estimates of relative velocity given in Eqs. (13.10), (13.12), and (13.14) are applicable even in LES. The important difference is that the stringent DNS requirement that $d_p \ll \eta$ is relaxed and replaced by $d_p \ll \mathcal{L}_c$. The above discussion clarifies the role of the stochastic contribution to particle motion. For small particles of time scale less than τ_c, the stochastic contribution will dominate and dictate the magnitude of the relative velocity. For larger particles, the stochastic contribution to the relative velocity will be sub-dominant to the deterministic contribution arising from the resolved-scale eddies. Nevertheless, for all particles the stochastic contribution is important in order to properly account for the role of subgrid scales in particle dispersion.

The EL approach becomes the method of choice provided the particle time scale is larger than the time scale of the smallest resolved eddy (i.e., $\tau_p > \tau_c$). The time-scale ratio can now be expressed as

$$\text{St}_c = \frac{\tau_p}{\tau_c} = \frac{1}{12\beta} \frac{1}{\phi(\text{Re})} \left(\frac{d}{\mathcal{L}_c}\right)^2 \left(\frac{\mathcal{L}_c}{\eta}\right)^{4/3} . \tag{13.23}$$

In Figure 13.5, τ_p/τ_c is plotted against d_p/η on log–log scale for different combinations of β and L/η, for the particular case where the LES filter is 100 times the Kolmogorov scale (i.e., for $\mathcal{L}_c/\eta = 100$). Thus, even if we limit to particle size an order of magnitude smaller than the smallest resolved eddy (i.e., $d_p/\mathcal{L}_c \lesssim 0.1$), provided the Reynolds number of carrier phase turbulence is sufficiently large and accordingly the cutoff length scale is much larger than the Kolmogorov scale ($\mathcal{L}_c/\eta \gg 1$), the time-scale ratio can be greater than 1 for a wide range of density ratios to make the EL approach very attractive.

In a given problem, particle size and β can be used with Figure 13.3 to obtain the corresponding particle Reynolds number, which can then be used to obtain a more precise estimate of the time-scale ratio. Again, the EL PU-LES approach can be pushed to consider larger particles of size comparable to the filter size (i.e., $d_p \sim \mathcal{L}_c$). In this case, even for bubbly flows ($\beta \rightarrow 3$), the EL approach becomes an attractive option. Note that in Figure 13.5, even for the $\beta = 3$ bubble case, the finite-Re correction applied corresponds to that of a rigid sphere. Instead, $\Phi(\text{Re})$ proposed by Mei (1994) for bubbles can be used, but the resulting change is minor.

14 Particle-Resolved Simulations

In this chapter we will discuss some of the numerical methodologies that are appropriate for particle-resolved simulations of multiphase flows. Our focus will be on PR-DNS, where all the flow scales of fluid motion are resolved along with the surface of the particles. PR-DNS simulations, however, come at a computational cost. The range of multiphase flow problems that can be simulated in a particle-resolved manner is limited. This limitation does not arise from the mathematical formulation. As discussed in Section 2.4, the mathematical formulation of PR-DNS is the easiest among all approaches to dispersed multiphase flows. All we need is to solve the Navier–Stokes equations in the continuous phase along with the rigid-body motion of the particle, with proper matching of velocity and stresses between the two phases at the interface. The complexity comes from the need to resolve all the scales in a fully resolved simulation. For example, with current computational capability, we can perform PR-DNS involving only a limited number of particles interacting with the continuous phase. Here too the particle Reynolds number must be limited since with increasing Re, small-scale flow features are generated whose resolution causes increasing demand. Full systems containing billions of particles will need to resort to other computational approaches that in one way or another rely upon averaging and thus on closure models. We shall consider such techniques in the coming chapters. The PR-DNS approach to be discussed in this chapter is thus best suited for the investigation of the dispersed multiphase flow at the microscale, which involves only a limited number of particles within the computational domain.

We will focus on numerical approaches that are designed to solve the governing equations within the different phases, the interaction between the phases at the interfaces, and the interaction of phases with the boundaries of the system, exactly from first principles. These approaches directly solve the fundamental equations of motion and accurately account for all the length and time scales of the problem, and thus avoid the need for any small-scale homogenization through spatial, temporal, or ensemble averaging. In particular, the interaction between the different phases is computed directly and any modeling of interfacial processes is thus avoided.

The emphasis will be on accurately simulating the details of the flow at the level of a limited number of interacting particles. We envision solving the following kinds of microscale problems with this approach: influence of turbulence on the settling velocity (or rise velocity) of an interacting swarm of particles (or bubbles); effect of particles or microbubbles on near-wall turbulent vortical structures; resuspension of sediment particles from a bed due to near-wall turbulence; close-range interaction and cluster formation of particles, and so on. If we extend the PR-DNS approach to consider Navier–Stokes equations within the dispersed phase (instead of rigid-body dynamics), we can also consider the dynamics of deformation of viscous droplets in straining flows, droplet motion through constricted tubes, to mention just a few. Details of the interface and the interaction processes will be fully resolved. Key information, such as force and torque on particles, or the deformation modes of droplets and bubbles, can be obtained from these simulations with a very high degree of confidence. Such microscale information abstracted from these PR-DNS can then be used in closure models that are invariably needed for investigating larger systems at the meso/macroscale.

There has been a rapidly growing body of literature on particle-resolved simulations over the past two decades. These advancements are both in terms of development of efficient numerical methodologies for computing the flow around an ever-growing number of particles/droplets/bubbles and in terms of fundamental microscale physics that enables the development of improved closure models from the particle-resolved simulations. In fact, many of the results presented in Chapters 5, 6, and 7 are obtained from PR-DNS of particles interacting among themselves, with ambient turbulence or with a nearby wall. It must be stressed that this chapter cannot do justice to all these developments. The purpose is only to introduce the concept of particle-resolved simulations with a few examples and motivate the reader to pursue the large body of literature.

14.1   Particle-Resolved Methods

There are two distinct possibilities for handling the geometric boundaries between the phases in a dispersed multiphase flow. The first approach is to define the phase boundaries and use a *body-fitted approach* in one or both phases, as necessary. Typically one starts with a surface mesh that defines the interface between the phases and then extends the surface mesh on either side to define curvilinear meshes into both the dispersed and the continuous phases. The curvilinear grid that conforms to the phase boundaries, greatly simplifies specification of interaction processes that occur across the interface. Furthermore, numerical methods on curvilinear grids are well developed and a desired level of accuracy can be maintained both within the different phases and along their interfaces. The main difficulty with this approach is grid generation. For example, for the case of flow in a pipe with several hundred embedded particles, obtaining an appropriate time-evolving body-fitted grid around all the particles is a nontrivial task. There are a few unstructured and structured grid-generation packages that are readily available in the market. Nevertheless, the computational cost of grid generation can increase rapidly as geometric complexity increases. This can be severely restrictive, particularly in cases where the phase boundaries are in time-dependent motion. The cost of grid generation can then overwhelm the cost of computing the flow.

The alternative to the body-fitted approach is the *body-free approach*, where the governing equations are solved on a grid that has been chosen *somewhat* independent of the boundaries between the phases. Here we distinguish two different approaches. The first uses a fixed grid that does not change over time and the boundaries between the phases are resolved within this fixed grid. The second approach is a dynamic grid that is typically chosen to be fine near the boundaries between the phases. This can be achieved using techniques such as adaptive mesh refinement (AMR). In both these approaches the mesh need not know where the interface between the phases is precisely, and as a result the mesh need not change every time step as the interface moves. The biggest advantage of the fixed grid is that it need not change as the particles move around within the computational domain. The fixed grid can be chosen based on the desired external and internal boundaries of the multiphase flow. The main point is that the grid is not decided based on the interface between the phases. In many applications, the fixed grid can be chosen to be a regular Cartesian mesh. But other structured and unstructured mesh arrangements may become more appropriate based on the external and internal boundaries of the multiphase flow. The interfaces between the continuous and dispersed phases are treated as embedded boundaries within the fixed grid. The two key issues to consider are: (1) how the interface is defined and advected on the fixed grid as the dispersed phase moves within the computational domain; and (2) how the interfacial interaction is accounted for in the motion of the different phases.

There are two different approaches for defining the interface: *front-tracking* and *front-capturing*.

In the front-tracking method, the interface is defined by a number of Lagrangian *marker points*. The interface dynamics is then captured by the motion of the Lagrangian marker points. In the front-capturing method, a global marker function is defined, whose particular value defines one or more disconnected surfaces in 3D, which are the desired interface between the phases. Thus, in the front-capturing method the interface is represented and followed in an Eulerian framework, while in the front-tracking method the interface is represented and tracked in a Lagrangian framework. Level-set methods (Sethian, 1999; Sethian and Smereka, 2003; Osher and Fedkiw, 2006), the volume-of-fluids (VOF) (Hirt and Nichols, 1981; Scardovelli and Zaleski, 1999), and the phase-field method (Jacqmin, 1999) are some of the popular variants of the front-capturing approach. Each of these methods have their pluses and minuses and they have also been used in combination to take advantage of their strengths.

In this chapter, we will primarily focus on the particle-resolved approach to dispersed multiphase flow containing rigid spherical particles. This considerably simplifies the problem of defining the interface between the phases. Information on the location of the particle centers along with their radii is sufficient to accurately reconstruct the interface between the phases. The problem of interface definition becomes complex in the case of droplets and bubbles due to their time-dependent deformation and processes such as breakup and agglomeration. By restricting attention to particle-laden multiphase flows, in this chapter we will not discuss the level-set, VOF, or phase-field approaches. The reader is referred to the many excellent works available in the literature.

We now come to the second issue pertaining to the interface, namely, how the velocity and stresses of the two phases are matched across the interface. Here again there are a few options and the most popular are the *sharp interface* and *immersed boundary methods*. In the sharp interface method, the sharpness of the interface between the phases is recognized both in the advection of the interface through the fixed grid and also in accounting for its influence on the bulk motion (Leveque and Li, 1994; Pember et al., 1995; Almgren et al., 1997; Udaykumar et al., 1997, 2001; Ye et al., 1999; Calhoun and LeVeque, 2000; Marella et al., 2005). In the computational cells cut through by the interface, the governing equations are solved by carefully computing the various local fluxes (or gradients). The above references have shown that by carefully modifying the computational stencil at the interface cells, a formal second-order accuracy can be maintained. In the immersed boundary method, the interface is defined by a large number of Lagrangian marker points. The influence of the interface on the adjacent phases is enforced as forcing, with the help of source terms in the governing equations, and in this process the interface is typically diffused over a few mesh cells. Different variants of feedback and direct forcing have been proposed for the implementation of the interface effect (Peskin, 1977; Goldstein et al., 1993; Saiki and Biringen, 1996; Fadlun et al., 2000; Kim et al., 2001; Mittal and Iaccarino, 2005). Thus, the jump conditions across the interface are enforced only in an integral sense. The advantage of the immersed boundary technique over the sharp interface approach is that the former is simpler to implement in a code than the latter. Furthermore, as the interface moves through a fixed Cartesian grid, the sharp interface approach requires recomputation of the discrete operators close to the interface. Accomplishing this in three dimensions in an efficient manner can be a challenge.

As an example of particle-resolved simulation methodology, in this chapter we will mainly discuss the immersed boundary methodology implemented with the direct forcing approach (Goldstein et al., 1993; Fadlun et al., 2000; Uhlmann, 2005). In both cases, this chapter only offers a glimpse. There are other particle-resolved simulation methodologies, such as lattice Boltzmann methods, discrete Lagrangian multiplier method, ghost fluid methods, etc. The reader should consult a recent, more advanced book that specializes in computational approaches to dispersed multiphase flows (Subramaniam and Balachandar, 2022). The reader can also consult Aref and Balachandar (2017) for further details on sharp interface methods and spectral methods for the high-resolution simulations of particle turbulence interaction.

14.2   Direct Forcing Immersed Boundary Methodology

In this section, the mathematical description of the fluid–particle system in the direct-forcing immersed boundary method (DF-IBM) framework is described. IBM was originally proposed by Peskin and his group at the Courant Institute (Peskin, 1977; McQueen and Peskin, 1989) but its application to multiphase flows was pioneered by Uhlmann (2005, 2008). The idea is to employ a regular fixed grid in which the governing Navier–Stokes equations are solved, but apply additional, appropriately distributed momentum and mass sources within the domain in order to satisfy the requisite interfacial conditions at the internal boundaries between the phases. Here we will briefly explain this methodology in the case of a fluid flow around a complex distribution of freely moving rigid solid particles. Of course, the immersed boundary method can be used for nonspherical and deforming particles, droplets, or bubbles. In fact, the original development by Peskin was for flow around a deforming heart valve.

The following incompressible Navier–Stokes equations are solved for the fluid in the Eulerian reference frame:

$$\nabla \cdot \mathbf{u} = 0, \tag{14.1}$$

$$\frac{\partial \mathbf{u}}{\partial t} = \underbrace{-(\mathbf{u} \cdot \nabla)\mathbf{u} = -\frac{\nabla p}{\rho_f} + \nu_f \nabla^2 \mathbf{u}}_{\mathbf{f}} + \mathbf{f}_{\mathrm{IB}}, \tag{14.2}$$

where $\mathbf{u}$ and p are the velocity and pressure fields, ρ_f and ν_f are the density and kinematic viscosity of the fluid. In the absence of an immersed boundary, the fluid velocity is advanced in time due to the combined action of advection, pressure, and diffusion, given by the term $\mathbf{f}$. The purpose of the additional immersed boundary forcing term $\mathbf{f}_{\mathrm{IB}}$ is to enforce the proper boundary condition at the boundary of the particles.

The first step of the DF-IBM methodology is to distribute N_l Lagrangian markers uniformly on the surface of the particle. You may be surprised to know that distributing N Lagrangian markers uniformly on the surface of a sphere has no analytic closed-form solution. One can use the algorithm of Saff and Kuijlaars (1997). The Lagrangian markers uniformly distributed using this algorithm are shown in Figure 14.1a. The interphase coupling between the fluid and the particle is realized by discrete volume forcing applied at the Lagrangian markers. It will be discussed below that for optimal performance of DF-IBM, the spacing between the Lagrangian markers must equal the smallest local grid size. Figure 14.1b shows a non-uniform distribution of Lagrangian markers, where the marker spacing increases from the bottom of the particle to the top. Such a Lagrangian marker distribution is needed in case of DF-IBM of a particle immersed in near-wall turbulence, with the particle extending from the wall layer into the buffer and log region. Since the background Eulerian grid resolution varies in the wall-normal direction, the corresponding Lagrangian marker spacing was also increased correspondingly (Akiki and Balachandar, 2016). The particle–fluid coupling force $\mathbf{f}_{\mathrm{IB}}$ in Eq. (14.2) is first calculated at the Lagrangian markers to ensure no-slip and no-penetration boundary conditions are satisfied on the solid surface. In the direct forcing method (Mohd-Yosof, 1997), the volume force acting at a Lagrangian marker is evaluated as

$$\mathbf{F}_{\mathrm{IB}}(X_l) = \frac{\mathbf{u}^d(\mathbf{X}_l) - \widetilde{\mathbf{u}}(\mathbf{X}_l)}{\Delta t}, \tag{14.3}$$

where $\mathbf{u}^d$ denotes the desired velocity of a Lagrangian marker prescribed by the rigid-body motion of the particle. $\widetilde{\mathbf{u}}(\mathbf{X}_l)$ is the intermediate fluid velocity obtained in the absence of the IB force term, interpolated to the Lagrangian marker from the nearby Eulerian grid points. The desired velocity $\mathbf{u}^d$ of a Lagrangian marker positioned at $\mathbf{X}_l$ on the particle whose centroid is located at $\mathbf{X}_p$ is

$$\mathbf{u}^d(\mathbf{X}_l) = \mathbf{V}_p + \mathbf{\Omega}_p \times (\mathbf{X}_l - \mathbf{X}_p), \tag{14.4}$$

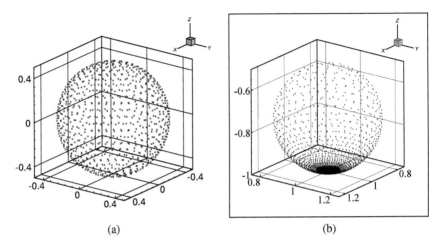

(a)   (b)

Figure 14.1 Evenly distributed Lagrangian markers on the surface of a spherical particle using the algorithm of Saff and Kuijlaars (1997). (a) The background Eulerian grid is uniform in all three directions and thus a uniform distribution of Lagrangian markers is employed. (b) The particle is close to a bottom wall and the Eulerian discretization is inhomogeneous in the vertical direction with a very fine grid near the bottom of the particle and a courser mesh near the top of the particle. A spiral distribution of Lagrangian markers is used (see Akiki and Balachandar, 2016) where the spacing between the markers is increased with increasing distance from the wall. In both cases, the Lagrangian marker spacing matches the local minimum Eulerian grid spacing. Reprinted from Akiki and Balachandar (2016), with permission from Elsevier.

where $\mathbf{V}_p$ and $\mathbf{\Omega}_p$ represent the translational and rotational velocities of the particle.

14.2.1   Interpolation and Projection

Interpolation is needed to convey data from the Eulerian grid to the Lagrangian marker, since the marker location will not coincide with the Eulerian grid points. In three dimensions, the simplest interpolation is tri-linear interpolation using the eight grid points that surround the Lagrangian marker. Such an interpolation is first order in accuracy and therefore consistent with a finite volume or a lower-order finite difference approach. In general, the spatial order of accuracy of interpolation must be consistent with the order of accuracy of the flow solver being used to solve the continuous phase. If a higher-order scheme is used to solve the governing equations (14.2), then interpolation must be carried out at the same level of accuracy. As an example, here we will follow the work presented in Yang and Balachandar (2020) and consider the governing equations being solved using a high-fidelity spectral element method (SEM) in the entire domain, including the volume occupied by the particles. In the spectral element method, the velocity is expanded using an Nth-order tensor product polynomial within each hexahedral element, while pressure is expanded using an $(N - 2)$th-order tensor product polynomial. The velocity field is collocated on the N^3 Gauss–Lobatto–Legendre (GLL) points within each spectral element (Patera, 1984; Deville et al., 2002). This spatial discretization method is known as PN/PN-2 (Maday and Patera, 1989). The governing equations are solved using a high-order, weighted residual technique that employs tensor-product polynomial bases. The resultant linear system is computed using the conjugate gradient solver with a suitable preconditioner. The time advancement of the fluid uses a backward difference formula (Aref and Balachandar, 2017). The semi-discretized equations yield a Poisson equation for pressure and a Helmholtz equation for each velocity component. The details are not the focus of this chapter and can be found in Yang and Balachandar (2020).

Barycentric Lagrange interpolation is adopted by evaluating the polynomial that collocates on the GLL grid points within an element at the location of a Lagrangian marker. The 3D barycentric Lagrange interpolation of a Lagrangian quantity $A(\mathbf{X})$ from an Eulerian quantity $a(\mathbf{x})$ is written as

$$A(\mathbf{X}) = \sum_{k=1}^{N} \sum_{j=1}^{N} \sum_{i=1}^{N} a(\mathbf{x}) \frac{w_i w_j w_k}{\bar{X}_{ijk}}, \qquad (14.5)$$

where $a(\mathbf{x})$ is an Eulerian property evaluated at the N^3 GLL grid points along each direction and $\bar{X}_{ijk} = (X - x_i)(Y - y_i)(Z - z_i)$. The barycentric weights w_i, w_j, w_k are computed as

$$w_j = \frac{1}{\prod_{k \neq j}(x_j - x_k)}, \quad j, k = 1, \dots, N. \qquad (14.6)$$

The barycentric Lagrange interpolation is computationally more efficient than the nodal Lagrange interpolation method. It requires $O(N^3)$ operations while the standard Lagrange interpolation requires $O(N^6)$ operations. It must be pointed out that the standard finite volume approach can be thought of as a subset of the above spectral element approach with $N = 2$. That is, each element is defined by the eight corners with the mass and momentum fluxes of any finite volume cell calculated at the six faces. In this case, tri-linear interpolation can be achieved with $O(2^3) = O(8)$ operations.

The immersed boundary force evaluated at the Lagrangian marker ($\mathbf{F}_{IB}$) needs to be projected back on the Eulerian grid. Here again there are different choices. The standard approach is to define a projection function g_M and carry out the projection as

$$\mathbf{f}_{IB}(\mathbf{x}_i) = \sum_{l=1}^{N_l} \mathbf{F}_{IB}(\mathbf{X}_l) \, g_M(|\mathbf{x}_i - \mathbf{X}_l|) \, W_l . \qquad (14.7)$$

Although the above summation is formally over all the N_l Lagrangian markers, the contribution to $\mathbf{f}_{IB}$ at any Eulerian grid point $\mathbf{x}_i$ comes only from the few Lagrangian markers that are nearby. In other words, for a nonzero contribution, $\mathbf{x}_i$ must lie within the compact support of the projection function g_M centered about $\mathbf{X}_l$. In the above, W_l is the volume weight associated with the lth Lagrangian marker. The value of this Lagrangian volume weight is often imprecisely defined. As we will see below, its precise value is not of primary importance.

There is considerable flexibility in the choice of the projection function. But the projection kernel must satisfy a few important physical constraints: g_M must decay monotonically with increasing distance between the Lagrangian marker $\mathbf{X}_l$ and the Eulerian grid point $\mathbf{x}_i$ and g_M must be properly normalized so that it integrates to unity. By normalizing the projection kernel, the Lagrangian weight is assigned sole responsibility of the amplitude of feedback forcing. Besides, the projection kernel typically has a compact support, which determines the number of Eulerian grid points influenced by a Lagrangian marker. Traditional IBM implementations have used regularized discrete delta functions (DDFs), which are developed for a uniform Cartesian grid (Roma et al., 1999; Peskin, 2002; Uhlmann, 2005). It is nontrivial to apply the DDFs in non-uniform and curvilinear grids (Pinelli et al., 2010; Akiki and Balachandar, 2016). Another simple approach to force projection is to use a Gaussian projection kernel. The one-dimensional Gaussian projection kernel can be expressed as

$$g_M(r) = \frac{1}{(\sigma\sqrt{2\pi})} e^{-\frac{r^2}{2\sigma^2}}, \tag{14.8}$$

where the kernel width $\delta_f = 2\sqrt{2\ln2}\sigma$. The ratio of $\delta_f/\overline{\Delta x}$ determines how many grid points the kernel spans on average, where $\overline{\Delta x}$ is the average grid size. For compact support, a cutoff distance r_c is defined beyond which the Gaussian function is taken to be zero, and the projection kernel is normalized to unity taking into account the cutoff distance; see Zwick and Balachandar (2019) for details of interpolation and projection.

14.2.2   The Motion of Particles and Their Interaction

The governing equations of the particles are the Newton–Euler equations of motion in the Lagrangian reference frame. The force and torque on a particle are composed of the fluid–particle coupling force/torque, particle–particle collisional force/torque, and external body force/torque. The fluid–particle coupling force $\mathbf{F}_{p,\text{hyd}}$ and torque $\tau_{p,\text{hyd}}$ acting on a particle can be obtained by integrating the tractional force over the particle surface. The translational and rotational momentum equations of a particle are

$$\rho_p m_p \frac{d\mathbf{V}_p}{dt} = \mathbf{F}_{p,\text{hyd}} + \mathbf{F}_{p,\text{col}} + (\rho_p - \rho_f)\mathcal{V}_p \mathbf{g}, \tag{14.9}$$

$$\mathbb{I}_p \frac{d\mathbf{\Omega}_p}{dt} = \tau_{p,\text{hyd}} + \tau_{p,\text{col}}, \tag{14.10}$$

where ρ_p is the mass density, $\mathcal{V}_p$ is the particle volume, m_p is the mass of a particle, and $\mathbb{I}_p$ is the moment of inertia of a particle. For a spherical particle of diameter d_p with uniform mass distribution, $m_p = \frac{1}{6}\pi\rho_p d_p^3$ and $\mathbb{I}_p = \frac{2}{5}\pi\rho_p d_p^5$.

The time integration of each particle position and velocity requires the evaluation of the net hydrodynamic force and torque on the particle. Typically, this will be accomplished by integrating the pressure and viscous stress distribution on the surface of the particle as given in Eq. (4.1). However, this is not the only way to proceed in the case of the immersed-boundary methodology, since the interface between the particle and the fluid has been somewhat smeared with use of the projection kernel. Following Uhlmann (2005), assuming rigid-body motion of the fluid within the immersed boundary, the hydrodynamic force and torque are obtained by integrals of the IBM force as

$$\mathbf{F}_{p,\text{hyd}} = -\int_{\Omega} \mathbf{F}_{\text{IB}} \, dV + \mathcal{V}_p \frac{d\mathbf{V}_p}{dt}, \tag{14.11}$$

$$\tau_{p,\text{hyd}} = -\int_{\Omega} (\mathbf{r} \times \mathbf{F}_{\text{IB}}) dV + \frac{\mathbb{I}_p}{\rho_p} \frac{d\mathbf{\Omega}_p}{dt}. \tag{14.12}$$

The Lagrangian quantities (particle position, angular orientation, translational velocity, and angular velocity) can be solved by integrating

$$\frac{d}{dt} \begin{pmatrix} \mathbf{X}_p \\ \boldsymbol{\theta}_p \\ \mathbf{V}_p \\ \mathbf{\Omega}_p \end{pmatrix} = \begin{pmatrix} \mathbf{V}_p \\ \mathbf{\Omega}_p \\ \frac{1}{(\rho_p - \rho_f)\mathcal{V}_p}\left(-\int_{\Omega_p} \mathbf{F}_{\text{IB}} \, dV + \mathbf{F}_{p,\text{col}} + (\rho_p - \rho_f)\mathcal{V}_p \mathbf{g} \right) \\ \frac{1}{\mathbb{I}_p}(-\int_{\Omega}(\mathbf{r} \times \mathbf{F}_{\text{IB}})dV + \tau_{p,\text{col}}) \end{pmatrix}. \tag{14.13}$$

When evolving the position of particles in response to force and torque acting on them, collisions among the particles become inevitable. In the above equations of motion, the collisional force and torque on a particle are represented by $\mathbf{F}_{p,\text{col}}$ and $\tau_{p,\text{col}}$. Typically, the soft-sphere discrete element method (DEM) model (Cundall and Strack, 1979) described in Section 9.1 is employed to deal with particle–particle and particle–wall collisions. The following variation of the soft-sphere collision model as described below was employed by Yang and Balachandar (2020). The normal collisional force on the ith particle due to its collision with the jth particle is modeled by a spring–dashpot combination as

$$\mathbf{F}_n^{ij} = -k_n|\delta|\mathbf{n}^{ij} - \eta_n \mathbf{W}_{cn}, \tag{14.14}$$

where k_n is the normal spring stiffness, $\mathbf{n}^{ij} = (\mathbf{X}_j - \mathbf{X}_i)/|\mathbf{X}_j - \mathbf{X}_i|$ the normal unit vector, η_n is the damping coefficient, and $\mathbf{W}_{cn}$ is the normal component of relative velocity. The overlapping distance δ is calculated as

$$\delta = |\mathbf{X}_i - \mathbf{X}_j| - \frac{d_{pi} + d_{pj}}{2} - \epsilon_n, \tag{14.15}$$

where ϵ_n is the threshold distance for collisional detection. The threshold distance is typically set at the order of the Eulerian grid size. The damping coefficient is computed as

$$\eta_n = \frac{1}{\sqrt{1/m_i + 1/m_j}} \times \frac{2\sqrt{k_n \ln(e_{\text{rest}})}}{\sqrt{\ln(e_{\text{rest}})^2 + \pi^2}}, \tag{14.16}$$

where e_{rest} is the dry restitution coefficient and m_i and m_j are the mass of the colliding particles.

The tangential collisional force is modeled as the contribution from two separate mechanisms. In the first, the tangential force is modeled by a tangential spring–dashpot system, which is proportional to the tangential displacement and relative velocity. The second tangential force is the Coulomb frictional force when two particles slide against each other, which is the product of the normal force exerted at the contacting point and the friction coefficient μ_c. The tangential force on the ith particle is computed as

$$\mathbf{F}_t^{ij} = -\min\left\{\mu_c|\mathbf{F}_n^{ij}|, \quad k_t\delta_t + \eta_t|\mathbf{W}_{cn}|\right\}\mathbf{n}_t^{ij}, \tag{14.17}$$

where k_t is the tangential spring stiffness, μ_c is the friction coefficient, and $\mathbf{n}_t^{ij}$ is the tangential unit vector along the direction of tangential slip velocity. For the given ith particle, the total collisional force is the sum of all binary collisions with all other overlapping particles:

$$\mathbf{F}_{p,\mathrm{col}}^i = \sum_{j=1}\mathbf{F}_n^{ij} + \mathbf{F}_t^{ij}. \tag{14.18}$$

Note that the tangential collisional force also produces a torque, thus affecting the rotational motion of a particle. The total torque on the ith particle due to all binary collisions is calculated as

$$\tau_{p,\mathrm{col}}^i = \sum_{j=1}^{m}(d_{pi}/2)\mathbf{n}^{ij} \times \mathbf{F}_t^{ij}. \tag{14.19}$$

14.3   Variations and Improvements of Direct Forcing

There have been many variations of immersed-boundary methodology other than the direct forcing technique described above. Even within the direct forcing technique there have been many advancements beyond the straightforward approach described above. All these variations strive to achieve the following goals. (i) Contain the feedback forcing to be identically zero in the entire fluid region, so that the standard Navier–Stokes equations are faithfully solved within the fluid. The forcing must therefore be localized along the interface and perhaps extended into the solid. (ii) The force distribution must be chosen appropriately such that the computed velocity field will satisfy the required boundary condition at the interface with the particles. There are no other restrictions on the shape of the interface or the boundary condition that can be satisfied along the interface.

Several different approaches to evaluating the feedback force have been advanced to satisfy the above two goals. These different approaches differ primarily in the manner in which the force field is computed and applied back on the fluid. Some of the forcing techniques in use are feedback forcing (Goldstein et al., 1993), direct forcing (Fadlun et al., 2000), and discrete mass and momentum forcing (Kim et al., 2001). The feedback-forcing approach is one of the earliest developments and it attempts to drive the fluid velocity at the particle surface to the desired velocity dictated by the particle's rigid-body motion as a damped oscillator. We desire the frequency of damping to be much larger than all relevant inverse time scales of fluid motion and the damping coefficient to be adequately large, in order for the velocity condition at the interface to be satisfied rapidly. Mohd-Yosof (1997) developed a very simple direct procedure for forcing the required velocity condition at the interface, which was later adapted and tested in Fadlun et al. (2000). In this approach, the forcing to be applied at each time is determined directly without solving an equation of damped oscillation. This method was further developed by Uhlmann (2005), as described in the previous section. One important difference is the shape of the projection kernel. The Gaussian projection kernel presented in Eq. (14.8) is simply motivated by the Gaussian filter function introduced when we derived the filtered equations in Chapter 11. But most DF-IBM methodologies do not use the Gaussian projection. A linear projection in the context of a uniform grid spacing of $\Delta x = \Delta y = \Delta z = h$ is defined as

$$g_M(r) = \begin{cases} |r|/h & \text{if} \quad |r| \le h, \\ 0 & \text{if} \quad |r| \le h. \end{cases} \tag{14.20}$$

Due to the compact support of the above linear function, each Lagrangian marker passes its force contribution to only the eight Eulerian grid points that surround it. In this sense, this projection is the counterpart of tri-linear interpolation. The above linear projection smears the interface force over one grid cell. This can be considered as an approximation to the ideal delta function forcing (Peskin, 1977) and it is first-order accurate. A formal second-order accurate approximation to the delta function is the most popular and can easily be constructed (Lai and Peskin, 2000):

$$g_M(r) = \begin{cases} \frac{1}{8}\left(3 - 2|r|/h + \sqrt{1 + 4|r|/h - 4(|r|/h)^2}\right) & \text{if} \quad |r| \le h, \\ \frac{1}{8}\left(5 - 2|r|/h - \sqrt{-7 + 12|r|/h - 4(|r|/h)^2}\right) & \text{if} \quad h \le |r| \le 2h, \\ 0 & \text{if} \quad 2h \le |r| \quad . \end{cases} \tag{14.21}$$

As discussed in Leveque and Li (1994) and Lai and Peskin (2000), in spite of the formal second-order accuracy of the projection, the use of the discrete delta function in feedback forcing results only in first-order accuracy. Furthermore, Kim et al. (2001) pointed out that the direct forcing employed by Fadlun et al. (2000) applies the forcing inside the fluid (i.e., even outside the solid region). Kim et al. proposed a modification to the direct-forcing strategy that strictly applies the momentum forcing only on the fluid–solid interface and inside the solid region. Thus, the Navier–Stokes equations are faithfully solved in the entire region occupied by the fluid. Their reformulation, however, requires that a mass source be included along the interface. In other words, the continuity equation is modified to

$$\nabla \cdot \mathbf{u} = q, \tag{14.22}$$

where q is strictly zero inside the fluid, but can take nonzero values along the interface and possibly in the solid region (where the velocity field is of no interest). Another important development is the particle-resolved uncontaminated-fluid reconcilable immersed boundary method (PUReIBM) of Subramaniam and co-workers (Tenneti et al., 2011). The principal advantage of this formulation is that the fluid stress at the particle surface is calculated directly from the flow solution (velocity and pressure fields), which – when integrated over the surface of a particle – yields the hydrodynamic force. Furthermore, the forcing is strictly applied only at points that are interior to the particle and the projection of the Lagrangian forcing is also only to neighboring grid nodes that do not reside inside the fluid phase. For more recent developments in the immersed-boundary methodology, the reader must refer to the following reviews and the references cited therein: Mittal and Iaccarino (2005); Kim and Choi (2019); Griffith and Patanker (2020).

14.3.1   Analysis of the Direct-Forcing IBM

The popularity of the direct-forcing IBM methodology is due to its simplicity and ability to obtain reasonably accurate answers. Despite its widespread usage, there are a few important questions that remain open regarding the implementation of the DF-IBM methodology that we discussed in the previous subsections. These include:

1. Is there an optimal number of Lagrangian markers to define the interface between a particle and the surrounding fluid?
2. The role of Lagrangian volume weight W_l in Eq. (14.7) is unclear. Although the common recommendation has been to set it equal to the local Eulerian cell volume (in dimensional terms), its role must be established and if there is an optimal value for the weight, it must be obtained.
3. Factors such as the translational and rotational velocity of the particle and the particle–fluid density ratio have been known to influence the stability of DF-IBM (Uhlmann, 2005). A fundamental explanation of these dependences is highly desired.
4. What are the spatial and temporal order of accuracy of DF-IBM as far as its ability to predict the net force and torque on the particle and in terms of its ability to enforce no-slip and no-penetration boundary conditions?

Towards answering these questions, Zhou and Balachandar (2021) theoretically analyzed DF-IBM regarding its accuracy, rate of convergence, and optimal distribution of Lagrangian markers and their volume weights. Without going through the detailed analysis presented in that paper, here we summarize the answers to the above questions:

1. The optimal distribution of Lagrangian markers on the surface of the particle is one in which the distance between the markers is equal to the local Eulerian grid spacing. If the spacing between the Lagrangian markers is larger than the Eulerian grid spacing, then the quality of the solution will be poor, with leakage of fluid into the particle between the markers. On the other hand, there is no advantage in increasing the number of markers and reducing the spacing between them to smaller than the Eulerian grid spacing. In fact, more than the optimal number of Lagrangian markers is detrimental since it decreases the stability limit.
2. The Lagrangian volume weight W_l serves the role of damping coefficient and thus the DF-IBM approach can yield the same answer for varying values of W_l. The larger the volume weight W_l, the faster is the convergence to the correct force coupling between the particle and the surrounding fluid. However, there is an upper stability limit, beyond which the method becomes unstable. Let

there be N_l Lagrangian markers uniformly distributed on the surface of the sphere, and let all the markers be given the same Lagrangian weight $W = W_l$ for $l = 1, \ldots, N_l$. In the limit where the particle is allowed to freely translate and rotate in response to the hydrodynamic force and torque acting on it, it has been shown that for stability (Zhou and Balachandar, 2021),

$$W \lesssim \frac{0.6\alpha}{\frac{1}{2} + \frac{N_l}{m} + \frac{2N_l}{3I_m}}, \tag{14.23}$$

where α is the stability limit of the explicit time-stepping algorithm used in DF-IBM. For example, in the case of explicit Euler time stepping, $\alpha = 1$ and in the case of the third-order Runge–Kutta scheme, $\alpha = 2.5$ (Aref and Balachandar, 2017). Here, $m = \mathcal{V}_p(\rho_p - \rho_f)/(\rho_f \mathcal{V}_{cell})$ is the nondimensional mass of the particle normalized by the mass of the fluid within the cell volume of $\mathcal{V}_{cell}$. The nondimensional moment of inertia $I_m = 4\mathbb{I}_p(\rho_p - \rho_f)/(d_p^2 \rho_p^2 \mathcal{V}_{cell})$. The additional empirical factor 0.6 has been added based on results obtained from DF-IBM simulations of a freely falling particle (Zhou and Balachandar, 2021). Below the stability limit, the choice of W_l matters in terms of the interface velocity error between the desired fluid velocity to be enforced at the Lagrangian markers and that which results in the DF-IBM calculation. The interface velocity error is proportional to $1/W$ and thus one should use the largest possible Lagrangian weight allowed by stability. The Lagrangian weight, however, does not influence the feedback forcing applied at the Lagrangian markers.

3. Equation (14.23) has been derived for a freely moving and rotating particle, with the three terms in the denominator coming from a stationary particle, translational motion, and rotational motion of the particle, respectively. In fact, the first term in the denominator is equal to $1/2$ only for an optimal distribution of more than 50 Lagrangian markers on the surface of the particle. In the limit where the Lagrangian markers are so far away from each other that they are uncorrelated, this term becomes $1/8$ (this limit is inappropriate because of the poor quality of the resulting solution). In the other limit where there are far more Lagrangian markers that are clustered very close to each other, this term becomes $N_l/8$. The second term arises from the translational motion of the particle and in the limit $m \to 0$ the stability limit decreases to zero. This happens either when the particle is neutrally buoyant (i.e., $\rho_p/\rho_f \to 0$) or when the particle becomes much smaller than the grid (i.e., $\mathcal{V}_p/\mathcal{V}_{cell} \to 0$). The factor $1/m$ implies that a large heavy particle in translational motion decreases the stability threshold only marginally compared to that of a stationary particle. The third term is from the rotational particle motion and it also contributes to the stability limit of the Lagrangian weight.

Uhlmann (2005) observed the DF-IBM simulation to become unstable when the particle-to-fluid density ratio is smaller than 1.2, for a Lagrangian volume-weight of $W = 1$. The theoretical prediction (14.23) also shows that for $W = 1$ the simulation is expected to be unstable for $\rho_p/\rho_f < 1.2$, consistent with the observations of Uhlmann. However, according to Eq. (14.23), stability can be recovered either by using a smaller volume-weight, or by increasing m with a smaller cell volume $\mathcal{V}_{cell}$. Reducing the volume-weight requires a smaller time step to maintain the precision. For both approaches, higher computational effort is needed to overcome the difficulty.

4. The DF-IBM is first-order accurate in time in terms of the fluid velocity imposed on the surface of the particle. In general, DF-IBM is only first-order accurate in space, especially close to the immersed boundary (Leveque and Li, 1994; Lai and Peskin, 2000). The error in enforcing the no-slip velocity boundary condition at the particle surface is proportional to Δt when the time step Δt is small.

Apart from answering the above important questions, the analysis of Zhou and Balachandar (2021) also provides other valuable insights into the accuracy and error behavior of DF-IBM. These can be enumerated as

5. The velocity error in a transient flow is also proportional to the flow oscillation frequency. Hence, a rapidly time-varying problem requires a very small time step. Low Reynolds number particulate flows require a small time step, which should be proportional to the Reynolds number. The usual practice of setting the time step according to the CFL number is not sufficient under such conditions. A diffusion-like time-step limit must additionally be satisfied.

6. Results of simple shear flow tests indicate that the velocity error is irrelevant to accurate prediction of the shear stress or the IBM force. This also means that the IBM force is quasi-constant, insensitive to the time step. In general, the shear force on the particle is overpredicted in a DF-IBM and the level of overprediction mainly depends on grid resolution and the shape of the projection kernel.

7. The largest permissible Lagrangian volume-weight is obtained when the Lagrangian markers are distributed farthest within the constraint that all the markers remain correlated with their neighbors. Thus, the optimal marker configuration uses the minimum number of Lagrangian markers to define the particle–fluid interface. This allows the largest time step in order to achieve the best computational efficiency. As the number of Lagrangian markers is increased above this minimum value, the volume-weight W must be decreased to stay within the stability limit and the time step needs to be reduced accordingly so as to limit the velocity error at the interface.

We now present the results of two simple examples of DF-IBM applied to flow over an isolated particle that were reported in Zhou and Balachandar (2021). In the first problem, a sphere of unit diameter is held fixed at the center of a large channel with an imposed uniform flow. The sphere was discretized with N_l evenly distributed markers. DF-IBM simulations were carried out for the following five different number of Lagrangian markers: $N_l = \{6, 24, 96, 384, 1284\}$. The channel is discretized with a uniform grid and five different grid resolutions of $\Delta x = \{0.72, 0.36, 0.18, 0.09, 0.05\}$ were considered. Different combinations of N_l and Δx have been used in the DF-

IBM simulations. The steady drag coefficient C_D obtained from DF-IBM is shown in Figure 14.2 for Re = 10 and 100. The solid line denotes the DF-IBM drag coefficient obtained with the "optimal" Lagrangian marker distribution, where Δx equals the marker separation. For Δx = 0.36, 0.18, and 0.09, three different number of markers were tested and these results for the same Δx are connected by a dashed line with symbols. The behavior of Re = 10 and 100 is quite similar. Also shown are results from a body-fitted simulation of very high resolution, which serves as the benchmark. The drag predicted by DF-IBM is always higher. With decreasing Δx and corresponding increase in N_l, the results approach the benchmark value. On the other hand, for a given Eulerian grid, varying the number of markers only has a small effect on the predicted C_D. The overall resolution of the flow field is determined by the Eulerian grid. The distribution of Lagrangian markers has no significant effect on the flow field as long as the adjacent markers are correlated. Therefore, it is better to adopt as coarse a marker distribution as possible within the constraint of correlation to save computational cost.

The second problem considered by Zhou and Balachandar (2021) is a sphere allowed to fall freely under gravity in a vertical channel with periodic side walls. In the DF-IBM implementation, the sphere surface is discretized with 528 equi-spaced Lagrangian markers. Two density ratios of 2.56 and 1.1 were considered. For ρ_p/ρ_f = 2.56, the sphere size and fluid viscosity are set to match the experimental settings in Mordant and Pinton (2000). Three different combinations of volume-weight and grid are investigated: (i) W = 1 and Δx equal to Lagrangian spacing; (ii) W = 2 and Δx equal to Lagrangian spacing; and (iii) W = 4 and $2\Delta x = \Delta y = \Delta z$ equal to Lagrangian spacing (i.e., the grid in the gravity direction is twice as dense as those in the other two directions). These combinations demonstrated that neither the weight nor the shape of the Eulerian grid cell affect the computed flow field. Figure 14.3 shows the simulated settling velocity, along with the experimental measurement, and the agreement is quite good. For ρ_p/ρ_f = 1.1, the volume-weight W is chosen well below unity to satisfy the stability requirement given in Eq. (14.23). As a consequence, the time step is chosen to be sufficiently small to make the velocity error small. Three different combinations of W and Δt are considered. All the cases are in agreement and despite the small density ratio a stable converged solution is obtained, since the stability criterion was satisfied.

14.4 Techniques Employing Body-Fitted Grids

There are a number of body-fitted approaches in terms of both grid generation, as well as solving the Navier–Stokes equations in the resulting curvilinear grids. There is a vast literature on both these fronts. Thus, our discussion in this section will be very brief and pertain strictly to simulations of particle-laden flows.

In the body-fitted grid approach, the surface of the particles must first be discretized, then extended outward into the region occupied by the fluid. The two major challenges with constructing body-fitted grids are as follows: (i) in the case of moving particles, the grid must be regenerated at each time step; and (ii) when two particles come very close to each other or when a particle approaches the boundaries of the computational

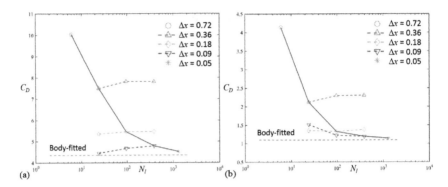

Figure 14.2 Plot of drag coefficient evaluated from the DF-IBM for varying number of Lagrangian markers. Results from different grid resolutions are shown. Simulation results of cases when the Eulerian grid resolution is the same as the Lagrangian marker spacing are connected by the solid line, which converges to the results of a highly resolved body-fitted grid. (a) Re = 10 and (b) Re = 100. Reprinted from Zhou and Balachandar (2021), with permission from Elsevier.

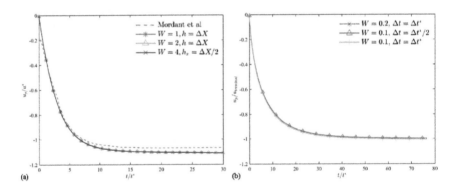

Figure 14.3 The settling velocity of a freely falling particle of density ratio (a) 2.56 and (b) 1.1. In these plots, the normalizing time $t^* = \sqrt{d_p/g}$, normalizing velocity $u^* = \sqrt{gd_p}$, and $\Delta t^* = 0.0038t^*$. Reprinted from Zhou and Balachandar (2021), with permission from Elsevier.

domain, it is very hard to generate a satisfactory grid that resolves the lubrication layer. Thus, in the case of moving particles, the body-fitted approach has generally been used for only an isolated particle moving in an unbounded domain, so that the governing equations are solved in a moving frame attached to the particle. In this moving frame of reference, the particle is stationary and thus re-gridding each time step is avoided. The moving frame is possible even in the presence of a nearby wall, as long as the particle motion is parallel to the wall. In the case of multiple particles, the body-fitted approach has generally been limited to investigation of flow over a stationary distribution of particles.

As examples of the body-fitted approach, we show three different discretizations in Figure 14.4. Each of these examples considers only idealized problems, with the focus on understanding a certain important aspect of multiphase-flow physics. Figure 14.4a shows a spectral discretization around an isolated particle in a large boundary-free computational domain. In this case the outer computational domain is chosen to be a much larger sphere concentric with the particle, so that a discretization based on spherical coordinates is used. The regular nature of the geometry also allows for the use of a highly accurate spectral methodology. The periodic $0 \leq \phi \leq 2\pi$ coordinate is treated with Fourier expansion, the $0 \leq \theta \leq \pi$ direction is not periodic, but allows sine or cosine expansion depending on the symmetry of the fluid quantity being discretized. In the radial direction that extends from the surface of the particle to the outer sphere, a Chebyshev discretization was used by Bagchi and Balachandar (2003b, 2004). It should be noted that the grid resolution is much finer on the right side than on the left. This is in order to better resolve the turbulent wake that forms on the leeward side of the particle and in the simulations, the flow is from left to right. The clustering of grid points was achieved with a non-uniform distribution of grid points along the θ direction with finer resolution near $\theta = 0$ and coarser resolution near $\theta = \pi$. The advantage of this simple geometry and the use of the accurate spectral methodology is that a highly complex flow at a large Reynolds number can be studied. Bagchi and Balachandar (2003b, 2004) used this body-fitted grid to study the effect of incoming turbulence on the force on an isolated particle and also the effect of the particle in modifying surrounding turbulence. The results from these simulations were discussed in Section 7.6.

Figure 14.4b shows a more complex body-fitted grid, which is needed because of the presence of a solid boundary in close proximity to the spherical particle. A spherical domain with a fully spectral approach is not appropriate. The figure shows a spectral element discretization (Deville et al., 2002) using code Nek5000. The computational domain is partitioned into hexahedral elements and within each element, velocity and pressure are represented in local Cartesian coordinates by tensor-product Lagrange polynomials of degree N_e and $N_e - 2$, respectively. Zeng et al. (2008) reported simulations that employed 3400 spectral elements, with each element resolved by $11 \times 11 \times 11$ Legendre–Gauss–Lobatto points, and thus employed an overall resolution of 3.4 million grid points. Except within a small bounding box around the particle, the hexahedral elements are rectangular in shape. Within the bounding box, the hexahedral elements deform from the outer box to the inner sphere. It is important to point out that the spherical surface of the particle is represented accurately as a tenth-order polynomial and not as a patchwork of flat surfaces. The isoparametric mapping helps in the transition from the spherical surface to the Cartesian mesh away from the sphere. Again the advantage of the body-fitted grid is that very small distances between the wall and the particle can be very accurately resolved and furthermore, the higher-order accuracy of the spectral element methodology is ideally suited for studying the interaction between wall turbulence and a nearby particle. Results from such spectral element simulations were discussed extensively in Chapter 8.

As the final example of a body-fitted grid, in Figure 14.4c a finite-volume discretization around a structured array of particles is shown. An unstructured tetrahedral grid is used to discretize the three-dimensional fluid volume around a face-centered-cubic (FCC) array of particles. The geometry and the flow are taken to be periodic in the transverse directions. Thus, the computational domain includes full particles embed-

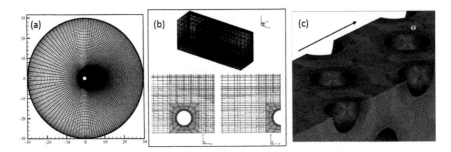

Figure 14.4 Three examples of a body-fitted grid. (a) A Fourier–Chebyshev spectral grid around a sphere that has been stretched to cluster the grid points near the particle and in the wake region. Taken from Bagchi and Balachandar (2002b). (b) A spectral element grid that has been designed to well resolve the narrow region between a particle and a flat wall. Taken from Zeng et al. (2008). A three-dimensional view as well as cut sections on two vertical planes passing through the center of the particle are presented. (c) A body-fitted grid around an FCC distribution of particles. GRIDPRO is used to first create the surface mesh on all the particle surfaces, which is then used as an input to TETGEN, which generates the body-conforming unstructured tetrahedral mesh inside the simulation domain. Taken from Mehta et al. (2016).

ded inside (these can be seen in the figure), half and quarter spheres intersected by the computational domain, which can be seen. Along the flow direction (marked by the arrow), the flow is not periodic and therefore a longer domain consisting of many particles is considered in the simulations by Mehta et al. (2016). GRIDPRO (Eiseman, 2016) is used to first create the surface mesh on all the particle surfaces that intersect the simulation domain and these include the full spheres and the half and quarter spheres that can be seen in the figure. This surface mesh is then used as an input to TETGEN (Si, 2015), which generates the body-conforming unstructured tetrahedral mesh inside the simulation domain. The quality and size of the tetrahedral elements in the domain determine the accuracy of the solution. Similarly, adequate mesh resolution on the particle surface is required to properly compute particle forces, which are determined by integrating flow properties on the surface of each particle. The quality and size of the elements are controlled by monitoring the element aspect ratio and maximum element volume. Surface mesh resolution is controlled by specifying the maximum element area. Mehta et al. (2016) performed a detailed study of surface and volume grid resolution required to accurately solve shock propagation over the FCC array of particles. This body-fitted grid generation was then extended in Mehta et al. (2018) to discretize the volume around a random distribution of particles.

15   Euler–Lagrange Approach

The Euler–Lagrange (EL) approach is also often referred to as the point-particle approach, since the particles are taken to be point masses, as far as their interactions with the surrounding continuous phase are concerned. In the particle-resolved approach, the presence of the particles was fed back to the surrounding continuous phase through the no-slip, no-penetration, isothermal or adiabatic, and other boundary conditions. These boundary conditions, without additional closure assumptions, directly controlled the mass, momentum, and energy exchanges between the particles and the surrounding fluid. Furthermore, these exchanges, which are in the form of tractional force, heat, and mass transfer, are properly distributed around the surfaces of the particles, and they accurately account for the presence of boundary layers, wakes, and other microscale features around the particles.

The significant computational simplicity of the EL approach is that the no-slip and no-penetration boundary conditions need not be explicitly applied at the interface between the particles and the surrounding fluid. Similarly, thermal boundary conditions (isothermal or adiabatic) need not be applied in the energy equation and species concentration boundary conditions in the mass balance. In fact, as stated earlier, in the EL approach, as far as the continuous phase is concerned, the particles are generally treated as just sources/sinks of mass, momentum, and energy that are appropriately projected back onto the continuous phase. With increasing particle volume fraction, one must also account for the volume effect of the particles and interparticle collisional interactions. Nevertheless, we do not need to identify the precise location of the particle–fluid interfaces and their influence on the flow at the microscale. The continuous-phase equations can be solved over the entire domain occupied by the multiphase flow, without demarcating regions occupied by the fluid versus regions occupied by the particles.

From a computational point of view, this offers a substantial twofold advantage. First, we avoid the need for either a complicated grid for the continuous phase that is body-fitted around all the particles, or the use of an immersed-boundary method with Lagrangian markers defining the interface. Second, and more importantly, the computational grid for the EL approach is not dictated to be finer than the particle size, but can be chosen to be larger than the particle diameter. Thus, the computational cost of the EL approach is typically orders of magnitude cheaper than the corresponding PR approach. Or, for the same computational cost, orders of magnitude more particles can be studied than with the PR approach.

The EL approach has a long history (Saffman, 1973; Riley and Patterson Jr., 1974; Maxey, 1987; Kallio and Reeks, 1989; McLaughlin, 1989; Elghobashi, 1991) and due to its computational simplicity its use continues to grow in many different engineering, environmental, and science applications. It has become a powerful and useful tool in the computation of turbulent multiphase flows in a variety of complex situations, from fundamental investigations (Squires and Eaton, 1991; Elghobashi and Truesdell, 1992, 1993; Wang and Maxey, 1993; Sundaram and Collins, 1997; Ferrante and Elghibashi, 2004) to applications such as fluidized beds and volcanic eruptions (Pepiot and Desjardins, 2012; Zwick and Balachandar, 2019; Lain and Sommerfield, 2020; Liu et al., 2021), to mention just a few examples.

There is of course a downside to the EL approach. The EL continuous-phase governing equations are not self-contained like the Navier–Stokes equations. The rigorously derived EL continuous-phase governing equations contain *secondary terms* that must be modeled with closure models. This difficulty arises from the fact that the EL continuous-phase governing equations are derived by applying a volume filter to the Navier–Stokes equations. In particular, a length scale $\mathscr{L}$ is chosen and only the dynamics of flow scales larger than $\mathscr{L}$ is explicitly retained in the EL equations; the effect of all subgrid scales appears as secondary terms and must be modeled. This brings to light two important limitations of the EL approach compared to the corresponding particle-resolved approach. First, in the EL approach, at any given instance in time, we only know the state of the unfiltered large scales. There are infinite possibilities for the state of the small scales that are filtered out, consistent with those of the known (or computed) large scales. But without knowledge of the precise state of the small scales, the best one can do is to predict the average or expected value of the effect of the small scales on the dynamics of the resolved large scales. In other words, the closure models of the secondary terms can only be statistical in nature. Second, the closure models in typical use are far from being perfect and therefore their predictions deviate from the true average or expected effect of the small scales. The above two limitations, respectively, contribute to uncertainty and error in the large-scale dynamics predicted by the EL approach. Thus, due to the filtering operation, the results of an EL simulation can never be as accurate as the underlying particle-resolved simulation.

The rigorously derived EL continuous-phase equations are far more complex than the standard Navier–Stokes equations, due to the presence of the secondary terms. It therefore becomes necessary to simplify the EL governing equations and the closure models. A quick survey of the literature will show that there is no single universal set of EL governing equations or a single universal EL formulation that has been agreed upon by researchers. A variety of EL formulations with different sets of governing equations have been solved. Much of the difference between the different EL formulations lies in the judicious simplifications made to the governing equations. This poses a great challenge not just for beginners, but even for seasoned researchers. Given a multiphase flow problem that is well suited for the EL approach, one is faced with the question: which among the many available EL approaches to pursue? Thus, the main focus of this chapter is to discuss the approximations often made in the different EL approaches and the advantages and disadvantages of the resulting simplified governing equations.

Table 15.1 The different limits of particle resolution, the associated ordering of length scales, and figure that represents the regime.

Resolution	Length scales			Figure
Particle-resolved (PR)	$\mathscr{L} \ll d_p$			13.4b, 13.4e
Particle-unresolved (PU) (small particle limit)	$\mathscr{L} \gg d_p$	$\eta \gg \mathscr{L} \gg d_p$	EL-DNS	13.4a
		$\mathscr{L} \gg d_p, \eta$	EL-LES	13.4d
Marginal resolution (MR) (large particle limit)	$\mathscr{L} \approx d_p$		EL-EXT	

A good understanding of the choices and their consequences will help the reader make informed decisions.

The EL approach was originally developed for applications where the following simplifying assumptions can be made: (i) the particles are of size much smaller than the computed flow scales (this corresponds to scenarios depicted in frames (a), (c), and (d) of Figure 13.4); (ii) the dispersed phase is sufficiently dilute that particle–particle interactions are rare; and (iii) of the five multiphase mechanisms (*volume*, *inertia*, *body-force*, *stress*, and *slip* effects) discussed in Section 2.2, not all are important. The advantage of these assumptions is that the resulting EL governing equations are considerably simplified. Furthermore, the closure models are well established and thus the EL approach is on a solid theoretical footing. Recent efforts have extended the EL approach to more complex applications where the above assumptions do not apply. Such applications involve dense flows of finite particle volume fraction and large particles of size comparable to the resolved flow scales. The purpose of this chapter is to obtain the EL governing equations and closure models that are applicable for these different scenarios.

15.1 Chapter Plan Based on Length Scales

The first step in applying the EL approach to a multiphase flow problem is to identify the filter length scale $\mathscr{L}$. This can be taken to be the cutoff length defined in Figure 13.4. The significance of $\mathscr{L}$ is that all the continuous-phase scales that are smaller will not be computed and only the dynamics of scales larger than $\mathscr{L}$ will be computed. By comparing $\mathscr{L}$ and the particle size d_p, we identify three different-resolution regimes that are shown in Table 15.1. In the limit $\mathscr{L} \ll d_p$, the particle is fully resolved and the PR computational methods have already been discussed in the previous chapter and therefore will not be considered here.

In the limit $\mathscr{L} \gg d_p$, the particle is unresolved and thus comes under the purview of the EL approach. We now bring in the third length scale: the Kolmogorov length scale η of the continuous-phase turbulence. As shown in Figure 13.4, three different particle unresolved regimes can be identified. If particles are smaller than the Kolmogorov scale, and if we can afford to resolve the entire range of turbulence, then $\eta > \mathscr{L} \gg d_p$ and we are in the situation depicted in Figure 13.4a. In this case, $\mathscr{L}$ can be chosen such that all turbulent eddies of size larger than the Kolmogorov scale are retained in the EL simulation. Because of scale separation between mesoscale turbulence and microscale pesudo turbulence, the precise choice of $\mathscr{L}$ is unimportant as long as it falls in between, as indicated in the figure. The advantage of this regime is that it is easier to develop accurately the closure models for the secondary terms presented in Table 11.1, since we need only account for the effects of pesudo turbulence.

In the scenarios depicted in Figures 13.4c and d, $\mathscr{L} \gg (\eta, d_p)$ and the entire range of turbulence is not resolved. The choice of $\mathscr{L}$ in the EL simulations is dictated by computational affordability. The EL approach applied to these latter scenarios will be referred to as **EL-LES**, while the EL approach applied to the scenario of frame (a) will be referred to as **EL-DNS**. In the context of EL-LES, the various closure models for the secondary terms presented in Table 11.1 must include the effects of both pesudo turbulence as well as small-scale turbulence that has been filtered out. In frame (c) the two contributions are well separated and therefore can be modeled perhaps independently. Closure modeling of the secondary terms is more complex in the scenario depicted in frame (d).

Here we must clarify the use of the terminology EL-DNS, which may be objectionable to many single-phase turbulence researchers. In classical fully resolved simulations of single-phase turbulence, the term "DNS" is reserved only for simulations where all the scales of flow are faithfully resolved. In this sense, only particle-resolved multiphase flow simulations qualify as DNS. The use of the term "DNS" in EL-DNS can be objected to, since flow variations on the scale of the particle are not resolved. Nevertheless, with this cautionary note we proceed to use the term EL-DNS (and in the next chapter EE-DNS). The reader should bear in mind that these are not first-principle simulations and as a result will require closure models, which will be the primary focus of this chapter.

It must be noted that if the energy contained within the filtered mesoscale turbulence is much larger than pesudo turbulence, then the latter can be ignored and the modeling of the former can perhaps be guided by subgrid closure models that are in common use in single-phase turbulence. In any case, under scenarios (c) and (d) the length scale $\mathscr{L}$ must be explicitly stated, since it influences the closure models of the secondary terms. A good understanding of the different length scales (and associated time scales) is essential in order to make judicious choices of $\mathscr{L}$ and the closure models, for a successful EL simulation.

For the purpose of this chapter, we refer to $\mathscr{L} \gg d_p$ as the "small particle limit" (see Table 15.1) and this limit will be considered first in this chapter. An increasing number of EL simulations have begun to consider $\mathscr{L} \approx d_p$, which will be referred to as the "large particle limit." Accordingly, the plan for the rest of this chapter is as follows.

- First, Section 15.2 will consider the small particle limit and obtain the governing continuous-phase and Lagrangian particle equations.
- In Section 15.3 we will consider the small particle limit and obtain the closure models of all the secondary variables. Here we will first consider the easier case of EL-DNS.
- Section 15.4 will discuss the more complex aspects of closure modeling for EL-LES.
- Then Section 15.5 will discuss how the EL governing equations capture the five

different multiphase flow mechanisms, in the small particle limit. This discussion will illustrate how simpler sets of EL equations can be obtained by ignoring the influence of unimportant multiphase mechanisms.

- Implementation details of the EL approach will be discussed in Section 15.6, where we will discuss additional complexities that arise at large-particle volume fraction.

- Finally, Section 15.7 will consider the complex "large particle limit" of the EL approach, where the particles are of size comparable to $\mathscr{L}$. This section will address how to correct for the self-induced velocity and temperature of a large particle whose size is comparable to the background grid in an EL simulation.

- This chapter includes one other section. In Section 15.8, we will address the force coupling method of Maxey and co-workers.

15.2   EL Governing Equations – Small Particle Limit

In this section we will develop the appropriate governing equations of the EL approach for the continuous and dispersed phases. In order to somewhat simplify the resulting equations, we will place a few restrictions on both the continuous and dispersed phases. Our attention will be limited to the regime where the continuous-phase flow can be considered incompressible, which leads to the simplification that the continuous-phase density is a constant given by ρ_f. We will also take the transport properties of the continuous phase, such as kinematic viscosity, thermal diffusivity, and specific heat capacity to be constant (i.e., the temperature and pressure dependence of these transport properties will be considered negligible). As for the dispersed phase, we will consider them to be nondeformable spheres, and to maintain their individuality. That is, each particle will remain distinct and processes such as merger or breakup of particles will be considered unimportant. We will also consider the particles to be homogeneously made up of the same material and the Biot number to be such that the particle can be considered isothermal. This assumption allows us to ignore the conjugate heat transfer problem within each particle. Each of these restrictions can be relaxed at the expense of making the governing equations somewhat more complicated. But they do not change the fundamental character of the EL equations and thus have been made to maintain focus on the essential multiphase physics.

Since the interfaces between the dispersed and the continuous phases will not be resolved, in the EL approach, the mathematical representation of the continuous phase will necessarily deviate from the true fluid velocity $\mathbf{u}(\mathbf{x}, t)$ and temperature $T_f(\mathbf{x}, t)$ fields that are defined only in regions occupied by the continuous phase. The formal way in which we avoid the need for explicitly tracking the interfaces between the phases is through the filtering operation defined in Chapter 11. We recall that in Figure 11.1 a function that was defined only in regions occupied by the fluid showed sharp discontinuities across the interfaces. But upon filtering with a Gaussian filter, the filtered function became both smooth and well defined over the entire domain. In Sections 11.2.1 to 11.2.3 this filtering strategy was rigorously applied to the continuous-phase governing equations to obtain filtered continuous-phase equations that can be applied over the entire volume of multiphase flow, without distinguishing regions occupied by the fluid and the particles.

15.2.1   Continuous-Phase Governing Equations

For the continuous-phase governing equations we start with the incompressible filtered mass, momentum, and temperature equations (11.99), (11.100), and (11.70). Recall that these equations assume (i) the continuous phase to be incompressible and (ii) the small particle limit of $\mathscr{L} \gg d_p$. We repeat here the filtered mass, momentum, and energy equations of the continuous phase:

$$\frac{\partial \phi_c}{\partial t} + \nabla \cdot (\phi_c \mathbf{u}_c) = - \sum_{l=1}^{N_p} G(\mathbf{x} - \mathbf{X}_l) \frac{\dot{m}_l}{\rho_f}, \tag{15.1}$$

$$\begin{aligned} \phi_c \rho_f \frac{D\mathbf{u}_c}{Dt} =& \nabla \cdot (\phi_c \boldsymbol{\tau}_{\text{sg}}) + \phi_c \rho_f \mathbf{g} + \phi_c \nabla \cdot \boldsymbol{\sigma}_c \\ &- \sum_{l=1}^{N_p} G(\mathbf{x} - \mathbf{X}_l) \left(\mathbf{F}'_l + \dot{\mathbf{M}}_l\right) + \mathbf{u}_c \left(\sum_{l=1}^{N_p} G(\mathbf{x} - \mathbf{X}_l)\dot{m}_l\right), \end{aligned} \tag{15.2}$$

$$\begin{aligned} \phi_c \rho_f C_{vf} \frac{DT_c}{Dt} =& - \nabla \cdot (\phi_c \mathbf{q}_{\text{sg}}) - \phi_c \nabla \cdot \mathbf{q}_c + \boldsymbol{\sigma}_c : \nabla \mathbf{u}_m \\ &- Q'_{\text{hyd}} - \left(W'_{\text{hyd}} - \mathbf{u}_c \cdot \mathbf{F}'_{\text{hyd}}\right) - \left(\dot{E} - \dot{m}E_c\right) + \mathbf{u}_c \cdot \left(\dot{\mathbf{M}} - \dot{m}\mathbf{u}_c\right), \end{aligned} \tag{15.3}$$

where in the temperature equation, Q'_{hyd}, W'_{hyd}, $\mathbf{F}'_{\text{hyd}}$, $\dot{m}$, $\dot{\mathbf{M}}$, and $\dot{E}$ must be replaced with the sum over contributions from the N_p particles as given in Eqs. (11.90), (11.91), (11.89), (11.85), and (11.86). The volume-weighted mixture velocity is defined as $\mathbf{u}_m = \phi_c \mathbf{u}_c + \phi_d \mathbf{u}_d$. The dynamic variables being solved are the filtered velocity $\mathbf{u}_c$, filtered pressure p_c (which is the isotropic part of $\boldsymbol{\sigma}_c$), and filtered temperature T_c. The continuous-phase volume fraction, which is defined as the complement of the particle volume fraction (i.e., $\phi_d = 1 - \phi_c$), must be obtained from the Lagrangian particle information. The formal definition of $\phi_d(\mathbf{x}, t)$ was given in Eq. (11.104) as the filter-weighted integral of the indicator function. In the small particle limit, the integral can be approximated as follows:

$$\phi_d(\mathbf{x}, t) = \sum V_{pl}\, G(\mathbf{x} - \mathbf{X}_l), \tag{15.4}$$

where V_{pl} is the volume of the lth particle. This approximation was employed in Section 11.2.4 when approximating the convolution integrals as a summation over all the particles, as they appear on the right-hand side of the above equations. We again note that to calculate ϕ_d at any point within the domain above, the summation has to be performed only for those particles that are within a few filter length scales of $\mathbf{x}$. Although the summation in the above equation is formally over all the particles within the domain, those particles that are well removed from **x** do not contribute to the summation.

The above multiphase governing equations differ from the standard Navier–Stokes equations in a few important ways: (i) the appearance of ϕ_c in the equations accounts for the fact that the continuous phase only occupies its local volume fraction, with the balance occupied by the particles; (ii) due to the filtering operation, the above equations include the residual stress τ_{sg} and the residual heat flux $\mathbf{q}_{\mathrm{sg}}$ contributions; (iii) back coupling of mass, momentum, and energy from the particles are represented by the summations on the right-hand side; and (iv) though the filtered macroscale stress tensor σ_c and the macroscale heat flux vector $\mathbf{q}_c$ appear as they do in the Navier–Stokes equations, as we will see below, their closure will be different from the single-phase limit.

15.2.2   Lagrangian Particle Equations

We now consider the Lagrangian governing equations of the particles. As noted before, we will limit attention to small particles and thus all the scales of the filtered macroscale continuous-phase flow are much larger than the particle size. The mass, momentum, angular momentum, and energy balances of each of the particles within the multiphase flow must be written following those given in Eq. (4.1). Let m_{pl} and I_{pl} be the mass and moment of inertia of the lth particle. For a rigid spherical particle

$$ m_{pl} = \frac{4\pi}{3} R_{pl}^3 \rho_p \quad \text{and} \quad I_{pl} = 2 m_{pl} R_{pl}^2 / 5, \tag{15.5} $$

where $R_{pl} = d_{pl}/2$ is the radius of the lth particle. Further, let the state of the lth particle be denoted by its position $\mathbf{X}_l$, translational velocity $\mathbf{V}_l$, angular velocity $\boldsymbol{\Omega}_l$, and temperature T_{pl}. In addition, let $V_{pl} = 4\pi R_{pl}^3/3$ be the volume of the lth particle, $m_{fl} = \rho_f V_{pl}$ the mass of fluid displaced by the particle, and C_p the constant heat capacity of the particles. Then the governing equations of the lth particle can be expressed as

$$
\begin{aligned}
\frac{dm_{pl}}{dt} &= \dot{m}_l, \\[4pt]
m_{pl}\frac{d\mathbf{V}_l}{dt} &= V_{pl}\left[\nabla \cdot \sigma_c\right]_{@l} + m_{pl}\mathbf{g} + \mathbf{F}'_l + \mathbf{F}_{cl} + \left(\dot{\mathbf{M}}_l - \dot{m}_l\mathbf{V}_l\right), \\[4pt]
I_{pl}\frac{d\boldsymbol{\Omega}_l}{dt} &= \frac{I_{fl}}{2}\left[\frac{D(\omega_c)}{Dt}\right]_{@l} + \mathbf{T}'_l + \mathbf{T}_{cl}, \\[4pt]
m_{pl}C_p\frac{dT_{pl}}{dt} &= -V_{pl}\left[\nabla \cdot \mathbf{q}_c\right]_{@l} + Q'_l + Q_{cl} + \left(W'_l - \mathbf{V}_l \cdot \mathbf{F}'_l - \boldsymbol{\Omega}_l \cdot \mathbf{T}'_l\right) \\[4pt]
&\quad + \left(\dot{E}_l - \dot{m}_l E_l\right) - \mathbf{V}_l \cdot \left(\dot{\mathbf{M}}_l - \dot{m}_l\mathbf{V}_l\right).
\end{aligned}
\tag{15.6}
$$

Let us examine each of the terms on the right-hand side. In the mass balance, $\dot{m}_l$ is the rate of mass gained by the lth particle by condensation and its value will be negative in case of evaporation. This term counterbalances the source term on the right-hand side of the continuous-phase mass balance (15.1), thus satisfying conservation of global mass. The exchanged mass carries with it momentum and energy as well, which are represented by $\dot{\mathbf{M}}_l$ and $\dot{E}_l$. These quantities need to be modeled and we shall discuss this in greater detail below. In the angular momentum balance, we have assumed the mass transfer to have no influence on the rate of change of angular velocity of the particle.

The first term on the right-hand side in the momentum, angular momentum, and energy equations corresponds to force, torque, and heat transfer, respectively, due to the undisturbed filtered ambient mesoscale flow at the lth particle. These undisturbed flow force, torque, and heat transfer are exerted even if the particle were replaced with the continuous phase. Here, $V_{pl} [\nabla \cdot \boldsymbol{\sigma}_c]_{@l}$ corresponds to the undisturbed flow force on the lth particle. In the angular momentum equation, I_{fl} is the moment of inertia of the fluid displaced by the particle and thus the first term corresponds to the undisturbed flow torque experienced by the particle. Similarly, the first term on the right-hand side in the energy equation corresponds to the undisturbed flow heat transfer to the lth particle. In the limit of small-sized particles being considered in this section, the variation of $\nabla \cdot \boldsymbol{\sigma}_c$, $D\omega_c/Dt$, and $\nabla \cdot \mathbf{q}_c$ over the volume of the particle has been ignored, and they are simply evaluated at the center of the particle.

The primed terms on the right-hand side are due to the perturbation flow induced by the presence of the particle, with $\mathbf{F}'_l$, $\mathbf{T}'_l$, and Q'_l representing the hydrodynamic force, torque, and heat transfer to the lth particle. They must be modeled in terms of the particle and filtered fluid properties, The availability of appropriate closure models for these interphase exchanges will be discussed in the next section in greater detail.

The terms with subscript cl on the right-hand side of the momentum, angular momentum, and energy equations correspond to force, torque, and heat transfer to the lth particle due to its contact with the other particles. These terms can be written as

$$\mathbf{F}_{cl} = \sum_{j \neq l} \mathbf{F}_{clj}, \quad \mathbf{T}_{cl} = \sum_{j \neq l} \mathbf{T}_{clj}, \quad Q_{cl} = \sum_{j \neq l} Q_{clj}, \tag{15.7}$$

where the sum is over all particles except the lth particle. Here, $\mathbf{F}_{clj}$ is the force on the lth particle due to collision with the jth particle; similarly, $\mathbf{T}_{clj}$ and Q_{clj} are the torque and heat transfer on the lth particle due to collision with the jth particle. The force and torque due to collision have been discussed in detail in Chapter 9. For example, the soft-sphere collision model can be used to calculate the collisional force and torque. The soft-sphere collision model allows for the simultaneous collision of the lth particle with more than one neighbor. The net collisional force and torque on the lth particle is the simple sum of all the individual contributions. Note that even though the sum is formally over all the other particles within the system, in practice simultaneous collision is likely with only a few neighbors. We have not discussed heat transfer due to collision, since it is not a significant effect in most applications of interest here. Accordingly, we can set $Q_{clj} \approx 0$.

In the momentum equation, the last two terms within parentheses can be interpreted in the following way. While $\dot{\mathbf{M}}_l$ is the momentum transferred to the lth particle due to mass transfer, only $\dot{m}_l \mathbf{V}_l$ actually resides in the transferred mass. So the reminder goes towards accelerating (or decelerating) the lth particle, depending on whether the balance is positive (or negative). Similar interpretations can be given to the pairs of terms within parentheses in the temperature equation. For example, $\dot{E}$ is the total energy transferred to the lth particle due to mass transfer. Only $\dot{m}_l E_l$ actually resides in the transferred mass. So the reminder goes toward increasing (or decreasing) the temperature of the lth particles depending on whether the balance is positive (or negative). As we will see below, W_l' and $\dot{\mathbf{M}}_l$ are modeled in such a way that the contribution to particle temperature from $(W_l' - \mathbf{V}_l \cdot \mathbf{F}_l' - \boldsymbol{\Omega}_l \cdot \mathbf{T}_l')$ and $\mathbf{V}_l \cdot (\dot{\mathbf{M}}_l - \dot{m}_l \mathbf{V}_l)$ will be set to zero. In other words, these work-related mechanisms do not alter the particle temperature.

15.3 Closure Relations of EL-DNS – Small Particle Limit

To complete the EL formulation, the continuous and dispersed-phase equations (15.1) to (15.6) must be supplemented with appropriate closures. In this section, we will consider the closure models of all the secondary variables presented in Table 11.1. The rest of this section will focus on modeling the following closure terms:

    I. Closure models of interphase momentum exchange $\mathbf{F}_l'$ and associated torque $\mathbf{T}_l'$.
    II. Closure model of work W_l' associated with the momentum exchange.
    III. Closure model of interphase heat transfer Q_l'.
    IV. Closure models of interphase mass exchange $\dot{m}_l$ and the associated $\dot{\mathbf{M}}_l$, $\dot{\mathbf{T}}_l$, and $\dot{E}_l$.
    V. Constitutive models of filtered continuous-phase stress tensor $\boldsymbol{\sigma}_c$ and heat flux $\mathbf{q}_c$.
    VI. Subgrid closure models of residual stress $\boldsymbol{\tau}_{\mathrm{sg}}$ and residual heat flux $\mathbf{q}_{\mathrm{sg}}$.

Even with the simplification of the "small particle" limit, obtaining accurate closure models remains a formidable challenge. Therefore, we will further simplify the problem with two important additional restrictions. First, our attention will be restricted to the EL-DNS regime of $\eta \gg \mathscr{L} \gg d_p$. In this regime, the filtered continuous-phase properties account for the entire range of mesoscale turbulence. Only the effect of pseudo turbulence needs to be modeled. This restriction considerably simplifies the development of the above-listed six closure problems, which are presented in the following six subsections. The added complexity of closure modeling in the EL-LES regime, where the effect of unresolved mesoscale turbulence must also be modeled, will be addressed after that.

    Second, the particle volume fraction will be considered to be modest. More specifically, in obtaining the closure models of force, torque, heat, and mass transfer of a particle, we ignore the direct perturbing influence of neighboring particles. But we account for the collective influence of all the neighbors through volume fraction dependence. It must be admitted that our understanding of even the collective influence of neighbors is limited. Reasonably accurate volume fraction dependence is known for only a few closure models. For example, volume fraction dependences of quasi-steady drag and mixture viscosity have been well studied. On the other hand, very little is known of the volume fraction dependences of torque or lift force. Thus, our presentation of the volume fraction effect on closure models will be incomplete in the following sections. We will present the volume fraction dependence of only those closure models that have been well established in the literature. Even these results are typically accurate only for modest values of particle volume fraction. Later subsections will address the additional challenges of closure modeling at large particle volume fraction.

15.3.1   Interphase Momentum Exchange

We start the discussion of closure relations with the modeling of interphase momentum coupling, which we have covered extensively in Chapter 4. By restricting to the EL-DNS regime of $\eta \gg \mathscr{L} \gg d_p$, we simply take the undisturbed flow at the lth particle to be the filtered continuous-phase quantities evaluated at $\mathbf{X}_l$. This allows the use of the finite-Re BBO equation (4.144) as the model for $\mathbf{F}'_l$. However, the following adaptations must be made in order to use Eq. (4.144) in the present context of the EL-DNS approach. (i) Since $\mathbf{F}'_l$ corresponds only to the force due to perturbation flow, the first term on the right-hand side of Eq. (4.144) must be excluded, since it corresponds to the undisturbed flow force. The undisturbed flow force has already been accounted for in the continuous and dispersed-phase momentum equations. (ii) All the continuous-phase quantities must be evaluated at the location of the lth particle. Therefore, variables such as $\mathbf{u}_c$ and $D\mathbf{u}_c/Dt$ in Eq. (4.144) must be replaced with $\mathbf{u}_{c@l}$ and $[D\mathbf{u}_c/Dt]_{@l}$, and so on. (iii) The particle-related quantities must be those of the lth particle and thus we replace the appropriate variables with $\mathbf{V}_l$ and $d\mathbf{V}_l/dt$. (iv) The BBO equation is for an isolated particle and finite volume corrections must be applied to quasi-steady, inviscid unsteady, and viscous unsteady forces. With these adaptations we obtain the following perturbation force model:

$$
\mathbf{F}'_l(t) = \underbrace{6\pi\mu_f R_{pl}[\mathbf{u}_{c@l} - \mathbf{V}_l]\Phi}_{\mathbf{F}'_{qsl}} + \underbrace{C_M\, m_{fl}\left[\left(\frac{D\mathbf{u}_c}{Dt}\right)_{@l} - \frac{d\mathbf{V}_l}{dt}\right]}_{\mathbf{F}'_{iul}}
$$

$$
+ \underbrace{6\pi\mu_f R_{pl}\int_{-\infty}^{t} K_{vu}\left[\frac{d\mathbf{u}_{c@l}}{dt} - \frac{d\mathbf{V}_l}{dt}\right]_{@\xi} d\xi}_{\mathbf{F}'_{vul}} + \mathbf{F}'_{Ll}.
$$

(15.8)

In the expression of the quasi-steady force $\mathbf{F}'_{qsl}$, the empirical correction factor $\Phi(\mathrm{Re}_l, \phi_{d@l})$ depends on both the Reynolds number and the particle volume fraction around the lth particle. The Reynolds number of the lth particle based on its relative velocity is $\mathrm{Re}_l = |\mathbf{u}_{c@l} - \mathbf{V}_l|d_{pl}/\nu_f$, and $\phi_{d@l}$ is the particle volume fraction field evaluated at $\mathbf{X}_l$. The volume fraction correction function given in Eq. (9.87) can be used as an accurate expression for $\Phi(\mathrm{Re}_l, \phi_{d@l})$ (Tenneti and Subramaniam, 2014). In the expression of the inviscid unsteady force $\mathbf{F}'_{iul}$, m_{fl} is the mass of fluid displaced by the lth particle. The added-mass coefficient as a function of local particle volume fraction can be taken to be $C_M(\phi_{d@l}) = (1 + 3.22\phi_{d@L})/2$ (see Eq. (9.96)). In the viscous unsteady force $\mathbf{F}'_{vul}$, the kernel $K_{vu}(t - \xi, \mathrm{Re}_l, \phi_{d@l})$ depends on the particle Reynolds number and local volume fraction, in addition to shifted time. The dependence on Re_l can be taken to be that given in Eq. (4.142) multiplied by the volume fraction dependence given in Eq. (9.97).

Depending on the situation, the above model for $\mathbf{F}'_l$ is often simplified. One commonly applied simplification is to ignore the added-mass and history force terms represented by the second and third terms on the right-hand side of Eq. (15.8). This simplification is typically made when the problem involves particles of density much larger than the surrounding fluid, as in the case of gas–solid flows, on the grounds that the particle acceleration is likely to be small due to its large inertia. However, caution is needed. dV_l/dt being small does not guarantee that the added-mass and history forces are small. Large values of continuous-phase acceleration, represented by the terms $D\mathbf{u}_c/Dt$ and $d\mathbf{u}_{c@l}/dt$, can contribute to substantial added-mass and history forces. See Ling et al. (2011a,b, 2013) for an estimate of the relative importance of the various forces.

If ambient shear effects become important, then lift force must be included with $\mathbf{F}'_{Ll}$ modeled as in Eq. (5.8). In addition, if the effect of particle rotation is important, then the combined effect of shear and particle rotation can be accounted for using the empirical relation given in Eq. (5.38). When a particle comes close to a wall, wall effects become important. In order to accurately predict processes such as deposition, resuspension, and wall impaction in an EL simulation, it is important to account for the additional wall effects in modeling $\mathbf{F}'_l$. The reader must refer to Chapters 4 and 7 and also consult other recent journal articles to arrive at an optimal model for $\mathbf{F}'_l$ that offers the best compromise between accuracy and simplicity.

In Eq. (15.6), $\mathbf{T}'_l$ represents the torque on the lth particle due to the net effect of the surface distribution of tractional forces on its surface. Under the EL-DNS assumption of this section, the torque on a particle can be modeled by the rotational BBO equation (5.33). Here, again, the first term on the right must be ignored, since $\mathbf{T}'_l$ accounts for only the effect of perturbation flow. Furthermore, Eq. (5.33) was derived in the zero-Re limit and therefore it must be extended to finite Re with empirical corrections. For example, the quasi-steady contribution must be replaced with the correlation given in Eq. (5.34). The wall effect on torque was discussed in Chapter 8. Thus, depending on the level of accuracy needed, the torque model to be employed in the EL approach can either be chosen to be simple or made more complex.

It should, however, be pointed out that there is no counterpart of particle torque in the continuous-phase governing equations. In Eq. (15.2), the feedback force $-\mathbf{F}'_l$ applied to the continuous phase with a symmetric Gaussian filter will not exert a torque on the fluid. This exposes an inconsistency, albeit perhaps a weak one, in the above EL formulation. This inconsistency in the continuous-phase momentum equation can be traced back to the assumption of very small particles. While the feedback torque was nonzero as defined in Eqs. (11.48) and (11.50), when approximating $\mathbf{F}'_{hyd}$ with the Gaussian sums as expressed in Eq. (11.89), the feedback torque from each particle was approximated to be zero. As a result of this inconsistency, global angular momentum is not conserved. Many EL simulations have computed particle-laden flows with particle rotation taken into account, while ignoring the feedback torque on the continuous phase. Their relative success can be used to argue that this inconsistency is not of great practical consequence. Nevertheless, either one should ignore the rotational motion of the particle altogether and be consistent with the filtered continuous-phase momentum equation (15.2), or consider more complex ways of applying the feedback force using multipole expansion that includes higher-order torque effects (see the force coupling method of Maxey and co-workers: Maxey and Patel, 2001; Lomholt et al., 2002; Lomholt and Maxey, 2003; Climent and Maxey, 2009; Yeo and Maxey, 2010), which will be discussed in Section 15.8.

15.3.2   Interphase Work Exchange

We now investigate the energy implications of the perturbation flow force $\mathbf{F}'_l$. As discussed above, the hydrodynamic force on the particle due to the perturbation flow is typically modeled as a superposition of quasi-steady, added-mass, viscous unsteady, and lift force contributions. Irrespective of how the hydrodynamic force is modeled, it can be seen in Eqs. (15.2) and (15.6) that the force applied to the particle and the force applied back on the surrounding fluid balance perfectly. Therefore, there is no net contribution to the total momentum of the system (continuous plus dispersed phases). Thus, as can be expected, the details of the interphase force models do not affect the global momentum balance. Similar conservation of the global energy between the fluid and the particles must be considered. This aspect was discussed earlier in Section 4.2.4 and here we will revisit the arguments in the present context of the EL approach.

In terms of energy,

$$\mathbf{V}_l \cdot \mathbf{F}'_l + \mathbf{\Omega}_l \cdot \mathbf{T}'_l \tag{15.9}$$

is the rate of work done on the lth particle and it contributes entirely to the rate of change of its translational and rotational kinetic energy. Thus, the term

$$W'_l - \mathbf{V}_l \cdot \mathbf{F}'_l - \mathbf{\Omega}_l \cdot \mathbf{T}'_l = 0 \tag{15.10}$$

in the particle temperature equation (15.6). In other words, the temperature of the particle is unaffected by the force and torque acting on its. This is due to the rigid-body nature of the particle. This will not be true in the case of a droplet, due to internal circulation induced by the relative motion with respect to the surrounding continuous phase. This internal circulation will lead to dissipation and conversion of kinetic energy to thermal energy. In which case, the equality given in (15.10) is not strictly correct. Nevertheless, the effect of dissipative heating of a droplet is often quite weak, so that Eq. (15.10) can be taken to apply for the dispersed phase in general.

The total work done of the continuous phase is exactly opposite to that for the dispersed phase, and thus the expression (15.10) as a model for W' may appear to apply for the continuous phase as well. However, this is not the case and further careful consideration is required, since we are concerned not with the original continuous-phase governing equations, but with the filtered version. We shall discuss this in this subsection.

The rate of change of kinetic energy of the fluid at the macroscale due to the feedback force from the lth particle can be obtained by dotting it with $\mathbf{u}_c$ and integrating over the domain:

$$-\mathbf{F}'_l \cdot \int G(\mathbf{x} - \mathbf{X}_l) \mathbf{u}_c(\mathbf{x}, t)\, d\mathbf{x} = -\mathbf{F}'_l \cdot \mathbf{u}_{cGl}, \tag{15.11}$$

where $\mathbf{u}_{cGl} = \int G(\mathbf{x} - \mathbf{x}_l) \mathbf{u}_c(\mathbf{x}, t)\, d\mathbf{x}$ is the Gaussian-weighted continuous-phase velocity at the lth particle. Though the integral is over the entire volume of the multiphase flow, its contribution is localized to be around the location of the lth particle due to the exponential decay of the Gaussian.[1] In general, the two kinetic energy changes will not balance. In other words, kinetic energy lost/gained by the particle is not fully balanced by the corresponding increase/decrease in kinetic energy of the surrounding filtered fluid motion. The imbalance is partly due to the symmetric spreading of the feedback force using the Gaussian filter. As mentioned earlier, the Gaussian filter results in zero feedback torque, and hence the contribution to kinetic energy of the surrounding fluid from torque is misrepresented.

Even if we ignore the torque and the rotational motion of the particle, $\mathbf{F}'_l \cdot \mathbf{V}_l$ will not balance the rate of change of kinetic energy of the fluid given in Eq. (15.11). The imbalance is due to the filtering operation. $\mathbf{F}'_l \cdot \mathbf{V}_l$ is precisely the rate of change of kinetic energy of the particle, as guaranteed by the rigid-body equation of motion. The negative of this must therefore be the rate of change of kinetic energy of the surrounding continuous phase, due to conservation of energy. The rate of change of kinetic energy given in Eq. (15.11) applies only to the filtered mesoscale. Equation (15.11) ignores the microscale flow around the particles, which has been filtered out and unaccounted for in the filtered velocity $\mathbf{u}_c$. Thus, for the total energy to be conserved, the difference $\mathbf{F}'_l \cdot (\mathbf{V}_l - \mathbf{u}_{cGl})$ must contribute to the rate of change of kinetic energy of the microscale flow that has been filtered out.

The contribution to the filtered microscale kinetic energy from each force component of $\mathbf{F}'_l$ can be investigated. For this we start with the following superposition:

$$\mathbf{F}'_l = \mathbf{F}'_{qsl} + \mathbf{F}'_{iul} + \mathbf{F}'_{vul} + \mathbf{F}'_{Ll}, \tag{15.12}$$

---

[1] The Gaussian weighted integral of the filtered velocity centered around the particle can be approximated by the filtered velocity at the center of the particle (i.e., it can be approximated as $\mathbf{u}_{cGl} \approx \mathbf{u}_{c@l}$). It should be noted that this approximation is unrelated to the size of the particle and the small-particle approximation made earlier in the simplification of the filtered continuous-phase governing equations. However, this approximation is likely to incur some error, since the Gaussian width is typically of the order of the grid resolution. Thus, the difference between $\mathbf{u}_{cGl}$ and $\mathbf{u}_{c@l}$ will depend on how rapidly $\mathbf{u}_c$ varies at the location of the particle. This approximation is, however, not necessary for the present discussion and therefore has not been pursued.

where the perturbation force on the lth particle has been written as the sum of quasi-steady, added-mass, viscous unsteady, and lift forces. From the definition of the quasi-steady force it can readily be seen that its contribution to the microscale kinetic energy is guaranteed to be positive definite. Due to the viscous nature of the quasi-steady force, this steady supply of microscale kinetic energy is balanced by an equal amount of energy dissipation. In other words, $\mathbf{F}'_{qsl} \cdot (\mathbf{V}_l - \mathbf{u}_{cGl})$ is guaranteed to be negative and is purely dissipative.

In contrast, from the definition of the inviscid-unsteady force given in Eq. (4.99), it can be seen that its contribution to perturbation flow kinetic energy $\left(\text{i.e., } \mathbf{F}'_{iul} \cdot (\mathbf{V}_l - \mathbf{u}_{cGl}) \right)$ can be positive or negative. As argued in Eq. (4.98), the kinetic energy of this inviscid perturbation flow can go up or down with increase or decrease in relatively velocity. This kinetic energy is associated with the potential flow component of the microscale motion and therefore will not be dissipated. This transfer of energy to the microscale motion is fully reversible, as can be expected from the inviscid nature of this force.

What about the other force contributions? In the case of viscous unsteady force, $\mathbf{F}'_{vul} \cdot (\mathbf{V}_l - \mathbf{u}_{cGl})$ is not guaranteed to be negative. But due to its viscous origin, the effect of this force is not reversible either. It can be expected that at least part of the corresponding microscale kinetic energy will be dissipated. In the case of lift force, $\mathbf{F}'_{Ll} \cdot (\mathbf{V}_l - \mathbf{u}_{cGl})$ is expected to be zero, since by definition, the lift force is normal to relative velocity.

Based on the above discussion, a good model for the rate of work associated with $\mathbf{F}'_l$ that appears in the filtered continuous-phase temperature equation is as follows:

$$W'_l = \left(\mathbf{F}'_{qsl} + \mathbf{F}'_{vul} \right) \cdot \mathbf{V}_l + \left(\mathbf{F}'_{iul} + \mathbf{F}'_{Ll} \right) \cdot \mathbf{u}_{cGl} . \qquad (15.13)$$

With this model, quasi-steady and viscous unsteady forces make a nonzero contribution to the continuous-phase energy equation through the term $W'_l - \mathbf{u}_{cGl} \cdot \mathbf{F}'_l$ in Eq. (15.3). This contribution is guaranteed to be positive in the case of quasi-steady force. But caution is needed, since the same does not apply for the viscous unsteady force. With the above definition of W'_l, the added-mass and lift forces do not make a net contribution to the filtered continuous-phase temperature equation. In essence, in Eq. (15.3) we can make the following substitution:

$$\left[W'_l - \mathbf{u}_{cGl} \cdot \mathbf{F}'_l \right] \approx \left[\left(\mathbf{F}'_{qsl} + \mathbf{F}'_{vul} \right) \cdot (\mathbf{V}_l - \mathbf{u}_{cGl}) \right] . \qquad (15.14)$$

This term then contributes to the dissipative heating of the continuous phase. Before leaving this topic, we remind the reader that the effect of dissipative heating on continuous-phase thermal evolution is generally weak in incompressible low-speed flows. If so, the above dissipative heating due to feedback force and the first term on the right-hand side of Eq. (15.3) (i.e., $\sigma_c : \nabla \mathbf{u}_m$) that represents the effect of dissipative heating due to filtered mesoscale continuous-phase motion can be neglected. For completeness we have retained these dissipative heating terms. In an EL application, these terms can be ignored if it can be established that the effect of dissipative heating on mesoscale flow is unimportant.

15.3.3 Interphase Energy Coupling

The definition of $Q'_{hyd}(\mathbf{x}, t)$ as the average thermal exchange from the dispersed to the continuous phase is given in Eq. (11.64). In the small particle limit, this can be approximated as the Gaussian-weighted sum of heat transfer from the particles in the neighborhood. The heat transfer Q'_l from the lth particle can be evaluated exactly in a particle-resolved simulation with knowledge of the continuous-phase thermal gradient around the surface of all the particles. The local heat flux along the surface of the particle can be expressed by the Fourier law of heat conduction as

$$\mathbf{q}'(\mathbf{x}', t) \cdot \mathbf{n} = -k_f \nabla T_f \cdot \mathbf{n}, \tag{15.15}$$

where k_f is the thermal conductivity of the fluid. Integration of the right-hand side over the surface of the lth particle with the filter function will then yield Q'_l.

However, such detailed information on the thermal field around the particles is unavailable in an EL simulation, since the filtered continuous-phase temperature T_c has lost all microscale details on the scale of the particle. Since T_c has been filtered to scales larger than a particle diameter, it is not possible to accurately evaluate the temperature gradient ∇T_f along the particle surface. The closure quest is to express Q'_l with a suitable model in terms of the filtered continuous-phase temperature and the temperature of the particle. This heat transfer problem was addressed in detail in Chapter 6 and for the present case of a small particle, the appropriate finite of BBO-like equation of heat transfer to the lth particle can be expressed as

$$
\begin{aligned}
Q'_l = m_{fl} C_{p,f} \left[\frac{DT_c}{Dt} \right]_{@l} + 2\pi k_f R_{pl} (T_{c@l} - T_{pl}) \, \mathrm{Nu}_l \\
+ 4\pi k_f R_{pl} \int_{-\infty}^{t} K_T(t - \xi) \left[\frac{dT_c}{dt} - \frac{dT_{pl}}{dt} \right]_{@l\xi} d\xi,
\end{aligned}
\tag{15.16}
$$

where $T_{c@l}$ is the filtered continuous-phase temperature at the location of the lth particle, which is taken to be the free-stream temperature of the particle. T_{pl} is the temperature of the lth particle, which is taken to be spatially uniform within the particle. Also, the subscript $@l\xi$ corresponds to the quantity being evaluated at the center of the lth particle at time ξ. The Nusselt number Nu_l is that of the lth particle and it can be calculated using the empirical correlation of Ranz and Marshall (1952) or Whitaker (1972): see Eq. (6.14).

The above expression for heat transfer adapted from those presented in Chapter 6 is for an isolated particle subjected to an ambient flow. The presence of neighboring particles will influence heat transfer, just as it increased the average drag coefficient (see Figure 9.10). The enhancement of mean heat transfer of a random array of particles as a function of particle volume fraction was discussed in Section 9.8. For example, the Nusselt number correlation of Sun et al. (2015) as a function of Reynolds number, Prandtl number, and local volume fraction given in Eq. (9.110) can be used. We also expect the thermal history kernel K_T to be influenced by the finite volume fraction. However, we lack any information on this dependence and without further research, we will have to settle for the thermal kernel of an isolated particle given in Eq. (6.18).

15.3.4   Interphase Mass Exchange

The definition of $\dot{m}(\mathbf{x}, t)$ as the filtered average of the interphase mass exchange from the dispersed to the continuous phase is given in Eq. (11.34). For each particle, this quantity can be precisely evaluated at the microscale with complete knowledge of the continuous and dispersed-phase properties around the surface of the particle. For example, in the case of mass exchange of species α due to evaporation, local mass exchange at a point on the surface of the particle can be expressed using Fick's law as

$$\rho_f(\mathbf{u} - \mathbf{u}_I) \cdot \mathbf{n} = -\mathcal{D}_\alpha \, \rho_f (\nabla Y_\alpha \cdot \mathbf{n}). \tag{15.17}$$

Here, $\mathcal{D}_\alpha$ is the mass diffusivity of species α, whose diffusion is the source of mass transfer, and Y_α is the mass fraction of this species in the continuous phase (Bird et al., 1960; Crowe et al., 2011).[2] According to Eq. (11.34), integration of the right-hand side of the above equation weighted by the filter function over the particle surface yields $\dot{m}$.

But the detailed information required for such surface integration is unavailable in an EL simulation. The filtered continuous-phase variables (i.e., those with subscript c) have lost all the microscale details on the scale of the particle. The closure quest is to express the mass exchange of each individual particle in terms of the filtered continuous-phase variables with a suitable model.

A closure model of $\dot{m}_l$ will also depend on the state of the particle. For the lth particle in a uniform free stream, mass transfer can be expressed as

$$\dot{m}_l = 4\pi R_{pl}^2 \rho_f \, h_{ml} \, (Y_{\alpha c@l} - Y_{\alpha l}), \tag{15.18}$$

where $Y_{\alpha c@l}$ is the species mass fraction of the free stream, which will be taken to be the value of the filtered mass fraction of the species in the continuous phase evaluated at the particle location. $Y_{\alpha l}$ is the species mass fraction in the continuous phase at the surface of the lth particle, which is microscale information and must be approximated. One approximation is as follows: by assuming the continuous phase that is in contact with the surface of the particle to be at the temperature of the particle T_{pl}, and assuming the evaporating species to be in saturation at that temperature, $Y_{\alpha l}$ can be evaluated. For an evaporating particle, $Y_{\alpha l} > Y_{\alpha c@l}$ and as a result $\dot{m}_l$ is negative.

In evaluating $\dot{m}_l$ using Eq. (15.18), the only other information needed is the nondimensional mass transfer coefficient $h_{ml} = \mathrm{Sh}_l \, \mathcal{D}_\alpha/(2R_{pl})$, where Sh_l is the Sherwood number of the lth particle. The Sherwood number is the key nondimensional quantity that models the net effect of the detailed mass exchange around the particle at the microscale. Just like the drag coefficient C_D of a particle in a uniform flow is modeled as a function of Reynolds number, the Sherwood number is modeled as a function of Reynolds number. The empirical Ranz–Marshal correlation for the Sherwood number is

$$\mathrm{Sh}_l = 2 + 0.6 \, \mathrm{Re}_l^{0.5} \, \mathrm{Sc}^{0.33}, \tag{15.19}$$

---

[2] Note that evaluation of mass flux using Fick's law requires additional calculation of the mass fraction field $Y_\alpha(\mathbf{x}, t)$ in the continuous phase. Thermodynamic and transport properties of the evaporating species will also be required.

where the Schmidt number $\text{Sc} = \nu_f/\mathcal{D}_\alpha$ is taken to be constant. The above model accounts for only the quasi-steady mass transfer. A more accurate model can be considered by appealing to heat and mass transfer analogy. An equation similar to Eq. (15.16) can be formulated to account for the unsteady effects of mass transfer. For now we leave such improvements for the reader to pursue.

Momentum and Energy Associated with Mass Exchange

In this subsection we shall consider momentum, internal, and kinetic energies associated with mass transfer. The exchange of mass and associated momentum and energy are precisely defined when the interface between the particle and the fluid is fully resolved. These definitions were presented in Eqs. (11.87) and (11.88) in terms of surface distributions of continuous-phase velocity, internal and total energy. When the details of the continuous phase are averaged with the filter operation and represented as filtered fluid velocity, internal and total energies (i.e., as $\mathbf{u}_{c@l}$, $e_{c@l}$, and $E_{c@l}$), the exchanged momentum and energy $\dot{\mathbf{M}}_l$, $\dot{e}_l$, and $\dot{k}_l$ need to be modeled in terms of these filtered continuous-phase quantities as well as the primary particle variables, $\mathbf{V}_l$, e_l, and E_l.

Although conceptually simple, consistent modeling of momentum and energy associated with mass transfer has been challenging. Before we discuss the model, let us look carefully at the exchange process as presented in Table 15.2, where the momentum, total, internal, and kinetic energy associated with the mass exchange $\dot{m}_l$ of the lth particle are presented in separate rows. The third column in the table presents these exchanges seen both from the perspective of the particle and the surrounding fluid. For example, as a result of the mass transfer, the lth particle gains a momentum of $\dot{\mathbf{M}}_l$, and the surrounding fluid correspondingly gains a momentum of $-\dot{\mathbf{M}}_l$. This exchange appeared in the respective momentum balances as

$$\frac{dm_{pl}\mathbf{V}_l}{dt} = \dot{\mathbf{M}}_l + \cdots \quad \text{and} \quad \frac{\partial \phi_c \rho_f \mathbf{u}_c}{\partial t} + \nabla \cdot (\phi_c \rho_f \mathbf{u}_c \mathbf{u}_c) = -G(\mathbf{x} - \mathbf{X}_l)\dot{\mathbf{M}}_l + \cdots .$$
(15.20)

In nonconservative form, the above can be rewritten as

$$m_{pl}\frac{d\mathbf{V}_l}{dt} = \left(\dot{\mathbf{M}}_l - \dot{m}_l\mathbf{V}_l\right) + \cdots \quad \text{and} \quad \phi_c \rho_f \frac{D\mathbf{u}_c}{Dt} = -G(\mathbf{x} - \mathbf{X}_l)\left(\dot{\mathbf{M}}_l - \dot{m}_l\mathbf{u}_c\right) + \cdots ,$$
(15.21)

where $\dot{m}_l\mathbf{V}_l$ is the momentum that resides in the mass $\dot{m}_l$ exchanged to the dispersed phase, while $-\int G(\mathbf{x} - \mathbf{X}_l)\dot{m}_l\mathbf{u}_c\, d\mathbf{x} = -\dot{m}_l\mathbf{u}_{cGl}$ is the momentum that resides in the mass $-\dot{m}_l$ exchanged to the filtered continuous phase. Thus, the adjusted right-hand sides contribute to changing the velocity of the dispersed and continuous phases. These are the contributions that appeared on the right-hand sides of Eqs. (15.6) and (15.2), which were in nonconservative form. The fourth column of Table 15.2 presents similarly adjusted sources of total, internal, and kinetic energies that contribute to $m_{pl}dE_l/dt$, $m_{pl}de_l/dt$, and $m_{pl}dk_l/dt$ in case of the particle and to $\phi_c\rho_f DE_c/Dt$, $\phi_c\rho_f De_c/Dt$, and $\phi_c\rho_f Dk_c/Dt$ for the filtered continuous phase.

Table 15.2 Detailed investigation of the momentum and energy associated with the mass exchange.

		Exchange	Adjusted Exchange	With Model (15.22)
Momentum	Particle	$\dot{\mathbf{M}}_l$	$\dot{\mathbf{M}}_l - \dot{m}_l \mathbf{V}_l$	$\dot{m}_l\left((\beta_{dM} - 1)\mathbf{V}_l + \beta_{cM}\mathbf{u}_{cGl}\right)$
	Fluid	$-\dot{\mathbf{M}}_l$	$-\left(\dot{\mathbf{M}}_l - \dot{m}_l \mathbf{u}_{cGl}\right)$	$-\dot{m}_l\left(\beta_{dM}\mathbf{V}_l + (\beta_{cM} - 1)\mathbf{u}_{cGl}\right)$
Total	Particle	$\dot{E}_l$	$\dot{E}_l - \dot{m}_l E_l$	$\dot{m}_l\left((\beta_{dE} - 1)E_l + \beta_{cE}E_{cGl}\right)$
Energy	Fluid	$-\dot{E}_l$	$-\left(\dot{E}_l - \dot{m}_l E_{cGl}\right)$	$-\dot{m}_l\left(\beta_{dE}E_l + (\beta_{cE} - 1)E_{cGl}\right)$
Internal	Particle	$\dot{e}_l$	$\dot{e}_l - \dot{m}_l e_l$	$\dot{m}_l\left((\beta_{dE} - 1)e_l + \beta_{cE}e_{cGl}\right)$
Energy	Fluid	$-\dot{e}_l$	$-(\dot{e}_l - \dot{m}_l e_{cGl})$	$-\dot{m}_l\left(\beta_{dE}e_l + (\beta_{cE} - 1)e_{cGl}\right)$
Kinetic	Particle	$\dot{k}_l$	$\dot{k}_l - \dot{m}_l k_l$	$\dot{m}_l\left((\beta_{dE} - 1)k_l + \beta_{cE}k_{cGl}\right)$
Energy	Fluid	$-\dot{k}_l$	$-\left(\dot{k}_l - \dot{m}_l k_{cGl}\right)$	$-\dot{m}_l\left(\beta_{dE}k_l + (\beta_{cE} - 1)k_{cGl}\right)$

We are now ready to consider appropriate models of $\dot{\mathbf{M}}_l$, $\dot{E}_l$, and so on. Simple models of these exchanges can be expressed as

$$\dot{\mathbf{M}}_l = \dot{m}(\beta_{dM}\mathbf{V}_l + \beta_{cM}\mathbf{u}_{cGl}), \qquad \dot{E}_l = \dot{m}(\beta_{dE}E_l + \beta_{cE}E_{cGl}),$$

$$\dot{e}_l = \dot{m}(\beta_{dE}e_l + \beta_{cE}e_{cGl}), \qquad \dot{k}_l = \dot{m}(\beta_{dE}k_l + \beta_{cE}k_{cGl}), \tag{15.22}$$

where β_{dM}, β_{cM}, β_{dE}, and β_{cE} are $O(1)$ constants, whose values can be determined from particle-resolved measurements or simulations. According to the above model, the velocity of the exchanged mass is at the weighted average of the particle and local filtered fluid velocities: $(\beta_{dM}\mathbf{V}_l + \beta_{cM}\mathbf{u}_{cGl})$. The total, internal, and kinetic energies of the exchanged mass are weighted similarly. This preserves the relations $\dot{E}_l = \dot{e}_l + \dot{k}_l$, $E_l = e_l + k_l$, and $E_{cGl} = e_{cGl} + k_{cGl}$. Note that the subscript cGl represents the filtered continuous-phase property integrated around the lth particle with Gaussian weight and it is expected to be somewhat different from that evaluated at the particle center (see Problem 15.2).

Adjusted exchanges with the above models are shown in column 5 of Table 15.2. As you will show in Problem 15.1, for a rigid isothermal particle we can make the approximation

$$\beta_{dM} = \beta_{dE} = 1 \qquad \text{and} \qquad \beta_{cM} = \beta_{cE} = 0, \tag{15.23}$$

which simply states that the velocity and energy of the exchanged mass is at the velocity and temperature of the rigid isothermal particle. The importance of this approximation becomes clear in the temperature equation of the particle. As can be seen in the fifth column of the table, the adjusted momentum and energy associated with the mass exchange does not contribute to altering the velocity or temperature of the particle.

The exchange models given in Eq. (15.23) will, however, have a nonzero influence on the velocity and temperature of the surrounding continuous phase. From the fifth column we can calculate the contribution to the continuous-phase acceleration to be $-\dot{m}_l (\mathbf{V}_l - \mathbf{u}_{cGl})$. Similarly, the effect of mass exchange on continuous-phase temperature is $-\dot{m}_l (e_l - e_{cGl})$, which can also be expressed as $-\dot{m}_l (C_p T_l - C_{vf} T_{cGl})$. If we assume the phase change to occur under the isothermal condition of the droplet, then the above can be separated into two parts:

$$\underbrace{-\dot{m}_l \left(C_p T_l - C_{vf}\right) T_l}_{\text{latent heat}} \underbrace{-\dot{m}_l C_{vf} \left(T_l - T_{cGl}\right)}_{\text{sensible heat}} . \tag{15.24}$$

According to the model, the latent and sensible heats contribute to changing the continuous-phase temperature, But they do not alter the particle temperature in a direct way. This is a consequence of assumption (15.23). If this does not correctly reflect the reality as captured in an experiment or in a particle-resolved simulation, then the approximation given in Eq. (15.23) must be modified appropriately to better capture the reality. For example, it should be noted that the assumption given in Eq. (15.23) becomes less accurate in the case of droplets, where there is strong internal circulation and the derivation presented below in Problem 15.1 does not apply. Further research is needed in order to develop better models of momentum and energy transfers associated with mass exchange.

Problem 15.1 The expressions given in Eqs. (11.87) and (11.88) have approached the mass, momentum, and energy exchange from the perspective of the continuous phase surrounding the lth particle. The same quantities can be considered from the perspective of the lth particle as well. After all, if the exchanges are positive (or negative) for the continuous phase, then they are negative (or positive) of equal magnitude for the dispersed phase. Write the mass exchange between the lth particle and the surrounding as

$$\dot{m}_l = \int_{S_l} \rho_p \left(\mathbf{u}_l(\mathbf{x}',t) - \mathbf{V}_l(t)\right) \cdot \mathbf{n}\, dA', \tag{15.25}$$

where $\mathbf{n}$ is the outward unit normal of the particle.

(a) By noting that the dispersed-phase density is orders of magnitude larger than the fluid-phase density, in case of evaporation or sublimation, compare the above with the integral within curly brackets in Eq. (11.85), and show that $|\mathbf{u}_l - \mathbf{u}| \gg |\mathbf{u}_l - \mathbf{V}|$. In other words, show that the relative motion of the interface with respect to the particle motion is quite small, while the fluid velocity at the particle surface can be different from the particle velocity in case of vigorous mass transfer.

(b) Obtain the following simple expression for the associated momentum exchange between the lth particle and the surrounding:

$$\dot{\mathbf{M}}_l = \int_{S_l} \rho_p \mathbf{V}(t) \left(\mathbf{u}_l(\mathbf{x}',t) - \mathbf{V}(t)\right) \cdot \mathbf{n}\, dA' = \dot{m}_l \mathbf{V}_l. \tag{15.26}$$

(c) Obtain similar expressions for kinetic and internal energies associated with the mass exchange as

$$\dot{k}_l = \dot{m}_l \frac{\mathbf{V}_l \cdot \mathbf{V}_l}{2}, \quad \dot{E}_l = \dot{m}_l E_l, \quad \text{and} \quad \dot{e}_l = \dot{m}_l e_l, \tag{15.27}$$

where $e_l = C_p T_l$ is the per-mass internal energy of the particle corresponding to temperature T_l and C_p is the specific heat of the dispersed phase, which is taken to be a constant. Also, $E_l = e_l + |\mathbf{V}_l|^2/2$. Substitute the above results for an individual particle in Eq. (11.86) and obtain

$$\dot{\mathbf{M}}(\mathbf{x},t) = \sum_{l=1}^{N_p} G(\mathbf{x}-\mathbf{X}_l)\,\dot{m}_l\mathbf{V}_l \quad \text{and} \quad \dot{E}(\mathbf{x},t) = \sum_{l=1}^{N_p} G(\mathbf{x}-\mathbf{X}_l)\,\dot{m}_l\left(e_l + \frac{1}{2}\mathbf{V}_l\cdot\mathbf{V}_l\right).$$

(15.28)

Problem 15.2 In this problem you will evaluate the difference between $\mathbf{u}_{cGl}$ and $\mathbf{u}_{c@l}$, where the former is the Gaussian-weighted integral of $\mathbf{u}_c$ centered around the lth particle, while the latter is $\mathbf{u}_c$ evaluated at the center of the lth particle. You will accomplish this by carrying out the integral after expanding $\mathbf{u}_c$ in a Taylor series as

$$\begin{aligned}
\mathbf{u}_{cGl} &= \int G(\mathbf{x}-\mathbf{X}_l)\,\mathbf{u}_c\,d\mathbf{x} \\
&= \int G(\mathbf{x}-\mathbf{X}_l)\left(\mathbf{u}_{c@l} + (\mathbf{x}-\mathbf{X}_l)\cdot(\nabla\mathbf{u}_c)_{@l}\right. \\
&\quad \left. + \frac{1}{2}(\mathbf{x}-\mathbf{X}_l)(\mathbf{x}-\mathbf{X}_l):(\nabla\nabla\mathbf{u}_c)_{@l} + \cdots\right)d\mathbf{x} \\
&= \mathbf{u}_{c@l} + \frac{1}{2}\sigma^2(\nabla^2\mathbf{u}_c)_{@l} + \cdots,
\end{aligned}$$

(15.29)

where we have used the definition of the Gaussian filter given in Eq. (11.4) and we recall that σ is related to the Gaussian filter width. Since $\mathbf{u}_c$ has already been filtered (or smoothened) by the Gaussian, we expect the contribution of the higher-order terms to be smaller.

15.3.5 Closure Models of σ_c and $\mathbf{q}_c$

We start with the closure model of the continuous-phase stress tensor, which from the discussion of Section 11.2.2 is given by

$$\sigma_c(\mathbf{x},t) = \frac{1}{\phi_c}\int \sigma(\mathbf{x}',t)\,I_c(\mathbf{x}',t)\,G(\mathbf{x}-\mathbf{x}')d\mathbf{x}'.$$

(15.30)

Thus, $\sigma_c(\mathbf{x},t)$ is the filtered continuous-phase stress around the point $\mathbf{x}$. We assume the fluid to be Newtonian and the stress tensor is given by

$$\sigma = -p\mathbf{I} + 2\mu_f\,\mathbf{S} + \lambda_f(\nabla\cdot\mathbf{u})\mathbf{I},$$

(15.31)

where the isotropic part is the fluid pressure p. The deviatoric part depends on the strain-rate tensor $\mathbf{S} = \left(\nabla\mathbf{u} + (\nabla\mathbf{u})^{\mathsf{T}}\right)/2$, with the constant of proportionality being the fluid viscosity μ_f. The last term corresponds to stress due to the compressibility of the fluid with λ_f being the bulk viscosity of the fluid. Substituting the above into Eq. (15.30) we obtain the closure model of the macroscale continuous-phase stress tensor as

$$\boldsymbol{\sigma}_c = -p_c \mathbf{I} + 2\mu_m \mathbf{S}_m + \lambda_m (\nabla \cdot \mathbf{u}_m)\mathbf{I}, \tag{15.32}$$

where $\mathbf{S}_m = \left(\nabla \mathbf{u}_m + (\nabla \mathbf{u}_m)^\mathrm{T}\right)/2$ is the strain rate of the volume-averaged mixture velocity $\mathbf{u}_m = \phi_c \mathbf{u}_c + \phi_d \mathbf{u}_d$. Several observations must be made. First, p_c is the filtered macroscale pressure given by

$$p_c(\mathbf{x}, t) = \frac{1}{\phi_c} \int p(\mathbf{x}', t)\, I_c(\mathbf{x}', t)\, G(\mathbf{x} - \mathbf{x}')d\mathbf{x}'. \tag{15.33}$$

Second, we have used the result of Problem 11.12 that the filtered strain rate of the continuous phase is the strain rate of the volume-averaged mixture velocity. Third, we recall that in a particle-laden flow, even when there is no relative velocity between the phases, the linear relation between average macroscale stress and macroscale strain rate changes. The increased fluid stress due to the no-slip and no-penetration boundary conditions imposed by the particles must be included with the Einstein-like viscosity correction. Therefore, in the above, one may consider using the mixture viscosity $\mu_m = \mu_f(1 + 2.5\phi_c)$ (see Eq. (4.59)) to account for the modification of the stress–strain relation. It must, however, be cautioned that in the presence of velocity slip between the particles and the surrounding fluid, effective viscosity may deviate from the above classic relation. In the last term, λ_m is the effective bulk viscosity of the mixture.

It must be emphasized that the closure presented in Eq. (15.32) is a model and thus subject to approximation error. Foremost among the approximations is the application of Einstein's viscosity correction, which is based on Stokes flow over a very dilute distribution of spherical particles that have no translational or rotational slip with the local fluid. When we apply the model to a finite volume fraction, finite Reynolds number flows, and nonzero slip velocity between the particles and the local fluid, the mixture viscosity relation must be appropriately modified. For example, with increasing volume fraction, interaction between the particles begins to play a role and must be taken into account in the stress–strain-rate relation. Also, changes in the two-particle (or radial) distribution function has an influence. The higher-order terms of effective mixture viscosity have been obtained as (Batchelor, 1974)

$$\mu_m = \mu_f \left(1 + 2.5\phi_d + 7.6\phi_d^2\right). \tag{15.34}$$

Starting from the Fourier law of heat conduction, the closure model of macroscale heat flux proceeds in an analogous manner and results in the following expression:

$$\mathbf{q}_c = k_m \nabla T_m. \tag{15.35}$$

Again, two aspects of the above expression must be considered. First, we have used the result of Problem 11.13 that the filtered thermal gradient of the continuous phase is the gradient of the volume-averaged mixture temperature $T_m = \phi_c T_c + \phi_d T_d$. Analogous to Einstein's viscosity, the effective thermal conductivity k_m of a multiphase mixture with embedded particles is higher than that of the pure fluid. In fact, this problem of effective thermal conductivity is much older than the effective viscosity problem, and was first solved by Maxwell (1873), who obtained the leading-order term that is proportional to the volume fraction of particles. Following the original derivation of Maxwell, higher-order terms have been obtained to account for higher volume fraction and other effects (Batchelor, 1974; Balachandar and Michaelides, 2022). An expression for effective thermal conductivity is given by

$$k_m = k_f + \frac{3(k_R - 1)}{k_R + 2}\phi_d + \frac{3(k_R - 1)^2}{(k_R + 2)^2}\phi_d^2, \tag{15.36}$$

where k_R is the ratio of particle to fluid thermal conductivity.

15.3.6 Closure Models of Subgrid Terms τ_{sg} and Q_{sg}

We start with the definition

$$\tau_{sg} = \frac{1}{\phi_c}\left(\phi_c \rho_c \tilde{\mathbf{u}}_c \tilde{\mathbf{u}}_c - \overline{(\rho_f \mathbf{u}\mathbf{u})}\right) \tag{15.37}$$

for the subgrid residual stress given in Eq. (11.40). As discussed in Problem 11.5, the residual stress can be decomposed into three contributions: Leonard stress, cross stress, and Reynolds stress. The Leonard stress accounts for the residual effect of the nonlinear interaction among the filtered large scales, and can be calculated explicitly in terms of the filtered variables, provided the filter G and the filtering operation are precisely defined. The cross stress arises from the residual interaction of the filtered large and small scales, while the Reynolds stress is often the dominant contribution and arises from the nonlinear interaction of the small scales.

Though closure modeling can be pursued for each term independently, here we will pursue modeling of τ_{sg} as the combined effect of all three contributions. However, for modeling purposes, we distinguish the two sources of unresolved velocity fluctuations, so that the effect of each can be modeled separately. The first source is "pseudo turbulence" due to the random location and motion of the particles. This accounts for velocity variation within the boundary layer and the wake around all the particles. The second source of subgrid fluctuation is due to a continuous-phase turbulence cascade. This second source is from turbulence that exists even in the absence of particles. We recognize the fact that in multiphase flows, the mesoscale turbulence cascade is modified by the presence of particles and therefore mesoscale turbulence is different from that of the corresponding single-phase limit. When a portion of this mesoscale turbulence is filtered out by the filtering operation, its contribution to subgrid stress must also be modeled.

We take this opportunity to introduce a modeling strategy that is applicable in all EL approaches. This strategy is illustrated in Table 15.3. In this section, we are concerned only with EL-DNS, where all the scales of mesoscale turbulence are resolved and therefore subgrid stress is entirely due to pseudo turbulence. In addition, if the particle volume fraction is very dilute, then the contribution from pseudo turbulence is small as well and one can assume $\tau_{sg} \approx 0$. In the case of EL-ELS, to be considered in the following subsection, in case of dilute particle distribution, the subgrid turbulence is mostly from the turbulence cascade and we can approximate $\tau_{sg} \approx \tau_{ct}$, where the

Table 15.3 The different modeling needs for subgrid stress and subgrid heat flux.

	$\phi_d \ll 1$ (dilute limit)	$\phi_d < 1$ (not dilute)
EL-DNS: $\eta > \mathcal{L} \gg d_p$	$\tau_{sg} \approx 0,\ \mathbf{q}_{sg} \approx 0$	$\tau_{sg} \approx \tau_{pt},\ \mathbf{q}_{sg} \approx \mathbf{q}_{pt}$
EL-LES: $\mathcal{L} \gg \eta, d_p$	$\tau_{sg} \approx \tau_{ct},\ \mathbf{q}_{sg} \approx \mathbf{q}_{ct}$	$\tau_{sg} \approx \tau_{ct} + \tau_{st},\ \mathbf{q}_{sg} \approx \mathbf{q}_{ct} + \mathbf{q}_{st}$
Large particle: $\mathcal{L} \sim d_p$	Complex	Complex

subscript ct corresponds to classical turbulence, as opposed to pseudo turbulence. In the case of EL-LES simulation of finite volume fraction flows, the modeling of τ_{sg} must include contributions from both filtered classical and pseudo turbulence. A simple addition is suggested in Table 15.3, although it is to be expected that the interaction between ambient and particle-generated turbulence is likely to be highly nonlinear and not simply additive. We will address EL-LES modeling of τ_{sg} in the next subsection. In the limit when particle size is comparable to filter size, even when the particle volume fraction is small, the filtered particle boundary layer and wake will affect subgrid stress at the grid points that are close to the particle. Thus, the closure model of τ_{sg} becomes complex, and the level of complexity increases with increasing particle volume fraction.

The rest of this subsection is concerned with EL-DNS modeling of pseudo turbulence. Two simple approaches to modeling pseudo turbulence were presented in Section 9.7.1. Both of them will be elaborated for application in an Euler–Lagrange methodology. The first approach is Eulerian based (see Eq. (9.100)), where τ_{pt} was expressed as a function of local particle Reynolds number and particle volume fraction. In the EL approach, the particle volume fraction $\phi_d(\mathbf{x}, t)$ was defined earlier in Eq. (15.4). The particle velocity field $\mathbf{u}_d(\mathbf{x}, t)$ and Reynolds number field $\text{Re}(\mathbf{x}, t)$ can similarly be defined by volume-weighting the individual particle contributions as (since the particle density is taken to be constant, this is tantamount to mass-weighting as well)

$$\mathbf{u}_d(\mathbf{x}, t) = \frac{1}{\phi_d} \sum V_{pl}\, \mathbf{V}_l\, G(\mathbf{x} - \mathbf{X}_l) \quad \text{and}$$

$$\text{Re}(\mathbf{x}, t) = \frac{1}{\phi_d} \sum V_{pl}\, \frac{|\mathbf{u}_c(\mathbf{X}_l, t) - \mathbf{V}_l|\, d_{pl}}{\nu}\, G(\mathbf{x} - \mathbf{X}_l). \tag{15.38}$$

Again the summation is formally over all the particles, however, only particles that are in the neighborhood of the point $\mathbf{x}$ will contribute. From the volume fraction and Reynolds number fields, the pseudo-turbulent kinetic energy and anisotropy coefficient can be obtained using Eqs. (9.101) and (9.102), which can then be substituted into Eq. (9.100) to obtain the pseudo-turbulent Reynolds stress tensor $\tau_{pt,axi}$. This Reynolds stress tensor was defined with the principle x-axis aligned with the direction of local relative velocity. However, in an EL simulation, at the point $(\mathbf{x}, t)$, the local relative velocity $\mathbf{u}_c(\mathbf{x}, t) - \mathbf{u}_d(\mathbf{x}, t)$ is not guaranteed to be along the x-axis and therefore the Reynolds stress tensor must be properly transformed to the computational coordinates. Let Q^{T} be the rotation matrix that transforms the x-coordinate to align with the direction of local relative velocity. Note that the definition of Q and its transpose Q^{T} is not unique, since owing to axisymmetry, the y and z-axes can transform to any two perpendicular directions on the plane normal to the local relative vector. With the rotation matrix defined, the pseudo-turbulent Reynolds stress to be used in the continuous-phase momentum equations can be written as

$$\boldsymbol{\tau}_{\mathrm{pt}} = \frac{1}{2}|\mathbf{u}_c - \mathbf{u}_d|^2 \left[\boldsymbol{Q}^{\mathrm{T}}\boldsymbol{\tau}_{\mathrm{pt,axi}}\boldsymbol{Q}\right], \qquad (15.39)$$

where we recognize the fact that in Section 9.7.1, $\boldsymbol{\tau}_{\mathrm{pt,axi}}$ was normalized by kinetic energy based on relative velocity.

The second approach is Lagrangian based, where the Reynolds stress of each individual particle is calculated, and thus can be applied directly to the EL approach. We recall Eq. (9.103) for the pseudo-turbulent Reynolds stress tensor of an individual particle $\mathscr{R}(\mathrm{Re}_l, \phi_{d@l})$ given in Section 9.7.1. From this, the Eulerian field of pseudo-turbulent subgrid stress is obtained by summing the contributions of all N particles within the system and filtering the sum as

$$\boldsymbol{\tau}_{\mathrm{pt}}(\mathbf{x}, t) = \sum_{l=1}^{N} \boldsymbol{Q}_l^{\mathrm{T}} \mathscr{R}(\mathrm{Re}_l, \phi_{d@l}) \boldsymbol{Q}_l \, G(\mathbf{x} - \mathbf{X}_l). \qquad (15.40)$$

Note that by definition, $\mathscr{R}(\mathrm{Re}_l, \phi_{d@l})$ is axisymmetric and oriented in the direction of relative velocity of the lth particle. The left and right multiplication with the rotational matrix $\boldsymbol{Q}_l$ and its transpose is to transform the contribution of the lth particle to the fixed coordinate in which $\boldsymbol{\tau}_{\mathrm{pt}}$ is defined. The transformed contribution of the lth particle is then multiplied by the Gaussian filter and summed over all N particles. Note that in this case, the pseudo-turbulent Reynolds stress of each particle has already been properly scaled by the square of relative velocity and therefore no further scaling is needed.

As the final step of the closure models, we now turn to models of the subgrid heat-flux term $\mathbf{q}_{\mathrm{sg}}$. In this section, we are concerned with EL-DNS and therefore the subgrid heat flux is entirely due to pseudo turbulence. Hence

$$\text{EL-DNS:} \quad \mathbf{q}_{\mathrm{sg}} = \mathbf{q}_{\mathrm{pt}}. \qquad (15.41)$$

For pseudo-turbulent heat flux the only available model is Eulerian based from the work of Sun et al. (2016), as discussed in Section 9.8.1. Based on the results presented there, a simple model of pseudo-turbulent heat flux is

$$\mathbf{q}_{\mathrm{pt}} = (1 - \phi_d)\left(0.2 + 1.2\phi_d - 1.24\phi_d^2\right)\exp(-0.002\,\mathrm{Re}_{su})\,(\mathbf{u}_c - \mathbf{u}_d)\,(T_c - T_d), \quad (15.42)$$

where $T_d(\mathbf{x}, t)$ is the Eulerian particle temperature field calculated by volume-weighting the individual Lagrangian particle temperatures as

$$T_d(\mathbf{x}, t) = \frac{1}{\phi_d}\sum V_{pl}\,T_{pl}\,G(\mathbf{x} - \mathbf{X}_l). \qquad (15.43)$$

15.4   Closure Relations of EL-LES – Small Particle Limit

In this section we will reconsider the closure models of all the secondary variables presented in Table 11.1, but now in the context of EL-LES, where the filtered continuous-phase properties do not account for the entire range of mesoscale turbulence. In other words, here we consider the limit $\mathcal{L} \gg (\eta, d_p)$. As a consequence of such a choice of filter length scale larger than the Kolmogorov scale, subgrid turbulence will consist of both unresolved mesoscale turbulence and pseudo turbulence. In what follows we will briefly consider each of the closure models that we discussed in the six subsections above. We will discuss modifications that must be made in order to account for unresolved mesoscale turbulence. As we will see, reasonable modifications can be made in some closure models, while in others we lack the necessary information to accurately account for unresolved mesoscale turbulence. As a result, the closure models of EL-LES are likely to be associated with larger errors and uncertainties than those presented for EL-DNS.

Closure Models of $\mathbf{F}'_l$ and $\mathbf{T}'_l$

The particle force model given in Eq. (15.8) still applies. However, the continuous-phase quantities that appear in the model must be properly redefined. For example, in the quasi-steady force, the computed EL velocity $\mathbf{u}_{c@l}$ now corresponds only to the large eddy portion of the fluid velocity evaluated at the location of the lth particle. The undisturbed fluid velocity seen by the lth particle must also include a contribution from the subgrid turbulence that has been filtered out. Since subgrid velocity fluctuation is not part of the EL solution, it must be modeled. For this, we resort to the Langevin model introduced in Section 7.5.6. Following that analysis, we express the fluid velocity seen by the particle as

$$\mathbf{u}_{@l} = \mathbf{u}_{c@l} + \mathbf{u}''_{@l}, \tag{15.44}$$

where the second term on the right is the stochastic component obtained from the Langevin model that accounts for the filtered subgrid component. According to the Langevin model given in Eq. (7.64), $\mathbf{u}''_{@l}$ at the lth particle is evaluated as a component parallel to the direction of relative velocity (i.e., in the direction of $\mathbf{u}_{c@l} - \mathbf{V}_l$) and a component perpendicular to the direction of relative velocity. Important adjustments must, however, be made. The Langevin model as developed in Section 7.5.6 accounts for the entire spectrum of turbulence, while in the present context, $\mathbf{u}''$ must only account for the effect of small eddies that have been filtered. Therefore, in the application of Eq. (7.64) and other related equations, quantities such as turbulent kinetic energy k and mean dissipation $\langle \epsilon \rangle$ must be correctly interpreted as those of subgrid turbulence. For further details on the application of the Langevin model in EL-LES, we refer the reader to the growing body of literature: Pozorski and Minier (1998); Minier et al. (2004); Dehbi (2008); Lattanzi et al. (2020, 2022).

Once obtained, $\mathbf{u}_{@l}$, which now includes the stochastic subgrid contribution, can be used to calculate the quasi-steady force on the lth particle. The inclusion of the stochastic component increases the level of particle-to-particle velocity variation. This increase in particle velocity fluctuation, if correctly captured with the Langevin model, plays an important role in accurately representing the longitudinal and transverse dispersion of the particles, by accounting for the effect of the small-scale eddies.

Similarly, in the evaluation of the added-mass and viscous unsteady forces in Eq. (15.8), $(D\mathbf{u}_c/Dt)_{@l}$ and $d\mathbf{u}_{c@l}/dt$ must be appropriately augmented with stochastic contributions to fluid acceleration following the fluid and the particle, respectively. Models for these subgrid accelerations are not as developed and well tested as those of subgrid velocity and therefore will not be discussed here. Furthermore, our understanding of the subgrid effect on torque on the lth particle is lacking and therefore its contribution is typically ignored.

Closure Model of W_l'

The dissipative nature of quasi-steady and viscous unsteady forces remains the same, even in the context of EL-LES, and therefore they make a nonzero contribution to the continuous-phase energy equation. Again, the added-mass and lift forces do not make a net contribution to the filtered continuous-phase thermal equation. In essence, instead of Eq. (15.14) we can now make the following substitution:

$$\left[W_l' - \mathbf{u}_{cGl} \cdot \mathbf{F}_l' \right] \approx \left[\left(\mathbf{F}_{qsl}' + \mathbf{F}_{vul}' \right) \cdot \left(\mathbf{V}_l - (\mathbf{u}_{cGl} + \mathbf{u}_{@l}'') \right) \right], \tag{15.45}$$

which accounts for the stochastic component of the subgrid fluid velocity seen by the lth particle. The effect of dissipative heating is generally weak in incompressible flows and therefore this contribution can often be ignored altogether.

Closure Model of Q_l'

The finite-Pe BBO-like equation of heat transfer of the lth particle given in Eq. (15.16) remains valid in the EL-LES regime as well. However, as with the velocity, the undisturbed fluid temperature at the particle location is only partially represented by $T_{c@l}$ in the expression for quasi-steady heat transfer. $T_{c@l}$ only corresponds to the fluid temperature due to the large eddies and the subgrid contribution from the filtered small-scale eddies must be included. A thermal Langevin model is required to model the stochastic contribution and the undisturbed fluid temperature at the particle location is given by

$$T_{@l} = T_{c@l} + T_{@l}''. \tag{15.46}$$

Ideally, subgrid contributions to DT_c/Dt and dT_c/dt must also be modeled if the undisturbed and history heat transfer contributions must include the effect of the subgrid contribution. Often these effects are ignored, mainly for lack of reliable models.

Closure Models of $\dot{m}_l$ and Associated Exchanges

The quasi-steady mass transfer model of Eq. (15.18) can still be used with the undisturbed mass fraction of the surrounding fluid given by $Y_{\alpha c@l} + Y_{\alpha@l}''$, with a Langevin model for the stochastic component. The Langevin model for mass transfer can be very similar to that for heat transfer, owing to the heat and mass transfer analogy. The momentum and energy exchanges associated with the mass transfer can simply be taken to be those given for EL-DNS without any major modification.

Modeling of σ_c and $\mathbf{q}_c$

Since these closure models pertain to only the filtered quantities, without the direct influence of the unresolved small scales, the modeling of σ_c and $\mathbf{q}_c$ remains the same as those given in Eqs. (15.32) and (15.35).

Modeling of Residual Stress $\boldsymbol{\tau}_{\mathrm{sg}}$ and Residual Heat Flux $\mathbf{q}_{\mathrm{sg}}$

The biggest difference between EL-DNS and EL-LES appears in the modeling of residual stress and flux. In EL-DNS, the subgrid eddies, whose effects on the resolved scales are modeled as $\boldsymbol{\tau}_{\mathrm{sg}}$ and $\mathbf{q}_{\mathrm{sg}}$, are entirely due to pseudo turbulence. In EL-LES, the subgrid eddies include filtered mesoscale turbulence, whose contribution to $\boldsymbol{\tau}_{\mathrm{sg}}$ and $\mathbf{q}_{\mathrm{sg}}$ will be considered here. As discussed in Section 9.7.1, the modeling of pseudo turbulence is directly related to the distribution of particles and their relative velocity with respect to the surrounding continuous phase. In contrast, in the case of subgrid turbulence, the turbulence cascade determines the back-effect of filtered subgrid scales on the large-scale dynamics. The influence of the suspended particles is indirect, only through the multiphase modification of mesoscale turbulence spectra from that of single-phase turbulence. As a result, the contributions to $\boldsymbol{\tau}_{\mathrm{sg}}$ and $\mathbf{q}_{\mathrm{sg}}$ from mesoscale turbulence have been modeled using traditional LES closures that have been well developed in the context of single-phase turbulence.

If we consider the regime where filtered mesoscale turbulence is the dominant portion of subgrid turbulence, the effect of pesudo turbulence becomes less important and can be ignored. In this regime, a simple gradient-diffusion model with eddy viscosity for subgrid residual stress can be written as

$$\boldsymbol{\tau}_{\mathrm{ct}} = 2\mu_t \mathbf{S}_a - \frac{2}{3}k\,\mathbf{I} \quad \text{with} \quad \mu_t = C_\mu \rho_c \phi_c \Delta^2 \, |(\mathbf{S}_a : \mathbf{S}_a)^{1/2}|, \tag{15.47}$$

where k is the subgrid kinetic energy, μ_t the eddy viscosity, $\mathbf{I}$ the identity tensor, Δ the mean size of the grid, and $\mathbf{S}_a = (\nabla \mathbf{u}_c + (\nabla \mathbf{u}_c)^{\mathrm{T}})/2 - (\nabla \cdot \mathbf{u}_c)\mathbf{I}/3$ is the anisotropic part of the filtered strain-rate tensor. The isotropic part of $\boldsymbol{\tau}_{\mathrm{sg}}$ is represented by $-2k\mathbf{I}/3$ and it need not be modeled, since this term can be combined with the filtered pressure p_c. The modeling of the anisotropic part reduces to evaluation of the model constant C_μ. A variety of approaches have been developed in the context of single-phase turbulence. The simplest among them being the Smagorinsky model, which assumes C_μ to be a constant that is traditionally set to 0.0256 (Lilly, 1967). A more sophisticated approach will be to use the dynamic Smagorinsky model (Germano et al., 1991; Moin et al., 1991; Pierce and Moin, 1998), where C_μ is dynamically evaluated at each grid point at each time, based on the local turbulence cascade. For stability, some spatial or temporal average is typically needed. The details of the dynamic Smagorinsky model and its variants is a rich topic in itself and can be found in books on LES.

A similar simple gradient-diffusion model with eddy diffusivity has been used for subgrid residual heat flux, yielding

$$\mathbf{q}_{\mathrm{ct}} = \frac{1}{\mathrm{Pr}_t}\mu_t C_{\mathrm{pf}}\,\nabla T_c\,, \tag{15.48}$$

where the turbulent Prandtl number Pr_t is typically taken to be an $O(1)$ constant. With the above model, since μ_t is already known from the residual stress, no additional effort is needed in computing the residual flux. It must be stressed that though the above-defined subgrid stress and heat flux models are the same as in the single-phase turbulence, the actual values of τ_{sg} and $\mathbf{q}_{\text{sg}}$ in a multiphase flow will be different from those of the corresponding single-phase turbulence, since the filtered multiphase velocity and thermal fields will differ substantially from their single-phase counterparts.

Almost all existing EL-LES simulations (Garcia et al., 2007; Arolla and Desjardins, 2015; Pakseresht and Apte, 2019; K. Liu et al., 2021; Taborda and Sommerfeld, 2021) have only used the above LES closures for subgrid residual stress and flux. This is generally justified if filtered mesoscale turbulence is much larger than the pseudo-turbulent contribution. However, in situations where filtered mesoscale turbulence and pesudo turbulence are of comparable strength, both must be included in the modeling of σ_{sg} and Q_{sg}. A naive approach is to add τ_{pt} and $\mathbf{q}_{\text{pt}}$ defined in Section 15.3.6 to the above LES closures. Further work is clearly needed to develop more accurate models that account for the nonlinear interaction between the two sources of subgrid turbulence.

15.5   Multiphase Effects and Further Simplifications

Despite the assumptions of incompressibility and constant fluid and particle properties, the governing equations of the continuous phase in Eqs. (15.1) to (15.3) remain more complicated than the standard single-phase Navier–Stokes equations. The difference accounts for the five multiphase mechanisms that were introduced in Section 2.2: *volume*, *inertia*, *body-force*, *thermodynamic/transport*, and *slip* effects. In this section, we will identify how each of these effects appear in the continuous-phase governing equations. This will allow us to appropriately simplify the governing equations if one or more of the multiphase mechanisms are known to be unimportant in a multiphase application. In the limit when all the multiphase mechanisms are absent, the flow becomes one-way coupled and we should recover the Navier–Stokes equations for the continuous phase.

Volume Effect

We recall that the volume effect simply accounts for the fact that the particles occupy a portion of the computational domain and the fluid fills only the remaining space. The volume effect directly influences the continuous-phase mass balance through the fact that the particle volume fraction is non-negligible (i.e., $\phi_d \neq 0$) and accordingly in Eq. (15.1), $\phi_c \neq 1$. If the volume occupied by the particles everywhere within the domain is negligibly small (i.e., if the volume effect is negligible), we can make the approximation $\phi_c \approx 1$, which simplifies the continuous-phase mass balance to

$$\nabla \cdot \mathbf{u}_c = -\sum_{l=1}^{N_P} G(\mathbf{x} - \mathbf{X}_l) \frac{\dot{m}_l}{\rho_f} . \tag{15.49}$$

In the absence of mass exchange between the phases, the above reduces to the standard incompressibility condition of $\nabla \cdot \mathbf{u}_c = 0$.

The volume effect also appears in other terms of the momentum and energy balances. In the momentum equation (15.2), only the fraction ϕ_c of the stress divergence force $\nabla \cdot \sigma_c$ is applied to the fluid and the balance $\phi_d \nabla \cdot \sigma_c$ contributes to particle momentum. Similarly, in the energy equation (15.3), in the second term on the right-hand side, only the fraction ϕ_c of the heat flux $\mathbf{q}_c$ is applied to the fluid. If the volume effect can be neglected, then in both these terms we can set $\phi_c \approx 1$. Another simplification that is appropriate is to approximate the volume-averaged mixture velocity to be the same as the continuous-phase velocity (i.e., $\mathbf{u}_m \approx \mathbf{u}_c$). The mixture velocity appears both in the definition of the stress tensor in Eq. (15.32) and in the viscous dissipation term on the right-hand side of Eq. (15.3), and therefore these terms can be simplified. Similarly in the closure of the subgrid heat transfer term in Eq. (15.35) we can make the following approximation for the mixture temperature: $T_m \approx T_c$. At low volume fraction we can also ignore the effect of pesudo turbulence.

Inertia Effect

In an incompressible multiphase flow, even though the fluid and the particle densities are constant, due to volume fraction variation, the density of the mixture $\rho_m = \phi_c \rho_f + \phi_d \rho_p$ varies over space and time. This variation in the mass of the mixture is the source of the inertia effect. The inertia effect appears on the left-hand sides of the continuous-phase momentum and energy equations (15.1) to (15.3) as the product $\phi_c \rho_f$, and the corresponding dispersed-phase inertia appears implicitly as the feedback force and heat transfer terms on the right-hand side. In applications where the density of the dispersed phase is of the order of the continuous phase, the inertia and volume effects are related. That is, if the volume effect can be ignored, then the inertia effect can be ignored as well by setting $\phi_c \rho_f \approx \rho_f$. In gas–solid flows, where the particle-to-gas density ratio is large, a small volume fraction does not necessarily guarantee a small particle mass loading. Only in the case $\phi_d \rho_p \ll \phi_c \rho_f$ can we ignore the inertia effect by setting $\phi_c \rho_f \approx \rho_f$ on the left-hand sides of the continuous-phase momentum and energy equations.

Body-Force Effect

The body-force effect is represented by the gravity term in the continuous and the particle momentum equations. As discussed in the introduction, in multiphase flows that are externally driven, if the flow and particle accelerations are much larger than the acceleration due to gravity, and if the particle and fluid velocities are much larger than the still-fluid settling velocity of the particles, then the effect of body force on multiphase dynamics becomes less important, and the body-force terms can be ignored in the momentum equations.

However, in other applications, especially in geophysical and environmental multiphase flows, the importance of body force must be carefully evaluated in order to properly decide whether to retain the gravity term or not in the momentum equations. In problems such as thermal convection or sediment-laden turbidity currents, where gravity is the sole source of multiphase dynamics, the body-force effect cannot be ignored even if the inertia effect can be. This leads to the important *Boussinesq approximation*, where we set $\phi_c = 1$ and $\rho_m = \rho_c$ in the governing equations except in the body-force term. The rationale for this approximation is as follows. If the particle mass loading is sufficiently small, we can ignore the volume and inertia effects and consider the fluid to occupy the entire volume. However, $\phi_c \neq 1$ becomes important when multiplied by $\mathbf{g}$ in the body-force term, which becomes the leading-order term in the momentum balance.

A subtle point regarding body force is worth noting. The body force on the continuous and dispersed phases can be combined to obtain the total body force per unit volume to be $\rho_m \mathbf{g}$. The total body force is then split into a horizontally averaged part that varies only along the vertical direction and a perturbation part as

$$\langle \rho_m \rangle_{xy}(z, t)\mathbf{g} + \rho_m' \mathbf{g}, \tag{15.50}$$

where $\langle \cdot \rangle_{xy}$ indicates an average over horizontal x–y planes. The first term can be written as a gradient and thus contributes only to the hydrostatic pressure variation. The second term contributes to multiphase dynamics. Thus, only the horizontal variation in mixture density can contribute to the body-force effect. In general, if there are other mechanisms driving the multiphase flow and if the inertia effect is small, then the body-force effect can also be deemed small.

Thermodynamic/Transport Effect
We now turn to the multiphase effect on the stress–strain-rate constitutive relation in the form of modified viscosity. At low volume fraction, this effect can be ignored by approximating the effective mixture viscosity as $\mu_m \approx \mu_f$, instead of Eq. (15.34). Similarly, at low volume fraction the effective thermal conductivity of the mixture can be taken to be that of the fluid itself, with the approximation $k_m \approx k_f$.

Slip Effect
The slip effect of nonzero relative velocity and nonzero relative temperature between the particle and the surrounding fluid is represented by the summation of $\mathbf{F}'$ and Q' over all the particles. In the limit where the slip effect is unimportant for the multiphase flow dynamics, these summation terms can be ignored in Eqs. (15.1) to (15.3). Also, the pesudo-turbulence terms $\boldsymbol{\tau}_{\mathrm{pt}}$ and $\mathbf{q}_{\mathrm{pt}}$ are related to the velocity and temperature difference and therefore they can be ignored in the absence of the slip effect. However, the most interesting aspect of multiphase flow is the slip effect. In fact, much of the book has been devoted to understanding and modeling the slip effect.

If any of the five multiphase mechanisms can be ignored, the corresponding terms in the continuous-phase governing equations can be simplified as outlined in the above paragraphs. In fact, in many of the multiphase flow applications one or more of the multiphase flow mechanisms are unimportant. This explains the varying forms of Euler–Lagrange governing equations used in the past investigations of multiphase flows. Researchers have cleverly simplified the governing equations by appropriately "turning-off" a combination of multiphase mechanisms as dictated by the problem at hand. When all the mechanisms are negligible, the governing equations reduce to the single-phase governing equations.

An example of the simplified formulation will be illustrated below. In the dilute limit, a common practice is to neglect the volume, inertia, and thermodynamic/transport effects. In addition, for illustration purposes, we will ignore any mass exchange between the phases (i.e., $\dot{m}_l = 0$). We will also neglect dissipative heating arising from both mesoscale and pesudo turbulence. The resulting simplified governing equations of the continuous phase are

$$\nabla \cdot \mathbf{u}_c = 0, \tag{15.51}$$

$$\rho_f \frac{D\mathbf{u}_c}{Dt} - \nabla \cdot \boldsymbol{\tau}_{\text{sg}} = -\nabla p_c + \phi_c \rho_f \mathbf{g} + \mu_f \nabla^2 \mathbf{u}_c - \sum_{l=1}^{N_p} G(\mathbf{x} - \mathbf{X}_l)\mathbf{F}_l', \tag{15.52}$$

$$\rho_f C_{vf} \frac{DT_c}{Dt} + \nabla \cdot \mathbf{q}_{\text{sg}} = -p_c(\nabla \cdot \mathbf{u}_c) + k_f \nabla^2 T_c - \sum_{l=1}^{N_p} G(\mathbf{x} - \mathbf{X}_l)Q_l'. \tag{15.53}$$

Only the body-force and slip effects are included in the above simplified set of continuous-phase equations. Note that in the case of EL-DNS, further simplifications can be made to the above equations by setting $\boldsymbol{\tau}_{\text{sg}} = 0$ and $Q_{\text{sg}} = 0$. The above equations must be solved with the Lagrangian equations of motion of the particles given in Eq. (15.6). The above EL-LES governing equations and their even simpler EL-DNS counterparts have been used to solve the problem of particle-laden isotropic turbulence, wall turbulence, and many more (Squires and Eaton, 1991; Elghobashi and Truesdell, 1992; Wang and Maxey, 1993; Portela and Oliemans, 2003; Ferrante and Elghibashi, 2004; Fox, 2012; Capecelatro and Desjardins, 2013; Kuerten, 2016; Park et al., 2017; Rahmani et al., 2018).

15.6   Implementation Details: Small Particle Limit

In this section we complete the mathematical formulation of the EL methodology by briefly describing how the governing equations are implemented and solved. Broadly speaking, this involves the following five important steps. An entire book could be devoted to the various numerical aspects of these five steps (see Prosperetti and Tryggvason (2007); Balachandar and Prosperetti (2007); Subramaniam and Balachandar (2022)). In this and the next two sections we will quickly review the salient aspects of the five steps.

I. Time Advancement of the Eulerian Continuous-Phase Quantities
The continuous-phase equations (15.1) to (15.3) are typically solved using finite-volume, finite-element, or finite-difference approaches. The computational domain is suitably divided into cells, elements, or grid points. The discretization of the left and right-hand sides of the governing equations at these cells or grid points is standard practice in CFD and therefore will not be elaborated. The primary variables of the continuous-phase governing equations are $\mathbf{u}_c$, p_c, and T_c, and all other secondary closure terms to be modeled have been presented in Section 15.3.

II. Time Advancement of the Lagrangian Particle Properties
The Lagrangian governing equations (15.6) are integrated in time to obtain the time evolution of each particle's position, mass, velocity, and temperature. Typically, the same time-integration scheme (e.g., Runge–Kutta) as that used for the continuous phase is employed in the integration of the Lagrangian equations as well.

III. Interpolation of the Eulerian Continuous-Phase Quantities to the Lagrangian Particle Locations
The interpolated continuous-phase information is needed in order to calculate the force and heat transfer on the particles. For example, in the evaluation of the quasi-steady force of the lth particle, the fluid velocity at the location of the particle denoted as $\mathbf{u}_{c@l} = \mathbf{u}_c(\mathbf{X}_l, t)$ is needed. There are many ways to interpolate the fluid velocity from the grid points to the particle location $\mathbf{X}_l$. The simplest being the zeroth-order accurate scheme of assigning the continuous-phase velocity of the finite volume cell. At the next level of accuracy, a tri-linear interpolation can be used to obtain the continuous-phase velocity from its value at the eight grid points surrounding the particle location. Higher-order Lagrange, spline, and Hermite interpolations have been tried to interpolate the fluid velocity to the particle location. The general rule of thumb is to maintain the same level of spatial accuracy used in the solution of the continuous-phase equations in the interpolation operation. The reader is referred to Balachandar and Maxey (1989), Kontomaris et al. (1992), Mancho et al. (2006), Garg et al. (2007), and Jacobs et al. (2007) for additional information on interpolation.

IV. Projection of the Lagrangian Particle Properties Onto the Eulerian Mesh
The projection step involves calculation of the summations over the particles on the right-hand side of the continuous-phase mass, momentum, and energy equations (15.1) to (15.3). In addition, the projection operation is employed in the definition of particle volume fraction in Eq. (15.4), particle velocity and thermal fields in Eq. (15.38) and (15.43). The important point to observe in these projection operations is that though the sum is formally defined over all the particles, to calculate the Eulerian field value at a point $\mathbf{x}$, only the contribution of those particles that are in the neighborhood matter, due to the exponential decay of the filter function.

V. Interparticle Collision Detection and Accounting for the Collisional Force
Note that even though the finite size of the particles was ignored in the continuous-phase governing equations (i.e., Eqs. (15.1) to (15.3) are solved over the entire computational domain), we account for the finite size when it comes to interparticle collisions. In the discrete element model (Cundall and Strack, 1979), the possibility of collision is detected between every pair of particles at each time step. Any pair of particles is taken to be in contact if the distance between their centers is less than the sum of their radii. For every such particle pair in contact, the soft-sphere collision algorithm outlined in Section 9.1.2 is used to calculate the repulsive force on the particles. In the implementation of the discrete element method, a particle can be in simultaneous contact with

a number of neighbors. The total collisional force, torque, and heat transfer due to all these contacts are added to obtain $\mathbf{F}_{cl}$, $\mathbf{T}_{cl}$, and Q_{cl} for application in Eq. (15.6).

It must be pointed out that the most expensive part of the collision algorithm is checking each particle pair for contact. In an N-particle system, a naive implementation will check all $N(N-1)/2$ particle pairs. However, such a complete check for collision is unnecessary. During each time step, the particles would have moved only by a short distance, typically less than a grid spacing, and therefore there is no chance of a particle colliding with others that are located far away. For each particle one needs to check contact only with those other particles in the neighborhood. This greatly reduces the collision detection to $O(N)$ operation. It must be recognized that under dilute conditions, the possibility of collision between particles is quite rare.

Two important numerical decisions are to be made: the level of spatial resolution will be controlled by the grid width Δx and the level of temporal resolution will be controlled by the time step Δt. The optimal grid width is dictated by the filter length scale $\mathscr{L}$. As a result of the filter operation, the macroscale governing equations (15.1) to (15.3) represents continuous-phase dynamics only at length scales larger than $\mathscr{L}$. Thus, a judicious choice of grid width is $\Delta x \approx \mathscr{L}/2$ to $\mathscr{L}/10$. The time step Δt is chosen to satisfy the following three conditions. (i) Explicit time-stepping of the advection and diffusion terms of the continuous-phase governing equation give rise to the familiar CFL and diffusion time-step limitations. For numerical stability, Δt must be chosen smaller than these stability limits. (ii) Explicit integration of the Lagrangian equations (15.6) requires the time step to be smaller than the smallest of the translational, rotational, and thermal time scales of all the particles (these time scales were defined in Section 2.3). (iii) In addition, to accurately calculate the collisional dynamics, the time step must be an order of magnitude smaller than the collisional time scale, which was defined in Problem 9.3.

It should be noted that the translational, rotational, and thermal time scales of a particle go as d_p^2. In other words, for very small particles we sometimes face the unpleasant situation of the Lagrangian time-step limitation being far most restrictive than the CFL and diffusion stability limits. This is somewhat counter-intuitive since the problem arises for very small particles whose motion is expected to closely follow that of the local fluid. There are different ways of avoiding the Lagrangian time-step limitation. The first option is to use an implicit time integration of the Lagrangian equations (which may not always be possible). The other option is to use analytical expressions of a particle's translational and rotational velocities and temperature, and thereby avoid numerical integration of the Lagrangian equations altogether. Since the Stokes number of such small particles is likely to be very small, it is far more computationally efficient to use equilibrium Eulerian expressions such as Eq. (12.27) for translational velocity, Eq. (12.30) for rotational velocity, and Eq. (12.28) for particle temperature. Examples of simpler simulations using the equilibrium Eulerian expressions were presented in Chapter 12.

Also, attention must be paid to the time-step limitation of the collision process. The material properties of the particle may be such that the collisional time scale, which represents the typical time duration over which a colliding pair is in contact, is much smaller than the CFL and diffusion time limits. Different approaches have been followed to alleviate the computational burden of very small collisional time step. The first approach is to substep only the collisional process. In this approach, the continuous-phase and Lagrangian time steps remain as dictated by CFL and diffusion time limits. Only the trajectory of particles under collisional motion is advanced at a much smaller time step, as required by the collisional time scale. An alternative is to suitably modify the material properties in order to allow larger collisional time scales on the order of CFL and diffusion time limits. This latter approach is advisable only if it can be ensured that this artificial change in collisional physics does not adversely affect the multiphase-flow physics that we are after.

15.6.1   Challenges at Finite Particle Volume Fraction

The continuous-phase governing equations (15.1) to (15.3) were obtained through a formal filtering process and involved the incompressibility and the constant thermodynamic and transport property assumptions. They, along with the Lagrangian governing equations (15.6), are formally appropriate for all volume fractions, from the very dilute to the close-packed limit. However, a number of complexities arise with increasing particle volume fraction. First, the volume, inertia, constitutive, and pesudo-turbulence effects become important with increasing volume fraction and therefore cannot be ignored. Furthermore, the simple models of effective mixture viscosity and thermal conductivity given in Eqs. (15.34) and (15.36) are accurate only for small values of ϕ_d. Therefore, in case of dense multiphase flows, appropriate expressions for effective transport properties of the mixture must be obtained to properly account for the constitutive effects.

Second, with increasing volume fraction, the frequency of collisions and the possibility of simultaneous collision between multiple particles increases. As the particle volume fraction approaches the close-packed limit, there can be enduring contact between particles. Under such conditions, collisional mechanics dominates and controls the overall momentum and energy balance. Fortunately, the soft-sphere model still remains applicable and can be used to account for particle–particle interaction even in the close-packed limit. However, the number of collisions and the associated force and torque calculations per time step increase greatly with increasing volume fraction. These collisional calculations can easily overwhelm all other calculations. Therefore, collision detection and collision force calculations must be implemented very efficiently when computing dense multiphase flows.

The most challenging aspect of the large volume fraction is fluid-mediated interactions among the particles. These interactions greatly affect the force on individual particles and their motion, and therefore must be accounted for in the EL simulation. At small values of ϕ_d, each particle's interaction with the surrounding fluid can be considered in isolation from all other neighbors. As a result, the force $\mathbf{F}'_l$ and heat transfer Q'_l on the lth particle are given only in terms of the particle properties and those of the undisturbed flow at the particle. With increasing volume fraction, to leading order, the effect of nearby particles is taken into account through the finite volume fraction corrections discussed in Section 9.4. For example, the volume fraction correction of Richardson and Zaki (1954) given in Eq. (9.86), or that of Tenneti and Subramaniam (2014) given in Eq. (9.87), should be used for the quasi-steady force. Similar finite volume fraction corrections given in Section 9.5 are needed for the added-mass and history forces. Finite volume fraction corrections must be considered for quasi-steady and unsteady contributions of heat transfer as well.

The above finite volume fraction corrections account for the collective effect of all the neighbors in an average statistical sense. It was discussed in Section 9.5.1 that in a random distribution of particles, the effect of neighbors on an individual particle varies substantially from the average effect predicted by the volume fraction correction. The fluid-mediated influence of neighbors induces substantial particle-to-particle variation, since each particle's neighborhood is unique in a random distribution. For instance, in the case of two particles approaching each other toward a collision, the force on the downstream particle drafting in the wake of the upstream particle cannot be accurately predicted based only on the average local volume fraction. However, the traditional approach has been to ignore such flow-mediated particle–particle interactions when the particles are not in contact. However, particle–particle interaction is instantly accounted for as soon as the particles come into contact. This practice of ignoring the influence of neighbors except when they are in physical contact is not physically consistent. Emerging techniques such as pairwise interaction extended the point-particle model (Akiki et al., 2016, 2017a,b; Balachandar et al., 2020) in an attempt to account for such particle–particle interaction effects.

15.7   EL Approach for Large Particles

We begin by clearly defining what we mean by small and large particles. It was discussed in Section 15.1 that the first step of an EL approach is to define the filter scale $\mathscr{L}$. Only the continuous-phase length scales larger then $\mathscr{L}$ are resolved in the EL simulation. The Eulerian grid spacing Δx is typically chosen to be smaller than $\mathscr{L}$. Here we define a small particle as one whose size is much smaller than the grid spacing or the filter width (i.e., $d_p \ll \Delta x$ or $\mathscr{L}$). A particle of size comparable to Δx or $\mathscr{L}$ will be considered large. In what follows, we will first address appropriate methods for accurately calculating hydrodynamic force and torque on a large particle, ignoring the fact that the fluid is two-way coupled (i.e., our initial treatment will be as if the problem is one-way coupled). The following subsection will then address the problem of self-induced perturbation flow due to the large particle and how to correct for it in the evaluation of hydrodynamic force and heat transfer. This correction is essential for accurate implementation of the EL methodology, since the self-induced error becomes large as the particle size approaches the grid size.

As long as the particle size is much smaller than the grid size (or equivalently much smaller than $\mathscr{L}$), the undisturbed fluid properties, such as fluid velocity, fluid pressure, and fluid temperature, at the particle location can be obtained by simply interpolating from the nearby Eulerian grid points onto the center of the particle.

The underlying assumption is that the fluid properties interpolated to any other point within the particle would yield nearly the same value, since the particle is much smaller than the scale of surrounding flow variation. This small-particle approximation was used in Eq. (15.6), and in all the closure models of Section 15.2, where the subscript $c@l$ indicates a continuous-phase property evaluated at the center of the lth particle through interpolation. As discussed earlier, in the implementation details in Section 15.4, a variety of options are available (e.g., tri-linear, Lagrange, Hermite, spectral interpolation) for performing the interpolation.

In the case of larger particles, the Lagrangian equations of motion, given in Eq. (15.6), must be rewritten as follows:

$$
\begin{aligned}
\frac{dm_{pl}}{dt} &= \dot{m}_l, \\
m_{pl}\frac{d\mathbf{V}_l}{dt} &= V_{pl}\overline{\nabla\cdot\boldsymbol{\sigma}_c}^v + m_{pl}\mathbf{g} + \mathbf{F}'_l + \mathbf{F}_{cl}, \\
I_{pl}\frac{d\boldsymbol{\Omega}_l}{dt} &= \frac{I_{fl}}{2}\overline{\frac{D(\boldsymbol{\omega}_c)}{Dt}}^v + \mathbf{T}'_l + \mathbf{T}_{cl}, \\
m_{pl}C_p\frac{dT_{pl}}{dt} &= -V_{pl}\overline{\nabla\cdot\mathbf{q}_c}^v + Q'_l + Q_{cl}.
\end{aligned}
\tag{15.54}
$$

In the above we have ignored the additional terms that were in Eq. (15.6). In the momentum, angular momentum, and energy equations, the first term on the right represents the undisturbed flow contribution. Here, the notation $\overline{(\cdot)}^v$ corresponds to an average of the argument over the volume occupied by the lth particle. Thus, $V_{pl}\overline{\nabla\cdot\boldsymbol{\sigma}_c}^v$ corresponds to the volume integral of $\nabla\cdot\boldsymbol{\sigma}_c$ over the volume of the lth particle and corresponds to the undisturbed flow force on the lth particle. Similarly, the first term on the right-hand side in the angular momentum and energy equations corresponds to the undisturbed flow contributions.

The evaluation of the volume averages is computationally complicated, since it requires integration of spatially varying quantities over the spherical volume of the particle. Using the result of Problem 4.16, the volume averages can be simplified as

$$
\begin{aligned}
\overline{\nabla\cdot\boldsymbol{\sigma}_c}^v &\approx [\nabla\cdot\boldsymbol{\sigma}_c]_{@l} + \frac{d_p^2}{40}\left[\nabla^2(\nabla\cdot\boldsymbol{\sigma}_c)\right]_{@l}, \\
\overline{\frac{D(\boldsymbol{\omega}_c)}{Dt}}^v &\approx \left[\frac{D(\boldsymbol{\omega}_c)}{Dt}\right]_{@l} + \frac{d_p^2}{40}\left[\frac{D(\nabla^2\boldsymbol{\omega}_c)}{Dt}\right]_{@l}, \\
\overline{\nabla\cdot\mathbf{q}_c}^v &\approx [\nabla\cdot\mathbf{q}_c]_{@l} + \frac{d_p^2}{40}\left[\nabla^2(\nabla\cdot\mathbf{q}_c)\right]_{@l},
\end{aligned}
\tag{15.55}
$$

where $[.]_{@l}$ stands for the argument being evaluated at $\mathbf{X}_l$, which is the center of the lth particle. Since the smallest length scale of the filtered macroscale quantities is $\mathscr{L}$, the Laplacian operator scales as $1/\mathscr{L}^2$. Thus, the ratio of the second to the leading-order term scales as $(d_p/\mathscr{L})^2$ and the neglected higher-order terms scale as $(d_p/\mathscr{L})^4$. For very small particles, by retaining only the first term on the right-hand side, we obtain the simplification given in Eq. (15.6). But with increasing particle size, higher-order terms must be retained in the approximation of the volume average.

It was argued earlier that the EL approach is on a sound theoretical footing when the particle size is smaller than the smallest scales of the continuous phase. The reason why the EL approach loses its accuracy for larger particles is due to our lack of understanding of the closure models of the secondary terms listed in Table 11.1. The closure relations presented in Section 15.2 are appropriate in the limit of small particles. These closures must be modified appropriately with increasing particle size.

For example, consider the closure model of $\mathbf{F}'_l$. In Section 15.3.1, the finite-Re BBO equation was presented as the closure model. This closure is appropriate for a small particle, since the ambient flow appears as a spatially uniform flow. Any spatial variation in the undisturbed ambient flow on the scale of the particle is small and can be ignored. For a large-sized particle, the ambient flow varies on the scale of the particle, and as discussed in Section 4.3, $\mathbf{F}'_l$ must be evaluated using the finite-Re MRG equation (4.145), along with additional lift forces as appropriate. The important difference between the finite-Re BBO and MRG equations is the use of surface and volume-averaged fluid quantities, instead of the quantities being evaluated at the particle center. All other closures (i.e., mass transfer, torque heat transfer, and Reynolds stress closures) must also be expressed in terms of surface and volume averages of macroscale-filtered continuous-phase quantities. While the use of the volume average in the undisturbed flow contributions (the first terms on the right-hand side of Eq. (15.54)) is exact, the same cannot be claimed of the other closures. As discussed in the derivation of the MRG equation, the use of surface and volume averages is well motivated by the Faxén correction, but nevertheless their use at finite Re is an approximation.

15.7.1   Self-induced Velocity Perturbation of Large Particles and its Correction

We will illustrate the importance of correcting for the self-induced perturbation flow with a simple example. Consider a two-way coupled EL simulation of an isolated particle in a steady uniform ambient flow. In the absence of the particle, all the grid points will have uniform ambient flow. With the particle in the flow, the hydrodynamic force on the particle will be fed back to the fluid. In the present framework, this force will be spread to the grid points surrounding the particle with the Gaussian filter function. Due to this feedback force, the flow field computed in the EL simulation will not be the uniform ambient flow in the neighborhood of the particle. The difference between this and the uniform ambient flow is the *self-induced perturbation* flow due to the presence of the particle. The self-induced perturbation flow attempts to capture the influence of the particle on the surrounding flow. However, in the EL simulation, there is a serious problem with the use of this perturbed velocity in the calculation of force on the particle.

If we consider the steady flow to be in the Stokes regime, the force on the particle will be predicted to be $3\pi\mu_f d_p u_f$, where u_f is the undisturbed ambient fluid velocity. In the EL simulation, this undisturbed velocity is recovered only far away from the

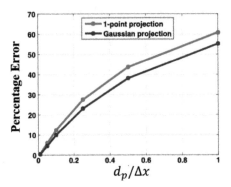

Figure 15.1 Percentage error in the calculation of Stokes drag on a particle in an EL simulation of uniform flow over an isolated particle. Results for Gaussian projection and a single-point projection of the force back on the fluid are presented. Reprinted from Horwitz and Mani (2016), with permission from Elsevier.

particle. The fluid velocity obtained at the center of the particle, or as an average over the surface of the particle, will deviate from u_f due to the self-induced perturbation. In general, since the feedback force is directed opposite to the ambient flow, the perturbed fluid velocity at the particle will be lower than u_f. Therefore, the Stokes drag predicted with the EL velocity will be lower and the difference will be substantial if the particle size is comparable to the grid size of the EL simulation. This point was nicely captured by Horwitz and Mani (2016), who presented the difference between the Stokes drag calculated with the true ambient velocity and that calculated with the perturbed EL velocity as a function of the ratio $d_p/\Delta x$. Their result is reproduced in Figure 15.1, where it can be seen that a particle of diameter one grid width results in almost 60% error. Even a particle 10 times smaller in size than the grid results in about 10% error. Only in the case of substantially small particles is the self-induced flow sufficiently small that its effect can be ignored in the calculation of the force.

The key lesson to be learnt is that large particles will result in significant self-induced perturbation flow, which must be accounted for in calculating the force on the particle. Naive application of the finite-Re MRG equation addressed in the previous section will result in erroneous force prediction. This problem has been recognized and numerical/analytical solutions have been proposed (Gualtier et al., 2015; Fukada et al., 2016, 2018; Horwitz and Mani, 2016, 2018; Ireland and Dejardins, 2017). The focus of this subsection is to present the analytical technique developed in Balachandar et al. (2019) and Balachandar and Liu (2023) to correct for the self-induced perturbation flow so that the hydrodynamic force on the particle can be evaluated accurately. The following subsection will present their analytical expression for evaluating the self-induced velocity disturbance in a steady flow. In the low-Re limit, they obtained an exact expression for the self-induced velocity due to a steady Gaussian feedback force using the Oseen approximation. Simulation results of a steady Gaussian feedback force are then used to obtain the self-induced velocity at finite Reynolds numbers. A comprehensive model for self-induced velocity correction is then presented to account for the general situation of time-dependent particle motion.

Self-induced Velocity in the Steady Regime

We will obtain a rigorous expression for the self-induced velocity of an isolated particle in an unbounded steady uniform ambient flow of velocity U. In the present situation of an isolated stationary particle in a vast expanse of fluid, the inertia, gravity, and constitutive effects can be ignored. We thus take into account only the volume and slip effects. Furthermore, let us start with the Stokes flow before considering the Oseen approximation and then the fully nonlinear flow. We start with the simplified governing equations (15.51) and (15.52), and use the fact that the isolated particle is located at the origin. We express the flow computed in an EL simulation as the sum $\mathbf{u}_c = U\mathbf{e}_x/\phi_c + \mathbf{u}'_c$, where the first term defines the filtered velocity in the absence of the feedback force, and accounts for the volume effect. The second term is the perturbation flow due to the feedback force, and thus represents the slip effect. The mass and momentum balance equations for the self-induced velocity perturbation are then

$$\nabla \cdot \left(\phi_c \mathbf{u}'_c\right) = 0 \quad \text{and} \quad 0 = -\nabla\left(\phi_c p_c\right) + \mu_f \nabla^2\left(\phi_c \mathbf{u}'_c\right) - F' G(\mathbf{x})\mathbf{e}_x, \qquad (15.56)$$

where F' is the steady streamwise hydrodynamic force on the particle, which with a negative sign is applied back on the fluid with the Gaussian filter function. The above EL equations of self-induced velocity perturbation are solved over the infinite space, including the volume occupied by the particle. The only boundary condition to be satisfied is the far-field requirement that $\mathbf{u}'_c \to 0$ as $|\mathbf{x}| \to \infty$. An exact solution of the above equations is sought by taking a 3D Fourier transform as carried out in Section 9.4.1, where the forward and backward 3D Fourier transforms were defined in Eq. (9.48). Note that the Fourier transform of the Gaussian feedback force becomes

$$-\frac{F'\mathbf{e}_x}{(2\pi)^3}\int G(\mathbf{x})\exp(-\iota\mathbf{k}\cdot\mathbf{x})\,d\mathbf{x} = -\frac{F'\mathbf{e}_x}{(2\pi)^3}\exp\left(-\frac{\sigma^2 k^2}{2}\right), \qquad (15.57)$$

where we have used the definition of the 3D Gaussian given in Eqs. (11.2) and (11.4). In the above equation, $k^2 = |\mathbf{k}|^2 = k_x^2 + k_y^2 + k_z^2$. Recall from Section 9.4.1 that the above momentum equation can be solved without the pressure term and the effect of pressure in enforcing incompressibility appears as the multiplicative projection tensor $\mathbf{P} = P_{lm}(\mathbf{k}) = (\delta_{lm} - k_l k_m/|\mathbf{k}|^2)$. We thus arrive at the following solution in Fourier space:

$$\widehat{\phi_c u'_{c,j}} = -\frac{F'}{(2\pi)^3 \mu_f k^2}\left[\delta_{jx} - \frac{k_x k_x}{k^2}\right]\exp\left(-\frac{\sigma^2 k^2}{2}\right) \quad \text{for} \quad j = x, y, z, \qquad (15.58)$$

where $\widehat{\phi_c u'_{c,j}}$ is the Fourier coefficient of the product $\phi_c \mathbf{u}'_c$ and the term within square brackets is $\mathbf{P}\cdot\mathbf{e}_x$.

We are interested in evaluating the self-induced velocity perturbation at the center of the particle. To obtain that, we take the inverse Fourier transform of the above Fourier-space solution and then substitute $\mathbf{x} = 0$ in order to evaluate the self-induced velocity perturbation at the origin where the particle center is located. Only the x-component is nonzero, and the other two components are zero by symmetry. The self-induced velocity perturbation at the particle center, denoted as $\mathbf{u}'_{c@}$, is then

$$\phi_{c@}\mathbf{u}'_{c@} = \int \overline{\phi_c u'_{c,j}}\, d\mathbf{k} = -\frac{F'}{6\pi\,\mu_f\sigma}\sqrt{\frac{2}{\pi}}\mathbf{e}_x\,. \tag{15.59}$$

The above is an important result with serious consequences for how the force on a particle is evaluated in an EL simulation. The true force on the stationary particle is given by the Stokes drag evaluated with the undisturbed ambient flow velocity $U\mathbf{e}_x$. However, in an EL simulation of the problem, the computed filtered fluid velocity at the center of the particle is not the true ambient velocity undisturbed by the particle. The true fluid velocity at the particle center is given by

$$\begin{matrix} \text{True fluid velocity} \\ \text{at the particle center} \end{matrix} = \underbrace{\phi_{1c@}\mathbf{u}_{c@}}_{\substack{\text{computed in} \\ \text{EL simulation}}} - \underbrace{\phi_{c@}\mathbf{u}'_{c@}}_{\substack{\text{self-induced} \\ \text{perturbation}}}\,. \tag{15.60}$$

Since $\phi_{c@}\mathbf{u}'_{c@}$ is expected to be negative, the true undisturbed ambient flow at the particle will be larger than that computed in the EL simulation.

In an EL simulation consisting of many particles, the filtered fluid velocity $\mathbf{u}_{c@}$ evaluated at a particle is influenced by its own presence and those of all other particles. We want to compensate only the effect of self-induced perturbation. In the first term, $\phi_{1c@}$ represents the fluid volume fraction evaluated at the particle center *only* due to the presence of that single particle (this is the reason for the inclusion of "1" in the subscript). Note that $\phi_{1c@}$ and $\phi_{c@}$ are the same in the case of an isolated particle, and hence they will not be distinguished in the self-induced perturbation. But in a typical EL simulation involving many particles, the two are not the same, since the latter includes the volume fraction effect of neighboring particles as well. Multiplication by $\phi_{1c@}$ in the first term thus accounts for the self-induced volume effect of the particle, while the second term on the right-hand side corrects for the self-induced slip effect of the feedback force.

From the above discussion, it is clear that the Stokes drag on the particle evaluated with the EL velocity $\mathbf{u}_{c@}$ will be lower than the actual force on the particle. If we substitute the Stokes drag expression for F', we obtain $|\phi_{c@}\mathbf{u}'_{c@}|/U = \left(1/\sqrt{2\pi}\right)(d_p/\sigma)$. Thus, the only parameter of importance is the size of the particle compared to the Gaussian filter width. Since the grid width Δx is typically chosen to be four to eight times smaller than σ, the size of the particle compared to the grid width plays an important role, as illustrated in Figure 15.1.

The value of the continuous-phase volume fraction at the center of an isolated particle can easily be obtained from the definition given in Eq. (11.17) as $\phi_{1c}(\mathbf{x} = 0) = 1 - \int G(\mathbf{x})d\mathbf{x}$. However, caution is required. In an EL simulation, if the particle volume fraction was simplified as given in Eq. (15.4), then $\phi_{1c}(\mathbf{x} = 0)$ must be correctly evaluated as $1 - V_p\, G(\mathbf{x} = 0)$. For a Gaussian filter function these expressions can easily be evaluated, and we get the following three different estimates:

$$\phi_{1c@} = \phi_{1c}(\mathbf{x} = 0)$$

$$= \begin{cases} 1 & \text{if volume effect ignored,} \\ 1 - \dfrac{1}{12\sqrt{2\pi}}\left(\dfrac{d_p}{\sigma}\right)^3 & \text{if } \phi_d \text{ approximated with Eq. (15.4),} \\ 1 - \mathrm{erf}\left[\dfrac{d_p}{2\sqrt{2}\sigma}\right] + \dfrac{d_p}{\sqrt{2\pi}\sigma}\exp\left[-\dfrac{d_p^2}{8\sigma^2}\right] & \text{if } \phi_d \text{ exactly evaluated with Eq. (11.104).} \end{cases}$$

$$(15.61)$$

Problem 15.3 *Oseen correction for* $\mathbf{u}'_{c@}$. In this problem you will extend the above analysis to small but finite Reynolds number with the Oseen approximation of the momentum equation (15.52). With this approximation, the nonlinear term on the left-hand side becomes $U\partial(\phi_c\mathbf{u}'_c)/\partial x$ and the governing equations are

$$\nabla \cdot (\phi_c \mathbf{u}'_c) = 0 \quad \text{and} \quad U\frac{\partial \phi_c \mathbf{u}'_c}{\partial x} = -\nabla\left(\frac{\phi_c p_c}{\rho_f}\right) + \nu_f \nabla^2\left(\phi_c \mathbf{u}'_c\right) - \frac{F'}{\rho_f}G(\mathbf{x})\mathbf{e}_x. \quad (15.62)$$

(a) Follow the solution procedure presented above, but with the additional term on the left-hand side, to obtain the solution in the Fourier space as

$$\widehat{\phi_c u'_{c,j}} = -\left\{\frac{1}{\left[\left(1 + \dfrac{\imath k_x U}{\nu_f k^2}\right)\right]}\right\} \frac{F'}{(2\pi)^3 \mu_f k^2}\left[\delta_{jx} - \frac{k_x k_x}{k^2}\right]\exp\left(-\frac{\sigma^2 k^2}{2}\right) \quad \text{for} \quad j = x, y, z,$$

$$(15.63)$$

where the term within curly brackets is the Oseen correction due to nonzero Reynolds number.

(b) Take the inverse Fourier transform of the above (consult Balachandar et al., 2019) to obtain the self-induced velocity perturbation at the particle center as

$$\phi_{c@}\mathbf{u}'_{c@} = \int \widehat{\phi_c u'_{c,j}}\, d\mathbf{k}$$

$$= -\frac{F'\mathbf{e}_x}{6\pi\,\mu_f \sigma}\sqrt{\frac{2}{\pi}}\underbrace{\frac{3}{\sqrt{2\pi}\,\mathrm{Re}_\sigma^3}\left(\pi - \sqrt{2\pi}\,\mathrm{Re}_\sigma + \frac{\pi}{2}\,\mathrm{Re}_\sigma^2 - \pi e^{\mathrm{Re}_\sigma^2/2}\mathrm{erfc}\left(\frac{\mathrm{Re}_\sigma}{\sqrt{2}}\right)\right)}_{\Psi_{\mathrm{os}}(\mathrm{Re}_\sigma)},$$

$$(15.64)$$

where the Reynolds number effect has been separated as the Oseen correction factor Ψ_{os}, which is only a function of $\mathrm{Re}_\sigma = \sigma U/\nu_f$.

(c) Show that this Oseen correction factor can be well approximated as $\Psi_{\mathrm{os}} \approx 1$ for $\mathrm{Re}_\sigma < 1$, while for a larger Reynolds number of $\mathrm{Re}_\sigma > 10$, the following offers an excellent approximation: $\Psi_{\mathrm{os}} \approx (3/2)\left(\sqrt{\pi/2}\right)/\mathrm{Re}_\sigma$.

Balachandar et al. (2019) extended the velocity correction results to finite particle Reynolds numbers using numerical simulations. They considered EL simulations

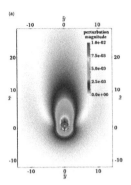 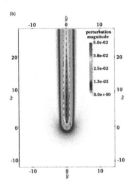

Figure 15.2 Contour plot of streamwise x velocity perturbation on the x–y plane passing through the center of the particle. (a) $\text{Re}_\sigma = 1$ and (b) $\text{Re}_\sigma = 100$. In both cases, $|\tilde{F}'| = 0.35$ and $\sigma/\Delta x = 1.7$. Superposed is the perturbation velocity vector plot. Also marked on the figures are the particle centered at the origin and a dashed circle of diameter 2.355σ. Reprinted from Balachandar et al. (2019), with permission from Elsevier.

where a stationary particle was subjected to uniform ambient flow in a very large domain. The particle force was fed back to the fluid using a Gaussian filter function. The three parameters of the numerical simulations are: (i) the Reynolds number Re_σ, which depends on the particle Reynolds number through the relation $\text{Re}_\sigma = (\sigma/d_p)\,\text{Re}$; (ii) the nondimensional force $\tilde{F}' = F'/(\rho_f U^2 \sigma^2)$; and (iii) the grid resolution given by the parameter $\sigma/\Delta x$. The simulations were performed for a range of all three parameters. Figure 15.2 shows the contour plot of streamwise velocity perturbation on the x–y plane passing through the center of the particle. The results for two different Reynolds numbers of $\text{Re}_\sigma = 1$ and $\text{Re}_\sigma = 100$ are shown for $|\tilde{F}'| = 0.35$. In both cases $\sigma/\Delta x = 1.7$. Superposed are the perturbation velocity vector plots. Also shown are the particle centered at the origin and a circle of diameter equal to the Gaussian width of 2.355σ (red circle).

These solutions can be compared to the Landau–Squires solution for a point momentum source discussed in Batchelor (2000). The only parameter of the Landau–Squires solution is the Reynolds number of the momentum source at the origin. In the limit of large Landau–Squires Reynolds number, the flow behaves like a narrow jet issuing from a small orifice located upstream of the origin, with a radial inward flow toward the origin outside of the jet. In the limit of small Reynolds number, the flow away from the origin approaches the Stokes flow. At distances $r \gg \sigma$ the numerical solution is well approximated by the Landau–Squires solution. However, in the near field over which the Gaussian forcing extends, departure can be significant. Most importantly, while the Landau–Squires solution is singular at the origin, the numerical solution is regularized with the Gaussian spreading. The numerical solution displays a narrow jet-like behavior in the higher Reynolds number wake, while the lower Reynolds number case displays an increased tendency toward fore–aft symmetry. The effect of the feedback force extends over a large region. At $\text{Re}_\sigma = 1$ the wake is somewhat broader and localized around the Gaussian force. But at the higher Reynolds number the wake does not decay rapidly in the streamwise direction and extends far downstream.

The strengths and weaknesses of the EL simulation are captured in the structure of these wake flows, as they can be compared with the actual wake behind a particle obtained from a particle-resolved simulation, such as those shown in Figures 7.1 and 7.3. The microscale details of the wake are clearly misrepresented in an EL simulation. On the other hand, the two-way coupling of a particle through the feedback force does create a perturbation flow which will influence its neighbors. Thus, fluid-mediated particle–particle interaction is included in a two-way coupled EL simulation. However, this two-way coupling influence is expected to be quite inaccurate in case of close-by particles, due to substantial differences in the near-field wake flow predicted by the Gaussian feedback force.

Here we are interested in the computed self-induced perturbation velocity at the origin. Not only the velocity at the center of the particle is modified, the velocity over the entire particle volume is affected and thus, even quantities such as surface-averaged fluid velocity will be influenced by the self-induced perturbation flow. Following the Oseen solution given in Eq. (15.64), the general solution of self-induced perturbation velocity can be expressed as

$$\phi_{c@} \mathbf{u}'_{c@} = -\frac{\mathbf{F}'}{6\pi \mu_f \sigma} \sqrt{\frac{2}{\pi}} \, \Psi, \qquad (15.65)$$

where Ψ is the correction function that accounts for the nonlinear effects. It depends on both Re_σ and $\tilde{F}'$ (i.e., $\Psi(\text{Re}_\sigma, \tilde{F}')$). The value of Ψ was obtained from a large number of simulations and the results were plotted against Re_σ for varying values of $\tilde{F}'$. This plot taken from Balachandar et al. (2019) is reproduced in Figure 15.3. The results for small values of nondimensional feedback force (i.e., for $\tilde{F}' = -0.0035$ and -0.035) are in excellent agreement with Ψ_{os} over the entire range of Re_σ. It is quite surprising that the results of Oseen's approximation remain accurate even for Reynolds numbers as large as 10^4, which was conjectured to be due to the simplicity of the flow arising from a Gaussian forcing. Some deviation from Ψ_{os} can be seen for $\tilde{F}' = 0.35$, and the deviation increases with further increase in $\tilde{F}'$. However, Balachandar et al. (2019) argued that the typical value of $\tilde{F}'$ is less than unity. So, for the present purposes we will assume $\Psi \approx \Psi_{os}$. As a result, the strength of the perturbation velocity is taken to increase linearly with feedback forcing.

Problem 15.4 *Estimation of true surface-averaged velocity.* Here you will extend the Oseen correction analysis of Problem 15.3 to obtain the true surface average of fluid velocity, with the intention that such a surface average can be used in the finite-Re MRG equation for an accurate evaluation of the particle force. We note that the surface average of fluid velocity obtained in a two-way coupled EL simulation will be

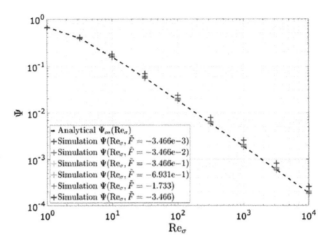

Figure 15.3 The self-induced velocity disturbance computed from the numerical simulations plotted as a function of Re_σ along with the analytical function Ψ_{os}. Reprinted from Balachandar et al. (2019), with permission from Elsevier.

corrupted by self-induced perturbation, which must be corrected to recover the true surface average. Following Eq. (15.60), we can write

True surface-averaged fluid velocity

$$= \overline{\phi_{1c}\mathbf{u}_c}^s - \overline{\phi_c\mathbf{u}_c'}^s$$

$$\approx \underbrace{\phi_{1c@}\mathbf{u}_{c@} + \frac{d_p^2}{24}\left[\nabla^2(\phi_{1c}\mathbf{u}_c)\right]_@}_{\substack{\text{computed in} \\ \text{EL simulation}}} - \underbrace{\left(\phi_{c@}\mathbf{u}_{c@}' + \frac{d_p^2}{24}\left[\nabla^2(\phi_c\mathbf{u}_c')\right]_@\right)}_{\substack{\text{self-induced} \\ \text{perturbation}}}. \qquad (15.66)$$

The first two terms on the right-hand side in the final line provide an excellent approximation of $\overline{\phi_{1c}\mathbf{u}_c}^s$, where it should be re-emphasized that ϕ_{1c} corresponds to the volume fraction field of an isolated particle. These first two terms can easily be calculated in an EL simulation from the computed filtered velocity field. The evaluation of the final two terms is the current focus. Since the third term has already been obtained in Eq. (15.65), we will now evaluate the Laplacian of self-induced perturbation velocity at the center of the particle, following the approach pursued earlier.

(a) In the Fourier space, the self-induced perturbation velocity was derived and presented in Eq. (15.63). From this, calculate the Laplacian as

$$\left[\nabla^2(\phi_c\mathbf{u}_c')\right]_@ = -\int k^2 \widehat{\phi_c\mathbf{u}_c'}\, d\mathbf{k} = \frac{F'\mathbf{e}_x}{6\pi\,\mu_f\sigma^3}\sqrt{\frac{2}{\pi}}\Psi_{\text{Lap}}, \qquad (15.67)$$

where

$$\Psi_{\text{Lap}}(\text{Re}_\sigma) = \frac{3}{\sqrt{2\pi}} \frac{\left[4\pi - 4\sqrt{2\pi}\,\text{Re}_\sigma + \pi\,\text{Re}_\sigma^2 + \pi\left(\text{Re}_\sigma^2 - 4\right) e^{\text{Re}_\sigma^2/2}\,\text{erfc}\left(\text{Re}_\sigma/\sqrt{2}\right)\right]}{\text{Re}_\sigma^3}.$$

(15.68)

For the above derivation you can consult Appendix B of Balachandar et al. (2019) to carry out the 3D integration.

(b) Show that the surface-averaged volume fraction-weighted self-induced velocity can then be approximated as

$$\overline{\phi_c \mathbf{u}_c'}^s = \phi_{c@} \mathbf{u}_{c@}' + \frac{d_p^2}{24} \left[\nabla^2(\phi_c \mathbf{u}_c')\right]_@ = -\frac{F' \mathbf{e}_x}{6\pi\mu_f\sigma} \sqrt{\frac{2}{\pi}} \underbrace{\left[\Psi_{\text{os}} - \frac{1}{24}\frac{d_p^2}{\sigma^2}\Psi_{\text{Lap}}\right]}_{\Psi_{\text{ef}}}.$$

(15.69)

As can be seen, the effect of the surface-averaging operation somewhat reduces the self-induced correction from that calculated at the particle center. This can be expected since the feedback Gaussian force has its peak at the location of the particle center and decays away. Also, Ψ_{Lap} is a function only of Re_σ.

(c) Show that Ψ_{Lap} can be well approximated by unity for $\text{Re}_\sigma < 1$, while for $\text{Re}_\sigma > 10$, $\Psi_{\text{Lap}} \approx 3\left(\sqrt{\pi/2}\right)/\text{Re}_\sigma$. We have defined Ψ_{ef} as the effective factor that accounts for the combined nonlinear and surface-average effects through the terms Ψ_{os} and Ψ_{Lap}, respectively. Note that Ψ_{ef} is only a function of Re_σ and (d_p/σ), whose explicit form, although somewhat complicated, is known.

Self-induced Correction Procedure

In this section, we address the important question: *In an EL simulation involving finite-sized particles, given the filtered fluid velocity field $\mathbf{u}_c$ and the particle motion, how can we accurately calculate the force on the particle using the finite-Re MRG equation by properly accounting for the self-induced perturbation flow?* We briefly summarize the work of Balachandar and Liu (2023) below.

Note that by adopting the MRG equation of force, we allow for the true fluid velocity to vary on the scale of the particle. However, it should be noted that even for a large particle of size comparable to the grid width, the variation in fluid velocity on the scale of the particle may not be very large, since the filtering operation would have smoothened the flow. In what follows we will describe the correction procedure for the steady finite-Re MRG force. If the computed EL fluid velocity is nearly uniform on the scale of the particle, one could simply use the finite-Re BBO equation instead of the finite-Re MRG equation and thereby avoid the need for surface averages.

In the quasi-steady limit, the force on a particle should be correctly expressed as (see Eq. (4.145))

$$\mathbf{F}' = 6\pi\mu_f R_p \left(\left[\overline{\phi_{1c}\mathbf{u}_c}^s - \overline{\phi_c\mathbf{u}_c'}^s\right] - \mathbf{V}\right)\Phi,$$

(15.70)

where the terms within square brackets can be recognized as the surface average of the true ambient fluid velocity. Again, multiplication by the single-particle volume fraction ϕ_{1c} accounts for the self-induced volume effect, while subtraction of $\overline{\phi_c \mathbf{u}'_c}^s$ accounts for the self-induced slip effect of the particle. The quest here is to rewrite the above relation to obtain a more explicit expression for $\mathbf{F}'$.

Substituting Eq. (15.69) for $\overline{\phi_c \mathbf{u}'_c}^s$ and rearranging, we obtain the *corrected quasi-steady drag* expression as

$$\mathbf{F}'_{qs} = \underbrace{\left(1 - \frac{d_p}{\sigma} \frac{\Phi \Psi_{ef}}{\sqrt{2\pi}}\right)^{-1}}_{\text{slip correction}} 6\pi \mu_f R_p \left(\overline{\phi_{1c}\mathbf{u}_c}^s - \mathbf{V}\right) \Phi. \qquad (15.71)$$

The subscript qs has been added to emphasize that this is only the quasi-steady force on the particle. The above can be substituted back to obtain the following expression for the actual relative velocity in terms of what is computed in the EL simulation:

$$\left(\left[\overline{\phi_{1c}\mathbf{u}_c}^s - \overline{\phi_c \mathbf{u}'_c}^s\right] - \mathbf{V}\right) = \underbrace{\left(1 - \frac{d_p}{\sigma} \frac{\Phi \Psi_{ef}}{\sqrt{2\pi}}\right)^{-1}}_{\text{slip correction}} \left(\overline{\phi_{1c}\mathbf{u}_c}^s - \mathbf{V}\right). \qquad (15.72)$$

Thus, the slip-correction function for relative velocity is identically the same as that for quasi-steady force. The correction increases the estimate of true relative velocity to be larger than that obtained from the EL filtered velocity field. Accordingly, the correction also increases the estimate of true quasi-steady force. In both cases the correction increases with increasing d_p/σ.

Despite the simple appearance of the finite-size correction function, it is implicit in nature due to the dependence of Φ on the particle Reynolds number Re and Ψ_{ef} on the filter Reynolds number Re_σ. These two Reynolds numbers are defined in terms of the true relative velocity. Therefore, as shown below, they must be expressed in terms of perturbed Reynolds numbers that can be computed in an EL simulation:

$$\begin{aligned}
\text{Re} &= \frac{|\overline{\phi_{1c}\mathbf{u}_c}^s - \overline{\phi_c \mathbf{u}'_c}^s - \mathbf{V}| d_p}{\nu_f} \\
&= \left(1 - \frac{d_p}{\sigma}\frac{\Phi\Psi_{ef}}{\sqrt{2\pi}}\right)^{-1} \frac{|\overline{\phi_{1c}\mathbf{u}_c}^s - \mathbf{V}| d_p}{\nu_f} = \left(1 - \frac{d_p}{\sigma}\frac{\Phi\Psi_{ef}}{\sqrt{2\pi}}\right)^{-1}\text{Re}_{EL}, \qquad (15.73) \\
\text{Re}_\sigma &= \frac{|\overline{\phi_{1c}\mathbf{u}_c}^s - \overline{\phi_c \mathbf{u}'_c}^s - \mathbf{V}| \sigma}{\nu_f} \\
&= \left(1 - \frac{d_p}{\sigma}\frac{\Phi\Psi_{ef}}{\sqrt{2\pi}}\right)^{-1} \frac{|\overline{\phi_{1c}\mathbf{u}_c}^s - \mathbf{V}| \sigma}{\nu_f} = \left(1 - \frac{d_p}{\sigma}\frac{\Phi\Psi_{ef}}{\sqrt{2\pi}}\right)^{-1}\text{Re}_{\sigma,EL}, \qquad (15.74)
\end{aligned}$$

where Re_{EL} and $\text{Re}_{\sigma,EL}$ are inaccurate Reynolds numbers computed with the EL velocity. Note that the two true Reynolds numbers are related as $\text{Re} = (d_p/\sigma)\,\text{Re}_\sigma$ and $\text{Re}_{EL} = (d_p/\sigma)\,\text{Re}_{\sigma,EL}$. The above are implicit expressions for Re and Re_σ, which can be solved for any value of Re_{EL} computed in an EL simulation, with the additional information of d_p/σ (see Figure 15.4).

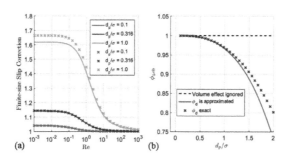

Figure 15.4 (a) Self-induced slip-correction function plotted against Re for three different values of d_p/σ. The lines are for the finite-Re MRG equation and the symbols are for the finite-Re BBO equation. (b) The self-induced volume correction $\phi_{1c@}$ plotted as a function of d_p/σ for the three different approximations given in Eq. (15.61). Reproduced from Balachandar and Liu (2023) with permission from Elsevier.

In summary, in the quasi-steady regime, the following steps should be pursued for each particle in order to properly correct for its self-induced perturbation flow in the calculation of particle force:

- As preprocessing step, compute and store ϕ_{1c} and the slip-correction functions Ψ_{os} and ψ_{Lap}.
- Compute the EL particle Reynolds number as $\mathrm{Re}_{EL} = |\overline{\phi_{1c}\mathbf{u}_c}^s - \mathbf{V}|d_p/\nu_f$.
- Solve Eq. (15.73) to obtain the true Reynolds number Re. Also obtain the filter Reynolds number as $\mathrm{Re}_\sigma = (\sigma/d_p)\,\mathrm{Re}$.
- From the values of Re, Re_σ, and d_p/σ, evaluate the value of the finite-size correction Ψ_{ef} using the definitions given in Eqs. (15.69), (15.64), and (15.68).
- Apply the correction given in Eq. (15.72) to calculate the true relative velocity, and the correction given in Eq. (15.71) to obtain the true quasi-steady force.

Unsteady Velocity Correction

So far we have developed an efficient methodology to correct for the self-induced perturbation velocity of a particle, when the feedback force of the particle on the surrounding fluid is dominated by the quasi-steady contribution. In this section, we will extend the correction procedure to conditions in which unsteady forces become important as well. Our discussion will be based on the analysis of Balachandar et al. (2019). Before proceeding further, let us first consider when unsteadiness becomes important. EL simulations of a freely falling particle were used to evaluate the accuracy of the quasi-steady model and based on the results, Balachandar et al. proposed that the unsteady model is important when $\tau_p^2 g/\sigma < 5$. From the definition of particle time scale τ_p, we rewrite this condition as

$$d_p^3 < \frac{1620\,\Phi^2(\mathrm{Re}_s)\nu_f^2}{\rho^2\,g}\frac{\sigma}{d_p}, \qquad (15.75)$$

where Re_s is the Reynolds number based on terminal velocity, which was presented in Figure 15.4 for different scenarios. From the above expression, with $\sigma/d_p = 10$, we obtain that unsteady effects play a role for water droplets of size smaller than about 70 μm falling in air and for sand particles of size smaller than about 1.1 mm in water. Unsteadiness in these examples is due to the acceleration of an initially stationary droplet or particle toward its terminal velocity. In many applications, unsteadiness may also arise from the time dependence of the fluid velocity seen by the particle. In which case, one must use the criterion $\tau_p/\tau_f < 5$, where τ_f is the time scale of fluid flow variation. Irrespective, it is clear that under a wide range of circumstances the effect of unsteadiness can become important.

A model for unsteady correction is to be employed only with an unsteady particle force model. Due to the finite size of the particle, we will consider the following MRG model for particle force:

$$\mathbf{F}'(t) = 6\pi \mu_f R_p \left[\overline{\phi_{1c}\mathbf{u}_c}^s - \overline{\phi_c\mathbf{u}_c'}^s - \mathbf{V} \right] \Phi + C_M m_f \left[\frac{D(\overline{\phi_{1c}\mathbf{u}_c}^v - \overline{\phi_c\mathbf{u}_c'}^v)}{Dt} - \frac{d\mathbf{V}}{dt} \right]$$
$$+ 6\pi \mu_f R_p \int_{-\infty}^{t} K_{vu} \left[\frac{d(\overline{\phi_{1c}\mathbf{u}_c}^s - \overline{\phi_c\mathbf{u}_c'}^s)}{dt} - \frac{d\mathbf{V}}{dt} \right] d\xi + \mathbf{F}'_{Ll}.$$

$$(15.76)$$

The fluid velocity computed in the EL simulation has been corrected so that $\overline{\phi_{1c}\mathbf{u}_c}^s - \overline{\phi_c\mathbf{u}_c'}^s$ is the undisturbed fluid velocity seen by the particle. Under unsteady conditions, Balachandar et al. (2019) obtained an expression for self-induced velocity perturbation, which can be adapted for the present discussion as

$$\overline{\phi_c\mathbf{u}_c'}^s = -\frac{1}{6\pi \mu_f \sigma}\sqrt{\frac{2}{\pi}} \left[\mathbf{F}'\Psi_{\text{ef}} - \int_{\infty}^{t} K_{u,\text{cor}}(t-\tau) \frac{d\mathbf{F}'\Psi_{\text{ef}}}{dt}\bigg|_{\tau} d\tau \right], \qquad (15.77)$$

where the first term on the right-hand side is the quasi-steady correction model derived earlier, now written for a general vectorial force. The second term is the unsteady correction written as a convolution integral where the kernel was given as (Balachandar et al., 2019)

$$K_{u,\text{cor}}(t-\tau) = \frac{1}{\sqrt{\exp\left[g\left(\frac{(t-\tau)|\mathbf{u}_r|}{\sigma} \right) \right] + \frac{2(t-\tau)\nu_f}{\sigma^2}}}, \qquad (15.78)$$

where the relative velocity vector is $\mathbf{u}_r = \overline{\phi_{1c}\mathbf{u}_c}^s - \overline{\phi_c\mathbf{u}_c'}^s - \mathbf{V}$ and the function g is given by

$$g(\xi) = [1 + \text{erf}(\xi - 8.08)] (1.71 + 0.193\xi) + [1 + \text{erf}(8.08 - \xi)] (0.4\xi). \qquad (15.79)$$

It can be observed that at small times the velocity correction kernel decays as one over the square root of the viscously scaled time, and at large times the decay is exponential.

Following the prescription given in Balachandar et al. (2019), the steps below can be implemented for each particle for unsteady correction of self-induced velocity:

- Compute the EL particle Reynolds number as $\text{Re}_{\text{EL}} = \left| \overline{\phi_{1c} \mathbf{u}_c}^s - \mathbf{V} \right| d_p / \nu_f$.
- Solve Eq. (15.73) to obtain the true Reynolds number. Also obtain the filter Reynolds number as $\text{Re}_\sigma = (\sigma/d_p)\,\text{Re}$.
- From the above, calculate Ψ_{ef} as given in Eq. (15.69).
- From the time history of $\mathbf{F}'$ and Ψ_{ef}, evaluate the current value of $d(\mathbf{F}'\Psi_{\text{ef}})/dt$.
- Compute the kernel $K_{u,\text{cor}}$ as a function of elapsed time using Eq. (15.78).
- From all the above information, compute the self-induced perturbation velocities $\overline{\phi_c \mathbf{u}'_c}^s$ and $\overline{\phi_c \mathbf{u}'_c}^v$ as given in Eq. (15.77).
- With the velocity corrections, calculate the force on the particle using Eq. (15.76).

Clearly the above correction procedure is computationally more involved than quasi-steady correction. It is needed only when the unsteady MRG (or BBO) model is used for force calculation. Calculation of the convolution integral for velocity correction is just as computationally expensive as calculation of the convolution integral in the unsteady force model. We close this discussion with a final note that while the unsteady force model in Eq. (15.76) involved an added-mass term, there is no such analog in the correction equation (15.77). This is because, in EL simulation, the Gaussian force does not impose a no-penetration condition on the flow. Thus, similar to the thermal problem considered in Section 6.2, there is only a convolution term for unsteady correction.

Role of Numerical Methodology

Until this point we have not discussed the role of numerical methodology in the correction procedure. The step-by-step procedures presented for the quasi-steady and unsteady corrections are generic and not specific to any numerical methodology. In this section, we will address the role of numerical methodology as dictated by: (i) the spatial discretization methodology, such as finite volume, finite difference, or spectral methodology and its order of accuracy; (ii) the grid width Δx in the region of the particle; (iii) the time advancement methodology, such as Runge–Kutta or backward difference scheme; and (iv) the time step Δt.

Even though there are a number of numerical parameters, they do not necessarily influence the self-induced perturbation flow of the feedback force. In many cases the correction procedure outlined above can be used irrespective of the numerical details. Numerical details become a factor only under conditions of inadequate resolution. So, the real question is how well the perturbation flow is numerically approximated in an implementation of the EL methodology. It must be noted that the steady and unsteady correction procedures presented above were based on analytical and numerical solutions that were well resolved. When the limit of grid independence is achieved, the numerical details do not matter. Irrespective of the specific spatial and temporal discretization used in the EL simulation, if Δx and Δt are sufficiently small, the resulting perturbation flow will have converged. Typically, a lower-order scheme will require a smaller Δx and Δt for adequate convergence than a higher-order scheme. In other words, provided the numerical methodology of the EL simulation is sufficiently converged to accurately capture the Gaussian-filtered forcing, its temporal variation, and the resulting perturbation flow, the above outlined correction procedures can be followed without modification. If the spatial or temporal resolution of the Gaussian forcing is marginal or insufficient, then numerical approximation will have an influence and the correction will depend on the discretization parameters $d_p/\Delta x$ and $\tau_p/\Delta t$ as well.

Another factor of importance is the use of the Gaussian filter in the above analysis. Other filter functions, such as the box filter given in Eq. (11.3), have been used in EL simulations. In the case of the Gaussian filter, its shape is dictated by the parameter σ and, as shown in Ireland et al. (2016) and Balachandar et al. (2019), four or more grid points per width of the Gaussian filter are sufficient for adequate resolution. In the case of the box filter, if the box size spans only the local finite volume cell and the feedback forcing is concentrated within that cell, then the perturbation flow will depend on the grid resolution and therefore the correction procedure must include $d_p/\Delta x$ as an important parameter. Other approaches to self-induced perturbation correction, such as those proposed by Horwitz and Mani (2016, 2018) and Fukada et al. (2016, 2018), can be pursued.

Additional Considerations

Here we present an important principle. *One must correct only those aspects of two-way coupling included in an EL simulation.* Based on this principle, different variants of the correction procedure can be considered, as displayed in Figure 15.5. (i) In a one-way coupled EL coupled simulation, there is no need for any correction, since there is no self-induced perturbation. The particle force can be calculated based on $\overline{\mathbf{u}}_c^{\,s}$. (ii) If the volume effect is ignored in a two-way coupled EL simulation, then the correction procedure must accordingly employ $\phi_{1c} = 1$, thus turning off only the self-influence of particle volume. (iii) In the EL simulation, if the particle force is calculated using the BBO equation instead of the MRG equation, then accordingly the Ψ_{ef} can be approximated by Ψ_{os}. (iv) The results of Balachandar et al. (2019) suggest that an unsteady correction is needed only when the Stokes number of the particle is less than 5. Under this condition, the quasi-steady algorithm can be extended to include the convolution integral of past history to account for the unsteady effects.

Two other factors can influence the correction procedure, as discussed in Balachandar and Liu (2023). First, if the force on the particle includes contributions other than the quasi-steady force, then the self-induced perturbation velocity will not align with the feedback force. The correction procedure must then be vectorial in nature and the correction function becomes a second-rank tensor that relates the velocity correction to the feedback force. Second, when a particle approaches a boundary of the fluid domain, then the wall effect will alter the self-induced perturbation correction. Both these aspects were considered in Balachandar and Liu (2023). The reader is recommended to consult this work in order to implement a more comprehensive correction procedure.

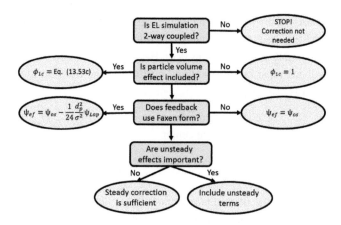

Figure 15.5 Flowchart showing decisions to be made in implementing the self-induced velocity correction procedure.

The correction function for both BBO and MRG approaches is plotted against Re for varying values of d_p/σ in Figure 15.4a. First and foremost, in all cases, the effect of slip correction is to increase the relative velocity, and thereby increase the force estimation to the true value. The difference between the BBO and MRG formulations is important only at small values of Re and for large particle size. Slip correction is negligible at large Reynolds numbers and particles much smaller than the filter width. At small and modest particle Reynolds numbers, even for particles as small as $d_p/\sigma = 0.1$, slip correction is of importance, without which the force on the particle will be under-predicted. Plotted in Figure 15.4b is the self-induced volume effect captured by the quantity $\phi_{1c@}$ plotted as a function of d_p/σ. All three evaluations of $\phi_{1c@}$ given in Eq. (15.61) are shown in the figure. It is clear that the volume effect is generally small. Even at $d_p/\sigma = 1$ its effect is only about 3%. Also, the approximate form is quite good for $d_p/\sigma < 1$.

15.7.2 Self-induced Thermal Perturbation of Large Particles and its Correction

All self-induced perturbation fields must be corrected in order to obtain the true undisturbed flow of a particle, which is what is needed in the point-particle models. In the case of heat transfer between the particulate and fluid phases, self-induced perturbation due to thermal slip or temperature difference between the particle and the undisturbed fluid must be properly accounted for in evaluating the heat transfer rate. The thermal correction of self-induced perturbation was considered by Liu et al. (2019) and here we will summarize their results. As with velocity correction, we will first present the correction procedure in the quasi-steady limit and then discuss the unsteady thermal correction procedure. Our discussion will be brief and the reader is referred to the paper for additional details on the derivation and interpretation of results.

Let Q' be the heat transferred to the particle and the negative of this is applied back to the fluid with the Gaussian smoothening function. This local source/sink of heat that is fed back to the fluid will in turn modify the local temperature field. However, the BBO and MRG heat-transfer models developed in Chapter 6 are based on the undisturbed temperature of the ambient flow. Hence, the self-induced thermal perturbation of the feedback must be obtained and corrected. The true fluid temperature at the particle center is given by

$$\underbrace{\begin{array}{c}\text{True fluid temperature} \\ \text{at the particle center}\end{array}}_{} = \underbrace{\phi_{1c@}T_{c@}}_{\substack{\text{computed in} \\ \text{EL simulation}}} - \underbrace{\phi_{c@}T'_{c@}}_{\substack{\text{self-induced} \\ \text{perturbation}}}, \tag{15.80}$$

where T_c is the filtered temperature field computed in an EL simulation and T'_c is the perturbation temperature field due to the feedback of Q'. As in velocity correction, ϕ_{1c} in the first term on the right-hand side represents the fluid volume fraction evaluated at the particle center due to the presence of that single particle, which is different from ϕ_c computed in an EL simulation with a distribution of particles. The self-induced perturbation field in the second term on the right-hand side is due to an isolated particle and we have used the fact that in this case $\phi_{1c@}$ and $\phi_{c@}$ are the same. As written above, the true undisturbed ambient temperature at the particle is different from that computed in the EL simulation. Multiplication by $\phi_{1c@}$ in the first term accounts for the self-induced volume effect of the particle, while the second term corrects for the self-induced thermal-slip effect.

The surface average of true undisturbed fluid temperature at the particle can be similarly defined as follows for use in the finite-Pe MRG thermal equation:

True surface-averaged temperature

$$= \overline{\phi_{1c}T_c}^s - \overline{\phi_cT'_c}^s$$

$$\approx \underbrace{\phi_{1c@}T_{c@} + \frac{d_p^2}{24}\left[\nabla^2(\phi_{1c}T_c)\right]_@}_{\substack{\text{computed in} \\ \text{EL simulation}}} - \underbrace{\left(\phi_{c@}T'_{c@} + \frac{d_p^2}{24}\left[\nabla^2(\phi_cT'_c)\right]_@\right)}_{\substack{\text{self-induced} \\ \text{perturbation}}}. \tag{15.81}$$

The first two terms in the last line can easily be calculated in an EL simulation from the computed filtered temperature field. The evaluation of the final two terms will be outlined in the problem below.

Problem 15.5 *Oseen correction for T'_c.* In this problem you will rework the thermal analysis in the Oseen limit following the presentation given in Liu et al. (2019). The governing equation of the self-induced thermal perturbation under Oseen approximation is

$$U\frac{\partial \phi_cT'_c}{\partial x} = \kappa_f\nabla^2(\phi_cT'_c) - \frac{Q'}{\rho_fC_{\text{pf}}}G(\mathbf{x}), \tag{15.82}$$

where κ_f is the thermal diffusivity of the fluid and C_{Pf} the specific heat capacity. The advantage of the Oseen approximation is that the above equation is decoupled from the mass and momentum equations (15.62). However, this is not true under full nonlinearity (i.e., at finite Peclect number), where the thermal equation will depend on the velocity field through the advection term.

(a) Follow the solution procedure presented for the velocity to obtain the solution of the temperature equation (15.82) in the Fourier space as

$$\widehat{\phi_c T_c'} = -\left\{ \frac{1}{\left(1 + \dfrac{\iota k_x U}{\kappa_f k^2}\right)} \right\} \frac{Q'}{(2\pi)^3 k_f \sigma^3 k^2} \exp\left(-\frac{\sigma^2 k^2}{2}\right), \qquad (15.83)$$

where the term within curly brackets is the Oseen correction.

(b) Take the inverse Fourier transform of $(1 - k^2)\widehat{\phi_c T_c'}$ (consult Liu et al., 2019) to obtain the surface-averaged volume fraction-weighted self-induced temperature as

$$\overline{\phi_c T_c'}^s = \phi_{c@} T_{c@}' + \frac{d_p^2}{24}\left[\nabla^2 (\phi_c T_c')\right]_@ = -\frac{Q'}{(2\pi)^{3/2}\, k_f \sigma}\underbrace{\left[\Psi_{\text{Tos}} - \frac{1}{24}\frac{d_p^2}{\sigma^2}\Psi_{\text{TLap}}\right]}_{\Psi_{\text{Tef}}},$$

$$(15.84)$$

where the dependence of Ψ_{Tos} and Ψ_{TLap} on the filter Péclet number $\mathrm{Pe}_\sigma = U\sigma/\kappa_f$ is given as

$$\Psi_{\text{Tos}}(\mathrm{Re}_\sigma) = \frac{\sqrt{\pi}}{\sqrt{2}\,\mathrm{Pe}_\sigma}\left[1 - \exp\left(\frac{1}{2}\mathrm{Pe}_\sigma^2\right)\mathrm{erfc}\left(\frac{1}{\sqrt{2}}\mathrm{Pe}_\sigma\right)\right], \qquad (15.85)$$

$$\Psi_{\text{TLap}}(\mathrm{Re}_\sigma) = \frac{1}{\mathrm{Pe}_\sigma}\left(\sqrt{2\pi} - \mathrm{Pe}_\sigma + \sqrt{\frac{\pi}{2}}\left(\mathrm{Pe}_\sigma^2 - 2\right)\exp\left(\frac{1}{2}\mathrm{Pe}_\sigma^2\right)\mathrm{erfc}\left(\frac{1}{\sqrt{2}}\mathrm{Pe}_\sigma\right)\right). \qquad (15.86)$$

The surface average reduces the self-induced thermal correction calculated at the particle center.

(c) Show that this Oseen correction factor can be well approximated as $\Psi_{\text{Tos}} \approx 1 - \sqrt{\pi}\,\mathrm{Pe}_\sigma/(2\sqrt{2})$ for small Pe_σ, while for a larger Peclet number the approximation is $\Psi_{\text{Tos}} \approx \left(\sqrt{\pi/2}\right)/\mathrm{Pe}_\sigma$.

(d) Also show that Ψ_{Lap} can be well approximated by the expansion $1 - \mathrm{Pe}_\sigma^2/3$ for small Pe_σ, while for large Pe_σ, $\Psi_{\text{TLap}} \approx \left(\sqrt{2\pi}\right)/\mathrm{Pe}_\sigma$.

Numerical simulations were performed (Liu et al., 2019) to obtain the self-induced thermal correction at finite Peclet number. The results showed that the finite-Re correction remained nearly the same as given by the Oseen correction, even at large Peclet number, provided the nondimensional feedback force is less than unity. This finding is the same as that for the velocity correction. Thus, in what follows we will ignore the modification for large feedback force and simply use the result presented in Eq. (15.84). The reader is referred to Liu et al. (2019) for a more accurate form if the nondimensional feedback force becomes large.

Quasi-steady Thermal Correction Procedure

Here we address the following question. *In an EL simulation, given the filtered fluid temperature field T_c, how can we accurately calculate the quasi-steady heat transfer by properly accounting for the self-induced thermal perturbation?* In the quasi-steady limit, the heat transfer to a particle is expressed as (see Eq. (6.10))

$$Q'_{qs} = 4\pi k_f R_p \left(\left[\overline{\phi_{1c} T_c}^s - \overline{\phi_c T_c'}^s \right] - T_p \right) \text{Nu} , \qquad (15.87)$$

where T_p is the uniform particle temperature. The terms within square brackets can be recognized as the surface average of the true ambient fluid temperature. Again, multiplication by the single particle volume fraction ϕ_{1c} accounts for the self-induced volume effect, while the subtraction of $\overline{\phi_c T_c'}^s$ accounts for the self-induced thermal slip of the particle. The quest here is to rewrite the above relation to obtain a more explicit expression for Q'_{qs}.

Substituting Eq. (15.84) for the perturbation $\overline{\phi_c T_c'}^s$ and rearranging, we obtain the *corrected quasi-steady heat transfer* expression as

$$Q'_{qs} = \underbrace{\left(1 - \frac{d_p}{\sigma} \frac{\text{Nu} \, \Psi_{\text{Tef}}}{\sqrt{2\pi}} \right)^{-1}}_{\text{thermal slip correction}} 4\pi k_f R_p \left(\overline{\phi_{1c} T_c}^s - T_p \right) \text{Nu} . \qquad (15.88)$$

The above can be substituted back to obtain the following expression for the actual relative temperature in terms of what is computed in the EL simulation:

$$\left(\left[\overline{\phi_{1c} T_c}^s - \overline{\phi_c T_c'}^s \right] - T_p \right) = \underbrace{\left(1 - \frac{d_p}{\sigma} \frac{\text{Nu} \, \Psi_{\text{Tef}}}{\sqrt{2\pi}} \right)^{-1}}_{\text{thermal slip correction}} \left(\overline{\phi_{1c} T_c}^s - T_p \right) . \qquad (15.89)$$

Thus, the slip-correction function for relative temperature is identically the same as that for heat transfer. The correction increases the estimate of true relative temperature to be larger than that obtained from the EL filtered thermal field. Accordingly, the correction also increases the estimate of true quasi-steady heat transfer.

Unlike the velocity correction function, the thermal correction function is an explicit expression, since it is only a function of Re_σ, Pe_σ, and d_p/σ, which do no depend on temperature difference. This simplicity is due to the linear nature of the energy equation as far as temperature is concerned. In summary, in the quasi-steady regime, the following steps should be pursued for each particle in order to properly correct for its self-induced thermal perturbation in the calculation of particle heat transfer:

- As preprocessing step, compute and store the thermal slip-correction functions Ψ_{Tos} and ψ_{TLap}.

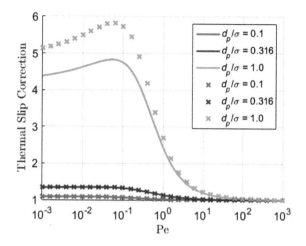

Figure 15.6 Self-induced slip-correction function plotted against Re for three different values of d_p/σ. The lines are for the finite-Re MRG equation and the symbols are for the finite-Re BBO equation. Taken from Liu et al. (2019).

- In the velocity correction steps, you would already have solved Eq. (15.73) to obtain the true Reynolds numbers. From Re_σ, obtain $\mathrm{Pe}_\sigma = \mathrm{Re}_\sigma \mathrm{Pr}$.
- From the values of Re_σ, Pe_σ, and d_p/σ, evaluate the value of the finite-size correction Ψ_{Tef}.
- Apply the correction given in Eq. (15.89) to calculate the true relative temperature, and the correction given in Eq. (15.88) to obtain the true quasi-steady heat transfer.

Once again, it must be emphasized that only those aspects of two-way coupling included in an EL simulation must be corrected. The proper variant of the correction procedure must be considered. (i) In a one-way coupled simulation there is no need for any correction. Particle heat transfer can be calculated based on $\overline{T}_c^{\,s}$. (ii) If the volume effect is ignored in a two-way coupled EL simulation, then the correction procedure must accordingly employ $\phi_{1c} = 1$. (iii) In the EL simulation, if the particle heat transfer is calculated using the BBO equation, instead of the MRG equation, then accordingly the Ψ_{Tef} can be approximated by Ψ_{Tos}. The correction functions for both BBO and MRG approaches are plotted against Pe for varying values of d_p/σ in Figure 15.6a. In obtaining these plots we have used $\mathrm{Pr} = 0.72$ corresponding to that of air and the Nusselt number correlation $\mathrm{Nu} = 2 + 0.6\,\mathrm{Re}^{1/2}\,\mathrm{Pr}^{1/3}$. First and foremost, in all cases, the effect of slip correction is to increase the relative temperature, and thereby increase the heat transfer estimation to the true value. The thermal correction is larger than the velocity correction. The difference between the BBO and MRG formulations is important only at small values of Pe and for large particle size. The thermal correction is negligible at large Peclet numbers and particles much smaller than the filter width. At small and modest particle Peclet numbers, even for particles as small as $d_p/\sigma = 0.1$, slip correction is of importance, without which heat transfer will be under-predicted.

Unsteady Thermal Correction

In this subsection, we will extend the thermal correction procedure to conditions in which unsteady heat transfer becomes important. Our discussion will be based on the analysis of Liu et al. (2019). The conditions of unsteady heat transfer were discussed in detail in Section 6.2. Similar conditions apply for the importance of thermal correction as well. From the results of EL simulations of an isolated particle undergoing thermal evolution, Liu et al. (2019) concluded that the quasi-steady model is sufficient only when $\tau_T/\tau_{fT} > 10$, where τ_T is the thermal time scale of the particle and τ_{fT} is the time scale on which the relative temperature of the fluid as seen by the particle varies. The above condition was observed to apply for $\mathrm{Pr}_\sigma < 1$, while for larger Pe_σ the importance of unsteadiness increases.

A model for unsteady thermal correction is needed only with an unsteady heat-transfer model. For a finite-sized particle, we will consider the following MRG model for particle heat transfer (see Eq. (6.10)):

$$
Q'(t) = 2\pi k_f R_p \left[\overline{\phi_{1c} T_c}^s - \overline{\phi_c T_c'}^s - T_p \right] \mathrm{Nu}
$$
$$
+ 4\pi k_f R_p \int_{-\infty}^t K_T \left[\frac{d(\overline{\phi_{1c} T_c}^s - \overline{\phi_c T_c'}^s)}{dt} - \frac{dT_p}{dt} \right] d\xi . \tag{15.90}
$$

The expression for unsteady self-induced thermal perturbation obtained by Balachandar et al. (2019) has been adapted as

$$
\overline{\phi_c T_c'}^s = -\frac{1}{\sqrt{2\pi}\, k_f \sigma} \left[Q' \Psi_{\mathrm{Tef}} - \int_{\infty}^t K_{T,\mathrm{cor}}(t-\tau) \left. \frac{dQ' \Psi_{\mathrm{Tef}}}{dt} \right|_\tau d\tau \right], \tag{15.91}
$$

where the first term on the right-hand side is the quasi-steady correction and the second term is the unsteady correction written as a convolution integral, with the kernel given as (Liu et al., 2019)

$$
K_{T,\mathrm{cor}}(t-\tau) = g_T(t-\tau) \frac{1}{\sqrt{1 + 2(t-\tau)/Pe_\sigma}} + (1 - g_T(t-\tau)) \exp(-2.356(t-\tau)) .
$$
$$
\tag{15.92}
$$

The function $g_T(\xi) = \mathrm{erfc}\left[c_1 \left(c_2^\xi - c_3 \right) \right]$, where the coefficients c_1, c_2, and c_3 as a function of Pe_σ are given in Table 15.4. As shown in Liu et al. (2019), the above composite kernel shows the correct analytical decay at short times and exponential decay at long times. We close the discussion with the following prescription that can be implemented for each particle for unsteady correction of self-induced temperature.

- Solve Eq. (15.73) to obtain the true Reynolds number. Also obtain the filter Reynolds and Péclet numbers as $\mathrm{Re}_\sigma = (\sigma/d_p)\mathrm{Re}$ and $\mathrm{Pe}_\sigma = \mathrm{Re}_\sigma\, \mathrm{Pr}$.
- From the above, calculate Ψ_{Tef}.
- From the time history of Q' and Ψ_{Tef}, evaluate the current value of $d(Q'\Psi_{\mathrm{Tef}})/dt$.
- Compute the kernel $K_{T,\mathrm{cor}}$ as a function of elapsed time using Eq. (15.92).
- From this information, compute the self-induced perturbation temperature $\overline{\phi_c T_c'}^s$ as given in Eq. (15.91).
- With the temperature correction, calculate the particle heat transfer using Eq. (15.90).

Table 15.4 Fitting parameters of the blending function for different Pe_σ.

Pe_σ	c_1	c_2	c_3
1	28.6521	1.0052	0.9869
$\sqrt{10}$	27.7924	1.0111	0.9880
10	8.92552	1.0506	0.9669
$10\sqrt{10}$	3.32093	1.1526	0.9123
100	2.55624	1.2051	0.8857
$100\sqrt{10}$	2.15369	1.2434	0.8617
1000	2.09725	1.2507	0.8579
$1000\sqrt{10}$	2.04844	1.2566	0.8542
10000	2.02633	1.2592	0.8525

Clearly, the above correction procedure is computationally more involved than quasi-steady correction. It is needed only when the unsteady MRG (or BBO) model is used for heat transfer.

15.8 Force Coupling Method

The force coupling methodology (FCM) was pioneered by Maxey and co-workers (Maxey and Patel, 2001; Lomholt et al., 2002; Lomholt and Maxey, 2003; Climent and Maxey, 2009; Yeo and Maxey, 2010). Strictly speaking, FCM is somewhere between Euler–Lagrange and particle-resolved methodologies, since it attempts to partially resolve the particles. We will discuss this methodology here since the interface between the particle and the surrounding fluid is not recognized in FCM. As in the EL methodology, the Navier–Stokes equations for the fluid are solved over the entire computational domain and nonslip and no-penetration boundary conditions are not explicitly enforced on the particle surface. Also, the influence of the particle is fed back to the fluid using back coupling of the force. There are two important departures from the EL approach. The back coupling involves both a monopole and a dipole source. Furthermore, the magnitudes of monopole and dipole are not determined by a drag or force law. They are decided internally in a self-consistent manner. For a comparative look at FCM and the other extended point-particle model using pairwise interaction, the reader can consult Balachandar and Maxey (2023).

The starting point of FCM does not involve application of a Gaussian filter function. The governing equations are thus the incompressible Navier–Stokes equations

$$\nabla \cdot \mathbf{u} = 0 \quad \text{and} \quad \rho_f \frac{D\mathbf{u}}{Dt} = -\nabla p + \mu \nabla^2 \mathbf{u} + \mathbf{f}(\mathbf{x}, t), \qquad (15.93)$$

where the added body-force distribution $\mathbf{f}(\mathbf{x}, t)$ accounts for the influence of the particles and is given as follows:

$$\mathbf{f}(\mathbf{x},t) = -\left(\sum_{l=1}^{N_p} \mathbf{F}'_l G_1(\mathbf{x} - \mathbf{X}_l) + \mathbf{D}'_l \cdot \nabla G_2(\mathbf{x} - \mathbf{X}_l)\right). \qquad (15.94)$$

The above represents the first two terms of the multipole expansion of the perturbation particle force. The first term represents the perturbation force $\mathbf{F}'$ the fluid exerts on the lth particle centered at $\mathbf{X}_l$. The second term is the dipole contribution whose magnitude is determined by the tensor $\mathbf{D}'_l$. The spreading functions G_1 and G_2 are chosen to be the 3D Gaussian introduced in Eqs. (11.2) and (11.4). While the Gaussian width was determined by the user in the EL approach based on the length scale $\mathcal{L}$ chosen, in FCM, the Gaussian widths σ_1 and σ_2 are uniquely determined by the requirement that the force and torque applied back to the fluid are consistent with the Stokes drag and torque.

As pointed out earlier, in FCM, the perturbation force on the lth particle is calculated from the particle motion using the following relation:

$$\mathbf{F}'_l = (m_p - m_f)\left(\frac{d\mathbf{V}_l}{dt} - \mathbf{g}\right). \qquad (15.95)$$

The fundamental difference between force coupling and EL methodologies becomes clear. In EL, $\mathbf{F}'$ was calculated as the sum of quasi-steady, added-mass, history, and lift forces, and then the particle motion is calculated as $m_p d\mathbf{V}_l/dt = m_f d\mathbf{u}_{@l}/dt + (m_p - m_f)\mathbf{g} + \mathbf{F}'_l$. In contrast, in FCM, the particle motion is calculated from the fluid velocity field as the volume average over the region occupied by the particle as

$$\mathbf{V}_l(t) = \int \mathbf{u}(\mathbf{x},t)\, G(\mathbf{x} - \mathbf{X}_l) d\mathbf{x}. \qquad (15.96)$$

The intention is that the fluid in the region of the particle behaves as the particle in its motion. From this supposition, the feedback force can be calculated as in Eq. (15.95) instead of the BBO or MRG force model. As demonstrated by Lomholt and Maxey (2003), if the Gaussian width is chosen as $\sigma_1 = R_p/\sqrt{\pi}$, then the feedback force will be captured exactly in the case of a uniform Stokes flow over an isolated particle. This choice was shown to work well even in case of small nonzero Reynolds numbers.

The innovative aspect of FCM is the contribution of the dipole term, where the tensor $\mathbf{D}'_l$ can be separated into symmetric and asymmetric parts as $\mathbf{D}'_l = \mathbf{D}'_{s,l} + \mathbf{D}'_{a,l}$, both of which must be determined from the flow field and the particle motion itself, without recourse to additional stress or torque modeling, just like the perturbation force $\mathbf{F}'_l$ was expressed in Eq. (15.95). The symmetric part accounts for the resistance a rigid particle offers to deformation. To approximate this behavior, FCM demands the average strain rate inside the particle to be zero:

$$\frac{1}{2} \int \left(\nabla \mathbf{u} + (\nabla \mathbf{u})^{\mathrm{T}}\right) G_2(\mathbf{x} - \mathbf{X}_l)\, d\mathbf{x} = 0. \qquad (15.97)$$

The above is a constraint and the symmetric part $\mathbf{D}'_{s,l}$ is chosen such that the above constraint is satisfied. This is implemented as an iterative procedure (Dance and Maxey,

2003), which typically converges within a few iterations. The role of the symmetric part is to increase the fluid stress outside the particle and thereby account for the enhanced viscosity effect discussed in Section 4.1.4.

The asymmetric part is related to the hydrodynamic torque $\mathbf{T}'_l$ acting on the particle, expressed in index notation as

$$\left(D'_{a,l}\right)_{ij} = -\frac{1}{2}\epsilon_{ijk}\left(T'_l\right)_k .\tag{15.98}$$

The hydrodynamic torque on the particle is then evaluated in terms of the angular velocity of the particle as

$$\mathbf{T}'_l = (I_p - I_f)\frac{d\mathbf{\Omega}_l}{dt},\tag{15.99}$$

where I_p is the rotational inertia of the particle and I_f is the rotational inertia of the displaced fluid. Again, the difference between force coupling and EL methodologies can be seen here. In EL, $\mathbf{T}'$ was calculated as the sum of quasi-steady history torques on the particle, and then the particle motion is calculated as $I_p(d\mathbf{\Omega}_l/dt) = m_f(d\omega_{@l}/dt) + \mathbf{T}'_l$. In contrast, in FCM, the angular velocity of the particle is obtained from the vorticity of the fluid averaged over the volume occupied by the particle as

$$\mathbf{\Omega}_l(t) = \frac{1}{2}\int \nabla \times \mathbf{u}(\mathbf{x},t)\,G(\mathbf{x}-\mathbf{X}_l)d\mathbf{x},\tag{15.100}$$

and then used to calculate the hydrodynamic torque on the particle. The Gaussian width σ_2 is obtained as $\sigma_2 = R_p/\left(6\sqrt{\pi}\right)^{1/3}$ by requiring that the hydrodynamic torque calculated as outlined above becomes exact in the case of Stokes flow.

An important aspect of the EL methodology is illustrated by the force coupling methodology. If the rotational motion of the particle along with the hydrodynamic torque on the particle is included in the EL methodology, then an appropriate way to include the feedback force within the framework of Gaussian filtering is to include an appropriate dipole source applied back on the fluid. The asymmetric part of the dipole can be obtained from the negative of the hydrodynamic torque applied to the particle. As far as the symmetric part is concerned, if the EL simulation includes a volume fraction-dependent viscosity correction, then there is no need for the symmetric part, since its effect has already been accounted for in the modified viscosity.

Problem 15.6 *Force-coupling solution in the Stokes limit.* In this problem you will solve the following steady Stokes equation with the Gaussian monopole and dipole forcing:

$$0 = -\nabla p + \mu_f \nabla^2 \mathbf{u} - (\mathbf{F}'G_1(\mathbf{x}-\mathbf{X}) + \mathbf{D}'\cdot\nabla G_2(\mathbf{x}-\mathbf{X})) .\tag{15.101}$$

(a) Use the 3D Fourier transform employed in the earlier problems of this chapter and obtain the following solutions that were given in Lomholt and Maxey (2003):

$$u_i = \left(A_1 \delta_{ij} + B_1 x_i x_j\right) F_j' + \left(\frac{dA_2}{dt} \frac{\delta_{ij} x_k}{r} + B_2(\delta_{ik} x_j + \delta_{jk} x_i) + \frac{dB_2}{dr} \frac{x_i x_j x_k}{r}\right) D_{jk}',$$
(15.102)

where $r = |\mathbf{x}|$ and the A and B functions are defined as follows:

$$A(r) = \frac{1}{8\pi\mu_f r} \left[\left(1 + \frac{\sigma^2}{r^2}\right) \mathrm{erf}\left(\frac{r}{\sqrt{2}\sigma}\right) - \frac{2\sigma}{\sqrt{2\pi}r} \exp\left(-\frac{r^2}{2\sigma^2}\right)\right],$$
(15.103)

$$B(r) = \frac{1}{8\pi\mu_f r^3} \left[\left(1 - \frac{3\sigma^2}{r^2}\right) \mathrm{erf}\left(\frac{r}{\sqrt{2}\sigma}\right) + \frac{6\sigma}{\sqrt{2\pi}r} \exp\left(-\frac{r^2}{2\sigma^2}\right)\right].$$
(15.104)

If we set $\sigma = \sigma_1$ or σ_2, then we obtain the definitions A_1 and B_1, and correspondingly A_2 and B_2. When integrated over the entire volume of the fluid, only the monopole term contributes to the net force on the fluid.

(b) By setting the net force on the fluid to be that given by Stokes drag (i.e., by letting $-\mathbf{F}' = -6\pi\mu_f R_p U \mathbf{e}_x$), show that only when $\sigma_1 = R_p/\sqrt{\pi}$:

$$\int \mathbf{u}(\mathbf{x}) G(\mathbf{x}) \, d\mathbf{x} = U\mathbf{e}_x.$$
(15.105)

This is how we get a consistent definition of σ_1. The velocity given by Eq. (15.102) was compared with the true Stokes flow around the particle by Lomholt and Maxey (2003) and the comparison is reproduced in Figure 15.7. It can be seen that while the velocity inside a real particle is a constant, the velocity inside the Gaussian force varies, but the volume average has been constrained to approach the particle velocity. As noted by Lomholt and Maxey (2003), Eq. (15.102) is in excellent agreement with the Stokes flow for $r > 1.25R_p$.

(c) Follow the steps presented in Lomholt and Maxey (2003) to obtain the following expression for the fluid velocity gradient averaged over the volume of the particle:

$$\int \frac{\partial u_i}{\partial x_j} G(\mathbf{x}) d\mathbf{x} = -\frac{1}{40\mu_f \pi^{3/2}\sigma_2^3} \left(D_{ij}' - \frac{1}{3}\delta_{ij} D_{kk}'\right) - \frac{1}{120\mu_f \pi^{3/2}\sigma_2^3} \left(D_{ij}' - D_{ji}'\right).$$
(15.106)

(d) Similar to force, by setting the net torque on the fluid to be that given by the Stokes solution (i.e., by setting $-\mathbf{T}' = -8\pi\mu_f R_p \Omega \mathbf{e}_z$), show that only when $\sigma_2 = R_p/(6\sqrt{\pi})^{1/3}$:

$$\frac{1}{2} \int \nabla \times \mathbf{u}(\mathbf{x}, t) G(\mathbf{x} - \mathbf{X}_l) d\mathbf{x} = \Omega \mathbf{e}_z.$$
(15.107)

This is how we get a consistent definition of σ_2. The velocity given by Eq. (15.102) for the asymmetric dipole was compared with the true Stokes flow around a spinning particle (Lomholt and Maxey, 2003) and the comparison is reproduced in Figure 15.8. It can be seen that while the circumferential velocity inside a real particle is linearly varying, the velocity inside the Gaussian force shows a complex variation. As noted by Lomholt and Maxey (2003), the FCM solution is in excellent agreement with the Stokes flow for $r > 1.25R_p$.

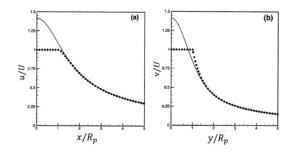

Figure 15.7 Velocity profile along (a) the x-axis and (b) the y-axis for an isolated sphere moving with velocity $U\mathbf{e}_x$. The exact Stokes solution is given by the dotted line and the continuous line is the FCM solution. Reprinted from Lomholt and Maxey (2003), with permission from Elsevier.

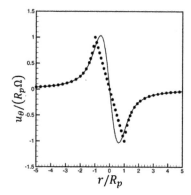

Figure 15.8 Circumferential velocity profile along the radial direction for an isolated sphere rotating with angular velocity $\Omega\mathbf{e}_z$. The exact Stokes solution is given by the dotted line and the continuous line is the FCM solution. Reprinted from Lomholt and Maxey (2003), with permission from Elsevier.

16 Euler–Euler Approach

The Euler–Euler (EE) approach derives its name from the fact that both the continuous and the dispersed phases are solved in the Eulerian frame of reference. For the fluid phase, the Eulerian frame is the natural choice and was pursued both in the particle-resolved (PR) and the Euler–Lagrange (EL) approaches. Particles are, however, inherently Lagrangian, and an Eulerian representation is possible only when the individual nature of the particles is erased. This requires that the particle-related Lagrangian quantities be suitably averaged, so that Eulerian fields of these quantities can be defined. The averaging process will allow particle volume fraction, particle velocity, and particle temperature fields to be defined as functions of space and time. The time evolution of these fields is then given by the governing equations of particulate mass, momentum, and energy, which must be obtained by correspondingly averaging the Lagrangian mass, momentum, and energy equations of individual particles.

In a polydisperse system, during the averaging process that leads to the Eulerian representation of the particle-related quantities, if the averaging is performed over all particles irrespective of their size, then the resulting particle volume fraction, velocity, and temperature fields account for the integrated behavior of all particle sizes. Such an indiscriminate average is often not desirable, since the behavior of small particles that respond well to turbulent eddies is substantially different from that of unresponsive large particles. Averaging these drastically different responses leads to serious difficulties in the proper interpretation of the resulting averaged quantities. The alternative is to divide the entire range of particle sizes into many narrow bins and average only those particles within each bin. This leads to particle volume fraction, velocity, and temperature fields for each particle size bin and associated equations of mass, momentum and energy conservation. The computational cost of solving the governing equations of many different particle size bins seriously limits the applicability of this approach. The EE method is thus often restricted to monodisperse systems or to few particle sizes. In contrast, the Lagrangian approach allows easy handling of a wide range of particle sizes.

When there are only a small number of particles within the system (typically smaller than the number of Eulerian grid points), the Lagrangian approach is preferred over the Eulerian approach. Partly because the computational cost of tracking a small number of particles in the Lagrangian frame is smaller than solving for the Eulerian particulate fields. Furthermore, when there are only a small number of particles distributed over a finely resolved Eulerian grid, it is difficult to define smooth Eulerian particulate fields. On the other hand, when there are billions and billions of particles within the particle-laden system, it is far more advantageous to adopt the EE approach, since it is impossible to track the motion of each and every particle.

The progression from the PR to EL and EE approaches can now be examined. The PR approach is fundamental and does not involve any approximation at the continuum level. PR results can be fully trusted. In the EL approach, while the rigid-body dynamics of the particles is fully resolved, the dynamics of the continuous phase is approximated only to scales that are larger than a filter length scale. That is, only the filtered (or averaged) governing equations of the continuous phase are solved. The filtering operation introduces Reynolds stress and other secondary terms in the fluid-phase governing equations, which must be substituted with their closure models. As discussed in the previous chapter, the filtering operation and the closure models introduce uncertainties and systematic errors in EL prediction.

The EE approach extends the filter operation to the particulate phase as well. The filter operation that was used to remove the small scales of the continuous phase is employed to smooth the particle-related quantities. As with the continuous-phase governing equations, as a result of the filtering operation, the particulate-phase Eulerian governing equations contain unknown secondary variables that are different from the primary variables being solved. Closure models are required in order to express the secondary variables empirically in terms of the primary variables (i.e., in terms of particle volume fraction, velocity, and temperature). Due to this additional layer of approximation, the results of the EE approach generally involve more uncertainties and larger errors than the corresponding EL results.

In the previous chapter, on the EL approach, we considered both the small-particle limit when the particle is much smaller than the filter width and the large-particle limit when the particle size is comparable to the grid size and the filter width. In the case of the EE approach, only the small-particle limit of $\mathcal{L} \gg d_p$ makes sense. In other words, since the particle equations are filtered, we will only consider situations where the particle is much smaller than the filter width, $\mathcal{L}$. Thus, in this chapter, we will first revisit the volume filtering technique of Chapter 11 applied to both the continuous and dispersed phases and obtain the governing equations of both the fluid and particulate phases in the small-particle limit. The definitions of particle volume fraction, velocity, and temperature fields will be presented in the process of obtaining the particle-phase governing equations.

Within the small-particle limit, we will consider two cases: (i) $\mathcal{L} > \eta$, when all the scales of fluid turbulence (not pseudo turbulence) are fully resolved; (ii) $\mathcal{L} < \eta$, when some of the fluid turbulence is filtered out in the filtering operation. Following the terminology of the previous chapter, these will be referred to as **EE-DNS** and **EE-LES**, respectively. In Section 16.2 we will discuss the closure models that are needed for the secondary terms that arise in the fluid and particulate phase governing equations, and compare these EE models with those of the corresponding EL closure models. Of particular importance to the particulate phase is the modeling of the interparticle collisional process. We will discuss closure modeling for both EE-DNS and EE-LES. In Section 16.3 we will consider the kinetic theory of granular media to: (i) obtain closure relations for particle pressure, frictional stress tensor, and collisional heat flux, which are additional secondary terms that appear in the particulate-phase momentum and energy equations; (ii) obtain an additional equation for the evolution of rms particle velocity fluctuation, which is commonly referred to as the granular temperature. In Sections 16.4 and 16.5 we will consider some important simplifications of the EE formulation that result in a simpler system of governing equations.

In the EE approach, where both the phases are being averaged, it is not necessary to restrict the averaging operations to spatial filtering. In Section 16.6, we will introduce ensemble averaging as an important statistical process and use it to obtain the average mass, momentum, and energy equations of both the continuous and dispersed phases. There we will carefully consider the fundamental differences between volume filtering and ensemble averaging. For now, it is sufficient to point out that the equations that result from volume filtering describe the large-scale dynamics of individual realizations. Only upon further averaging over a large number of EL realizations do the averaged continuous and dispersed-phase properties approach their corresponding ensemble average. On the other hand, the ensemble-averaged equations provide direct access to ensemble-averaged statistics. However, an ensemble-averaged flow will not provide insight into the dynamics of individual realizations.

16.1   EE Governing Equations from Volume Filtering

This section will present the governing equations of the continuous and dispersed phases. To simplify the discussion and the resulting equations, we will place a few restrictions on both the continuous and dispersed phases, as was done in the description of the EL approach. The continuous phase will be considered incompressible, and therefore ρ_f is a constant. We will also take transport properties, such as kinematic viscosity, thermal diffusivity, and specific heat capacity to be constant. As for the dispersed phase, we will consider the particles to be spherical and ignore processes such as the merger or breakup of particles. We will also consider the particles to be homogeneously made up of the same material and to be isothermal. Each of these restrictions can be relaxed at the expense of making the governing equations somewhat more complicated.

16.1.1   Continuous-Phase Governing Equations

The filtered continuous-phase mass, momentum, and temperature equations were derived in Chapter 11. Here we start from Eqs. (11.99), (11.100), and (11.70), which already assume the continuous phase to be incompressible, and proceed to apply the other simplifications listed above. In the EE approach, since the particulate phase is also averaged, the length scale of the filter is much larger than the particle diameter and in fact the filter width needs to be larger than the interparticle distance in order for the resulting particle-related fields to be smooth and continuous. These conditions satisfy the additional assumptions involved in the derivation of the simplified governing equations. In the EE approach, the filtered mass, momentum, and energy equations of the continuous phase are

$$\rho_f \left(\frac{\partial \phi_c}{\partial t} + \nabla \cdot (\phi_c \mathbf{u}_c) \right) = -\dot{m},$$

(16.1)

$$\phi_c \rho_f \frac{D\mathbf{u}_c}{Dt} = \nabla \cdot (\phi_c \boldsymbol{\tau}_{\text{sg}}) + \phi_c \rho_f \mathbf{g} + \phi_c \nabla \cdot \boldsymbol{\sigma}_c - \mathbf{F}'_{\text{hyd}} - \left(\dot{\mathbf{M}} - \dot{m} \mathbf{u}_c \right),$$

(16.2)

$$\phi_c \rho_f C_{\text{vf}} \frac{DT_c}{Dt} = - \nabla \cdot (\phi_c \mathbf{q}_{\text{sg}}) - \phi_c \nabla \cdot \mathbf{q}_c + \boldsymbol{\sigma}_c : \nabla \mathbf{u}_m$$
$$- Q'_{\text{hyd}} - \left(W'_{\text{hyd}} - \mathbf{u}_c \cdot \mathbf{F}'_{\text{hyd}} \right) - \left(\dot{E} - \dot{m} E_c \right) + \mathbf{u}_c \cdot \left(\dot{\mathbf{M}} - \dot{m} \mathbf{u}_c \right).$$

(16.3)

Note that the volume-weighted mixture velocity was previously defined as $\mathbf{u}_m = \phi_c \mathbf{u}_c + \phi_d \mathbf{u}_d$. The above governing equations are the same as those of the EL approach, given as Eqs. (15.1), (15.2), and (15.3), with the only difference that in the EL approach the mass, momentum, and energy feedback to the continuous phase was summed over all the particles around the point $\mathbf{x}$ weighted by the Gaussian filter, whereas in the EE approach the feedback terms are calculated based on the Eulerian particulate fields, as will be elaborated below. The differences between the above governing equations and the unfiltered Navier–Stokes equations of the particle-resolved approach were discussed in Chapter 15.

16.1.2 Dispersed-Phase Governing Equations

In Chapter 11, we derived the dispersed-phase governing equations in the Eulerian frame with a rigorous application of the filtering operation. These equations, given as Eqs. (11.122), (11.123), and (11.124), are rewritten here for completeness as

$$\rho_p \left(\frac{\partial \phi_d}{\partial t} + \nabla \cdot (\phi_d \mathbf{u}_d) \right) = \dot{m},$$

(16.4)

$$\phi_d \rho_p \frac{d\mathbf{u}_d}{dt} = \nabla \cdot (\phi_d \boldsymbol{\tau}_{d,\text{sg}}) + \underbrace{\nabla \cdot (\phi_d \boldsymbol{\sigma}_{d,\text{co}})}_{\substack{\text{collisional} \\ \text{effect}}} + \phi_d \rho_p \mathbf{g} + \phi_d \nabla \cdot \boldsymbol{\sigma}_c + \mathbf{F}'_{\text{hyd}} + \left(\dot{\mathbf{M}} - \dot{m} \mathbf{u}_d \right),$$

(16.5)

$$\phi_d \rho_p C_p \frac{dT_d}{dt} = - \nabla \cdot (\phi_d \mathbf{q}_{d,\text{sg}}) - \phi_d \nabla \cdot \mathbf{q}_c$$
$$+ \underbrace{\Gamma}_{\substack{\text{collisional} \\ \text{effect}}} + Q'_{\text{hyd}} + (W'_{\text{hyd}} - \mathbf{u}_d \cdot \mathbf{F}'_{\text{hyd}})$$
$$+ (\dot{E} - \dot{m} E_d) - \mathbf{u}_d \cdot \left(\dot{\mathbf{M}} - \dot{m} \mathbf{u}_d \right).$$

(16.6)

The above equations reflect the fact that in the EE approach, individual particle trajectories are not followed. Only the collective motion represented by the dispersed-phase primary variables ϕ_d, $\mathbf{u}_d$, and T_d are solved. While the filtered continuous-phase equations have been discussed in detail in the context of the EL approach, the above filtered particulate-phase equations have not been discussed before in this book and therefore we will discuss them here in greater detail.

In Eq. (16.4), the right-hand side corresponds to the mass source of the particulate phase. It is positive in case of processes such as condensation or reverse sublimation, where mass goes from the fluid to the particulate phase, and negative in case of evaporation or sublimation. This term precisely balances the continuous-phase source/sink that appears on the right-hand side of Eq. (16.1), with the net effect that the total mass of the continuous and the dispersed phases is conserved.

In the momentum equation (16.5), the first term on the right-hand side is the Reynolds stress term that arises because the velocity of particles within the filtering volume deviates from the mean value and it is analogous to the familiar fluid Reynolds stress term that appears in Eq. (16.2). The definition of particle Reynolds stress can be found in Eq. (11.114). On the right-hand side, the second term accounts for the inter-particle collisional effect, which we shall discuss in greater detail in this chapter. The third term corresponds to the gravitational body force that acts on the particles. The fourth term on the right-hand side is the undisturbed flow force that acts on the particles within the filtering volume. This term, along with the corresponding force on the fluid given by the second term on the right-hand side of Eq. (16.2), yields the total force $\nabla \cdot \boldsymbol{\sigma}_c$ that acts on the fluid–particle mixture. Note that the fluid and particles share this pressure and viscous stress force in the proportion of their volume. The fifth term on the right-hand side accounts for all other hydrodynamic forces that are exerted on the particles within the filtering volume due to the surrounding fluid. It excludes the undisturbed fluid force, which appeared as the fourth term. The last term on the right-hand side is the adjusted contribution to particulate phase acceleration by the the mass exchange between the phases.

In the temperature equation (16.6), the first term on the right-hand side is the Reynolds heat flux term that arises from an individual particle's temperature deviating from the mean dispersed-phase temperature and it is analogous to the fluid Reynolds heat flux term that appears in Eq. (16.3). The second term on the right-hand side is the undisturbed flow heat transfer. This term, along with the corresponding term in the fluid energy equation given by the second term on the right-hand side of Eq. (16.3), yields the total conductive heat flux $\nabla \cdot \mathbf{q}_c$ that acts on the mixture. Note that the fluid and particles share this total diffusive heat flux in proportion to their volume. The third term on the right-hand side accounts for the dissipative heating of particles due to interparticle collisions and we shall discuss this contribution in greater detail in this chapter. The fourth term on the right-hand side accounts for the direct heat transfer contribution to the particles within the filtering volume from the surrounding fluid. In the second line of Eq. (16.6), the pair of terms within the first set of parentheses correspond to the net effect of work done on the particles due to $\mathbf{F}'_{\text{hyd}}$. The rest of the terms on the right-hand side are the net increase in the temperature of the particles as a result of the mass exchange between the phases.

16.2 Closure Relations of the EE Approach

The continuous and dispersed-phase governing equations must be supplemented with appropriate closures. In this section we will consider the closure models of all the secondary variables that appear in the governing equations. We have already considered the modeling of these secondary variables in the context of the EL approach. Their modeling in the EE approach is similar and therefore we will focus mainly on aspects that are different. As in the EL approach, we will address modeling of the following closure terms:

I. Closure models of interphase momentum exchange $\mathbf{F}'_{\text{hyd}}$.

II. Closure model of work W'_{hyd} associated with the momentum exchange.

III. Closure model of interphase heat transfer Q'_{hyd}.

IV. Closure models of interphase mass exchange $\dot{m}$ and the associated $\dot{\mathbf{M}}$ and $\dot{e}$.

V. Constitutive models of filtered continuous-phase stress tensor $\boldsymbol{\sigma}_c$ and heat flux $\mathbf{q}_c$. These models remain exactly the same as those discussed in Section 15.3.5 in the context of the EL approach and therefore will not be repeated here.

VI. Subgrid closure models of fluid residual stress $\boldsymbol{\tau}_{\text{sg}}$ and residual flux $\mathbf{q}_{\text{sg}}$.

VII. Subgrid closure models of particle residual stress $\boldsymbol{\tau}_{d,\text{sg}}$ and residual flux $\mathbf{q}_{d,\text{sg}}$.

VIII. Closure models of collisional stress $\boldsymbol{\sigma}_{d,\text{co}}$ and inelastic collisional dissipation Γ.

Comparing the above list with that of the EL approach presented in Chapter 15, we note four important differences. (i) In the EL closure, mass, momentum, and energy coupling between the phases were in terms of individual particles. As a result, quantities to be modeled, such as $\mathbf{F}'_l$, were Lagrangian in nature and they included subscript l denoting that they correspond to the lth particle. In contrast, in the EE approach, quantities such as $\mathbf{F}'_{\text{hyd}}$ are Eulerian fields that are averaged over particles located around the point $\mathbf{x}$. (ii) In the EE approach, we ignore the rotational motion of the particles. (iii) Since the particle quantities are filtered in the EE approach, we additionally require closure models for particle residual stress and heat flux (i.e., $\boldsymbol{\tau}_{d,\text{sg}}$ and $\mathbf{q}_{d,\text{sg}}$). (iv) Furthermore, the effect of interparticle collisions must now be accounted for through the terms $\boldsymbol{\sigma}_{d,\text{co}}$ and Γ, since interparticle collisions are not computed directly using DEM as soft collisions.

As discussed in the context of the EL approach, the closure models of the secondary terms become more complicated as the particle volume fraction increases. In obtaining the closure models of force, heat, and mass transfer of a particle, we will ignore the direct perturbing influence of neighboring particles. We only attempt to account for the collective influence of all the neighbors through the volume fraction dependence of the models. Once again we admit that our current understanding of the collective influence of neighbors is limited. Thus, our presentation of volume fraction effect on closure models will be simple and the reader should consult recent research developments.

Before proceeding to consider the modeling of each of the to-be-closed secondary terms, we repeat the cautionary note given at the end of Section 11.3. The nature of the filtered equations depends on the closure models being used for the secondary terms. In particular, ill-posed closures have been known to result in a nonhyperbolic nature of the resulting system (in the inviscid limit). This problem has been identified and several simple correction procedures have been advanced (Stuhmiller, 1977; Lhuillier, 1982; Lhuillier et al., 2013; Theofanous et al., 2018; Balakrishnan and Bellan, 2021). More recent formulations of the closure models that accurately preserve hyperbolicity have been proposed (Fox, 2019; Fox et al., 2020). In the following sections, the emphasis will be on the physics of the different closure terms. The suggested closures are not intended to be the best available. The reader should always consult the current literature as models are constantly updated as more information becomes available.

16.2.1   Interphase Momentum Exchange

In the EE approach, the particles are guaranteed to be much smaller than the filter length scale. As a result, the undisturbed flow of particles located in the neighborhood of the point $\mathbf{x}$ can be taken to be the filtered continuous-phase quantities evaluated at the point. This allows the use of the finite-Re BBO equation (4.144) as the model for $\mathbf{F}'_{\text{hyd}}$ with appropriate finite volume fraction corrections applied to quasi-steady, inviscid unsteady, and viscous unsteady forces. The resulting perturbation force model, which does not include the undisturbed flow force, is given as

$$
\mathbf{F}'_{\text{hyd}} = \frac{\phi_d}{\mathcal{V}_p} \left\{ \underbrace{6\pi\mu_f R_p [\mathbf{u}_c - \mathbf{u}_d]\Phi}_{\mathbf{F}'_{\text{qs}}} + \underbrace{C_M\, m_f \left[\frac{D\mathbf{u}_c}{Dt} - \frac{d\mathbf{u}_d}{dt} \right]}_{\mathbf{F}'_{\text{fu}}} \right.
$$

$$
\left. + \underbrace{6\pi\mu_f R_p \int_{-\infty}^{t} K_{\text{vu}} \left[\frac{d\mathbf{u}_c}{dt} - \frac{d\mathbf{u}_d}{dt} \right]_{@\xi} d\xi}_{\mathbf{F}'_{vu}} + \mathbf{F}'_L \right\} .
$$

(16.7)

As defined in the above equation, the total force $\mathbf{F}'_{\text{hyd}}$ and all its components are functions of $\mathbf{x}$ and t. In contrast, their EL counterparts were defined for each individual particle. Note that the terms within curly brackets correspond to the force on a single particle and the multiplicative factor $\phi_d/\mathcal{V}_p$ accounts for the number of particles within a unit volume. In writing the above force expression, we have assumed the particles to be monodispersed and the particle radius to be R_p, particle volume $\mathcal{V}_p$, and the mass of displaced fluid m_f. If the system is polydisperse, then the particles must be divided into size bins, and the force on each size bin is given by the above expression, with appropriate values of R_p, $\mathcal{V}_p$, and m_f. This force is applied in the filtered momentum equation (16.5) of each particle size bin. But their sum is applied in the filtered continuous-phase momentum equation (16.2).

The quasi-steady force $\mathbf{F}'_{qs}$ must be correctly interpreted as the average quasi-steady force on particles that are within the filtering volume located around $\mathbf{x}$ and weighted by the filter function. In general, $\mathbf{F}'_{qs}$ represents the drag force on many particles, each of which may be moving at a velocity that is different from $\mathbf{u}_d$. As discussed in the EL approach, therefore, the correction Φ can be expected to be a function of (i) the particle Reynolds number $\mathrm{Re} = |\mathbf{u}_c - \mathbf{u}_d| d_p / \nu_f$ based on the local relative velocity and (ii) the local particle volume fraction ϕ_d to account for the clustering effect of particles. We have already seen that the dependence of Φ on Re and ϕ_d can be well approximated with empirical correlations such as that obtained by Tenneti and Subramaniam (2014): see Eq. (9.87). In terms of Φ as a function of Re and ϕ_d, it can be argued that empirical correlations such as that given in Eq. (9.87) are better suited as a model for the EE approach than for the EL approach. This is because these empirical correlations were developed by averaging the drag on a random distribution of particles, and thus better approximate the average quasi-steady drag of the EE approach as defined above. They are not intended to be used as models of drag on individual particles, as employed in the EL approach. For additional discussion on this distinction, the reader should refer to Balachandar (2020).

The added-mass coefficient C_M and the viscous unsteady kernel K_{vu} can be taken to be those defined for the EL approach, although now they are for the average force on all the particles in the neighborhood of a spatial point. In the limit of large particle-to-fluid density ratio, the added-mass and history forces represented by the second and third terms on the right-hand side can be neglected, thereby considerably simplifying the force expression. However, we repeat the caution that was made earlier in the context of the EL method, that a large density ratio only guarantees $d\mathbf{u}_d/dt$ to be small. Therefore, the added-mass and history forces can be large even in a particle-laden gas flow if the value of fluid acceleration represented by $D\mathbf{u}_c/Dt$ and $d\mathbf{u}_c/dt$ is large. Here, $\mathbf{F}'_L$ corresponds to the lift force on the particle, which can be retained if its contribution is important to dispersed-phase motion. Other forces as appropriate can be added to the above force expression.

One other difference between the EL and EE formulations in terms of particle rotation must be discussed briefly. In the EE approach, the vorticity of the particle velocity field can be calculated as $\nabla \times \mathbf{u}_d$. However, it is not the same as twice the angular velocity of individual particles as calculated in the EL approach. In the EL approach, the angular velocity of each particle can be calculated directly, which corresponds to the "rotation" of the particle about its own axis. The particles' velocities computed in the EL approach can be projected onto an Eulerian field, whose curl is similar to $\nabla \times \mathbf{u}_d$ of the EE approach. The angular velocity of the individual particles and the angular velocity of the particle velocity field are not the same. For example, we can consider a contrived situation where all the particles are spinning about their center, but translate along parallel paths at constant velocity (i.e., $\mathbf{u}_d$ = constant vector) with no angular velocity of the particle velocity field. In practical situations of freely moving small suspended particles, the angular velocity of individual particles and that of the particle velocity field are somewhat related, since both are dictated by local fluid vorticity.

EE-DNS

We now address an important difference between EE and EL methodologies in the modeling of the different force contributions. We will address this difference first in the context of EE-DNS where $\eta \gg \mathscr{L} \gg d_p$. Model quantities such as Φ, C_M, and K_{vu} that embody the essence of quasi-steady, added-mass, and viscous unsteady forces must depend not only on Re and ϕ_d, but also on the rms particle velocity variation v_{rms} within the filtering volume. This dependence was not necessary in the case of the EL approach, since the drag of each particle is calculated with full understanding of particle velocity. In contrast, in the EE approach, since the velocity of individual particles deviates from their average value of $\mathbf{u}_d$, a measure of this local variation of particle velocity must be included as an additional parameter in the EE force model. The quasi-steady drag based on the average Reynolds number is not the same as the average quasi-steady drag calculated in terms of the Gaussian-weighted average of the drag of all the particles within the filtering volume. The difference is due to the nonlinear dependence between drag and Re. The difference depends on the level of particle velocity variation. The subgrid particle velocity variations arise both from direct interactions between the particles in the form of interparticle collisions and due to fluid-mediated particle–particle interactions, such as drafting and tumbling.

The additional dependence on the rms particle velocity variation raises two complications. First, the rms particle velocity variation v_{rms} as a function of $(\mathbf{x}, t)$ is not among the primary variables being solved. This must then be considered as an additional secondary quantity, which must be modeled in terms of the primary variables. As we will see over the rest of this chapter, the rms particle velocity fluctuation plays a fundamental role in the modeling of other closure terms as well, and is often referred to as the granular temperature. In this chapter we will derive an equation for this evolution. The second complication is to express the additional dependence of Φ, C_M, and K_{vu} on v_{rms}. Only limited information exists on the dependence of Φ on v_{rms} (Huang et al., 2017, 2019; Tavanashad et al., 2019). Very little is known about the dependence of other force contributions on particle velocity variation.

EE-LES

We now briefly comment on the modeling of $\mathbf{F}'_{\mathrm{hyd}}$ in the case of EE-LES (i.e., $\mathscr{L} \gg \eta \gg d_p$), where $\mathbf{u}_c$ corresponds to only the large eddy portion of the fluid velocity. As a result, Re $= |\mathbf{u}_c - \mathbf{u}_d| d_p / \nu_f$ now accounts only for the difference between the filtered large-scale fluid velocity and the average particle velocity. The quasi-steady drag correction will therefore depend additionally on the subgrid fluid velocity variation. In the EL formulation, since individual particles are tracked, we used the Langevin model to predict the subgrid fluid velocity seen by each particle. This approach is not possible for the collective motion of the dispersed phase being considered in the EE approach. Thus, subgrid fluid velocity fluctuations (e.g., quantified as u_{rms}) must also be included as a parameter in the evaluation of the different force contributions. For example, in EE-LES, at a minimum one must consider Φ to be a function of Re, ϕ_d, v_{rms}, and u_{rms}. Of course, the rms particle velocity fluctuation now depends not only on interparticle collisions and fluid-mediated close interactions, but also on subgrid fluid velocity variation within the filter volume. These added complexities are often ignored in most practical implementations of the EE methodology for lack of reliable models. It is sufficient to say that active research is needed in order to develop drag models that systematically account for subgrid fluid and particle velocity variations, to extend beyond the current knowledge of $\phi(\mathrm{Re}, \phi_d)$. Similar improvements are needed in the modeling of unsteady drag and lift contributions to interphase force coupling.

Problem 16.1 In preparation for the discussion to follow on interfacial work done by the hydrodynamic force, in this problem, we will expose an important consequence of averaging over the dispersed phase. In this problem you will examine $\mathbf{F}'_{\mathrm{hyd}} \cdot \mathbf{u}_d$.

(a) Show that this 3D field is not the same as the Gaussian-weighted sum of $\mathbf{F}'_l \cdot \mathbf{V}_l$ over all the particles. Using the definitions given in Eqs. (11.89) and (11.110), derive the following expression:

$$\mathbf{F}'_{\mathrm{hyd}} \cdot \mathbf{u}_d = -\left(\sum_{l=1}^{N_p} G(\mathbf{x} - \mathbf{X}_l) \mathbf{F}'_l \right) \frac{\sum_{l=1}^{N_p} G(\mathbf{x} - \mathbf{X}_l) \mathscr{V}_l \mathbf{V}_l}{\sum_{l=1}^{N_p} G(\mathbf{x} - \mathbf{X}_l) \mathscr{V}_l}. \tag{16.8}$$

(b) Show that the above in general is not equal to $\sum_{l=1}^{N_p} G(\mathbf{x} - \mathbf{X}_l) \mathbf{F}'_l \cdot \mathbf{V}_l$. Only in the limit where the velocity of all the particles contributing to the Gaussian-weighted summation are the same, are the two equal.

16.2.2 Interphase Work Exchange

We now investigate energy implications of the interphase coupling force $\mathbf{F}'_{\mathrm{hyd}}$. In particular, in this subsection we are interested in developing a suitable model for the term $W'_{\mathrm{hyd}} - \mathbf{u}_c \cdot \mathbf{F}'_{\mathrm{hyd}}$ that appears in the continuous-phase temperature equation (16.3) and a suitable model for the term $W'_{\mathrm{hyd}} - \mathbf{u}_d \cdot \mathbf{F}'_{\mathrm{hyd}}$ that appears in the dispersed-phase temperature equation (16.6). In developing these models we will use the knowledge that they contribute toward changing the continuous and dispersed-phase temperatures, respectively.

First let us consider modeling of the term $W'_{\mathrm{hyd}} - \mathbf{u}_d \cdot \mathbf{F}'_{\mathrm{hyd}}$ in the filtered dispersed-phase equation (16.6). As discussed in Problem 16.1, $\mathbf{u}_d \cdot \mathbf{F}'_{\mathrm{hyd}}$ only approximately accounts for the rate of work done on the particles that contributes to the rate of change of particulate kinetic energy. Nevertheless, we will insist on the fact that in case of rigid particles, the hydrodynamic force on the particles should only contribute to the change in particulate kinetic energy and should not contribute to altering the internal energy. Accordingly, we will set

$$\text{In Eq. (16.6): } \quad W'_{\mathrm{hyd}} - \mathbf{u}_d \cdot \mathbf{F}'_{\mathrm{hyd}} = 0. \tag{16.9}$$

It is important to note that the above relation *should not* be considered as a model for W'_{hyd}.

Similarly, in the filtered continuous-phase equation (16.3), $-\mathbf{F}'_{\text{hyd}} \cdot \mathbf{u}_c$ is the rate of change of kinetic energy of the filtered large-scale fluid motion. The two kinetic energy changes do not balance if $\mathbf{u}_c \neq \mathbf{u}_d$. In other words, kinetic energy lost/gained by the particle is not fully gained/lost by the surrounding filtered fluid motion. The imbalance is due to the filtering operation.

We recall that $\mathbf{u}_c$ accounts only for the filtered macroscale fluid motion and ignores the microscale flow around the particles. Similarly for the particles, $\mathbf{u}_d$ is only their filtered macroscale motion and ignores the deviation of individual particle velocity from the average. Thus, for the total energy to be conserved, the difference $\mathbf{F}'_{\text{hyd}} \cdot (\mathbf{u}_d - \mathbf{u}_c)$ must contribute to the rate of change of kinetic energy of the microscale motion that has been filtered out.

The contribution to kinetic energy from each force component of $\mathbf{F}'_{\text{hyd}}$ was investigated in Section 15.3.2 for the EL formulation. That analysis applies for the present EE formulation as well, but with the quasi-steady, added-mass, viscous unsteady, and lift forces evaluated as functions of $\mathbf{x}$ and t. Only the quasi-steady force is strictly dissipative. By definition, the inviscid and lift forces are nondissipative. The viscous unsteady force is at least partly dissipative. Based on these arguments, a good model for the rate of work associated with $\mathbf{F}'_{\text{hyd}}$ on the continuous phase is as follows:

$$\left(\mathbf{F}'_{\text{qs}} + \mathbf{F}'_{vu}\right) \cdot \mathbf{u}_d + \left(\mathbf{F}'_{iu} + \mathbf{F}'_{Ll}\right) \cdot \mathbf{u}_c . \tag{16.10}$$

With this model, quasi-steady and viscous unsteady forces make a nonzero contribution to the continuous-phase internal energy equation through the term $W'_{\text{hyd}} - \mathbf{u}_c \cdot \mathbf{F}'_{\text{hyd}}$. This contribution is guaranteed to be positive in the case of quasi-steady force. But caution is needed, since the same does not apply for the viscous unsteady force. With the above definition, the added-mass and lift forces do not make a net contribution to the filtered continuous-phase thermal equation. In essence, in the continuous-phase temperature equation we can make the following substitution:

$$\text{In Eq. (16.3):} \quad \left(W'_{\text{hyd}} - \mathbf{u}_c \cdot \mathbf{F}'_{\text{hyd}}\right) \approx \frac{\phi_d}{\mathscr{V}_p}\left((\mathbf{F}'_{\text{qs}} + \mathbf{F}'_{vu}) \cdot (\mathbf{u}_d - \mathbf{u}_c)\right) . \tag{16.11}$$

This term then contributes to dissipative heating of the continuous phase. In most low-speed applications, this term is likely to make only a negligible contribution to the thermal evolution of the multiphase flow and therefore can be ignored.

16.2.3   Interphase Energy Coupling

The definition of $Q'_{\text{hyd}}(\mathbf{x}, t)$ as the average thermal exchange from the dispersed to the continuous phase is given in Eq. (11.64). The closure quest is to express $Q'_{\text{hyd}}(\mathbf{x}, t)$ in terms of the filtered continuous and dispersed-phase temperatures. For the present case of the small-particle limit, the appropriate finite-Pe BBO-like equation of heat transfer is

$$
\begin{aligned}
Q'_{\text{hyd}} = \frac{\phi_d}{\mathscr{V}_p} \Bigg\{ & m_f C_{p,f} \frac{DT_c}{Dt} + 2\pi k_f R_p (T_c - T_d)\, \text{Nu} \\
& + 4\pi k_f R_p \int_{-\infty}^{t} K_T(t-\xi) \left[\frac{dT_c}{dt} - \frac{dT_p}{dt} \right]_\xi d\xi \Bigg\},
\end{aligned}
\tag{16.12}
$$

where T_c and T_d are the filtered continuous and particulate-phase temperatures. In the above equation, the Nusselt number Nu plays the role of Φ in quasi-steady force modeling. In the EL-DNS approach, Nu was modeled as a function of particle Reynolds number, Prandtl number, and local volume fraction. In the present case of EE-DNS, due to spatial filtering of particle velocity and temperature, the rms particle velocity v_{rms} and rms particle temperature $T_{d,\text{rms}}$ will additionally determine the average quasi-steady heat transfer. Thus, Nu must conceptually be a function of Re, Pr, ϕ_d, v_{rms}, and $T_{d,\text{rms}}$. In case of EE-LES, subgrid fluid velocity and temperature fluctuations, quantified in terms of u_{rms} and $T_{c,\text{rms}}$, will contribute to Nusselt number modeling as well. Clearly, closure modeling rapidly becomes a very complex undertaking. In general, Nu is greatly simplified by using the empirical correlation of Ranz and Marshall (1952) or Whitaker (1972) (see Eq. (6.14)). The effect of finite particle volume fraction is taken into account with correlations such as the one developed by Sun et al. (2015) given in Eq. (9.110). The dependence on other parameters remains to be explored.

16.2.4 Interphase Mass Exchange

The definition of $\dot{m}(\mathbf{x}, t)$ as the filtered interphase mass exchange from the continuous to the dispersed phase is given in Eq. (11.34). The closure quest is to express the mass exchange in terms of the filtered continuous and particulate-phase variables. A suitable model of this closure is shown as

$$
\dot{m} = \frac{\phi_d}{\mathscr{V}_p} 4\pi R_p^2 \rho_f\, h_m \left(Y_{\alpha c} - Y_{\alpha d} \right),
\tag{16.13}
$$

where $Y_{\alpha c}$ is the species mass fraction of the continuous phase at $\mathbf{x}$. $Y_{\alpha d}$ is the species mass fraction in the continuous phase at the surface of the particle, which is evaluated by assuming the continuous phase that is in contact with the surface of the particle to be at the temperature of the particle T_d, and assuming the evaporating species to be in saturation at that temperature. The nondimensional mass transfer coefficient is given by $h_m = \text{Sh}\, \mathcal{D}_\alpha / (2R_p)$, where Sh is the Sherwood number. The Sherwood number for mass transfer plays the same role as Nu for heat transfer and thus the Nusselt number closure discussed in the previous subsection can be used for Sherwood number closure. Thus, the Ranz–Marshal correlation given in Eq. (15.19) can be used for quasi-steady mass transfer. To account for unsteady effects, Eq. (16.12) can be used with mass transfer replacing heat transfer. The effect of finite particle volume fraction can be taken into account with correlations given in Eq. (9.110).

We now consider momentum and energy exchanges between the phases associated with the mass exchange. As we did in the EL approach, $\dot{\mathbf{M}}$, $\dot{E}$, $\dot{e}$, and $\dot{k}$ can be written as a product of $\dot{m}$ and an add-mixture of local continuous and dispersed-phase properties. While Eq. (15.22) was for an individual particle, the corresponding EE model is

$$\dot{\mathbf{M}} = \dot{m}(\beta_{dM}\mathbf{u}_d + \beta_{cM}\mathbf{u}_c), \qquad \dot{E} = \dot{m}(\beta_{dE}E_d + \beta_{cE}E_c),$$

$$\dot{e} = \dot{m}(\beta_{dE}e_d + \beta_{cE}e_c), \qquad \dot{k} = \dot{m}(\beta_{dE}k_d + \beta_{cE}k_c), \qquad (16.14)$$

where E_d, e_d, and k_d are the total, internal, and kinetic energy averaged over the dispersed phase and their definitions are similar to those given in Eq. (11.110). The above more general definition, with proper choice of the coefficients β_{dM}, β_{cM}, β_{dE}, and β_{cE}, can offer flexibility in partitioning dissipative heating between the dispersed and continuous phases. Such flexibility may be desired in case of droplets with internal circulation. However, such models can be developed only with fundamental information from particle-resolved simulations. For now, following what we did in the EL approach, we will assume the mass exchange not to directly influence the dispersed-phase temperature, and arrive at the following closure:

$$\dot{\mathbf{M}} = \dot{m}\mathbf{u}_d \quad \text{and} \quad \dot{e} = \dot{m}e_d = \dot{m}C_pT_d. \qquad (16.15)$$

The above equations imply that the velocity and internal energy of the exchanged fluid are the same as those of the locally averaged dispersed phase. With these assumptions we see that the last terms on the right-hand side of Eqs. (16.5) and (16.6) become zero. On the other hand, the mass exchange contributes to the right-hand side of the continuous-phase momentum and energy equations.

The models presented above are arbitrary choices. They assume the mass added or removed from the continuous phase does not affect the filtered velocity and temperature equations of the dispersed phase. A more accurate model may add or remove the right amount of energy from the dispersed phase. One must rely on results from appropriately designed experiments or particle-resolved simulations to develop more accurate models. Corresponding changes must be made to the momentum and energy equations of both phases. Finally, before we leave this topic, in the following three problems, you can explore the further consequences of averaging over the dispersed phase in the EE approach.

Problem 16.2 In Eq. (15.28), the momentum associated with mass exchange was expressed as a weighted sum of the mass exchange times the velocity of the lth particle. However, in an EE simulation where the particulate phase is filtered, there is no direct access to individual particle quantities such as $\dot{m}_l$ and $\mathbf{V}_l$. Therefore, the momentum associated with mass exchange must be expressed equivalently as $\dot{\mathbf{M}} \approx \dot{m}\mathbf{u}_d$. In this problem you will show that assuming $\dot{\mathbf{M}} = \dot{m}\mathbf{u}_d$ is not the same as the Lagrangian definition $\dot{\mathbf{M}} = \sum_{l=1}^{N_p} G(\mathbf{x} - \mathbf{X}_l)\dot{m}_l\mathbf{V}_l$.

(a) Obtain the following expression:

$$\dot{m}\mathbf{u}_d = \left(\sum_{l=1}^{N_p} G(\mathbf{x}-\mathbf{X}_l)\dot{m}_l\right) \frac{\sum_{l=1}^{N_p} G(\mathbf{x}-\mathbf{X}_l)\mathscr{V}_l\mathbf{V}_l}{\sum_{l=1}^{N_p} G(\mathbf{x}-\mathbf{X}_l)\mathscr{V}_l}. \qquad (16.16)$$

(b) Thereby show that it is different from $\sum_{l=1}^{N_p} G(\mathbf{x} - \mathbf{X}_l)\dot{m}_l \mathbf{V}_l$. If all the particles in the neighborhood of the point $\mathbf{x}$ have identical particle velocity, then show that the two are equal.

Problem 16.3 In this problem, we will consider the kinetic energy exchange associated with mass transfer. In the filtered particulate formulation, the kinetic energy associated with mass exchange is expressed as $\dot{k} \approx \dot{m}\mathbf{u}_d \cdot \mathbf{u}_d/2$.

(a) Derive the following expression:

$$\dot{m}\frac{\mathbf{u}_d \cdot \mathbf{u}_d}{2} = \frac{1}{2}\left(\sum_{l=1}^{N_p} G(\mathbf{x} - \mathbf{X}_l)\dot{m}_l\right)\left|\frac{\sum_{l=1}^{N_p} G(\mathbf{x} - \mathbf{X}_l)\mathcal{V}_l\mathbf{V}_l}{\sum_{l=1}^{N_p} G(\mathbf{x} - \mathbf{X}_l)\mathcal{V}_l}\right|^2. \tag{16.17}$$

(b) Show that the above is the same as $\sum_{l=1}^{N_p} G(\mathbf{x} - \mathbf{X}_l)\dot{m}_l \mathbf{V}_l \cdot \mathbf{V}_l/2$ only when the velocity of all the particles is the same as the average $\mathbf{u}_d$.

Problem 16.4 In this problem, you will consider the internal energy exchange associated with mass transfer. In the filtered particulate formulation, the internal energy associated with mass exchange is expressed as $\dot{e} \approx \dot{m}e_d$.

(a) Derive the following expression:

$$\dot{m}e_d = \left(\sum_{l=1}^{N_p} G(\mathbf{x} - \mathbf{X}_l)\dot{m}_l\right)\frac{\sum_{l=1}^{N_p} G(\mathbf{x} - \mathbf{X}_l)\mathcal{V}_l e_l}{\sum_{l=1}^{N_p} G(\mathbf{x} - \mathbf{X}_l)\mathcal{V}_l}. \tag{16.18}$$

(b) Show that the right-hand side is equal to $\sum_{l=1}^{N_p} G(\mathbf{x} - \mathbf{X}_l)\dot{m}_l e_l$ only if all the particles in the neighborhood of the point $\mathbf{x}$ have identical particle temperature.

16.2.5 Closure Models of Subgrid Terms τ_{sg} and $\mathbf{q}_{sg}$

We recall that subgrid residual stress and residual flux account for the effect of small-scale fluid velocity and thermal fluctuations on the large-scale flow dynamics. As discussed under the EL approach, subgrid fluctuations arise both from classical turbulence cascade and from pseudo turbulence generated by the particles. We refer to Table 15.3 and point out that the different approximations to τ_{sg} and $\mathbf{q}_{sg}$ presented there remain applicable even under the EE approach.

In the case of EE-DNS, subgrid stress is entirely due to pseudo turbulence. Two approaches to modeling pseudo turbulence were presented in Section 9.7.1, but only the Eulerian-based approach is applicable in an EE simulation. In Eq. (9.100), τ_{pt} was expressed as a function of local particle Reynolds number and particle volume fraction, both of which can easily be calculated as 3D time-dependent fields in an

EE simulation. Pseudo-turbulent kinetic energy and anisotropy coefficients can be obtained using Eqs. (9.101) and (9.102), which can then be substituted into Eq. (9.100) to obtain the pseudo-turbulent Reynolds stress tensor $\tau_{pt,axi}$. This Reynolds stress tensor was defined with the principle axis aligned with the direction of local relative velocity. As discussed in Section 15.3.6, the pseudo-turbulent Reynolds stress tensor in the computational coordinates can be obtained using the relation (15.39). Similarly, the pseudo-turbulent heat flux can readily be obtained using the Eulerian-based model of Sun et al. (2016). The details of this evaluation have already been discussed in Section 15.3.6 for the EL approach.

The difference between EE-DNS and EE-LES lies in the modeling of the filtered mesoscale turbulence cascade τ_{ct} and $\mathbf{q}_{ct}$. The details of EE-LES are the same as those presented in Chapter 15 for the EL-LES approach, and therefore will not be repeated here. In EL-LES, the subgrid eddies were not limited to pseudo turbulence, as they also include filtered mesoscale turbulence. As stated in the context of EL-LES, contributions to τ_{ct} and $\mathbf{q}_{ct}$ from mesoscale turbulence can be modeled using traditional LES closures that have been developed for single-phase turbulence.

16.2.6   Closure Models of Subgrid Terms $\tau_{d,\text{sg}}$ and $\mathbf{q}_{d,\text{sg}}$

The particle residual stress arises from variations in the velocity of individual particles within the filtering volume. Similarly, particle residual heat flux is due to variations in the temperature of individual particles from the mean value. In this sense, $\tau_{d,\text{sg}}$ and $\mathbf{q}_{d,\text{sg}}$ are analogous to τ_{sg} and $\mathbf{q}_{\text{sq}}$. While the latter pair quantifies variations in fluid velocity and temperature within the filter volume, the former pair quantifies particle-to-particle variation. Two sources of subgrid particle velocity variation can be identified. (i) Due to fluid velocity variations within the filter volume (which is the source of τ_{sg}), the velocity of particles within the filter volume will also vary as different particles are moved differently by the different subgrid eddies. The fluid velocity variation within the filter volume can in turn be either at the mesoscale due to classical turbulence cascade or at the microscale due to pseudo turbulence generated by the particles. Particle-to-particle velocity variation within the filter volume arises both in response to classical and pseudo turbulence. While classical turbulence is the only source of particle velocity fluctuation at low volume fraction, with increasing volume fraction, fluid-mediated particle–particle interactions are responsible for subgrid particle velocity variation. (ii) The second source of particle velocity variation is due to direct particle–particle interaction in the form of collisions. Each collision imparts a significant change in the velocity of the colliding particles and thus contributes to particle velocity fluctuation from the mean. These two subgrid stresses will be referred to as "turbulent stress" $\tau_{d,\text{t}}$ due to fluid velocity fluctuation and "kinetic stress" $\tau_{d,\text{ki}}$ due to collisional velocity fluctuation. Although these two mechanisms influence each other, a simple approximation will be to assume

$$\tau_{d,\text{sg}} = \tau_{d,\text{t}} + \tau_{d,\text{ki}}.$$

(16.19)

In this section we are primarily concerned with the modeling of $\tau_{d,\mathrm{t}}$ and in Section 16.3.4 we will consider a simple definition for the kinetic stress $\tau_{d,\mathrm{ki}}$.

When the volume fraction is low, collisional interactions between the particles are rare and particle residual stress is nearly entirely driven by surrounding fluid turbulence. This is the limit that will be considered in this section. Since particle velocity variations are driven by fluid velocity variation, in this section, we will develop closures that will express $\tau_{d,\mathrm{t}}$ in terms of the corresponding continuous-phase quantity τ_{sg}. In the other extreme, when interparticle collisions are frequent, collisions may be the dominant mechanism of particle-to-particle velocity variation. In fact, in this limit, particle velocity variation due to collision could be driving local pseudo turbulence of the fluid, which could dominate classical turbulence. Hence, any attempt to express $\tau_{d,\mathrm{ki}}$ in terms of τ_{sg} is inappropriate. The collision-dominated regime, where $\tau_{d,\mathrm{ki}}$ must be modeled in terms of collisional particle velocity fluctuation, will be considered in the next section of this chapter. In the intermediate regime, contributions to particle residual stress and heat flux may arise both from particle–turbulence and particle–particle collisional interactions. Note that interparticle collisions contribute primarily to momentum change of the individual particles. Here we ignore heat transfer effects of collisions and thus thermal variation of the particles within the filter volume is entirely due to variation in fluid temperature. In other words, $\mathbf{q}_{d,\mathrm{sg}} = \mathbf{q}_{d,\mathrm{t}}$.

Similar to Table 15.3 that showed the different contributions to fluid residual stress and heat flux, for particles we present Table 16.1, where the different contributions to subgrid particle residual stress and heat flux are highlighted. Three different particle volume fraction regimes are considered: the small volume fraction regime, where particle–particle encounters are rare and can be ignored, the moderate volume fraction regime, where particle–particle interactions are mostly mediated by fluid; and the large volume fraction regime, where collisional effects dominate. In the case of EE-DNS at small volume fraction, both $\tau_{d,\mathrm{sg}}$ and $\mathbf{q}_{d,\mathrm{sg}}$ are small and can be ignored. At moderate volume fraction, the contribution to $\tau_{d,\mathrm{sg}}$ and $\mathbf{q}_{d,\mathrm{sg}}$ is primarily from pseudo turbulence and therefore has been denoted as $\tau_{d,\mathrm{pt}}$ and $\mathbf{q}_{d,\mathrm{pt}}$. In the case of EE-LES, at small volume fraction, the subgrid turbulence is primarily classical turbulence and hence the turbulence contributions have been denoted as $\tau_{d,\mathrm{ct}}$ and $\mathbf{q}_{d,\mathrm{ct}}$. At moderate volume fraction, both pseudo and classical turbulences contribute to $\tau_{d,\mathrm{sg}}$ and $\mathbf{q}_{d,\mathrm{sg}}$. At large volume fraction, both in EE-DNS and EE-LES, particle stress is primarily from a collisional process, which we denote $\tau_{d,\mathrm{ki}}$ and will be considered in Section 16.3.4.

We now proceed to consider dilute and moderate volume fraction regimes to get appropriate expressions for $\tau_{d,\mathrm{sg}}$ and $\mathbf{q}_{d,\mathrm{sg}}$ by relating them to τ_{sg} and $\mathbf{q}_{\mathrm{sq}}$. In doing so, we rely on the fact that turbulent fluctuations of particle velocity and temperature are related to turbulent fluctuations of fluid velocity seen by the particles in their trajectories. In the previous subsection (which in turn referred to Section 15.3.6), we presented explicit expressions for the residual fluid stress and flux. In this section the plan is to express $\tau_{d,\mathrm{t}}$ and $\mathbf{q}_{d,\mathrm{t}}$ in terms of the already obtained subgrid fluid stress and flux. Such a relation between subgrid particle velocity variation and subgrid fluid velocity variation is appropriate in case of particles of small Stokes number, since the motion of such particles is controlled by the local fluid velocity. Furthermore, limiting attention only to particles of small Stokes number is appropriate in the present context of the EE approach.

Toward the goal of obtaining a relation between $\tau_{d,\mathrm{sg}}$ and τ_{sg}, we start by revisiting the following definitions:

$$\phi_d \tau_{d,\mathrm{sg}} = \rho_p \left(\phi_d \mathbf{u}_d \mathbf{u}_d - \overline{\mathbf{V}\mathbf{V}}^d \right) \quad \text{and} \quad \phi_c \tau_{\mathrm{sg}} = \rho_f \left(\phi_c \mathbf{u}_c \mathbf{u}_c - \overline{\mathbf{u}\mathbf{u}} \right), \quad (16.20)$$

where $\overline{()}^d/\phi_d$ and $\overline{()}/\phi_c$ indicate the filter-weighted average over the region occupied by the particle and the fluid, respectively. We then make use of the following equilibrium Eulerian expansion for particle velocity derived in Section 12.3 (ignoring gravitational settling):

$$\mathbf{v} = \mathbf{u} - (1 - \beta)\tau_p \frac{D\mathbf{u}}{Dt}, \quad (16.21)$$

where we note that $\mathbf{v}$ is the particle velocity field that is the Eulerian counterpart of the Lagrangian particle velocity $\mathbf{V}$. With the above equations we can obtain the following relation between the particle and fluid residual stresses:

$$\phi_d \tau_{d,\mathrm{sg}} = \phi_d \frac{\rho_p}{\rho_f} \tau_{\mathrm{sg}} + \underbrace{\rho_p \phi_d \left(\frac{\overline{\mathbf{u}\mathbf{u}}}{\phi_c} - \frac{\overline{\mathbf{u}}}{\phi_c}\frac{\overline{\mathbf{u}}}{\phi_c} \right) - \rho_p \phi_d \left(\frac{\overline{\mathbf{u}\mathbf{u}}^d}{\phi_d} - \frac{\overline{\mathbf{u}}^d}{\phi_d}\frac{\overline{\mathbf{u}}^d}{\phi_d} \right)}_{\text{effect of preferential accumulation}}$$

$$+ \underbrace{2(1-\beta)\tau_p \rho_p \phi_d \left(\frac{1}{\phi_d} \overline{\mathbf{u}\frac{D\mathbf{u}}{Dt}}^d - \frac{\overline{\mathbf{u}}^d}{\phi_d}\frac{1}{\phi_d}\overline{\frac{D\mathbf{u}}{Dt}}^d \right)}_{\text{effect of velocity–acceleration correlation}} + O(\tau_p^2). \quad (16.22)$$

To leading order, $\tau_{d,\mathrm{sg}}$ and τ_{sg} are related by the particle-to-fluid density ratio. This is simply because of their definitions. To leading order, particle and fluid velocity fluctuations are equal, in the absence of any particle inertia effects. Although τ_p does not appear explicitly in the "effect of preferential accumulation," this term is identically zero in the absence of particle inertia. In this term, we note that $\overline{\mathbf{u}}^d$ refers to fluid velocity averaged over the region occupied by the particles, and can be referred to as the "Lagrangian average of fluid velocity." Therefore, in the definition of $\overline{\mathbf{u}}^d$, the fluid velocity is extended to the region occupied by the particles as well, which is appropriate in the EE approach. Note that if particles are uniformly distributed within the filter volume, then $\overline{\mathbf{u}}^d/\phi_d \approx \overline{\mathbf{u}}/\phi_c$. Similar equalities apply to the other quantities as well, and the entire term goes to zero. Only at finite particle inertia, due to preferential accumulation of particles, when particles are non-uniformly distributed within the filter volume, will the fluid velocity averaged over the particle volume be different from that averaged over the fluid volume. This is the reason why this contribution is termed the effect of preferential accumulation of particles. The last term on the right-hand side is the $O(\tau_p)$ contribution due to velocity–acceleration correlation. Again, the velocity and acceleration are those of the fluid, but evaluated at the particle location. The velocity–acceleration correlation has been shown to play an important role in turbulent closure models (Sawford et al., 2003; Pope, 2014).

Table 16.1 The different modeling needs for subgrid particle residual stress and residual heat flux.

	$\phi_d \ll 1$ (very dilute)	Moderate ϕ_d	Dense
EL-DNS: $\eta > \mathcal{L} \gg d_p$	$\tau_{d,\text{sg}} \approx 0$	$\tau_{d,\text{sg}} \approx \tau_{d,\text{t}} \approx \tau_{d,\text{pt}}$	$\tau_{d,\text{sg}} \approx \tau_{d,\text{ki}}$
	$\mathbf{q}_{d,\text{sg}} \approx 0$	$\mathbf{q}_{d,\text{sg}} \approx \mathbf{q}_{d,\text{t}} \approx \mathbf{q}_{d,\text{pt}}$	
EL-LES: $\mathcal{L} \gg \eta, d_p$	$\tau_{d,\text{sg}} \approx \tau_{d,\text{t}} \approx \tau_{d,\text{ct}}$	$\tau_{d,\text{sg}} \approx \tau_{d,\text{t}}$	$\tau_{d,\text{sg}} \approx \tau_{d,\text{ki}}$
	$\mathbf{q}_{d,\text{sg}} \approx \mathbf{q}_{d,\text{t}} \approx \mathbf{q}_{d,\text{ct}}$	$\mathbf{q}_{d,\text{sg}} \approx \mathbf{q}_{d,\text{t}}$	

Problem 16.5 Start from the equilibrium Eulerian expression given in Eq. (16.21) to obtain the following $O(\tau_p)$ accurate expression:

$$\overline{\mathbf{vv}}^d = \overline{\mathbf{uu}}^d - 2(1 - \beta)\tau_p \overline{\mathbf{u}\frac{D\mathbf{u}}{Dt}}^d + O\left(\tau_p^2\right) . \tag{16.23}$$

Substitute this in the expression for $\tau_{d,\text{sg}}$ in Eq. (16.20) to obtain the relation given in Eq. (16.22).

Problem 16.6 In this problem you will include the effect of gravitational settling to the above analysis. Start from the following equilibrium Eulerian expression given in Section 12.3 that includes the effects of both settling and fluid acceleration:

$$\mathbf{v} = \mathbf{u} - (1 - \beta)\tau_p \left(\frac{D\mathbf{u}}{Dt} - \mathbf{g}\right) . \tag{16.24}$$

Show that the relation (16.22) remains the same even with the inclusion of gravitational settling. The fact that the expression remains the same simply indicates that there are only two mechanisms responsible for the difference between particle and fluid residual stresses. These two mechanisms are the effect of preferential accumulation and velocity–acceleration correlation. However, the quantitative impact of both these effects is modified in the presence of gravitational settling.

The above approach can be duplicated to relate the particle residual heat flux to that of the fluid, using the equilibrium Eulerian approximation of particle temperature. We start by proposing the following simpler definitions of the particle and fluid heat flux that are analogous to the definitions of residual stress:

$$\phi_d \mathbf{q}_{d,\text{sg}} = \rho_p \left(\overline{\mathbf{V}T_p}^d - \phi_d \mathbf{u}_d T_d\right) \quad \text{and} \quad \phi_c \mathbf{q}_{\text{sg}} = \rho_f \left(\overline{\mathbf{u}T_f} - \phi_c \mathbf{u}_c T_c\right) . \tag{16.25}$$

We then make use of the following equilibrium Eulerian expansion for the particle temperature derived in Section 12.3:

$$T_p = T_f - \tau_T \frac{DT_f}{Dt} . \tag{16.26}$$

With the above equations, we can obtain the following relation between the particle and fluid residual stresses:

$$
\phi_d \mathbf{q}_{d,\mathrm{sg}} = \phi_d \frac{\rho_p}{\rho_f} \mathbf{q}_{\mathrm{sg}} - \rho_p \phi_d \left(\frac{\overline{\mathbf{u} T_f}}{\phi_c} - \frac{\overline{\mathbf{u}}}{\phi_c} \frac{\overline{T_f}}{\phi_c} \right) + \rho_p \phi_d \left(\frac{\overline{\mathbf{u} T_f}^d}{\phi_d} - \frac{\overline{\mathbf{u}}^d}{\phi_d} \frac{\overline{T_f}^d}{\phi_d} \right)
$$

$$
\underbrace{\qquad}_{\text{effect of preferential accumulation}}
$$

$$
- (1-\beta)\tau_p \rho_p \phi_d \left(\frac{1}{\phi_d} \overline{T_f \frac{D\mathbf{u}}{Dt}}^d - \frac{\overline{T_f}^d}{\phi_d} \frac{1}{\phi_d} \overline{\frac{D\mathbf{u}}{Dt}}^d \right) \tag{16.27}
$$

$$
\underbrace{\qquad}_{\text{effect of temperature–acceleration correlation}}
$$

$$
- \tau_T \rho_p \phi_d \left(\frac{1}{\phi_d} \overline{\mathbf{u} \frac{DT_f}{Dt}}^d - \frac{\overline{\mathbf{u}}^d}{\phi_d} \frac{1}{\phi_d} \overline{\frac{DT_f}{Dt}}^d \right) + O(\tau_p^2, \tau_T^2, \tau_p \tau_T).
$$

$$
\underbrace{\qquad}_{\text{effect of velocity–thermal–acceleration correlation}}
$$

Again, to leading order, $\mathbf{q}_{d,\mathrm{sg}}$ and $\mathbf{q}_{\mathrm{sg}}$ are related by the particle-to-fluid density ratio. The "effect of preferential accumulation" is identically zero in the absence of particle inertia, since particle distribution within the filter volume is then uniform and the average over the particle volume is the same as the average over the fluid volume. Only at finite particle inertia, due to non-uniform distribution of particles within the filter volume, does the contribution from the second term on the right-hand side become nonzero. The last two terms on the right-hand side are the $O(\tau_p, \tau_T)$ contribution due to temperature–acceleration and velocity–thermal–acceleration correlations.

Problem 16.7 Obtain the relation (16.27).

The expression given in Eq. (16.22) is useful in understanding the mechanisms that are responsible for the difference between the residual stresses of the particle and those of the fluid. However, the relation is not very useful in evaluating $\tau_{d,\mathrm{sg}}$ in an EE simulation in terms of the local value of τ_{sg}, since the other two terms in the equation are unknown. We know the qualitative behavior that these additional terms go to zero in the limit of zero particle inertia (or as the particle Stokes number becomes very small). We now present a very simple approach to quantitatively account for the effect of particle inertia and obtain an explicit expression for particle residual stress. For this, we recall the discussion of Section 7.5.5 where we obtained expressions for rms particle velocity fluctuation in terms of corresponding local rms fluid velocity fluctuation. We exploit the results of Section 7.5.5 in the present context of EE modeling by first assuming the particle residual stress to be isotropic (i.e., particle velocity fluctuation without any preferred direction). The rms subgrid fluid velocity fluctuation is defined as

$$
u_{\mathrm{rms}}^2 = \frac{1}{3\rho_f} \mathrm{tr}(\boldsymbol{\tau}_{\mathrm{sg}}). \tag{16.28}
$$

We apply the theory of Wang and Stock (1993) to the present EE context of solving filtered fluid and particle governing equations. The particle Stokes number St is evaluated in terms of the fluid time scale at the filter length scale of $\mathcal{L}$. Assuming the filter length scale to be in the inertial range, the particle Stokes number can be expressed as

$$St = \frac{\tau_p}{(\mathcal{L}^{2/3}/\epsilon^{1/3})}, \tag{16.29}$$

where ϵ is the local dissipation rate that can be obtained from the EE simulation. With the assumption of isotropy, the expression for particle residual stress and rms particle velocity fluctuation becomes

$$\tau_{d,t} = -\rho_p \frac{u_{rms}^2}{1+St} \mathbf{I} \quad \text{and} \quad v_{rms}^2 = \frac{1}{3\rho_p} tr(\tau_{d,t}) = \frac{u_{rms}^2}{1+St}. \tag{16.30}$$

Here, $\mathbf{I}$ is the identity tensor. In the limit $St \to 0$, the particle residual stress becomes equal to the isotropic part of the fluid residual stress, scaled by the particle-to-fluid density ratio. As the particle time scale increases relative to the time scale of subgrid eddies, the particle motion differs from that of the local fluid and the general effect is for the particle residual stress to decrease with increasing Stokes number.

The particle residual heat flux can similarly be modeled in terms of fluid residual heat flux as

$$\mathbf{q}_{d,t} = -\frac{\rho_p}{\rho_f} \frac{\mathbf{q}_{sg}}{1+St_{Th}}, \tag{16.31}$$

where St_{Th} is the thermal Stokes number appropriately defined as the ratio between the particle thermal time scale and the fluid time scale at the filter length scale. It should be pointed out that the models (16.30) and (16.31) can be used for particle fluctuations arising from both classical turbulence and pseudo turbulence. The difference lies in how the inputs τ_{sg} and $\mathbf{q}_{sg}$ are approximated. In case of pseudo turbulence, both τ_{sg} and $\mathbf{q}_{sg}$ will be taken to be fluid velocity and thermal fluctuations arising from the wake effect of the particles. In the case of classical turbulence, τ_{sg} and $\mathbf{q}_{sg}$ will be taken to be the subgrid LES closure. In situations where both contributions are important, τ_{sg} and $\mathbf{q}_{sg}$ from the two sources must be appropriately combined.

The effect of anisotropy can easily be built into the above models following the results of Section 7.5.5. In order to do that, we define the relative velocity between the particles and the local fluid to be the parallel direction: $\mathbf{e}_\parallel = (\mathbf{u}_c - \mathbf{u}_d)/|\mathbf{u}_c - \mathbf{u}_d|$. According to theory, the particle rms velocity (and accordingly the particle residual stress component) along this direction will be different from that along the two perpendicular directions. We still assume the fluid residual stress to be isotropic and u_{rms} to be given by Eq. (16.28). Equations (7.51) and (7.52) are then used to obtain the rms particle velocity fluctuation along the parallel and perpendicular directions. There we referred to particle velocity fluctuations along the parallel and perpendicular directions as $\langle V_3'^2 \rangle^{1/2}$ and $\langle V_1'^2 \rangle^{1/2} = \langle V_2'^2 \rangle^{1/2}$. In their evaluation, care must be taken to properly define the relative velocity parameter γ and the inertia parameter St_T in terms of the filter scale turbulence. Based on this, we can define the following subgrid particle Reynolds stress and rms particle velocity fluctuation:

$$\tau_{d,t} = -\rho_p Q^{\mathrm{T}} \begin{bmatrix} \langle V_3'^2 \rangle & 0 & 0 \\ 0 & \langle V_1'^2 \rangle & c \\ 0 & 0 & \langle V_2'^2 \rangle \end{bmatrix} Q \quad \text{and} \quad v_{\mathrm{rms}}^2 = \frac{1}{3} \left(\langle V_1'^2 \rangle + \langle V_2'^2 \rangle + \langle V_3'^2 \rangle \right).$$

$$(16.32)$$

In the above, Q^{T} is the rotation matrix that transforms the x-coordinate to align with the direction of local relative velocity, and T means transpose. We leave the details to the reader.

We close this section by stressing the approximations involved in the above model. The above closure models assume the flow turbulence to be isotropic. Further research is needed to generalize the results to anisotropic and inhomogeneous conditions. Since the isotropic part can be combined with pressure, modeling of the anisotropic part plays an important role.

Problem 16.8 Let us now consider a simple idealized problem that can illustrate an important facet of multiphase flow. Consider a volume of fluid–particle mixture acted upon by an external pressure gradient dp/dx that is spatially uniform. Let there be no spatial variation and as a result, the fluid velocity, particle velocity, and particle volume fraction are all spatially homogeneous. Furthermore, assume the flow to be one-dimensional with only the x-component of fluid and particle velocities.

(a) Make use of spatial homogeneity and obtain the following continuous and dispersed-phase momentum equations from their general form given in Eqs. (16.2) and (16.5). In deriving these equations, assume the flow to be laminar and therefore the residual stresses are zero. Also, since there is no velocity gradient, σ_c has only a pressure contribution. We will also assume the gravitation effect to be negligible and no mass exchange between the phases (i.e., $\dot{m} = 0$):

$$\phi_c \rho_f \frac{du_c}{dt} = -\phi_c \frac{dp}{dx} - F'_{x,\mathrm{hyd}},$$

$$(16.33)$$

$$\phi_d \rho_p \frac{du_d}{dt} = -\phi_d \frac{dp}{dx} + F'_{x,\mathrm{hyd}},$$

$$(16.34)$$

where, due to the absence of spatial gradients, $\partial/\partial t = D/Dt = d/dt$. Let us model the x-component of the hydrodynamic force on the particles as the sum of only quasi-steady and added-mass forces as

$$F'_{x,\mathrm{hyd}} = \frac{\phi_d \rho_p}{\tau_p}(u_c - u_d) + \phi_d \rho_f C_M \left(\frac{du_c}{dt} - \frac{du_d}{dt} \right).$$

$$(16.35)$$

(b) Combine the two equations to obtain a single equation for the velocity difference $u_c - u_d$ and solve it to obtain the result

$$u_c - u_d = u_{\mathrm{rel}} \left(1 - \exp\left[-t/t_m \right] \right),$$

$$(16.36)$$

where

$$u_{\mathrm{rel}} = -\left(\frac{\rho_p - \rho_f}{\rho_p} \right) \left(\frac{\phi_c \tau_p}{\rho_m} \right) \left(\frac{dp}{dx} \right) \quad \text{and} \quad t_m = \left(\frac{\rho_p + \phi_c \rho_f}{\rho_p} \right) \frac{\rho_f \tau_p}{\rho_m}.$$

$$(16.37)$$

The solution indicates that the relative velocity approaches a constant value of u_{rel} on a time scale of t_m. While both the fluid and particles accelerate as $-(1/\rho_m)dp/dx$, their relative velocity remains a constant after the initial transient.

16.3   Kinetic Granular Theory

At the beginning of Section 16.2 we introduced collision-related secondary terms in the dispersed-phase governing equations, but deferred their formal derivation. They will now be covered in this section. The description of pressure, viscous stress, and thermal flux in terms of molecular collisional processes is well described by the kinetic theory of gases (Irving and Kirkwood, 1950; Chapman and Cowling, 1990). The earliest effort to consider particulate systems analogous to molecular collisions in gases goes back to Maxwell (1873), who used this approach to describe the rings of Saturn. A more modern application of the kinetic theory of gases to particulate systems is credited to Bagnold (1954), who studied dense granular flows that occur in a geophysical context. In the 1980s, Jenkins and Savage (1983) and Lun et al. (1984) pioneered the application of kinetic theory to model the collisional process between particles and the development of rigorous expressions of collisional stress and thermal conduction. In this section, we will follow the approach outlined in Gidaspow (1994). Our discussion will be brief and focused on the end goal of developing appropriate models for collision-related secondary quantities. Clearly, the topic is deep and entire books have been devoted to kinetic granular theory. The reader is referred to these books.

16.3.1   Preliminaries

We start with the notion that properties such as particle velocity and temperature change abruptly after each collision. This change can be quantified by considering the property ψ of the two colliding particles just before collision and just after collision. If we label the two colliding particles as "1" and "2," and their pre- and post-collision states with superscripts "in" and "out," the change in ψ can be written as

$$\Delta\psi = \Delta\psi_1 + \Delta\psi_2 = \left(\psi_1^{out} - \psi_1^{in}\right) + \left(\psi_2^{out} - \psi_2^{in}\right), \tag{16.38}$$

where $\Delta\psi_1$ and $\Delta\psi_2$ are the collisional property change in particles "1" and "2," while $\Delta\Psi$ is the total property change of the two-particle system. We will later consider ψ to be particle momentum and particle energy to derive quantities such as collisional pressure, viscosity, and thermal conductivity. Our interest lies in obtaining an expression for the average rate of change of the particle property ψ due to collisions. Since we are interested in the average rate of change, we must consider all possible collisions between the particles. In the analysis to follow, we will use the field representation of particle velocity given in Eq. (11.107) as $\mathbf{v}$, which correctly reduces to the Lagrangian particle velocity in regions occupied by the particles.

Granular Temperature

Before we proceed with the formal derivation of the average collisional effect of particle momentum and energy, let us first consider a simple intuitive approach that is credited to Haff (1983). The intuitive approach surprisingly will yield accurate functional forms of collisional pressure, viscosity, and thermal conductivity, whose formal and more accurate derivations will follow in the subsequent subsections. For the intuitive approach, we start with the notion of granular temperature, Θ, introduced by Ogawa (1978). Granular temperature measures the average fluctuating velocity of particles about their mean value. Let $\langle \mathbf{v} \rangle(\mathbf{x}, t)$ be the ensemble-averaged particle velocity, then by definition

$$\Theta(\mathbf{x}, t) = \frac{1}{3} \langle |\mathbf{v} - \langle \mathbf{v} \rangle|^2 \rangle, \tag{16.39}$$

where the angle brackets can be chosen to be the ensemble average. Since collisions between the particles are due to their fluctuating motion, Θ plays an important role in deciding the frequency of collisions as well as the net effect of collisions that appear as collisional pressure, viscosity, and thermal conductivity in the governing equations. The term "granular temperature" is used to draw analogy to the fact that the thermodynamic definition of temperature is a measure of molecular velocity fluctuations about their mean motion. However, while T_d is the particle-phase thermodynamic temperature which measures molecular velocity fluctuations within each particle, Θ is a measure of particle-to-particle velocity fluctuation, and thus they must be clearly differentiated. It should be noted that Θ has units of (length/time)^2 and not degree Kelvin. There is another major difference between molecular velocity fluctuations and granular velocity fluctuations. In the case of molecular velocity, the magnitude of fluctuations is many orders larger than the mean value. In the case of granular velocity, fluctuations and mean are often of the same order. Furthermore, molecular collisions are elastic and energy conserving, while granular collisions are inelastic and frictional and thus are dissipative.

Granular Pressure

Let us now consider a particulate system where the particles are undergoing collisions due to their fluctuating velocity $\mathbf{v} - \langle \mathbf{v} \rangle$ (see Figure 16.1a). Also, the average inter-particle distance can be expressed in terms of the local volume fraction as $l \propto d_p \phi_d^{-1/3}$. This means that the average distance a particle has to travel to collide with a neighbor goes as $\propto d_p(\phi_d^{-1/3} - \phi_{\mathrm{cl}}^{-1/3})$, where ϕ_{cl} is the maximum particle volume fraction at close packing when the mean interparticle distance becomes zero. We then consider the collisional pressure p_d to be the rate of momentum change due to collisions (corresponding to the force of collision) divided by the cross-section area of collision. Mathematically, this translates to

$$p_d \propto \frac{\left(\rho_p \mathscr{V}_p \sqrt{\Theta}\right)\left(\sqrt{\Theta}/(d_p(\phi_d^{-1/3} - \phi_{\mathrm{cl}}^{-1/3}))\right)}{d_p^2} \propto \frac{\rho_p \Theta}{\phi_d^{-1/3} - \phi_{\mathrm{cl}}^{-1/3}}, \tag{16.40}$$

where in the numerator of the first proportionality, the first term is proportional to the momentum change associated with one collision and the second term is proportional to the rate of collisions (velocity divided by average distance to next collision), and the denominator is proportional to the area of collision. In the second proportionality we have used the fact that the particle volume $\mathcal{V}_p \propto d_p^3$. We thus have recovered the familiar result that pressure is proportional to density times the squared velocity. The collisional pressure also scales inversely as the mean gap between the particles. However, this scaling is accurate only for small values of ϕ_d and is inappropriate as ϕ_d approaches the close-packing limit.

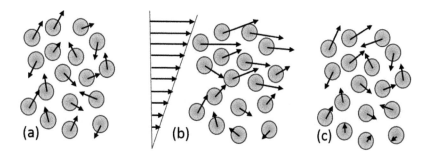

Figure 16.1   (a) A schematic of particles in random motion. In this case there is no mean velocity and the level of random particle velocity fluctuation is statistically uniform. (b) In this schematic, the level of random particle velocity fluctuation is statistically uniform, but there is a mean streamwise motion that increases with y and the mean velocity is shown on the left. (c) In this case there is no mean velocity and the level of random particle velocity fluctuation increases in y.

Granular Shear Viscosity

We follow a similar argument to obtain an estimate of collisional shear viscosity. The statistical symmetries of the collisional process must be considered. If $\langle \mathbf{v} \rangle$ and Θ are spatially uniform, then the collisional process is spatially isotropic and there is only collisional pressure. For shear viscosity, consider an inhomogeneous scenario, where the mean particle x-velocity varies along the y-direction (i.e., only $d\langle v_x \rangle/dy \neq 0$). This scenario is depicted in Figure 16.1b. If we consider Θ to be homogeneous, then the rate of collisions of a particle located within the middle of the bed with its neighbors remains the same as the previous case without mean shear. Also, the collisional cross-section remains proportional to d_p^2. We focus on the differential effect of collisions due to mean shear. Although the effect of fluctuating velocity on all collisions remains the same, the mean velocity of collisions for the slow-moving particles located at negative y differs from the mean velocity of collisions for the fast-moving particles in the upper layer. The differential x-momentum associated with the collision can be estimated to be $\rho_p \mathcal{V}_p \Delta v_x$, where $\Delta v_x = d_p(d\langle v_x \rangle/dy)$ is the mean streamwise velocity difference across one particle diameter. The collisional shear stress can then be expressed as the net rate of x-momentum change due to collisions divided by the cross-section area of collision:

$$\sigma_{d,xy} \propto \frac{(\rho_p \mathscr{V}_p d_p (d\langle v_x \rangle / dy))(\sqrt{\Theta}/(d_p(\phi_d^{-1/3} - \phi_{cl}^{-1/3})))}{d_p^2} . \tag{16.41}$$

If particle shear viscosity is defined as $\mu_d = \sigma_{d,xy}/(d\langle v_x \rangle / dy)$, we obtain the following scaling relation:

$$\mu_d \propto \frac{\rho_p d_p \sqrt{\Theta}}{\phi_d^{-1/3} - \phi_{cl}^{-1/3}} , \tag{16.42}$$

which predicts shear viscosity to scale as the square root of granular temperature.

Granular Conductivity

In a similar manner, we can obtain the scaling relation of granular conductivity. To do so, we consider a homogeneous flow of uniform mean velocity (i.e., $\langle \mathbf{u} \rangle$ constant), while the granular temperature varies in the y-direction as shown in Figure 16.1c. Again, the rate of collisions and cross-sectional area of collision remain the same. However, the kinetic energy of particle velocity fluctuations in the lower layer is lower than the kinetic energy of particle velocity fluctuations in the upper layer. Thus, collisions have the net effect of transporting the kinetic energy of particle velocity fluctuations in the negative y-direction. This mean kinetic energy transport scales as $\rho_p \mathscr{V}_p (\Theta(y - dy/2) - \Theta(y + dy/2))/2 \sim -\rho \mathscr{V}_p d_p (d\Theta/dy)/2$. With this, the collisional flux of granular temperature can be expressed as

$$q_{\Theta,y} \propto \frac{\left(-\rho_p \mathscr{V}_p d_p (d\Theta/dy)\right)\left(\sqrt{\Theta}/\left(d_p(\phi_d^{-1/3} - \phi_{cl}^{-1/3})\right)\right)}{d_p^2} . \tag{16.43}$$

If the particle granular conductivity is defined as $k_\Theta = -q_{\Theta,y}/(d\Theta/dy)$, we obtain the following scaling relation:

$$k_\Theta \propto \frac{\rho_p d_p \sqrt{\Theta}}{\phi_d^{-1/3} - \phi_{cl}^{-1/3}} . \tag{16.44}$$

Granular Dissipation

As mentioned, there are fundamental differences between molecular collisions and granular collisions. Important among them is the dissipative nature of particulate collisions. In molecular collisions, energy is conserved and as a result, the temperature of an insulated volume of gas will remain the same over time. Meanwhile, the granular kinetic energy within an insulated particulate system will not be preserved. In other words, consider a triply periodic cubic box with a random initial distribution of particles within the volume at the desired volume fraction. Let the particles be given a random initial velocity such that the mean velocity is zero and a nonzero granular temperature that is uniform within the box. If we let the system evolve from this initial state, the particles will start to move within the triply periodic box and collide. The level of particle velocity fluctuation measured in terms of granular temperature will decrease over time due to two dissipative mechanisms. The first dissipative mechanism is the drag force between the particle and the surrounding fluid and the second dissipative mechanism is the inelastic and frictional nature of collision between the particles. Thus, even in the absence of surrounding fluid, due to interparticle collisions, the granular temperature will steadily decrease over time and this is known as granular cooling. The lost kinetic energy of particle velocity fluctuation clearly goes to heating or thermal energy of the particles.

Later we will derive an equation of balance for the granular temperature, where the dissipative effect will appear as granular dissipation. In preparation, here we will get an estimate of the granular dissipation following the scaling argument presented above. Again, the frequency of collisions and cross-section area remain the same. From the discussion of particle–particle collision presented in Section 9.1, the average kinetic energy lost in collision can be estimated as $\rho_p \mathcal{V}_p (1 - \epsilon_c^2)\Theta$. Granular dissipation can now be expressed as

$$\Gamma \propto \frac{\left(\rho_p \mathcal{V}_p (1 - \epsilon_c^2)\Theta\right)\left(\sqrt{\Theta}/(d_p(\phi_d^{-1/3} - \phi_{cl}^{-1/3}))\right)}{d_p^2} \propto \frac{\rho_p(1 - \epsilon_c^2)\Theta^{3/2}}{\phi_d^{-1/3} - \phi_{cl}^{-1/3}} . \tag{16.45}$$

In summary, we have derived scaling relations of p_d, μ_d, k_Θ, and Γ in terms of their dependence on particle volume fraction and granular temperature. Explicit expressions for these quantities using granular kinetic theory will be obtained in the following sections.

16.3.2 Inelastic Boltzmann–Enskog Equation

In this section we revisit the one-particle and two-particle distribution functions that were introduced in Chapter 3 in the description of the dispersed phase. We will briefly discuss equations that govern these distribution functions as advanced by Boltzmann, and later extended by Enskog and others. Our emphasis will be on the statistical mechanic description of the collisional process. We start with $f_1(\mathbf{x}, \mathbf{v}, t)$ as the one-particle distribution function. By definition, the number of particles in an elemental volume $d\mathbf{x}$ centered at $\mathbf{x}$ with velocity within the range $d\mathbf{v}$ around $\mathbf{v}$ at time t is given by $f_1(\mathbf{x}, \mathbf{v}, t)d\mathbf{v}d\mathbf{x}$. With this definition, the number density of particles at $(\mathbf{x}, t)$ or the number of particles per unit volume is given by

$$n(\mathbf{x}, t) = \int_{\mathbf{v}} f_1(\mathbf{x}, \mathbf{v}, t)d\mathbf{v} . \tag{16.46}$$

The average particle velocity and the granular temperature are given by

$$\langle \mathbf{v} \rangle(\mathbf{x}, t) = \int_{\mathbf{v}} \mathbf{v} f_1(\mathbf{x}, \mathbf{v}, t)d\mathbf{v} \quad \text{and} \quad \Theta(\mathbf{x}, t) = \int_{\mathbf{v}} (\mathbf{v} - \langle \mathbf{v} \rangle)^2 f_1(\mathbf{x}, \mathbf{v}, t)d\mathbf{v} . \tag{16.47}$$

The time evolution of the one-particle distribution function is given by the Boltzmann equation

$$\frac{\partial f_1}{\partial t} + \mathbf{v} \cdot \nabla f_1 + \mathbf{A} \cdot \nabla_{\mathbf{v}} f_1 = \dot{f}_{col} , \tag{16.48}$$

where the second term on the left-hand side is the net flux of particles into an elemental volume around $(\mathbf{x}, \mathbf{v}, t)$ from neighboring spatial locations $(\mathbf{x} \pm d\mathbf{x}, \mathbf{v}, t)$. The third term on the left-hand side is the net flux of particles from neighboring velocity values (i.e., $(\mathbf{x}, \mathbf{v} \pm d\mathbf{v}, t)$). The right-hand side accounts for the collisional contribution to the rate of change of f_1. In the above, $\mathbf{A}(\mathbf{x}, t)$ is particle acceleration and $\nabla_{\mathbf{v}}$ is gradient with respect to velocity as the variable. The left-hand side can be interpreted as the total derivative of f_1 following a characteristic that goes from $(\mathbf{x}, \mathbf{v}, t)$ to $(\mathbf{x} + \mathbf{v}\Delta t, \mathbf{v} + \mathbf{A}\Delta t, t + \Delta t)$. In the absence of collisions, f_1 is constant along the characteristics. This is simply a recognition of the fact that particle position $\mathbf{x}$ advances based on velocity $\mathbf{v}$, while the velocity of a particle advances based on its acceleration. Thus, as position and velocity change, the particle distribution function is preserved along the characteristic. The change in f_1 along the characteristic depends only on the collisional process. Due to collisions, over a small time increment Δt, a particle with initial state $(\mathbf{x}, \mathbf{v}, t)$ may no longer arrive at $(\mathbf{x} + \mathbf{v}\Delta t, \mathbf{v} + \mathbf{A}\Delta t, t + \Delta t)$, and similarly a particle that was initially not at $(\mathbf{x}, \mathbf{v}, t)$ may end up arriving at $(\mathbf{x} + \mathbf{v}\Delta t, \mathbf{v} + \mathbf{A}\Delta t, t + \Delta t)$. The contributions of these two mechanisms are denoted as $\dot{f}_-$ and $\dot{f}_+$, and together $\dot{f}_{\text{col}} = \dot{f}_+ - \dot{f}_1$ account for the change in f_1 due to collisions. Note that $\dot{f}_-$ contributes to a loss and is therefore denoted with a negative sign.

We now proceed to derive an expression for $\dot{f}_{\text{col}}$. To do that, consider particle "1" to be the reference particle located at $\mathbf{x}$ and having velocity $\mathbf{v}$ (see Figure 16.2a). For simplicity, we will consider a monodisperse system where all the particles are of diameter d_p. Let the second particle be at location $\mathbf{x}_2$ and have velocity $\mathbf{v}_2$. The conditions for a collision to occur in an infinitesimal time are (i) $\mathbf{x}_2 = \mathbf{x} + d_p \mathbf{n}$ and (ii) $(\mathbf{v} - \mathbf{v}_2) \cdot \mathbf{n} > 0$, where the first condition states that the second particle's surface is infinitely close to the surface of the first particle along the direction $\mathbf{n}$ (where the unit vector $\mathbf{n}$ is directed out from the center of the first particle) and the second condition requires that the two particles are approaching each other along their line of separation (as opposed to moving apart).

The probability of simultaneously finding a particle located at $\mathbf{x}$ with velocity $\mathbf{v}$ and another located at $\mathbf{x} + d_p \mathbf{n}$ with velocity $\mathbf{v}_2$ is given by the two-particle distribution function (also known as the *pair distribution function*) $f_2(\mathbf{x}, \mathbf{v}, \mathbf{x} + d_p \mathbf{n}, \mathbf{v}_2, t)$. More precisely, the probability of finding a particle located within an elemental volume $d\mathbf{x}_1$ centered at $\mathbf{x}$ with velocity within the range $\mathbf{v} + d\mathbf{v}$ and another within an elemental volume $d\mathbf{x}_2$ centered at $\mathbf{x} + d_p \mathbf{n}$ with velocity within the range $\mathbf{v}_2 + d\mathbf{v}_2$ is given by $f_2(\mathbf{x}, \mathbf{v}, \mathbf{x} + d_p \mathbf{n}, \mathbf{v}_2, t) d\mathbf{x}_1 \, d\mathbf{x}_2 \, d\mathbf{v} \, d\mathbf{v}_2$. In general, both the one-particle and the pair distribution functions are functions of time. Only in a statistically stationary flow can their time dependence be dropped. An important point must be noted. As a result of collisions, Eq. (16.48) for a one-particle distribution function now involves a two-particle distribution function. If one were to write an equation for the rate of change of the two-particle distribution function, it will involve a three-particle distribution function, and so on. In what follows, we will use the idea of *molecular chaos* to close this hierarchy and express f_2 back in terms of f_1.

In the frame of reference attached to the first particle, consider a sphere of radius d_p on which the center of the colliding second sphere must lie. Consider an infinitesimal

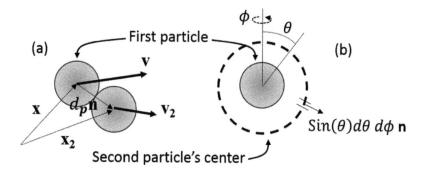

Figure 16.2 (a) A schematic of two particles undergoing collision. The condition for collision is that the distance between the centers is equal to the sum of their radius and the relative velocity projected along the line of separation is such that the two particles approach each other. (b) Schematic showing all possible locations of the second particle in the frame centered on the first particle.

surface patch of solid angle $d\mathbf{n} = \sin\theta d\theta d\phi\, \mathbf{n}$ (see Figure 16.2b, where θ and ϕ denote the spherical coordinates centered about the first particle). The cross-sectional area of this surface patch is $d_p^2 d\mathbf{n}$. The flux of the second particle toward the first particle in unit time is then given by $(\delta\mathbf{v} \cdot \mathbf{n})d_p^2 d\mathbf{n}$, which must be multiplied by $f_2(\mathbf{x}, \mathbf{v}, \mathbf{x} + d_p\mathbf{n}, \mathbf{v}_2, t)$ to account for the likelihood that the two particles are at locations $\mathbf{x}$ and $\mathbf{x} + d\mathbf{x}$ with the right velocities. In the above, $\delta\mathbf{v} = \mathbf{v} - \mathbf{v}_2$, which must be positive when projected along $\mathbf{n}$. Otherwise, the flux of the second particles will be directed away from the first particle and cannot contribute to collision. Each such collision contributes to the reference particle with initial state $(\mathbf{x}, \mathbf{v}, t)$ not arriving at $(\mathbf{x} + \mathbf{v}\Delta t, \mathbf{v} + \mathbf{A}\Delta t, t + \Delta t)$. This provides the following expression:

$$\dot{f}_- = d_p^2 \iint\limits_{\delta\mathbf{v}\cdot\mathbf{n}>0} (\delta\mathbf{v} \cdot \mathbf{n})f_2(\mathbf{x}, \mathbf{v}, \mathbf{x} + d_p\mathbf{n}, \mathbf{v}_2, t)\, d\mathbf{n}\, d\mathbf{v}_2, \qquad (16.49)$$

where the integrals are over all solid angles $\mathbf{n}$ and over all possible velocities $\mathbf{v}_2$ of the colliding particle. Similarly, we now consider the reference particle whose pre-collision velocity is $\mathbf{v}_{\mathrm{pre}}$ colliding with a second particle located at $\mathbf{x} - d_p\mathbf{n}$ with a pre-collision velocity of $\mathbf{v}_{2\mathrm{pre}}$. Their post-collision velocities are $\mathbf{v}$ and $\mathbf{v}_2$. From the collisional physics considered in Section 9.1, we have the relation

$$\mathbf{v} = \mathbf{v}_{\mathrm{pre}} - \frac{(1 + \epsilon_c)}{2}(\mathbf{v}_{\mathrm{pre}} - \mathbf{v}_{2\mathrm{pre}}) \cdot \mathbf{n}, \qquad (16.50)$$

$$\mathbf{v}_2 = \mathbf{v}_{2\mathrm{pre}} + \frac{(1 + \epsilon_c)}{2}(\mathbf{v}_{\mathrm{pre}} - \mathbf{v}_{2\mathrm{pre}}) \cdot \mathbf{n}. \qquad (16.51)$$

With such collisions, we can express

$$\dot{f}_+ = d_p^2 \iint_{\delta\mathbf{v}_{\text{pre}} \cdot \mathbf{n} < 0} (\delta\mathbf{v}_{\text{pre}} \cdot \mathbf{n}) \, f_2(\mathbf{x}, \mathbf{v}_{\text{pre}}, \mathbf{x} - d_p\mathbf{n}, \mathbf{v}_{2\text{pre}}, t) \, d\mathbf{n} \, d\mathbf{v}_{2\text{pre}} \tag{16.52}$$

$$= \frac{d_p^2}{\epsilon_c^2} \iint_{\delta\mathbf{v} \cdot \mathbf{n} > 0} (\delta\mathbf{v} \cdot \mathbf{n}) \, f_2(\mathbf{x}, \mathbf{v}_{\text{pre}}, \mathbf{x} - d_p\mathbf{n}, \mathbf{v}_{2\text{pre}}, t) \, d\mathbf{n} \, d\mathbf{v}_2 , \tag{16.53}$$

where factor $1/\epsilon_c$ comes from the replacement of $\delta\mathbf{v}_{\text{pre}} \cdot \mathbf{n}$ by $\delta\mathbf{v} \cdot \mathbf{n}$ and another factor $1/\epsilon_c$ comes from replacing $d\mathbf{v}_{2\text{pre}}$ by $d\mathbf{v}_2$. Substituting the above expressions in $\dot{f}_{\text{col}} = \dot{f}_+ - \dot{f}_1$, we obtain the final expression of the right-hand side of Eq. (16.48) as

$$\dot{f}_{\text{col}} = d_p^2 \iint_{\delta\mathbf{v} \cdot \mathbf{n} > 0} (\delta\mathbf{v} \cdot \mathbf{n}) \left[\frac{1}{\epsilon_c^2} f_2(\mathbf{x}, \mathbf{v}_{\text{pre}}, \mathbf{x} - d_p\mathbf{n}, \mathbf{v}_{2\text{pre}}, t) - f_2(\mathbf{x}, \mathbf{v}, \mathbf{x} + d_p\mathbf{n}, \mathbf{v}_2, t) \right] d\mathbf{n} \, d\mathbf{v}_2 . \tag{16.54}$$

16.3.3 Effect of Collisions on Particle Properties

We now proceed to express the rate of change of a particle-related quantity ψ due to collisions. The process to be followed is similar to that outlined above. But we will also take into account the change in ψ due to each kind of collision and its associated probability. We again restrict attention only to binary collisions. Simultaneous collisions between more than two particles are not considered, and this is appropriate except at very high volume fractions. Within the framework of binary collisions between the two particles shown in Figure 16.2b, let us first consider the change in ψ with the first particle as reference, whose position and velocity are $\mathbf{x}$ and $\mathbf{v}_1$. The second particle is at $\mathbf{x} + d_p\mathbf{n}$ with velocity $\mathbf{v}_2$. The change in ψ of the reference particle is given by (Andreotti et al., 2013)

$$\dot{\psi}_{\text{col}} = d_p^2 \iiint_{\mathbf{v}_{12} \cdot \mathbf{n} > 0} \Delta\psi_1 (\mathbf{v}_{12} \cdot \mathbf{n}) f_2(\mathbf{x}, \mathbf{v}_1, \mathbf{x} + d_p\mathbf{n}, \mathbf{v}_2, t) \, d\mathbf{n} \, d\mathbf{v}_1 \, d\mathbf{v}_2 , \tag{16.55}$$

where the integrals are over all orientations of $\mathbf{n}$, and all possible values of $\mathbf{v}_1$ and $\mathbf{v}_2$, but under the collisional condition that $\mathbf{v}_{12} \cdot \mathbf{n} > 0$, where $\mathbf{v}_{12} = \mathbf{v}_1 - \mathbf{v}_2$. The rate of change of ψ due to collisions can also be considered with the second particle as the reference, now located at $\mathbf{x}$ with velocity $\mathbf{v}_2$, while the first particle's location with respect to the reference particle becomes $\mathbf{x} - d_p\mathbf{n}$. This yields the alternate definition

$$\dot{\psi}_{\text{col}} = d_p^2 \iiint_{\mathbf{v}_{12} \cdot \mathbf{n} > 0} \Delta\psi_2 (\mathbf{v}_{12} \cdot \mathbf{n}) f_2(\mathbf{x} - d_p\mathbf{n}, \mathbf{v}_1, \mathbf{x}, \mathbf{v}_2, t) \, d\mathbf{n} \, d\mathbf{v}_1 \, d\mathbf{v}_2 . \tag{16.56}$$

We now take the average of the above two expressions and further write it as a symmetric and an asymmetric contribution:

$$
\begin{aligned}
\dot{\psi}_{\text{col}} = \; & \frac{d_p^2}{4} \iiint_{\mathbf{v}_{12}\cdot\mathbf{n}>0} (\Delta\psi_1 + \Delta\psi_2)(\mathbf{v}_{12}\cdot\mathbf{n})(f_2(\mathbf{x},\mathbf{v}_1,\mathbf{x}+d_p\mathbf{n},\mathbf{v}_2,t) \\
& + f_2(\mathbf{x}-d_p\mathbf{n},\mathbf{v}_1,\mathbf{x},\mathbf{v}_2,t))\,d\mathbf{n}\,d\mathbf{v}_1\,d\mathbf{v}_2 \\
& + \frac{d_p^2}{4} \iiint_{\mathbf{v}_{12}\cdot\mathbf{n}>0} (\Delta\psi_1 - \Delta\psi_2)(\mathbf{v}_{12}\cdot\mathbf{n})(f_2(\mathbf{x},\mathbf{v}_1,\mathbf{x}+d_p\mathbf{n},\mathbf{v}_2,t) \\
& - f_2(\mathbf{x}-d_p\mathbf{n},\mathbf{v}_1,\mathbf{x},\mathbf{v}_2,t))\,d\mathbf{n}\,d\mathbf{v}_1\,d\mathbf{v}_2\,.
\end{aligned}
\tag{16.57}
$$

In the first term on the right-hand side, we recognize the fact that to leading order

$$
f_2(\mathbf{x},\mathbf{v}_1,\mathbf{x}+d_p\mathbf{n},\mathbf{v}_2,t) \approx f_2(\mathbf{x}-d_p\mathbf{n},\mathbf{v}_1,\mathbf{x},\mathbf{v}_2,t),
\tag{16.58}
$$

since this only involves a shift by a small distance d_p. The same leading-order expansion is inappropriate for the second term, since it will render it zero. To obtain the first non-zero contribution from the second term, we Taylor series expand both the two-particle distribution functions about the point of contact between the two particles. This yields

$$
\begin{aligned}
& f_2(\mathbf{x},\mathbf{v}_1,\mathbf{x}+d_p\mathbf{n},\mathbf{v}_2,t) - f_2(\mathbf{x}-d_p\mathbf{n},\mathbf{v}_1,\mathbf{x},\mathbf{v}_2,t) \\
& = d_p \nabla f_2(\mathbf{x}-(d_p/2)\mathbf{n},\mathbf{v}_1,\mathbf{x}+(d_p/2)\mathbf{n},\mathbf{v}_2,t) + O\left(d_p^2\right).
\end{aligned}
\tag{16.59}
$$

Substituting the above into Eq. (16.57), we obtain the result

$$
\dot{\psi}_{\text{col}} = \nabla \cdot (\phi_d \boldsymbol{\sigma}[\psi]) + S[\psi],
\tag{16.60}
$$

where $\boldsymbol{\sigma}[\psi]$ should be interpreted as the collisional stress of the quantity ψ and $S[\psi]$ as the source term of the quantity ψ. The leading-order collisional stress term is

$$
\phi_d \boldsymbol{\sigma}[\psi] = \frac{d_p^3}{4} \iiint_{\mathbf{v}_{12}\cdot\mathbf{n}>0} (\Delta\psi_1 - \Delta\psi_2)(\mathbf{v}_{12}\cdot\mathbf{n})\,\mathbf{n}\,f_2(\mathbf{x}-(d_p/2)\mathbf{n},\mathbf{v}_1,\mathbf{x}
$$
$$
+ (d_p/2)\mathbf{n},\mathbf{v}_2,t)\,d\mathbf{n}\,d\mathbf{v}_1\,d\mathbf{v}_2,
\tag{16.61}
$$

and the collisional source term is given by

$$
S[\psi] = \frac{d_p^2}{2} \iiint_{\mathbf{v}_{12}\cdot\mathbf{n}>0} (\Delta\psi_1 + \Delta\psi_2)(\mathbf{v}_{12}\cdot\mathbf{n})f_2(\mathbf{x}-d_p\mathbf{n},\mathbf{v}_1,\mathbf{x},\mathbf{v}_2,t)\,d\mathbf{n}\,d\mathbf{v}_1\,d\mathbf{v}_2\,.
\tag{16.62}
$$

We will soon choose the quantity ψ to be particle momentum or kinetic energy, and in order to evaluate the effect of collisions on these quantities as given by the above equations, we first need to obtain expressions for the two-particle distributions functions. This will be considered in the following subsection.

16.3.4   Maxwellian Distribution and Kinetic Stress

We make an important assumption that the velocities of the particles fluctuate about the local average (or the filtered value) following a Gaussian or normal distribution. The normal velocity distribution is also known as the Maxwellian distribution. With this assumption, the one-particle distribution function can be written as

$$f_1(\mathbf{x}, \mathbf{v}) = n(\mathbf{x}, t) \left(\frac{1}{2\pi\Theta(\mathbf{x}, t)} \right)^{3/2} \exp\left(-\frac{(\mathbf{v} - \langle \mathbf{v} \rangle)^2}{2\Theta(\mathbf{x}, t)} \right), \tag{16.63}$$

where $n(\mathbf{x}, t)$ is the average number density of particles at $\mathbf{x}$ and t. The granular temperature can be expressed in terms of the one-particle distribution function as

$$\frac{1}{3} \int_{-\infty}^{\infty} |\mathbf{v} - \langle \mathbf{v} \rangle|^2 f_1(\mathbf{x}, t) \, d\mathbf{v} = \Theta. \tag{16.64}$$

Note that the $(\mathbf{x}, t)$ dependence of the one-particle distribution function comes only from the dependence of $\langle \mathbf{v} \rangle$, n, and Θ on space and time.

In Section 16.2.6 we considered particle residual stress arising from particle-to-particle variation in particle velocity. There we focused on fluid turbulence as the source of particle velocity variation in modeling $\tau_{d,t}$. In this subsection we will consider particle velocity variation arising from interparticle collisions in the case of a not-so-dilute suspension of particles. The kinetic contribution to particle residual stress can be expressed as

$$\phi_d \tau_{d,\mathrm{ki}} = -m_p \int \mathbf{v}'\mathbf{v}' \, f_1(\mathbf{x}, \mathbf{v}) \, d\mathbf{v}, \tag{16.65}$$

where f_1 is taken to be the Maxwellian distribution given above. In the above, $\mathbf{v}' = \mathbf{v} - \langle \mathbf{v} \rangle$ is the velocity fluctuation about the mean. Substituting the Maxwellian distribution and carrying out the integration, it can be shown that

$$\phi_d \tau_{d,\mathrm{ki}} = -\rho_p \Theta \mathbf{I}. \tag{16.66}$$

This particle stress due to collision-induced particle velocity fluctuation is also known as the *kinetic stress*. It can be seen that to leading order the kinetic stress is isotropic and behaves as particle pressure. As obtained, the kinetic stress is linearly related to granular temperature.

16.3.5   Molecular Chaos and Radial Distribution Function

Earlier, in Section 3.4, we discussed the two-particle distribution function, in the restricted context of an instantaneous statistical description of a random distribution of particles, without regard to particle velocity. We expressed the two-particle distribution function in terms of a one-particle distribution function using the radial distribution function. In the present context, in the previous subsection, we have defined the two-particle distribution function more generally by including the velocity of the two particles as additional descriptors. Using the same approach as in Eq. (3.33), we can express

$$f_2(\mathbf{x}_1, \mathbf{v}_1, \mathbf{x}_2, \mathbf{v}_2, t) = g(\mathbf{x}_1, \mathbf{x}_2, t) f_1(\mathbf{x}_1, \mathbf{v}_1, t) f_2(\mathbf{x}, \mathbf{v}_2, t), \qquad (16.67)$$

where $f_1(\mathbf{x}_1, \mathbf{v}_1, t)$ and $f_1(\mathbf{x}_2, \mathbf{v}_2, t)$ are the one-particle distribution functions for the first and second particle and $g(\mathbf{x}_1, \mathbf{x}_2, t)$ is the radial distribution function. Justification for the above approximation is as follows. In the limit where the two particles are far away (i.e., $|\mathbf{x}_2 - \mathbf{x}_1| \gg d_p$), it is reasonable to assume that the two particles are quite independent and therefore $f_2(\mathbf{x}_1, \mathbf{v}_1, \mathbf{x}_2, \mathbf{v}_2, t) \simeq f_1(\mathbf{x}_1, \mathbf{v}_1, t) f_2(\mathbf{x}, \mathbf{v}_2, t)$, which is consistent with the definition that the radial distribution function becomes unity at large distances. In the present context of collisions, we are concerned with the limit $|\mathbf{x}_2 - \mathbf{x}_1| = d_p$. Even in this limit, if the particle volume fraction is low, it is reasonable to assume $f_2(\mathbf{x}_1, \mathbf{v}_1, \mathbf{x}_2, \mathbf{v}_2, t) \simeq f_1(\mathbf{x}_1, \mathbf{v}_1, t) f_2(\mathbf{x}, \mathbf{v}_2, t)$, since the distribution of each particle is nearly independent of each other. Only when the particle volume fraction increases, due to the finite size of the particle compared to the mean interparticle separation, does the probability of finding a second particle close to the first particle increase, thus increasing the probability of collision. This increase is accounted for by the radial distribution function. The above expression implicitly employs the concept of molecular chaos advanced by Boltzmann. Note that the two colliding particles are required to be close only in their position and not in their velocity. Thus, it is assumed that the velocity of the two colliding particles is uncorrelated and as a result the radial distribution function has no dependence on particle velocity.

If we assume the radial distribution function to be homogeneous and isotropic, then g is only a function of the separation vector. Furthermore, since $|\mathbf{x}2 - \mathbf{x}_1| = d_p$, in Eq. (16.67), the radial distribution function can be taken to be its value at close contact between monodispersed particles, which is a function only of the local volume fraction and will be denoted as g_0. Several different approximations of $g_0(\phi_d)$ have been advanced in the literature. The most prominent among them is the following Carnahan–Starling formula (Carnahan and Starling, 1969), which has been verified experimentally to be accurate at low to moderate volume fractions:

$$g_0(\phi_d) = \frac{2 - \phi_d}{2(1 - \phi_d)^3}. \qquad (16.68)$$

Other similar expressions that are more accurate when the volume fraction approaches the close-packing limit have been proposed in the literature.

Substituting for the one-particle Maxwellian distribution function, the two-particle radial distribution function that appears in the integral in Eq. (16.61) can be rewritten as

$$f_2(\mathbf{x} - (d_p/2)\mathbf{n}, \mathbf{v}_1, \mathbf{x} + (d_p/2)\mathbf{n}, \mathbf{v}_2, t)$$
$$= g_0 f_1(\mathbf{x} - (d_p/2)\mathbf{n}, \mathbf{v}_1, t) f_1(\mathbf{x} + (d_p/2)\mathbf{n}, \mathbf{v}_2, t)$$
$$= n_1 n_2 g_0 \left(\frac{1}{4\pi^2 \Theta_1 \Theta_2} \right)^{3/2} \exp \left(-\frac{(\mathbf{v}_1 - \langle \mathbf{v}_1 \rangle)^2}{2\Theta_1} - \frac{(\mathbf{v}_2 - \langle \mathbf{v}_2 \rangle)^2}{2\Theta_2} \right), \qquad (16.69)$$

where subscripts 1 and 2 for $\langle \mathbf{v} \rangle$, n, and Θ indicate these quantities being evaluated at the first and second particle, respectively. This can be simplified by identifying the centers of the two particles to be only slightly displaced from their contact point. Taylor series expansion of the two, one-particle distribution functions about the contact point yields

$$f_2(\mathbf{x} - (d_p/2)\mathbf{n}, \mathbf{v}_1, \mathbf{x} + (d_p/2)\mathbf{n}, \mathbf{v}_2, t)$$

$$= g_0 f_1(\mathbf{x}, \mathbf{v}_1, t) f_1(\mathbf{x}, \mathbf{v}_2, t) + g_0 \frac{d_p}{2} f_1(\mathbf{x}, \mathbf{v}_1, t) f_1(\mathbf{x}, \mathbf{v}_2, t) \nabla \ln \left(\frac{f_1(\mathbf{x}, \mathbf{v}_2, t)}{f_1(\mathbf{x}, \mathbf{v}_1, t)} \right) + O(d_p^2).$$

(16.70)

With the above derivation we are now ready to consider specific examples of ψ and calculate important collisional properties. Before that, as a preparatory step, you will need to complete the derivations given in the following problem.

Problem 16.9 Assume a Maxwellian distribution of one-particle velocity fluctuation and the definition of granular temperature given in Eq. (16.47) to obtain the following relations:

$$\iint v_{12i} v_{12j} f_1(\mathbf{x}, \mathbf{v}_1, t) f_1(\mathbf{x}, \mathbf{v}_2, t) \, d\mathbf{v}_1 d\mathbf{v}_2 = 2n^2 \Theta \delta_{ij}, \qquad (16.71)$$

$$\iint v_{12i} v_{12i} f_1(\mathbf{x}, \mathbf{v}_1, t) f_1(\mathbf{x}, \mathbf{v}_2, t) \, d\mathbf{v}_1 d\mathbf{v}_2 = 6n^2 \Theta. \qquad (16.72)$$

In obtaining the above relations, it will be assumed that the mean velocity remains the same for the two colliding particles due to their proximity (i.e., $\langle \mathbf{v}_1 \rangle = \langle \mathbf{v}_2 \rangle$). As a result, in obtaining the above relation, without loss of generality it can be assumed that the mean velocity is zero.

16.3.6   Collisional Effect on Momentum Balance

We are now ready to calculate the effect of collision on the particulate momentum balance equation. Specifically, we want to obtain the term(s) that appear in Eq. (16.5) accounting for the effect of interparticle collisions. To do so, we set $\psi = \rho_p \mathscr{V}_p \mathbf{v}$ in Eqs. (16.61) and (16.62). Since momentum is conserved in each collision, $\Delta\psi_1 + \Delta\psi_2 = 0$ and as a result there is no collisional source term in the particle-phase momentum equation. To calculate the collisional stress expression, we revisit the hard-sphere collisional model considered in Section 9.1. We adapt the general expression given in Eq. (9.11) for the present discussion of monodispersed particles. Furthermore, if the frictional effect of collision is neglected, then

$$\Delta(\rho_p \mathscr{V}_p \mathbf{v}_1) = -\Delta(\rho_p \mathscr{V}_p \mathbf{v}_2) = -\frac{\rho_p \mathscr{V}_p}{2}(1 + \epsilon_c)(\mathbf{v}_{12} \cdot \mathbf{n})\mathbf{n}. \qquad (16.73)$$

Substituting Eq. (16.73) for $\Delta \psi$ in Eq. (16.61), along with Eq. (16.70) for the two-particle distribution function, we obtain the following expression for particle stress:

$$\phi_d \boldsymbol{\sigma}_{d,\text{co}} =$$

$$\underbrace{-\frac{d_p^3}{4} g_0 \rho_p \mathscr{V}_p (1 + \epsilon_c) \iiint_{\mathbf{v}_{12} \cdot \mathbf{n} > 0} (\mathbf{v}_{12} \cdot \mathbf{n})^2 \mathbf{n}\mathbf{n} f_1(\mathbf{x}, \mathbf{v}_1, t) f_1(\mathbf{x}, \mathbf{v}_2, t) \, d\mathbf{n} \, d\mathbf{v}_1 \, d\mathbf{v}_2}_{\phi_d \boldsymbol{\sigma}_{d1}}$$

$$\underbrace{-\frac{d_p^4}{8} g_0 \rho_p \mathscr{V}_p (1 + \epsilon_c) \iiint_{\mathbf{v}_{12} \cdot \mathbf{n} > 0} (\mathbf{v}_{12} \cdot \mathbf{n})^2 \mathbf{n}\mathbf{n} f_1(\mathbf{x}, \mathbf{v}_1, t) f_1(\mathbf{x}, \mathbf{v}_2, t) \nabla \ln \left(\frac{f_1(\mathbf{x}, \mathbf{v}_2, t)}{f_1(\mathbf{x}, \mathbf{v}_1, t)} \right) d\mathbf{n} \, d\mathbf{v}_1 \, d\mathbf{v}_2}_{\phi_d \boldsymbol{\sigma}_{d2}},$$

$$\tag{16.74}$$

where $\boldsymbol{\sigma}_{d,\text{co}} = \boldsymbol{\sigma}[\rho_p \mathscr{V}_p \mathbf{v}]$.

Problem 16.10 Consult the original derivations presented in Chapman and Cowling (1990) and obtain the following identity:

$$\int (\mathbf{v}_{12} \cdot \mathbf{n})^2 \mathbf{n}\mathbf{n} \, d\mathbf{n} = \frac{2\pi}{15} (2\mathbf{v}_{12}\mathbf{v}_{12} + |\mathbf{v}_{12}|^2 \mathbf{I}). \tag{16.75}$$

The above integral is over the entire solid angle for all orientations of $\mathbf{n}$.

With the above identity, the first collision stress can be expressed as

$$\phi_d \boldsymbol{\sigma}_{d1} = -\frac{\pi d_p^3}{30} g_0 \rho_p \mathscr{V}_p (1 + \epsilon_c) \iint (2\mathbf{v}_{12}\mathbf{v}_{12} + |\mathbf{v}_{12}|^2 \mathbf{I}) f_1(\mathbf{x}, \mathbf{v}_1, t) f_1(\mathbf{x}, \mathbf{v}_2, t) \, d\mathbf{v}_1 \, d\mathbf{v}_2. \tag{16.76}$$

Now we use the results of Problem 16.9 to carry out the integrals and obtain the following expression for particle pressure:

$$\phi_d \boldsymbol{\sigma}_{d1} = -\phi_d p_{d1} \mathbf{I} = -2 g_0 \rho_p \phi_d^2 (1 + \epsilon_c) \Theta \, \mathbf{I}, \tag{16.77}$$

where we have made the following substitution: $\pi d_p^3 \mathscr{V}_p n^2 / 6 = \phi_d^2$. Similar steps can be followed in obtaining an explicit expression for the second collision stress, whose derivation is more involved. We refer the reader to the discussion presented in Gidaspow's (1994) book and simply provide the final result as

$$\phi_d \boldsymbol{\sigma}_{d2} = \frac{4}{3\sqrt{\pi}} g_0 \rho_p d_p \phi_d^2 (1 + \epsilon_c) \Theta^{1/2} \left(\frac{3}{5} (\nabla \mathbf{u}_d + (\nabla \mathbf{u}_d)^{\mathsf{T}}) + \nabla \cdot \mathbf{u}_d \mathbf{I} \right), \tag{16.78}$$

where we have used the fact that the mean particle velocity in the context of the EE methodology is $\mathbf{u}_d$. From the above, we obtain the collisional dynamic and bulk viscosities of the particulate phase as

$$\mu_d = \frac{4}{5\sqrt{\pi}} g_0 \rho_p d_p \phi_d (1 + \epsilon_c) \Theta^{1/2} \quad \text{and} \quad \lambda_d = \frac{4}{3\sqrt{\pi}} g_0 \rho_p d_p \phi_d (1 + \epsilon_c) \Theta^{1/2}. \tag{16.79}$$

The final expression for particle stress becomes

$$\phi_d \boldsymbol{\sigma}_{d,\mathrm{co}} = (-\phi_d p_{d1}\mathbf{I} + \lambda_d(\nabla \cdot \mathbf{u}_d))\phi_d \,\mathbf{I} + \mu_d \phi_d \left(\nabla \mathbf{u}_d + (\nabla \mathbf{u}_d)^{\mathrm{T}}\right) . \qquad (16.80)$$

16.3.7   Boltzmann–Maxwell Transport Theorem

In the previous sections, granular temperature Θ emerged as an important quantity that determined a number of other important quantities such as collisional pressure and collisional viscosity. However, Θ is not part of the primary variables being solved in an EE simulation. Therefore, the goal of this section is to derive a fundamental equation for granular temperature (i.e., fluctuating particle kinetic energy equation). The first step in this process is to derive Maxwell's transport theorem. Consider a property $\psi(\mathbf{v})$ being carried by the particles that is primarily only a function of $\mathbf{v}$ (and not t or x). Its transport equation can be written from the Boltzmann equation (16.48) as

$$\int \psi \left(\frac{\partial f_1}{\partial t} + \mathbf{v} \cdot \nabla f_1 + \mathbf{A} \cdot \nabla_\mathbf{v} f_1\right) d\mathbf{v} = \int \psi \dot{f}_{\mathrm{col}} \, d\mathbf{v} . \qquad (16.81)$$

We then rewrite each term on the left-hand side in the following manner:

$$\int \psi \frac{\partial f_1}{\partial t} d\mathbf{v} = \frac{\partial}{\partial t} \int \psi f_1 \, d\mathbf{v} = \frac{\partial n\langle\psi\rangle}{\partial t} , \qquad (16.82)$$

$$\int \psi \mathbf{v} \cdot \nabla f_1 \, d\mathbf{v} = \nabla \cdot \int \psi \mathbf{v} f_1 \, d\mathbf{v} = \nabla \cdot (n\langle\psi\mathbf{v}\rangle), \qquad (16.83)$$

$$\int \psi \mathbf{A} \cdot \nabla_\mathbf{v} f_1 \, d\mathbf{v} = \int \nabla_\mathbf{v} \cdot (\psi \mathbf{A} f_1) \, d\mathbf{v} - \int f_1 \mathbf{A} \cdot \nabla_\mathbf{v} \psi \, d\mathbf{v} = -n\frac{\mathbf{F}}{m_p} \cdot \langle\nabla_\mathbf{v}\psi\rangle . \qquad (16.84)$$

In obtaining the above equations we have used a number of simple facts. (i) In the first equality of the first equation, we have commuted the time derivative with the integral over $\mathbf{v}$. (ii) In the first equality of the second equation, we have commuted the spatial derivative with the integral over $\mathbf{v}$. (iii) In the first two equations, we have used the fact that ψ is not a function of t or x. (iv) With the application of the Gauss theorem in the first term of the third equation (i.e., $\int \nabla_\mathbf{v} \cdot (\psi \mathbf{A} f_1) \, d\mathbf{v}$), the volume integral becomes the surface integral for very large $\mathbf{v}$, whose vanishing probability makes this term go to zero. (v) In the third equation, $\mathbf{A}$'s dependence on $\mathbf{v}$ can be ignored. (v) Finally, in the last equation, we have replaced acceleration by $\mathbf{F}/m_p$, where m_p is the mass of the particle.

We substitute the above three equations on the left-hand side of Eq. (16.81) and for the right-hand side we make use of the collisional model derived in Eq. (16.60) to obtain

$$\frac{\partial n\langle\psi\rangle}{\partial t} + \nabla \cdot (n\langle\psi\mathbf{v}\rangle) - n\frac{\mathbf{F}}{m_p} \cdot \langle\nabla_\mathbf{v}\psi\rangle = \nabla \cdot (\phi_d \boldsymbol{\sigma}[\psi]) + S[\psi] . \qquad (16.85)$$

The above master equation can be used to obtain the particle mass balance equation (16.4) by setting $\psi = m_p$ (i.e., choosing the particle property ψ to be the mass of the monodisperse particle). If we ignore mass transfer between the phases, it can be noted that $\langle m_p \rangle = m_p$, since the mass of the particle remains the same. Furthermore, $n(\mathbf{x},t)m_p = \rho_p\phi_d(\mathbf{x},t)$, $n\langle m_p\mathbf{v}\rangle = \rho_p\phi_d\mathbf{u}_d$, and $\nabla_{\mathbf{v}}m_p = 0$. The collisional contribution to particle mass is identically zero (i.e., $\sigma[m_p] = S[m_p] = 0$). With these reductions, we obtain

$$\frac{\partial \rho_p\phi_d}{\partial t} + \nabla \cdot (\rho_p\phi_d\mathbf{u}_d) = 0. \qquad (16.86)$$

Thus, the particle mass conservation equation that we obtained from the filtering operation can be obtained from the Boltzmann equation as well. This is the foundation of the kinetic theory of gases.

The above process can be repeated with $\psi = m_p\mathbf{V}$ to obtain the momentum balance equation for the particles. In obtaining this equation, we make the following substitutions: $n\langle m_p\mathbf{v}\rangle = \rho_p\phi_d\mathbf{u}_d$, $n\langle m_p\mathbf{v}\mathbf{v}\rangle = \rho_p\phi_d\langle\mathbf{v}\mathbf{v}\rangle = \rho_p\phi_d\mathbf{u}_d\mathbf{u}_d - \phi_d\tau_{d,\mathrm{ki}}$, and $\nabla_{\mathbf{v}}(m_p\mathbf{v}) = m_p\mathbf{I}$. Furthermore, we note that collisions conserve momentum and as a result the collisional source term $S[m_p\mathbf{v}] = 0$. From these substitutions we obtain

$$\frac{\partial \rho_p\phi_d\mathbf{u}_d}{\partial t} + \nabla \cdot (\rho_p\phi_d\mathbf{u}_d\mathbf{u}_d) = \nabla \cdot (\phi_d\tau_{d,\mathrm{ki}}) + n\mathbf{F} + \nabla \cdot (\phi_d\sigma_{d,\mathrm{co}}). \qquad (16.87)$$

In the above equation, we have recognized the collisional stress contribution to particle momentum $\sigma[m_p\mathbf{v}]$ to be $\sigma_{d,\mathrm{co}}$, obtained earlier in Eq. (16.80).

The above momentum balance can be compared with the complete particle-phase momentum balance equation (16.5). The fundamental difference between the two is as follows. The above momentum balance ignores the effect of the fluid surrounding the particles, but rigorously accounts for the effects of interparticle collisions, whereas the complete momentum equation (16.5) was derived from a rigorous filtering operation that accounts for the interaction between the particles and the surrounding fluid. But the filtering approach did not account for the effect of interparticle collisions, and the collisional terms were simply added to Eq. (16.5). Since the fluid effects are ignored in Eq. (16.87), the residual stress is only from the kinetic contribution from interparticle collisions and thus $\nabla \cdot (\phi_d\tau_{d,\mathrm{sg}})$ appears as $\nabla \cdot (\phi_d\tau_{d,\mathrm{ki}})$, which can be expressed as $-\rho_p\nabla\Theta$ with the use of the closure relation given in Eq. (16.66). In the above equation, $\mathbf{F}$ is the constant external force on a particle and thus $n\mathbf{F}$ accounts for the external force on all the particles within a unit volume and corresponds to the gravitational body-force term in Eq. (16.5). The complete momentum equation (16.5) also includes the force terms $\phi_d \nabla \cdot \sigma_c$ and $\mathbf{F}'_{\mathrm{hyd}}$, which are the undisturbed and perturbation flow forces exerted on the particles from the surrounding fluid. They are absent in Eq. (16.87) due to the neglect of the surrounding fluid. Or we could interpret $n\mathbf{F}$ as the total force acting on the particle including those from the surrounding fluid.

16.3.8 Granular Temperature Equation

We now turn our attention to obtaining the granular temperature. This is accomplished by setting $\psi = m_p|\mathbf{v}|^2/2$ in Eq. (16.85) to obtain

$$\frac{1}{2}\frac{\partial \rho_p\phi_d\langle\mathbf{v}\cdot\mathbf{v}\rangle}{\partial t} + \frac{1}{2}\nabla\cdot(\rho_p\phi_d\langle(\mathbf{v}\cdot\mathbf{v})\mathbf{v}\rangle) - n\mathbf{F}\cdot\langle\mathbf{v}\rangle = \nabla\cdot\left(\phi_d\sigma\left[\frac{m_p}{2}|\mathbf{v}|^2\right]\right) + S\left[\frac{m_p}{2}|\mathbf{v}|^2\right].$$
$$(16.88)$$

Each term of the above equation must now be expanded to its final form, which we shall do below. First, we separate the particle velocity into a mean and a fluctuating component as $\mathbf{v} = \langle \mathbf{v} \rangle + \mathbf{v}' = \mathbf{u}_d + \mathbf{v}'$. Then, we make use of the relations given in Problem 16.11, which the reader should derive. Substituting for the different terms we obtain the following:

$$
\frac{1}{2}\frac{\partial \rho_p \phi_d (|\mathbf{u}_d|^2 + 3\Theta)}{\partial t} + \frac{1}{2}\nabla \cdot \left(\rho_p \phi_d \mathbf{u}_d (|\mathbf{u}_d|^2 + 3\Theta) \right)
$$
$$
= n\mathbf{F} \cdot \mathbf{u}_d + \nabla \cdot \left(\phi_d \mathbf{u}_d \cdot (\boldsymbol{\tau}_{d,\mathrm{ki}} + \boldsymbol{\sigma}_{d,\mathrm{co}}) \right) - \nabla \cdot \left(\phi_d (\mathbf{q}_{d,\mathrm{ki}} + \mathbf{q}_{d,\mathrm{co}}) \right) + S\left[\frac{m_p}{2}|\mathbf{v}|^2\right].
$$
$$(16.89)$$

The above equation includes both the fluctuation kinetic energy (i.e., granular temperature) as well as the mean kinetic energy. By subtracting the latter we can get the final desired equation for the granular temperature. An equation for the mean kinetic energy is obtained in Problem 16.13. By subtracting Eq. (16.95) from the above equation of total kinetic energy, we obtain the following final expression for granular temperature:

Granular temperature equation

$$
\boxed{\frac{3}{2}\rho_p \phi_d \frac{d\Theta}{dt} = \phi_d (\boldsymbol{\tau}_{d,\mathrm{ki}} + \boldsymbol{\sigma}_{d,\mathrm{co}}) : \nabla \mathbf{u}_d - \nabla \cdot \left(\phi_d (\mathbf{q}_{d,\mathrm{ki}} + \mathbf{q}_{d,\mathrm{co}}) \right) - \Gamma.}
$$
$$(16.90)$$

We are now ready to interpret the different terms of the above granular temperature equation. The term on the left-hand side is the total derivative of granular temperature following the mean particle velocity $\mathbf{u}_d$ and in obtaining this term we have made use of the particle mass balance. The first term on the right-hand side corresponds to the production of granular temperature by mean shear, where both kinetic and collisional stresses participate. The second term on the right is the conduction of granular temperature from regions of large Θ. This term also receives contributions from kinetic and collisional mechanisms. The last term on the right (i.e., $\Gamma = -S\left[\frac{m_p}{2}|\mathbf{v}|^2\right]$) is dissipation of granular temperature due to inelastic collisions, and this loss of particle fluctuation kinetic energy appears as the source term in the particle temperature equation (16.6). The general expression for collisional source has been given in Eq. (16.62), where we can substitute $\psi = m_p|\mathbf{v}|^2/2$ to obtain an explicit expression for Γ. The resulting derivation is somewhat involved and we refer the reader to Jenkins and Savage (1983), Ding and Gidaspow (1990), or Gidaspow (1994). The final expression for inelastic collisional dissipation is

$$
\Gamma = 3\left(1 - \epsilon_c^2\right)\rho_p \phi_d^2 g_0 \Theta \left(\frac{4}{d_p}\left(\frac{\Theta}{\pi}\right)^{1/2} - \nabla \cdot \mathbf{u}_d \right),
$$
$$(16.91)$$

where ϵ_c is the coefficient of restitution. The granular temperature equation should be solved in addition to the particle-phase mass, momentum, and energy equations (16.4), (16.5), and (16.6).

The above granular temperature equation has been derived only accounting for interparticle collisions, without any consideration of particles' interactions with the surrounding fluid. Since the influence of the surrounding fluid was ignored, the force **F** acting on the particles was taken to be only a constant external force, such as gravity. With the inclusion of hydrodynamic force on the particles, **F** must be interpreted as the total force on the particles, which will not be a constant and therefore cannot be taken out of the ensemble average. This will result in additional terms to the granular temperature equation. On the one hand, external fluid turbulence will enhance the level of particle velocity fluctuation, while on the other, the drag force of the surrounding fluid will have the effect of damping particle velocity fluctuation.

Problem 16.11 Use the separation of particle velocity into the mean and fluctuating components, along with the definition of granular temperature and the properties of ensemble average over all possible particle velocities, to obtain the following two relations:

$$\frac{1}{2}\langle \mathbf{v} \cdot \mathbf{v} \rangle = \frac{1}{2}\mathbf{u}_d \cdot \mathbf{u}_d + \frac{3}{2}\Theta, \tag{16.92}$$

$$\frac{1}{2}\langle (\mathbf{v} \cdot \mathbf{v})\mathbf{v} \rangle = \mathbf{u}_d \left(\frac{1}{2}\mathbf{u}_d \cdot \mathbf{u}_d + \frac{3}{2}\Theta \right) - \frac{1}{\rho_p}\mathbf{u}_d \cdot \tau_{d,\text{ki}} + \frac{1}{\rho_p}\mathbf{q}_{d,\text{ki}}, \tag{16.93}$$

where kinetic stress was earlier defined as $\tau_{d,\text{ki}} = \rho_p \mathbf{u}_d \mathbf{u}_d - \rho_p \langle \mathbf{vv} \rangle = -\rho_p \langle \mathbf{v}'\mathbf{v}' \rangle$ and its closure was presented in Eq. (16.66). The kinetic energy flux is similarly defined as $\mathbf{q}_{d,\text{ki}} = \rho_p \langle (\mathbf{v}' \cdot \mathbf{v}')\mathbf{v}' \rangle$.

Problem 16.12 Show that the collision contribution can be separated into a mean portion and a contribution from the fluctuating component

$$\sigma \left[\frac{m_p}{2}|\mathbf{v}|^2 \right] = \underbrace{\sigma \left[\frac{m_p}{2}|\mathbf{v}'|^2 \right]}_{-\phi_d \mathbf{q}_{d,\text{co}}} + \mathbf{u}_d \cdot \sigma_{d,\text{co}}, \tag{16.94}$$

where $\mathbf{q}_{d,\text{co}}$ is the collisional flux of granular temperature.

Problem 16.13 By dotting the momentum equation (16.87) with the mean velocity $\mathbf{u}_d$, obtain the following expression:

$$\frac{1}{2}\frac{\partial \rho_p \phi_d |\mathbf{u}_d|^2}{\partial t} + \frac{1}{2}\nabla \cdot \left(\rho_p \phi_d \mathbf{u}_d |\mathbf{u}_d|^2 \right)$$
$$= n\mathbf{F} \cdot \mathbf{u}_d - \phi_d(\tau_{d,\text{co}} + \sigma_{d,\text{co}}) : \nabla \mathbf{u}_d + \nabla \cdot (\phi_d \mathbf{u}_d \cdot (\tau_{d,\text{co}} + \sigma_{d,\text{co}})). \tag{16.95}$$

Granular Thermal Conduction

The granular temperature equation (16.90) involved five secondary variables that must be modeled. They are $\tau_{d,ki}$, $\sigma_{d,co}$, Γ, $\mathbf{q}_{d,ki}$, and $\mathbf{q}_{d,co}$. Of these, we have already considered the modeling of the first three, with their closure equations given as Eqs. (16.66), (16.80), and (16.91). In this subsection, we will consider modeling of the kinetic and collision parts of granular conduction represented by $\mathbf{q}_{d,ki}$, and $\mathbf{q}_{d,co}$. We start with the kinetic component, which from its definition given in Problem 16.11 can be expressed as

$$\phi_d \mathbf{q}_{d,ki} = m_p \int f_1 \, (\mathbf{v}' \cdot \mathbf{v}')\mathbf{v}' dv' \, . \tag{16.96}$$

If we assume the one-particle distribution function f_1 to be given by a Maxwellian distribution, then the above integral is identically zero due to the odd power of $\mathbf{v}'$. A nonzero contribution of the kinetic term arises only at the next level, when the non-Maxwellian nature of the one-particle distribution function is taken into account. We refer the reader to Gidaspow (1994), where the following closure is presented:

$$\phi_d \mathbf{q}_{d,ki} = -\frac{75\sqrt{\pi}}{384}\rho_p d_p \Theta^{1/2}\nabla\Theta \, . \tag{16.97}$$

We now turn to the closure model of $\mathbf{q}_{d,co}$. From Eq. (16.95) we recognize this term to be the collisional stress term associated with $\psi = m_p|\mathbf{v}|^2/2$. From the definition given in Eq. (16.61), following the process that led to Eq. (16.74), we obtain the following expression:

$$\phi_d \mathbf{q}_{d,co}$$
$$= \underbrace{\frac{d_p^3}{4} g_0 \rho_p \mathcal{V}_p \iiint_{\mathbf{v}_{12} \cdot \mathbf{n}>0} \Delta \mathbf{v}_1^2(\mathbf{v}_{12} \cdot \mathbf{n})\mathbf{n} f_1(\mathbf{x},\mathbf{v}_1,t)f_1(\mathbf{x},\mathbf{v}_2,t) \, d\mathbf{n} \, d\mathbf{v}_1 \, d\mathbf{v}_2}_{\phi_d \mathbf{q}_{d1}}$$
$$+ \underbrace{\frac{d_p^4}{8} g_0 \rho_p \mathcal{V}_p \iiint_{\mathbf{v}_{12} \cdot \mathbf{n}>0} \Delta \mathbf{v}_1^2(\mathbf{v}_{12} \cdot \mathbf{n})\mathbf{n} f_1(\mathbf{x},\mathbf{v}_1,t)f_1(\mathbf{x},\mathbf{v}_2,t)\nabla \ln\left(\frac{f_1(\mathbf{x},\mathbf{v}_2,t)}{f_1(\mathbf{x},\mathbf{v}_1,t)}\right) \, d\mathbf{n} \, d\mathbf{v}_1 \, d\mathbf{v}_2}_{\phi_d \mathbf{q}_{d2}} \, . \tag{16.98}$$

For a Maxwellian distribution, $\mathbf{q}_{d1}$ is zero. Taking into account the first correction to such a distribution, the following expression can be obtained (Gidaspow, 1994):

$$\phi_d \mathbf{q}_{d1} = -\frac{15\sqrt{\pi}}{32}\rho_p \phi_d d_p \Theta^{1/2}\left(1 + \frac{6}{5}\phi_d g_0(1 + \epsilon_c)\right)\nabla\Theta \, . \tag{16.99}$$

The second integral can also be evaluated:

$$\phi_d \mathbf{q}_{d2} = -\frac{2}{\sqrt{\pi}}\rho_p \phi_d^2 d_p \Theta^{1/2} g_0(1 + \epsilon_c)\nabla\Theta \, . \tag{16.100}$$

This completes our brief discussion of granular kinetic theory. Before we leave the topic, it must be pointed out that the closure models presented here for collisional stress, heat flux, and dissipative heating are not unique. Alternate closure models have been derived using different assumptions (Lun et al., 1984; Andreotti et al., 2013). The reader is encouraged to consult other books on granular media for a deeper understanding.

We conclude this section by pointing out that the complete set of EE governing equations for the continuous phase are given as Eqs. (16.1) to (16.3) and the corresponding governing equations for the dispersed phase are given as Eqs. (16.4) to (16.6) supplemented with the granular temperature equation given as Eq. (16.90). These equations involve many secondary terms that must be closed or expressed in terms of the primary variables. These closure models were divided into those arising from turbulent fluctuations and those arising from collisional processes. The turbulence-related closure models were presented in Section 16.2 and collision-related closure models were systematically derived in Section 16.3. From this extensive list of closure terms it is clear that the complete EE formulation is quite complicated. In the following section, we will describe some traditional attempts at simplifying the EE approach. However, we caution that such simplifications are not always appropriate and there are many multiphase flow applications where the complete set of EE equations, as presented above, is needed.

16.4   Simplified Formulation – Mixture Approach

In this and the subsequent section we will consider different simplifications of the EE formulation that result in a considerably simpler set of governing equations. This simpler set of equations arose from judicious assumptions and in problems where these assumptions are valid, the advantage of having to deal only with a simpler set of governing equations can be substantial.

Consider the situation where the particles are very small in size compared to the Kolmogorov scale. As a result, let the particle Stokes number with respect to the entire range of turbulent scales be quite small. An immediate consequence of this small Stokes number approximation is that the particles' inertia with respect to the surrounding fluid can be ignored and they behave like tracer particles. Furthermore, let us consider the particles to be sufficiently small that their settling velocity is negligible compared to the surrounding fluid velocity, during the time period of interest. We now recall the equilibrium expansion for particle velocity in terms of the surrounding fluid velocity given in Eq. (12.27). With the above assumptions, we are interested in the limit of very small Stokes number and nondimensional settling velocity (i.e., St, $|\tilde{\mathbf{V}}_s| \to 0$). In this limit, the particle velocity is simply the local fluid velocity. The corresponding thermal Stokes number of the particle is also expected to be small, with the consequence that the particle temperature is the same as that of the surrounding fluid. Thus, we have the following two important simplifications:

$$\mathbf{u}_m = \mathbf{u} = \mathbf{v} \quad \text{and} \quad T_m = T_f = T_p, \tag{16.101}$$

where we have denoted the mixture velocity and temperature to be $\mathbf{u}_m$ and T_m. Additionally, from Problem 12.3 we note that the angular velocity of such particles can be taken to be the same as that of the surrounding fluid.

There are several applications where the above assumptions are appropriate. For example, consider the problem of a slurry flow through a pipe where the suspended particles are very fine and powder-like. There are other modern applications where nanoparticles are suspended in fluid to enhance the thermal properties of the mixture. Another class of applications is in the area of dusty gas pioneered by Saffman, Carrier and others (Carrier, 1958; Saffman, 1962; Marble, 1970). They considered the suspended dust particles to be so small that their relative motion with respect to the surrounding gas can be ignored and the mixture can be treated as a single fluid with appropriate local properties of the mixture. Therefore, the mixture approach to be discussed below is sometimes also referred to as the *dusty gas* approach. It is also sometimes referred to as the "one-fluid" or "single-fluid" formulation of the multiphase flow. The mixture framework has been used successfully in the past to study multiphase turbulent jets and plumes using the terminology of locally homogeneous flows (Faeth, 1983, 1987).

Although the material properties, such as density, viscosity, and thermal conductivity, of both the continuous and the dispersed phases are constants, the corresponding properties of the mixture will vary over space and time, due to the fact that the volume fraction of each phase will vary over space and time. The governing equations of the mixture will thus resemble those of a variable-property fluid. As an example, we first introduce the mixture density as

$$\rho_m(\mathbf{x}, t) = \rho_f \, \phi_c(\mathbf{x}, t) + \rho_p \, \phi_d(\mathbf{x}, t) \,. \tag{16.102}$$

Thus, the space-time dependence of ρ_m arises from the space-time dependence of the volume fraction variation.

The mass balance of the mixture can be obtained by summing Eqs. (16.1) and (16.4), where the mass-exchange term cancels between the phases. Similarly, the momentum and energy equations of the two phases are added together to obtain the mixture equations. The dusty gas assumption leads to the following, considerably simpler set of governing equations for the mixture:

$$\boxed{\frac{\partial \rho_m}{\partial t} + \mathbf{u}_m \cdot \nabla \rho_m = 0, \quad \nabla \cdot \mathbf{u}_m = 0,} \tag{16.103}$$

$$\boxed{\rho_m \frac{D\mathbf{u}_m}{Dt} = \rho_m \mathbf{g} + \nabla \cdot \boldsymbol{\sigma}_m,} \tag{16.104}$$

$$\boxed{\rho_m C_m \frac{DT_m}{Dt} = \boldsymbol{\sigma}_m : \nabla \mathbf{u}_m - \nabla \cdot \mathbf{q}_m.} \tag{16.105}$$

In the above, we have assumed the fluid and particle residual stresses to be small and neglected them. This is often a good assumption since the above mixture formulation requires the particles to be very small. The filter size can therefore be chosen to be much larger than the particle size, but smaller than the Kolmogorov scale. If the filter size is chosen to be larger than the Kolmogorov scale then the residual stress $\nabla \cdot \boldsymbol{\tau}_{\text{sg}}$

must be included in the momentum equation and the residual heat flux $\nabla \cdot \mathbf{q}_{sg}$ must be added to the temperature equation and these come from mesoscale turbulence. In Eq. (16.105), the specific heat of the mixture is given by $\rho_m C_m = \rho_f \, \phi_c C_{vf} + \rho_p \, \phi_d C_p$.

Note that when the fluid is incompressible, the mixture also remains incompressible, since the particles move with the fluid. The mass balance of the mixture takes the form $\nabla \cdot \mathbf{u}_m = -(D\rho_m/Dt)/\rho_m = 0$, as given in Eq. (16.103). In the above, $D/Dt = \partial/\partial t + \mathbf{u}_m \cdot \nabla$ is the total derivative following the mixture velocity. The momentum and energy equations can be closed with the following constitutive relations:

$$\sigma_m = -p_m \mathbf{I} + \mu_m \left(\nabla \mathbf{u}_m + (\nabla \mathbf{u}_m)^{\mathrm{T}} \right) \quad \text{and} \quad \mathbf{q}_m = -k_m \nabla T_m, \tag{16.106}$$

where μ_m and k_m are mixture dynamic viscosity and thermal conductivity, respectively, which are in general functions of particle volume fraction, and their expressions can be found in Eqs. (15.34) and (15.36). With these closure relations, we have a complete system of equations for the mixture that can be solved using conventional CFD codes.

We can now investigate which multiphase flow effects are included in this mixture approach. The volume effect becomes irrelevant since the particles are precisely moving lock-step with the surrounding fluid. Also, due to the assumption that the particle velocity and temperature are the same as those of the surrounding fluid, the slip effect is taken to be zero. The inertia effect is retained due to the appearance of ρ_m on the left-hand side of the momentum and energy equations. Similarly, the use of ρ_m in the buoyancy term accounts for the body-force effect of the suspended particulates. Finally, by modeling mixture viscosity and thermal conductivity to be dependent on the local particle volume fraction, we account for the thermodynamic/transport effect.

This formulation places no restriction on what the particle volume fraction can be. In other words, the above governing equations apply for very small particles even at finite values of particle volume fraction. That is the reason why inertia, body-force, and constitutive effects are retained in the governing equations. If the particle volume fraction is small, then depending on the problem, one or more of these effects can be turned off. For example, the inertial effect can be neglected by settling the mixture density to be simply ρ_f on the left-hand side of the above equations. On the other hand, the body-force effect can be ignored by setting the mixture density to be ρ_f in the buoyancy term. The constitutive effects can be ignored by taking the mixture viscosity and conductivity to be those of the fluid.

An important advantage of the mixture formulation is that it is not restricted to monodisperse systems. The above governing equations readily apply to a polydisperse system as well, provided all the particles are of sufficiently small size that their Stokes number is small. All the particles, irrespective of their size, follow the fluid in their velocity and temperature. As a result, particle size does not play a role. In calculating the particle volume fraction and the mixture density, all the different-sized particles are taken into account.

16.5   Simplified Formulation – Sedimentation Approximation

There are a number of environmental flows where suspended sediments play an important role. The suspended sediments often drive the flow and therefore must be studied as multiphase flow. These environmental flows present a very wide range of scales, with their length on the order of kilometers, depth on the order of meters, and Kolmogorov eddies on the order of millimeter or smaller. In comparison, the suspended sediment particles are typically a few hundred microns or less. Since the dispersed-phase is much smaller than the continuous-phase scale, the mixture formulation of the previous section applies. However, two modifications are warranted. First, the concentration or the volume fraction of suspended sediments is generally very low and thus the inertial and the thermodynamic/transport effects are not important. However, the body-force effect is of primary importance, since the excess weight of the suspended sediments drives the flow. Second, although the settling velocity of the suspended sediments is generally small compared to the flow velocity, on the global scale of the flow, sedimentation plays an important role in dictating the vertical variation of sediment concentration. Therefore, the velocity of the dispersed phase cannot be considered the same as the surrounding fluid. The effect of gravitational settling must be included. There are other engineering and technological applications that present a similar scenario. In these applications, the gravitational settling of particles and the buoyancy effect are of primary importance. In such applications, the governing equations to be obtained here are of relevance.

Consider the case of a sediment-laden river flow, where the sediment is quartz-based sand or silt particles of specific gravity 2.65. If the volume fraction of suspended sediment is 1%, then the mixture density is $\rho_m = 1016.5 \ \text{kg/m}^3$, which is only 1.65% larger than the fluid (water) density of $\rho_f = 1000 \ \text{kg/m}^3$. Thus, in terms of fluid inertia, the sediment-laden water is not much heavier than clear water. But the sustained action of the excess weight due to this small density difference is sufficient to generate intense flows. As an example, consider a lock separating a tank of well-mixed sediment-laden fluid from a tank of clear fluid. Let the fluid in both tanks be of height 1 m. If the lock is released, the heavier sediment-laden water will flow along the bottom into the clear-water tank. This is due to the difference in the hydrostatic pressure at the bottom of the sediment-laden tank compared to the clear-fluid tank. The pressure difference between the bottoms of the two tanks is $(\rho_m - \rho_f)gH$, where H is the height of the tank. The velocity of the resulting flow scales as $\sqrt{(\rho_m - \rho_f)gH/\rho_f}$, which can be estimated to be 0.4 m/s in the present example. It is clear that strong multiphase flows can be driven by the buoyancy effect of even modest density differences.

In the mass balance, if we set ρ_m to be a constant approximately equal to the fluid density, we arrive at the standard incompressibility condition. Similarly, on the left-hand side of the mixture momentum and energy equations we set $\rho_m = \rho_f$. We also set the mixture properties to be those of the fluid (i.e., $\mu_m = \mu_f$ and $k_m = k_f$). With these substitutions we can rewrite the governing equations as

$$\boxed{\nabla \cdot \mathbf{u}_c = 0,} \qquad (16.107)$$

$$\boxed{\rho_f \frac{D\mathbf{u}_c}{Dt} = \rho_m \mathbf{g} - \nabla p_c + \mu_f \nabla^2 \cdot \mathbf{u}_c,} \qquad (16.108)$$

$$\boxed{\rho_f C_{vf} \frac{DT_c}{Dt} = k_f \nabla^2 \cdot T_c,} \qquad (16.109)$$

$$\boxed{\frac{\partial \phi_d}{\partial t} + \nabla \cdot ((\mathbf{u}_c + V_s \mathbf{e}_g)\phi_d) = \nabla \cdot \left(D_\phi \nabla \phi_d\right),} \qquad (16.110)$$

where the mixture density ρ_m is given by Eq. (16.102), $\mathbf{e}_g$ is the gravitational direction, and D_ϕ is the diffusion coefficient of particles. In the energy equation, viscous and pressure heating terms are neglected as they are quite small in most applications. In fact, in the sediment-laden problem addressed above, the temperature equation decouples from the continuity and momentum equations and therefore is not of importance.

Three differences between the above equations and Eqs. (16.103) to (16.105) must be observed. First, in the above, we use the fluid velocity $\mathbf{u}_c$ instead of the mixture velocity, since the particle velocity is only slightly different from the fluid velocity due to settling. Second, as a consequence, in the particle volume fraction equation, the advection velocity of particles is taken to be a simple sum of the local fluid velocity plus the still-fluid particle settling velocity. The addition of settling velocity V_s in Eq. (16.110) has two important consequences – one physical and one technical. As a result of the settling process, the sediment volume fraction will tend to shift toward the bottom boundary. If not countered, all the particles will eventually deposit at the bottom boundary. At a particular elevation z within the domain, $V_s \overline{\phi_d}^{H}$ corresponds to downward flux of sediment due to settling, where the overbar with the superscript H corresponds to an average over the horizontal plane. $\overline{v_c' \phi_d'}^{H}$ corresponds to turbulent flux of particles across the plane, where v_c' and ϕ_d' correspond to vertical velocity and particle concentration fluctuation from their horizontal average. To achieve a steady state, the downward settling flux must be balanced by an upward turbulent flux of particles. In other words, since $V_s \overline{\phi_d}^{H}$ is guaranteed to point down, the turbulent flux $\overline{v_c' \phi_d'}^{H}$ must point up. This in turn means positive (and negative) vertical velocity fluctuations must be correlated with higher (and lower) than average sediment concentrations. So, sediment particles will naturally concentrate in regions of upwelling (positive vertical velocity fluctuation) to positively contribute to $\overline{v_c' \phi_d'}^{H}$. The nature of wall turbulence is such that a sediment volume fraction gradient will develop with higher sediment concentration closer to the bottom boundary and gradual reduction with vertical distance. This sediment concentration gradient will increase with increasing settling velocity V_s.

The second technical aspect of settling velocity is that it varies with sediment size. While the mixture equations (16.103) to (16.105) apply to polydisperse systems, the particle volume fraction equation (16.110) must be solved for each particle size separately, since their V_s will be different. Thus, in a polydisperse system, the entire size spectrum of particles must be divided into a number of bins, each characterized by its settling velocity V_s, and Eq. (16.110) must be solved for each size bin to calculate the volume fraction of all particles of that size class. If the particle size spectra is divided into N_B bins, then there are N_B volume fraction fields denoted by $\phi_{d,i}$ for $i = 1, \ldots, N_B$. Each must be obtained by solving Eq. (16.110) with the appropriate value of $V_{s,i}$. The total particle volume fraction is then given by $\phi_d = \sum_1^{N_B} \phi_{d,i}$, which is used to calculate the mixture density and applied in the buoyancy term.

The third difference is the addition of the diffusion term to the right-hand side of the sediment volume fraction equation, where the diffusion coefficient of particles D_ϕ accounts for the effect of subgrid velocity fluctuation that has not be accounted for in the particle velocity field $\mathbf{u} + V_s \mathbf{e}_g$. The need for this additional term can be motivated by considering the simple example of a spatially uniform flow in which particles are initially distributed only within a spherical volume and outside of this volume there are no particles. In other words, at $t = 0$, the volume fraction is non-zero within that spherical volume, and zero outside. For simplicity of discussion, we will ignore additional effects such as gravitational settling and thus the particles are traveling with the fluid. Even in the absence of any influence of external turbulence, the pseudo turbulence generated by the particles will influence each other's motion. The resulting particle-to-particle velocity fluctuation is the source of particle diffusion. In other words, as a result of particle velocity variation, particles will diffuse from regions of high to low concentration. Strictly speaking, this addition to the right-hand side of the particle volume fraction equation must be balanced by the counter-diffusion of fluid that flows into the high particle concentration region to replace the volume vacated by the particles that diffuse out. This corresponds to a term $\nabla \cdot \left(D_\phi \nabla \phi_c \right)$ that must be added to the right-hand side of the continuous-phase mass balance equation. In the above, we have ignored this term with the assumption that the volume fraction of particles is quite low everywhere that, to leading order, $\phi_c \approx 1$.

The curious reader may rightfully ask the question: why add this term to the simplified EE equations presented in this section in an *ad hoc* manner? Should this term be added to the right-hand side of the rigorously derived dispersed-phase mass balance equation (16.4)? If not, how is this effect of particles diffusing down the concentration gradient taken into account in Eq. (16.4)? Toward answering these questions, we first note that the added term can be rewritten as $-\nabla \cdot (\mathbf{u}_{dif} \phi_d)$, where the diffusion velocity is defined as $\mathbf{u}_{dif} = -D_\phi \nabla \phi_d / \phi_d$. This diffusion velocity can be added to the other two velocities and the second term on the left-hand side could have been written as $\nabla \cdot ((\mathbf{u}_c + V_s \mathbf{e}_g + \mathbf{u}_{dif}) \phi_d)$. In the present simplified EE formulation, since we assumed the particles to move with the fluid velocity $\mathbf{u}_c$ owing to their small inertia, we had to explicitly add the other two effects that are of importance, namely, the effect of sustained particle settling through $V_s \mathbf{e}_g$ and the effect of particles diffusing down the concentration gradient due to their fluid-mediated mutual interaction at the microscale through the diffusion velocity $\mathbf{u}_{dif}$. Note that the simplified EE formulation given in Eqs. (16.107) to (16.110) does not include a separate equation for the particle velocity field. As a result, we had to use the following effective model for particle velocity:

$$\mathbf{u}_d = \mathbf{u}_c + V_s \mathbf{e}_g - D_\phi \frac{\nabla \phi_d}{\phi_d} \,. \qquad (16.111)$$

The added diffusion term is not needed in a properly formulated complete set of EE equations since the diffusion contribution to particle velocity is naturally taken into account in the dispersed-phase momentum equation through the residual stress $\tau_{d,\mathrm{sg}}$ in Eq. (16.5) (recall that the residual stress also includes the kinetic stress $\tau_{d,\mathrm{ki}}$).

Despite its relative simplicity, the governing equations (16.107) to (16.110), even in the monodisperse particle limit, can give rise to very complex multiphase flows. Turbidity currents are great examples of such flows. Figure 16.3 shows three different turbidity currents where a sediment-laden current enters the domain on the left and is allowed to leave on the right. The current flows at the bottom of the domain submerged below a deep body of nearly stagnant fluid. The only difference between the three cases shown is the slope of the bottom boundary: 2.86° (Case 1), 0.72° (Case 2), and 0.29° (Case 3), respectively. These slopes give rise to vastly different turbulent flows that are characterized as: (i) supercritical, where the interface between the current and the ambient is vigorously turbulent (Salinas et al., 2021b); (ii) transcritical, where the interface between the current and the ambient alternates between quiescent and turbulent states, separated by internal hydraulic jumps; and (iii) subcritical, where the interface between the current and the ambient is laminar (Salinas et al., 2021a). Note that the results presented in Figure 16.3 are from highly resolved 3D turbulent simulations, whose 3D vortical structure for Case 1 is shown in Figure 16.4. A very complex pattern of three families of vortex structures can be seen (see Salinas et al., 2021b for further details). The 2D views shown in Figure 16.3 are span-averaged sediment concentration (upper frame) and gradient Richardson number (bottom frame) in each case (also see Salinas et al. (2018, 2019a,b, 2022); Zúñiga et al. (2022); Salinas et al. (2023)).

It is important to emphasize that these are truly multiphase flows, since without the suspended sediments there will be no flow. Furthermore, the relation between sediments and turbulence in these flows is very complex. Turbulence, at least near the bottom boundary, is essential to keep the sediments in suspension – or else the sediment will settle to the bottom and the flow will cease to exist. Turbulence is needed to keep the sediment in suspension, and suspended sediment is needed to sustain turbulence – this is the symbiotic relation between the two. On the other hand, sediments self-stratify due to their settling and create stable vertical density stratification; the effect of such stable stratification is to damp turbulence (Shringarpure et al., 2012). This delicate balance between turbulence production and turbulence damping is the source of strong difference in the behavior of subcritical, supercritical, and transcritical currents.

Before we leave this topic, an important turbulence modulating effect of particles, which has often been ignored, must be emphasized. In multiphase flows, the question of turbulence modulation can be studied at three different levels: (i) at the microscale of individual particles, their unsteady wakes at higher particle Reynolds numbers can contribute to turbulence augmentation, while wake dissipation can contribute to turbulence damping; (ii) at the mesoscale of collections of particles, instabilities can be amplified or damped, thereby promoting or suppressing turbulence; (iii) at the

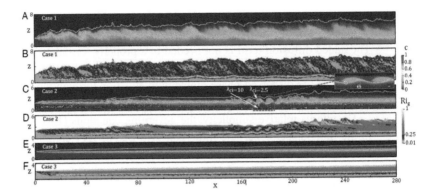

Figure 16.3 (A, C, and E) Spanwise averaged concentration field at one instant in time in the statistically steady-state regime for (A) supercritical, (C) transcritical, and (E) subcritical currents. White contours correspond to sediment volume fraction of 1%. (B, D, and F) Spanwise averaged gradient Richardson number at one instant in time in the statistically steady-state regime for the supercritical, transcritical, and subcritical cases. The wall-normal axis is stretched for better visualization. Taken from Salinas et al. (2020).

macroscale, as seen in the turbidity current example, system-level stratification can greatly modify the nature of turbulence. This large-scale effect of particle–turbulence interaction is not limited to geophysical flows. For example, consider a turbulent bubbly flow in a horizontal pipe. Due to buoyancy there will be a sustained upward flux of bubbles to the top of the pipe, which will be countered by a turbulent flux that tends to move the bubbles toward the bottom. The resulting density stratification, even at small bubble volume fractions, will have a profound effect on turbulence distribution within the pipe. This large-scale effect of turbulence modulation by the dispersed phase is far less studied and understood than the micro and mesoscale modulation of turbulence by particles.

16.6 Ensemble-Averaged Equations

In this section, we will consider the ensemble average as an important approach in the understanding and analysis of multiphase flows. The ensemble average implies an average over an ensemble of realizations. That is, the multiphase problem must be repeated countless number of times, together forming the ensemble of realizations. Due to the turbulent nature of the flow and the random distribution and motion of the particles, no two realization will be identical, although all the realizations are statistically the same. An average over all the realizations of the ensemble yields the ensemble average. Since infinite repetitions of the multiphase flow are clearly out of the question, the ensemble average is typically approximated by an average over a finite number of realizations. In a complex turbulent multiphase flow, even repeating the experiment a few times can be a daunting task. If the turbulence being considered is statistically stationary, then an average over time can be used to approximate the ensemble average. If the turbulence is

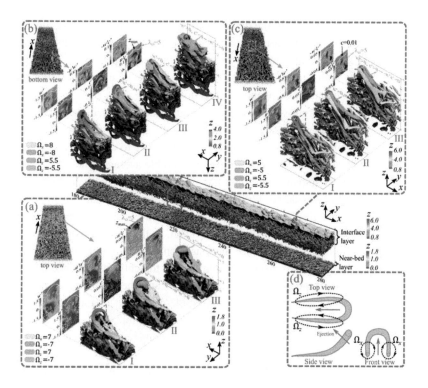

Figure 16.4 Isosurface of swirling strength colored by the bed-normal location, together with an isosurface of ϕ_d for half the domain. Three/four close-up views of an example hairpin structure within the (a) near-bed, (b) lower-interface, and (c) upper-interface regions are shown. In frames (a)–(c) we also show the contours of velocity and concentration perturbations and cross-correlations on a $y-z$ plane going through the middle of the hairpin (dashed red planes), together with contours of constant swirling strength. Also, we show isosurfaces of $-u'w'$ (light blue), $w'\phi'_d$ (light violet and brown), bed-normal vorticity (light green and cyan), and streamwise vorticity (light yellow and pink). (d) Schematic of a hairpin vortex. Reprinted from Salinas et al. (2021b). Copyright 2021 by the American Physical Society.

nearly homogeneous along one or more directions, then again the averages along those directions can be used in place of the ensemble average. From the ensemble average, one can obtain a wealth of statistical information about the multiphase flow, including the mean, standard deviation, two-point correlation, two-time correlation, etc.

The approach to be pursued in this section must be clearly contrasted from the approaches that we have pursued up until now in the past few chapters and sections. In the particle-resolved approach, each simulation only considers an individual realization of the multiphase flow. The same applies for the EL approach as well, since individual particles are being tracked within the flow. The EE approaches discussed in the previous sections also pertain only to individual realizations. Although the filtering operation filters out all details smaller than the filter length scale and erases the Lagrangian nature of the particles, the resulting filtered continuous and particulate fields are those of an individual realization. Thus, the results of a PR, EL, or EE simulation provides deeper insight into the dynamics of an individual realization, albeit at differing levels of resolution. Ensemble-averaged statistical information can be obtained from the PR, EL, and EE approaches by repeating the simulation many times and averaging over these realizations, averaging over time in case of statistical stationarity, and averaging over spatially homogeneous directions.

The alternative is to obtain the ensemble-averaged statistical information directly by solving a set of ensemble-averaged equations that govern the dynamics of the ensemble-averaged quantities, without having to calculate the dynamics of individual realizations and then perform the ensemble average. Such direct access to ensemble-averaged statistics certainly has its advantages. However, two important limitations must be considered. First, the ensemble-averaged equations involve secondary terms, similar to those encountered in the EL and EE approaches, whose reliable modeling still remains a major challenge. In fact, in the EL and EE approaches, only the effect of a portion of turbulence that has been filtered out needs to be modeled, whereas in the ensemble-averaged equations, since the entire turbulence has been averaged out, the importance of the secondary terms and the challenge of modeling them greatly increases. Second, from the ensemble average it is not possible to deterministically predict the dynamics of an individual realization.

As we will see in this section, the ensemble-averaged equations to be derived below will appear to be similar in form to the EE equations of the previous sections. In the filtered EE equations, only the small-scale dynamics has been filtered out, whereas in the ensemble-averaged equations the entire range of turbulence and the random distribution of particles have been averaged out. As a result, the ensemble-averaged flow is likely to be far smoother than those obtained in an EE simulation. Hence we reserve the term *two-fluid formulation* for the ensemble-averaged equations.

The above discussions can be summarized with an analogy to the single-phase turbulence simulation approaches. Particle-resolved simulations are equivalent to direct numerical simulations (DNS), while EL and EE approaches correspond to large eddy simulation (LES). The ensemble-averaged multiphase flow equations to be discussed can be compared to the Reynolds-averaged Navier–Stokes (RANS) approach.

16.6.1   Governing Equation

Ensemble-averaged governing equations for multiphase flow have a rich history starting from the mixture theories of multiphase flow as an application of rational mechanics (Drew, 1983; Gidaspow, 1994; Drew and Passman, 2006). The reader should also consider the works of Zhang and Prosperetti (1994, 1997). The derivation to be presented below will closely follow the elegant approach of Joseph et al. (1990). The notation will, however, closely follow that employed in this book.

We start with the indicator function I_c defined in Eq. (11.1) at the beginning of the filter operation, which is unity in regions occupied by the fluid and zero in regions occupied by the particles. We also introduce the notation $\langle \cdot \rangle$ to denote the ensemble average, which corresponds to an average over all the realizations that make up the ensemble. The process of deriving the governing equations using the ensemble average is similar to that followed in Chapter 11 for the filtered governing equation. Analogous to Eqs. (11.17) and (11.104), ensemble averages of the fluid and solid indicator functions yield the ensemble-averaged definitions of fluid and particle volume fraction as

$$\langle \phi_c \rangle = \langle I_c \rangle \quad \text{and} \quad \langle \phi_d \rangle = \langle (1 - I_c) \rangle , \tag{16.112}$$

where the angle brackets in the volume fraction on the left-hand side are simply a notation to distinguish them from the volume fractions ϕ_c and ϕ_d defined with the spatial filter. The difference between the ensemble average and the spatial filter operations can be explored with the particle volume fraction as an example. In case of $\langle \phi_d \rangle(\mathbf{x}, t)$, this quantity is obtained by averaging over an ensemble of realizations. Therefore, $\langle \phi_d \rangle(\mathbf{x}, t)$ indicates the fraction of realizations where the position $\mathbf{x}$ at time t is occupied by a particle, whereas the quantity $\phi_d(\mathbf{x}, t)$ corresponds to the fractional volume of space around the point $\mathbf{x}$ at time t that is occupied by particles, in one particular realization. While $\phi_d(\mathbf{x}, t)$ pertains to a single realization, $\langle \phi_d \rangle(\mathbf{x}, t)$ is an ensemble-averaged quantity. Clearly these two quantities are not the same.

The ensemble-averaged fluid and particle velocity fields are then defined as

$$\langle \mathbf{u}_c \rangle = \frac{\langle I_c \mathbf{u} \rangle}{\langle \phi_c \rangle} \quad \text{and} \quad \langle \mathbf{u}_d \rangle = \frac{\langle (1 - I_c) \mathbf{v} \rangle}{\langle \phi_d \rangle} , \tag{16.113}$$

where $\mathbf{v}$ was defined in Eq. (11.107) as the field representation of particle velocity. Other ensemble-averaged fluid and particle quantities are defined similarly. In the above, and the discussion to follow, we assume the fluid density ρ_f and particle density ρ_p to be constants. The mass balance equations are obtained by observing that the fluid and particle indicator functions are material quantities (i.e., the values of these functions do not change following the material – clearly a particle remains a particle when followed). Mathematically, this property translates to the material (or total) derivative of the indicator function being zero, as written below:

$$\frac{\partial I_c}{\partial t} + \mathbf{u} \cdot \nabla I_c = 0 \quad \text{and} \quad \frac{\partial (1 - I_c)}{\partial t} + \mathbf{v} \cdot \nabla (1 - I_c) = 0. \tag{16.114}$$

We then use the fact that the fluid and particle velocities are incompressible. In particular, $\nabla \cdot \mathbf{u} = 0$ can be combined with the second term of the first equation to obtain $\partial I_c / \partial t + \nabla \cdot (I_c \mathbf{u}) = 0$. Now taking the ensemble average we obtain

$$\frac{\partial \langle I_c \rangle}{\partial t} + \nabla \cdot \langle I_c \mathbf{u} \rangle = 0 \quad \Rightarrow \quad \frac{\partial \langle \phi_c \rangle}{\partial t} + \nabla \cdot (\langle \phi_c \rangle \langle \mathbf{u}_c \rangle) = 0 . \tag{16.115}$$

The important difference between the ensemble average and the spatial filter operation is that the ensemble average, which only involves averaging over an ensemble realization, commutes with both the time and space derivatives. In other words, Gauss and Leibniz rules are not needed. This property greatly simplifies the derivations to be presented in this section. The same steps as above can be followed to obtain

$$\frac{\partial \langle \phi_d \rangle}{\partial t} + \nabla \cdot (\langle \phi_d \rangle \langle \mathbf{u}_d \rangle) = 0 . \tag{16.116}$$

The above two ensemble-averaged mass balance equations are of the same form as the mass balance equations (16.1) and (16.4) obtained with the filter approach, except for the fact that the above equations are in the absence of mass exchange between the phases. Additionally, the above equations describe the mass conservation of the ensemble-averaged flow, while the filtered equations describe the mass balance of an individual realization at scales larger than the filter length scale.

The ensemble-averaged continuous-phase momentum equation is obtained by multiplying the Navier–Stokes equation by I_c and taking the ensemble average. This leads to the following expression:

$$\rho_f \left(\frac{\partial \langle I_c \mathbf{u} \rangle}{\partial t} + \nabla \cdot \langle I_c \mathbf{uu} \rangle \right) = \langle I_c \nabla \cdot \boldsymbol{\sigma} \rangle + \rho_f \langle I_c \mathbf{g} \rangle, \tag{16.117}$$

where to obtain the left-hand side we have added $\mathbf{u}(\partial I_c / \partial t + \nabla \cdot (I_c \mathbf{u})) = 0$ to the Navier–Stokes equation before taking the ensemble average. The stress term on the right-hand side can be written as

$$\langle I_c \nabla \cdot \boldsymbol{\sigma} \rangle = \nabla \cdot \langle I_c \boldsymbol{\sigma} \rangle - \langle \nabla I_c \cdot \boldsymbol{\sigma} \rangle = \nabla \cdot (\langle \phi_c \rangle \langle \boldsymbol{\sigma} \rangle) - \langle \mathbf{F} \rangle_{\text{hyd}}, \tag{16.118}$$

where in deriving the second equality we have used the fact that ∇I_c is nonzero only at the interface between the particles and the surrounding fluid and on the particle surface it is equal to the outward unit normal vector $\mathbf{n}$. We then recognize $\mathbf{n} \cdot \boldsymbol{\sigma} = \mathbf{t}$ as the tractional force. Therefore, $\langle \mathbf{F} \rangle_{\text{hyd}}(\mathbf{x}, t) = \langle \mathbf{t} \rangle$ is the ensemble average of the local hydrodynamic force applied on the particle by the fluid over all realizations in which the particle–fluid interface lies at the point $\mathbf{x}$ at time t. This ensemble-averaged quantity is different from $\mathbf{F}'_{\text{hyd}}$, which corresponds to the volume average of the hydrodynamic force on all the particles located around the point $\mathbf{x}$ at time t, weighted by the Gaussian filter, in one realization. Furthermore, the perturbation force $\mathbf{F}'_{\text{hyd}}$ does not account for the undisturbed flow force. Thus, $\langle \mathbf{F} \rangle_{\text{hyd}}(\mathbf{x}, t)$ is different from $\mathbf{F}'_{\text{hyd}}$, although the two are related.

We define the ensemble-averaged Reynolds stress as

$$\nabla \cdot (\langle \phi_c \rangle \langle \boldsymbol{\tau} \rangle_{\text{sg}}) = \rho_f \nabla \cdot (\langle \phi_c \rangle \langle \mathbf{u}_c \rangle \langle \mathbf{u}_c \rangle) - \rho_f \nabla \cdot \langle I_c \mathbf{uu} \rangle. \tag{16.119}$$

Substituting these into Eq. (16.117), we obtain the final continuous-phase ensemble-averaged momentum equation as

$$\rho_f \left(\frac{\partial \langle \phi_c \rangle \langle \mathbf{u}_c \rangle}{\partial t} + \nabla \cdot (\langle \phi_c \rangle \langle \mathbf{u}_c \rangle \langle \mathbf{u}_c \rangle) \right) - \nabla \cdot (\langle \phi_c \rangle \langle \boldsymbol{\tau} \rangle_{\text{sg}})$$
$$= \rho_f \langle \phi_c \rangle \mathbf{g} + \nabla \cdot (\langle \phi_c \rangle \langle \boldsymbol{\sigma} \rangle) - \langle \mathbf{F} \rangle_{\text{hyd}}. \tag{16.120}$$

The above assumes zero mass transfer between the phases and therefore there is no momentum exchange associated with the mass transfer. A similar derivation for the particulate phase will lead to the following particulate phase momentum equation (see Joseph et al., 1990):

$$\rho_p \left(\frac{\partial \langle \phi_d \rangle \langle \mathbf{u}_d \rangle}{\partial t} + \nabla \cdot (\langle \phi_d \rangle \langle \mathbf{u}_d \rangle \langle \mathbf{u}_d \rangle) \right) - \nabla \cdot (\langle \phi_d \rangle \langle \boldsymbol{\tau} \rangle_{d,\text{sg}})$$
$$= \nabla \cdot (\langle \phi_d \rangle \langle \boldsymbol{\sigma}_d \rangle) + \langle \mathbf{F} \rangle_{\text{hyd}} + \rho_p \langle \phi_d \rangle \mathbf{g}, \tag{16.121}$$

where the particle Reynolds stress is defined as

$$\nabla \cdot (\langle \phi_d \rangle \langle \boldsymbol{\tau} \rangle_{d,\text{sg}}) = \rho_f \nabla \cdot (\langle \phi_d \rangle \langle \mathbf{u}_d \rangle \langle \mathbf{u}_d \rangle) - \rho_p \nabla \cdot \langle (1 - I_c) \mathbf{v v} \rangle. \qquad (16.122)$$

Problem 16.14 *An expression for ensemble-averaged macroscale strain rate.* In this problem you should follow the steps of Problem 11.15 and derive the following result on the ensemble-averaged strain rate (also refer to Joseph et al., 1990):

$$\overline{\frac{1}{2} \langle \nabla \mathbf{u} + (\nabla \mathbf{u})^{\mathrm{T}} \rangle} = \frac{1}{2} \left(\nabla \langle \mathbf{u}_m \rangle + (\nabla \langle \mathbf{u}_m \rangle)^{\mathrm{T}} \right). \qquad (16.123)$$

We have thus obtained the fundamental result that the ensemble-averaged strain rate of the continuous phase is the strain rate of the ensemble-averaged mixture velocity.

Problem 16.15 *Ensemble-averaged energy equations.* In this problem you will first derive the ensemble-averaged fluid temperature equation. Start with the following (unfiltered/unaveraged) thermal equation for the fluid that applies in the region outside the particles:

$$\rho_f C_{\text{vf}} \left(\frac{\partial T_f}{\partial t} + \mathbf{u} \cdot \nabla T_f \right) = \nabla \cdot \mathbf{q}, \qquad (16.124)$$

where we have neglected viscous heating and compressional heating/cooling terms as they are generally small in incompressible flows. Multiply the above equation by I_f and take the ensemble average.

(a) Using the first of the identities given in Eq. (16.114), obtain the following equation:

$$\rho_f C_{\text{vf}} \left(\frac{\partial \langle \phi_c \rangle \langle T_c \rangle}{\partial t} + \nabla \cdot (\langle \phi_c \rangle \langle T_c \rangle \langle \mathbf{u}_c \rangle) \right) + \nabla \cdot (\langle \phi_c \rangle \langle Q \rangle_{\text{sg}}) = \nabla \cdot (\langle \phi_c \rangle \langle \mathbf{q} \rangle) - \langle Q \rangle_{\text{hyd}}, \qquad (16.125)$$

where the Reynolds flux term is given by $\nabla \cdot (\langle \phi_c \rangle \langle Q \rangle_{\text{sg}}) = \rho_f C_{\text{vf}} \nabla \cdot \langle I_c T_f \mathbf{u} \rangle - \rho_f C_{\text{vf}} \nabla \cdot (\langle \phi_c \rangle \langle T_c \rangle \langle \mathbf{u}_c \rangle)$. In obtaining the right-hand side, you need to use the following relation, which is similar to that given in Eq. (16.118):

$$\langle I_c \nabla \cdot \mathbf{q} \rangle = \nabla \cdot \langle I_c \mathbf{q} \rangle - \langle \nabla I_c \cdot \mathbf{q} \rangle = \nabla \cdot (\langle \phi_c \rangle \langle \mathbf{q} \rangle) - \langle Q \rangle_{\text{hyd}}, \qquad (16.126)$$

where we have used the fact that ∇I_c is zero if the point $\mathbf{x}$ at time t is fully inside the fluid or the particle. Only in realizations where the particle–fluid interface is at the point $\mathbf{x}$ at time t, is ∇I_c equal to the outward normal pointing into the fluid. Thus, $\langle \nabla I_c \cdot \mathbf{q} \rangle$ corresponds to the ensemble average of heat flux to the particle over all instances where the particle–fluid interface happens to be located at $\mathbf{x}$. As discussed in the context of interphase force coupling, Q'_{hyd} and $\langle Q \rangle_{\text{hyd}}$ are related quantities, but are distinctly different as their definitions are not the same.

(b) Following a similar process, derive the following ensemble-averaged particle temperature equation:

$$\rho_p C_p \left(\frac{\partial \langle \phi_d \rangle \langle T \rangle_d}{\partial t} + \nabla \cdot (\langle \phi_d \rangle \langle T_d \rangle \langle \mathbf{u}_d \rangle) \right) + \nabla \cdot (\langle \phi_d \rangle \langle Q \rangle_{d,\mathrm{sg}}) = \nabla \cdot (\langle \phi_d \rangle \langle \mathbf{q}_d \rangle) + \langle Q \rangle_{\mathrm{hyd}} \,. \tag{16.127}$$

16.6.2   Some Examples

We now consider a simple particle-laden flow problem whose results from ensemble-averaged simulations will be compared against the results of the corresponding EE simulations. This comparison on the one hand will illustrate the computational simplicity of the ensemble-averaged approach, while on the other it will demonstrate the difficulty of arriving at accurate closure models of the ensemble-averaged equations. The problem to be considered is gravity-driven particle-laden flow within an inclined channel of slope θ. The schematic of a possible experimental setup is shown in Figure 16.5. The flow within the channel is driven by the excess density of the water–sediment mixture over that of clear water. We are interested in studying the nature of the flow sufficiently downstream of the entrance where the flow can be considered fully developed. The flow represents an idealized case of an environmental turbidity current in which ambient water entrainment is not allowed with the presence of a top boundary.

Let us consider a length L of the channel in the fully developed regime. Let the height of the channel be $2H$ and the width be W. The excess weight of the suspended particles within this volume along the flow direction is given by $(\rho_p - \rho_f)(2HLW)\phi_{d0} \, g \sin(\theta)$, where ϕ_{d0} is the average particle volume fraction over the entire channel. This streamwise force is counterbalanced by friction along the walls whose magnitude is given by $(2LW)\rho_f u_*^2$, where $2LW$ is the area of the top and bottom walls of the channel and u_* is the friction velocity. Balancing these two forces, we can estimate the friction velocity to be

$$u_* = \sqrt{\frac{\rho_p - \rho_f}{\rho_f} g H \phi_{d0} \sin(\theta)} \,. \tag{16.128}$$

The maximum velocity of the particle-laden fluid within the channel is typically an order of magnitude larger than u_*.

The suspended particles will be considered to be much smaller than the Kolmogorov scale of turbulence. Furthermore, the volume fraction of suspended particles ϕ_{d0} will be considered to be quite small and the flow is considered to be quite dilute. With these considerations, the volume, inertia, slip, and thermodynamic/transport effects can be considered negligible. The only multiphase mechanism of importance is the buoyancy effect. The settling effect of particles, although small, plays an important role in establishing the vertical particle concentration gradient. The resulting stable stratification has a strong damping influence on turbulence. Based on these assumptions, we observe that the simpler EE governing equations (16.107), (16.108), and (16.110) are well suited for this problem. The temperature equation is not important in this problem. The two nondimensional parameters of importance in this flow are the shear

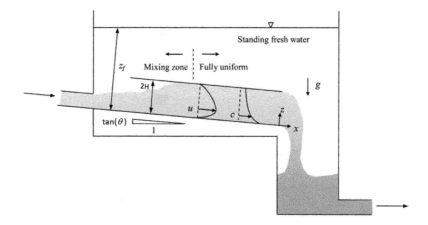

Figure 16.5 Schematic of the fully developed particle-laden flow in a channel that is inclined at an angle θ. Taken with permission from Yeh et al. (2013).

Reynolds number, defined as $\mathrm{Re}_* = u_* H / \nu$, and the nondimensional settling velocity of the sediment, V_s / u_*. The Reynolds number dictates the intensity of turbulence and controls the size of the smallest eddies compared to the channel half-height H. The nondimensional settling velocity controls the stratification effect.

EE-DNS simulations of this flow at a Reynolds number of $\mathrm{Re}_* = 180$ were considered by Cantero et al. (2009). At this modest value of the Reynolds number, the turbulence cascade was fully resolved. The particles are much smaller than the grid and therefore they were neither tracked individually nor the flow around them resolved. Hence, these simulations qualify as EE-DNS and each simulation captured the time evolution of one realization of the particle-laden turbulent flow within the channel. In the limit of zero particle settling velocity, the particles are uniformly distributed within the channel and the flow is classical turbulent channel flow. As the settling velocity increases, the concentration of particles is large near the bottom wall and decreases with vertical distance. Thus, the streamwise driving is skewed toward the bottom, and stable stratification tends to dampen turbulence. The most striking aspect of this flow, discovered by Cantero et al. (2009, 2012), is that the flow completely laminarizes when V_s / u_* increases above a certain threshold that depends on Re_*.

The ensemble-averaged flow of the same problem was considered by Yeh et al. (2013), and in particular they focused on a rigorous comparison of the results against those obtained with EE-DNS. In the fully developed regime, the turbulent flow is homogeneous along the streamwise x and spanwise z directions. Furthermore, the flow is statistically stationary. Thus, the ensemble-averaged velocity and particle volume fraction fields are only functions of the wall-normal y-coordinate. In addition, ensemble-averaged spanwise and wall-normal velocities are identically zero. These symmetries greatly simplify the ensemble-averaged equations. For example, the streamwise velocity is only a function of y (i.e., $\langle u_c \rangle (y)$). Also, $\langle v_c \rangle = \langle w_c \rangle = 0$ and together they automatically satisfy the continuity equation. In the ensemble-averaged equations, only the body-force or buoyancy effect is retained and all other multiphase mechanisms (volume, inertia, slip, and thermodynamic/transport effects) are ignored. With this, it is assumed that $\phi_c \approx 1$, consistent with the low particle volume fraction assumption. Furthermore, the particle momentum equation is not solved separately and the particle velocity field is taken to be given by Eq. (16.111). With these approximations, the x-component of the momentum equation (16.120) becomes

$$0 = \frac{d\tau_{xy,\text{sg}}}{dy} + \mu_f \frac{d^2 \langle u_c \rangle}{dy^2} + (\rho_p - \rho_f)g\langle \phi_d \rangle, \qquad (16.129)$$

where $\tau_{xy,\text{sg}}$ is the nonzero Reynolds stress component. The only other equation that is needed is the particle volume fraction equation (16.116). In the present problem, with the particle velocity field given by Eq. (16.111), we obtain

$$-V_s \frac{d\langle \phi_d \rangle}{dy} + \frac{df_{d,\text{sg}}}{dy} - \frac{d}{dy}\left(D_\phi \frac{d\langle \phi_d \rangle}{dy}\right) = 0, \qquad (16.130)$$

where $f_{d,\text{sg}} = \langle v_c \rangle_d \langle \phi_d \rangle$ is the volume-fraction-weighted, ensemble-averaged vertical flux of particles due to turbulence. In this, $\langle v_c \rangle_d$ is the average y-component of fluid velocity as seen by the particles. Note that even though the ensemble-averaged fluid velocity is zero, the same is not true for the particle (i.e., $\langle v_c \rangle \approx 0$, but $\langle v_c \rangle_d = 0$). This is due to the positive correlation between particle concentration and upward fluid velocity implied in the fluid velocity averaged at the particle locations. In a more traditional notation, $f_{d,\text{sg}}$ would have been denoted as Reynolds particle flux $\overline{v'_c \phi'_d}^H$ (see Yeh et al., 2013).

The above two equations are for the primary variables $\langle u_c \rangle(y)$ and $\langle \phi_d \rangle(y)$. Thus, $\tau_{xy,\text{sg}}$ and $f_{d,\text{sg}}$ must be interpreted as secondary variables that must be modeled in terms of the primary variables. The accuracy of the solution to the ensemble-averaged equations depends on how good the closure models are. The advantage of the ensemble-averaged governing equations over those of the EE-DNS approach given in Eqs. (16.107) to (16.110) is their simplicity. The ensemble-averaged approach requires only the solution of two ODEs. In fact, they can be solved sequentially in an uncoupled manner – the second equation can be solved first to obtain $\langle \phi_d \rangle$, which can then be used to obtain $\langle u_c \rangle$. In contrast, the EE-DNS approach requires the solution of five coupled time-dependent 3D PDEs. Since all the scales of turbulent motion are resolved, these simulations were computationally demanding. But the price to be paid in the ensemble-averaged approach is quite steep. Its accuracy is often limited because of our inability to model the secondary variables in a universal manner.

The results of three different cases taken from Yeh et al. (2013) will be presented below: Case 1, $\text{Re}_* = 180$ and $V_s/u_* = 0.005$; Case 2, $\text{Re}_* = 180$ and $V_s/u_* = 0.175$; and Case 3, $\text{Re}_* = 180$ and $V_s/u_* = 0.2125$. Thus, the effect of settling velocity increases from Case 1 to Case 3. In the case of the ensemble-averaged approach, three different Reynolds-averaged (RANS) models were considered: Mellor–Yamada, standard $k - \epsilon$, and quasi-equilibrium $k - \epsilon$. These are well-established RANS models and we will not present them in any detail. The reader is referred to Yeh et al. (2013) and the sources cited therein. The RANS prediction, along with the ensemble average of the EE-DNS results for the fluid velocity, is shown in Figure 16.6 for the three different cases. It should be pointed out that all three models are well calibrated to perform well for a turbulent channel flow, which will be the limit of $V_s/u_* = 0$. The results of Case 1 with weak settling velocity show that the RANS results are in good agreement with the EE-DNS results. With increasing V_s/u_*, the differences between the three models increase and they do not agree well with the behavior predicted by the EE-DNS simulation. In particular, as the particles settle toward the bottom boundary, the driving force increases more in the lower half of the channel than in the upper half, and this leads the peak of the velocity profile to shift toward the bottom boundary. The ensemble-averaged simulations do not capture this trend accurately for the highest settling velocity considered.

The corresponding results for $\langle \phi_d \rangle$ normalized by the average over the entire channel ϕ_{d0} are shown in Figure 16.7. In the limit $V_s/u_* = 0$, the normalized particle volume fraction is a constant equal to unity everywhere. For Case 1 with weak settling velocity, away from the boundaries, $\langle \phi_d \rangle$ follows a linear profile that is well predicted by the ensemble-averaged formulation. Again, with increasing settling velocity, the ensemble-averaged approach without accurate closures is unable to recover the correct ensemble average. This point was discussed at greater length by Yeh et al. (2013). In fact, they point out that with further increase in settling velocity, the EE-DNS simulation shows a total suppression of turbulence (Cantero et al., 2009), whereas this important physics cannot be predicted by the three RANS models. It is interesting to note that the closure models that performed well at smaller settling velocity are not suitable for capturing the subgrid physics at higher settling velocity. This lack of universality in the closure of the secondary terms has been one of the biggest challenges.

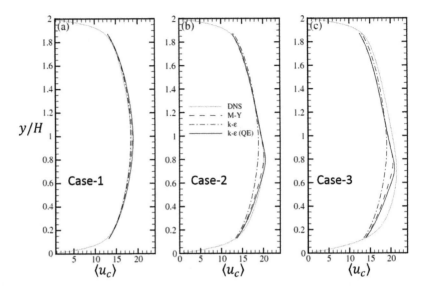

Figure 16.6 Comparison of the vertical profile of the ensemble-averaged streamwise velocity obtained from the EE-DNS simulation with those obtained from ensemble-averaged simulation using three different RANS models. Taken with permission from Yeh et al. (2013).

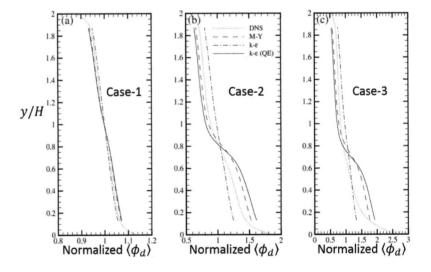

Figure 16.7 Comparison of the vertical profile of the ensemble-averaged normalized particle volume fraction obtained from the EE-DNS simulation with those obtained from ensemble-averaged simulation using three different RANS models. Taken with permission from Yeh et al. (2013).

Appendix A Index Notation

In this book we use the standard notation that a bold variable denotes a vector. Some examples are the fluid velocity $\mathbf{u}(\mathbf{x}, y)$ or the particle velocity $\mathbf{V}(t)$. The fluid velocity is a vector field (i.e., it is dependent on the space variable $\mathbf{x}$), whereas the Lagrangian particle velocity is just a time-dependent velocity vector. In index notation they will be denoted as $u_i(\mathbf{x}, t)$ and V_i, where the index $i = 1, 2, 3$.

In this appendix, we will very briefly go over some of the index notation definitions and identities that will be used in Chapter 4 and in the rest of the book. We caution that the treatment is neither extensive nor rigorous. We are assuming that you already know much of this material from your graduate-level fluid mechanics course. This appendix is only to serve as a quick refresher. If it is not familiar, you must take an applied mathematics course or read a book on mathematics for physicists and engineers.

There are some simple standard rules of index notation that we will follow: (i) any index that appears only once is called a *free index*; (ii) if an index repeats twice in a term then it implies a sum over 1, 2, and 3 of that repeated index; and (iii) if an index repeats more than twice in a term then you must have made an error.

According to the rule, (i) if there is no free index in a term, then that term is a scalar. An example will be pressure, which will be denoted by the variable p, where it can be noted that it comes with no index of any sort. (ii) If there is one free index, then the term is a vector. Examples are position $\mathbf{x}$ and velocities $\mathbf{u}$ and $\mathbf{V}$, which in index notation become x_i, u_i, and V_i. Here, u_i indicates the vector $(u_1, u_2, u_3)^{\mathrm{T}}$, which includes the three components of velocity. (iii) If there are two free indices, then the quantity is a second-rank tensor. An example is the stress tensor σ, which in index notation is written as σ_{ij}, where i and j are both free indices. As i and j take values 1, 2, and 3, we can see that the second-rank tensor has nine components (just as a vector has three components), which can be written in 3×3 matrix form as

$$\sigma = \sigma_{ij} = \begin{bmatrix} \sigma_{11} & \sigma_{12} & \sigma_{13} \\ \sigma_{21} & \sigma_{22} & \sigma_{23} \\ \sigma_{31} & \sigma_{32} & \sigma_{33} \end{bmatrix}. \tag{A.1}$$

For physical quantities, the indices have a definite meaning. For example, in the case of σ_{ij}, the first index gives the direction of force per area (i.e., the direction of the stress component) acting on a surface whose normal is given by the second index. (iv) If there are three free indices, then the quantity is a third-rank tensor and has $3^3 = 27$

components. This list goes on. Note that a scalar is a zeroth-rank tensor and a vector is a first-rank tensor.

We now introduce two important constant tensors. The first is the *Kronecker delta*, written as

$$\delta_{ij} = \begin{cases} 1 & \text{if } i = j, \\ 0 & \text{if } i \neq j. \end{cases} \tag{A.2}$$

The Kronecker delta is a constant second-rank tensor and, when written as a 3×3 matrix, it is an identity matrix. The second useful tensor is the *Levi-Civita symbol*, which can be expressed as

$$\epsilon_{ij} = \begin{cases} 1 & \text{if } (i, j, k) = (1, 2, 3), \ (2, 3, 1), \text{ or } (3, 1, 2), \\ -1 & \text{if } (i, j, k) = (1, 3, 2), \ (2, 1, 3), \text{ or } (3, 2, 1), \\ 0 & \text{otherwise}. \end{cases} \tag{A.3}$$

The Levi-Civita symbol is a constant third-rank tensor, since it has three free indices i, j, and k. Both the Kronecker delta and the Levi-Civita symbols are handy tools in index notation, as we will soon see.

The scalar, vector, second- and higher-order tensors obey standard mathematical operations such as addition, subtraction, multiplication, and so on, in specific ways, which we shall go over quickly. You can add or subtract a scalar with a scalar, a vector with a vector, a tensor with a tensor, and so on, and the sum or difference will in turn result in a scalar, vector, tensor, ... In physical terms, you can write velocity as a sum of two other velocities – all three are vectors and all are of dimension length/time. Certainly, it does not make sense to add pressure with velocity.

As far as multiplication, we define three kinds of multiplication: (i) outer product; (ii) inner product; and (iii) cross product. Let us apply these multiplications between two vectors **a** and **b**. In index notation, their definitions are given below:

$$\begin{aligned} \text{Outer product:} \quad & \mathbf{a}\,\mathbf{b} && = a_i\,b_j, \\ \text{Inner product:} \quad & \mathbf{a} \cdot \mathbf{b} && = a_i\,b_i, \\ \text{Cross product:} \quad & \mathbf{a} \times \mathbf{b} && = \epsilon_{ijk}\,a_j\,b_k. \end{aligned} \tag{A.4}$$

From the above definitions, we see that the outer product of two vectors results in a second-rank tensor, since there are two free indices at the end. The inner product of two vectors results in a scalar, since the repeated index i indicates a sum over $i = 1, 2, 3$. The inner product of two vectors can thus be expanded as $a_1 b_1 + a_2 b_2 + a_3 b_3$. The cross product of two vectors results in another vector. This can be seen in the index notation, where i is the only free (unmatched) index, while j and k are repeated. In an expanded form, the ith component of the cross product can be written as

$$\begin{aligned} (\mathbf{a} \times \mathbf{b})_i = \ & \epsilon_{i11}\,a_1\,b_1 + \epsilon_{i12}\,a_1\,b_2 + \epsilon_{i13}\,a_1\,b_3 \\ & \epsilon_{i21}\,a_2\,b_1 + \epsilon_{i22}\,a_2\,b_2 + \epsilon_{i23}\,a_2\,b_3 \\ & \epsilon_{i31}\,a_3\,b_1 + \epsilon_{i32}\,a_3\,b_2 + \epsilon_{i33}\,a_3\,b_3 \quad \text{for } i = 1, 2, 3. \end{aligned} \tag{A.5}$$

The above definitions of outer, inner, and cross product can be generalized when **a** is an nth-rank tensor and **b** is an mth-rank tensor (i.e., **a** has n free indices and **b** has m free indices). Their outer product will be a more complex tensor of rank $n + m$, whereas the rank of their inner product will be smaller. If there is only one free index that is common between **a** and **b**, then the rank of $\mathbf{a} \cdot \mathbf{b}$ is $n + m - 2$. In the case of a cross product, one free index of **a** combines with one free index of **b** to yield a free index in the cross product, and thus the rank of $\mathbf{a} \times \mathbf{b}$ is $n + m - 1$.

Problem A.1 With the above definitions, prove the following important properties of index notation.

- a_i for $i = 1, 2, 3$ is equal to a_j for $j = 1, 2, 3$.
- $a_i\, b_j = a_k\, b_l$ for $i, j, k, l = 1, 2, 3$.
- $a_i\, b_i = a_j\, b_j$.

The first important property is that the indices are dummy variables. You can replace one dummy index with another, since they all run 1, 2, 3. In case of a repeated index, if you decide to replace one of them, you must do so for both. After replacement there should still be a repeated index.

Problem A.2 With the above definitions, prove the following identities of the Kronecker delta and Levi-Civita symbols. All these are very useful identities that will be needed repeatedly over the rest of the book. It is not enough for you to prove these identities. You must remember them and start using them with ease as and when needed.

- $\delta_{ii} = 3$.
- $\delta_{ij} = \delta_{ji}$ for all i, j. That is, the order of the two indices does not matter in Kronecker delta. This also means Kronecker delta is symmetric.
- $\delta_{ij}\delta_{il} = \delta_{jl}$.
- $A_i\delta_{ij} = A_j$. This shows that Kronecker delta has the property of switching the free index of a tensor, through the use of a repeated index.
- $\epsilon_{ijk} = \epsilon_{kij} = \epsilon_{jki}$ for all i, j, k. This important property shows that the indices of the Levi-Civita symbol can be cyclically varied.
- $\epsilon_{ijk} = -\epsilon_{ikj} = -\epsilon_{jik} = -\epsilon_{kji}$ for all i, j, k. This important property shows that the indices of the Levi-Civita symbol can be acyclically varied, but with a negative sign.
- $\epsilon_{ijk}\delta_{jk} = 0$ for all i.
- $\epsilon_{ijk}\epsilon_{ilm} = \delta_{jl}\delta_{km} - \delta_{jm}\delta_{kl}$.
- If **A** is a matrix whose rows are A_{1i}, A_{2j}, A_{3k}, then $\det[\mathbf{A}] = \epsilon_{ijk}A_{1i}A_{2j}A_{3k}$. Show that the right-hand side can also be written as a cross product of two rows of the matrix dotted with the third row.

Before leaving this appendix, we point out that we will deal with tensor quantities that are products of other tensor quantities. For example, the outer product of the fluid velocity vector with itself is **u u**, which is also sometimes written as **u** ⊗ **u**. It is a second-rank tensor and in index notation will be denoted as $u_i u_j$. Similarly, $x_i x_j x_k$ is a third-rank tensor. On the other hand, $u_i u_j x_j$ is a vector, since a sum over j is implied and i is the only remaining free index.

Appendix B Vector Calculus

In this appendix we will very briefly go over some of the vector calculus identities that are used in Chapter 4 and in the rest of the book. We again caution that the treatment is neither extensive nor rigorous. This appendix should only be used as a refresher.

In vector calculus, the important quantity is the gradient operator ∇, which is a vector, and its three components are the derivatives along three orthogonal directions. In Cartesian coordinates, the three components are $(\partial/\partial x, \partial/\partial y, \partial/\partial z)^{\mathrm{T}}$. Thus, in index notation, we have

$$\nabla = \frac{\partial}{\partial x_i} \quad \text{for } i = 1, 2, 3, \tag{B.1}$$

where x_1, x_2, and x_3 are also referred to as the x, y, and z-coordinates. The gradient operator behaves as any other vector and can be used in outer, inner, and cross products. However, since it is a derivative operator, it needs to operate on another quantity. That is, it needs to be followed by another fluid mechanical quantity to the right of it and this quantity must be a field variable (i.e., it must be a function of $\mathbf{x}$) in order for the gradient to be nonzero. For example, ∇ operating on a constant scalar, vector, or tensor (that is not a function of $\mathbf{x}$) is identically zero.

Let us now consider the three different products of ∇ and the velocity vector field $\mathbf{u}(\mathbf{x}, t)$. Since the field $\mathbf{u}$ is a function of space and time, its gradient is in general nonzero. The three products are

$$\text{Outer product: } \nabla\mathbf{u} = \frac{\partial u_j}{\partial x_i},$$

$$\text{Inner product: } \nabla \cdot \mathbf{u} = \frac{\partial u_1}{\partial x_1} + \frac{\partial u_2}{\partial x_2} + \frac{\partial u_3}{\partial x_3}, \tag{B.2}$$

$$\text{Cross product: } \nabla \times \mathbf{u} = \epsilon_{ijk} \frac{\partial u_k}{\partial x_j},$$

where $\nabla\mathbf{u}$ is known as the *gradient of* $\mathbf{u}$, $\nabla \cdot \mathbf{u}$ is known as the *divergence of* $\mathbf{u}$, and $\nabla \times \mathbf{u}$ is known as the *curl of* $\mathbf{u}$, which yields the vorticity field. All these definitions involve x_1, x_2, and x_3 derivatives of the three components of velocity. All the rules of index notation discussed in the previous appendix apply to the gradient operator as well. Note that ∇ can operate repeatedly. For example, $\nabla\nabla\mathbf{u}$ corresponds to a gradient operator outer product with a gradient operator outer product with velocity. Due to the two outer products, the result will be a third-rank tensor, which in index notation can be written as $\partial^2 u_k/(\partial x_i \partial x_j)$. As i, j, and k run over 1, 2, and 3, $\nabla\nabla\mathbf{u}$ can be seen to have 27 components. Of particular importance is the Laplace operator, which is defined as

$$\nabla^2 = \nabla \cdot \nabla = \frac{\partial^2}{\partial x_i \partial x_i} = \frac{\partial^2}{\partial x_1^2} + \frac{\partial^2}{\partial x_2^2} + \frac{\partial^2}{\partial x_3^2}, \tag{B.3}$$

where the last relation comes from summing over the repeated index i.

Problem B.1 Prove the following identities where p is a scalar field, $\mathbf{A}$, $\mathbf{B}$, and $\mathbf{u}$ are vector fields.

- $\nabla \mathbf{x}$ is equal to the identity matrix or, in index notation, $\partial x_j / \partial x_i = \delta_{ij}$.
- $\nabla \times (\nabla p) \equiv 0$.
- $\nabla \cdot (\nabla p) = \nabla^2 p$.
- $\nabla \cdot (\nabla \times \mathbf{A}) = 0$.
- $\nabla \times (\mathbf{A} \times \mathbf{B}) = \mathbf{A}(\nabla \cdot \mathbf{B}) + (\mathbf{B} \cdot \nabla)\mathbf{A} - \mathbf{B}(\nabla \cdot \mathbf{A}) - (\mathbf{A} \cdot \nabla)\mathbf{B}$.
- $\mathbf{u} \cdot \nabla \mathbf{u} = (\nabla \times \mathbf{u}) \times \mathbf{u} + \frac{1}{2}\nabla(\mathbf{u} \cdot \mathbf{u})$.
- $\nabla^2 \mathbf{u} = \nabla(\nabla \cdot \mathbf{u}) - \nabla \times (\nabla \times \mathbf{u})$.

We now consider some useful results that will be of great help in following the derivations presented in Chapter 4. We start with the definition that r is the length of the position vector $\mathbf{x}$. That is, $r = (\mathbf{x} \cdot \mathbf{x})^{1/2}$. As an example, we derive the following relations that are used in obtaining the growing and decaying solutions of Section 4.1.1:

$$\frac{\partial r}{\partial x_i} = \frac{\partial (x_j x_j)^{1/2}}{\partial x_i} = \frac{1}{2(x_j x_j)^{1/2}} 2x_j \frac{\partial x_j}{\partial x_i} = \frac{1}{r} x_j \delta_{ij} = \frac{x_i}{r} \, ;$$

$$\frac{\partial r^k}{\partial x_i} = kr^{k-1} \frac{\partial r}{\partial x_i} = kr^{k-2} x_i \, . \tag{B.4}$$

Problem B.2 Follow the steps above to obtain these two relations:

$$\frac{\partial^2}{\partial x_i \partial x_j}\left(\frac{1}{r}\right) = \frac{x_i x_j}{r^5} - \frac{\delta_{ij}}{3r^3} \quad \text{and} \quad \frac{\partial^3}{\partial x_i \partial x_j x_k}\left(\frac{1}{r}\right) = \frac{x_i x_j x_k}{r^7} - \frac{x_i \delta_{jk} + x_j \delta_{ki} + x_k \delta_{ij}}{5r^5}.$$

Problem B.3 We first show that the Laplacian of $1/r$ is zero:

$$\frac{\partial}{\partial x_j}\left(\frac{1}{r}\right) = -\frac{x_j}{r^3} \quad \Rightarrow \quad \frac{\partial^2}{\partial x_j \partial x_j}\left(\frac{1}{r}\right) = \frac{\partial}{\partial x_j}\left(-\frac{x_j}{r^3}\right) = -\frac{\delta_{jj}}{r^3} + \frac{3x_j x_j}{r^5} = 0, \tag{B.5}$$

where we use the identities $\delta_{jj} = 3$ and $x_j x_j = r^2$. You will follow these steps to prove

$$\frac{\partial^2}{\partial x_j \partial x_j}\left(\frac{x_i}{r^3}\right) = 0 \, . \tag{B.6}$$

Example B.4 We present here as an example some manipulations that you will need to obtain the relations presented in Chapter 4. In this example, we start with the following results for pressure and particular velocity given in Eqs. (4.10) and (4.14):

$$p(\mathbf{x}) = a_1 u_{cj} \frac{x_j}{r^3} \quad \text{and} \quad \mathbf{u} = \frac{a_1}{2\mu_f} u_{cj} \frac{x_i x_j}{r^3},$$ (B.7)

where we note that a_1 and μ_f are scalar constants and u_{cj} is a vector constant. Our goal is to show that the above pressure and velocity satisfy the Stokes equation $0 = -\nabla p + \mu_f \nabla^2 \mathbf{u}$. To do that, we first obtain the pressure gradient as

$$-\nabla p = -\frac{\partial}{\partial x_i}\left(a_1 u_{cj} \frac{x_j}{r^3}\right) = -a_1 u_{cj}\left(\frac{\delta_{ij}}{r^3} - \frac{3x_i x_j}{r^5}\right),$$ (B.8)

where we have used the result obtained in Eq. (B.4). Toward obtaining the Laplacian of velocity, let us first calculate the gradient of velocity as

$$\frac{\partial u_i}{\partial x_k} = \frac{\partial}{\partial x_k}\left(\frac{a_1}{2\mu_f} u_{cj} \frac{x_i x_j}{r^3}\right) = \frac{a_1}{2\mu_f} u_{cj}\left(\delta_{ik}\frac{x_j}{r^3} + \delta_{jk}\frac{x_i}{r^3} - \frac{3x_i x_j x_k}{r^5}\right).$$ (B.9)

Follow this procedure and take another derivative of the above expression to obtain

$$\mu_f \nabla^2 \mathbf{u} = a_1 u_{cj}\left(\frac{\delta_{ij}}{r^3} - \frac{3x_i x_j}{r^5}\right).$$ (B.10)

This, combined with the expression for the pressure gradient, shows that they satisfy the Stokes equation.

The final section of this appendix recollects Gauss's theorem from vector calculus, which we write as

$$\int_\Omega \nabla \cdot \mathbf{u}\, d\Omega = \int_S \mathbf{u} \cdot \mathbf{n}\, dS,$$ (B.11)

where on the left-hand side we have the volume integral over the volume Ω, and on the right-hand side the integral over the surface S of that volume. Here, $\mathbf{u}$ is any vector field and $\mathbf{n}$ is the outward-pointing surface normal vector. It is very important to refresh your understanding of this theorem.

Problem B.5 In this problem you will derive two variants of Gauss's theorem. Let $\mathbf{u} = \mathbf{a}\phi(\mathbf{x})$, where $\mathbf{a}$ is a constant vector independent of $\mathbf{x}$, whereas the scalar field ϕ is a function of space. Derive the following generalization of the Gauss theorem:

$$\int_\Omega \nabla\phi\, d\Omega = \int_S \phi\mathbf{n}\, dS.$$ (B.12)

Instead, let $\mathbf{u} = \mathbf{a} \times \mathbf{w}(\mathbf{x})$, where $\mathbf{a}$ is a constant vector and $\mathbf{w}$ is a vector field. Derive the following generalization of Gauss's theorem:

$$\int_\Omega \nabla \times \mathbf{w}\, d\Omega = \int_S \mathbf{n} \times \mathbf{w}\, dS.$$ (B.13)

Appendix C Added Dissipation of an Isolated Particle

Consider a large domain D_∞ around the particle whose surface will be denoted ∂D_∞. We will eventually consider the effect of a random distribution of non-interacting rigid particles within this volume, but for now in this appendix we consider an isolated particle located at the origin. We impose the undisturbed flow (4.47) on this surface, and thus the velocity perturbation goes to zero on this large outer surface. We now proceed to calculate the rate of work that needs to be done at this surface in order to maintain the straining flow. For this analysis we will work only with the deviatoric stress that excludes the isotropic pressure component, since we are only interested in the viscous dissipation part. The imposed velocity at the outer surface is given by Eq. (4.47), while the tractional force due to total deviatoric stress can be expressed as $(2\mu_f S_{cik} + \tau_{ik})n_k$, where τ_{ik} is the deviatoric stress due to the perturbation flow induced by the particle. Now, the local rate of work (or power) can be calculated as the dot product of the velocity $S_{cij}x_j$ with the tractional force. It then follows that the rate of work done on the fluid at the outer surface ∂D_∞ is equal to

$$\int_{\partial D_\infty} S_{cij}x_j(2\mu_f S_{cik} + \tau_{ik})n_k\,dA = 2\mu_f S_{cij}S_{cij}D_\infty + S_{cij}\int_{\partial D_\infty} \tau_{ik}x_jn_k\,dA, \quad \text{(C.1)}$$

where D_∞ is the volume of the entire space within the large domain including the particle. Note that S_{cij} is a constant over the entire volume. Also, we have used the divergence theorem to convert the surface integral of the first term into a volume integral. Note that the rate of work done on the fluid goes toward balancing viscous dissipation. The first term on the right is the dissipation of the undisturbed ambient straining flow. The second term on the right is of particular interest since it represents the added dissipation due to the inclusion of the particle.

To evaluate the second term of Eq. (C.1), we consider the following application of the divergence theorem:

$$S_{cij}\int_{D_\infty - V_P} \frac{\partial \tau_{ik}x_j}{\partial x_k}\,dV = S_{cij}\int_{\partial D_\infty} \tau_{ik}x_jn_k\,dA - S_{cij}\int_S \tau_{ik}x_jn_k\,dA, \quad \text{(C.2)}$$

where V_p is the volume of the particle whose surface is denoted by S. Thus, the above volume integral is over the large domain D_∞, but outside of the particle. Note that n_k is the outward normal of a closed surface. For the argument of the left-hand side we can use the relation $\partial(\tau_{ik}x_j)/\partial x_k = x_j(\partial\tau_{ik}/\partial x_k) + \tau_{ij}$ and furthermore we make use of the fact that the perturbation flow satisfies the Stokes equation (i.e., $\partial\tau_{ik}/\partial x_k = 0$). Using these we can simplify the left-hand side of the above equation to $S_{cij} \int_{D_\infty - V_P} \tau_{ij} dV$. This term can be converted to a surface integral over the particle as follows:

$$
S_{cij} \int_{D_\infty - V_P} \tau_{ij} dV = S_{cij} \int_{D_\infty - V_p} 2\mu_f \frac{\partial u_i}{\partial x_j} dV
$$

$$
= S_{cij} \int_{\partial D_\infty} 2\mu_f u_i n_j dA - S_{cij} \int_S 2\mu_f u_i n_j dA
$$

$$
= -S_{cij} \int_S 2\mu_f u_i n_j dA , \tag{C.3}
$$

where we note that the perturbation flow is nonvortical and therefore $\tau_{ij} = 2\mu_f(\partial u_i / \partial x_j)$. Also, we have used the divergence theorem to obtain the second equality and the fact that the perturbation velocity is zero on ∂D_∞ in obtaining the last simplification.

Substituting Eq. (C.3) into Eq. (C.2) and the resulting expression into the right-hand side of Eq. (C.1), we obtain the rate of work done on the fluid as

$$
2\mu_f S_{cij} S_{cij} D_\infty + S_{cij} \int_S \left(\tau_{ik} x_j n_k - 2\mu_f u_i n_j \right) dA . \tag{C.4}
$$

The second term on the right-hand side is again the added dissipation due to the presence of the particle and therefore is identical to the second term on the right-hand side of Eq. (C.1). However, the difference in the integral is now over the surface of the particle, as opposed to the integral over the arbitrary outer surface. This term depends on the perturbation velocity and the corresponding deviatoric stress, which can be evaluated from the solution given in Eq. (4.52). Performing this calculation, we obtain

$$
\tau_{ik} x_j n_k - 2\mu_f u_i n_j
$$

$$
= 5\mu_f S_{cik} x_j n_k \left(2\frac{R_P^5}{r^5} - \frac{R_P^3}{r^3} \right) + \frac{25}{r^2} \mu_f S_{ckl} x_i x_j x_l n_k \left(\frac{R_P^3}{r^3} - \frac{R_P^5}{r^5} \right) . \tag{C.5}
$$

To evaluate the integral of the above expression over the surface of the sphere, we use the following identities:

$$
\int_S n_j n_k dA = \frac{4\pi}{3} \delta_{jk} \quad \text{and} \quad \int_S n_i n_j n_k n_l dA = \frac{4\pi}{15} \left(\delta_{ij} \delta_{kl} + \delta_{ik} \delta_{jl} + \delta_{il} \delta_{jk} \right) . \tag{C.6}
$$

Further, we note that on the surface of the sphere, $x_j = R_p n_j$ in the evaluation of the integral. Carrying out the integration, we arrive at

$$
S_{cij} \int_S \left(\tau_{ik} x_j n_k - 2\mu_f u_i n_j \right) dA = \frac{20\pi}{3} \mu_f R_p^3 S_{cij} S_{cij} . \tag{C.7}
$$

Appendix D Solution of the Helmholtz Equation

Here we seek to solve the scalar Helmholtz equation

$$\nabla^2 \psi + k^2 \psi = 0, \tag{D.1}$$

where k is the wavenumber. Consider a spherical coordinate system (r, θ, φ), where r is the radial coordinate, $(0 \le \theta \le \pi)$ is the zenith angle from the positive x-axis, and $(0 \le \varphi \le 2\pi)$ is the azimuthal angle. In the spherical coordinate we define the gradient and Laplace operators as

$$\nabla = \left(\frac{\partial}{\partial r}, \frac{1}{r} \frac{\partial}{\partial \theta}, \frac{1}{r \sin \theta} \frac{\partial}{\partial \varphi} \right),$$
$$\nabla^2 = \frac{1}{r^2} \frac{\partial}{\partial r} r^2 \frac{\partial}{\partial r} + \frac{1}{r^2 \sin \theta} \frac{\partial}{\partial \theta} \sin \theta \frac{\partial}{\partial \theta} + \frac{1}{r^2 \sin^2 \theta} \frac{\partial^2}{\partial \varphi^2}. \tag{D.2}$$

The divergence of a vector field $\mathbf{A} = (A_r, A_\theta, A_\varphi)$ can be expressed as

$$\nabla \cdot \mathbf{A} = \frac{1}{r^2} \frac{\partial}{\partial r} r^2 A_r + \frac{1}{r \sin \theta} \frac{\partial}{\partial \theta} \sin \theta A_\theta + \frac{1}{r \sin \theta} \frac{\partial}{\partial \varphi} A_\varphi. \tag{D.3}$$

We seek an axisymmetric solution for ψ and as a result, $\partial/\partial \varphi = 0$. Furthermore, using separation of variables, we can write

$$\psi = R(r)\, \Theta(\theta). \tag{D.4}$$

Substituting into Eq. (D.1) and separating into an equation for R and an equation for Θ, we obtain

$$\frac{1}{R} \frac{\partial}{\partial r} r^2 \frac{\partial R}{\partial r} + k^2 r^2 = n(n+1),$$
$$\frac{1}{\Theta} \frac{1}{\sin \theta} \frac{\partial}{\partial \theta} \sin \theta \frac{\partial \Theta}{\partial \theta} = -n(n+1), \tag{D.5}$$

where $n(n+1)$ is the constant of separation. The substitutions $x = kr$ and $Z = x^{1/2} R$ are made and, after some manipulation, we obtain

$$x^2 \frac{\partial^2 Z}{\partial x^2} + x \frac{\partial Z}{\partial x} + \left[x^2 - \left(n + \frac{1}{2} \right)^2 \right] Z = 0. \tag{D.6}$$

Solutions to this equation are Bessel functions of half-integral order. In the original variables the solution can be written as

$$R(r) = \begin{cases} h^{(1)}(kr) & \text{for} \quad \text{Im}(kr) > 0, \\ h^{(2)}(kr) & \text{for} \quad \text{Im}(kr) < 0, \end{cases} \tag{D.7}$$

where $h^{(1)}$ and $h^{(2)}$ are spherical Hankel functions of the first and second kind, and here they represent outgoing and incoming disturbance waves, respectively.

We rearrange the second line of Eq. (D.5) and make the substitution $y = \cos\theta$ to obtain

$$(1 - y^2)\frac{\partial^2\Theta}{\partial y^2} - 2y\frac{\partial\Theta}{\partial y} + n(n+1)\Theta = 0 . \tag{D.8}$$

Solutions to the above are associated Legendre polynomials $P_n^0(\cos\theta)$. So the final solutions of the Helmholtz equation (D.1) in spherical coordinates are

$$\psi(r,\theta) = \begin{cases} h^{(1)}(kr)P_n^0(\cos\theta) & \text{for} \quad \text{Im}(kr) > 0 \text{ (outgoing)}, \\ h^{(2)}(kr)P_n^0(\cos\theta) & \text{for} \quad \text{Im}(kr) < 0 \text{ (incoming)} . \end{cases} \tag{D.9}$$

Appendix E Derivation of the Perturbation Force of the BBO Equation

Due to the axisymmetric nature of the flow, the tractional force on the surface of the particle can be expressed as

$$\sigma \cdot \mathbf{n} = \sigma_{rr}\, \mathbf{e}_r + \sigma_{r\theta}\, \mathbf{e}_\theta\,, \tag{E.1}$$

where $\mathbf{e}_r$ and $\mathbf{e}_\theta$ are the unit vectors along the radial and tangential directions. The expressions for the components of the stress tensor in terms of ϕ and ψ are given by (Guz, 2009)

$$\sigma_{rr} = \rho_f \frac{\partial \phi}{\partial t} + 2\mu_f \left[\left(-\nabla^2 + \frac{\partial^2}{\partial r^2} \right)\phi + \left(r\frac{\partial^3}{\partial r^3} + 3\frac{\partial^2}{\partial r^2} - r\frac{\partial}{\partial r}\nabla^2 - \nabla^2 \right)\psi \right],$$

$$\sigma_{r\theta} = 2\mu_f \left[\left(\frac{\partial^2}{\partial r \partial \theta}\frac{1}{r} \right)\phi + \left\{ \frac{\partial}{\partial \theta}\left(\frac{\partial^2}{\partial r^2} + \frac{1}{r}\frac{\partial}{\partial r} - \frac{1}{r^2} - \frac{1}{2}\nabla^2 \right) \right\}\psi \right]. \tag{E.2}$$

Here we note an important point that was implicit in the above discussion. As the direction of relative velocity $(\mathbf{v} - \mathbf{u}_c)$ changes over time, so does the spherical coordinate system used in the above analysis. The zenith angle θ is defined with respect to the relative velocity vector. Thus, in evaluating the tractional force it is far more advantageous to represent it in terms of the relative velocity vector and surface normal $\mathbf{n}$, instead of $\mathbf{e}_r$ and $\mathbf{e}_\theta$. Following Maxey and Riley (1983) and Gatignol (1983), we write the tractional force on the surface of the sphere as

$$[\sigma \cdot \mathbf{n}]_{r=R_p} = \rho_f \left(\frac{A}{R_p^2} - B\, h_1 \right)\left(\frac{d(\mathbf{V} - \mathbf{u}_c)}{dt} \cdot \mathbf{n} \right)\mathbf{n} + 3\rho_f B\, h_1 \frac{d(\mathbf{V} - \mathbf{u}_c)}{dt}, \tag{E.3}$$

where we have substituted the expressions for ϕ and ψ from Eqs. (4.80), (4.81), and (4.85) into Eq. (E.2) and simplified. In the above, the first term on the right is the inviscid contribution to the tractional force, which at every elemental surface area on the sphere is normal to the sphere. In Eq. (E.1), the viscous contribution to the tractional force has both radial and tangential components. But most interestingly, when combined, the viscous part of the tractional force on any elemental surface area on the sphere is in the direction of relative velocity, as indicated by the second term on the right in the above equation.

Substituting the expressions for A and B and integrating the tractional force over the surface of the sphere, we can obtain the following expression for the force on the particle due to the perturbation flow:

$$\mathbf{F}_{\text{pert}}(t) = \oint_{S_p} [\sigma \cdot \mathbf{n}]_{r=R_p} \, dA$$

$$= \frac{4}{3} \pi R_p^2 \rho_f \frac{d(\mathbf{V} - \mathbf{u}_c)}{dt} \left[\frac{A}{R_p^2} + 2B \, h_1 \right]$$

$$= m_f \frac{d(\mathbf{V} - \mathbf{u}_c)}{dt} \left(\frac{9}{2 k_2^2 R_p^2} - \frac{9\iota}{2 k_2 R_p} - \frac{1}{2} \right), \qquad (E.4)$$

where $\iota = \sqrt{-1}$. In obtaining the third expression, in addition to Eq. (4.85) for the constants A and B, we have also used the property of the Hankel function that $h_1(k_2 R_p)/h_0(k_2 R_p) = (1 - \iota k_2 R_p)/(k_2 R_p)$.

Now that we have an expression for the perturbation force, we implement our original plan to replace $k_2^2 = -(1/\nu_f)(\partial/\partial t)$ with its Laplace transform. We make the following replacements:

$$k_2^2 \to -\frac{s}{\nu_f} \quad \text{and} \quad k_2 \to \frac{\iota \sqrt{s}}{\sqrt{\nu_f}}. \qquad (E.5)$$

Furthermore, we make use of the Laplace transform relation $\mathscr{L}[d(\mathbf{V} - \mathbf{u}_c)/dt] = s\mathscr{L}[(\mathbf{V} - \mathbf{u}_c)]$, where $\mathscr{L}$ stands for the Laplace transform. Substituting, we obtain the perturbation force in the Laplace space as

$$\mathscr{L}[\mathbf{F}_{\text{pert}}] = -m_f \left(\frac{9\nu_f}{2R_p^2} + \frac{s}{2} + \frac{9\sqrt{s}\,\nu_f}{2R_p} \right) \mathscr{L}[\mathbf{V} - \mathbf{u}_c]. \qquad (E.6)$$

The Laplace inverse of each of these terms can readily be obtained. That of the first term yields $6\pi \mu_f R_p (\mathbf{u}_c - \mathbf{V})$, which is the component of force proportional to the velocity difference. The Laplace transform of the second term yields

$$\frac{1}{2} m_f \left[\frac{d\mathbf{u}_c}{dt} - \frac{d\mathbf{V}}{dt} \right]. \qquad (E.7)$$

By carefully rewriting the last term, its Laplace inverse can be written as a convolution integral

$$\mathscr{L}^{-1} \left[\left(-m_f \frac{9\sqrt{\nu_f}}{2\sqrt{s}R_p} \right) (s\mathscr{L}[\mathbf{V} - \mathbf{u}_c]) \right]$$

$$= 6\pi \mu_f R_p \int_{-\infty}^{t} \frac{R_p}{\sqrt{\pi \nu_f (t - \xi)}} \left[\frac{d\mathbf{u}_c}{dt} - \frac{d\mathbf{V}}{dt} \right]_{@\xi} d\xi, \qquad (E.8)$$

where we have used the fact $\mathscr{L}^{-1} \left[1/\sqrt{(s)} \right] = 1/\sqrt{\pi t}$. Now the above three contributions to the perturbation force can be added to the undisturbed flow force given in Eq. (4.87) to write

$$\mathbf{F}_{\text{pert}}(t) = 6\pi\mu_f R_p(\mathbf{u}_c - \mathbf{V}) + \frac{1}{2}m_f\left[\frac{d\mathbf{u}_c}{dt} - \frac{d\mathbf{V}}{dt}\right]$$

$$+ 6\pi\mu_f R_p \int_{-\infty}^t \frac{R_p}{\sqrt{\pi\nu_f(t-\xi)}}\left[\frac{d\mathbf{u}_c}{dt} - \frac{d\mathbf{V}}{dt}\right]_{@\xi} d\xi. \qquad \text{(E.9)}$$

Appendix F Derivation of the MRG Equation with Reciprocal Theorem

We employ the *Lorentz reciprocal theorem* (Happel and Brenner, 1965; Kaneda, 1980) which allows us to evaluate the force on a particle in a Stokes flow without actually solving for the details of the perturbation flow. In Section 4.1.1, we obtained the solution for the homogeneous ambient flow problem (i.e., when the ambient flow was spatially uniform). This solution will be used in the reciprocal theorem to obtain an expression for the perturbation force in the more complex situation of an inhomogeneous (i.e., spatially varying) ambient flow. Read Peres (1929), who used the reciprocal theorem to obtain a simple and elegant proof of Faxén's theorem for steady incompressible flows. Maxey and Riley (1983) and Gatignol (1983) also made use of the reciprocal theorem in their derivations.

The reciprocal theorem considers two zero-Re Stokes flow fields around a spherical particle. The two fields will be denoted as $(\mathbf{u}_1, p_1)$ and $(\mathbf{u}_2, p_2)$ and both of them satisfy Eq. (4.119). Here, the subscript $i = \{1, 2\}$ indicates the two flow fields. Let the far-field boundary conditions of these two velocity fields vanish far away from the sphere, being zero for $t \leq 0$. Then, the reciprocal relation between the two flows can be cast as an integral relation,

$$\oint_{S_p} \mathscr{L}(\mathbf{u}_2) \cdot (\mathscr{L}(\boldsymbol{\sigma}_1) \cdot \mathbf{n}) \, dA = \oint_S \mathscr{L}(\mathbf{u}_1) \cdot (\mathscr{L}(\boldsymbol{\sigma}_2) \cdot \mathbf{n}) \, dA, \tag{F.1}$$

where $\mathscr{L}(\cdot)$ is the Laplace transform. The integration is over the surface of the sphere, and $\boldsymbol{\sigma}_i$ are the stress fields created by the respective flows. A version of the above relation was derived in Eq. (4.64).

We choose the unknown complex flow field arising from the inhomogeneous boundary condition on the surface of the sphere to be flow field 1: $(\mathbf{u}_1, p_1)$. Its inhomogeneous boundary condition is then given by the first line of Eq. (4.121) as

$$\mathbf{u}_1(\mathbf{x}, t) = \mathbf{V} - \mathbf{u}_c \quad \text{for} \quad |\mathbf{x}| = R_p. \tag{F.2}$$

For now we will simply refer to this boundary condition as $\mathbf{u}_{1,p}(\mathbf{x}, t)$ and later make the substitution $\mathbf{u}_{1,p}(\mathbf{x}, t) = \mathbf{V} - \mathbf{u}_c$. The desired quantity of interest is the net time-dependent force on the particle, which can be expressed in Laplace space as

$$\mathscr{L}(\mathbf{F}_1) = \oint_{S_p} \mathscr{L}(\boldsymbol{\sigma}_1) \cdot \mathbf{n} \, dA. \tag{F.3}$$

We let flow field 2: $(\mathbf{u}_2, p_2)$ be that due to the unsteady motion of a sphere in a spatially uniform time-dependent ambient fluid. This solution was obtained in Section 4.1.5. Let the uniform boundary condition for $\mathbf{u}_2$ be

$$\mathbf{u}_2(\mathbf{x}, t) = \mathbf{u}_{2,p}(t) \quad \text{for} \quad |\mathbf{x}| = R_p, \tag{F.4}$$

which only depends on time and is independent of $\mathbf{x}$. Using Eq. (F.1) and the boundary conditions given by Eqs. (F.2) and (F.4) leads to

$$\mathscr{L}(\mathbf{u}_{2,p}) \cdot \mathscr{L}(\mathbf{F}_1) = \oint_{S_p} \mathscr{L}(\mathbf{u}_{1,p}(\mathbf{x}, t)) \cdot (\mathscr{L}(\sigma_2) \cdot \mathbf{n}) \, dA, \tag{F.5}$$

where we have used the fact that $\mathscr{L}(\mathbf{u}_{2,p})$ is a constant and therefore can be taken out of the surface integral. The surface integral of $\mathscr{L}(\sigma_1) \cdot \mathbf{n}$ is then given by Eq. (F.3). To further simplify this relation, we observe from Eq. (E.3) that in the present case of axisymmetric flow, the traction vector $\mathscr{L}(\sigma_2) \cdot \mathbf{n}$ on the surface of the sphere can be written as a linear combination of components in the radial direction and along the direction of particle motion:

$$\mathscr{L}(\sigma_2) \cdot \mathbf{n} = H_1(s)(\mathscr{L}(\mathbf{u}_{2,p}) \cdot \mathbf{n})\mathbf{n} + H_2(s)\mathscr{L}(\mathbf{u}_{2,p}), \tag{F.6}$$

where s is the Laplace variable. Comparing the above to Eq. (E.3), we obtain

$$H_1(s) = \rho_f \left(\frac{A}{R_p^2} - B\, h_1 \right) s \quad \text{and} \quad H_2(s) = 3\rho_f B\, h_1\, s. \tag{F.7}$$

The important property that we will exploit is that $H_1(s)$ and $H_2(s)$ are independent of the details of the motion and are invariant on the surface of the sphere (i.e., independent of the spherical coordinates θ and φ). Substituting Eq. (F.6) in Eq. (F.5) and rearranging,

$$\mathscr{L}(\mathbf{u}_{2,p}) \cdot \mathscr{L}(\mathbf{F}_1)$$
$$= H_1(s) \oint_{S_p} (\mathscr{L}(\mathbf{u}_{2,p}) \cdot \mathbf{n})(\mathscr{L}(\mathbf{u}_{1,p}(\mathbf{x})) \cdot \mathbf{n}) dA + H_2(s) \oint_{S_p} \mathscr{L}(\mathbf{u}_{1,p}(\mathbf{x})) \cdot \mathscr{L}(\mathbf{u}_{2,p}) dA, \tag{F.8}$$

where for notational convenience we have suppressed the time dependence of $\mathbf{u}_{1,p}$. We now note that the spatially uniform $\mathscr{L}(\mathbf{u}_{2,p})$ can be taken out of the second integral and if we make the substitution $\mathbf{n} = \mathbf{x}/R_p$, where $\mathbf{x}$ is the position vector from the center of the sphere, we obtain

$$\frac{H_1(s)}{R_p} \oint_{S_p} \left[(\mathbf{x} \cdot \mathscr{L}(\mathbf{u}_{2,p}))\, \mathscr{L}(\mathbf{u}_{1,p}(\mathbf{x})) \right] \cdot \mathbf{n}\, dA + H_2(s) \left(\oint_{S_p} \mathscr{L}(\mathbf{u}_{1,p}(\mathbf{x}))\, dA \right) \cdot \mathscr{L}(\mathbf{u}_{2,p}). \tag{F.9}$$

We now use Gauss's theorem to convert the first surface integral to a volume integral and obtain

$$\frac{H_1(s)}{R_p} \int_{V_p} \nabla \cdot \left[(\mathbf{x} \cdot \mathscr{L}(\mathbf{u}_{2,p}))\, \mathscr{L}(\mathbf{u}_{1,p}(\mathbf{x})) \right] dV + H_2(s) \left(\oint_{S_p} \mathscr{L}(\mathbf{u}_{1,p}(\mathbf{x}))\, dA \right) \cdot \mathscr{L}(\mathbf{u}_{2,p}). \tag{F.10}$$

As the next step we use the following relations to further simplify the equation. First, due to incompressibility, $\nabla \cdot \mathscr{L}(\mathbf{u}_{1,p}(\mathbf{x})) = \mathscr{L}(\nabla \cdot \mathbf{u}_{1,p}(\mathbf{x})) = 0$, where we have used the fact that the Laplace operator commutes with the divergence operator. In addition, we note $\nabla \left(\mathbf{x} \cdot \mathcal{L}(\mathbf{u}_{2,p}) \right) = (\nabla \mathbf{x}) \cdot \mathcal{L}(\mathbf{u}_{2,p} = \mathcal{L}(\mathbf{u}_{2,p})$, where we have again used the fact that $\mathcal{L}(\mathbf{u}_{2,p})$ is not a function of $\mathbf{x}$. Making these substitutions, we get

$$\mathcal{L}(\mathbf{u}_{2,p}) \cdot \mathcal{L}(\mathbf{F}_1)$$

$$= \mathcal{L}(\mathbf{u}_{2,p}) \cdot \left\{ \frac{H_1(s)}{R_p} \int_{V_p} \mathcal{L}(\mathbf{u}_{1,p}(\mathbf{x})) \, dV + H_2(s) \oint_{S_p} \mathcal{L}(\mathbf{u}_{1,p}(\mathbf{x})) \, dA \right\}. \qquad \text{(F.11)}$$

As the final step we define the following volume and surface averages:

$$\overline{f(\mathbf{x},t)}^V = \frac{3}{4\pi R_p^3} \int_{V_p} f(\mathbf{x} - \mathbf{x}',t) \, d\mathbf{x}',$$

$$\overline{f(\mathbf{x},t)}^S = \frac{1}{4\pi R_p^2} \oint_{S} f(\mathbf{x} - \mathbf{x}',t) \, d\mathbf{x}'. \qquad \text{(F.12)}$$

This allows us to rewrite Eq. (F.11) as

$$\mathcal{L}(\mathbf{F}_1) = \frac{4}{3}\pi R_p^2 H_1(s) \mathcal{L}\left(\overline{\mathbf{u}_{1,p}(\mathbf{x},t)}^V \right) + 4\pi R_p^2 H_2(s) \mathcal{L}\left(\overline{\mathbf{u}_{1,p}(\mathbf{x},t)}^S \right). \qquad \text{(F.13)}$$

Thus, we have obtained an expression for the force on a spherical particle due to the inhomogeneous velocity boundary condition on its surface in terms of surface and volume averages of the undisturbed ambient flow that are the inhomogeneous boundary condition.

F.1   Force Expression in the Time Domain

The result from the reciprocal theorem can be specialized to the problem at hand. Since flow field 1 was chosen for the desired perturbation flow due to the inhomogeneous (spatially varying) ambient flow, $\mathbf{F}_{\text{pert}} = \mathbf{F}_1$. Also, we make the substitution $\mathbf{u}_{1,p} = \mathbf{V} - \mathbf{u}_c$ and obtain the following result from the reciprocal theorem:

$$\mathcal{L}(\mathbf{F}_{\text{pert}}) = m_f \left[G_v \mathcal{L}\left(\overline{\mathbf{u}_c(\mathbf{x},t) - \mathbf{v}}^V \right) + G_s \mathcal{L}\left(\overline{\mathbf{u}_c(\mathbf{x},t) - \mathbf{v}}^S \right) \right], \qquad \text{(F.14)}$$

where G_v is the nondimensional resistance in the Laplace space that acts on the volume-averaged relative velocity and G_s is the nondimensional resistance in the Laplace space that acts on the surface-averaged relative velocity. From Eq. (F.7) and the definitions of A and B given in Eq. (4.85), these resistances can be expressed as

$$G_v = -\frac{H_1(s)}{\rho_f R_p} = \frac{s}{2},$$

$$G_s = -\frac{H_2(s)}{-\rho_f R_p} = \left(\frac{9\nu_f}{2R_p^2} + \frac{9\sqrt{s\,\nu_f}}{2R_p} \right). \qquad \text{(F.15)}$$

By comparing the above definitions with those of Eq. (E.6), we can see that the operators G_v and G_s together are identical to the presentation in Eq. (E.6). However, here the operator G_v operates only over the volume-averaged relative velocity and the operator G_s operates only over the surface-averaged relative velocity. In the limit when the undisturbed ambient flow $\mathbf{u}_c$ does not vary spatially, its average over the surface or the volume of the particle will yield the same constant value. In which case, the resistances G_v and G_s can be added and we recover the result of Eq. (E.6). Finally, the Laplace inverse of Eq. (F.14) can be taken following the steps of Section 4.1.5, yielding

$$\mathbf{F}_{\text{pert}}(t) = 6\pi\mu_f R_p(\overline{\mathbf{u}_c}^s - \mathbf{V}) + \frac{1}{2}m_f \left[\frac{d\overline{\mathbf{u}_c}^v}{dt} - \frac{d\mathbf{V}}{dt}\right]$$
$$+ 6\pi\mu_f R_p \int_{-\infty}^t \frac{R_p}{\sqrt{\pi\nu_f(t-\xi)}} \left[\frac{d\overline{\mathbf{u}_c}^s}{dt} - \frac{d\mathbf{V}}{dt}\right]_{@\xi} d\xi .$$

(F.16)

References

Abgrall, R., and Saurel, R. (2003). Discrete equations for physical and numerical compressible multiphase mixtures. *Journal of Computational Physics*, **186**(2), 361–396.

Acrivos, A., and Taylor, T. D. (1962). Heat and mass transfer from single spheres in Stokes flow. *Physics of Fluids*, **5**(4), 387–394.

Akiki, G., and Balachandar, S. (2016). Immersed boundary method with non-uniform distribution of Lagrangian markers for a non-uniform Eulerian mesh. *Journal of Computational Physics*, **307**, 34–59.

Akiki, G., Jackson, T. L., and Balachandar, S. (2016). Force variation within arrays of monodisperse spherical particles. *Physical Review Fluids*, **1**(4), 044202.

Akiki, G., Jackson, T. L., and Balachandar, S. (2017a). Pairwise interaction extended point-particle model for a random array of monodisperse spheres. *Journal of Fluid Mechanics*, **813**, 882–928.

Akiki, G., Moore, W. C., and Balachandar, S. (2017b). Pairwise-interaction extended point-particle model for particle-laden flows. *Journal of Computational Physics*, **351**, 329–357.

Aliseda, A., Cartellier, A., Hainaux, F., and Lasheras, J. C. (2002). Effect of preferential concentration on the settling velocity of heavy particles in homogeneous isotropic turbulence. *Journal of Fluid Mechanics*, **468**, 77–105.

Almgren, A. S., Bell, J. B., Colella, P. and Marthaler, T. (1997). A Cartesian grid projection method for the incompressible Euler equations in complex geometries. *SIAM Journal of Scientific Computing*, **18**, 1289.

Anderson, T. B., and Jackson, R. (1967). Fluid mechanical description of fluidized beds. Equations of motion. *Industrial and Engineering Chemistry Fundamentals*, **6**(4), 527–539.

Andreotti, B., Forterre, Y., and Pouliquen, O. (2013). *Granular Media: Between Fluid and Solid*. Cambridge University Press.

Annamalai, S., and Balachandar, S. (2017). Faxén form of time-domain force on a sphere in unsteady spatially varying viscous compressible flows. *Journal of Fluid Mechanics*, **816**, 381–411.

Ardekani, A. M., and Rangel, R. H. (2008). Numerical investigation of particle–particle and particle–wall collisions in a viscous fluid. *Journal of Fluid Mechanics*, **596**, 437–466.

Aref, H., and Balachandar, S. (2017). *A First Course in Computational Fluid Dynamics.* Cambridge University Press.

Arolla, S. K., and Desjardins, O. (2015). Transport modeling of sedimenting particles in a turbulent pipe flow using Euler–Lagrange large eddy simulation. *International Journal of Multiphase Flow*, **75**, 1–11.

Asmolov, E. S., and McLaughlin, J. B. (1999). The inertial lift on an oscillating sphere in a linear shear flow. *International Journal of Multiphase Flow*, **25**(4), 739–751.

Auton, T. R. (1987). The lift force on a spherical body in a rotational flow. *Journal of Fluid Mechanics*, **183**, 199–218.

Auton, T. R., Hunt, J. C. R., and Prud'Homme, M. (1988). The force exerted on a body in inviscid unsteady non-uniform rotational flow. *Journal of Fluid Mechanics*, **197**, 241–257.

Ayala, O., Rosa, B., and Wang, L. P. (2008). Effects of turbulence on the geometric collision rate of sedimenting droplets. Part 2. Theory and parameterization. *New Journal of Physics*, **10**(7), 075016.

Baer, M. R., and Nunziato, J. W. (1986). A two-phase mixture theory for the deflagration-to-detonation transition (DDT) in reactive granular materials. *International Journal of Multiphase Flow*, **12**(6), 861–889.

Bagchi, P., and Balachandar, S. (2002a). Effect of free rotation on the motion of a solid sphere in linear shear flow at moderate Re. *Physics of Fluids*, **14**, 2719–2737.

Bagchi, P., and Balachandar, S. (2002b). Steady planar straining flow past a rigid sphere at moderate Reynolds number. *Journal of Fluid Mechanics*, **466**, 365–407.

Bagchi, P., and Balachandar, S. (2002c). Shear versus vortex-induced lift force on a rigid sphere at moderate Re. *Journal of Fluid Mechanics*, **473**, 379–388.

Bagchi, P., and Balachandar, S. (2003a). Inertial and viscous forces on a rigid sphere in straining flows at moderate Reynolds numbers. *Journal of Fluid Mechanics*, **481**, 105–148.

Bagchi, P., and Balachandar, S. (2003b). Effect of turbulence on the drag and lift of a particle. *Physics of Fluids*, **15**(11), 3496–3513.

Bagchi, P., and Balachandar, S. (2004). Response of the wake of an isolated particle to an isotropic turbulent flow. *Journal of Fluid Mechanics*, **518**, 95–123.

Bagchi, P., and Kottam, K. (2008). Effect of freestream isotropic turbulence on heat transfer from a sphere. *Physics of Fluids*, **20**(7), 073305.

Bagnold, R. A. (1954). Experiments on a gravity-free dispersion of large solid spheres in a Newtonian fluid under shear. *Proceedings of the Royal Society of London, Series A*, **225**(1160), 49–63.

Balachandar, S. (2003). Parameterization of force on a particle/drop/bubble. *Multiphase Science and Technology*, **15**, 157–171.

Balachandar, S. (2009). A scaling analysis for point-particle approaches to turbulent multiphase flows. *International Journal of Multiphase Flow*, **35**(9), 801–810.

Balachandar, S. (2020). Lagrangian and Eulerian drag models that are consistent between Euler–Lagrange and Euler–Euler (two-fluid) approaches for homogeneous systems. *Physical Review Fluids*, **5**(8), 084302.

Balachandar, S., and Eaton, J. K. (2010). Turbulent dispersed multiphase flow. *Annual Review of Fluid Mechanics*, **42**, 111–133.

Balachandar, S., and Ha, M. Y. (2001). Unsteady heat transfer from a sphere in a uniform cross-flow. *Physics of Fluids*, **13**(12), 3714–3728.

Balachandar, S., and Ha, M. Y. (2021). A note on thermal history kernel for unsteady heat transfer of a spherical particle. *International Journal of Heat and Mass Transfer*, **180**, 121781.

Balachandar, S., and Liu, K. (2023). A correction procedure for self-induced velocity of a finite-sized particle in two-way coupled Euler–Lagrange simulations. *International Journal of Multiphase Flow*, **159**, 104316.

Balachandar, S., and Maxey, M. R. (1989). Methods for evaluating fluid velocities in spectral simulations of turbulence. *Journal of Computational Physics*, **83**(1), 96–125.

Balachandar, S., and Maxey, M. R. (2023). Deterministic extended point-particle models. In Subramaniam, S., and Balachandar, S. (eds), *Modeling Approaches and Computational Methods for Particle-Laden Turbulent Flows*. Elsevier, pp. 299–330.

Balachandar, S., and Michaelides, E. E. (2022). Dispersed multiphase heat and mass transfer. *Annual Review of Heat Transfer*, **24**, 173–215.

Balachandar, S., and Prosperetti, A. (2007). *IUTAM Symposium on Computational Approaches to Multiphase Flow: Proceedings of an IUTAM Symposium Held at Argonne National Laboratory, October 4–7, 2004*. Springer Science & Business Media.

Balachandar, S., Liu, K., and Lakhote, M. (2019). Self-induced velocity correction for improved drag estimation in Euler–Lagrange point-particle simulations. *Journal of Computational Physics*, **376**, 160–185.

Balachandar, S., Moore, W. C., Akiki, G., and Liu, K. (2020). Toward particle-resolved accuracy in Euler–Lagrange simulations of multiphase flow using machine learning and pairwise interaction extended point-particle (PIEP) approximation. *Theoretical and Computational Fluid Dynamics*, **34**(4), 401–428.

Balachandar, S., Peng, C., and Wang, L.-P. (2023). Turbulence modulation by suspended finite-sized particles – Towards physics-based multiphase subgrid modelling, preprint, arXiv: 2311.13493.

Balakrishnan, K., and Bellan, J. (2018). High-fidelity modeling and numerical simulation of cratering induced by the interaction of a supersonic jet with a granular bed of solid particles. *International Journal of Multiphase Flow*, **99**, 1–29.

Balakrishnan, K., and Bellan, J. (2021). Fluid density effects in supersonic jet-induced cratering in a granular bed on a planetary body having an atmosphere in the continuum regime. *Journal of Fluid Mechanics*, **915**, A29.

Barnea, E., and Mizrahi, J. (1973). A generalized approach to the fluid dynamics of particulate systems: Part 1. General correlation for fluidization and sedimentation in solid multiparticle systems. *Chemical Engineering Journal*, **5**(2), 171–189.

Basset, A. (1888). *A Treatise on Hydrodynamics*. Deighton, Bell & Company.

Batchelor, G. K. (1970). The stress system in a suspension of force-free particles. *Journal of Fluid Mechanics*, **41**(3), 545–570.

Batchelor, G. K. (1972). Sedimentation in a dilute dispersion of spheres. *Journal of Fluid Mechanics*, **52**(2), 245–268.

Batchelor, G. K. (1974). Transport properties of two-phase materials with random structure. *Annual Review of Fluid Mechanics*, **6**(1), 227–255.

Batchelor, G. K. (2000). *An Introduction to Fluid Dynamics*. Cambridge University Press.

Bayewitz, M. H., Yerushalmi, J., Katz, S., and Shinnar, R. (1974). The extent of correlations in a stochastic coalescence process. *Journal of Atmospheric Sciences*, **31**(6), 1604–1614.

Beard, K. V., and Ochs III, H. T. (1983). Measured collection efficiencies for cloud drops. *Journal of Atmospheric Sciences*, **40**(1), 146–153.

Bec, J., Homann, H., and Ray, S. S. (2014). Gravity-driven enhancement of heavy particle clustering in turbulent flow. *Physical Review Letters*, **112**(18), 184501.

Berrouk, A. S., Laurence, D., Riley, J. J., and Stock, D. E. (2007). Stochastic modeling of inertial particle dispersion by subgrid motion for LES of high Reynolds number pipe flow. *Journal of Turbulence*, **8**, N50.

Beetstra, R., Van der Hoef, M. A., and Kuipers, J. A. M. (2007). Drag force of intermediate Reynolds number flow past mono- and bi-disperse arrays of spheres. *AIChE Journal*, **53**(2), 489–501.

Beylkin, G., and Monzón, L. (2005). On approximation of functions by exponential sums. *Applied and Computational Harmonic Analysis*, **19**(1), 17–48.

Biesheuvel, A., and Spoelstra, S. (1989). The added mass coefficient of a dispersion of spherical gas bubbles in liquid. *International Journal of Multiphase Flow*, **15**(6), 911–924.

Bird, R. B., Stewart, W. E., and Lightfoot, E. N. (1960). *Transport Phenomena*. Wiley.

Birkner, F. B., and Morgan, J. J. (1968). Polymer flocculation kinetics of dilute colloidal suspensions. *Journal of the American Water Works Association*, **60**(2), 175–191.

Birouk, M., and Gokalp, I. (2006). Current status of droplet evaporation in turbulent flows. *Progress in Energy and Combustion Science*, **32**(4), 408–423.

Bluemink, J. J., Lohse, D., Prosperetti, A., and Van Wijngaarden, L. (2008). A sphere in a uniformly rotating or shearing flow. *Journal of Fluid Mechanics*, **600**, 201–233.

Bogner, S., Mohanty, S., and Rüde, U. (2015). Drag correlation for dilute and moderately dense fluid-particle systems using the lattice Boltzmann method. *International Journal of Multiphase Flow*, **68**, 71–79.

Bordas, R., Roloff, C., Thevenin, D., and Shaw, R. A. (2013). Experimental determination of droplet collision rates in turbulence. *New Journal of Physics*, **15**(4), 045010.

Bosse, T., Kleiser, L., and Meiburg, E. (2006). Small particles in homogeneous turbulence: Settling velocity enhancement by two-way coupling. *Physics of Fluids*, **18**(2), 027102.

Botto, L., and Prosperetti, A. (2012). A fully resolved numerical simulation of turbulent flow past one or several spherical particles. *Physics of Fluids*, **24**(1), 013303.

Bouchet, G., Mebarek, M., and Dušek, J. (2006). Hydrodynamic forces acting on a rigid fixed sphere in early transitional regimes. *European Journal of Mechanics B/Fluids*, **25**(3), 321–336.

Boussinesq, J. (1885). Sur la résistance qu'oppose un liquide indéfini au repos au mouvement varié d'une sphére solide. *Comptes rendus de l'Académie des Sciences, Paris*, **100**, 935–937.

Brady, J. F., and Bossis, G. (1988). Stokesian dynamics. *Annual Review of Fluid Mechanics*, **20**, 111–157.

Braeunig, J. P., Desjardins, B., and Ghidaglia, J. M. (2007). A pure Eulerian scheme for multimaterial fluid flows. Paper presented at the workshop "Numerical Methods for Multi-material Fluid Flows," Czech Technical University in Prague.

Brandt, L., and Coletti, F. (2022). Particle-laden turbulence: progress and perspectives. *Annual Review of Fluid Mechanics*, **54**, 159–189.

Brennen, C. E. (2005). *Fundamentals of Multiphase Flow*. Cambridge University Press.

Brenner, H. (1961). The slow motion of a sphere through a viscous fluid towards a plane surface. *Chemical Engineering Science*, **16**(3–4), 242–251.

Brilliantov, N. V., and Pöschel, T. (2010). *Kinetic Theory of Granular Gases*. Oxford University Press

Brinkman, H. C. (1947). Fluid flow in a porous medium. *Applied Science Research A*, **27**(143149), 42.

Brucato, A., Grisafi, F., and Montante, G. (1998). Particle drag coefficients in turbulent fluids. *Chemical Engineering Science*, **53**, 3295–3314.

Brush Jr., L. M., Ho, H.-W., and Yen, B.-C. (1964). Accelerated motion of a sphere in a viscous fluid. *Journal of the Hydraulics Division*, **90**(1), 149–160.

Burgers, J. M. (1995). Hydrodynamics – On the influence of the concentration of a suspension upon the sedimentation velocity (in particular for a suspension of spherical particles). In Nieuwstadt, F. T. M., and Steketee, J. A. (eds), *Selected Papers of J. M. Burgers*. Springer, pp. 452–477.

Burton, T. M., and Eaton, J. K. (2005). Fully resolved simulations of particle–turbulence interaction. *Journal of Fluid Mechanics*, **545**, 67.

Buscall, R., Goodwin, J. W., Ottewill, R. H., and Tadros, T. F. (1982). The settling of particles through Newtonian and non-Newtonian media. *Journal of Colloid and Interface Science*, **85**(1), 78–86.

Calhoun, D., and LeVeque, R. J. (2000). A Cartesian grid finite-volume method for the advection–diffusion equation in irregular geometries. *Journal of Computational Physics*, **157**, 143–180.

Candelier, F., and Souhar, M. (2007). Time-dependent lift force acting on a particle moving arbitrarily in a pure shear flow, at small Reynolds number. *Physical Review E*, **76**(6), 067301.

Cantero, M. I., Lee, J. R., Balachandar, S., and Garcia, M. H. (2007). On the front velocity of gravity currents. *Journal of Fluid Mechanics*, **586**, 1–39.

Cantero, M. I., Balachandar, S., Cantelli, A., Pirmez, C., and Parker, G. (2009). Turbidity current with a roof: Direct numerical simulation of self-stratified turbulent channel flow driven by suspended sediment. *Journal of Geophysical Research: Oceans*, **114**, C3.

Cantero, M. I., Cantelli, A., Pirmez, C., et al. (2012). Emplacement of massive turbidites linked to extinction of turbulence in turbidity currents. *Nature Geoscience*, **5**(1), 42–45.

Capecelatro, J., and Desjardins, O. (2013). An Euler–Lagrange strategy for simulating particle-laden flows. *Journal of Computational Physics*, **238**, 1–31.

Capecelatro, J., and Desjardins, O. (2022). Volume-filtered Euler–Lagrange method for strongly coupled fluid–particle flows. In Subramaniam, S., and Balachandar, S. (eds), *Modeling Approaches and Computational Methods for Particle-Laden Turbulent Flows*. Elsevier, pp. 383–418.

Caporaloni, M., Tampieri, F., Trombetti, F., and Vittori, O. (1975). Transfer of particles in nonisotropic air turbulence. *Journal of the Atmospheric Sciences*, **32**(3), 565–568.

Carnahan, N. F., and Starling, K. E. (1969). Equation of state for nonattracting rigid spheres. *Journal of Chemical Physics*, **51**(2), 635–636.

Carrier, G. F. (1958). Shock waves in a dusty gas. *Journal of Fluid Mechanics*, **4**(4), 376–382.

Carter, D., Petersen, A., Amili, O., and Coletti, F. (2016). Generating and controlling homogeneous air turbulence using random jet arrays. *Experiments in Fluids*, **57**(12), 189.

Chakraborty, P., Balachandar, S., and Adrian, R. J. (2005). On the relationships between local vortex identification schemes. *Journal of Fluid Mechanics*, **535**, 189–214.

Chandrasekhar, S. (1943). Stochastic problems in physics and astronomy. *Reviews of Modern Physics*, **15**(1), 1.

Chang, E. J., and Maxey, M. R. (1994). Unsteady flow about a sphere at low to moderate Reynolds number. Part 1. Oscillatory motion. *Journal of Fluid Mechanics*. **277**, 347–379.

Chang, E. J., and Maxey, M. R. (1995). Unsteady flow about a sphere at low to moderate Reynolds number. Part 2. Accelerated motion. *Journal of Fluid Mechanics*, **303**, 133–153.

Chapman, S., and Cowling, T. G. (1990). *The Mathematical Theory of Non-Uniform Gases: An Account of the Kinetic Theory of Viscosity, Thermal Conduction and Diffusion in Gases*. Cambridge University Press.

Chen, R. C., and Wu, J. L. (2000). The flow characteristics between two interactive spheres. *Chemical Engineering Science*, **55**(6), 1143–1158.

Cheng, L. Y., Drew, D. A., and Lahey Jr., R. T. (1978). Virtual mass effects in two-phase flow (No. NUREG/CR-0020). Rensselaer Polytechnic Inst., Troy, NY. Dept. of Nuclear Engineering.

Cherukat, P., and McLaughlin, J. B. (1990). Wall-induced lift on a sphere. *International Journal of Multiphase Flow*, **16**(5), 899–907.

Cherukat, P., and McLaughlin, J. B. (1994). The inertial lift on a rigid sphere in a linear shear flow field near a flat wall. *Journal of Fluid Mechanics*, **263**, 1–18. Corrigendum: *Journal of Fluid Mechanics*, **285**, 407.

Chester, W., Breach, D. R., and Proudman, I. (1969). On the flow past a sphere at low Reynolds number. *Journal of Fluid Mechanics*, **37**(04), 751–760.

Childress, S. (1972). Viscous flow past a random array of spheres. *Journal of Chemical Physics*, **56**(6), 2527–2539.

Chouippe, A., and Uhlmann, M. (2015). Forcing homogeneous turbulence in direct numerical simulation of particulate flow with interface resolution and gravity. *Physics of Fluids*, **27**(12), 123301.

Chouippe, A., and Uhlmann, M. (2019). On the influence of forced homogeneous-isotropic turbulence on the settling and clustering of finite-size particles. *Acta Mechanica*, **230**(2), 387–412.

Chung, J. N., and Troutt, T. R. (1988). Simulation of particle dispersion in an axisymmetric jet. *Journal of Fluid Mechanics*, **186**, 199–222.

Clamen, A., and Gauvin, W. H. (1969). Effects of turbulence on the drag coefficients of spheres in a supercritical flow regime. *AIChE Journal*, **15**(2), 184–189.

Clift, R., Grace, J. R., and Weber, M. E. (1978). *Bubbles, Drops and Particles*. Academic Press. Reprinted 2005, Courier Corporation.

Climent, E., and Maxey, M. R. (2009). The force coupling method: A flexible approach for the simulation of particulate flows. In Feuillebois, F., and Sellier, A. (eds), *Methods for Creeping Flows*. Ressign Press.

Coimbra, C. F. M., L'Esperance, D., Lambert, R. A., Trolinger, J .D., and Rangel, R. H. (2004). An experimental study on stationary history effects in high-frequency Stokes flows. *Journal of Fluid Mechanics*, **504**, 353–363.

Constantinescu, G. S., and Squires, K. D. (2003). LES and DES investigations of turbulent flow over a sphere at Re = 10,000. *Flow, Turbulence and Combustion*, **70**(1–4), 267–298.

Cooley, M. D. A., and O'Neill, M. E. (1969). On the slow motion generated in a viscous fluid by the approach of a sphere to a plane wall or stationary sphere. *Mathematika*, **16**(1), 37–49.

Corrsin, S., and Lumley, J. (1956). On the equation of motion for a particle in turbulent fluid. *Applied Scientific Research*, **6**(2), 114–116.

Costa, P., Picano, F., Brandt, L., and Breugem, W. P. (2018). Effects of the finite particle size in turbulent wall-bounded flows of dense suspensions. *Journal of Fluid Mechanics*, **843**, 450–478.

Cox, R. G., and Brenner, H. (1968). The lateral migration of solid particles in Poiseuille flow: I Theory. *Chemical Engineering Science*, **23**(2), 147–173.

Cox, R. G., and Hsu, S. K. (1977). The lateral migration of solid particles in a laminar flow near a plane. *International Journal of Multiphase Flow*, **3**(3), 201–222.

Crowe, C. T., Troutt, T. R., and Chung, J. N. (1996). Numerical models for two-phase turbulent flows. *Annual Review of Fluid Mechanics*, **28**, 11–43.

Crowe, C. T., Schwarzkopf, J. D., Sommerfeld, M., and Tsuji, Y. (2011). *Multiphase Flows with Droplets and Particles*. CRC Press.

Csanady, G. T. (1963). Turbulent diffusion of heavy particles in the atmosphere. *Journal of the Atmospheric Sciences*, **20**(3), 201–208.

Cundall, P. A., and Strack, O. D. (1979). A discrete numerical model for granular assemblies. *Geotechnique*, **29**(1), 47–65.

Cuzzi, J. N., Hogan, R. C., Paque, J. M., and Dobrovolskis, A. R. (2001). Size-selective concentration of chondrules and other small particles in protoplanetary nebula turbulence. *Astrophysical Journal*, **546**(1), 496.

Daley, D., and Vere-Jones, D. (2003). *An Introduction to the Theory of Point Processes*. Springer.

Dance, S. L., and Maxey, M. R. (2003). Incorporation of lubrication effects into the force-coupling method for particulate two-phase flow. *Journal of Computational Physics*, **189**(1), 212–238

Davies, C. (1966). Deposition from moving aerosols. In Davies, C. N. (ed.), *Aerosol Science*. Academic Press, pp. 393–446.

Davila, J., and Hunt, J. C. R. (2001). Settling of small particles near vortices and in turbulence. *Journal of Fluid Mechanics*, **440**, 117–145.

de Almeida, F. C. (1976). The collisional problem of cloud droplets moving in a turbulent environment: Part I. A method of solution. *Journal of the Atmospheric Sciences*, **33**(8), 1571–1578.

de Almeida, F. C. (1979a). The collisional problem of cloud droplets moving in a turbulent environment: Part II. Turbulent collision efficiencies. *Journal of the Atmospheric Sciences*, **36**(8), 1564–1576.

de Almeida, F. C. (1979b). The effects of small-scale turbulent motions on the growth of a cloud droplet spectrum. *Journal of the Atmospheric Sciences*, **36**(8), 1557–1563.

Dean, W., and O'Neill, M. E. (1964). A slow motion of viscous liquid caused by the rotation of a solid sphere. *Mathematika*, **11**, 67–74.

Deen, N. G., Peters, E. A. J. F., Padding, J. T., and Kuipers, J. A. M. (2014). Review of direct numerical simulation of fluid–particle mass, momentum and heat transfer in dense gas–solid flows. *Chemical Engineering Science*, **116**, 710–724.

Dehbi, A. (2008). Turbulent particle dispersion in arbitrary wall-bounded geometries: A coupled CFD-Langevin-equation based approach. *International Journal of Multiphase Flow*, **34**(9), 819–828.

Dehbi, A. (2010). Validation against DNS statistics of the normalized Langevin model for particle transport in turbulent channel flows. *Powder Technology*, **200**(1–2), 60–68.

Delichatsios, M. A., and Probstein, R. F. (1975). Coagulation in turbulent flow: Theory and experiment. *Journal of Colloid and Interface Science*, **51**(3), 394–405.

Dennis, S. C. R., and Walker, J. D. A. (1971) Calculation of the steady flow past a sphere at low and moderate Reynolds numbers. *Journal of Fluid Mechanics*, **48**, 771–789.

Deville, M. O., Fischer, P. F., and Mund, E. H. (2002). *High-Order Methods for Incompressible Fluid Flow*. Cambridge University Press.

Ding, J., and Gidaspow, D. (1990). A bubbling fluidization model using kinetic theory of granular flow. *AIChE Journal*, **36**(4), 523–538.

Dorgan, A. J., and Loth, E. (2007). Efficient calculation of the history force at finite Reynolds numbers. *International Journal of Multiphase Flow*, **33**(8), 833–848.

Drew, D. A. (1983). Mathematical modeling of two-phase flow. *Annual Review of Fluid Mechanics*, **15**(1), 261–291.

Drew, D. A., and Passman, S. L. (2006). *Theory of Multicomponent Fluids*. Springer.

Druzhinin, O. A. (1995). On the two-way interaction in two-dimensional particle-laden flows: The accumulation of particles and flow modification. *Journal of Fluid Mechanics*, **297**, 49–76.

Druzhinin, O. A., and Elghobashi, S. (1998). Direct numerical simulations of bubble-laden turbulent flows using the two-fluid formulation. *Physics of Fluids*, **10**(3), 685–697.

Druzhinin, O. A., and Elghobashi, S. (1999). On the decay rate of isotropic turbulence laden with microparticles. *Physics of Fluids*, **11**, 602–610.

Duru, P., Nicolas, M., Hinch, J., and Guazzelli, E. (2002). Constitutive laws in liquid-fluidized beds. *Journal of Fluid Mechanics*, **452**, 371–404.

Duru, P., Koch, D. L., and Cohen, C. (2007). Experimental study of turbulence-induced coalescence in aerosols. *International Journal of Multiphase Flow*, **33**(9), 987–1005.

Eaton, J. K., and Fessler, J. R. (1994). Preferential concentration of particles by turbulence. *International Journal of Multiphase Flow*, **20**, 169–209.

Einstein, A. (1906). Eine neue bestimmung der moleküldimensionen. *Annalen der Physik*, **19**, 289–306. Correction in *Annalen der Physik*, **34**, 591 (1911).

Eiseman, P. R. (2016). GRIDPRO v6.1, Topology Input Language Manual. Program Development Company, White Plains, NY.

Elghannay, H. A., and Tafti, D. K. (2016). Development and validation of a reduced order history force model. *International Journal of Multiphase Flow*, **85**, 284–297.

Elghobashi, S. (1991). Particle-laden turbulent flows: Direct simulation and closure models. *Applied Scientific Research*, **48**(3–4), 301–314.

Elghobashi, S. (1994). On predicting particle-laden turbulent flows. *Applied Scientific Research*, **52**(4), 309–329.

Elghobashi, S., and Truesdell, G. C. (1992). Direct simulation of particle dispersion in a decaying isotropic turbulence. *Journal of Fluid Mechanics*, **242**, 655–700.

Elghobashi S., and Truesdell, G. C. (1993). On the 2-way interaction between homogeneous turbulence and dispersed solid particles. 1: Turbulence modification. *Physics of Fluids*, **5**, 1790–1801.

Esmaily, M., and Mani, A. (2020). Modal analysis of the behavior of inertial particles in turbulence subjected to Stokes drag. *Physical Review Fluids*, **5**(8), 084303.

Esmaily-Moghadam, M., and Mani, A. (2016). Analysis of the clustering of inertial particles in turbulent flows. *Physical Review Fluids*, **1**(8), 084202.

Evrard, F., Denner, F., and van Wachem, B. (2019). A multi-scale approach to simulate atomisation processes. *International Journal of Multiphase Flow*, **119**, 194–216.

Fadlun, E. A., Verzicco, R., Orlandi, P., and Mohd-Yusof, J. (2000). Combined immersed-boundary finite-difference methods for three-dimensional complex flow simulations. *Journal of Computational Physics*, **161**, 35–60.

Faeth, G. M. (1983). Evaporation and combustion of sprays. *Progress in Energy and Combustion Science*, **9**(1–2), 1–76.

Faeth, G. M. (1987). Mixing, transport and combustion in sprays. *Progress in Energy and Combustion Science*, **13**(4), 293–345.

Falkovich, G., Fouxon, A., and Stepanov, M. G. (2002). Acceleration of rain initiation by cloud turbulence. *Nature*, **419**(6903), 151–154.

Faxén, H. (1923). Die Bewegung einer starren Kugel längs der Achse eines mit zäher Flüssigkeit gefüllten Rohres. *Arkiv foer Matematik, Astronomi och Fysik*, **17**(27), 1–28.

Fedkiw, R. P., Aslam, T., Merriman, B., and Osher, S. (1999). A non-oscillatory Eulerian approach to interfaces in multimaterial flows (the ghost fluid method). *Journal of Computational Physics*, **152**(2), 457–492.

Feng, Z. G., and Michaelides, E. E. (2000). A numerical study on the transient heat transfer from a sphere at high Reynolds and Peclet numbers. *International Journal of Heat and Mass Transfer*, **43**(2), 219–229.

Ferenc, J. S., and Néda, Z. (2007). On the size distribution of Poisson Voronoi cells. *Physica A: Statistical Mechanics and its Applications*, **385**(2), 518–526.

Ferrante, A., and Elghobashi, S. (2004). On the physical mechanisms of drag reduction in a spatially developing turbulent boundary layer laden with microbubbles. *Journal of Fluid Mechanics*, **503**, 345–355.

Ferry, J., and Balachandar, S. (2001). A fast Eulerian method for disperse two-phase flow. *International Journal of Multiphase Flow*, **27**(7), 1199–1226.

Ferry, J., and Balachandar, S. (2002). Equilibrium expansion for the Eulerian velocity of small particles. *Powder Technology*, **125**(2–3), 131–139.

Ferry, J., and Balachandar, S. (2005). Equilibrium Eulerian approach for predicting the thermal field of a dispersion of small particles. *International Journal of Heat and Mass Transfer*, **48**(3–4), 681–689.

Ferry, J., Rani, S. L., and Balachandar, S. (2003). A locally implicit improvement of the equilibrium Eulerian method. *International Journal of Multiphase Flow*, **29**(6), 869–891.

Fessler, J. R., Kulick, J. D., and Eaton, J. K. (1994). Preferential concentration of heavy particles in a turbulent channel flow. *Physics of Fluids*, **6**(11), 3742–3749.

Feuillebois, F., and Lasek, A. (1978). On the rotational historic term in nonstationary Stokes flow. *Quarterly Journal of Mechanics and Applied Mathematics*, **31**(4), 435–443.

Fevrier, P., Simonin, O., and Squires, K. D. (2005). Partitioning of particle velocities in gas–solid turbulent flows into a continuous field and a spatially uncorrelated random distribution: Theoretical formalism and numerical study. *Journal of Fluid Mechanics*, **533**, 1–46.

Finn, J. R., Li, M., and Apte, S. V. (2016). Particle based modelling and simulation of natural sand dynamics in the wave bottom boundary layer. *Journal of Fluid Mechanics*, **796**, 340–385.

Fornari, W., Picano, F., Sardina, G., and Brandt, L. (2016). Reduced particle settling speed in turbulence. *Journal of Fluid Mechanics*, **808**, 153–167.

Fornari, W., Zade, S., Brandt, L., and Picano, F. (2019). Settling of finite-size particles in turbulence at different volume fractions. *Acta Mechanica*, **230**(2), 413–430.

Fox, R. O. (2012). Large-eddy-simulation tools for multiphase flows. *Annual Review of Fluid Mechanics*, **44**, 47–76

Fox, R. O. (2019). A kinetic-based hyperbolic two-fluid model for binary hard-sphere mixtures. *Journal of Fluid Mechanics*, **877**, 282–329.

Fox, R. O., Laurent, F., and Vié, A. (2020). A hyperbolic two-fluid model for compressible flows with arbitrary material-density ratios. *Journal of Fluid Mechanics*, **903**, A5.

Friedlander, S. K. (1960a). On the particle-size spectrum of atmospheric aerosols. *Journal of Meteorology*, **17**(3), 373–374.

Friedlander, S. K. (1960b). Similarity considerations for the particle-size spectrum of a coagulating, sedimenting aerosol. *Journal of the Atmospheric Sciences*, **17**(5), 479–483.

Friedlander, S. K., and Johnstone, H. F. (1957). Deposition of suspended particles from turbulent gas streams. *Industrial and Engineering Chemistry*, **49**(7), 1151–1156.

Frost, D. L., Gregoire, Y., Petel, O., Goroshin, S., and Zhang, F. (2012). Particle jet formation during explosive dispersal of solid particles. *Physics of Fluids*, **24**(9), 091109.

Fukada, T., Takeuchi, S., and Kajishima, T. (2016). Interaction force and residual stress models for volume-averaged momentum equation for flow laden with particles of comparable diameter to computational grid width. *International Journal of Multiphase Flow*, **85**, 298–313.

Fukada, T., Fornari, W., Brandt, L., Takeuchi, S., and Kajishima, T. (2018). A numerical approach for particle–vortex interactions based on volume-averaged equations. *International Journal of Multiphase Flow*, **104**, 188–205.

Fung, J. C. H. (1997). Gravitational settling of small spherical particles in unsteady cellular flow fields. *Journal of Aerosol Science*, **28**(5), 753–787.

Garcia, M., Riber, E., Simonin, O., and Poinsot, T. (2007). Comparison between Euler/Euler and Euler/Lagrange LES approaches for confined bluff-body gas–solid flow prediction. In *Proceedings of the 6th International Conference on Multiphase Flow, ICMF 2007*.

Garcia-Villalba, M., Kidanemariam, A. G., and Uhlmann, M. (2012). DNS of vertical plane channel flow with finite-size particles: Voronoi analysis, acceleration statistics and particle-conditioned averaging. *International Journal of Multiphase Flow*, **46**, 54–74.

Garg, R., Narayanan, C., Lakehal, D., and Subramaniam, S. (2007). Accurate numerical estimation of interphase momentum transfer in Lagrangian–Eulerian simulations of dispersed two-phase flows. *International Journal of Multiphase Flow*, **33**(12), 1337–1364.

Gatignol, R. (1983). The Faxén formulae for a rigid particle in an unsteady non-uniform Stokes flow. *Journal de Mecanique Theorique et Appliquee*, **1**, 143–160.

Geier, M., Pasquali, A., and Schonherr, M. (2017). Parametrization of the cumulant lattice Boltzmann method for fourth order accurate diffusion Part II: Application to flow around a sphere at drag crisis. *Journal of Computational Physics*, **348**, 889–898.

Germano, M., Piomelli, U., Moin, P., and Cabot, W. H. (1991). A dynamic subgrid-scale eddy viscosity model. *Physics of Fluids A: Fluid Dynamics*, **3**(7), 1760–1765.

Ghosal, S., and Moin, P. (1995). The basic equations for the large eddy simulation of turbulent flows in complex geometry. *Journal of Computational Physics*, **118**(1), 24–37.

Gidaspow, D. (1994). *Multiphase Flow and Fluidization: Continuum and Kinetic Theory Descriptions*. Academic Press.

Gillespie, D. T. (1972). The stochastic coalescence model for cloud droplet growth. *Journal of the Atmospheric Sciences*, **29**(8), 1496–1510.

Gillespie, D. T. (1975). Three models for the coalescence growth of cloud drops. *Journal of the Atmospheric Sciences*, **32**(3), 600–607.

Guildenbecher, D. R., López-Rivera, C., and Sojka, P. E. (2009). Secondary atomization. *Experiments in Fluids*, **46**(3), 371–402.

Glawe, D. D., and Samimy, M. (1993). Dispersion of solid particles in compressible mixing layers. *Journal of Propulsion and Power*, **9**(1), 83–89.

Goldhirsch, I. and Zanetti, G. (1993). Clustering instability in dissipative gases. *Physical Review Letters*, **70**(11), 1619.

Goldman, A. J., Cox, R. G., and Brenner, H. (1966). The slow motion of two identical arbitrarily oriented spheres through a viscous fluid. *Chemical Engineering Science*, **21**(12), 1151–1170.

Goldman, A. J., Cox, R. G., and Brenner, H. (1967a). Slow viscous motion of a sphere parallel to a plane wall: I Motion through a quiescent fluid. *Chemical Engineering Science*, **22**(4), 637–651.

Goldman, A. J., Cox, R. G., and Brenner, H. (1967b). Slow viscous motion of a sphere parallel to a plane wall: II Couette flow. *Chemical Engineering Science*, **22**(4), 653–660.

Goldstein, D., Handler, R., and Sirovich, L. (1993). Modeling a no-slip flow boundary with an external force field, *Journal of Computational Physics*, **105**, 354–366.

Good, G. H., Ireland, P. J., Bewley, G. P., Bodenschatz, E., Collins, L. R., and Warhaft, Z. (2014). Settling regimes of inertial particles in isotropic turbulence. *Journal of Fluid Mechanics*, **759**, R3.

Gore, R. A., and Crowe, C. T. (1990). Discussion of particle drag in a dilute turbulent two-phase suspension flow. *International Journal of Multiphase Flow*, **16**, 359–361.

Grabowski, W. W., and Wang, L. P. (2013). Growth of cloud droplets in a turbulent environment. *Annual Review of Fluid Mechanics*, **45**, 293–324.

Griffith, B. E., and Patankar, N. A. (2020). Immersed methods for fluid–structure interaction. *Annual Review of Fluid Mechanics*, **52**, 421–448.

Grover, S. N., and Pruppacher, H. R. (1985). The effect of vertical turbulent fluctuations in the atmosphere on the collection of aerosol particles by cloud drops. *Journal of the Atmospheric Sciences*, **42**(21), 2305–2318.

Gualtieri, P., Picano, F., Sardina, G., and Casciola, C. M. (2015). Exact regularized point particle method for multiphase flows in the two-way coupled regime. *Journal of Fluid Mechanics*, **773**, 520–561.

Gunn, D. J. (1978). Transfer of heat or mass to particles in fixed and fluidized beds. *International Journal of Heat and Mass Transfer*, **21**(4), 467–476.

Gurvich, A. S., and Yaglom, A. M. (1967). Breakdown of eddies and probability distributions for small-scale turbulence. *Physics of Fluids*, **10**(9), S59–S65.

Guz, A. N. (2009). *Dynamics of Compressible Viscous Fluid*. Cambridge Scientific Publishers.

Haff, P. K. (1983). Grain flow as a fluid-mechanical phenomenon. *Journal of Fluid Mechanics*, **134**, 401–430.

Hall, D. (1988). Measurements of the mean force on a particle near a boundary in turbulent flow. *Journal of Fluid Mechanics*, **187**, 451–466.

Happel, J., and Brenner, H. (1965). *Low Reynolds Number Hydrodynamics*. Prentice-Hall.

Hasimoto, H. (1959). On the periodic fundamental solutions of the Stokes equations and their application to viscous flow past a cubic array of spheres. *Journal of Fluid Mechanics*, **5**(2), 317–328.

Haworth, D. C., and Pope, S. B. (1986). A generalized Langevin model for turbulent flows. *Physics of Fluids*, **29**(2), 387–405.

He, L., Tafti, D. K., and Nagendra, K. (2017). Evaluation of drag correlations using particle resolved simulations of spheres and ellipsoids in assembly. *Powder Technology*, **313**, 332–343.

Helfinstine, R. A., and Dalton, C. (1974). Unsteady potential flow past a group of spheres. *Computers and Fluids*, **2**(1), 99–112.

Herron, I. H., Davis, S. H., and Bretherton, F. P. (1975) On the sedimentation of a sphere in a centrifuge. *Journal of Fluid Mechanics*, **68**, 209–234

Hertz, H. (1882). Über die Berührung fester elastische Körper. *J. Reine und Angewandte Mathematik*, **92**, 156–171. English translation: On the contact of rigid elastic solids and on hardness: Chapter 6 of *Miscellaneous Papers by H. Hertz*, Macmillan, 1896.

Hetsroni, G. (1989). Particle–turbulence interaction. *International Journal of Multiphase Flow*, **15**(5), 735–746.

Hill, R. J. (2002). Scaling of acceleration in locally isotropic turbulence. *Journal of Fluid Mechanics*, **452**, 361–370.

Hill, R. J., Koch, D. L., and Ladd, A. J. (2001). Moderate-Reynolds-number flows in ordered and random arrays of spheres. *Journal of Fluid Mechanics*, **448**, 243–278.

Hinze, J. O. (1959). *Turbulence*. McGraw-Hill.

Hirt, C. W., and Nichols, B. D. (1981). Volume of fluid (VOF) method for the dynamics of free boundaries. *Journal of Computational Physics*, **39**(1), 201–225.

Ho, B. P., and Leal, L. G. (1974). Inertial migration of rigid spheres in two-dimensional unidirectional flows. *Journal of Fluid Mechanics*, **65**(2), 365–400.

Hocking, L. M., and Jonas, P. R. (1970). The collision eficiency of small drops. *Quarterly Journal of the Royal Meteorological Society*, **96**, 722–720.

Homann, H., Bec, J., and Grauer, R. (2013). Effect of turbulent fluctuations on the drag and lift forces on a towed sphere and its boundary layer. *Journal of Fluid Mechanics*, **721**, 155–179.

Horowitz, M., and Williamson, C. H. K. (2010). The effect of Reynolds number on the dynamics and wakes of freely rising and falling spheres. *Journal of Fluid Mechanics*, **651**, 251–294.

Horwitz, J. A. K., and Mani, A. (2016). Accurate calculation of Stokes drag for point-particle tracking in two-way coupled flows. *Journal of Computational Physics*, **318**, 85–109.

Horwitz, J. A. K., and Mani, A. (2018). Correction scheme for point-particle models applied to a nonlinear drag law in simulations of particle–fluid interaction. *International Journal of Multiphase Flow*, **101**, 74–84.

Houim, R. W., and Oran, E. S. (2016). A multiphase model for compressible granular–gaseous flows: Formulation and initial tests. *Journal of Fluid Mechanics*, **789**, 166–220.

Hu, K. C., and Mei, R. (1998). Particle collision rate in fluid flows. *Physics of Fluids*, **10**(4), 1028–1030.

Huang, Z., Wang, H., Zhou, Q., and Li, T. (2017). Effects of granular temperature on inter-phase drag in gas–solid flows. *Powder Technology*, **321**, 435–443.

Huang, Z., Zhang, C., Jiang, M., Wang, H., and Zhou, Q. (2019). Effects of particle velocity fluctuations on inter-phase heat transfer in gas–solid flows. *Chemical Engineering Science*, **206**, 375–386.

Humphrey, J. A. C., and Murata, H. (1992). On the motion of solid spheres falling through viscous fluids in vertical and inclined tubes. *ASME Journal of Fluids Engineering*, **114**, 2–11.

Hunt, J. C. R. (1982). Self-similar particle-size distributions during coagulation: Theory and experimental verification. *Journal of Fluid Mechanics*, **122**, 169–185.

Ireland, P. J., and Desjardins, O. (2017). Improving particle drag predictions in Euler-Lagrange simulations with two-way coupling. *Journal of Computational Physics*, **338**, 405–430.

Ireland, P. J., Bragg, A. D., and Collins, L. R. (2016). The effect of Reynolds number on inertial particle dynamics in isotropic turbulence. Part 2. Simulations with gravitational effects. *Journal of Fluid Mechanics*, **796**, 659–711.

Irving, J. H., and Kirkwood, J. G. (1950). The statistical mechanical theory of transport processes. IV. The equations of hydrodynamics. *Journal of Chemical Physics*, **18**(6), 817–829.

Ishii, M. (1975). Thermo-fluid dynamic theory of two-phase flow. NASA Sti/Recon Technical Report A, 75, 29657.

Jacobs, G. B., Kopriva, D. A., and Mashayek, F. (2007). Towards efficient tracking of inertial particles with high-order multidomain methods. *Journal of Computational and Applied Mathematics*, **206**(1), 392–408.

Jacqmin, D. (1999). Calculation of two-phase Navier–Stokes flows using phase-field modeling. *Journal of Computational Physics*, **155**(1), 96–127.

Jebakumar, A. S., Magi, V., and Abraham, J. (2019). Lattice-Boltzmann simulations of flow past stationary particles in a channel. *Numerical Heat Transfer, Part A: Applications*, **76**(5), 281–300.

Jeffrey, D. J., and Onishi, Y. (1984). Calculation of the resistance and mobility functions for two unequal rigid spheres in low-Reynolds-number flow. *Journal of Fluid Mechanics*, **139**, 261–290.

Jeffrey, G. B. (1915). On the steady rotation of a solid of revolution in a viscous fluid. *Proceedings of the London Mathematical Society*, **2**(1), 327–338.

Jenkins, J. T., and Savage, S. B. (1983). A theory for the rapid flow of identical, smooth, nearly elastic, spherical particles. *Journal of Fluid Mechanics*, **130**, 187–202.

Jenny, M., Dušek, J., and Bouchet, G. (2004). Instabilities and transition of a sphere falling or ascending freely in a Newtonian fluid. *Journal of Fluid Mechanics*, **508**, 201–239.

Jin, C., Potts, I., and Reeks, M. W. (2015). A simple stochastic quadrant model for the transport and deposition of particles in turbulent boundary layers. *Physics of Fluids*, **27**(5), 053305.

Jin, C., Potts, I., and Reeks, M. W. (2016). The effects of near wall corrections to hydrodynamic forces on particle deposition and transport in vertical turbulent boundary layers. *International Journal of Multiphase Flow*, **79**, 62–73.

Johnson, T. A. and Patel, V. C. (1999). Flow past a sphere up to a Reynolds number of 300. *Journal of Fluid Mechanics*, **378**, 19–70.

Jonas, P. R., and Goldsmith, P. (1972). The collection efficiencies of small droplets falling through a sheared air flow. *Journal of Fluid Mechanics*, **52**(3), 593–608.

Joseph, D. D., Lundgren, T. S., Jackson, R., and Saville, D. A. (1990). Ensemble averaged and mixture theory equations for incompressible fluid–particle suspensions. *International Journal of Multiphase Flow*, **16**(1), 35–42.

Kajishima, T., Takiguchi, S., Hamasaki, H., and Miyake, Y. (2001). Turbulence structure of particle-laden flow in a vertical plane channel due to vortex shedding. *JSME International Journal Series B Fluids and Thermal Engineering*, **44**, 526–535.

Kallio, G. A., and Reeks, M. W. (1989). A numerical simulation of particle deposition in turbulent boundary layers. *International Journal of Multiphase Flow*, **15**(3), 433–446.

Kaneda, Y. (1980). A generalization of Faxén's theorem to nonsteady motion of an almost spherical drop in an arbitrary flow of a compressible fluid. *Physica A*, **101**, 407–422.

Kapila, A. K., Menikoff, R., Bdzil, J. B., Son, S. F., and Stewart, D. S. (2001). Two-phase modeling of deflagration-to-detonation transition in granular materials: Reduced equations. *Physics of Fluids*, **13**(10), 3002–3024.

Kawanisi, K., and Shiozaki, R. (2008). Turbulent effects on the settling velocity of suspended sediment. *Journal of Hydraulic Engineering*, **134**(2), 261–266.

Kendall, K. (1971). The adhesion and surface energy of elastic solids. *Journal of Physics D: Applied Physics*, **4**(8), 1186–1195.

Kim, I., Elghobashi, S., and Sirignano, W. A. (1993). Three-dimensional flow over two spheres placed side by side. *Journal of Fluid Mechanics*, **246**, 465–488.

Kim, I., Elghobashi, S., and Sirignano, W. A. (1998). On the equation for spherical-particle motion: Effect of Reynolds and acceleration numbers. *Journal of Fluid Mechanics*, **367**, 221–253.

Kim, J., and Balachandar, S. (2012). Mean and fluctuating components of drag and lift forces on an isolated finite-sized particle in turbulence. *Theoretical and Computational Fluid Dynamics*, **26**(1–4), 185–204.

Kim, J., Kim, D., and Choi, H. (2001). An immersed-boundary finite-volume method for simulations of flow in complex geometries. *Journal of Computational Physics*, **171**, 132–150.

Kim, S., and Karrila, S. J. (2013). *Microhydrodynamics: Principles and Selected Applications*. Courier Corporation.

Kim, W., and Choi, H. (2019). Immersed boundary methods for fluid–structure interaction: A review. *International Journal of Heat and Fluid Flow*, **75**, 301–309.

King, M. R., and Leighton Jr., D. T. (1997). Measurement of the inertial lift on a moving sphere in contact with a plane wall in a shear flow. *Physics of Fluids*, **9**(5), 1248–1255.

Klett, J. D., and Davis, M. H. (1973). Theoretical collision efficiencies of cloud droplets at small Reynolds numbers. *Journal of the Atmospheric Sciences*, **30**(1), 107–117.

Kline, S. J., Reynolds, W. C., Schraub, F. A., and Runstadler, P. W. (1967). The structure of turbulent boundary layers. *Journal of Fluid Mechanics*, **30**(4), 741–773.

Kontomaris, K., Hanratty, T. J., and McLaughlin, J. B. (1992). An algorithm for tracking fluid particles in a spectral simulation of turbulent channel flow. *Journal of Computational Physics*, **103**(2), 231–242.

Kops-Werkhoven, M. M., and Fijnaut, H. M. (1981). Dynamic light scattering and sedimentation experiments on silica dispersions at finite concentrations. *Journal of Chemical Physics*, **74**(3), 1618–1625.

Krishnan, G. P., and Leighton Jr., D. T. (1995). Inertial lift on a moving sphere in contact with a plane wall in a shear flow. *Physics of Fluids*, **7**(11), 2538–2545.

Kuerten, J. G. M. (2016). Point-particle DNS and LES of particle-laden turbulent flow – a state-of-the-art review. *Flow, Turbulence and Combustion*, **97**(3), 689–713.

Kumar, V. S., and Kumaran, V. (2005). Voronoi cell volume distribution and configurational entropy of hard-spheres. *Journal of Chemical Physics*, **123**(11), 114501.

Kurose, R., and Komori, S. (1999). Drag and lift forces on a rotating sphere in a linear shear flow. *Journal of Fluid Mechanics*, **384**, 183–206.

Lai, M.-C., and Peskin, C. S. (2000). An immersed boundary method with formal second-order accuracy and reduced numerical viscosity. *Journal of Computational Physics*, **160**, 705–719.

Lain, S., and Sommerfeld, M. (2020). Influence of droplet collision modelling in Euler/Lagrange calculations of spray evolution. *International Journal of Multiphase Flow*, **132**, 103392.

Lamb, H. (1932). *Hydrodynamics*. Cambridge University Press.

Lance, M., and Bataille, J. (1991). Turbulence in the liquid phase of a uniform bubbly air–water flow. *Journal of Fluid Mechanics*, **222**, 95–118.

Landau, L. D., and Lifshitz, E. M. (1987). *Fluid Mechanics*, 2nd ed. Pergamon Press.

Landweber, L., and Miloh, T. (1980). Unsteady Lagally theorem for multipoles and deformable bodies. *Journal of Fluid Mechanics*, **96**(1), 33–46.

Lattanzi, A. M., Tavanashad, V., Subramaniam, S., and Capecelatro, J. (2020). Stochastic models for capturing dispersion in particle-laden flows. *Journal of Fluid Mechanics*, **903**, A7.

Lattanzi, A. M., Tavanashad, V., Subramaniam, S., and Capecelatro, J. (2022). Stochastic model for the hydrodynamic force in Euler–Lagrange simulations of particle-laden flows. *Physical Review Fluids*, **7**(1), 014301.

Lazar, E. A., Mason, J. K., MacPherson, R. D., and Srolovitz, D. J. (2013). Statistical topology of three-dimensional Poisson–Voronoi cells and cell boundary networks. *Physical Review E*, **88**(6), 063309.

Lazaro, B. J., and Lasheras, J. C. (1989). Particle dispersion in a turbulent, plane, free shear layer. *Physics of Fluids A*, **1**(6), 1035–1044.

Lee, L., and Leveque, R. J. (2003). An immersed interface method for incompressible Navier–Stokes equations. *SIAM Journal on Scientific Computing*, **25**(3), 832–856.

Lee, H., and Balachandar, S. (2010). Drag and lift forces on a spherical particle moving on a wall in a shear flow at finite Re. *Journal of Fluid Mechanics*, **657**, 89–125.

Lee, H., and Balachandar, S. (2017). Effects of wall roughness on drag and lift forces of a particle at finite Reynolds number. *International Journal of Multiphase Flow*, **88**, 116–132.

Lee, P. S., Cheng, M. T., and Shaw, D. T. (1981). The influence of hydrodynamic turbulence on acoustic turbulent agglomeration. *Aerosol Science and Technology*, **1**(1), 47–58.

Legendre, D., and Magnaudet, J. (1998). The lift force on a spherical bubble in a viscous linear shear flow. *Journal of Fluid Mechanics*, **368**, 81–126.

Leveque, R. J., and Li, Z. (1994). The immersed interface method for elliptic equations with discontinuous coefficients and singular sources, *SIAM Journal of Numerical Analysis*, **31**, 1019–1044.

Lhuillier, D. (1982). Forces d'inertie sur une bulle en expansion se déplaçant dans un fluide. *Comptes rendus de l'Académie des Sciences, Paris*, **295**(11), 95–98.

Lhuillier, D., Chang, C. H., and Theofanous, T. G. (2013). On the quest for a hyperbolic effective-field model of disperse flows. *Journal of Fluid Mechanics*, **731**, 184–194.

Li, X., Balachandar, S., Lee, H., and Bai, B. (2019). Fully resolved simulations of a stationary finite-sized particle in wall turbulence over a rough bed. *Physical Review Fluids*, **4**(9), 094302.

Liang, S. C., Hong, T., and Fan, L. S. (1996). Effects of particle arrangements on the drag force of a particle in the intermediate flow regime. *International Journal of Multiphase Flow*, **22**(2), 285–306.

Lilly, D. K. (1967). The representation of small-scale turbulence in numerical simulation experiments. *Proceedings of the IBM Scientific Computing Symposium on Environmental Sciences*, pp. 195–210.

Lin, C. L., and Lee, S. C. (1975). Collision efficiency of water drops in the atmosphere. *Journal of the Atmospheric Sciences*, **32**(7), 1412–1418.

Lin, C. J., Lee, K. J., and Sather, N. F. (1970a). Slow motion of two spheres in a shear field. *Journal of Fluid Mechanics*, **43**(1), 35–47; **32**(7), 1412–1418.

Lin, C. J., Perry, J. H., and Schowalter, W. R. (1970b). Simple shear flow round a rigid sphere: Inertial effects and suspension rheology. *Journal of Fluid Mechanics*, **44**(1), 1–17.

Ling, C. H. (1995). Criteria for incipient motion of spherical sediment particles. *Journal of Hydraulic Engineering*, **121**(6), 472–478

Ling, Y., Haselbacher, A., and Balachandar, S. (2011a). Importance of unsteady contributions to force and heating for particles in compressible flows: Part 1: Modeling and analysis for shock–particle interaction. *International Journal of Multiphase Flow*, **37**(9), 1026–1044.

Ling, Y., Haselbacher, A., and Balachandar, S. (2011b). Importance of unsteady contributions to force and heating for particles in compressible flows. Part 2: Application to particle dispersal by blast waves. *International Journal of Multiphase Flow*, **37**(9), 1013–1025.

Ling, Y., Parmar, M., and Balachandar, S. (2013). A scaling analysis of added-mass and history forces and their coupling in dispersed multiphase flows. *International Journal of Multiphase Flow*, **57**, 102–114.

Liu, B. Y., and Agarwal, J. K. (1974). Experimental observation of aerosol deposition in turbulent flow. *Journal of Aerosol Science*, **5**(2), 145–155.

Liu, B. Y., and Ilori, T. A. (1974). Aerosol deposition in turbulent pipe flow. *Environmental Science and Technology*, **8**(4), 351–356.

Liu, K., Lakhote, M., and Balachandar, S. (2019). Self-induced temperature correction for inter-phase heat transfer in Euler–Lagrange point-particle simulation. *Journal of Computational Physics*, **396**, 596–615.

Liu, K., Huck, P. D., Aliseda, A., and Balachandar, S. (2021). Investigation of turbulent inflow specification in Euler–Lagrange simulations of mid-field spray. *Physics of Fluids*, **33**(3), 033313.

Liu, Y. J., Nelson, J., Feng, J., and Joseph, D. D. (1993). Anomalous rolling of spheres down an inclined plane, *Journal of Non-Newtonian Fluid Mechanics*, **50**, 305–329.

Lomholt, S., and Maxey, M. R. (2003). Force-coupling method for particulate two-phase flow: Stokes flow. *Journal of Computational Physics*, **184**(2), 381–405.

Lomholt, S., Stenum, B., and Maxey, M. R. (2002). Experimental verification of the force coupling method for particulate flows. *International Journal of Multiphase Flow*, **28**(2), 225–246.

Longmire, E. K., and Eaton, J. K. (1992). Structure of a particle-laden round jet. *Journal of Fluid Mechanics*, **236**, 217–257.

Lorentz, H. A. (1896). A general theorem concerning the motion of a viscous fluid and a few consequences derived from it. *Zittingsverlag Akademie van Wetenschappen Amsterdam*, **5**, 168–175.

Loth, E. (2016). Overview of numerical approaches. In Michaelides, E., Crowe, C. T., and Schwarzkopf, J. D. (eds), *Multiphase Flow Handbook*, 2nd ed. CRC Press, pp. 79–94.

Lovalenti, P. M., and Brady, J. F. (1993a). The hydrodynamic force on a rigid particle undergoing arbitrary time-dependent motion at small Reynolds number. *Journal of Fluid Mechanics*, **256**, 561–605.

Lovalenti, P. M., and Brady, J. F. (1993b). The force on a sphere in a uniform flow with small-amplitude oscillations at finite Reynolds number. *Journal of Fluid Mechanics*, **256**, 607–614.

Lu, J., Das, S., Peters, E. A. J. F., and Kuipers, J. A. M. (2018). Direct numerical simulation of fluid flow and mass transfer in dense fluid-particle systems with surface reactions. *Chemical Engineering Science*, **176**, 1–18.

Lu, J. C., and Tryggvason, G. (2006). Numerical study of turbulent bubbly downflows in a vertical channel. *Physics of Fluids*, **18**, 103302.

Luding, S., and Herrmann, H. J. (1999). Cluster-growth in freely cooling granular media. *Chaos: An Interdisciplinary Journal of Nonlinear Science*, **9**(3), 673–681.

Lun, C. K. K., Savage, S. B., Jeffrey, D. J., and Chepurniy, N. (1984). Kinetic theories for granular flow: Inelastic particles in Couette flow and slightly inelastic particles in a general flowfield. *Journal of Fluid Mechanics*, **140**, 223–256.

Lundgren, T. S. (1972). Slow flow through stationary random beds and suspensions of spheres. *Journal of Fluid Mechanics*, **51**(2), 273–299.

Maday, Y., and Patera, A. T. (1989). Spectral element methods for the incompressible Navier–Stokes equations. *State-of-the-Art Surveys on Computational Mechanics* (A90–47176 21–64). American Society of Mechanical Engineers, pp. 71–143.

Magnaudet, J., Rivero, M., and Fabre, J. (1995). Accelerated flows past a rigid sphere or a spherical bubble. Part 1. Steady straining flow. *Journal of Fluid Mechanics*, **284**, 97–135.

Mancho, A. M., Small, D., and Wiggins, S. (2006). A comparison of methods for interpolating chaotic flows from discrete velocity data. *Computers and Fluids*, **35**(4), 416–428.

Manton, M. J. (1977). On the coalescence of water droplets in turbulent clouds. *Tellus*, **29**(1), 1–7.

Marble, F. E. (1970). Dynamics of dusty gases. *Annual Review of Fluid Mechanics*, **2**, 397–446.

Marchioli, C., Picciotto, M., and Soldati, A. (2007). Influence of gravity and lift on particle velocity statistics and transfer rates in turbulent vertical channel flow. *International Journal of Multiphase Flow*, **33**(3), 227–251.

Marchisio, D. L., and Fox, R. O. (eds). (2007). *Multiphase Reacting Flows: Modelling and Simulation*. Springer.

Marella, S., Krishnan, S., Liu, H., and Udaykumar, H. S. (2005). Sharp interface Cartesian grid method I: An easily implemented technique for 3D moving boundary computations. *Journal of Computational Physics*, **210**(1), 1–31.

Mason, B. J. (1969). Some outstanding problems in cloud physics–the interaction of microphysical and dynamical processes. *Quarterly Journal of the Royal Meteorological Society*, **95**(405), 449–485.

Maude, A. D. (1961). End effects in a falling-sphere viscometer. *British Journal of Applied Physics*, **12**(6), 293–295.

Maxey, M. R. (1987). The gravitational settling of aerosol particles in homogeneous turbulence and random flow fields. *Journal of Fluid Mechanics*, **174**, 441–465.

Maxey, M. R., and Patel, B. K. (2001). Localized force representations for particles sedimenting in Stokes flow. *International Journal of Multiphase Flow*, **27**(9), 1603–1626.

Maxey, M. R., and Riley, J. J. (1983). Equation of motion for a small rigid sphere in a nonuniform flow. *Physics of Fluids*, **26**(4), 883–889.

Maxwell, J. C. (1873). *A Treatise on Electricity and Magnetism*, Vol. 1. Clarendon Press.

Mazzuoli, M., Blondeaux, P., Vittori, G., Uhlmann, M., Simeonov, J., and Calantoni, J. (2020). Interface-resolved direct numerical simulations of sediment transport in a turbulent oscillatory boundary layer. *Journal of Fluid Mechanics*, **885**, A28.

McLaughlin, J. B. (1989). Aerosol particle deposition in numerically simulated channel flow. *Physics of Fluids*, **1**(7), 1211–1224.

McLaughlin, J. B. (1991). Inertial migration of a small sphere in linear shear flows. *Journal of Fluid Mechanics*, **224**, 261–274.

McLaughlin, J. B. (1993). The lift on a small sphere in wall-bounded linear shear flows. *Journal of Fluid Mechanics*, **246**, 249–265.

McQueen, D. M., and Peskin, C. S. (1989). A three-dimensional computational method for blood flow in the heart. II. Contractile fibers. *Journal of Computational Physics*, **82**, 289–297.

Mehrabadi, M., Tenneti, S., Garg, R., and Subramaniam, S. (2015). Pseudo-turbulent gas-phase velocity fluctuations in homogeneous gas–solid flow: Fixed particle assemblies and freely evolving suspensions. *Journal of Fluid Mechanics*, **770**, 210–246.

Mehta, R. D. (1985). Aerodynamics of sports balls. *Annual Review of Fluid Mechanics*, **17**(1), 151–189.

Mehta, Y., Neal, C., Jackson, T. L., Balachandar, S., and Thakur, S. (2016). Shock interaction with three-dimensional face centered cubic array of particles. *Physical Review Fluids*, **1**(5), 054202.

Mehta, Y., Neal, C., Salari, K., Jackson, T. L., Balachandar, S., and Thakur, S. (2018). Propagation of a strong shock over a random bed of spherical particles. *Journal of Fluid Mechanics*, **839**, 157–197.

Mei, R. (1992). An approximate expression for the shear lift force on a spherical particle at finite Reynolds number. *International Journal of Multiphase Flow*, **18**(1), 145–147.

Mei, R. (1994). Effect of turbulence on the particle settling velocity in the nonlinear drag range. *International Journal of Multiphase Flow*, **20**, 273–284.

Mei, R., and Adrian, R. J. (1992). Flow past a sphere with an oscillation in the free-stream velocity and unsteady drag at finite Reynolds number. *Journal of Fluid Mechanics*, **237**, 323–341.

Meiburg, E., and Kneller, B. (2010). Turbidity currents and their deposits. *Annual Review of Fluid Mechanics*, **42**, 135–156.

Meiburg, E., Wallner, E., Pagella, A., Riaz, A., Härtel, C., and Necker, F. (2000). Vorticity dynamics of dilute two-way-coupled particle-laden mixing layers. *Journal of Fluid Mechanics*, **421**, 185-227.

Merle, A., Legendre, D., and Magnaudet, J. (2005). Forces on a high-Reynolds-number spherical bubble in a turbulent flow. *Journal of Fluid Mechanics*, **532**, 53–62.

Michaelides, E. E. (1992). A novel way of computing the Basset term in unsteady multiphase flow computations. *Physics of Fluids A: Fluid Dynamics*, **4**(7), 1579–1582.

Michaelides, E. E. (2003). Hydrodynamic force and heat/mass transfer from particles, bubbles, and drops – the Freeman scholar lecture. *Journal of Fluids Engineering*, **125**(2), 209–238.

Michaelides, E. E., and Feng, Z. G. (1994). Heat transfer from a rigid sphere in a non-uniform flow and temperature field. *International Journal of Heat and Mass Transfer*, **37**, 2069–2076.

Minier, J. P. (2015). On Lagrangian stochastic methods for turbulent polydisperse two-phase reactive flows. *Progress in Energy and Combustion Science*, **50**, 1–62.

Minier, J. P., Peirano, E., and Chibbaro, S. (2004). PDF model based on Langevin equation for polydispersed two-phase flows applied to bluff-body gas–solid flows. *Physics of Fluids*, **16**, 2419–2431.

Minier, J. P., Chibbaro, S., and Pope, S. B. (2014). Guidelines for the formulation of Lagrangian stochastic models for particle simulations of single-phase and dispersed two-phase turbulent flows. *Physics of Fluids*, **26**(11), 113303.

Mittal, R. (1999). Planar symmetry in the unsteady wake of a sphere. *AIAA Journal*, **37**(3), 388–390.

Mittal, R., and Iaccarino, G. (2005). Immersed boundary methods. *Annual Reviews in Fluid Mechanics*, **37**, 239–261.

Mittal, R., Wilson, J. J., and Najjar, F. M. (2002). Symmetry properties of the transitional sphere wake. *AIAA Journal*, **40**(3), 579–582.

Miyazaki, K., Bedeaux, D., and Avalos, J. B. (1995). Drag on a sphere in slow shear flow. *Journal of Fluid Mechanics*, **296**, 373–390.

Mizukami, M., Parthasarathy, R. N., and Faeth, G. M. (1992). Particle-generated turbulence in homogeneous dilute dispersed flows. *International Journal of Multiphase Flow*, **18**(3), 397–412.

Mohd-Yusof, J. (1996). Interaction of massive particles with turbulence. PhD thesis, Department of Mechanical Engineering, Cornell University, Ithaca, NY.

Mohd-Yusof, J. (1997). Combined immersed boundaries B-Spline methods for simulations of flows in complex geometries in complex geometries. *CTR Annual Research Briefs*, NASA Ames/Stanford University.

Moin, P., Squires, K., Cabot, W., and Lee, S. (1991). A dynamic subgrid-scale model for compressible turbulence and scalar transport. *Physics of Fluids*, **3**(11), 2746–2757.

Mollinger, A. M., and Nieuwstadt, F. T. M. (1996). Measurement of the lift force on a particle fixed to the wall in the viscous sublayer of a fully developed turbulent boundary layer. *Journal of Fluid Mechanics*, **316**, 285–306.

Monchaux, R., Bourgoin, M., and Cartellier, A. (2010). Preferential concentration of heavy particles: A Voronoi analysis. *Physics of Fluids*, **22**, 103304.

Mordant, N., and Pinton, J. F. (2000). Velocity measurement of a settling sphere. *European Physical Journal B – Condensed Matter and Complex Systems*, **18**(2), 343–352.

Moore, W. C., and Balachandar, S. (2019). Lagrangian investigation of pseudo-turbulence in multiphase flow using superposable wakes. *Physical Review Fluids*, **4**(11), 114301.

Moore, W. C., Balachandar, S., and Akiki, G. (2019). A hybrid point-particle force model that combines physical and data-driven approaches. *Journal of Computational Physics*, **385**, 187–208.

Morris, J. F., and Brady, J. F. (1998). Pressure-driven flow of a suspension: Buoyancy effects. *International Journal of Multiphase Flow*, **24**(1), 105–130.

Mougin, G., and Magnaudet, J. (2002). The generalized Kirchhoff equations and their application to the interaction between a rigid body and an arbitrary time-dependent viscous flow. *International Journal of Multiphase Flow*, **28**(11), 1837–1851.

Murray, J. (1965). On the mathematics of fluidization Part 1. Fundamental equations and wave propagation. *Journal of Fluid Mechanics*, **21**(3), 465–493.

Najjar, F. M., and Balachandar, S. (1998). Low-frequency unsteadiness in the wake of a normal flat plate. *Journal of Fluid Mechanics*, **370**, 101–147.

Naso, A., and Prosperetti, A. (2010). The interaction between a solid particle and a turbulent flow. *New Journal of Physics*, **12**(3), 033040.

Natarajan, R., and Acrivos, A. (1993). The instability of the steady flow past spheres and disks. *Journal of Fluid Mechanics*, **254**, 323–344.

Necker, F., Härtel, C., Kleiser, L., and Meiburg, E. (2002). High-resolution simulations of particle-driven gravity currents. *International Journal of Multiphase Flow*, **28**(2), 279–300.

Nielsen P. (1984). On the motion of suspended sand particles. *Journal of Geophysical Research*, **89**, 616–626.

Nielsen P. (1993). Turbulence effects on the settling of suspended particles. *Journal of Sedimentary Petrology*, **63**, 835–838.

Nott, P. R., and Brady, J. F. (1994). Pressure-driven flow of suspensions: Simulation and theory. *Journal of Fluid Mechanics*, **275**, 157–199.

Nott, P. R., Guazzelli, E., and Pouliquen, O. (2011). The suspension balance model revisited. *Physics of Fluids*, **23**(4), 043304.

Odar, F. (1966). Verification of the proposed equation for calculation of the forces on a sphere accelerating in a viscous fluid. *Journal of Fluid Mechanics*, **25**(3), 591–592.

Odar, F., and Hamilton, W. S. (1964). Forces on a sphere accelerating in a viscous fluid. *Journal of Fluid Mechanics*, **18**(2), 302–314.

Ogawa, S. (1978). Multitemperature theory of granular materials. *Proceedings of the US–Japan Seminar on Continuum Mechanical and Statistical Approaches in the Mechanics of Granular Materials*, Gakajutsu Bunken Fukyu-Kai, pp. 208–217.

Ohl, C. D., Tijink, A., and Prosperetti, A. (2003). The added mass of an expanding bubble. *Journal of Fluid Mechanics*, **482**, 271–290.

O'Neill, M. E. (1964). A slow motion of viscous liquid caused by a slowly moving solid sphere. *Mathematika*, **11**(1), 67–74.

O'Neill, M. E., and Majumdar, S. R. (1970a). Asymmetrical slow viscous fluid motions caused by the translation or rotation of two spheres. Part I: The determination of exact solutions for any values of the ratio of radii and separation parameters. *Zeitschrift für angewandte Mathematik und Physik*, **21**(2), 164–179.

O'Neill, M. E., and Majumdar, S. R. (1970b). Asymmetrical slow viscous fluid motions caused by the translation or rotation of two spheres. Part II: Asymptotic forms of the solutions when the minimum clearance between the spheres approaches zero. *Zeitschrift für angewandte Mathematik und Physik*, **21**(2), 180–187.

Okuyama, K., Kousaka, Y., and Yoshida, T. (1978). Turbulent coagulation of aerosols in a pipe flow. *Journal of Aerosol Science*, **9**(5), 399–410.

Oseen, C. W. (1927). *Hydrodynamik*. Akademische Verlagsgesellschaft.

Osher, S., and Fedkiw, R. (2006). *Level Set Methods and Dynamic Implicit Surfaces*. Springer.

Pakseresht, P., and Apte, S. V. (2019). Volumetric displacement effects in Euler–Lagrange LES of particle-laden jet flows. *International Journal of Multiphase Flow*, **113**, 16–32.

Pan, Y., and Banerjee, S. (1997). Numerical investigation of the effects of large particles on wall turbulence. *Physics of Fluids*, **9**, 3786–3807.

Park, G. I., Bassenne, M., Urzay, J., and Moin, P. (2017). A simple dynamic subgrid-scale model for LES of particle-laden turbulence. *Physical Review Fluids*, **2**(4), 044301.

Parmar, M. (2010). Unsteady forces on a particle in compressible flows. PhD thesis, University of Florida, Gainesville, FL.

Parmar, M., Annamalai, S., Balachandar, S., and Prosperetti, A. (2018). Differential formulation of the viscous history force on a particle for efficient and accurate computation. *Journal of Fluid Mechanics*, **844**, 970–993.

Parmar, M., Haselbacher, A., and Balachandar, S. (2008). On the unsteady inviscid force on cylinders and spheres in subcritical compressible flow. *Philosophical Transactions of the Royal Society of London, Series A*, **366**(1873), 2161–2175.

Parmar, M., Haselbacher, A., and Balachandar, S. (2010). Improved drag correlation for spheres and application to shock-tube experiments. *AIAA Journal*, **48**(6), 1273–1276.

Parmar, M., Haselbacher, A., and Balachandar, S. (2011). Generalized Basset–Boussinesq–Oseen equation for unsteady forces on a sphere in a compressible flow. *Physical Review Letters*, **106**(8), 084501.

Parthasarathy, R. N. (1990). Homogeneous dilute turbulent particle-laden water flows. PhD thesis, University of Michigan, Ann Arbor, MI.

Parthasarathy, R. N., and Faeth, G. M. (1990). Turbulence modulation in homogeneous dilute particle-laden flows. *Journal of Fluid Mechanics*, **220**, 485–514.

Patera, A. T. (1984). A spectral element method for fluid dynamics: Laminar flow in a channel expansion. *Journal of Computational Physics*, **54**(3), 468–488.

Peakall, J., McCaffrey, B., and Kneller, B. (2000). A process model for the evolution, morphology, and architecture of sinuous submarine channels. *Journal of Sedimentary Research*, **70**(3), 434–448.

Pearson, H. J., Valioulis, I. A., and List, E. J. (1984). Monte Carlo simulation of coagulation in discrete particle-size distributions. Part 1. Brownian motion and fluid shearing. *Journal of Fluid Mechanics*, **143**, 367–385.

Pedinotti, S., Mariotti, G., and Banerjee, S. (1992). Direct numerical simulation of particle behaviour in the wall region of turbulent flows in horizontal channels. *International Journal of Multiphase Flow*, **18**(6), 927–941.

Pember, R. B., Bell, J. B., Colella, P., Crutchfield, W. Y., and Welcome, M. L. (1995). An adaptive Cartesian grid method for unsteady compressible flow in irregular regions. *Journal of Computational Physics*, **120**, 278–304.

Peng, C., Ayala, O. M., and Wang, L. P. (2020). Flow modulation by a few fixed spherical particles in a turbulent channel flow. *Journal of Fluid Mechanics*, **884**, A15.

Pepiot, P., and Desjardins, O. (2012). Numerical analysis of the dynamics of two- and three-dimensional fluidized bed reactors using an Euler–Lagrange approach. *Powder Technology*, **220**, 104–121.

Peres, J. (1929). Action of an obstacle on a viscous fluid. Simple proof of Faxén's equations. *Comptes rendus de l'Académie des Sciences, Paris*, **188**, 310–312.

Perez-Madrid, A., Rubí, J. M., and Bedeaux, D. (1990) Motion of a sphere through a fluid in stationary homogeneous flow. *Physica A*, **163**, 778–790.

Peskin, C. S. (1977). Numerical analysis of blood flow in the heart. *Journal of Computational Physics*, **25**, 220–252.

Peskin, C. S. (2002). The immersed boundary method. *Acta Numerica*, **11**, 479–517.

Pierce, C. D., and Moin, P. (1998). A dynamic model for subgrid-scale variance and dissipation rate of a conserved scalar. *Physics of Fluids*, **10**(12), 3041–3044.

Pinelli, A., Naqavi, I. Z., Piomelli, U., and Favier, J. (2010). Immersed-boundary methods for general finite-difference and finite-volume Navier–Stokes solvers. *Journal of Computational Physics*, **229**(24), 9073–9091.

Pismen, L. M., and Nir, A. (1978). On the motion of suspended particles in stationary homogeneous turbulence. *Journal of Fluid Mechanics*, **84**(1), 193–206.

Poe, G. G., and Acrivos, A. (1975). Closed-streamline flows past rotating single cylinders and spheres: Inertia effects. *Journal of Fluid Mechanics*, **72**, 605–623.

Pope, S. B. (1994). Lagrangian PDF methods for turbulent flows. *Annual Review of Fluid Mechanics*, **26**(1), 23–63.

Pope, S. B. (2001). *Turbulent Flows*. Cambridge University Press.

Pope, S. B. (2014). The determination of turbulence-model statistics from the velocity–acceleration correlation. *Journal of Fluid Mechanics*, **757**, R1.

Portela, L. M., and Oliemans, R. V. (2003). Eulerian–Lagrangian DNS/LES of particle–turbulence interactions in wall-bounded flows. *International Journal for Numerical Methods in Fluids*, **43**(9), 1045–1065.

Pozorski, J., and Minier, J. P. (1998). On the Lagrangian turbulent dispersion models based on the Langevin equation. *International Journal of Multiphase Flow*, **24**(6), 913–945.

Prahl, L., Holzer, A., Arlov, D., Revstedt, J., Sommerfeld, M., and Fuchs, L. (2007). On the interaction between two fixed spherical particles. *International Journal of Multiphase Flow*, **33**(7), 707–725.

Prosperetti, A., and Tryggvason, G. (eds). (2007). *Computational Methods for Multiphase Flow*. Cambridge University Press.

Proudman, I., and Pearson, J. R. A. (1957). Expansions at small Reynolds numbers for the flow past a sphere and a circular cylinder. *Journal of Fluid Mechanics*, **2**(03), 237–262.

Pushkin, D. O., and Aref, H. (2002). Self-similarity theory of stationary coagulation. *Physics of Fluids*, **14**(2), 694–703.

Rahmani, M., Geraci, G., Iaccarino, G., and Mani, A. (2018). Effects of particle polydispersity on radiative heat transfer in particle-laden turbulent flows. *International Journal of Multiphase Flow*, **104**, 42–59.

Raithby, G. D., and Eckert, E. R. G. (1968). The effect of turbulence parameters and support position on the heat transfer from spheres. *International Journal of Heat and Mass Transfer*, **11**(8), 1233–1252.

Ramkrishna, D. (2000). *Population Balances: Theory and Applications to Particulate Systems in Engineering*. Academic Press.

Rani, S. L., and Balachandar, S. (2003). Evaluation of the equilibrium Eulerian approach for the evolution of particle concentration in isotropic turbulence. *International Journal of Multiphase Flow*, **29**(12), 1793–1816.

Rani, S. L., Winkler, C. M., and Vanka, S. P. (2004). Numerical simulations of turbulence modulation by dense particles in a fully developed pipe flow. *Powder Technology*, **141**(1–2), 80–99.

Ranz, W. E., and Marshall, W. R. (1952). Evaporation from drops. *Chemical Engineering Progress*, **48**, 141–146.

Rao, K. K., Nott, P. R., and Sundaresan, S. (2008). *An Introduction to Granular Flow*. Cambridge University Press.

Rashidi, M., Hetsroni, G., and Banerjee, S. (1990). Particle–turbulence interaction in a boundary layer. *International Journal of Multiphase Flow*, **16**(6), 935–949.

Reeks, M. W. (1977). On the dispersion of small particles suspended in an isotropic turbulent fluid. *Journal of Fluid Mechanics*, **83**(3), 529–546.

Reeks, M. W. (1983). The transport of discrete particles in inhomogeneous turbulence. *Journal of Aerosol Science*, **14**(6), 729–739.

Riboux, G., Risso, F., and Legendre, D. (2010). Experimental characterization of the agitation generated by bubbles rising at high Reynolds number. *Journal of Fluid Mechanics*, **643**, 509–539.

Richardson, J., and Zaki, W. (1954). Fluidization and sedimentation – Part I. *Transactions of the Institution of Chemical Engineers*, **32**, 38–58.

Richter, D. H. (2015). Turbulence modification by inertial particles and its influence on the spectral energy budget in planar Couette flow. *Physics of Fluids*, **27**, 063304.

Richter, D. H., and Sullivan, P. P. (2014). Modification of near-wall coherent structures by inertial particles. *Physics of Fluids*, **26**, 103304.

Riley, J. J., and Patterson Jr., G. S. (1974). Diffusion experiments with numerically integrated isotropic turbulence. *Physics of Fluids*, **17**(2), 292–297.

Risso, F. (2016). Physical interpretation of probability density functions of bubble-induced agitation. *Journal of Fluid Mechanics*, **809**, 240–263.

Rivero, M., Magnaudet, J., and Fabre, J. (1991). New results on the forces exerted on a spherical body by an accelerated flow. *Comptes Rendus de l'Academie des Sciences, Serie II*, **312**(13), 1499–1506.

Roma, A. M., Peskin, C. S., and Berger, M. J. (1999). An adaptive version of the immersed boundary method. *Journal of Computational Physics*, **153**(2), 509–534.

Rosa, B., Wang, L. P., Maxey, M. R., and Grabowski, W. W. (2011). An accurate and efficient method for treating aerodynamic interactions of cloud droplets. *Journal of Computational Physics*, **230**(22), 8109–8133.

Rouson, D. W., and Eaton, J. K. (2001). On the preferential concentration of solid particles in turbulent channel flow. *Journal of Fluid Mechanics*, **428**, 149–169.

Rubinow, S. I., and Keller, J. B. (1961). The transverse force on a spinning sphere moving in a viscous fluid. *Journal of Fluid Mechanics*, **11**(3), 447–459.

Rudinger, G. (2012). *Fundamentals of Gas Particle Flow*, Vol. 2. Elsevier.

Rudoff, R. R., and Bachalo, W. D. (1988). Measurement of droplet drag coefficients in polydispersed turbulent flow fields. *Proceedings of the AIAA Aerospace Meeting*, Reno, NV.

Saff, E. B., and Kuijlaars, A. B. (1997). Distributing many points on a sphere. *The Mathematical Intelligencer*, **19**(1), 5–11.

Saffman, P. G. (1962). On the stability of laminar flow of a dusty gas. *Journal of Fluid Mechanics*, **13**, 120–128.

Saffman, P. G. (1965). The lift on a small sphere in a slow shear flow. *Journal of Fluid Mechanics*, **22**(2), 385–400.

Saffman, P. G. (1973). On the settling speed of free and fixed suspensions. *Studies in Applied Mathematics*, **52**(2), 115–127.

Saffman, P. G. F., and Turner, J. S. (1956). On the collision of drops in turbulent clouds. *Journal of Fluid Mechanics*, **1**(1), 16–30.

Saiki, E. M., and Biringen, S. (1996). Numerical simulation of a cylinder in uniform flow: Application of a virtual boundary method. *Journal of Computational Physics*, **123**, 450–465.

Salinas, J. S., Shringarpure, M., Cantero, M. I., and Balachandar, S. (2018). Mixing at a sediment concentration interface in turbulent open channel flow. *Environmental Fluid Mechanics*, **18**, 173–200.

Salinas, J. S., Cantero, M. I., Shringarpure, M., and Balachandar, S. (2019a). Properties of the body of a turbidity current at near-normal conditions: 1. Effect of bed slope. *Journal of Geophysical Research: Oceans*, **124**(11), 7989–8016.

Salinas, J. S., Cantero, M. I., Shringarpure, M., and Balachandar, S. (2019b). Properties of the body of a turbidity current at near-normal conditions: 2. Effect of settling. *Journal of Geophysical Research: Oceans*, **124**(11), 8017–8035.

Salinas, J., Balachandar, S., Shringarpure, M., Fedele, J., Hoyal, D., and Cantero, M. (2020). Soft transition between subcritical and supercritical currents through intermittent cascading interfacial instabilities. *Proceedings of the National Academy of Sciences*, **117**(31), 18278–18284.

Salinas, J. S., Balachandar, S., Shringarpure, M., et al. (2021a). Anatomy of subcritical submarine flows with a lutocline and an intermediate destruction layer. *Nature Communications*, **12**(1), 1–11.

Salinas, J. S., Balachandar, S., and Cantero, M. I. (2021b). Control of turbulent transport in supercritical currents by three families of hairpin vortices. *Physical Review Fluids*, **6**(6), 063801.

Salinas, J. S., Zúñiga, S., Cantero, M. I., Shringarpure, M., Fedele, J., Hoyal, D., and Balachandar, S. (2022). Slope dependence of self-similar structure and entrainment in gravity currents. *Journal of Fluid Mechanics*, **934**, R4.

Salinas, J. S., Balachandar, S., Zúñiga, S. L., Shringarpure, M., Fedele, J., Hoyal, D., and Cantero, M. I. (2023). On the definition, evolution, and properties of the outer edge of gravity currents: A direct-numerical and large-eddy simulation study. *Physics of Fluids*, **35**, 016610.

Sankagiri, S., and Ruff, G. A. (1997). Measurement of sphere drag in high turbulent intensity flows. *Proceedings of the ASME Fluids Engineering Division*, **244**, 277–282.

Sangani, A. S., Zhang, D. Z., and Prosperetti, A. (1991). The added mass, Basset, and viscous drag coefficients in nondilute bubbly liquids undergoing small-amplitude oscillatory motion. *Physics of Fluids*, **3**(12), 2955–2970.

Saurel, R., and Abgrall, R. (1999). A multiphase Godunov method for compressible multifluid and multiphase flows. *Journal of Computational Physics*, **150**(2), 425–467.

Sawford, B. L., Yeung, P. K., Borgas, M. S., et al. (2003). Conditional and unconditional acceleration statistics in turbulence. *Physics of Fluids*, **15**(11), 3478–3489.

Saye, R. I., and Sethian, J. A. (2011). The Voronoi implicit interface method for computing multiphase physics. *Proceedings of the National Academy of Sciences*, **108**(49), 19498–19503.

Scardovelli, R., and Zaleski, S. (1999). Direct numerical simulation of free-surface and interfacial flow. *Annual Review of Fluid Mechanics*, **31**(1), 567–603.

Schiller, L., and Naumann, A. (1933). Über die grundlegenden Berechnungen bei der Schwerkraftaufbereitung. *Zeitschrift des Vereines Deutscher Ingenieure*, **77**(12), 318–320.

Schöneborn, P. R. (1975). The interaction between a single particle and an oscillating fluid. *International Journal of Multiphase Flow*, **2**(3), 307–317.

Segre, G., and Silberberg, A. J. (1962a). Behaviour of macroscopic rigid spheres in Poiseuille flow Part 1. Determination of local concentration by statistical analysis of particle passages through crossed light beams. *Journal of Fluid Mechanics*, **14**(1), 115–135.

Segre, G., and Silberberg, A. (1962b). Behaviour of macroscopic rigid spheres in Poiseuille flow Part 2. Experimental results and interpretation. *Journal of Fluid Mechanics*, **14**(1), 136–157.

Sethian, J. A., and Smereka, P. (2003). Level set methods for fluid interfaces. *Annual Review of Fluid Mechanics*, **35**(1), 341–372.

Sequeiros, O. E., Naruse, H., Endo, N., Garcia, M. H., and Parker, G. (2009). Experimental study on self-accelerating turbidity currents. *Journal of Geophysical Research*, **114**, C05025

Sethian, J. A. (1999). *Level Set Methods and Fast Marching Methods: Evolving Interfaces in Computational Geometry, Fluid Mechanics, Computer Vision, and Materials Science*. Cambridge University Press.

Seyed-Ahmadi, A., and Wachs, A. (2020). Microstructure-informed probability-driven point-particle model for hydrodynamic forces and torques in particle-laden flows. *Journal of Fluid Mechanics*, **900**, A21.

Shaffer, F., Gopalan, B., Breault, R. W., et al. (2013). High speed imaging of particle flow fields in CFB risers. *Powder Technology*, **242**, 86–99.

Shaw, R. A. (2003). Particle–turbulence interactions in atmospheric clouds. *Annual Review of Fluid Mechanics*, **35**(1), 183–227.

Shaw, R. A., Reade, W. C., Collins, L. R., and Verlinde, J. (1998). Preferential concentration of cloud droplets by turbulence: Effects on the early evolution of cumulus cloud droplet spectra. *Journal of the Atmospheric Sciences*, **55**(11), 1965–1976.

Shotorban, B., and Balachandar, S. (2006). Particle concentration in homogeneous shear turbulence simulated via Lagrangian and equilibrium Eulerian approaches. *Physics of Fluids*, **18**(6), 065105.

Shotorban, B., and Balachandar, S. (2007). A Eulerian model for large-eddy simulation of concentration of particles with small Stokes numbers. *Physics of Fluids*, **19**, 118107.

Shotorban, B., and Balachandar, S. (2009). Two-fluid approach for direct numerical simulation of particle-laden turbulent flows at small Stokes numbers. *Physical Review E*, **79**(5), 056703.

Shotorban, B., and Mashayek, F. (2006). A stochastic model for particle motion in large-eddy simulation. *Journal of Turbulence*, **7**, 18.

Shringarpure, M., Cantero, M. I., and Balachandar, S. (2012). Dynamics of complete turbulence suppression in turbidity currents driven by monodisperse suspensions of sediment. *Journal of Fluid Mechanics*, **712**, 384–417.

Si, H. (2015). TetGen, a Delaunay-based quality tetrahedral mesh generator. *ACM Transactions on Mathematical Software*, **41**(2), 11.

Siddani, B., and Balachandar, S. (2023). Point-particle drag, lift, and torque closure models using machine learning: Hierarchical approach and interpretability. *Physical Review Fluids*, **8**, 014303.

Siddani, B., Balachandar, S., and Fang, R. (2021a). Rotational and reflectional equivariant convolutional neural network for data-limited applications: Multiphase flow demonstration. *Physics of Fluids*, **33**, 103323.

Siddani, B., Balachandar, S., Moore, W. C., Yang, Y., and Fang, R. (2021b). Machine learning for physics-informed generation of dispersed multiphase flow using generative adversarial networks. *Theoretical and Computational Fluid Dynamics*, **35**, 807–830.

Sierakowski, A. J., and Prosperetti, A. (2016). Resolved-particle simulation by the Physalis method: Enhancements and new capabilities. *Journal of Computational Physics*, **309**, 164–184.

Simcik, M., and Ruzicka, M. C. (2013). Added mass of dispersed particles by CFD: Further results. *Chemical Engineering Science*, **97**, 366–375.

Simeonov, J. A., and Calantoni, J. (2012). Modeling mechanical contact and lubrication in direct numerical simulations of colliding particles. *International Journal of Multiphase Flow*, **46**, 38–53.

Smith, G. F. (1971). On isotropic functions of symmetric tensors, skew-symmetric tensors and vectors. *International Journal of Engineering Science*, **9**, 899–916.

Smoluchowski, M. V. (1911). On the mutual action of spheres which move in a viscous liquid. *Bulletin of the Academy of Science Cracovie A*, **1**, 28–39.

Smoluchowski, M. V. (1918). Versuch einer mathematischen Theorie der Koagulationskinetik kolloider Lösungen. *Zeitschrift für Physikalische Chemie*, **92**(1), 129–168.

Son, M., and Hsu, T. J. (2008). Flocculation model of cohesive sediment using variable fractal dimension. *Environmental Fluid Mechanics*, **8**, 55–71.

Soo, S. L. (1956). Statistical properties of momentum transfer in two-phase flow. *Chemical Engineering Science*, **5**(2), 57–67.

Soo, S. L. (1975). Equation of motion of a solid particle suspended in a fluid. *Physics of Fluids*, **18**(2), 263–264.

Squires, K. D., and Eaton J. (1991). Preferential concentration of particles by turbulence. *Physics of Fluids*, **3**, 1169–1179.

Sridhar, G., and Katz, J. (1995). Drag and lift forces on microscopic bubbles entrained by a vortex. *Physics of Fluids*, **7**, 389–399.

Stewart, C., McGrath, T., and Balachandar, S. (2018). Soft-sphere simulations of a planar shock interaction with a granular bed. *Physical Review Fluids*, **3**, 034308,

Stimson, M., and Jeffery, G. B. (1926). The motion of two spheres in a viscous fluid. *Proceedings of the Royal Society London, Series A*, **111**(757), 110–116.

Stommel, H. (1949). Trajectories of small bodies sinking slowly through convection cells. *Journal of Marine Research*, **8**(11), 24–29.

Stoyan, D., and Stoyan, H. (1994). *Fractals, Random Shapes, and Point Fields*. Wiley.

Stoyan, D., Kendall, W. S., and Mecke, J. (1995). *Stochastic Geometry and its Applications*. Akademie-Verlag.

Stuhmiller, J. H. (1977). The influence of interfacial pressure forces on the character of two-phase flow model equations. *International Journal of Multiphase Flow*, **3**(6), 551–560.

Subramaniam, S. (2000). Statistical representation of a spray as a point process. *Physics of Fluids*, **12**(10), 2413–2431.

Subramaniam, S. (2001). Statistical modeling of sprays using the droplet distribution function. *Physics of Fluids*, **13**(3), 624–642.

Subramaniam, S. (2013). Lagrangian–Eulerian methods for multiphase flows. *Progress in Energy and Combustion Science*, **39**(2), 215–245.

Subramaniam, S., and Balachandar, S. (2022). *Modeling Approaches and Computational Methods for Particle-Laden Turbulent Flows*. Elsevier.

Subrahmanyam, T. V., and Forssberg, K. E. (1990). Fine particles processing: Shear-flocculation and carrier flotation: A review. *International Journal of Mineral Processing*, **30**(3–4), 265–286.

Sun, B., Tenneti, S., and Subramaniam, S. (2015). Modeling average gas–solid heat transfer using particle-resolved direct numerical simulation. *International Journal of Heat and Mass Transfer*, **86**, 898–913.

Sun, B., Tenneti, S., Subramaniam, S., and Koch, D. L. (2016). Pseudo-turbulent heat flux and average gas–phase conduction during gas–solid heat transfer: Flow past random fixed particle assemblies. *Journal of Fluid Mechanics*, **798**, 299–349.

Sundaram, S., and Collins, L. R. (1997). Collision statistics in an isotropic particle-laden turbulent suspension. Part 1. Direct numerical simulations. *Journal of Fluid Mechanics*, **335**, 75–109.

Taborda, M. A., and Sommerfeld, M. (2021). Reactive LES–Euler/Lagrange modelling of bubble columns considering effects of bubble dynamics. *Chemical Engineering Journal*, **407**, 127222.

Tackie, E., Bowen, B. D., and Epstein, N. (1983). Hindered settling of uncharged and charged submicrometer spheres. *Annals of the New York Academy of Sciences*, **404**(1), 366–367.

Takemura, F., and Magnaudet, J. (2003). The transverse force on clean and contaminated bubbles rising near a vertical wall at moderate Reynolds number. *Journal of Fluid Mechanics*, **495**, 235–253.

Tanaka, T., Yamagata, K., and Tsuji, Y. (1990). Experiment on fluid forces on a rotating sphere and spheroid. *Proceedings of the 2nd KSME–JSME Fluids Engineering Conference*. The Korean Society of Mechanical Engineers, pp. 366–369.

Tanemura, M. (2003). Statistical distributions of Poisson Voronoi cells in two and three dimensions. *FORMA*, **18**(4), 221–247.

Tang, Y., Peters, E. A. J. F., Kuipers, J. A. M., Kriebitzsch, S. H. L., and Hoef, M. A. (2015). A new drag correlation from fully resolved simulations of flow past monodisperse static arrays of spheres. *AIChE Journal*, **61**(2), 688–698.

Tavanashad, V., Passalacqua, A., Fox, R. O., and Subramaniam, S. (2019). Effect of density ratio on velocity fluctuations in dispersed multiphase flow from simulations of finite-size particles. *Acta Mechanica*, **230**(2), 469–484.

Tavassoli, H., Kriebitzsch, S. H. L., Van der Hoef, M. A., Peters, E. A. J. F., and Kuipers, J. A. M. (2013). Direct numerical simulation of particulate flow with heat transfer. *International Journal of Multiphase Flow*, **57**, 29–37.

Tavanashad, V., Passalacqua, A., and Subramaniam, S. (2021). Particle-resolved simulation of freely evolving particle suspensions: Flow physics and modeling. *International Journal of Multiphase Flow*, **135**, 103533

Taylor, G. I. (1922). Diffusion by continuous movements. *Proceedings of the London Mathematical Society, Series 2*, **2**(1), 196–212.

Taylor, G. I. (1928). The forces on a body placed in a curved or converging stream of fluid. *Proceedings of the Royal Society of London, Series A*, **120**(785), 260–283.

Tchen, C. M. (1947). Mean value and correlation problems connected with the motion of small particles suspended in a turbulent fluid. PhD thesis, Delft University of Technology, Delft.

Tee, Y. H., Barros, D. C., and Longmire, E. K. (2020). Motion of finite-size spheres released in a turbulent boundary layer. *International Journal of Multiphase Flow*, **133**, 103462.

Ten Cate, A., Derksen, J. K., Portela, L. M. and Van den Akker, H. E. A. (2004). Fully resolved simulations of colliding monodisperse spheres in forced isotropic turbulence. *Journal of Fluid Mechanics*, **519**, 233–271.

Tennekes, H., and Lumley, J. L. (2018). *A First Course in Turbulence*. MIT Press.

Tenneti, S., and Subramaniam, S. (2014). Particle-resolved direct numerical simulation for gas–solid flow model development. *Annual Review of Fluid Mechanics*, **46**, 199–230.

Tenneti, S., Garg, R., and Subramaniam, S. (2011). Drag law for monodisperse gas–solid systems using particle-resolved direct numerical simulation of flow past fixed assemblies of spheres. *International Journal of Multiphase Flow*, **37**(9), 1072–1092.

Theofanous, T. G., Mitkin, V., and Chang, C. H. (2018). Shock dispersal of dilute particle clouds. *Journal of Fluid Mechanics*, **841**, 732–745.

Tiwari, S. S., Pal, E., Bale, S., et al. (2020a). Flow past a single stationary sphere, 1. Experimental and numerical techniques. *Powder Technology*, **365**, 115–148.

Tiwari, S. S., Pal, E., Bale, S., et al. (2020b). Flow past a single stationary sphere, 2. Regime mapping and effect of external disturbances. *Powder Technology*, **365**, 215–243.

Tollmien, W. (1938). Uber krafte und momente in schwach gekrummten oder konvergenten stromungen. *Ingenieur-Archiv*, **9**(4), 308–326.

Tomboulides, A., Orszag, S., and Karniadakis, G. (1993). Direct and large-eddy simulations of axisymmetric wakes. *Proceedings of the 31st AIAA Meeting*, Reno, NV.

Torobin, L. B., and Gauvin, W. H. (1961). Fundamental aspects of solids–gas flow: Part VI: Multiparticle behavior in turbulent fluids. *Canadian Journal of Chemical Engineering*, **39**(3), 113–120.

Torquato, S. (2013). *Random Heterogeneous Materials: Microstructure and Macroscopic Properties*. Springer.

Tri, B. D., Oesterle, B., and Deneu, F. (1990). Premiers résultats sur la portance d'une sphere en rotation aux nombres de Reynolds intermediaires. *Comptes rendus de l'Académie des Sciences, Series II*, **311**, 27–31.

Tryggvason, G., Scardovelli, R., and Zaleski, S. (2011). *Direct Numerical Simulations of Gas–Liquid Multiphase Flows*. Cambridge University Press.

Tsuji, Y., Morikawa, Y., and Terashima, K. (1982). Fluid-dynamic interaction between two spheres. *International Journal of Multiphase Flow*, **8**(1), 71–82.

Tsuji, Y., Morikawa, Y., and Shiomi, H. (1984). LDV measurements of an air–solid two-phase flow in a vertical pipe. *Journal of Fluid Mechanics*, **139**, 417–434.

Tsuji, Y., Tanaka, T., and Ishida, T. (1992). Lagrangian numerical simulation of plug flow of cohesionless particles in a horizontal pipe. *Powder Technology*, **71**(3), 239–250.

Tunstall, E. B., and Houghton, G. (1968). Retardation of falling spheres by hydrody-namic oscillations. *Chemical Engineering Science*, **23**(9), 1067–1081.

Turner, J. S. (1979). *Buoyancy Effects in Fluids*. Cambridge University Press.

Udaykumar, H. S., Kan, H.-C., Shyy, W., and Tran-son-Tay, R. (1997). Multiphase dynamics in arbitrary geometries on fixed Cartesian grids. *Journal of Computational Physics*, **137**, 366–405.

Udaykumar, H. S., Mittal, R., Rampunggoon, P., and Khanna, A. (2001). A sharp inter-face Cartesian grid method for simulating flows with complex moving boundaries. *Journal of Computational Physics*, **174**, 345–380.

Uhlherr, P. H. T., and Sinclair, C. G. (1970). The effect of freestream turbulence on the drag coefficients of spheres, *Proc. Chemca.*, **1**, 1–12.

Uhlmann, M. (2005). An immersed boundary method with direct forcing for the simulation of particulate flows. *Journal of Computational Physics*, **209**(2), 448–476.

Uhlmann, M. (2008). Interface-resolved direct numerical simulation of vertical partic-ulate channel flow in the turbulent regime. *Physics of Fluids*, **20**, 053305.

Uhlmann, M., and Chouippe, A. (2017). Clustering and preferential concentration of finite-size particles in forced homogeneous-isotropic turbulence. *Journal of Fluid Mechanics*, **812**, 991–1023.

Valioulis, I. A., and List, E. J. (1984). A numerical evaluation of the stochastic com-pleteness of the kinetic coagulation equation. *Journal of the Atmospheric Sciences*, **41**(16), 2516–2530.

Van Hout, R. (2011). Time-resolved PIV measurements of the interaction of polystyrene beads with near-wall-coherent structures in a turbulent channel flow. *International Journal of Multiphase Flow*, **37**(4), 346–357.

Van Hout, R. (2013). Spatially and temporally resolved measurements of bead resus-pension and saltation in a turbulent water channel flow. *Journal of Fluid Mechanics*, **715**, 389–423.

Van Hout, R., Eisma, J., Elsinga, G. E., and Westerweel, J. (2018). Experimental study of the flow in the wake of a stationary sphere immersed in a turbulent boundary layer. *Physical Review Fluids*, **3**(2), 024601.

Variano, E. A., and Cowen, E. A. (2008). A random-jet-stirred turbulence tank. *Journal of Fluid Mechanics*, **604**, 1–32.

Vasseur, P., and Cox, R. G. (1977). The lateral migration of spherical particles sedi-menting in a stagnant bounded fluid. *Journal of Fluid Mechanics*, **80**(3), 561–591.

Veldhuis, C. H. J., and Biesheuvel, A. (2007). An experimental study of the regimes of motion of spheres falling or ascending freely in a Newtonian fluid. *International Journal of Multiphase Flow*, **33**(10), 1074–1087.

Voinov, V. V., Voinov, O. V., and Petrov, A. G. (1973). Hydrodynamic interaction between bodies in a perfect incompressible fluid and their motion in nonuniform streams (in Russian). *Prikladnaya Matematika i Mekhanika*, **37**(4), 680–689. English translation, *Journal of Applied Mathematics and Mechanics*, **37**(4), 642–651.

Wakaba, L., and Balachandar, S. (2005). History force on a sphere in a weak linear shear flow. *International Journal of Multiphase Flow*, **31**(9), 996–1014.

Wakaba, L., and Balachandar, S. (2007). On the added mass force at finite Reynolds and acceleration numbers. *Theoretical and Computational Fluid Dynamics*, **21**(2), 147–153.

Wang, C. C. (1970). A new representation theorem for isotropic functions. *Archive for Rational Mechanics and Analysis*, **36**, 166–197.

Wang, L. P. (2022). Coagulation in turbulent particle-laden flows. In Subramaniam, S., and Balachandar, S. (eds), *Modeling Approaches and Computational Methods for Particle-Laden Turbulent Flows* Elsevier.

Wang, L. P., and Grabowski, W. W. (2009). The role of air turbulence in warm rain initiation. *Atmospheric Science Letters*, **10**(1), 1–8.

Wang, L. P., and Maxey, M. R. (1993). Settling velocity and concentration distribution of heavy particles in homogeneous isotropic turbulence. *Journal of Fluid Mechanics*, **256**, 27–68.

Wang, L. P., and Stock, D. E. (1993). Dispersion of heavy particles by turbulent motion. *Journal of the Atmospheric Sciences*, **50**(13), 1897–1913.

Wang, L. P., Wexler, A. S., and Zhou, Y. (2000). Statistical mechanical description and modelling of turbulent collision of inertial particles. *Journal of Fluid Mechanics*, **415**, 117–153.

Wang, Q., Squires, K. D., Chen, M., and McLaughlin, J. B. (1997). On the role of the lift force in turbulence simulations of particle deposition. *International Journal of Multiphase Flow*, **23**(4), 749–763.

Warnica, W. D., Renksizbulut, M., and Strong, A. B. (1994). Drag coefficient of spherical liquid droplets. *Experiments in Fluids*, **18**, 265–270.

Wen, F., Kamalu, N., Chung, J. N., Crowe, C. T., and Troutt, T. R. (1992). Particle dispersion by vortex structures in plane mixing layers. *Journal of Fluids Engineering*, **114**(4), 657–666.

Whitaker, S. (1972). Forced convection heat transfer correlations for flow in pipes, past flat plates, single spheres, and for flow in packed beds and tube bundles. *AIChE Journal*, **18**, 361–371.

Wicker, R. B., and Eaton, J. K. (2001). Structure of a swirling, recirculating coaxial free jet and its effect on particle motion. *International Journal of Multiphase Flow*, **27**(6), 949–970.

Willen, D. P., and Prosperetti, A. (2019). Resolved simulations of sedimenting suspensions of spheres. *Physical Review Fluids*, **4**(1), 014304.

Williams, F. A. (1958). Spray combustion and atomization. *Physics of Fluids*, **1**(6), 541–545.

Winkler, C. M., Rani, S. L., and Vanka, S. P. (2004). Preferential concentration of particles in a fully developed turbulent square duct flow. *International Journal of Multiphase Flow*, **30**(1), 27–50.

Winterwerp, J. C. (1998). A simple model for turbulence induced flocculation of cohesive sediment. *Journal of Hydraulic Research*, **36**(3), 309–326.

Woodcock, L. V. (1976). Glass transition in the hard-sphere model. *Journal of the Chemical Society, Faraday Transactions 2: Molecular and Chemical Physics*, **72**, 1667–1672.

Wu, F. C., and Chou, Y. J. (2003). Rolling and lifting probabilities for sediment entrainment. *Journal of Hydraulic Engineering*, **129**(2), 110–119.

Wu, J. S., and Faeth, G. M. (1993). Sphere wakes in still surroundings at intermediate Reynolds numbers. *AIAA Journal*, **31**(8), 1448–1455.

Wu, J.-S., and Faeth, G. M. (1994a). Sphere wakes at moderate Reynolds numbers in aturbulent environment. *AIAA Journal*, **32**(3), 535–541.

Wu, J.-S., and Faeth, G. M. (1994b). Effect of ambient turbulence intensity on sphere wakes at intermediate Reynolds numbers. *AIAA Journal*, **33**, 171–173.

Yang, F. L. (2010). A formula for the wall-amplified added mass coefficient for a solid sphere in normal approach to a wall and its application for such motion at low Reynolds number. *Physics of Fluids*, **22**(12), 123303.

Yang, T. S., and Shy, S. S. (2003). The settling velocity of heavy particles in an aqueous near-isotropic turbulence. *Physics of Fluids*, **15**(4), 868–880.

Yang, T. S., and Shy, S. S. (2005). Two-way interaction between solid particles and homogeneous air turbulence: Particle settling rate and turbulence modification measurements. *Journal of Fluid Mechanics*, **526**, 171–216.

Yang, Y., and Balachandar, S. (2021). A scalable parallel algorithm for direct-forcing immersed boundary method for multiphase flow simulation on spectral elements. *Journal of Supercomputing*, **77**(3), 2897–2927.

Yau, M. K., and Rogers, R. R. (1996). *A Short Course in Cloud Physics*. Elsevier.

Ye, T., Mittal, R., Udaykumar, H. S., and Shyy, W. (1999). An accurate Cartesian grid method for viscous incompressible flows with complex immersed boundaries. *Journal of Computational Physics*, **156**, 209–240.

Yearling, P. R., and Gould, R. D. (1995). Convective heat and mass transfer from single evaporating water, methanol and ethanol droplets (No. CONF-951135–). American Society of Mechanical Engineers.

Yeh, T. H., Cantero, M., Cantelli, A., Pirmez, C., and Parker, G. (2013). Turbidity current with a roof: Success and failure of RANS modeling for turbidity currents under strongly stratified conditions. *Journal of Geophysical Research: Earth Surface*, **118**(3), 1975–1998.

Yeo, K., and Maxey, M. R. (2010). Simulation of concentrated suspensions using the force-coupling method. *Journal of Computational Physics*, **229**(6), 2401–2421.

Yu, M., Yu, X., and Balachandar, S. (2021). Flocculation dynamics of cohesive sediment in isotropic turbulence. Submitted to *Water Resources*.

Yu, M., Yu, X., Balachandar, S., Mehta, A., and Manning, A. (2022). Restructuring of cohesive sediment flocs by turbulence. In preparation.

Yu, X., Hsu, T. J., and Balachandar, S. (2013). Convective instability in sedimentation: Linear stability analysis. *Journal of Geophysical Research: Oceans*, **118**(1), 256–272.

Yu, X., Hsu, T. J., and Balachandar, S. (2014). Convective instability in sedimentation: 3D numerical study. *Journal of Geophysical Research: Oceans*, **119**(11), 8141–8161.

Yudine, M. I. (1959). Physical considerations on heavy-particle diffusion. *Advances in Geophysics*, Vol. 6. Elsevier, pp. 185–191.

Yun, G., Kim, D., and Choi, H. (2006). Vortical structures behind a sphere at subcritical Reynolds numbers. *Physics of Fluids*, **18**(1), 015102.

Zaichik, L. I., Simonin, O., and Alipchenkov, V. M. (2003). Two statistical models for predicting collision rates of inertial particles in homogeneous isotropic turbulence. *Physics of Fluids*, **15**(10), 2995–3005.

Zaidi, A. A. (2018). Study of particle inertia effects on drag force of finite sized particles in settling process. *Chemical Engineering Research and Design*, **132**, 714–728.

Zaidi, A. A., Tsuji, T., and Tanaka, T. (2014). A new relation of drag force for high Stokes number monodisperse spheres by direct numerical simulation. *Advanced Powder Technology*, **25**(6), 1860–1871.

Zarin, N. A., and Nicholls, J. A. (1971). Sphere drag in solid rockets – non-continuum and turbulence effects. *Combustion Science and Technology*, **3**, 273–280.

Zeng, L., Balachandar, S., and Fischer, P. (2005). Wall-induced forces on a rigid sphere at finite Re. *Journal of Fluid Mechanics*, **536**, 1–25.

Zeng, L., Balachandar, S., Fischer, P., and Najjar, F. (2008). Interactions of a stationary finite-sized particle with wall turbulence. *Journal of Fluid Mechanics*, **594**, 271–305.

Zeng, L., Najjar, F., Balachandar, S., and Fischer, P. (2009). Forces on a finite-sized particle located close to a wall in a linear shear flow. *Physics of Fluids*, **21**(3), 033302.

Zeng, L., Balachandar, S., and Najjar, F. M. (2010). Wake response of a stationary finite-sized particle in a turbulent channel flow. *International Journal of Multiphase Flow*, **36**(5), 406–422.

Zhang, D. Z., and Prosperetti, A. (1994). Averaged equations for inviscid disperse two-phase flow. *Journal of Fluid Mechanics*, **267**, 185–219.

Zhang, D. Z., and Prosperetti, A. (1997). Momentum and energy equations for disperse two-phase flows and their closure for dilute suspensions. *International Journal of Multiphase Flow*, **23**(3), 425–453.

Zhang, J., Zhang, J. P., and Fan, L. S. (2005). Effect of particle size ratio on the drag force of an interactive particle. *Chemical Engineering Research and Design*, **83**(4), 339–343.

Zhang, J. F., and Zhang, Q. H. (2011). Lattice Boltzmann simulation of the flocculation process of cohesive sediment due to differential settling. *Continental Shelf Research*, **31**(10), S94–S105.

Zhou, J., Adrian, R. J., Balachandar, S., and Kendall, T. M. (1999). Mechanisms for generating coherent packets of hairpin vortices in channel flow. *Journal of Fluid Mechanics*, **387**, 353–396.

Zhou, K., and Balachandar, S. (2021). An analysis of the spatio-temporal resolution of the immersed boundary method with direct forcing. *Journal of Computational Physics*, **424**, 109862.

Zhou, W., and Dušek, J. (2015). Chaotic states and order in the chaos of the paths of freely falling and ascending spheres. *International Journal of Multiphase Flow*, **75**, 205–223.

Zhou, Z., Jin, G., Tian, B., and Ren, J. (2017). Hydrodynamic force and torque models for a particle moving near a wall at finite particle Reynolds numbers. *International Journal of Multiphase Flow*, **92**, 1–19.

Zhu, C., Liang, S. C., and Fan, L. S. (1994). Particle wake effects on the drag force of an interactive particle. *International Journal of Multiphase Flow*, **20**(1), 117–129.

Zhu, C., Fan, L. S., and Yu, Z. (2021). *Dynamics of Multiphase Flows*. Cambridge University Press.

Zinchenko, A. Z. (1994). Algorithm for random close packing of spheres with periodic boundary conditions. *Journal of Computational Physics*, **114**(2), 298–307.

Zou, J. F., Ren, A. L., and Deng, J. (2005). Study on flow past two spheres in tandem arrangement using a local mesh refinement virtual boundary method. *International Journal for Numerical Methods in Fluids*, **49**(5), 465–488.

Zuber, N. (1964). On the dispersed two-phase flow in the laminar flow regime. *Chemical Engineering Science*, **19**(11), 897–917.

Zúñiga, S. L., Salinas, J. S., Balachandar, S., and Cantero, M. I. (2022). Universal nature of rapid evolution of conservative gravity and turbidity currents perturbed from their self-similar state. *Physical Review Fluids*, **7**, 043801.

Zwick, D., and Balachandar, S. (2019). Dynamics of rapidly depressurized multiphase shock tubes. *Journal of Fluid Mechanics*, **880**, 441–477.

Index

3Γ distribution, 335

absolute instability, 185, 186
added-mass force, 119
Archimedes force, 116
averaging, 51
  volume, 51

back coupling, 4
baseball, 189
Basset history kernel, 119
Basset–Boussinesq–Oseen (BBO) equation, 90
body-centered cubic, 70
body-fitted approach, 470
body-free approach, 470
Boltzmann constant, 228
Boussinesq approximation, 25, 516
box filter, 388
Brinkman's equation, 324
Brownian constant, 228
Brownian motion, 228

closure problem, 398
coagulation efficiency, 354
collision efficiency, 354
continuity effect, 222
convolution integral, 389
cricket ball, 189
critical Reynolds number, 193
cross stress, 401, 508
crossing trajectory effect, 218
curve ball, 189

diffusivity, 222
  fluid parcel, 222
  particle, 222
discrete element method, 382
discrete element model (DEM), 303, 518
dispersed multiphase flows, 2
  continuous phase, 2
  dispersed phase, 2
double-threaded wake, 185, 273, 277
doublet, 94

drag crisis, 189
drag law
  Hadamard–Rybczynski, 30
  standard, 29
droplet density function, 76
dusty gas, 20, 455
dusty-gas approximation, 434, 588

eddy-decay effect, 218
Einstein's viscosity correction, 104
ensemble, 47
  joint, 81
  macro, 82
  meso, 82
  micro, 81
equation of state, 407
equilibrium, 434
equilibrium approximation, 434
Euler–Euler formulation, 83
Euler–Lagrange formulation, 83
Euler–Lagrange-DNS (EL-DNS), 490
Euler–Lagrange-LES (EL-LES), 490

face-centered cubic, 70
fast-tracking, 207
Favre filtered energy, 401
Faxén theorem, 124
Fick's law, 89, 502
form drag, 97
forward coupling, 3
Fourier law, 89
front-capturing method, 471
  level-set, 471
  volume of fluid (VOF), 471
front-tracking method, 471
  marker points, 471
fully resolved, *see* particle-resolved simulation
fully resolved simulation, 82

Gaussian filter, 388
geometric collision kernel, 354
golf ball, 189
granular cooling, 573

granular kinetic theory, 5
granular temperature, 422, 550, 570
granular temperature equation, 586

Hadamard–Rybczynski drag, 100
Helmholtz decomposition, 112
Hill's spherical vortex, 101
hindered settling, 330
Hopf bifurcation, 185
hyperbolicity, 419
hypopicnal flows, 5

immersed boundary method, 472
imperfect bifurcation, 277
inertial clustering, *see* preferential accumulation
inertial memory effect, 219
inviscid-unsteady force, 118

kinetic stress, 420, 562, 578
Kronecker delta, 606

Lagrangian stochastic method, 228
laminar wake, 192, 203
Langevin model, 230, 291
Leibniz rule, 395
Leonard stress, 401, 508
Levi-Civita symbol, 606
Liouville densities, 76
locally homogeneous flows, 588
lognormal distribution, 85
loitering, 208
Lorentz reciprocal theorem, 312, 619

Magnus lift force, 161
Matérn hard-core process, 75
Maxey–Riley–Gatignol (MRG) equation, 90
Maxwellian distribution, 578
mesoscale instabilities
    double-diffusive instability, 16
    Holmboe instability, 16
    Kelvin–Helmholtz instability, 16
    lobe and cleft instability, 16
    Rayleigh–Taylor instability, 16
    Richtmeyer–Meshkov instability, 17
method of reflections, 248, 326
mobility matrix, 314
molecular chaos, 574, 579
multimaterial flows, 3
multiphase flow, 1
    gas–liquid, 1
    gas–solid, 1
    liquid–liquid, 1
    liquid–solid, 1
    solid–gas, 1
multiphase mechanisms, 20
    body-force effect, 20

inertia effect, 20
slip effect, 20
thermodynamic/transport effect, 20
volume effect, 20

nozzling effect, 399, 402
number density function (NDF), 76
Nusselt number, 29

one-sided vortex shedding, 186

Péclet number, 169
pair distribution function, 61, 574
pair probability function, 66
particle accumulation effect, 374
particle diffusivity tensor, 215
particle size distribution, 85
particle time scale, 28
    thermal, 29
    translational, 28
particle-resolved DNS (PR-DNS), 463
particle-resolved LES (PR-LES), 466
particle-resolved simulation, 50
particle-unresolved DNS (PU-DNS), 464
particle-unresolved LES (PU-LES), 466
perfect bifurcation, 277
phase, 1
point-particle, 321
Poisson distribution, 56
Poisson Voronoi distribution, 71
polydispersity, 78, 84, 85
population balance equation (PBE), 76, 86, 381
preferential accumulation, 9
pressure-gradient force, 115
probability density function
    generic, 60
    specific, 59
pseudo turbulence, 8, 47, 182, 508
pseudo vector, 286

quasi-steady force, 117

radial distribution function, 58, 61, 66, 328, 374,
    578
random sequential addition process, 68
Ranz–Marshal correlation, 502, 559
realization, 47
    joint, 81
    macro, 82
    meso, 82
    micro, 81
reciprocal relation, 109, 249
reciprocal theorem, 108, 126
reflectional invariance, 54
regimes of multiphase flow, 4
    collisional, 5, 45

compaction, 5
contact, 5
four-way coupled, 4
one-way coupled, 4
three-way coupled, 4
two-way coupled, 4
renormalization, 203, 328
resistance matrix, 312
reverse swing, 189
Reynolds stress, 508
    ensemble-averaged, 598
    particle, 552, 599
Rosin–Rammler distribution, 85
rotational invariance, 54
rotlet, 104

Saffman length, 147
Saffman length scale, 262
Schmidt number, 503
sediment-laden flows, 2
separated multiphase flows, 2
settling velocity, 11
    fast-tracking, 12
    loitering effect, 11, 204
    nonlinear effect, 12, 204
    particle trapping, 12, 204
    trajectory bias, 12, 204
    two-way coupling effect, 12, 204
sharp interface method, 471
Sherwood number, 502, 559
side-by-side arrangement, 310
single-phase limit, 18
single-threaded wake, 184
skin friction drag, 97
Smoluchowski coagulation equation, 360
statistically complete collision equation, 358
statistics
    axisymmetry, 62
    homogeneous, 54
    isotropy, 54
    stationary, 53
stochastic point process, 74
Stokes drag, 91

Stokes length, 147
Stokes length scale, 262
Stokes number, 9, 12
Stokesian dynamics, 5
Stokeslet, 94
Stommel retention zone, 11
stresslet, 106
Strouhal number, 194
superficial velocity, 345
suspension balance model, 5
suspensions, 5
swelling algorithm, 70

tandem arrangement, 310
3Γ distribution, 71
three different scales, 7
    macroscale, 7
    mesoscale, 7
    microscale, 7
trajectory bias, 207
translational invariance, 54
true vector, 286
turbophoresis, 10, 448
turbophoretic migration, 233, 419
turbulent stress, 562
turbulent wake, 192, 203
two-fluid formulation, 83, 596
two-phase flow, 1

virtual-mass force, 119
Voronoi tessellation, 70
vortex shedding, 196
    chaotic, 196
    low-frequency oblique, 196
    quasi periodic, 196
    steady oblique, 196
    zig-zag, 196

Weber number, 352
Wiener white-noise process, 228
Wigner–Seitz radius, 58
Williams spray equation, 76, 381